The Practice of Social Research

Earl Babbie

Chapman University

CENGAGE

Australia • Brazil • Canada • Mexico • Singapore • United Kingdom • United States

CENGAGE

The Practice of Social Research,
Fifteenth Edition
Earl Babbie

Product Director: Laura Ross

Associate Product Manager: Kori Alexander

Product Assistant: Shelby Blakey

Learning Designer: Emma Guiton

Senior Content Manager: Kathy Sands-Boehmer

Digital Design Lead: Matt Altieri

IP Analyst: Deanna Ettinger

Marketing Manager: Tricia L. Salata

Senior Marketing Development Manager: Nicole Hurst

Art Director: Nadine Ballard

Manufacturing Planner: Karen Hunt

Production Service: MPS Limited

Cover Designer: Nadine Ballard

Cover Image: cigdem/ShutterStock.com

For product information and technology assistance, contact us at
Cengage Customer & Sales Support, 1-800-354-9706
or **support.cengage.com.**
For permission to use material from this text or product, submit all requests online at **www.cengage.com/permissions.**
Further permissions questions can be e-mailed to
permissionrequest@cengage.com.

Library of Congress Control Number: 2020932559

Student Edition:
ISBN 13: 978-0-357-36076-7
ISBN 10: 0-357-36076-1

Loose-leaf Edition:
ISBN 13: 978-0-357-36081-1
ISBN 10: 0-357-36081-8

Cengage
200 Pier 4 Boulevard
Boston, MA 02210
USA

Cengage is a leading provider of customized learning solutions with employees residing in nearly 40 different countries and sales in more than 125 countries around the world. Find your local representative at **www.cengage.com.**

To learn more about Cengage platforms and services, register or access your online learning solution, or purchase materials for your course, visit **www.cengage.com.**

Printed at CLDPC, USA, 01-22

Dedication

Suzanne Babbie

A Note from the Author

Writing is my joy, sociology my passion. I delight in putting words together in a way that makes people learn or laugh or both. Sociology shows up as a set of words, also. It represents our last, best hope for planet-training our race and finding ways for us to live together. I feel a special excitement at being present when sociology, at last, comes into focus as an idea whose time has come.

I grew up in small-town Vermont and New Hampshire. When I announced that I wanted to be an auto-body mechanic, my teacher, like my dad, told me I should go to college instead. When young Malcolm Little announced he wanted to be a lawyer, his teacher told him a "colored boy" should be something more like a carpenter. The difference in our experiences says something powerful about the idea of a level playing field. The inequalities among ethnic groups run deep, as Malcolm X would go on to point out.

I ventured into the outer world by way of Harvard, the U.S. Marine Corps, UC Berkeley, and 12 years teaching at the University of Hawaii. I resigned from teaching in 1980 and wrote full time for seven years, until the call of the classroom became too loud to ignore. For me, teaching is like playing jazz. Even if you perform the same number over and over, it never comes out the same way twice and you don't know exactly what it'll sound like until you hear it. Teaching is like writing with your voice.

After some 20 years of teaching at Chapman University in southern California, I have now shifted my venue by moving to Arkansas and getting a direct experience of southern/midwestern life. When that's balanced by periodic returns to my roots in Vermont, I feel well rounded in my sociological experiences.

While no longer in residence at Chapman University, I maintain an active participation there as an Emeritus Professor. Also, upon my retirement, Chapman established an undergraduate research center, which they generously named the Earl Babbie Research Center. The Center keeps me actively participating with friends and colleagues at Chapman.

Contents in Brief

Contents in Detail

Part 3   Modes of Observation

CHAPTER 8

Experiments   228

CHAPTER 9

Survey Research   249

Boxed Features

Preface

The book in your hands has been about four decades in the making. It began in the classroom, when I was asked to teach a seminar in survey research. Frustrated with the lack of good textbooks on the subject, I began to dream up something I called "A Survey Research Cookbook and Other Fables," which was published in 1973 with a more sober title: *Survey Research Methods.*

The book was an immediate success. However, there were few courses limited to survey research. Several instructors around the country asked if "the same guy" could write a more general methods book, and *The Practice of Social Research* appeared two years later. The latter book has become a fixture in social research instruction with this Fifteenth Edition. The official two-volume Chinese edition was published in Beijing in 2000, and there are numerous other non-English versions around the world.

Over the life of this book, successive revisions have been based in large part on suggestions, comments, requests, and corrections from my colleagues around the country and, increasingly, around the world. Many also requested a shorter book with a more applied orientation.

Whereas the third quarter of the twentieth century saw a greater emphasis on quantitative, pure research, the century ended with a renaissance of concern for applied sociological research (sometimes called *sociological practice*) and also a renewed interest in qualitative research.

Changes in the Fifteenth Edition

A revision like this depends heavily on the input from students and faculty, who have been using earlier editions. Some of those suggestions resulted in two new features that have been added to every chapter:

General Changes

- Each chapter begins with a list of numbered learning objectives that are keyed to the relevant discussion in that chapter.

- Each chapter begins with a "What Do You Think?" puzzle for students to think about. Later in the chapter, there are answers included in "What Do You Think... Revisited."

- As with each edition, I have included illustrative data (from the U.S. Census, opinion polls, observational studies) wherever possible. This doesn't change the methodological purposes for using the data but it keeps the reader in closer touch with the real world.

- A letter from the book to the student, something I developed awhile back in *The Basics of Social Research* (a shorter version of this book), and I thought *Practice* readers would enjoy it and benefit from it, too.

- Added a new feature—"What Do You Think?"—which poses a puzzle at the outset of each chapter and answers it later, using the materials covered in the chapter.

Chapter Changes

In addition to those bookwide changes, here are some of the additional updates you'll find in specific chapters of the book. Many of these changes were made in response to comments and requests from students and faculty.

Part One: An Introduction to Inquiry

1 Human Inquiry and Science

- Shifted from birth to fertility rates and updated Table 1-1
- Expanded discussion of probabilistic causation
- Added an introduction to the Ethics of Human Inquiry
- Updated GSS table on attitudes toward gays and lesbians to 2014
- Deleted box on "The Hardest Hit"
- Added box on "Social Research Making a Difference"
- Added new Figure 1-1
- Added box on "Fertility Rate Implications"
- Moved "Wheel of Science" figure to Chapter 2
- Clarified "dark side of the moon" comment
- Added box on the role of women in early social research

2 Paradigms, Theory, and Research

- Added some bibliographic citations for classic references
- Added introductory discussion of logic and rationality
- Added box on "The Power of Paradigms"
- Added box on "Framing a Hypothesis"
- Added box on "Church Involvement"
- Moved "Wheel of Science" figure from Chapter 1
- Replaced section on "Links between Theory and Research" with "The Importance of Theory in the 'Real World'"

3 The Ethics and Politics of Social Research

- Added example of Facebook 2012 study violating informed consent and Facebook's creation of an ethics review process
- Added an example on confidentiality and anonymity in qualitative studies
- Pointed readers to the 2015 revision of the AAPOR Code of Ethics

Part Two: The Structuring of Inquiry

4 Research Design

- Added box reporting a graduate student's experience in the field
- Expanded the discussion of Figure 4-1
- Added new figure comparing time variable and different designs
- Cited Peter Lynn book on longitudinal surveys
- Added new section on mixed modes
- Cited Akerlof and Kennedy on the evaluation of environmental degradation studies
- Introduced new trend study of American fears (2014 and 2018)
- Discussed panel attrition in qualitative panel studies
- Updated marijuana approval to 2016 GSS
- Updated states legalizing recreational marijuana
- Introduced the concept of statistical significance
- Made comparison of idiographic/Nomothetic less repetitive
- Expanded the explanation of necessary and sufficient causes in Main Points
- Improved the explanation of the ecological fallacy in Main Points
- Dropped an older example to trim section on longitudinal design

5 Conceptualization, Operationalization, and Measurement

- Added discussion of measuring ethnicity in Cornwall County, Britain
- Added discussion of cognitive interviewing
- Added example of bullying in the workplace
- Added test of whether the terms *baby* or *fetus* affected abortion attitudes
- Added discussion of definition of rape and other variables
- Added section on use of cognitive interviewing in Korean language
- Integrated Gender and Race box into text
- Added a box on "Conceptualization"
- Added box "On to Hollywood"
- Added box on "Pregnant Chads and Voter Intentions"

- Added example of biased questions with regard to gun control

6   Indexes, Scales, and Typologies

- Updated the abortion example of a Guttman scale to 2012 GSS
- Cited Vision of Humanity's global peace index
- Cited the World Economic Forum's "Global Competitiveness Index" for rating 142 economies
- Updated data on State ratings on health
- Added discussion of research on various Likert formats online
- Added box on "Best College in the United States"
- Added box on "Assessing Women's Status"

7   The Logic of Sampling

- Updated presidential election polling
- Introduced term *chain referral*
- Added Michael Brick's prediction of a rebirth of quota sampling
- Discussed FCC rules on calling cell phones
- Expanded discussion of sampling for online surveys
- Revised box on selecting random numbers because of new table in Appendix B
- Added related box on sampling in Iran to sampling in the United States (or anywhere)
- Cited Nate Silver's FiveThirtyEight.com rating of pollsters
- Added study citing problems with snowball sampling
- Critiqued use of polls to pick GOP presidential candidates for 2015–2016 debates and the Democrats in the run-up to their 2020 presidential primary
- Added box on "Representative Sampling"
- Updated data on the extent of wireless phones

Part Three: Modes of Observation

8   Experiments

- Expanded on the introduction to online experiments
- Added example of an online experiment raising ethical questions

9   Survey Research

- Quoted from AAPOR report on mobile devices
- Cited an article on tablet-based surveys
- Cited Nate Silver's 538 rating of pollsters
- Added Pew Center tips on Web surveys
- Added a comparison of low-response telephone surveys and nonprobability online surveys
- Added study of nonresponse impact on univariate and multivariate analyses
- Mentioned Google Forms as an alternative to Survey Monkey

10   Qualitative Field Research

- Added example of participatory research in South Africa
- Added citation on uses of video for data collection
- Added box of evaluation of NGOs
- Added participant comment from Bangladeshi PAR study
- Added Van Cleave study of Crook County courthouse
- Reported an interview study of white supremacists
- Added study using online forums to study menopausal women

11   Unobtrusive Research

- Introduced Google Public Data
- Introduced Topsy Social Analytics
- Introduced the Association of Religious Data Archives and Their Measurement Wizard
- Discussed Tyler Vigen's work on spurious correlations among big data
- Added study comparing Mother's and Father's Day cards
- Updated comparison of male and female earnings
- Added *Sociological Perspectives* special issue on ethnography

12   Evaluation Research

- Added box on evaluation research by non-profit organizations
- Added box on "Solutions without Problems"
- Updated murder rates in states with and without death penalty

Part Four: Analysis of Data

13 Qualitative Data Analysis

- Updated election maps to 2016
- Added dedoose to QDA list

14 Quantitative Data Analysis

- Updated GSS data on church attendance
- Updated table on marijuana use and age
- Updated table on gender and church attendance
- Updated table exploring explanations for gender differences in pay

15 The Logic of Multivariate Analysis

- No changes

16 Social Statistics

- Deleted section on "Discriminant Analysis"
- Updated figure on election results to 2016
- Added section on Demographic Analyses

17 Reading and Writing Social Research

- Added citation to my e-book, *Avoiding Plagiarism*
- Gave an example of URL reference citation

Pedagogical Features

Although students and instructors alike have told me that the past editions of this book were effective tools for learning research methods, I see this edition as an opportunity to review the book from a pedagogical standpoint—fine-tuning some elements and adding others. Here's the resulting package for the Fifteenth Edition.

- **Learning Objectives:** Each chapter includes learning objectives to guide the student's understanding and comprehension of the chapter materials.

- **Chapter Introduction:** Each chapter opens with an introduction that lays out the main ideas in that chapter and, importantly, relates them to the content of other chapters in the book.

- **Clear and Provocative Examples:** Students often tell me that the examples—real and hypothetical—have helped them grasp difficult and/or abstract ideas, and this edition

has many new examples as well as some that have proved particularly valuable in earlier editions.

- **Full-Color Graphics:** From the first time I took a course in research methods, most of the key concepts have made sense to me in graphical form. Whereas my task here has been to translate those mental pictures into words, I've also included some illustrations. Advances in computer graphics have helped me communicate to the Cengage Learning artists what I see in my head and would like to share with students. I'm delighted with the new graphics in this edition.

- **Boxed Examples and Discussions:** Students tell me they like the boxed materials that highlight particular ideas and studies as well as vary the format of the book. In this edition, I've updated *Issues and Insights* boxed features to elaborate on the logic of research elements, *How to Do It* boxes to provide practical guidance, and *Applying Concepts in Everyday Life* features to help students see how the ideas they're reading about apply to real research projects, as well as to their lives.

- **Running Glossary:** There is a running glossary throughout the text. Key terms are highlighted in the text, and the definition for each term is listed at the bottom of the page where it first appears. This makes it easier for students learn the definitions of these terms and to locate them in each chapter so they can review them in context.

- **Main Points:** At the end of each chapter, a concise list of main points provides both a brief chapter summary and a useful review. The main points let students know exactly what ideas they should focus on in each chapter.

- **Key Terms:** A list of key terms follows the main points. These lists reinforce the students' acquisition of necessary vocabulary. The new vocabulary in these lists is defined in context within the chapters. The terms are boldfaced in the text, are defined in the running glossary that appears at the bottom of the page throughout the text, and are included in the glossary at the back of the book.

- **Proposing Social Research:** This series of linked exercises invites students to apply what they've learned in each chapter to the development of their own research proposal.

- **Review Questions:** This review aid allows students to test their understanding of the chapter concepts and apply what they've learned.

- **Appendixes:** As in previous editions, a set of appendixes provides students with some research tools, such as a guide to the library, a table of random numbers, and more.

- **Clear and Accessible Writing:** This is perhaps the most important "pedagogical aid" of all. I know that all authors strive to write texts that are clear and accessible, and I take some pride in the fact that this "feature" of the book has been one of its most highly praised attributes through fourteen previous editions. It's the one thing most often mentioned by the students who write to me. For the Fifteenth Edition, the editors and I have taken special care to reexamine literally every line in the book—pruning, polishing, embellishing, and occasionally restructuring for a maximally "reader-friendly" text. Whether you're new to this book or intimately familiar with previous editions, I invite you to open to any chapter and evaluate the writing for yourself.

Supplements

The Practice of Social Research, Fifteenth Edition, is accompanied by a wide array of supplements prepared for both the instructor and student to create the best learning environment inside as well as outside the classroom. All the continuing supplements have been thoroughly revised and updated, and several are new to this edition. I invite you to examine and take full advantage of the teaching and learning tools available to you.

The Practice of Social Research MindTap represents a new approach to a highly personalized, online learning platform. It combines all of a student's learning tools—the chapter reading, Practice activities within APLIA, Explore activities and new Create activities—into a Learning Path that guides the student through the research methods course. Instructors personalize the experience by customizing the presentation of these learning tools for their students, even seamlessly introducing their own content into the Learning Path.

Digital Resources

The MindTap digital platform offers:

- An interactive eBook, in which students can highlight key text, add notes, and create custom flashcards.

- Practice activities that empower students toward authentic and thoughtful learning experiences.

- Explore activities that allow students to interact with hypothetical social experiments in a scenario-based learning activity.

- Create activities asking students to piece together the different stages of the research process in each chapter and that will help them build up to a capstone project at the end of the course.

- A capstone project that can serve as a term paper or final project.

- A digital test bank, which includes multiple-choice, true/false, and essay questions for each chapter.

- A fully mobile experience via the MindTap mobile app, so students can read or listen to textbooks and study with the aid of instructor notifications and flashcards.

Instructor's Manual

This supplement offers the instructor chapter outlines, lecture outlines, behavioral objectives, teaching suggestions and resources, video suggestions, and questions/activities to guide a research project.

Cengage Learning Testing powered by Cognero®

Cengage Learning Testing powered by Cognero is a flexible online system that allows instructors to author, edit, and manage test bank content and quickly create multiple test versions. You can deliver tests from your LMS, your classroom—or wherever you want.

PowerPoint® Slides

Helping make your lectures more engaging, PowerPoint slides outline the chapters of the main text in a classroom-ready presentation, making it easy for instructors to assemble, edit, publish, and present custom PowerPoint slides.

Acknowledgments

It would be impossible to acknowledge adequately all the people who have influenced this book. My earlier methods text, *Survey Research Methods*, was dedicated to Samuel Stouffer, Paul Lazarsfeld, and Charles Glock. I again acknowledge my debt to them.

Many colleagues helped me through the several editions of *The Practice of Social Research* and my shorter text, *The Basics of Social Research.* Their contributions are still present in this edition of *Practice*, as are the end results from unsolicited comments and suggestions from students and faculty around the world.

Over the years, I have become more and more impressed by the important role played by editors in books like this. Since 1973, I've worked with varied sociology editors at Wadsworth, which has involved the kinds of adjustments you might need to make in as many successive marriages.

Although my name appears on the spine of this book and elsewhere in it, I want you to know that a volume like this can only appear by virtue of the work of a team of publishing professionals. Among the many whose hands have stroked this book along the way, I especially want to name Product Manager Kori Alexander, Learning Designer Emma Guiton, Content Manager Kathy Sands-Boehmer, Art Director Nadine Ballard, Digital Lead Matt Altieri, Product Assistant Shelby Blakey and Copy Editor Debbie Stone. Without their time and thoughtful attention, this book would remain a dream residing in my computer. Thank you.

I have dedicated this book to my soulmate, best friend, and wife, Suzanne Babbie. I see in Suze those things I am most proud of in myself, except I see purer versions of those qualities in her. She ennobles what is possible in a human being, and I become a better person because of her example.

I've asked my author and your instructor to chat among themselves so you and I can have a private conversation. Before you start reading this book, I want to let you in on something: I know you may not want me. You may not have chosen to take this course. My guess is that you're reading me because I've been assigned in a required research methods class. In that case, it's a bit like an arranged marriage.

I also know that you likely have some concerns about this course, especially its potential difficulty. If you do, you're not alone. I certainly don't want to *create* such concerns. However, I know from years of personal experience that many students feel anxious at the beginning of a social research course. In this short chat, I want to reassure you that it will not be as bad as you think. You may even enjoy this course. You see, a great many students from all over the world have written to my author to say just that: They were worried about the course at the beginning, but they ended up truly enjoying it.

So, to be clear, I'm not Freddy Krueger or Chucky—some monster plotting to make your college years miserable. I'm not even a dean. It's a little early in our relationship to call myself your *friend*, of course, but I do get called that a lot. I'm confident we can work together.

Benjamin Spock, the renowned author and pediatrician, began his books on child care by assuring new parents that they already knew more about caring for children than they thought they did. I want to begin on a similar note. Before you've read very far, you'll see that you already know a great deal about the practice of social research. In fact, you've been conducting social research all your life. From that perspective,

this book aims at helping you sharpen skills you already have and perhaps show you some tricks that may not have occurred to you.

If you're worried about *statistics* in a course like this, I must tell you something. There *are* some statistics. But it's not what you think. It's not just an evil swarm of numbers. Statistics has a logic that allows us to do amazing things. Did you know that questioning around 2,000 people, properly selected, can let us forecast the results of an election in which over 100 million people vote? I think you might find it's worth learning some statistics in order to understand how that sort of thing works. (In all my years as a textbook, I've never gotten tired of that example.)

Chapters 14 to 16 contain quite a bit of statistics, because they deal with quantitative (numerical) data analysis. Frankly, my author has never found a way of teaching students how to do statistical analyses without using some statistics. However, you'll find more emphasis on the *logic* of statistics than on mathematical calculations.

Maybe I should let you in on a little secret: My author never took a basic statistics course!

In his undergraduate major, statistics wasn't required. When he arrived at graduate school, a simple misunderstanding (really, you can't blame him for this) led him to indicate he had already taken introductory statistics when that wasn't, well, *technically* true. He only got an A in the advanced graduate statistics course because it focused on the logic of statistics more than on calculations. Statistics made *sense* to him, even without memorizing the calculations.

Here's a more embarrassing secret that he probably wouldn't want you to know. When he published his first research methods textbook in 1973, his chapter on statistics had only three calculations—*and he got two of them wrong.* (He's gotten much better, by the way. However, if you find any mistakes, please write him. I'm much happier when everything between my covers is in good order.)

The purpose of these confessions is not to downplay the importance of statistical analyses: I shall present them to you with the highest respect. My purpose is to let you know that statistics is not a mystical world that only math wizards enter. Statistics is a powerful tool that will help you understand the world around you. My author and I merely want to help you learn enough of it to wield that tool to your advantage.

What can you do if you come across something in this book or in class that you simply don't understand? You have several options:

1. Assume that it will never matter that much and go on with your life.

2. Decide that you are too stupid to understand such sophisticated ideas.

3. Ask someone else in the course if they understand it.

4. Ask your instructor to clarify it.

5. In case of emergency: e-mail my author at ebabbie@mac.com.

Options (1) and (2) are *not* good choices. Try (3), (4), and (5)—in that order.

With regard to number (5), by the way, please realize that tens of thousands of students around the world are using this book, in many languages, every semester, so it may take my author a little while to get back to you. He doesn't have a workshop of methodology elves helping him. Here's a hint: Do not frame your question in the form of a take-home exam, as in "What are three advantages of qualitative research over quantitative research?" My author doesn't answer those sorts of questions. *You* are the one taking the exam. He's taken enough exams already. Besides, he would give answers that leave out all the great material your instructor brings to the course.

Speaking of your instructor, by the way, please know that this is not the easiest course to teach. Even if the statistics are not as heavy as you thought, you'll be asked to open yourself up to new ways of seeing and understanding. That's

not necessarily comfortable, and your instructor has taken on the task of guiding you through whatever confusion and/or discomfort you may experience. So, give 'em a break.

Instructors know that this course typically produces lower-than-average teacher evaluations. Personally, I think it's because of the subject matter as well as the fears students bring to the course. So when it's time for evaluations, please separate your instructor's performance from any concerns you may have had about the material. Of course, you might find yourself thoroughly enjoying the subject of social science research. My author and I do, and so does your instructor. We plan to do everything possible to share that enjoyment with you.

If you're at all concerned about the state of the world (and I think you should be), it's worth knowing that social research is a key to solving most major problems. No joke. Consider the problem of overpopulation, for example. My author is fond of calling it the "mother of all social problems." (You'll get used to his sense of humor as you make your way through my pages. Be sure to check the glossary, by the way.)

Anyway, back to overpopulation. Most simply put, there are more people on the planet than it can sustain, even at the impoverished standard of living many of those people suffer. If everyone were living like those in the most developed countries, our resources would last about a week and a half and our carbon footprint would crush us like bugs. And the world's population is growing by about 80 million people a year. That's another United States every four years.

Where would you go for an answer to a problem like that? My author is fond of saying that at first people asked, "What causes all the babies?" and they turned to the biologists for help. But when they learned what was causing the babies, that didn't solve the problem. Frankly, they weren't willing to give up sex. So they turned to the rubber industry for help. That made some difference, but the population continued to grow. Finally, people turned to the chemical industry: "Can't we just take a pill and be able to have sex without producing babies?" Soon the pills were developed and they made some difference, but the population still continued to grow.

As I've learned from my author, the key to population growth lies in the social structures that lead people to have more babies than is needed to perpetuate the human species (roughly two babies per couple). Consider, for example, the social belief that a woman is not "really a woman" until she has given birth, or the complementary belief that a man is not "really a man" until he has sired young. Some people feel that they should produce children to take care of them when they are old, or to perpetuate their name (the father's name in most cases). Many other social perspectives promote the production of more than enough babies.

The biologists, chemists, and rubber manufacturers can't address those causes of overpopulation. That is precisely where social researchers come in. Social researchers can discover the most powerful causes of social problems like overpopulation, prejudice, war, and climate change (yes, even climate change) and explore ways of combating them.

The pressing need for well-trained social researchers is what motivates my author and your instructor to do what they do. It also explains why you may be required to take this course—even against your will. We're arming you to make a powerful difference in the world around you. What you do with that new ability is up to you, but we hope you will use it only for the good.

I'll turn you over to my author now. I'll do everything I can to make this a fun and useful course for you.

CHAPTER 1

Human Inquiry and Science

CHAPTER OVERVIEW

All of us try to understand and predict the social world. Scientific inquiries—and social research in particular—are designed to avoid the pitfalls of ordinary human inquiry.

© Monkey Business Images/Shutterstock.com

Introduction

This book is about knowing things—not so much *what* we know as *how* we know it. Let's start by examining a few things you probably know already.

You know the world is round. You probably also know it's cold on the dark side of the moon (the side facing away from the sun), and you know people speak Japanese in Japan. You know that vitamin C can prevent colds and that unprotected sex can result in AIDS.

How do you know? If you think for a minute, you'll see you know these things because somebody told them to you, and you believed them. You may have read in *National Geographic* that people speak Japanese in Japan, and that made sense to you, so you didn't question it. Perhaps your physics or astronomy instructor told you it was cold on the dark side of the moon, or maybe you heard it on the news.

Some of the things you know seem obvious to you. If I asked you how you know the world is round, you'd probably say, "Everybody knows that." There are a lot of things everybody knows. Of course, at one time, everyone "knew" the world was flat.

Most of what you know is a matter of agreement and belief. Little of it is based on personal experience and discovery. A big part of growing up in any society, in fact, is the process of learning to accept that what everybody around you "knows" is so. If you don't know those same things, you can't really be a part of the group. If you were to question seriously that the world *is* round, you'd quickly find yourself set apart from other people. You might be sent to live in a hospital with others who ask questions like that.

So, most of what you know is a matter of believing what you've been told. Understand that there's nothing wrong with you in that respect. That's simply the way human societies are structured. The basis of knowledge is agreement. Because you can't learn all you need to know through personal experience and discovery alone, things are set up so you can simply believe what others tell you. You know some things through tradition and others from "experts." I'm not saying you shouldn't question this received knowledge; I'm just drawing your attention to the way you and society normally get along regarding what is so.

There are other ways of knowing things, however. In contrast to knowing things through agreement, you can know them through direct experience—through observation. If you dive into a glacial stream flowing through the Canadian Rockies, you don't need anyone to tell you it's cold.

When your experience conflicts with what everyone else knows, though, there's a good chance you'll surrender your experience in favor of agreement. For example, imagine you've come to a party at my house. It's a high-class affair, and the drinks and food are excellent. In particular, you're taken by one of the appetizers I bring around on a tray: a breaded, deep-fried tidbit that's especially zesty. You have a couple—they're so delicious! You have more. Soon you're subtly moving around the room to be wherever I am when I arrive with a tray of these nibblies.

Finally, you can contain yourself no longer. "What are they?" you ask. I let you in on the secret: "You've been eating breaded, deep-fried worms!" Your response is dramatic: Your stomach rebels, and you promptly throw up all over

What do you think?

The decision to have a baby is deeply personal. No one is in charge of who will have babies in the United States in any given year or of how many will be born. Although you must get a license to marry or go fishing, you do not need a license to have a baby. Many couples delay pregnancy, some pregnancies happen by accident, and some pregnancies are planned. Given all these uncertainties and idiosyncrasies, how can baby-food and diaper manufacturers know how much inventory to produce from year to year? By the end of this chapter, you should be able to answer this question.

See the *What do you think? . . . Revisited* box toward the end of the chapter.

Earl Babbie

the living room rug. What a terrible thing to serve guests!

The point of the story is that *both* of your feelings about the appetizer were quite real. Your initial liking for them was certainly real, but so was the feeling you had when you found out what you'd been eating. It should be evident, however, that the disgust you felt was strictly a product of the agreements you have with those around you that worms aren't fit to eat. That's an agreement you began the first time your parents found you sitting in a pile of dirt with half of a wriggling worm dangling from your lips. When they pried your mouth open and reached down your throat for the other half of the worm, you learned that worms are not acceptable food in our society.

Aside from these agreements, what's wrong with worms? They're probably high in protein and low in calories. Bite-sized and easily packaged, they're a distributor's dream. They are also a delicacy for some people who live in

Earl Babbie

We learn some things by experience, others by agreement. This young man seems to be learning by personal experience.

societies that lack our agreement that worms are disgusting. Some people might love the worms but be turned off by the deep-fried breading.

Here's a question to consider: "Are worms *really* good or *really* bad to eat?" And here's a more interesting question: "*How could you know* which was really so?" This book is about answering the second question.

Looking for Reality

Reality is a tricky business. You've probably long suspected that some of the things you "know" may not be true, but how can you actually know what's real? People have grappled with this question for thousands of years.

Knowledge from Agreement Reality

One answer that has arisen out of that grappling is science, which offers an approach to both agreement reality and experiential reality. Scientists have certain criteria that must be met before they'll accept the reality of something they haven't personally experienced. In general, an assertion must have both *logical* and *empirical* support: It must make sense, and it must not contradict actual observation. Why do earthbound scientists accept the assertion that it's cold on the dark side of the moon (away from the sun)? First, it makes sense, because the surface heat of the moon comes from the sun's rays. Second, the scientific measurements made on the moon's dark side confirm the expectation. So, scientists accept the reality of things they don't personally experience—they accept an **agreement reality**—but they have special standards for doing so.

More to the point of this book, however, science offers a special approach to the discovery of reality through personal experience—that is, to the business of inquiry. **Epistemology** is the science of knowing; **methodology** (a subfield of epistemology) might be called the science of finding out. This book is an examination and presentation of social science methodology, or how social scientists find out about human social life. You'll see that some of the methods coincide with the traditional image of science but others have been specially geared to sociological concerns.

In the rest of this chapter, we'll look at inquiry as an activity. We'll begin by examining inquiry as a natural human activity, something you and I have engaged in every day of our lives. Next, we'll look at some kinds of errors we make in normal inquiry, and we'll conclude by examining what makes science different. We'll see some of the ways science guards against common human errors in inquiry.

"Issues and Insights: Social Research Making a Difference" gives an example of controlled social research challenging what "everybody knows."

Ordinary Human Inquiry

Practically all people exhibit a desire to predict their future circumstances. We seem quite willing, moreover, to undertake this task using *causal* and *probabilistic* reasoning. First, we generally recognize that future circumstances are somehow caused or conditioned by present ones. We learn that swimming beyond the reef may bring an unhappy encounter with a shark. As students we learn that studying hard will result in better grades. Second, we also learn that such patterns of cause and effect are *probabilistic* in nature: The effects occur more often when the causes occur than when the causes are absent—but not always. Thus, students learn that studying hard produces good grades in most instances, but not every time. We recognize the danger of swimming beyond the reef, without believing that every such swim will be fatal.

As we'll see throughout the book, science makes these concepts of causality and probability more explicit and provides techniques for dealing with them more rigorously than does casual human inquiry. It sharpens the skills we already have by making us more conscious, rigorous, and explicit in our inquiries.

In looking at ordinary human inquiry, we need to distinguish between prediction and understanding. Often, we can make predictions without understanding—perhaps you can predict rain when your trick knee aches. And often, even if we don't understand why, we're willing to act on the basis of a demonstrated predictive ability. The racetrack buff who finds that the third-ranked horse in the third race of the day always wins will probably keep betting without knowing, or caring, why it works out that way.

Whatever primitive drives or instincts motivate human beings, satisfying these urges depends heavily on the ability to predict future circumstances. However, the attempt to predict is often placed in a context of knowledge and understanding. If we can understand *why* things are related to one another, why certain regular patterns occur, we can predict even better than if we simply observe and remember those patterns. Thus, human inquiry aims at answering both "what" and "why" questions, and we pursue these goals by observing and figuring out.

As I suggested earlier, our attempts to learn about the world are only partly linked to direct, personal inquiry or experience. Another, much larger, part comes from the agreed-on knowledge that others give us. This agreement reality both assists and hinders our attempts to find out for ourselves. To see how, consider two important sources of our secondhand knowledge—tradition and authority.

---

agreement reality Those things we "know" as part and parcel of the culture we share with those around us.

epistemology The science of knowing; systems of knowledge.

methodology The science of finding out; procedures for scientific investigation.

Issues and Insights

Social Research Making a Difference

Medication errors in U.S. hospitals kill or injure about 770,000 patients each year, and the newly developed Computerized Physician Order Entry (CPOE) systems have been widely acclaimed as the solution to this enormous problem, which stems in part from the traditional system of using handwritten prescriptions.

Medical science research has generally supported the new technology, but an article in the *Journal of the American Medical Association* in March 2005 sent a shock wave through the medical community. The sociologist Ross Koppel and his colleagues used several of the research techniques you'll be learning in this book to test the effectiveness of the new technology. Their conclusion: CPOE was not nearly as effective as claimed; it did not prevent errors in medication (Koppel et al., 2005).

As you can imagine, those manufacturing and selling the equipment were not thrilled by the research, and it has generated an ongoing discussion within the health-care community. At last count, the study had been cited over 20,000 times in other articles, and Koppel has become a sought-after expert in this regard.

Source: Kathryn Goldman Schuyler, Medical Errors: Sociological Research Makes News, *Sociological Practice Newsletter* (American Sociological Association, Section on Sociological Practice), Winter 2006, p. 1.

Tradition

Each of us inherits a culture made up, in part, of firmly accepted knowledge about the workings of the world and the values that guide our participation in it. We may learn from others that eating too much candy will decay our teeth, that the circumference of a circle is approximately twenty-two sevenths of its diameter, or that masturbation will make you blind. Ideas about gender, race, religion, and different nations that you learned as you were growing up would fit in this category. We may test a few of these "truths" on our own, but we simply accept the great majority of them, the things that "everybody knows."

Tradition, in this sense of the term, offers some clear advantages to human inquiry. By accepting what everybody knows, we avoid the overwhelming task of starting from scratch in our search for regularities and understanding. Knowledge is cumulative, and an inherited body of knowledge is the jumping-off point for developing more of it. We often speak of "standing on the shoulders of giants"—that is, starting with the knowledge base of previous generations.

At the same time, tradition may be detrimental to human inquiry. If we seek a fresh understanding of something that everybody already understands and has always understood, we may be marked as fools for our efforts. More to the point, however, most of us rarely even think of seeking a different understanding of something we all "know" to be true.

Authority

Despite the power of tradition, new knowledge appears every day. Aside from our personal inquiries, we benefit throughout life from new discoveries and understandings produced by others. Often, acceptance of these new acquisitions depends on the status of the discoverer. You're more likely to believe the epidemiologist who declares that the common cold can be transmitted through kissing, for example, than to believe your Uncle Pete saying the same thing.

Like tradition, authority can both assist and hinder human inquiry. We do well to trust the judgment of the person who has special training, expertise, and credentials in a given matter, especially in the face of controversy. At the same time, inquiry can be greatly hindered by a legitimate authority who errs within his or her own special province. Biologists, after all, do make mistakes in the field of biology.

Inquiry is also hindered when we depend on the authority of experts speaking outside their realm of expertise. For example, consider the political or religious leader with no biochemical expertise who declares that marijuana is a dangerous drug. The advertising industry plays heavily on this misuse of authority by, for example, having popular athletes discuss the nutritional value of breakfast cereals or movie actors evaluate the performance of automobiles.

Both tradition and authority, then, are double-edged swords in the search for knowledge about the world. Simply put, they provide us with a starting point for our own inquiry, but they can lead us to start at the wrong point and can push us off in the wrong direction.

Errors in Inquiry and Some Solutions

Quite aside from the potential dangers of tradition and authority, we often stumble and fall when we set out to learn for ourselves. Let's look at some of the common errors we make in our casual inquiries and the ways science guards against those errors.

Inaccurate Observations

Quite frequently, we make mistakes in our observations. For example, what was your methodology instructor wearing on the first day of class? If you have to guess, that's because most of our daily observations are casual and semiconscious. That's why we often disagree about "what really happened."

In contrast to casual human inquiry, scientific observation is a conscious activity. Simply making observation more deliberate can reduce error. If you had to guess what your instructor was wearing the first day of class, you'd probably make a mistake. If you had gone to the first class meeting with a conscious plan to observe and record what your instructor was wearing, however, you'd likely be more accurate. (You might also need a hobby.)

In many cases, both simple and complex measurement devices help guard against inaccurate observations. Moreover, they add a degree of precision well beyond the capacity of the unassisted human senses. Suppose, for example, that you had taken color photographs of your instructor that day. (See earlier comment about needing a hobby.)

Overgeneralization

When we look for patterns among the specific things we observe around us, we often assume that a few similar events are evidence of a general pattern. That is, we tend to overgeneralize on the basis of limited observations. This can misdirect or impede inquiry.

Imagine that you're a reporter covering an animal-rights demonstration. You have just two hours to turn in your story. Rushing to the scene, you start interviewing people, asking them why

they're demonstrating. If the first two demonstrators you interview give you essentially the same reason, you might simply assume that the other 3,000 would agree. Unfortunately, when your story appears, your editor could get scores of letters from protesters who were there for an entirely different reason.

Realize, of course, that we must generalize to some extent in order to survive. It's probably not a good idea to keep asking whether *this* rattlesnake is poisonous. Assume they all are. At the same time, we have a tendency to overgeneralize.

Scientists guard against overgeneralization by seeking a sufficiently large sample of observations. The **replication** of inquiry provides another safeguard. Basically, this means repeating a study and checking to see if the same results occur each time. Then, as a further test, the study can be repeated under slightly varied conditions.

Selective Observation

One danger of overgeneralization is that it can lead to selective observation. Once you have concluded that a particular pattern exists and have developed a general understanding of why it does, you'll tend to focus on future events and situations that fit the pattern, and you'll ignore those that don't. Racial and ethnic prejudices depend heavily on selective observation for their persistence.

In another example, here's how Lewis Hill recalls growing up in rural Vermont:

> Haying began right after the Fourth of July. The farmers in our neighborhood believed that anyone who started earlier was sure to suffer all the storms of late June in addition to those following the holiday which the old-timers said were caused by all the noise and smoke of gunpowder burning. My mother told me that my grandfather and other Civil War veterans claimed it always rained hard after a big battle. Things didn't always work out the way the older residents promised, of course, but everyone remembered only the times they did.
>
> (Hill, 2000: 35)

Sometimes a research design will specify in advance the number and kind of observations to be made, as a basis for reaching a conclusion. If you

---

replication Repeating an experiment to expose or reduce error.

and I wanted to learn whether women were more likely than men to support the legality of abortion, we'd commit ourselves to making a specified number of observations on that question in a research project. We might select a thousand people to be interviewed on the issue. Alternatively, when making direct observations of an event, such as an animal-rights demonstration, social scientists make a special effort to find "deviant cases"—those who do not fit into the general pattern.

Illogical Reasoning

There are other ways in which we often deal with observations that contradict our understanding of the way things are in daily life. Surely one of the most remarkable creations of the human mind is "the exception that proves the rule." That idea doesn't make any sense at all. An exception can draw attention to a rule or to a supposed rule (in its original meaning, "prove" meant "test"), but in no system of logic can it validate the rule it contradicts. Even so, we often use this pithy saying to brush away contradictions with a simple stroke of illogic. This is particularly common in relation to group stereotypes. When a person of color, a woman, or a gay violates the stereotype someone holds for that group, it somehow "proves" that, aside from this one exception, the stereotype remains "valid" for all the rest. For example, a woman business executive who is kind and feminine is taken as "proof" that all other female executives are mean and masculine.

What statisticians have called the *gambler's fallacy* is another illustration of illogic in day-to-day reasoning. A consistent run of either good or bad luck is presumed to foreshadow its opposite. An evening of bad luck at poker may kindle the belief that a winning hand is just around the corner; many a poker player has stayed in a game much too long because of that mistaken belief. (A more reasonable conclusion is that they are not very good at poker.)

Although all of us sometimes fall into embarrassingly illogical reasoning in daily life, scientists avoid this pitfall by using systems of logic consciously and explicitly. Chapter 2 will examine the logic of science in more depth. For now, it's enough to note that logical reasoning is a conscious activity for scientists, who have colleagues around to keep them honest.

Science, then, attempts to protect us from the common pitfalls of ordinary inquiry. Accurately observing and understanding reality is not an obvious or trivial matter, as we'll see throughout this chapter and this book.

Before moving on, I should caution you that scientific understandings of things are also constantly changing. Any review of the history of science will provide numerous examples of old "knowledge" being supplanted by new "knowledge." It's easy to feel superior to the scientists of a hundred or a thousand years ago, but I fear there is a tendency to think those changes are all behind us. Now, we know the way things are.

In *The Half-Life of Facts* (2012), Samuel Arbesman addresses the question of how long today's scientific "facts" survive reconceptualization, retesting, and new discoveries. For example, half of what medical science knew about hepatitis and cirrhosis of the liver was replaced in 45 years.

The fact that scientific knowledge is constantly changing actually points to a strength of scientific scholarship. Whereas cultural beliefs and superstitions may survive unchallenged for centuries, scientists are committed to achieving an ever better understanding of the world. My purpose in this book is to prepare you to join that undertaking.

The Foundations of Social Science

The two pillars of science are logic and observation. A scientific understanding of the world must (1) make sense and (2) correspond with what we observe. Both elements are essential to science and relate to three major aspects of the overall scientific enterprise: theory, data collection, and data analysis.

In the most general terms, scientific theory deals with logic, data collection with observation, and data analysis with patterns in what is observed and, where appropriate, the comparison of what is logically expected with what is actually observed. Though most of this textbook deals with data collection and data analysis—demonstrating how to conduct empirical research—recognize that social *science* involves all three elements. As such, Chapter 2 of this book concerns the theoretical context of research; Parts 2 and 3 focus on data collection; and Part 4 offers an introduction to the analysis of data. Figure 1-1 offers a schematic view of how this book addresses these three aspects of social science.

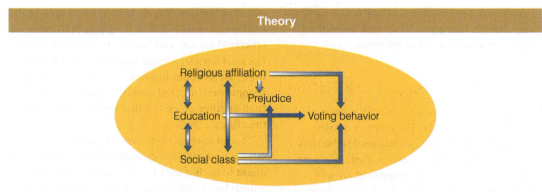

Chapters 2–3

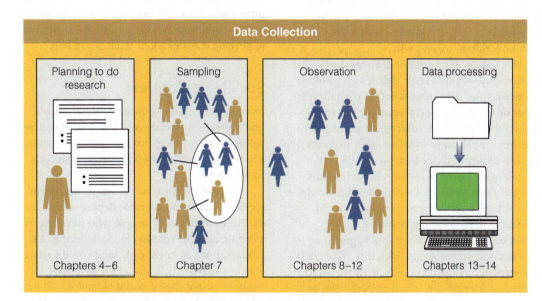

Part 4

FIGURE 1-1

Social Science = Theory + Data Collection + Data Analysis. This figure offers a schematic overview of the major stages of social research, indicating where each is discussed in this book.

Let's turn now to some of the fundamental issues that distinguish social science from other ways of looking at social phenomena.

Theory, Not Philosophy or Belief

Social science theory has to do with what is, not with what *should* be. For many centuries, however, social theory has combined these two orientations. Social philosophers liberally mixed their observations of what happened around them, their speculations about why, and their *ideas* about how things ought to be. Although modern social scientists may do the same from time to time, realize that social *science* has to do with how things are and why.

This means that scientific **theory**—and science itself—cannot settle debates on value. Science cannot determine whether capitalism is better or worse than socialism except in terms of agreed-on criteria. To determine scientifically whether capitalism or socialism most supports human dignity and freedom, we would first have to agree on some measurable definitions of dignity and freedom. Our conclusions would depend totally on this agreement and would have no general meaning beyond it.

By the same token, if we could agree that suicide rates, say, or giving to charity were good measures of a religion's quality, then we could determine scientifically whether Buddhism or Christianity is the better religion. Again, our conclusion would be inextricably tied to the given criterion. As a practical matter, people seldom agree on criteria for determining issues of value, so science is seldom useful in settling such debates. In fact, questions like these are so much a matter of opinion and belief that scientific inquiry is often viewed as a threat to what is "already known."

We'll consider this issue in more detail in Chapter 12, when we look at evaluation research. As you'll see, social scientists have become increasingly involved in studying programs that reflect ideological points of view, such as affirmative action or welfare reform. One of the biggest problems researchers face is getting people to agree on criteria of success and failure. Yet such criteria are essential if social science research is to tell us anything useful about matters of value. By analogy, a stopwatch can't tell us if one sprinter is better than another unless we first agree that speed is the critical criterion.

Social science, then, can help us know only what is and why. We can use it to determine what ought to be, but only when people agree on the criteria for deciding what's better than something else—an agreement that seldom occurs. With that understood, let's turn now to some of the fundamental bases upon which social science allows us to develop theories about what is and why.

Social Regularities

In large part, social science theory aims to find patterns in social life. That aim, of course, applies to all science, but it sometimes presents a barrier to people when they first approach social science.

Actually, the vast number of formal norms in society create a considerable degree of regularity. For example, only people who have reached a certain age can vote in elections. In the U.S. military, until recently, only men could participate in combat. Such formal prescriptions, then, regulate, or regularize, social behavior.

Aside from formal prescriptions, we can observe other social norms that create more regularities. Republicans are more likely than Democrats to vote for Republican candidates. University professors tend to earn more money than do unskilled laborers. Men earn more than do women. (We'll look at this pattern in more depth later in the book.) The list of regularities could go on and on.

Three objections are sometimes raised in regard to such social regularities. First, some of the regularities may seem trivial. For example, Republicans vote for Republicans; everyone knows that. Second, contradictory cases may be cited, indicating that the "regularity" isn't totally regular. Some laborers make more money than some professors do. Third, it may be argued that the people involved in the regularity could upset the whole thing if they wanted to.

Let's deal with each of these objections in turn.

theory A systematic explanation for the observations that relate to a particular aspect of life: juvenile delinquency, for example, or perhaps social stratification or political revolution.

The Charge of Triviality

During World War II, Samuel Stouffer, one of the greatest social science researchers, organized a research branch in the U.S. Army to conduct studies in support of the war effort (Stouffer et al. 1949–1950). Many of the studies focused on the morale among soldiers. Stouffer and his colleagues found that there was a great deal of "common wisdom" regarding the bases of military morale. Much of the research undertaken by this organization was devoted to testing these "obvious" truths.

For example, people had long recognized that promotions affect morale in the military. When military personnel get promotions and the promotion system seems fair, morale rises. Moreover, it makes sense that people who are getting promoted will tend to think the system is fair, whereas those passed over will likely think the system is unfair. By extension, it seems sensible that soldiers in units with slow promotion rates will tend to think the system is unfair, and those in units with rapid rates will think the system is fair. But was this the way they really felt?

Stouffer and his colleagues focused their studies on two units: the Military Police (MPs), which had the slowest promotion rate in the Army, and the Army Air Corps (forerunner of the U.S. Air Force), which had the fastest promotion rate. It stood to reason that MPs would say the promotion system was unfair and that the air corpsmen would say it was fair. The studies, however, showed just the opposite.

Notice the dilemma faced by a researcher in a situation such as this. On the one hand, the observations don't seem to make sense. On the other hand, an explanation that makes obvious good sense isn't supported by the facts.

A lesser scientist would have set the problem aside "for further study." Stouffer, however, looked for an explanation for his observations, and eventually he found it. Robert Merton, Alice Kitt (1950), and other sociologists at Columbia University had begun thinking and writing about something they called *reference group theory*. This theory says that people judge their lot in life less by objective conditions than by comparing themselves with others around them—their reference group. For example, if you lived among poor people, a salary of $50,000 a year would make you feel like a millionaire. But if you lived among people who earned $500,000 a year, that same $50,000 salary would make you feel impoverished.

Stouffer applied this line of reasoning to the soldiers he had studied. Even if a particular MP had not been promoted for a long time, it was unlikely that he knew some less-deserving person who had gotten promoted more quickly. Nobody got promoted in the MPs. Had he been in the Air Corps—even if he had gotten several promotions in rapid succession—he would probably have been able to point to someone less deserving who had gotten even faster promotions. An MP's reference group, then, was his fellow MPs, and the air corpsman compared himself with fellow corpsmen. Ultimately, then, Stouffer reached an understanding of soldiers' attitudes toward the promotion system that (1) made sense and (2) corresponded to the facts.

This story shows that documenting the obvious is a valuable function of any science, physical or social. Charles Darwin coined the phrase *fool's experiment* to describe much of his own research—research in which he tested things that everyone else "already knew." As Darwin understood, the obvious all too often turns out to be wrong; thus, apparent triviality is not a legitimate objection to any scientific endeavor.

What about Exceptions?

The objection that there are always exceptions to any social regularity does not mean that the regularity itself is unreal or unimportant. A particular woman may well earn more money than most men, but that provides small consolation to the majority of women, who earn less—the pattern still exists. Social regularities, in other words, are probabilistic patterns, and they are no less real simply because some cases don't fit the general pattern.

This point applies in physical science as well as social science. Subatomic physics, for example, is a science of probabilities. In genetics, the mating of a blue-eyed person with a brown-eyed person will probably result in a brown-eyed offspring. The birth of a blue-eyed child does not destroy the observed regularity, because the geneticist states only that a brown-eyed offspring is more likely and, further, that brown-eyed offspring will be born in a certain percentage of the cases. The social scientist makes a similar, probabilistic prediction—that women overall are likely to earn less than men. Once a pattern like this is observed, the social scientist has grounds for asking why it exists.

People Could Interfere

Finally, the objection that the conscious will of the actors could upset observed social regularities does not pose a serious challenge to social science. This is true even though a parallel situation does not appear to exist in the physical sciences. (Presumably, physical objects cannot violate the laws of physics, although the probabilistic nature of subatomic physics once led some observers to postulate that electrons had free will.) There is no denying that a religious, right-wing bigot could go to the polls and vote for an agnostic, left-wing African American if he wanted to upset political scientists studying the election. All voters in an election could suddenly switch to the underdog just to frustrate the pollsters. Similarly, workers could go to work early or stay home from work and thereby prevent the expected rush-hour traffic. But these things do not happen often enough to seriously threaten the observation of social regularities.

Social regularities, then, do exist, and social scientists can detect them and observe their effects. When these regularities change over time, social scientists can observe and explain those changes.

There is a slightly different form of human interference that makes social research particularly challenging. Social research has a *recursive* quality, in that what we learn about society can end up changing things so that what we learned is no longer true. For example, every now and then you may come across a study reporting "The Ten Best Places to Live," or something like that. The touted communities aren't too crowded, yet they have all the stores you'd ever want; the schools and other public facilities are great, crime is low, the ratio of doctors per capita is high, and the list goes on. What happens when this information is publicized? People move there, the towns become overcrowded, and eventually, they are not such nice places to live. More simply, imagine what results from a study that culminates in a published list of the least-crowded beaches or fishing spots.

In 2001, the Enron Corporation was fast approaching bankruptcy and some of its top executives were quietly selling their shares in the company. During this period, those very executives were reassuring employees of the corporation's financial solvency and recommending that workers keep their own retirement funds invested in the company. As a consequence of this deception, those employees lost most of their retirement funds at the same time that they were becoming unemployed.

The events at Enron led two Stanford business-school faculty, David Larcker and Anastasia Zakolyukina (2010), to see if it would be possible to detect when business executives are lying. Their study analyzed tens of thousands of conference-call transcripts, identified instances of executives fibbing, and looked for speech patterns associated with those departures from the truth. For example, Larcker and Zakolyukina found that when the executives lied, they tended to use exaggerated emotions, for instance, calling business prospects "fantastic" instead of "good." The research found other tip-offs that executives were lying, such as fewer references to shareholders and fewer references to themselves. Given the type of information derived from this study—uncovering identifiable characteristics of lying—who do you suppose will profit most from it? Probably the findings will benefit business executives and those people who coach them on how to communicate. There is every reason to believe that a follow-up study of top executives in, say, ten years will find very different speech patterns from those used today.

Aggregates, Not Individuals

Social regularities do exist, then, and are worthy of theoretical and empirical study. As such, social scientists study primarily social patterns rather than individual ones. These patterns reflect the *aggregate* or collective actions and situations of many individuals. Although social scientists often study motivations and actions that affect individuals, they seldom study the individual per se. That is, they create theories about the nature of group, rather than individual, life. Whereas psychologists focus on what happens *inside* individuals, social scientists study what goes on *between* them: examining everything from couples, to small groups and organizations, on up to whole societies—and even interactions between societies.

Sometimes the collective regularities are amazing. Consider the birth rate, for example. People have babies for an incredibly wide range of personal reasons. Some do it because their parents want them to. Some think of it as a way of completing their womanhood or manhood. Others want to hold their marriages together. Still others have babies by accident.

If you have had a baby, you could probably tell a much more detailed, idiosyncratic story. Why did you have the baby when you did, rather than a year earlier or later? Maybe your house burned down and you had to delay a year before you could afford to have the baby. Maybe you felt that being a family person would demonstrate maturity, which would support a promotion at work.

Everyone who had a baby last year had a different set of reasons for doing so. Yet, despite this vast diversity, despite the idiosyncrasy of each individual's reasons, the General Fertility Rate in a society (the number of live births per 1,000 women 15 to 50 years of age) is remarkably consistent from year to year. See Table 1-1 for some fertility rates in the United States.

If the U.S. fertility rates were 30, 20, 70, 55, and 80 in five successive years, demographers would begin dropping like flies. As you can see, however, social life is far more orderly than that. Moreover, this regularity occurs without society-wide regulation. As mentioned earlier, no one plans how many babies will be born or determines who will have them. (See

TABLE 1-1

Fertility Rates in the United States: 2006–2013

Year	Fertility Rate per 1,000 Women Ages 15–50
2006	54.9
2007	55.0
2008	58.5
2009	57.0
2010	54.6
2011	54.0
2012	54.1
2013	51.6

Source: U.S. Bureau of the Census, Historical Table 3. Births in the past year per 1,000 women, by Age: ACS, 2006–2013 [XLSX], accessed July 15, 2016, at http://www.census.gov/hhes/fertility/data/cps/historical.html.

"Applying Concepts in Everyday Life: Fertility-Rate Implications" for a look at how the analysis of fertility rates can serve many purposes.)

Social science theories try to explain why aggregated patterns of behavior are so regular, even when the individuals participating in them may change over time. We could say that social scientists don't seek to explain people per se. They try instead to understand the *systems* in which people operate, which in turn explain why people do what they do. The elements in such a system are not people but *variables*.

Concepts and Variables

Our most natural attempts at understanding are usually concrete and idiosyncratic. That's just the way we think.

Imagine that someone says to you, "Women ought to get back into the kitchen where they belong." You're likely to hear that comment in terms of what you know about the speaker. If it's your old Uncle Harry who is also strongly opposed to daylight saving time, ZIP Codes, and personal computers, you're likely to think that his latest pronouncement simply fits into his rather dated point of view about things in general.

If, on the other hand, the statement issues forth from a politician who is trailing a female challenger and who has also begun making statements about women being emotionally unfit for public office and not understanding politics, you may hear his latest comment in the context of this political challenge.

In both examples, you're trying to understand the thoughts of a particular individual. In social science, researchers go beyond that level of understanding to seek insights into classes or types of individuals. Regarding the two examples just described, they might use terms such as *old-fashioned* or *bigot* to describe the kind of person who made the comment. In other words, they try to place the individual in a set of similar individuals, according to a particular, defined concept.

By examining an individual in this way, social scientists can make sense out of more than one person. In understanding what makes the bigoted politician think the way he does, they'll also learn about other people who are "like him." In other words, they have not been studying bigots as much as *bigotry*.

Bigotry here is spoken of as a *variable* because it varies. Some people are more bigoted than others.

Applying Concepts in Everyday Life

Fertility-Rate Implications

Take a minute to reflect on the practical implications of the data you've just seen. The *What Do You Think?* box for this chapter asked how baby-food and diaper manufacturers could plan production from year to year. The consistency of U.S. fertility rates suggests that this is not the problem it might have seemed.

Who else might benefit from this kind of analysis? What about health-care workers and educators? Can you think of anyone else?

What if we analyzed fertility rates by region of the country, by ethnicity, by income level, and so forth? Clearly, these additional analyses could make the data even more useful. As you learn about the options available to social researchers, I think you'll gain an appreciation for the practical value that research can have for the whole society.

Social scientists are interested in understanding the system of variables that causes bigotry to be high in one instance and low in another.

The idea of a system composed of variables may seem rather strange, so let's look at an analogy. The subject of a physician's attention is the patient. If the patient is ill, the physician's purpose is to help that patient get well. By contrast, a medical researcher's subject matter is different: the variables that cause a disease, for example. The medical researcher may study the physician's patient, but only as a carrier of the disease.

Of course, medical researchers care about real people, but in the actual research, patients are directly relevant only for what they reveal about the disease under study. In fact, when researchers can study a disease meaningfully without involving actual patients, they do so.

Social research involves the study of variables and the *attributes* that compose them. Social science theories are written in a language of variables, and people become involved only as "carriers" of those variables. Here's a closer look at what social scientists mean by variables and attributes.

Attributes, or values, are characteristics or qualities that describe an object—in this case, a person. Examples include female, Asian, alienated, conservative, dishonest, intelligent, and farmer. Anything you might say to describe yourself or someone else involves an attribute.

Variables, on the other hand, are logical sets of attributes. The variable *occupation* is composed of attributes such as farmer, professor, and truck driver. *Social class* is a variable composed of a set of attributes such as upper class, middle class, and lower class. Sometimes it helps to think of attributes as the categories that make up a variable. See Figure 1-2 for a schematic review of what social scientists mean by variables and attributes.

Sex and *gender* are examples of variables. These two variables are not synonymous, but distinguishing them can be complicated. I will try to simplify the matter here and abide by that distinction throughout this book.

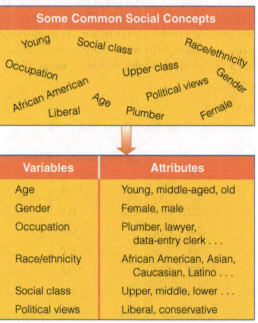

Some Common Social Concepts

Young Social class Race/ethnicity
Occupation Upper class Gender
African American Age Political views
Liberal Plumber Female

Variables	Attributes
Age	Young, middle-aged, old
Gender	Female, male
Occupation	Plumber, lawyer, data-entry clerk . . .
Race/ethnicity	African American, Asian, Caucasian, Latino . . .
Social class	Upper, middle, lower . . .
Political views	Liberal, conservative

FIGURE 1-2

Variables and Attributes. In social research and theory, both variables and attributes represent social concepts. Variables are sets of related attributes (categories, values).

---

attribute A characteristic of a person or a thing.

variable A logical set of attributes. The variable *sex* is made up of the attributes *male* and *female*.

Most simply put, *sex* refers to biological/ physiological differences, and the attributes comprising this variable are *male* and *female*, *men* and *women*, or *boys* and *girls*.

Gender, on the other hand, is a social distinction, referring to what is generally expected of men and women. Notice that these "general expectations" can vary from culture to culture and over time. Note also that some men will exhibit feminine behaviors and characteristics, while some women will exhibit masculine behaviors and characteristics. One set of attributes comprising gender is *masculine* and *feminine*.

However, the real complication comes when women as a class are treated differently from men as a class, but not because of their physical differences. A good example is gender discrimination in income. As we'll see later in this book, American women overall earn less than men, even when they do the same job and have the same credentials. It has nothing to do with being feminine or masculine, but it is not logically based on their different plumbing, either. The pattern of differential pay for women and men is based, instead, on established social patterns regarding women and men. Traditionally in America, for example, men have been the main breadwinners for their family, whereas women typically worked outside the home to provide the family with some supplemental income. Even though this work pattern has changed a good deal, and women's earnings are often an essential share of the family income, the pattern of monetary compensation—that of men earning more than women—has been slower to change.

Thus, we shall use the term *sex* whenever the distinction between men and women is relevant to biological differences. For example, there is a correlation between sex and height in that men are, on average, taller than women. This is not a social distinction but a physiological one. Most of the times we distinguish men and women in this book, however, will be in reference to social distinctions, such as the example of women being paid less than men or women being underrepresented in elected political offices. In those cases, we shall use the term *gender*. The attributes *men* and *women* will often be used for both *sex* and *gender*.

The relationship between attributes and variables lies at the heart of both description and explanation in science. For example, we might describe a college class in terms of the variable *sex* by reporting the observed frequencies of the attributes *male* and *female*: "The class is 60 percent men and 40 percent women." An unemployment rate can be thought of as a description of the variable *employment status* of a labor force in terms of the attributes *employed* and *unemployed*. Even the report of annual family income for a city is a summary of attributes composing that variable: *$13,124*, *$30,980*, *$55,000*, and so forth. Sometimes the meanings of the concepts that lie behind social science concepts are fairly clear. Other times they aren't.

The relationship between attributes and variables is more complicated when we move from description to explanation and it gets to the heart of the variable language of scientific theory. Here's a simple example, involving two variables, *education* and *prejudice*. For the sake of simplicity, let's assume that the variable *education* has only two attributes: *educated* and *uneducated*. (Chapter 5 will address the issue of how such things are defined and measured.) Similarly, let's give the variable *prejudice* two attributes: *prejudiced* and *unprejudiced*.

Now let's suppose that 90 percent of the uneducated are prejudiced, and the other 10 percent are unprejudiced. And let's suppose that 30 percent of the educated people are prejudiced, and the other 70 percent are unprejudiced. This is illustrated graphically in Figure 1-3a.

Figure 1-3a illustrates a relationship or association between the variables *education* and *prejudice*. This relationship can be seen in terms of the pairings of attributes on the two variables. There are two predominant pairings: (1) those who are educated and unprejudiced and (2) those who are uneducated and prejudiced. Here are two other useful ways of viewing that relationship.

First, let's suppose that we play a game in which we bet on your ability to guess whether a person is prejudiced or unprejudiced. I'll pick the people one at a time (not telling you which ones I've picked), and you have to guess whether each person is prejudiced. We'll do it for all 20 people in Figure 1-3a. Your best strategy in this case would be to guess prejudiced each time, because 12 out of the 20 are categorized that way. Thus, you'll get 12 right and 8 wrong, for a net success of 4.

a. There is an apparent relationship between education and prejudice.

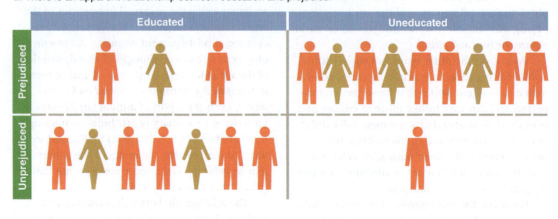

b. There is *no* apparent relationship between education and prejudice.

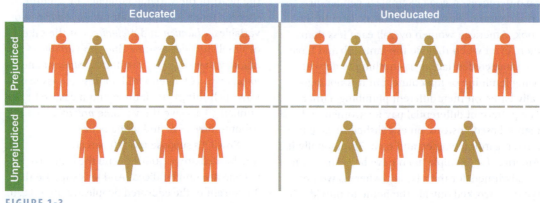

FIGURE 1-3

Illustration of Relationship between Two Variables (Two Possibilities). Variables such as *education* and *prejudice* and their attributes (*educated/uneducated, prejudiced/unprejudiced*) are the foundation for the examination of causal relationships in social research.

Now let's suppose that when I pick a person from the figure, I have to tell you whether the person is educated or uneducated. Your best strategy now would be to guess "prejudiced" for each uneducated person and "unprejudiced" for each educated person. If you follow that strategy, you'll get 16 right and 4 wrong. Your improvement in guessing "prejudiced" by knowing education illustrates what it means to say that variables are related.

Second, by contrast, let's consider how the 20 people would be distributed if education and prejudice were unrelated to each other. This is illustrated in Figure 1-3b. Notice that half the people are educated, and half are uneducated. Also notice that 12 of the 20 (60 percent) are prejudiced. Given that 6 of the 10 people in each group are prejudiced, we conclude that the two variables are unrelated to each other. Knowing a person's education

would not be of any value to you in guessing whether that person was prejudiced.

We'll be looking at the nature of relationships among variables in some depth in Part 4 of this book. In particular, we'll see some of the ways relationships can be discovered and interpreted in research analysis. A general understanding of relationships now, however, will help you appreciate the logic of social science theories.

Theories describe the relationships we might logically expect among variables. Often, the expectation involves the idea of *causation*. A person's attributes on one variable are expected to cause, predispose, or encourage a particular attribute on another variable. In Figure 1-3a, something about being educated apparently leads people to be less prejudiced than if they are uneducated.

Applying Concepts in Everyday Life

Independent and Dependent Variables

Let's talk about dating. Some dates are great and some are awful; others are somewhere in between. So the *quality of dates* is a variable and "great," "OK," and "awful" might be the attributes making up that variable. (If dating isn't a relevant activity for you right now, perhaps you can pretend or substitute something similar.)

Now, have you noticed something that seems to affect the quality of different dates? (If you are not dating, perhaps you can recall prior dating or simply imagine it.) Perhaps it will have something to do with

the kind of person you dated, your activities on the date, something about your behavior, the amount of money spent, or the like. Can you give it a name that enables you to identify that factor as a variable (e.g., physical attractiveness, punctuality)? Can you identify a set of attributes comprising that variable?

Consider the *quality* or the *characteristics* of the dates: Which is the independent variable and which is the dependent variable? (When we get to Chapter 12, "Evaluation Research," you'll learn ways of determining whether the variable you identified really matters.)

As I'll further discuss later in the book, *education* and *prejudice* in this example would be regarded as **independent** and **dependent variables**, respectively. Because *prejudice* depends on something, we call it the dependent variable, which depends on an independent variable, in this case *education*. Although the educational levels of the people being studied vary, that variation is independent of prejudice.

Notice, at the same time, that educational variations can be found to depend on something else—such as the educational level of our subjects' parents. People whose parents have a lot of education are more likely to get a lot of education than are those whose parents have little education. In this relationship, the subject's education is the dependent variable and the parents' education the independent variable. We can say that the independent variable is the cause and the dependent variable the effect. (See "Applying Concepts in Everyday Life: Independent and Dependent Variables" for more.)

At this point, we can see that our discussion of Figure 1-3 involved the interpretation of data. We looked at the distribution of the 20 people in terms of the two variables. In constructing a social science theory, we would derive an expectation regarding the relationship between the two variables, based on what we know about each. We know, for example, that education exposes people to a wide range of cultural variation and to diverse points of view—in short, it broadens their perspectives. Prejudice, on the other hand, represents a narrower perspective. Logically, then, we might expect education and prejudice to be somewhat incompatible. We might therefore arrive at an expectation that increasing education would

reduce the occurrence of prejudice, an expectation that our observations would support.

Because Figure 1-3 has illustrated two possibilities—that education reduces the likelihood of prejudice or that it has no effect—you might be interested in knowing what is actually the case. There are, of course, many types of prejudice. For this illustration, let's consider prejudice against gays and lesbians. Over the years, the General Social Survey (GSS) has asked respondents whether a homosexual relationship between two adults is "always wrong, almost always wrong, sometimes wrong, or not wrong at all." In 2014, 40 percent of those interviewed said that homosexuality was always wrong. However, this response is strongly related to the respondents' education, as Table 1-2 indicates.

Notice that the theory has to do with the two variables *education* and *prejudice*, not with people as

---

independent variable A variable with values that are not problematical in an analysis but are taken as simply given. An independent variable is presumed to cause or determine a dependent variable. If we discover that religiosity is partly a function of sex—women are more religious than are men—sex is the independent variable and *religiosity* is the dependent variable. Note that any given variable might be treated as independent in one part of an analysis and as dependent in another part of it. *Religiosity* might become an independent variable in an explanation of crime rates.

dependent variable A variable assumed to depend on or be caused by another (independent variable). If you find that *income* is partly a function of *amount of formal education, income* is being treated as a dependent variable.

TABLE 1-2

Education and Antigay Prejudice

Level of Education	Percent Saying Homosexuality Is Always Wrong
Less than high school graduate	60
High school graduate	43
Junior college	34
Bachelor's degree	27
Graduate degree	27

such. People are the carriers of those two variables, so we can see the relationship between the variables only when we observe people. Ultimately, however, the theory uses a language of variables. It describes the associations that we might logically expect to exist between particular attributes of different variables. You can do this data analysis for yourself with nothing more than a connection to the Internet. See "How to Do It: Analyzing Data Online with the General Social Survey (GSS)."

The Purposes of Social Research

Chapter 4 will examine the various purposes of social research in some detail, but previewing them here will be useful. To begin, sometimes social research is a vehicle for *exploring* something—that is, mapping out a topic that may warrant further study later. This could involve looking into a new political or religious group, learning something about the use of a new street drug, and so forth. The methods vary greatly and the conclusions are usually suggestive rather than definitive. Still, careful exploratory social research can dispel some misconceptions and help focus future research.

Some social research is done for the purpose of *describing* the state of social affairs: What is the unemployment rate? What is the racial composition of a city? What percentage of the population holds a particular political view or plans to vote for a certain candidate? Careful empirical description takes the place of speculation and impressions.

Often, social research aims at *explaining* something—providing reasons for phenomena, in terms of causal relationships. Why do some cities have higher unemployment rates than others? Why are some people more prejudiced than others? Why are women likely to earn less than

men for doing the same job? Ordinary, everyday discourse offers an abundance of answers to such questions, but some of those answers are simply wrong. Explanatory social research provides reasons that are more trustworthy.

While some studies focus on one of these three purposes, a given study often has elements of all three. For example, when Kathleen A. Bogle (2008) undertook in-depth interviews of college students to study the phenomenon of "hooking up," she uncovered some aspects that might not have been expected, fulfilling an exploratory purpose. When two people "hook up," does that mean they have sex? Bogle found substantial ambiguities in that regard; some students felt that sex was part of the definition of that dating form, whereas others did not.

Her study also provides excellent descriptions of the students' various experiences of hooking up. While her in-depth interviews with 76 students at two universities in one region of the country do not allow us to draw quantitative conclusions about all college students in the United States, they provide an excellent qualitative description of the phenomenon—not just norms but wild variations as well. Not everyone will have interviewee Stephen's experience of his partner throwing up on him during sex, or having her call him Anthony instead of Stephen at a critical moment. (You'll learn more about the difference between "qualitative" and "quantitative" research later.)

Bogle's interviews also point to some of the causes, or explanations, of different kinds of hooking up. For example, the students' *beliefs* about their peers' behavior strongly influenced how they hooked up. Thus, it would be difficult to categorize this study as exploratory, descriptive, or explanatory, as it has elements of all three types.

It's worth noting here that the purpose of some research is limited to understanding, whereas other research efforts are deliberately intended to bring about social change, creating a more workable or a more just society.

The Ethics of Human Inquiry

Most of this book is devoted to the logic and techniques of doing social research, but you'll soon discover an ethical dimension running throughout the discussion. You'll learn that medical, social, and other studies of human beings have often used methods later condemned as unethical. In Chapter 3 and throughout the book, we examine the various concerns that distinguish ethical from unethical research.

The ethical concerns will make more sense to you as you learn more about the actual techniques of doing research. Be sure to consider this important issue as you read each chapter.

Some Dialectics of Social Research

There is no one way to do social research. (If there were, this would be a much shorter book.) In fact, much of the power and potential of social research lies in the many valid approaches it comprises.

Four broad and interrelated distinctions underlie these approaches. Though these distinctions can be seen as competing choices, a good social researcher thoroughly learns each. This is what I mean by the "dialectics" of social research: a fruitful tension between these complementary concepts.

Idiographic and Nomothetic Explanation

All of us go through life explaining things. We do it every day. You explain why you did poorly or well on an exam, why your favorite team is winning or losing, why you may be having trouble getting dates. In our everyday explanations, we engage in two distinct forms of causal reasoning, though we do not ordinarily distinguish them.

Sometimes we attempt to explain a single situation in idiosyncratic detail. Thus, for example, you may have done poorly on an exam because (1) you had forgotten there was an exam that day, (2) it was in your worst subject, (3) a traffic jam made you late for class, (4) your roommate had kept you up the night before the exam with loud music, (5) the police kept you until dawn demanding to know what you had done with your roommate's stereo—and with your roommate, for that matter, and (6) a band of coyotes ate your textbook. Given all these circumstances, it is no wonder that you did poorly.

This type of causal reasoning is called an **idiographic** explanation. *Idio* in this context means unique, separate, peculiar, or distinct, as in the word *idiosyncrasy*. When we have completed an idiographic explanation, we feel that we fully understand the causes of what happened in this particular instance. At the same time, the scope of our explanation is limited to the case at hand. Although parts of the idiographic explanation might apply to other situations, our intention is to explain one case fully.

Now consider a different kind of explanation. Every time you study with a group, you do better on an exam than when you study alone. Your favorite team does better at home than on the road. Athletes get more dates than do members of the biology club. Notice that this type of explanation is more general, covering a wider range of experience or observation. It speaks implicitly of the relationship between variables: for example, (1) whether or not you study in a group and (2) how well you do on the exam. This type of explanation—labeled **nomothetic**—seeks to explain a class of situations or events rather than a single one. Moreover, it seeks to explain "economically," using only one or just a few explanatory factors. Finally, it settles for a partial rather than a full explanation.

In each of these examples, you might qualify your causal statements with *on the whole, usually, all else being equal,* and the like. Thus, you usually do better on exams when you've studied in a group, but not always. Similarly, your team has won some games on the road and lost some at home. And the gorgeous head of the biology club may get lots of dates, while the defensive lineman Pigpen-the-Terminator may spend a lot of Saturday nights alone punching heavy farm equipment. Such exceptions are acceptable within a broader range of overall explanation.

---

idiographic An approach to explanation in which we seek to exhaust the idiosyncratic causes of a particular condition or event. Imagine trying to list all the reasons why you chose to attend your particular college. Given all those reasons, it's difficult to imagine your making any other choice.

nomothetic An approach to explanation in which we seek to identify a few causal factors that generally impact a class of conditions or events. Imagine the two or three key factors that determine which colleges students choose, such as proximity, reputation, and so forth.

How to Do It

Analyzing Data Online with the General Social Survey (GSS)

You can test the relationship between prejudice and education for yourself if you have a connection to the Internet. We'll come back to this method for analyzing data later, in Chapter 14, but here's a quick peek in case you're interested.

If you go to http://sda.berkeley.edu/sdaweb/analysis/?dataset=gss14, you will find yourself at a web page like the one that follows. As you can see, the page is divided into two sections: a column listing variables on the left, and a form containing a variety of filters, options, and fields on the right. I've indicated how you would work your way into the hierarchical list of variables to locate questionnaire items dealing with attitudes about homosexuality. For this example, I've selected HOMOSEX.

In the form on the right, I've indicated that we want to analyze differences in attitudes for different educational levels, measured in this case by the variable called "DEGREE." By typing YEAR(2012) into the Selection Filter field, I've indicated that we want to do this analysis using the GSS survey conducted in 2012.

If you are interested in trying this yourself, fill out the form as I have done. Then, click the button marked "Run the Table" at the bottom of the form, and you'll get a colorful table with the results. Once you've done that, try substituting other variables you might be interested in. Or see if the relationship between HOMOSEX and DEGREE was pretty much the same in, say, 1996.

The National Opinion Research Center (NORC) at the University of Chicago conducts a periodic national survey of American public opinion for the purpose of making such data available for analysis by the social research community. This comprehensive project is called the General Social Survey (GSS).

Beginning in 1972, large national samples were surveyed annually in face-to-face interviews; that frequency was reduced to every other year starting in 1994. Though conducted less often, the GSS interviews are lengthy and each takes over an hour to complete, making it possible to obtain a wide range of information about the demography and the opinions of the American population. The number of topics covered in a given survey is further increased by presenting different questions to different subsets of the overall sample. In the successive surveys, some questions are always asked while others are repeated only from time to time. Thus, it is possible to track changes in things such as political orientations, attendance at religious services, or attitudes toward abortion.

The GSS is a powerful resource for social scientists, since everyone from undergraduates through faculty members has access to a vast data set that would otherwise be available to only a few. In the early years of the GSS, data were made available to the research community by mailing physical data sets (cards or tapes) to researchers. Many data examples in this book come from that source. You can learn more about the GSS at the official website maintained by the University of Michigan.

As we noted earlier, patterns are real and important even when they are not perfect.

Both the idiographic and the nomothetic approaches to understanding can serve you in your daily life. The nomothetic patterns you discover might offer a good guide for planning your study habits, but the idiographic explanation is more convincing to your parole officer.

By the same token, both idiographic and nomothetic reasoning are powerful tools for social research. Researchers who seek an exhaustive understanding of the inner workings of a particular juvenile gang or the corporate leadership of a particular multinational conglomerate engage in idiographic research: They try to understand that particular group as fully as possible.

A. Libin and J. Cohen-Mansfield (2000) have contrasted the way these two approaches are used in studies of the elderly (gerontology). Some studies focus on the experiences of individuals in the totality of their life situations, whereas other studies look for statistical patterns describing the elderly in general. The authors then suggest ways to combine idiographic and nomothetic approaches in gerontology.

SDA 4.0   Selected Study: **GSS 1972-2012 Cumulative Datafile**

Analysis Create Variables Download Custom Subset Search   Standard Codebook   Codebook by Year of Interview

Variable Selection

Selected: [   ]   [View]

Copy to: Row   Col   Ctrl   Filter

Mode: ○ Append ● Replace

| Tables | Means | Correl. matrix | Comp. correl. | Regression | Logit/Probit | List values |

SDA Frequencies/Crosstabulation Program
Help: General / Recoding Variables

Row: HOMOSEX   (Required)
Column: DEGREE
Control: [   ]
Selection Filter(s): YEAR(2012)
Weight: COMPWT - Composite weight: WTSSALL * OVERSAMP * FC ▼

▼ Output Options

Cell contents:
Percentaging: ☑ Column ☐ Row ☐ Total

Sample design: ● Complex ○ SRS

☐ Confidence intervals - Level: 95 percent ◊
☐ Standard error of each percent
☐ Design effect (deft) for each percent   ☐ Z-statistic
☐ Unweighted N   ☑ Weighted N

Other options:
☐ Summary statistics   ☐ Question text   ☑ Color coding
☐ Suppress table   ☐ Include missing-data values

Title: [   ]

▶ Chart Options

▶ Decimal Options

[Run the Table] [Clear Fields]

Variable tree (left panel):
- ▶ 🖿 CASE IDENTIFICATION AND YEAR
- ▶ 🖿 RESPONDENT BACKGROUND VARIABLES
- ▶ 🖿 PERSONAL AND FAMILY INFORMATION
- ▶ 🖿 ATTITUDINAL MEASURES - NATIONAL PROBLEMS
- ▶ 🖿 PERSONAL CONCERNS
- ▶ 🖿 SOCIETAL CONCERNS
- ▶ 🖿 WORKPLACE AND ECONOMIC CONCERNS
- ▼ 🖿 CONTROVERSIAL SOCIAL ISSUES
  - ▶ 🖿 Gender Issues
  - ▶ 🖿 Abortion
  - ▼ 🖿 Family Planning, Sex, and Contraception
    - CHLDIDEL - IDEAL NUMBER OF CHILDREN
    - CHLDMORE - EXPECT MORE CHILDREN
    - CHLDNUM - HOW MANY CHILDREN EXPECTED
    - CHLDSOON - CHILDREN EXPECTED IN 5 YEAR
    - PILL - BIRTH CONTROL INFORMATION
    - TEENPILL - BIRTH CONTROL INFORMATION TO
    - PILLOKY - BIRTH CONTROL TO TEENAGERS 14
    - PILLOK - BIRTH CONTROL TO TEENAGERS 14-
    - SEXEDUC - SEX EDUCATION IN PUBLIC SCHO
    - DIVLAW - DIVORCE LAWS
    - DIVLAWY - DIVORCE LAWS-VERSION Y
    - SPDUE - EVER ENTITLED TO ALIMONY OR CH
    - SPPAID - REGULARLY RECEIVED ALIMONY - C
    - PREMARSX - SEX BEFORE MARRIAGE
    - TEENSEX - SEX BEFORE MARRIAGE -- TEENS
    - XMARSEX - SEX WITH PERSON OTHER THAN
    - HOMOSEX - HOMOSEXUAL SEX RELATIONS
    - HOMOCHNG - HOMOSEXUALITY: INHERENT O
  - ▶ 🖿 Pornography
  - ▶ 🖿 Child Discipline
  - ▶ 🖿 Suicide
  - ▶ 🖿 Activism
  - ▶ 🖿 Violent Experiences
  - ▶ 🖿 Media Exposure
  - ▶ 🖿 Interviewer Observations
  - ▶ 🖿 Experimental Variables
  - ▶ 🖿 Abortion Part Two
  - ▶ 🖿 Working Mothers
  - ▶ 🖿 Women's Rights
  - ▶ 🖿 Race Part Two
  - ▶ 🖿 How Often Think About Topics
  - ▶ 🖿 Recent Traumatic Events
  - ▶ 🖿 Social Issues Scales
  - ▶ 🖿 Important Life Aspects
  - ▶ 🖿 Personal Concerns
- ▶ 🖿 MILITARY ISSUES
- ▶ 🖿 OBLIGATIONS AND RESPONSIBILITIES

sda.berkeley.edu

Much social research involves the analysis of masses of statistical data. As valuable as the examination of overall patterns can be, it can come at the risk of losing sight of the individual men and women those data represent. Both the "macro" and the "micro" are important to our grasp of social dynamics, and some social research focuses specifically on the detailed particulars of real lives at the ground level of society. Throughout this book, I'll highlight recent studies that reflect this approach to understanding social life.

Statistically, unwed childbirth, especially among the poor in America, is likely to lead to a host of problems in the years that follow. Both the child and the mother are likely to struggle and suffer. The children are less likely to do well in school and later in life, and the mothers are likely to struggle in low-paying jobs or may reconcile themselves to living on welfare. The trend toward unwed births has increased dramatically in recent decades, especially among the poor. As a reaction to these problems, in 2005 the Bush administration launched a "Healthy Marriage Initiative," aimed at encouraging childbearing couples to marry. Voices for and against the program were raised with vigor.

Women in Social Research

At present, women are equal partners with men in social research—with women currently earning substantially more graduate degrees in the social sciences than men—but it has not always been that way. Early on, men clearly predominated, but there were actually women social researchers from the beginning, though their contributions have generally been ignored in recounting the history of social research.

For example, Auguste Comte (1798–1857) is generally regarded as the Father of Sociology, creating the French term, *Sociologie*, for example. His writings on positivism laid the groundwork for the new science. His works so impressed Harriet Martineau (1802–1876) in Britain that she translated them into English. Before long, Comte was advising students to read Martineau's English translations rather than the French originals. In her own right, Martineau pioneered social research into the family, children, religion, and race relations.

Or consider Florence Nightingale (1820–1910), most famous for professionalizing the field of nursing during and following the Crimean War. Less well known today, she was also an active quantitative researcher regarding sanitation, health, gender, and related fields. Moreover, she

was a pioneer in the use of infographics, such as pie charts and graphs, to present statistical results in easily graspable forms.

In the United States, Jane Addams (1860–1935) is best known for her social service contributions through Hull House in Chicago, but she was also an active researcher in connection with that work, publishing articles in the *American Journal of Sociology*, for example. Addams was a force for social activism in sociology, an orientation that has risen and fallen repeatedly since. She is the only sociologist to have earned a Nobel Prize (1931).

These are but a few of the women who were active in the creation and evolution of social research. This is not the only field in which women's contributions have been ignored by history, but it's time to set the record straight in social research.

Sources: Mark J. Perry, *Women earned majority of doctoral degrees in 2017 for 9th straight year and outnumber men in grad school 137 to 100*, AEIdeas blog, October 3, 2018, http://www.aei.org/publication/women-earned-majority-of-doctoral-degrees-in-2017-for-9th-straight-year-and-outnumber-men-in-grad-school-137-to-100-2/. Fiona Armstrong, *Celebrating the impact of women in social science*, Economic and Social Research Council blog, August 3, 2018, https://blog.esrc.ac.uk/2018/03/08/celebrating-the-impact-of-women-in-social-science/.

In *Promises I Can Keep: Why Poor Women Put Motherhood before Marriage* (Berkeley: University of California Press, 2005), Kathryn Edin and Maria Kefalas raise a question that, perhaps, should have been asked before a solution to the perceived problem was promoted: Why do poor women bear children outside of wedlock? The two social scientists spent five years speaking one-on-one with many young women who had borne children out of wedlock. Some of the things the researchers learned dramatically contradicted various common assumptions. Whereas many Americans have bemoaned the abandonment of marriage among the poor; for example, the women interviewed tended to speak highly of the institution, indicating that they hoped to be married one day. Many, however, were willing to settle down only with someone trustworthy and stable. Better to remain unmarried than to enter a bad marriage.

At the same time, these young women felt strongly that their ultimate worth as women centered on their bearing children. Most preferred being an unmarried mother to being a childless woman, the real tragedy in their eyes. This was only one finding among many that

contradicts common assumptions, perhaps even some of your own.

The box "Women in Social Research" indicates that women are not just the subjects of social research but are also the researchers.

As you can see, social scientists can access two distinct kinds of explanations. Just as physicists treat light as a particle in some experiments and as a wave in others, social scientists can search for relatively superficial universals today and probe the narrowly particular tomorrow. Both are good science, both are rewarding, and both can be fun.

Inductive and Deductive Theory

Like idiographic and nomothetic forms of explanation, inductive and deductive thinking both play a role in our daily lives. They, too, represent an important variation in social research.

There are two routes to the conclusion that you do better on exams if you study with others. On the one hand, you might find yourself puzzling, halfway through your college career, about why you do so well on exams sometimes but so poorly at other times. You might list all the exams you've taken, noting how well you did on each. Then you might try to recall any circumstances shared by all the good exams and all the poor ones. Did you do better on multiple-choice exams or essay exams? Morning exams or afternoon exams? Exams in the natural sciences, the humanities, or the social sciences? Times when you studied alone or…BAM! It occurs to you that you have almost always done best on exams when you studied with others. This mode of inquiry is known as **induction**.

Inductive reasoning moves from the particular to the general, from a set of specific observations to the discovery of a pattern that represents some degree of order among all the given events. Notice, incidentally, that your discovery doesn't necessarily tell you *why* the pattern exists—just that it does.

Here's a very different way you might have arrived at the same conclusion about studying for exams. Imagine approaching your first set of exams in college. You wonder about the best ways to study—how much to review, how much to focus on class notes. You learn that some students prepare by rewriting their notes in an orderly fashion. Then you consider whether to study at a measured pace or pull an all-nighter just before the exam. Among these musings, you might ask whether you should get together with other students in the class or just study on your own. You could evaluate the pros and cons of both options.

Studying with others might not be as efficient, because a lot of time might be spent on things you already understand. On the other hand, you can understand something better when you've explained it to someone else. And other students might understand parts of the course that you haven't grasped yet. Several minds can reveal perspectives that might have escaped you. Also, your commitment to study with others makes it more likely that you'll study rather than watch the special retrospective on TV.

In this fashion, you might add up the pros and cons and conclude, logically, that you'd benefit from studying with others. It seems reasonable to you, the way it seems reasonable that you'll do better if you study rather than not. Sometimes we say things like this are true "in theory." To complete the process, we test whether they're true in practice. For a complete

test, you might study alone for half your exams and study with others for the rest. This procedure would test your logical reasoning.

This second mode of inquiry, **deduction**, moves from the general to the specific. It moves from (1) a pattern that might be logically or theoretically expected to (2) observations that test whether the expected pattern actually occurs. Notice that deduction begins with "why" and moves to "whether," whereas induction moves in the opposite direction.

As you'll see later in this book, these two very different approaches present equally valid avenues for science. Each can stimulate the research process, prompting the researcher to take on specific questions and to frame the manner in which they are addressed. Moreover, you'll see how induction and deduction work together to provide ever more powerful and complete understandings.

Notice, by the way, that the distinction between the deductive and inductive is not necessarily linked to the nomothetic and idiographic modes. For example, idiographically and deductively, you might prepare for a particular date by taking into account everything you know about the person you're dating, trying to anticipate logically how you can prepare—what kinds of clothing, behavior, hairstyle, oral hygiene, and so forth will likely produce a successful date. Or, idiographically and inductively, you might try to figure out what it was exactly that caused your last date to call 911. A nomothetic, deductive approach arises when you coach others on your "rules of dating," wisely explaining why their dates will be impressed to hear them expound on the dangers of satanic messages concealed in

---

induction The logical model in which general principles are developed from specific observations. Having noted that Jews and Catholics are more likely to vote Democratic than are Protestants, you might conclude that religious minorities in the United States are more affiliated with the Democratic party, and then your task is to explain why.

deduction The logical model in which specific expectations of hypotheses are developed on the basis of general principles. Starting from the general principle that all deans are meanies, you might anticipate that this one won't let you change courses. This anticipation would be the result of deduction.

rock-and-roll lyrics. When you later review your life and wonder why you didn't date more musicians, you might engage in nomothetic induction. Thus, there are four possible approaches, which are used as much in life as in research.

We'll return to induction and deduction later in the book. At this point, let's turn to a third broad distinction that generates rich variations in social research.

Determinism versus Agency

The two preceding sections are based implicitly on a more fundamental issue. As you pursue your studies of social research methods, particularly when you examine causation and explanation in data analysis, you will come face to face with one of the most nagging dilemmas in the territory bridging social research and social philosophy: determinism versus agency. As you explore examples of causal social research, this issue comes to a head.

Imagine that you have a research grant to study the causes of racial prejudice. Having created a reasonable measure of prejudice so that you can distinguish those with higher or lower degrees of prejudice, you will be able to explore its causes. You may find, for example, that people living in certain regions of the country are, overall, more prejudiced than those living in other regions. Certain political orientations seem to promote prejudice, as do certain religious orientations. Economic insecurities may increase prejudice and result in the search for scapegoats. Or, if you are able to determine something about your subjects' upbringing—the degree of prejudice expressed by their parents, for example—you may discover more causes of prejudice.

Typically, none of these "causes" will be definitive, but each adds to the likelihood of a subject being prejudiced. Imagine, for example, a woman who was raised in a generally prejudiced region by prejudiced parents. She now holds political and religious views that support such prejudice, and she feels at risk of losing her job. When you put all those causes together, the likelihood of such a person being prejudiced is very high.

Notice the significance of the word *likelihood* in this discussion. As indicated earlier in this chapter, social researchers deal with a probabilistic causation. Thus, the convergence of all the causes of prejudice just mentioned would produce a high probability that the person

in question would appear prejudiced in our measurements. Even though the determinism involved in this approach is not perfect, it is deterministic all the same.

Missing in this analysis is what is variously called "choice," "free will," or, as social researchers tend to prefer, "agency." What happened to the individual? How do you feel about the prospect of being a subject in such an analysis? Let's say you consider yourself an unprejudiced person; are you willing to say you were destined to turn out that way because of forces and factors beyond your control? Probably not, and yet that's the implicit logic behind the causal analyses that social researchers so often engage in.

The philosophical question here is whether human behaviors are determined by their particular environment or whether they feel and act out of their personal choice or agency. I cannot pretend to offer an ultimate answer to this question, which has challenged philosophers and others throughout the history of human consciousness. But I can share the working conclusion I have reached as a result of observing and analyzing human behavior over a few decades.

I've tentatively concluded that (1) each of us possesses considerable free choice or agency, but (2) we readily allow ourselves to be controlled by environmental forces and factors, such as those described earlier in the example of prejudice. As you explore the many examples of causal analysis in this book and elsewhere in the social research literature, this giving away of agency will become obvious.

More shocking, if you pay attention to the conversations of daily life—yours as well as those of others—you will find that we constantly deny having choice or agency. Consider these few examples:

"I couldn't date someone who smokes."

"I couldn't tell my mother that."

"I couldn't work in an industry that manufactures nuclear weapons."

The list could go on for pages, but I hope this makes the point. In terms of human agency, you *could* do any of these things, although you might *choose* not to. However, you rarely explain your behavior or feeling on the basis of choice. If your classmates suggest you join them at a party or the movies and you reply, "I can't. I have an exam tomorrow," in fact, you could blow off the exam and join them; but you choose not to. (Right?) However, you rarely take responsibility for such a decision. You blame it on external forces: Why did the professor have to give an exam the day after the big party?

This situation is very clear in the case of love. Which of us ever *chooses* to love someone, or to be in love? Instead, we speak of "falling in love," sort of like catching a cold or falling in a ditch. The iconic anthem for this point of view is the set of 1913 lyrics, courtesy of songwriter, Joseph McCarthy:

> You made me love you.
> I didn't want to do it.

As I said at the outset of this discussion, the dilemma of determinism versus agency continues to bedevil philosophers, and you will find its head poking up from time to time throughout this book. I can't give you an ultimate answer to it, but I wanted to alert you to its presence.

The question of *responsibility* is an important aspect of this issue. Although it lies outside the realm of this book, I would like to bring it up briefly. Social research occurs in the context of a sociopolitical debate concerning who is responsible for a person's situation and their experiences in life. If you are poor, for example, are you responsible for your low socioeconomic status or does the responsibility lie with other people, organizations, or institutions?

Social research typically looks for ways that social structures (from interaction patterns to whole societies), affect the experiences and situations of individual members of society. Thus, your poverty might be a consequence of being born into a very poor family and having little opportunity for advancement. Or the closing of a business, exporting jobs overseas, or a global recession might lie at the root of your poverty.

Notice that this approach works against the notion of agency that we have discussed. Moreover, while social scientists tend to feel social problems should be solved at the societal level—through legislation, for example—this is a disempowering view for an individual. If you take the point of view that your poverty, bad grade, or rejected job application is the result of forces beyond your control, then you are conceding that you have no power. There is more power in assuming you have it than in assuming you are the helpless victim of circumstances. You can do

this without denying the power of social forces around you. In fact, you may exercise your individual responsibility by setting out to change the social forces that have an impact on your life. This complex view calls for a healthy **tolerance for ambiguity**, which is an important ability in the world of social research.

Qualitative and Quantitative Data

The distinction between quantitative and qualitative data in social research is essentially the distinction between numerical and nonnumerical data. When we say someone is intelligent, we've made a qualitative assertion. When psychologists and others measure intelligence by IQ scores, they are attempting to quantify such a qualitative assessment. For example, a psychologist might say that a person has an IQ of 120.

Every observation is qualitative at the outset, whether it be your experience of someone's intelligence, the location of a pointer on a measuring scale, or a check mark entered in a questionnaire. None of these things is inherently numerical or quantitative, but converting them to a numerical form is useful at times. (Chapter 14 deals specifically with the quantification of data.)

Quantification often makes our observations more explicit. It can also make aggregating and summarizing data easier. Further, it opens up the possibility of statistical analyses, ranging from simple averages to complex formulas and mathematical models. Thus, a social researcher might ask whether you tend to date people older or younger than yourself. A quantitative answer to this seems easily attained. The researcher asks how old each of your dates has been and calculates an average. Case closed.

Or is it? Although "age" here represents the number of years people have been alive, sometimes people use the term differently; perhaps for some people "age" really means "maturity." Though your dates may tend to be a little older than you, they may act more immaturely and thus represent the same "age." Or someone might see "age" as how young or old your dates look or

---

tolerance for ambiguity The ability to hold conflicting ideas in your mind simultaneously, without denying or dismissing any of them.

maybe the degree of variation in their life experiences, their worldliness. These latter meanings would be lost in the quantitative calculation of average age. Qualitative data are richer in meaning and detail than are quantitative data. This is implicit in the cliché, "He is older than his years." The poetic meaning of this expression would be lost in attempts to specify how much older.

This richness of meaning stems in part from ambiguity. If the expression means something to you when you read it, that particular meaning arises from your own experiences, from people you've known who might fit the description of being "older than their years" or perhaps the times you've heard others use that expression. Two things about this phrase are certain: (1) You and I probably don't mean exactly the same thing when we say it, and (2) if I say it, you don't know exactly what I mean, and vice versa.

It might be possible to quantify this concept, however. For example, we might establish a list of life experiences that would contribute to what we mean by *worldliness*:

Getting married

Getting divorced

Having a parent die

Seeing a murder committed

Being arrested

Being exiled

Being fired from a job

Running away with the circus

We might quantify people's worldliness as the number of such experiences they've had: the more they have experienced, the more worldly we'd say they were. If we thought of some experiences as more powerful than others, we could give those experiences more points. Once we had made our list and point system, scoring people and comparing their worldliness would be pretty straightforward. We would have no difficulty agreeing on who had more points than whom.

To quantify a concept like worldliness, we need to be explicit about what we mean. By focusing specifically on what we'll include in our measurement of the concept, however, we also exclude any other meanings. Inevitably, then, we face a trade-off: Any explicated, quantitative measure will be more superficial than the corresponding qualitative description.

What a dilemma! Which approach should we choose? Which is more appropriate to social research?

The good news is that we don't need to choose. In fact, we shouldn't. Both qualitative and quantitative methods are useful and legitimate in social research. Some research situations and topics are amenable mostly to qualitative examination, others mostly to quantification. We need both.

However, because these two approaches call for different skills and procedures, you may feel more comfortable with and become more adept in one mode than the other. You'll be a stronger researcher, however, to the extent that you can learn both approaches. At the very least, you should recognize the legitimacy of both.

Finally, you may have noticed that the qualitative approach seems more aligned with idiographic explanations, whereas nomothetic explanations are more easily achieved through quantification. Though this is true, these relationships are not absolute. Moreover, both approaches present considerable "gray area." Recognizing the distinction between qualitative and quantitative research doesn't mean that you must identify your research activities with one to the exclusion of the other. A complete understanding of a topic often requires both techniques.

The contributions of these two approaches are widely recognized today. For example, when Stuart Biddle and his colleagues (2001) at the University of Wales set out to review the status of research in the field of sport and exercise psychology, they were careful to examine the uses of both quantitative and qualitative techniques, drawing attention to those they felt were underused.

The apparent conflict between these two fundamental approaches has been neatly summarized by Paul Thompson (2004: 238–9):

> Only a few sociologists would openly deny the logic of combining the strengths of both quantitative and qualitative methods in social research.... In practice, however, despite such wider methodological aspirations in principle, social researchers have regrettably become increasingly divided into two camps, many of whose members know little of each other even if they are not explicitly hostile.

In reviewing the frequent disputes over the superiority of qualitative or quantitative methods, Anthony Onwuegbuzie and Nancy Leech

What do you think?...Revisited

This chapter opened with a question regarding uncontrolled variations in society—specifically, giving birth. We noted that there is no apparent control over who will or will not have a baby during a given year. Indeed, many babies are unplanned and thus are conceived "by accident." For the most part, the women who have babies differ from one year to the next, and each baby results from idiosyncratic, deeply personal reasons.

As the data introduced in this chapter indicate, however, aggregate social life operates differently from individual experiences of living in society. Although predicting whether a specific person or couple will decide to have a child at a given time is difficult, a greater regularity exists at the level of groups, organizations, and societies. This regularity is produced by social structure, culture, and other forces that individuals may or may not be aware of. Reflect, for example, on the impact of a housing industry that provides too few residences to accommodate large families, in contrast to one in which accommodation is the norm. Whereas that single factor would not absolutely determine the childbearing choices of a particular person or couple, it would have a predictable, overall effect across the whole society. And *social* researchers are chiefly interested in describing and understanding social patterns, not individual behaviors. This book will share with you some of the logic and tools social researchers use in that quest.

(2005) suggest that the two approaches have more similarities than differences. They further argue that using both approaches strengthens social research. My intention in this book is to focus on the complementarity of these two approaches rather than on any apparent competition between them.

Now that you've learned about the foundations of social research, I hope you can see how vibrant and exciting such research is. All we need is an open mind and a sense of adventure—and a good grounding in the basics of social research.

The Research Proposal

I conclude this chapter by introducing a practical learning feature that will run throughout the book: the preparation of a research proposal. Most organized research begins with a description of what is planned in the project: what questions it will raise and how it will answer them. Often such proposals are created for the purpose of getting the resources needed to conduct the research envisioned.

One way to learn the topics of this course is to use them in writing a research proposal. Each chapter ends with an exercise describing a step in this process. Even if you will not actually conduct a major research project, you can lay out a plan for doing so. Your instructor may use this as a course requirement. If not, you can still use the exercises to test your mastery of each chapter.

SAGrader is a computer program designed to assist you with this sort of exercise. It will accept a draft submission and critique it, pointing to elements that are missing, for example.

There are many organizational structures for research proposals. I've created a fairly typical one for you to use with this book. Here is the proposal outline, indicating which chapters in the book most directly deal with each topic:

Introduction (Chapter 1)

Review of the Literature (Chapters 2, 17; Appendix A)

Specifying the Problem/Question/Topic (Chapters 5, 6, 12)

Research Design (Chapter 4)

Data-Collection Method (Chapters 4, 8–11)

Selection of Subjects (Chapter 7)

Ethical Issues (Chapter 3)

Review of Literature (Chapters 2, 17)

Data Analysis (Chapters 13–16)

Bibliography (Chapter 17)

I'll have more to say about each of these topics as we move through the book, beginning with this chapter's exercise, where we'll discuss what might go into the introduction. Chapter 4 will have an extended section on the research proposal, and Chapter 17 will help you pull all the parts of the proposal into a coherent whole.

MAIN POINTS

Introduction

- The subject of this book is how we find out about social reality.

Looking for Reality

- Much of what we know, we know by agreement rather than by experience. Scientists accept an agreement reality but have special standards for doing so.

- The science of knowing is epistemology; the science of finding out is methodology.

- Inquiry is a natural human activity. Much of ordinary human inquiry seeks to explain events and predict future events.

- When we understand through direct experience, we make observations and seek patterns or regularities in what we observe.

- Two important sources of agreed-on knowledge are tradition and authority. However, these useful sources of knowledge can also lead us astray.

- Whereas we often observe inaccurately in day-to-day inquiry, researchers seek to avoid such errors by making observation a careful and deliberate activity.

- We sometimes jump to general conclusions on the basis of only a few observations, so scientists seek to avoid overgeneralization by committing to a sufficient number of observations and by replicating studies.

- There is a risk that we will pay attention only to observations that fit with our expectations; this is selective observation.

- In everyday life we sometimes reason illogically. Researchers seek to avoid illogical reasoning by being as careful and deliberate in their reasoning as in their observations. Moreover, the public nature of science means that others can always challenge faulty reasoning.

The Foundations of Social Science

- Social theory attempts to discuss and explain what is, not what should be. Theory should not be confused with philosophy or belief.

- Social science looks for regularities in social life.

- Social scientists are interested in explaining human aggregates, not individuals.

- Theories are written in the language of variables.

- A variable is a logical set of attributes. An attribute is a characteristic. *Sex*, for example, is a variable made up of the attributes *male* and *female*. So is *gender* when those attributes refer to social rather than biological distinctions.

- In causal explanation, the presumed cause is the independent variable, and the affected variable is the dependent variable.

- Social research has three main purposes: exploring, describing, and explaining social

phenomena. Many research projects reflect more than one of these purposes.

- Ethics plays a key role in the practice of social research.

Some Dialectics of Social Research

- Whereas idiographic explanations seek to present a full understanding of specific cases, nomothetic explanations seek to present a generalized account of many cases.

- Inductive theories reason from specific observations to general patterns. Deductive theories start from general statements and predict specific observations.

- The underlying logic of traditional science implicitly suggests a deterministic cause-and-effect model in which individuals have no choice, although researchers do not say, or necessarily believe, that.

- Some researchers are intent on focusing attention on the "agency" by which the subjects of study are active, choice-making agents. The issue of free will versus determinism is an old one in philosophy, and people exhibit conflicting orientations in their daily behavior, sometimes proclaiming their freedom and other times denying it.

- Quantitative data are numerical; qualitative data are not. Both types of data are useful for different research purposes.

KEY TERMS

agreement reality	induction
attribute	methodology
deduction	nomothetic
dependent variable	replication
epistemology	theory
idiographic	tolerance for ambiguity
independent variable	variable

REVIEW QUESTIONS

1. How would the discussion of variables and attributes apply to physics, chemistry, and biology?

2. Identify a social problem that you feel ought to be addressed and solved. What are the variables represented in your description of the problem? Which of those variables would you monitor in determining whether the problem was solved?

3. Suppose you were interested in studying the quality of life among elderly people. What quantitative and qualitative indicators might you examine?

4. How might social research be useful to such professionals as physicians, attorneys, business executives, police officers, and newspaper reporters?

CHAPTER 2

Paradigms, Theory, and Research

CHAPTER OVERVIEW

Social scientific inquiry is an interplay of theory and research, logic and observation, induction and deduction—and of the fundamental frames of reference known as paradigms.

Thomas La Mela / Shutterstock.com

Introduction

This chapter explores some specific ways theory and research work hand in hand during the adventure of inquiry into social life. We'll begin by looking at several fundamental frames of reference, called **paradigms**, that underlie social theories and inquiry. Whereas theories seek to explain, paradigms provide ways of looking. In and of themselves, paradigms don't explain anything, but they provide logical frameworks within which theories are created. As you'll see in this chapter, theories and paradigms intertwine throughout the search for meaning in social life.

Since I'll be using the term, "logic," several times in this chapter, I should clarify its meaning. The words *logic* and *rationality* are sometimes used interchangeably, but that's a mistake. Logic is a system of principles for determining the validity of an explanation. Thus, a set of religious beliefs can provide a theological system of explanation: for example, God rewarding or punishing good or bad behavior. Astrology provides another logic for explaining things.

Rationality is a logical system of explanation that is typically associated with science, though it is difficult to define in the space

available here. Rational logic operates in the natural realm, excluding supernatural explanations. It is often described as objective, unaffected by the scientist's personal beliefs or emotions. Wherever possible, rational logic is allied with empirical observations. If we reason logically that men and women will differ in some regard, we can then make observations to see if it's true. The testing of logical expectations must include the possibility of failure as well as success.

In a sense, much of this book will illustrate the use of rational principles and practices in social research. For example, we'll see that a requirement for causal relationships is that the cause needs to happen before the effect. I'll conclude this discussion with one more important clarification. Rational social science explanations do not have to assume that human behavior itself is rational. Sometimes it is and sometimes it isn't, but even nonrational or irrational human behavior can be susceptible to rational explanation.

Some restaurants in the United States are fond of conducting political polls among their customers before an upcoming election. Some people take these polls very seriously because of their uncanny history of predicting winners. By the same token, some movie theaters have achieved similar success by offering popcorn in bags picturing either donkeys or elephants. Years ago, granaries in the Midwest offered farmers a chance to indicate their political preferences through the bags of grain they selected.

Such oddities offer some interest. They all present the same pattern over time, however:

paradigm A model or framework for observation and understanding that shapes both what we see and how we understand it. The conflict paradigm causes us to see social behavior one way, the interactionist paradigm causes us to see it differently.

What do you think?

Scholars such as George Herbert Mead make a powerful argument that social life is really a matter of interactions and their residue. You and I meet each other for the first time, feel each other out, and mutually create rules for dealing with each other. The next time we meet, we'll probably fall back on these rules, which tend to stay with us. Think about your first encounters with a new professor or making a new friend. Mead suggests that all the social patterns and structures that we experience are created in this fashion.

Other scholars, such as Karl Marx, argue that social life is fundamentally a struggle among social groups. According to Marx, society is a class struggle in which the "haves" and the "have-nots" are pitted against each other in an attempt to dominate others and to avoid being dominated. He claims that, rather than being mutually created by individuals, rules for behavior grow out of the economic structure of a society.

Which of these very different views of society is true? Or does the truth lie somewhere else?

See the *What do you think?...Revisited* box toward the end of the chapter.

They work for a while, but then they fail. Moreover, we can't predict when or why they will fail.

These unusual polling techniques point to the shortcomings of "research findings" based only on the observation of patterns. Unless we can offer logical explanations for such patterns, the regularities we've observed may be mere flukes, chance occurrences. If you flip coins long enough, you'll get ten heads in a row. Scientists could adapt a street expression to describe this situation: "Patterns happen."

Logical explanations are what theories seek to provide. Further, theory functions three ways in research. First, it prevents our being taken in by flukes. If we can't explain why Ma's Diner has predicted elections so successfully, we run the risk of supporting a fluke. If we know why it has happened, however, we can anticipate whether it will work in the future.

Second, theories make sense of observed patterns in ways that can suggest other possibilities. If we understand the reasons why broken homes produce more juvenile delinquents than do intact homes—lack of supervision, for example— we can take effective action, such as establishing after-school youth programs.

Third, theories can shape and direct research efforts, pointing toward likely discoveries through empirical observation. If you were looking for your lost keys on a dark street, you could whip your flashlight around randomly—or you could use your memory of where you had been to limit your

search to more likely areas. Theory, by analogy, directs researchers' flashlights where they will most likely observe interesting patterns of social life.

This is not to say that all social science research is tightly intertwined with social theory. Sometimes social scientists undertake investigations simply to discover the state of affairs, such as an evaluation of whether an innovative social program is working or a poll to determine which candidate is winning a political race. Similarly, descriptive ethnographies, such as anthropological accounts of preliterate societies, produce valuable information and insights in and of themselves. However, even studies such as these often go beyond pure description to ask, "Why?" Theory is directly relevant to "why" questions.

There are many routes to understanding social life.

Some Social Science Paradigms

There is usually more than one way to make sense of things. In daily life, for example, liberals and conservatives often explain the same phenomenon—teenagers using guns at school, for example—quite differently. So might the parents and teenagers themselves. But underlying each of these different explanations, or theories, is a paradigm—one of the fundamental models or frames of reference we use to organize our observations and reasoning.

Paradigms are often difficult to recognize as such because they are so implicit, assumed, taken for granted. They seem more like "the way things are" than like one possible point of view among many. Here's an illustration of what I mean.

Where do you stand on the issue of human rights? Do you feel that individual human beings are sacred? Are they "endowed by their creator with certain inalienable rights," as asserted by the U.S. Declaration of Independence? Are there some things that no government should do to or ask of its citizens?

Consider that many other cultures today regard the Western (and particularly U.S.) commitment to the sanctity of the individual as bizarre. Historically, it is decidedly a minority viewpoint. For example, although many Asian countries now subscribe to some "rights" that belong to individuals, those are balanced against the "rights" of families, organizations, and society at large. When criticized for violating human rights, Asian leaders often point to high crime rates and social disorganization in Western societies as the cost of what they see as our radical "cult of the individual."

No matter what our beliefs, it's useful to recognize that our views and feelings in this matter are the result of the paradigm into which we have been socialized. The sanctity of the individual is not an objective fact of nature; it is a point of view, a paradigm. All of us operate within many such paradigms.

When we recognize that we are operating within a paradigm, two benefits accrue. First, we are better able to understand the seemingly bizarre views and actions of others who are operating from within a different paradigm. Second, at times we can profit from stepping outside our paradigm. We can see new ways of seeing and explaining things. We can't do that as long as we mistake our paradigm for reality.

Paradigms play a fundamental role in science, just as they do in daily life. Thomas Kuhn (1970) drew attention to the role of paradigms in the history of the natural sciences. Major scientific paradigms have included such fundamental viewpoints as Copernicus's conception of the earth moving around the sun (instead of the reverse), Darwin's theory of evolution, Newton's mechanics, and Einstein's relativity. Which scientific theories "make sense" depends on which paradigm scientists maintain.

Although we sometimes think of science as developing gradually over time, marked by important discoveries and inventions, Kuhn says that, historically, one paradigm would become entrenched, resisting substantial change. Eventually, however, the shortcomings of that paradigm would become obvious in the form of observations that violated the expectations suggested by the paradigm. These are often referred to as *anomalies*, events that fall outside expected or standard patterns. For a long time in American society as elsewhere, a fundamental belief system regarding sex and gender held that only men were capable of higher learning. In that situation, every demonstrably learned woman was an anomalous challenge to the traditional view. When the old paradigm was sufficiently challenged, Kuhn suggested, a new paradigm would emerge and supplant the old one. Kuhn's classic book on this subject is titled, appropriately enough, *The Structure of Scientific Revolutions.*

Social scientists have developed several paradigms for understanding social behavior. The fate of supplanted paradigms in the social sciences, however, differs from what Kuhn observed in the natural sciences. Natural scientists generally believe that the succession of paradigms represents progress from false views to true ones. No modern astronomer believes that the sun revolves around the earth, for example.

In the social sciences, on the other hand, theoretical paradigms may gain or lose popularity, but they're seldom discarded. Social science paradigms represent a variety of views, each of which offers insights the others lack while ignoring aspects of social life that the others reveal.

Each of the paradigms we're about to examine offers a different way of looking at human social life. Each makes certain assumptions about the nature of social reality. Ultimately, paradigms cannot be true or false; as ways of looking, they can only be more or less useful. Rather than deciding which paradigms are true or false, try to find ways they might be useful to you. As we'll see, each can open up new understandings, suggest different kinds of theories, and inspire different kinds of research.

Macrotheory and Microtheory

Let's begin with a discussion that encompasses many of the paradigms to be discussed. Some theorists focus their attention on society at large or at least on large portions of it. **Macrotheory** deals with large, aggregate entities of society or even whole societies. Topics of study for macrotheory include the struggle among economic classes in a society, international relations, and the interrelations among major institutions in society, such as government, religion, and family.

Some scholars have taken a more intimate view of social life. **Microtheory** deals with issues of social life at the level of individuals and small groups. Dating behavior, jury deliberations, and student–faculty interactions provide apt subjects for a microtheoretical perspective. Such studies often come close to the realm of psychology, but whereas psychologists typically focus on what goes on inside humans, social scientists study what goes on among them. Some researchers prefer to limit *macrotheory* to the study of whole societies. In that case, the intermediate level between macrotheory and microtheory is called *mesotheory*: studying organizations, communities, and perhaps social categories such as gender.

The basic distinction between macrotheory and microtheory cuts across the paradigms we'll examine next. Whereas some of them, such as symbolic interactionism and ethnomethodology, often work best at the microlevel, others, such as the conflict paradigm, can be pursued at either the microlevel or the macrolevel.

Early Positivism

When the French philosopher Auguste Comte (1798–1857) coined the term *sociologie* in 1822,

he launched an intellectual adventure that is still unfolding today. Comte was arguably the first to identify society as a phenomenon that can be studied scientifically. (Initially, he wanted to label his enterprise *social physics*, but another scholar preempted that term.)

Prior to Comte's time, society simply *was*. To the extent that people recognized different kinds of societies or changes in society over time, religious paradigms predominantly explained these differences. The state of social affairs was often seen as a reflection of God's will. Alternatively, people were challenged to create a "City of God" on earth to replace sin and godlessness.

Comte separated his inquiry from religion, replacing religious belief with scientific objectivity. His "positive philosophy" postulated three stages of history. A "theological stage" predominated throughout the world until about 1300 C.E. During the next 500 years, a "metaphysical stage" replaced God with ideas such as "nature" and "natural law." Finally, Comte felt that he was launching the third stage of history, in which science would replace religion and metaphysics; knowledge would be based on observations through the five senses rather than on belief. Again, Comte felt that society could be studied and understood rationally and that sociology could be as scientific as biology or physics.

Comte's view came to form the foundation for subsequent development of the social sciences. In his optimism for the future, he coined the term *positivism* to describe this scientific approach, believing that scientific truths could be positively verified through empirical observations and the logical analysis of what was observed. In recent decades, the idea of positivism has come under serious challenge, as we'll see later in this discussion.

---

macrotheory A theory aimed at understanding the "big picture" of institutions, whole societies, and the interactions among societies. Karl Marx's examination of the class struggle is an example of macrotheory.

microtheory A theory aimed at understanding social life at the level of individuals and their interactions. Explaining how the play behavior of girls differs from that of boys is an example of microtheory.

Conflict Paradigm

Karl Marx (1818–1883) suggested that social be-havior could best be seen as the process of con-flict: the attempt to dominate others and to avoid being dominated. Marx focused primarily on the struggle among economic classes. Specifically, he examined the way capitalism produced the op-pression of workers by the owners of industry. Marx's interest in this topic did not end with analytical study: He was also ideologically com-mitted to restructuring economic relations to end the oppression he observed.

The conflict paradigm is not limited to economic analyses. Georg Simmel (1858–1918) was particularly interested in small-scale conflict, in contrast to the class struggle that interested Marx. Simmel noted, for example, that conflicts among members of a tightly knit group tended to be more intense than those among people who did not share feelings of belonging and intimacy.

In a more recent application of the conflict paradigm, when Michel Chossudovsky's (1997) analysis of the International Monetary Fund (IMF) and World Bank suggested that these two international organizations were increas-ing global poverty rather than eradicating it, he directed his attention to the competing interests involved in the process. In theory, the chief in-terest being served should be that of the poor people of the world or perhaps the impoverished Third World nations. The researcher's inquiry, however, identified many other interested parties who benefited: the commercial lending institu-tions who made loans in conjunction with the IMF and World Bank and multinational corpora-tions seeking cheap labor and markets for their goods, to name two. Chossudovsky's analysis concluded that the interests of the banks and corporations tended to take precedence over those of the poor people, who were the intended beneficiaries. Moreover, he found that many policies were weakening national economies in developing nations, as well as undermining democratic governments.

Although applications of the conflict paradigm often focus on class, gender, and ethnic struggles, it would be appropriate to apply it whenever different groups have competing in-terests. For example, we could fruitfully apply it to understanding relations among different departments in an organization, fraternity and sorority rush weeks, or student–faculty–administrative relations, to name just a few.

These examples should illustrate some of the ways you might view social life if you were taking your lead from the conflict paradigm. To explore the applicability of this paradigm, you might take a minute to skim through a daily newspaper or news magazine and identify events you could interpret in terms of individuals and groups attempting to dominate each other and avoid being dominated. The theoretical concepts and premises of the conflict paradigm might help you make sense out of these events.

Symbolic Interactionism

As we have seen, whereas Marx chiefly ad-dressed macrotheoretical issues—large institu-tions and whole societies in their evolution through the course of history—Simmel was more interested in the ways individuals interacted with one another, or the "micro" aspects of society; he began by examining dyads (groups of two people) and triads (groups of three), for example. Similarly, he wrote about "the web of group affiliations" (Wolff 1950).

Simmel was one of the first European so-ciologists to influence the development of U.S. sociology. His focus on the nature of interac-tions in particular influenced George Herbert Mead (1863–1931), Charles Horton Cooley (1864–1929), and others, who took up the cause and developed it into a powerful paradigm for research.

Cooley, for example, introduced the idea of the "primary group," those intimate associates with whom we share a sense of belonging, such as our family, friends, and so forth. Cooley also wrote of the "looking-glass self" we form by looking at the reactions of people around us. If everyone treats us as beautiful, for example, we conclude that we are. See how fundamentally this paradigm differs from the society-level con-cerns of Marx.

Similarly, Mead emphasized the impor-tance of our human ability to "take the role of the other," imagining how others feel and how they might behave in certain circumstances. As we gain an idea of how people in general see things, we develop a sense of what Mead called the "generalized other" (Strauss 1977). Mead also felt that most interactions revolved around individuals' reaching a common understanding through language and other symbolic systems, hence the term *symbolic interactionism.*

Here's one way you might apply this paradigm to an examination of your own life. The next time you meet someone new, watch how your knowledge of each other unfolds through the process of interaction. Notice also any attempts you make to manage the image you are creating in the other person's mind.

Clearly this paradigm can lend insights into the nature of interactions in ordinary social life, but it can also help us understand unusual forms of interaction, as when researchers Robert Emerson, Kerry Ferris, and Carol Brooks Gardner (1998), in a study still relevant today, set out to understand the nature of "stalking." Through interviews with numerous stalking victims, they came to identify different motivations among stalkers, stages in the development of a stalking scenario, how people can recognize whether they are being stalked, and what they can do about it.

Ethnomethodology

Whereas some social science paradigms emphasize the impact of social structure (such as norms, values, and control agents) on human behavior, others do not. Harold Garfinkel (1917–2011), a contemporary sociologist, took the point of view that people are continually creating social structure through their actions and interactions—that they are, in fact, creating their realities. Thus, when you and your instructor meet to discuss your term paper, even though there are myriad expectations about how you should act, the conversation will somewhat differ from any of those that have occurred before, and how you both act will somewhat modify your future expectations. That is, discussing your term paper will impact your future interactions with other professors and students.

Given the tentativeness of reality in this view, Garfinkel suggested that people are continuously trying to make sense of the life they experience. In a way, he suggested that everyone is acting like a social scientist: hence the term *ethnomethodology,* or "methodology of the people."

How would you approach learning about people's expectations and how they make sense out of their world? One technique ethnomethodologists use is to break the rules, to violate people's expectations. If you try to talk to me about your term paper, but I keep talking about football, any expectations you had for my behavior might become apparent. We might also see how you make sense out of my behavior. ("Maybe he's using football as an analogy for understanding social systems theory.")

In another example of ethnomethodology, John Heritage and David Greatbatch (1992) examined the role of applause in British political speeches: How did the speakers evoke applause, and what function did it serve (for example, to complete a topic)? Research within the ethnomethodological paradigm often focuses on communication.

You can find many interesting opportunities to try the ethnomethodological paradigm. For instance, the next time you get on an elevator, don't face the front (that's the norm, or expected behavior). Instead, just stand quietly facing the rear of the elevator. See how others react to this behavior. Just as important, notice how you feel about it. If you do this experiment a few times, you should begin to develop a feel for the ethnomethodological paradigm.* See "Applying Concepts in Everyday Life: The Power of Paradigms," for more on this topic.

We'll return to ethnomethodology in Chapter 10, when we discuss field research. For now, let's turn to a very different paradigm.

Structural Functionalism

Structural functionalism, sometimes also known as "social systems theory," grows out of a notion introduced by Comte and others: A social entity, such as an organization or a whole society, can be viewed as an organism. Like organisms, a social system is made up of parts, each of which contributes to the functioning of the whole.

*I am grateful to my colleague, Bernard McGrane, for this experiment. Barney also has his students eat dinner with their hands, watch TV without turning it on, and engage in other strangely enlightening behavior (McGrane 1994).

Applying Concepts in Everyday Life

The Power of Paradigms

In this chapter, we are looking at some of the social science paradigms used to organize and make sense out of social life, and we are seeing the impact those paradigms have on what is observed and how it is interpreted. The power of paradigms, however, extends well beyond the scientific realm. You can look almost anywhere in the world and see conflicts among religious, ethnic, political, and other cultural paradigms.

Consider the September 11, 2001, attacks on the World Trade Center and the Pentagon. Widely varied interpretations, reflecting radically different paradigms, blamed the attacks on Osama bin Laden, Saddam Hussein, Israel, the Bush administration, God, homosexuals, and feminists. Some of these explanations may strike you as bizarre, but they made perfectly good sense within the worldviews of those espousing them. That's the power that paradigms have in all areas of life.

By analogy, consider the human body. Each component—such as the heart, lungs, kidneys, skin, and brain—has a particular job to do. The body as a whole cannot survive unless each of these parts does its job, and none of the parts can survive except as a part of the whole body. Or consider an automobile, composed of tires, steering wheel, gas tank, spark plugs, and so forth. Each of the parts serves a function for the whole; taken together, that system can get us across town. None of the individual parts would be of much use to us by itself, however.

The view of society as a social system, then, looks for the "functions" served by its various components. We might consider a football team as a social system—one in which the quarterback, running backs, offensive linemen, and others have their own jobs to do for the team as a whole. Or, we could look at a symphony orchestra and examine the functions served by the conductor, the first violinist, and the other musicians.

Social scientists using the functionalist paradigm might note that the function of the police, for example, is to exercise social control—encouraging people to abide by the norms of society and bringing to justice those who do not. We could just as reasonably ask what functions criminals serve in society. Within the functionalist paradigm, we'd see that criminals serve as job security for the police. In a related observation, Emile Durkheim (1858–1917) suggested that crimes and their punishment provided an opportunity for the reaffirmation of a society's values. By catching and punishing a thief, we reaffirm our collective respect for private property.

To get a sense of the functionalist paradigm, thumb through your college or university catalog and assemble a list of the administrators (such as president, deans, registrar, campus security, maintenance personnel). Figure out what each of them does. To what extent do these roles relate to the chief functions of your college or university, such as teaching or research? Suppose you were studying some other kind of organization. How many of the school administrators' functions would also be needed in, say, an insurance company?

In applying the functionalist paradigm to everyday life, people sometimes make the mistake of thinking that functionality, stability, and integration are necessarily good, or that the functionalist paradigm makes that assumption. However, when social researchers look for the "functions" served by poverty, racial discrimination, or the oppression of women, they are not justifying such things. Rather, they seek to understand the roles such things play in the larger society, as a way of understanding why they persist and how they could be eliminated.

Earl Babbie

The functionalist paradigm is based on the assumption that all elements of society serve some function for the operation of the whole, even "negative" elements. Crime, for example, serves the function of providing employment for police. This doesn't justify crime, but it explains part of how crime is woven into the social whole.

Feminist Paradigms

When Ralph Linton concluded his anthropological classic, *The Study of Man* (1937: 490), by speaking of "a store of knowledge that promises to give man a better life than any he has known," no one complained that he had left women out. Linton was using the linguistic conventions of his time; he implicitly included women in all his references to men. Or did he?

When feminists (of both genders) first began questioning the use of masculine nouns and pronouns whenever gender was ambiguous, their concerns were often viewed as petty. Many felt the issue was one of women having their feelings hurt, their egos bruised. But be honest: When you read Linton's words, what did you picture? An amorphous, genderless human being, or a masculine persona?

In a similar way, researchers looking at the social world from a feminist paradigm have called attention to aspects of social life that other paradigms do not reveal. In fact, feminism has established important theoretical paradigms for social research. In part it focuses on gender differences and how they relate to the rest of social organization. These paradigms draw attention to the oppression of women in many societies, which in turn sheds light on oppression in general.

Feminist paradigms not only reveal the treatment of women or the experience of oppression but they often also point to limitations in how other aspects of social life are examined and understood. For example, feminist perspectives are often related to a concern for the environment. As Greta Gard suggests,

> The way in which women and nature have been conceptualized historically in Western intellectual tradition has resulted in devaluing whatever is associated with women, emotion, animals, nature, and the body, while simultaneously elevating in value those things associated with men, reason, humans, culture, and the mind. One task of ecofeminism has been to expose these dualisms and the ways in which feminizing nature and naturalizing or animalizing women has served as justification for the domination of women, animals and the earth.

(1993: 5)

Feminist paradigms have also challenged the prevailing notions concerning consensus in society. Most descriptions of the predominant beliefs, values, and norms of a society are written by people representing only portions of society.

In the United States, for example, such analyses have typically been written by middle-class white men—and, not surprisingly, they have written about the beliefs, values, and norms they themselves share. Though George Herbert Mead spoke of the "generalized other" that each of us becomes aware of and can "take the role of," feminist paradigms question whether such a generalized other even exists.

Mead used the example of learning to play baseball to illustrate how we learn about the generalized other. As shown here, Janet Lever's research suggests that understanding the experience of boys may tell us little about girls.

> Girls' play and games are very different. They are mostly spontaneous, imaginative, and free of structure or rules. Turn-taking activities like jump rope may be played without setting explicit goals. Girls have far less experience with interpersonal competition. The style of their competition is indirect, rather than face to face, individual rather than team affiliated. Leadership roles are either missing or randomly filled.

(1986: 86)

Social researchers' growing recognition of the intellectual differences between men and women led the psychologist Mary Field Belenky and her colleagues to speak of *Women's Ways of Knowing* (1986). In-depth interviews with 45 women enabled the researchers to distinguish five perspectives on knowing that challenge the view of inquiry as obvious and straightforward:

- *Silence:* Some women, especially early in life, feel themselves isolated from the world of knowledge, their lives largely determined by external authorities.
- *Received knowledge:* From this perspective, women feel themselves capable of taking in and holding knowledge originating with external authorities.
- *Subjective knowledge:* This perspective opens up the possibility of personal, subjective knowledge, including intuition.
- *Procedural knowledge:* Some women feel they have learned the ways of gaining knowledge through objective procedures.
- *Constructed knowledge:* The authors describe this perspective as "a position in which women view all knowledge as contextual, experience themselves as creators of knowledge, and value both subjective and objective strategies for knowing."

(Belenky et al. 1986: 15)

"Constructed knowledge" is particularly interesting in the context of our previous discussions. The positivistic paradigm of Comte would have a place neither for "subjective knowledge" nor for the idea that truth might vary according to its context. The ethnomethodological paradigm, on the other hand, would accommodate these ideas.

Introduced by Nancy Hartsock in 1983, the term *feminist standpoint theory* refers to the idea that women have knowledge about their status and experience that is not available to men. This viewpoint has evolved over time. For example, scholars have come to recognize that there is no single female experience. Further, different kinds of women (varying by wealth, ethnicity, or age, for example) have quite different experiences of life in society, all the while sharing things in common because of their gender. This sensitivity to variations in the female experience is also a main element in *third-wave feminism*, which began in the 1990s.

To try out feminist paradigms, you might want to look into the possibility of discrimination against women at your college or university. Are the top administrative positions held equally by men and women? How about secretarial and clerical positions? Are men's and women's sports supported equally? Read through the official history of your school. Does it include men and women equally? (If you attend an all-male or all-female school, of course, some of these questions won't apply.)

As we just saw, feminist paradigms reflect not only a concern for the unequal treatment of women but also an epistemological recognition that men and women perceive and understand society differently. Social theories created solely by men, which has been the norm, run the risk of having an unrecognized bias. A similar case can be made for theories created almost exclusively by white people.

Critical Race Theory

As a general rule, whenever you find the word *critical* in the name of a paradigm or theory, it will likely refer to a nontraditional view, one that may be at odds with the prevailing paradigms of an academic discipline or with the mainstream structure of society.

The roots of critical race theory are generally associated with the civil rights movement of the mid-1950s and race-related legislation of the 1960s. By the mid-1970s, with fears that the strides toward equality were beginning to bog down, civil rights activists and social scientists began the codification of a paradigm based on an awareness of race and a commitment to racial justice.

This was not the first time sociologists had paid attention to the status of nonwhites in American society. Perhaps the best-known African American sociologist in the history of the discipline was W. E. B. DuBois, who published *The Souls of Black Folk* in 1903. Among other things, DuBois pointed out that African Americans lived their lives through a "dual consciousness": as Americans and as black people. By contrast, white Americans have seldom reflected on being white. If you were American, white was simply assumed. If you were not white, you have been seen as, and made to feel like, the exception. So imagine the difference between an African American sociologist and a white sociologist creating a theory of social identity. Their theories of identity would likely differ in some fundamental ways, even if they were not limiting their analyses to their own race.

Much of the contemporary scholarship in critical race theory has to do with the role of race in politics and government, studies often undertaken by legal scholars as well as social scientists. Thus, for example, Derrick Bell (1980) critiqued the U.S. Supreme Court's landmark *Brown v. Board of Education* decision, which struck down the "separate but equal" system of school segregation. He suggested that the Court was motivated by the economic and political interests of the white majority, not by the objective of educational equality for African American students. In his analysis, Bell introduced the concept of **interest convergence**, suggesting that laws will only be changed to benefit African Americans if and when those changes are seen to further the interests of whites. Richard Delgado (2002) provides an excellent overview of how subsequent critical race theorists have pursued Bell's reasoning.

interest convergence The thesis that majority-group members will only support the interests of minorities when those actions also support the interests of the majority group.

Rational Objectivity Reconsidered

We began with Comte's assertion that we can study society rationally and objectively. Since his time, the growth of science, the decline of superstition, and the rise of bureaucratic structures have put rationality more and more at the center of social life. As fundamental as rationality is to most of us, however, some contemporary scholars have raised questions about it.

For example, positivistic social scientists have sometimes erred in assuming that humans will always act rationally. I'm sure your own experience offers ample evidence to the contrary. Many modern economic models also assume that people will make rational choices in the economic sector: They will choose the highest-paying job, pay the lowest price, and so forth. This assumption, however, ignores the power of such matters as tradition, loyalty, and image that compete with reason in determining human behavior.

A more sophisticated positivism would assert that we can rationally understand even nonrational human behavior. Here's an example. In the famous "Asch experiment" (Asch 1958), a group of subjects is presented with a set of lines on a screen and asked to identify the two lines of equal length.

Imagine yourself a subject in such an experiment. You're sitting in the front row of a classroom among a group of six subjects. A set of lines is projected on the wall in front of you (see Figure 2-1). The experimenter asks the subjects, one at a time, to identify the line to the right (A, B, or C) that matches the length of line X.

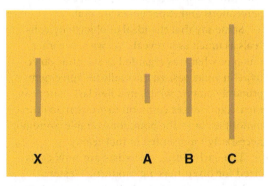

FIGURE 2-1

The Asch Experiment. Subjects in the Asch experiment have a seemingly easy task: to determine whether A, B, or C is the same length as X. But there's more here than meets the eye.

The correct answer (B) is obvious to you. To your surprise, you find that all the other subjects agree on a different answer!

The experimenter announces that all but one of the group has gotten the correct answer; that is, you've gotten it wrong. Then a new set of lines is presented, and you have the same experience. The obviously correct answer is wrong, and everyone but you seems to understand that.

As it turns out, of course, you're the only real subject in the experiment—all the others are working with the experimenter. The purpose is to see whether you would be swayed by public pressure and go along with the incorrect answer. In one-third of the initial experiments, Asch found that his subjects did just that.

Choosing an obviously wrong answer in a simple experiment is an example of nonrational behavior. But as Asch went on to show, experimenters can examine the circumstances that lead more or fewer subjects to go along with the incorrect answer. For example, in subsequent studies, Asch varied the size of one group and the number of "dissenters" who chose the "wrong" (that is, the correct) answer. Thus, it is possible to study nonrational behavior rationally and scientifically.

More radically, we can question whether social life abides by rational principles at all. In the physical sciences, developments such as chaos theory, fuzzy logic, and complexity have suggested that we may need to fundamentally rethink the orderliness of physical events.

The contemporary challenge to positivism, however, goes beyond the question of whether people behave rationally. In part, the criticism of positivism challenges the idea that scientists can be as objective as the scientific ideal assumes. Most scientists would agree that personal feelings can and do influence the problems scientists choose to study, their choice of what to observe, and the conclusions they draw from their observations.

As with rationality, there is a radical critique of objectivity. Whereas scientific objectivity has long stood as an unquestioned ideal, some contemporary researchers suggest that subjectivity might actually be preferred in some situations, as we glimpsed in the discussions of feminism and ethnomethodology. Let's take a moment to return to the dialectic of subjectivity and objectivity.

To begin with, all our experiences are inescapably subjective. There is no way out. We can see only through our own eyes, and anything peculiar to our eyes will shape what we see. We can hear things only the way our particular ears and brain transmit and interpret sound waves. You and I, to some extent, hear and see different realities. And both of us experience quite different physical "realities" than do bats, for example. In what to us is total darkness, a bat "sees" things such as flying insects by emitting a sound we humans can't hear. Echolocation, or the reflection of the bat's sound, creates a "sound picture" precise enough for the bat to home in on the moving insect and snatch it up. In a similar vein, scientists on the planet Xandu might develop theories of the physical world based on a sensory apparatus that we humans can't even imagine. Maybe they have x-ray vision or hear colors.

Despite the inescapable subjectivity of our experience, we humans seem to be wired to seek an agreement on what is "really real," what is objectively so. Objectivity is a conceptual attempt to get beyond our individual views. It is ultimately a matter of communication, as you and I attempt to find a common ground in our subjective experiences. Whenever we succeed in our search, we say we are dealing with objective reality. This is the agreement reality discussed in Chapter 1.

Perhaps the most significant studies in the history of social science were conducted in the 1930s by a Turkish American social psychologist, Muzafer Sherif (1935), who slyly said he wanted to study "autokinetic effects." To do this, he put small groups in totally darkened rooms, save for a single point of light in the center of the wall in front of the participants. Sherif explained that the light would soon begin to move about, and the subjects were to determine how far it was moving—a difficult task with nothing else visible as a gauge of length or distance.

Amazingly, the subjects in each group agreed on the distance the point of light moved about. Oddly, however, the groups arrived at quite different conclusions. Strangest of all—as you may have guessed—the point of light had remained stationary. If you stare at a fixed point of light long enough, it will seem to move (Sherif's "autokinetic effect"). Notice, however, that each of the groups agreed on a specific delusion. The movement of the light was real to them, but it was a reality created out of nothing: a socially constructed reality.

Whereas our subjectivity is individual, our search for objectivity is social. This is true in all aspects of life, not just in science. While you and I prefer different foods, we must agree to some extent on what is fit to eat and what is not, or else there could be no restaurants, no grocery stores, no food industry. The same argument could be made regarding every other form of consumption. There could be no movies or television, no sports.

Social scientists as well have found benefits in the concept of objective reality. As people seek to impose order on their experience of life, they find it useful to pursue this goal as a collective venture. What are the causes and cures of prejudice? Working together, social researchers have uncovered some answers that hold up to intersubjective scrutiny. Whatever your subjective experience of things, for example, you can discover for yourself that as education increases, prejudice tends to decrease. Because each of us can discover this independently, we say it is objectively true.

From the seventeenth century through the middle of the twentieth, the belief in an objective reality—that people could see and measure ever more clearly—predominated in science. For the most part, it was held not simply as a useful paradigm but as The Truth. The term *positivism* generally represents the belief in a logically ordered, objective reality that we can come to know. This is the view challenged today by postmodernists and others who suggest that perhaps only our perceptions and experiences are real.

Some say that the ideal of objectivity conceals as much as it reveals. As we saw earlier, much of what was regarded as scientific objectivity in years past was actually an agreement primarily among white, middle-class, European men. Experiences common to women, to ethnic minorities, or to the poor, for example, were not necessarily represented in that reality.

The early anthropologists are now criticized for often making modern, Westernized "sense" out of the beliefs and practices of nonliterate tribes around the world—sometimes portraying their subjects as superstitious savages. We often call orally transmitted beliefs about the distant past "creation myth," whereas we speak of our own beliefs as "history." Increasingly today, there is a demand to find the native logic by which various peoples make sense out of life.

Ultimately, we'll never know whether there is an objective reality that we experience subjectively or whether our concepts of an objective reality are illusory. So desperate is our need to know just what is going on, however, that both the positivists and the postmodernists are sometimes drawn into the belief that their view is real and true. There is a dual irony in this. On the one hand, the positivist's belief in the reality of the objective world must ultimately be based on faith; it cannot be proved by "objective" science, because that's precisely what is at issue. And the postmodernists, who say nothing is objectively so, do at least feel the absence of objective reality is *really* the way things are.

Postmodernism is often portrayed as a denial of the possibility of social science, but that's not the point I am making. This textbook makes no assumption about the existence or absence of an objective reality. At the same time, human beings demonstrate an extensive and robust ability to establish agreements as to what's "real." This appears in regard to rocks and trees, as well as ghosts and gods, and even more-elusive ideas such as loyalty and treason. Whether or not something like "prejudice" really exists, research into its nature can take place, because enough people agree that prejudice does exist, and researchers can use techniques of inquiry to study it.

Another social science paradigm, **critical realism**, suggests that we define "reality" as that which can be seen to have an effect. Since prejudice clearly has an observable effect in our lives, it must be judged "real" in terms of this point of view. This paradigm fits interestingly with the statement attributed to the early American sociologist W. I. Thomas: "If men define situations as real, they are real in their consequences" (1928: 571–72).

This book will not require or even encourage you to choose among positivism, postmodernism, or any of the other paradigms discussed in this chapter. In fact, I invite you to look for value in any and all of them as you seek to understand the world that may or may not exist around you.

Similarly, as social researchers, we are not forced to align ourselves entirely with either of these approaches. Instead, we can treat them as two distinct arrows in our quiver. Each approach compensates for the weaknesses of the other by suggesting complementary perspectives that can produce useful lines of inquiry.

The renowned British physicist, Stephen Hawking, has elegantly described the appealing simplicity of the positivistic model but tempers his remarks with a recognition of the way science is practiced.

> *According to this way of thinking, a scientific theory is a mathematical model that describes and codifies the observations we make. A good theory will describe a large range of phenomena on the basis of a few simple postulates and will make definite predictions that can be tested. If the predictions agree with the observations, the theory survives that test, though it can never be proved to be correct. On the other hand, if the observations disagree with the predictions, one has to discard or modify the theory. (At least, that is what is supposed to happen. In practice, people often question the accuracy of the observations and the reliability and moral character of those making the observations.)*

> *(2001: 31)*

In summary, a rich variety of theoretical paradigms can be brought to bear on the study of social life. With each of these fundamental frames of reference, useful theories can be constructed. We turn now to some of the issues involved in theory construction, which are of interest and use to all social researchers, from positivists to postmodernists—and all those in between. Now let's look at some other fundamental options for organizing social research.

Elements of Social Theory

As we have seen, paradigms are general frameworks or viewpoints: literally "points from which to view." They provide ways of looking at life and are grounded in sets of assumptions about the nature of reality.

---

critical realism A paradigm that holds that things are real insofar as they produce effects.

Whereas a paradigm offers a way of looking, a theory aims to explain what we see. Theories are systematic sets of interrelated statements intended to explain some aspect of social life. Thus, theories flesh out and specify paradigms. Recall from Chapter 1 that social scientists engage in both idiographic and nomothetic explanations. Idiographic explanations seek to explain a limited phenomenon as completely as possible—explaining why a particular woman voted as she did, for example—whereas nomothetic explanations attempt to explain a broad range of phenomena at least partially: identifying a few factors that account for much voting behavior in general.

Let's look a little more deliberately now at some of the elements of a theory. As I mentioned in Chapter 1, science is based on observation. In social research, *observation* typically refers to seeing, hearing, and (less commonly) touching. A corresponding idea is *fact*. Although for philosophers "fact" is as complex a notion as "reality," social scientists generally use the term to refer to some phenomenon that has been observed. It is a fact, for example, that Donald Trump defeated Hillary Clinton in the 2016 presidential election in the Electoral College.

Scientists aspire to organize many facts under "rules" called *laws*. Abraham Kaplan (1964: 91) defines *laws* as universal generalizations about classes of facts. The law of gravity is a classic example: Bodies are attracted to each other in proportion to their masses and in inverse proportion to the distance separating them.

Laws must be truly universal, however, not merely accidental patterns found among a specific set of facts. It is a fact, Kaplan points out (1964: 92), that in each of the U.S. presidential elections from 1920 to 1960, the major candidate with the longest name won. That is not a law, however, as shown by elections since. The earlier pattern was a coincidence.

Sometimes called "principles," laws are important statements about what is so. We speak of them as being "discovered," granting, of course, that our paradigms affect what we choose to look for and what we see. Laws in and of themselves do not explain anything. They just summarize the way things are. Explanation is a function of theory, as we'll see shortly.

There are no social science laws that claim the universal certainty of those of the natural sciences. Social scientists debate among themselves whether such laws will ever be discovered. Perhaps social life essentially does not abide by invariant laws. This does not mean that social life is so chaotic as to defy prediction and explanation. As we saw in Chapter 1, social behavior falls into patterns, and those patterns quite often make perfect sense, although we may have to look below the surface to find the logic.

As I just indicated, laws should not be confused with theories. Whereas a law is an observed regularity, a *theory* is a systematic explanation for observations that relate to a particular aspect of life. For example, someone might offer a theory of juvenile delinquency, prejudice, or political revolution.

Theories explain observations by means of concepts. Jonathan Turner (1989: 5) calls concepts the "basic building blocks of theory." *Concepts* are abstract elements representing classes of phenomena within the field of study. The concepts relevant to a theory of juvenile delinquency, for example, include "juvenile" and "delinquency," for starters. A "peer group"—the people you hang around with and identify with—is another relevant concept. "Social class" and "ethnicity" are undoubtedly relevant concepts in a theory of juvenile delinquency. "School performance" might also be relevant.

A *variable* is a special kind of concept. Some of the concepts just mentioned refer to things, and others refer to sets of things. As we saw in Chapter 1, each variable comprises a set of attributes; thus, *delinquency*, in the simplest case, is made up of *delinquent* and *not delinquent*. A theory of delinquency would aim at explaining why some juveniles are delinquent and others are not.

Axioms or *postulates* are fundamental assertions, taken to be true, on which a theory is grounded. In a theory of juvenile delinquency, we might begin with axioms such as "Everyone desires material comforts" and "The ability to obtain material comforts legally is greater for the wealthy than for the poor." From these we might proceed to *propositions*: specific conclusions, derived from the axiomatic groundwork, about the relationships among concepts. From our beginning axioms about juvenile delinquency, for example, we might reasonably formulate the proposition that poor youths are more likely to break the law to gain material comforts than are rich youths.

This proposition, incidentally, accords with Robert Merton's classic attempt to account for deviance in society. Merton (1957: 139–57) spoke of the agreed-on means and ends of a society. In Merton's model, nondeviants are those who share the societal agreement as to desired ends (such as a new car) and the means prescribed for achieving them (such as to buy it). One type of deviant—Merton called this type the "innovator"—agrees on the desired end but does not have access to the prescribed means for achieving it. Innovators find another method, such as crime, of attaining the desired end.

From propositions, in turn, we can derive hypotheses. A **hypothesis** is a specified testable expectation about empirical reality that follows from a more general proposition. Thus, a researcher might formulate the hypothesis, "Poor youths have higher delinquency rates than do rich youths." Research is designed to test hypotheses. In other words, research will support (or fail to support) a theory only indirectly—by testing specific hypotheses that are derived from theories and propositions.

Let's look more clearly at how theory and research come together.

Two Logical Systems Revisited

In Chapter 1, I introduced deductive and inductive theory with a promise that we would return to them later. It's later.

The Traditional Model of Science

Most of us have a somewhat idealized picture of "the scientific method." It is a view gained as a result of the physical-science education we've received ever since our elementary school days. Although this traditional model of science tells only a part of the story, it's helpful to understand its logic.

There are three main elements in the traditional model of science, typically presented in the order in which they are implemented: theory, operationalization, and observation. Let's look at each in turn.

Theory

At this point we're already well acquainted with the idea of theory. According to the traditional model of science, scientists begin with a theory, from which they derive a *hypothesis* that they can test. (See "How to Do It: Framing a Hypothesis," for more.) So, for example, as social scientists we might have a theory about the causes of juvenile delinquency. Let's assume that we've arrived at the hypothesis that delinquency is inversely related to social class. That is, as social class goes up, delinquency goes down.

Operationalization

To test any hypothesis, we must specify the meanings of all the variables involved in it: *social class* and *delinquency* in the present case. For example, *delinquency* might be specified as "being arrested for a crime," or "being convicted of a crime," and so forth. For this particular study, *social class* might be specified by family income.

Next, we need to specify how we'll measure the variables we have defined. **Operationalization** literally means the operations involved in measuring a variable. There are many ways we can pursue this topic, each of which allows for different ways of measuring our variables.

For simplicity, let's assume we're planning to conduct a survey of high school students. We might operationalize *delinquency* in the form of the question "Have you ever stolen anything?" Those who answer "yes" will be classified as delinquents in our study; those who say "no" will be classified as nondelinquents. Similarly, we might operationalize *social class* by asking respondents, "What was your family's income last year?" and providing them with a set of family income categories: under $25,000; $25,000–$49,999; $50,000–$99,999; and $100,000 or above.

---

hypothesis A specified testable expectation about empirical reality that follows from a more general proposition; more generally, an expectation about the nature of things derived from a theory. It is a statement of something that ought to be observed in the real world if the theory is correct.

operationalization One step beyond conceptualization, operationalization is the process of developing operational definitions, or specifying the exact operations involved in measuring a variable.

How to Do It

Framing a Hypothesis

As we have seen, the deductive method of research typically focuses on the testing of a hypothesis. Let's take a minute to look at how to create a hypothesis for testing.

Hypotheses state an expected causal relationship between two (or more) variables. Let's suppose you're interested in student political orientations, and your review of the literature and your own reasoning suggest to you that college major will play some part in determining students' political views. Already, we have two variables: *major* and *political orientation*. Moreover, *political orientation* is the dependent variable; you believe it depends on something else: the independent variable, *major*.

Now we need to specify the attributes composing each of those variables. For simplicity's sake, let's assume two political orientations: liberal and conservative. To simplify the matter of major, let's suppose your research interests center on the presumed differences between business students and those in the social sciences.

Even with these simplifications, you would need to specify more concretely how you'll recognize a liberal or a conservative when you come across them in your study. This process of specification will be discussed at length in Chapter 5. For now, let's assume you'll ask student-subjects whether they consider themselves liberal or conservative, letting the students decide what the terms mean to them. (As we'll see later, this simple dichotomy is unlikely to work in practice, as some students want to identify themselves as independents and other students as something else.)

Identifying students' majors isn't as straightforward as you might think. For example, what disciplines make up the social sciences in your study? Also, must students have declared their majors already, or can they just be planning to major in one of the relevant fields?

Once these issues have been settled, you're ready to state your hypothesis. For example, it might be "Students majoring in the social sciences will be more likely to identify themselves as liberals than will those majoring in business."

In addition to this basic expectation, you may wish to specify "more likely" in terms of how *much* more likely. Chapter 14 will provide some options in this regard.

At this point someone might object that "delinquency" can mean something more or different from having stolen something at one time or another or that social class isn't necessarily the same as family income. Some parents might think body piercing is a sign of delinquency even if their children don't steal, and to some "social class" might include an element of prestige or community standing as well as how much money a family has. For the researcher testing a hypothesis, however, the meaning of variables is exactly and only what the **operational definition** specifies.

In this respect, scientists are very much like Humpty Dumpty in Lewis Carroll's *Through the Looking Glass*. "When I use a word," Humpty Dumpty tells Alice, "it means just what I choose it to mean—neither more nor less."

"The question is," Alice replies, "whether you can make words mean so many different things."

> **operational definition** The concrete and specific definition of something in terms of the operations by which observations are to be categorized. The operational definition of "earning an A in this course" might be "correctly answering at least 90 percent of the final-exam questions."

To which Humpty Dumpty responds, "The question is, which is to be master—that's all" ([1895] 2009: 190).

Scientists have to be "masters" of their operational definitions for the sake of precision in observation, measurement, and communication. Otherwise, we would never know whether a study that contradicted ours did so only because it used a different set of procedures to measure one of the variables and thus changed the meaning of the hypothesis being tested. Of course, this also means that to evaluate a study's conclusions about juvenile delinquency and social class, or any other variables, we need to know how those variables were operationalized.

The way we have operationalized the variables in our imaginary study could be open to other problems, however. Perhaps some respondents will lie about having stolen anything, in which case we'll misclassify them as nondelinquent. Some respondents will not know their family incomes and will give wrong answers; others may be embarrassed and lie. We'll consider such issues in detail in Part 2.

Our operationalized hypothesis now is that the highest incidence of delinquents will be found among respondents who select the lowest family income category (under $25,000), a lower percentage of delinquents will be found in the $25,000–$49,999 category, still fewer delinquents will be found in the $50,000–$99,999 category, and the lowest percentage of delinquents will be found in the $100,000 or above category.

Observation

The final step in the traditional model of science involves actual observation, looking at the world and making measurements of what is seen. Having developed theoretical clarity and expectations and having created a strategy for looking, all that remains is to look at the way things actually appear.

Let's suppose our survey produced the following data:

	Percent Delinquent
Under $25,000	20
$25,000–$49,999	15
$50,000–$99,999	10
$100,000 or above	5

Observations producing such data would confirm our hypothesis. But suppose our findings were as follows:

	Percent Delinquent
Under $25,000	15
$25,000–$49,999	15
$50,000–$99,999	15
$100,000 or above	15

These findings would disconfirm our hypothesis regarding family income and delinquency. Disconfirmability or the possibility of falsification is an essential quality in any hypothesis. In other words, if there is no chance that our hypothesis will be disconfirmed, it hasn't said anything meaningful. You can't test whether a hypothesis is true unless your test contains the possibility of deciding it's false.

For example, the hypothesis that "juvenile delinquents" commit more crimes than do "nondelinquents" cannot possibly be disconfirmed, because criminal behavior is intrinsic to the notion of delinquency. Even if we recognize that some young people commit crimes without being caught and labeled as delinquents, they couldn't threaten our hypothesis, because our

observations would lead us to conclude they were law-abiding nondelinquents.

Figure 2-2 provides a schematic diagram of the traditional model of scientific inquiry. In it, we see the researcher beginning with an interest in a phenomenon (such as juvenile delinquency). Next comes the development of a theoretical understanding, in this case, that a single concept (such as social class) might explain others. The theoretical considerations result in an expectation about what should be observed if the theory is correct. The notation $Y = f(X)$ is a conventional way of saying that Y (for example, *delinquency*) is a function of (depends on) X (for example, *social class*). At that level, however, X and Y still have rather general meanings that could give rise to quite different observations and measurements. Operationalization specifies the procedures that will be used to measure the variables. The lowercase y in Figure 2-2, for example, is a precisely measurable indicator of capital Y. This operationalization process results in the formation of a testable hypothesis: For example, self-reported theft is a function of

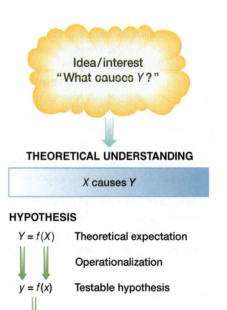

FIGURE 2-2

The Traditional Model of Science. The deductive model of scientific inquiry begins with a sometimes vague or general question, which is subjected to a process of specification, resulting in hypotheses that can be tested through empirical observations.

Issues and Insights

Hints for Stating Hypotheses

by Riley E. Dunlap

Department of Sociology, Oklahoma State University.

A hypothesis is the basic statement that is tested in research. Typically a hypothesis states a relationship between two variables. (Although it is possible to use more than two variables, you should stick to two for now.) Because a hypothesis makes a prediction about the relationship between the two variables, it must be testable so you can determine whether the prediction is right or wrong when you examine the results obtained in your study. A hypothesis must be stated in an unambiguous manner to be clearly testable. What follows are suggestions for developing testable hypotheses.

Assume you have an interest in trying to predict some phenomenon, such as "attitudes toward women's liberation," and that you can measure such attitudes on a continuum ranging from "opposed to women's liberation" to "neutral" to "supportive of women's liberation." Also assume that, lacking a theory, you'll rely on "hunches" to come up with variables that might be related to attitudes toward women's liberation.

In a sense, you can think of hypothesis construction as a case of filling in the blank: "_____ is related to attitudes toward women's liberation." Your job is to think of a variable that might plausibly be related to such attitudes and then to word a hypothesis that states a relationship between the two variables (the one that fills in the "blank" and "attitudes toward women's liberation"). You need to do so in a precise manner so that you can determine clearly whether the hypothesis is supported or not when you examine the results (in this case, most likely the results of a survey).

The key is to word the hypothesis carefully so that the prediction it makes is quite clear to you as well as to others. If you use age, note that saying "Age is related to attitudes toward women's liberation" does not say precisely how you think the two are related. (In fact, the only way this hypothesis could be falsified is if you fail to find a statistically significant relationship of any type between age and attitudes toward women's liberation). In this case, a couple of steps are necessary. You have two options:

1. "Age is related to attitudes toward women's liberation, with younger adults being more supportive than older adults." (Or, you could state the opposite, if you believed older people are likely to be more supportive.)
2. "Age is negatively related to support for women's liberation." Note here that I specify "support" for women's liberation (SWL) and then predict a negative relationship—that is, as age goes up, I predict that SWL will go down.

In this hypothesis, note that both of the variables (*age*, the independent variable or likely "cause," and *SWL*, the dependent variable or likely "effect") range from low to high. This feature of the two variables is what allows you to use "negatively" (or "positively") to describe the relationship.

Notice what happens if you hypothesize a relationship between *gender* and SWL. Because gender is a nominal variable (as you'll learn in Chapter 5), it does not range from low to high—people are either male or female (the two attributes of the variable *gender*). Consequently, you must be careful in stating the hypothesis unambiguously:

1. "Gender is positively (or negatively) related to SWL" is not an adequate hypothesis, because it doesn't specify how you expect gender to be related to SWL—that is, whether you think men or women will be more supportive of women's liberation.
2. It is tempting to say something like "Women are positively related to SWL," but this really doesn't work because female is only an attribute, not a full variable (*gender* is the variable).
3. "Gender is related to SWL, with women being more supportive than men" would be my recommendation. Or, you could say, "with men being less supportive than women," which makes the identical prediction. (Of course, you could also make the opposite prediction, that men are more supportive than women are, if you wished.)
4. Equally legitimate would be "Women are more likely to support women's liberation than are men." (Note the need for the second "are," or you could be construed as hypothesizing that women support women's liberation more than they support men—not quite the same idea.)

The above examples hypothesized relationships between a "characteristic" (age or gender) and an "orientation" (attitudes toward women's liberation). Because the causal order is pretty clear (obviously age and gender come before attitudes and are less alterable), we could state the hypotheses as I've done, and everyone would assume that we were stating causal hypotheses.

Finally, you may run across references to the **null hypothesis**, especially in statistics. Such a hypothesis predicts no relationship (technically, no statistically significant relationship) between the two variables, and it is always implicit in testing hypotheses. Basically, if you have hypothesized a positive (or negative) relationship, you are hoping that the results will allow you to reject the null hypothesis and verify your hypothesized relationship.

---

null hypothesis In connection with hypothesis testing and tests of statistical significance, the hypothesis that suggests there is no relationship among the variables under study. You may conclude that the variables are related after having statistically rejected the null hypothesis.

family income. Observations aimed at finding out whether this statement accurately describes reality are part of what is typically called *hypothesis testing*. (See "Issues and Insights: Hints for Stating Hypotheses," for more on the process of formulating hypotheses.)

This operationalization process results in the formation of a testable hypothesis: For example, increasing family income reduces self-reported theft. Observations aimed at finding out whether this is true are part of what is typically called "hypothesis testing."

Deduction and Induction Compared

The traditional model of science uses deductive logic (see Chapter 1). In this section, we're going to see how deductive logic fits into social science research and contrast it with inductive logic. W. I. B. Beveridge, a philosopher of science, describes these two systems of logic as follows:

> *Logicians distinguish between inductive reasoning (from particular instances to general principles, from facts to theories) and deductive reasoning (from the general to the particular, applying a theory to a particular case). In induction one starts from observed data and develops a generalization which explains the relationships between the objects observed. On the other hand, in deductive reasoning one starts from some general law and applies it to a particular instance.*

> *(1950: 113)*

The classical illustration of deductive logic is the familiar syllogism "All men are mortal; Socrates is a man; therefore, Socrates is mortal." This syllogism presents a theory and its operationalization. To prove it, you might then perform an empirical test of Socrates' mortality. That is essentially the approach discussed as the traditional model.

Using inductive logic, you might begin by noting that Socrates is mortal and by observing several other men as well. You might then note that all the observed men were mortals, thereby arriving at the tentative conclusion that all men are mortal.

Let's consider an actual research project as a vehicle for comparing the roles of deductive and inductive logic in theory and research.

A Case Illustration

Years ago, Charles Glock, Benjamin Ringer, and I (1967) set out to discover what caused differing levels of church involvement among U.S. Episcopalians. Several theoretical or quasi-theoretical positions suggested possible answers. I'll focus on only one here—what we came to call the "Comfort hypothesis."

In part, we took our lead from the Christian injunction to care for "the halt, the lame, and the blind" and those who are "weary and heavy laden." At the same time, ironically, we noted the Marxist assertion that religion is "the opium of the people." Given both, it made sense to expect the following, which was our hypothesis: "Parishioners whose life situations most deprive them of satisfaction and fulfillment in the secular society turn to the church for comfort and substitute rewards" (Glock, Ringer, and Babbie 1967: 107–8).

Having framed this general hypothesis, we set about testing it. Were those deprived of satisfaction in the secular society in fact more religious than those who received more satisfaction from the secular society? To answer this, we needed to determine who was deprived. Our questionnaire included items that intended to indicate whether parishioners were relatively deprived or gratified in secular society.

To start, we reasoned that men enjoy higher status than do women in our generally male-dominated society. It followed that, if our hypothesis were correct, women should appear more religious than men. Once the survey data had been collected and analyzed, our expectation about gender and religion was clearly confirmed. On three separate measures of religious involvement—ritual (for example, church attendance), organizational (for example, belonging to church organizations), and intellectual (for example, reading church publications)—women were more religious than men. On our overall measure, women scored 50 percent higher than men.

In another test of the Comfort hypothesis, we reasoned that in a youth-oriented society, old people would be more deprived of secular gratification than the young would be. Once again, the data confirmed our expectation. The oldest parishioners were more religious than were the middle-aged, who were more religious than were the young adults.

Social class—measured by education and income—afforded another test, which was successful in proving our hypothesis. Those with low social status were more involved in the church than were those with high social status.

The hypothesis was even confirmed in a test that went against everyone's commonsense expectations. Despite church posters showing worshipful young families and bearing the slogan "The Family That Prays Together Stays Together," the Comfort hypothesis suggested that parishioners who were married and had children—the clear U.S. ideal at that time—would enjoy secular gratification in that regard. As a consequence, they should be less religious than those who lacked one or both family components. Thus, we hypothesized that parishioners who were both single and childless should be the most religious; those with either spouse or child should be somewhat less religious; and those who were married and had children—representing the ideal pictured on all those posters—should be least religious of all. That's exactly what we found.

Finally, the Comfort hypothesis suggested that the various kinds of secular deprivation would be cumulative: Those with all the characteristics associated with deprivation should be the most religious; those with none should be the least. When we combined the four individual measures of deprivation into a composite measure (see Chapter 6 for methods of doing this), the theoretical expectation was exactly confirmed. Comparing the two extremes, we found that single, childless, older, lower-class, female parishioners scored more than three times as high on the measure of church involvement than did young, married, upper-class fathers.

This research example clearly illustrates the logic of the deductive model. Beginning with general, theoretical expectations about the impact of social deprivation on church involvement, we derived concrete hypotheses linking specific measurable variables, such as *age* and *church attendance*. We then analyzed the actual empirical data to determine whether empirical reality supported the deductive expectations. Sounds good, right?

Alas, I've been fibbing a little bit just now. To tell the truth, although we began with an interest in discovering what caused variations in church involvement among Episcopalians, we didn't actually begin with a Comfort hypothesis, or any other hypothesis for that matter. (In the interest of further honesty, Glock and Ringer initiated the study, and I joined it years after the data had been collected.)

A questionnaire was designed to collect information from parishioners that might shed a bit of light on why some participated in the church more than others, but construction of the questionnaire was not guided by any precise, deductive theory. Once the data were collected, the task of explaining differences in religiosity began with an analysis of variables that have a wide impact on people's lives, including *gender, age, social class,* and *family status.* Each of these four variables was found to relate strongly to church involvement in the ways already described. Rather than being good news, this presented a dilemma.

Glock recalls discussing his findings with colleagues over lunch at the Columbia faculty club. Once he had displayed the tables illustrating the impact of the variables and their cumulative effect, a colleague asked, "What does it all mean, Charlie?" Glock was at a loss. Why were those variables so strongly related to church involvement?

That question launched a process of reasoning about what the several variables had in common, aside from their impact on religiosity. (The composite index was originally labeled "Predisposition to Church Involvement.") Eventually, we saw that each of the four variables also reflected differential status in the secular society, and then we had the thought that perhaps the issue of comfort was involved. Thus, the inductive process had moved from concrete observations to a general theoretical explanation. See "Applying Concepts in Everyday Life: Church Involvement," for more on the Glock study.

A Graphic Contrast

As the preceding case illustration shows, theory and research can be accomplished both inductively and deductively. Figure 2-3 shows a graphic comparison of the deductive and inductive methods. In both cases, we are interested in the relationship between the number of hours spent studying for an exam and the grade earned on that exam. Using the deductive method, we would begin by examining the matter logically. Doing well on an exam reflects a student's ability to recall and to manipulate information. Both of these abilities should be increased by exposure to the information before the exam.

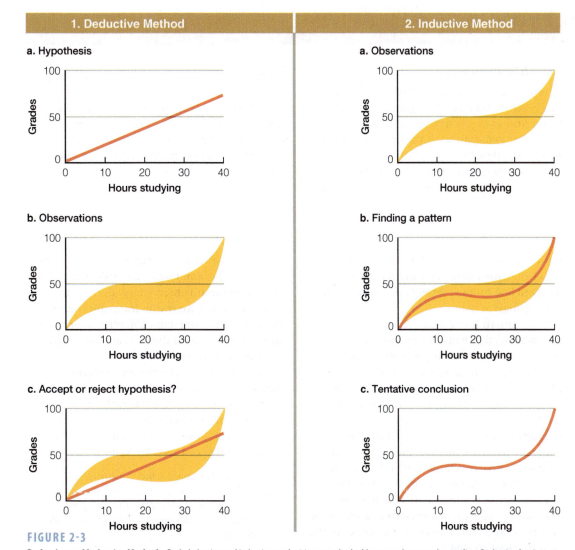

FIGURE 2-3

Deductive and Inductive Methods. Both deduction and induction are legitimate and valuable approaches to understanding. Deduction begins with an expected pattern that is tested against observations, whereas induction begins with observations and seeks to find a pattern within them.

In this fashion, we would arrive at a hypothesis suggesting a positive relationship between the number of hours spent studying and the grade earned on the exam. That is, we expect grades to increase as the hours of studying increase. If increased study hours produced decreased grades, we would call it a negative relationship. The hypothesis is represented by the line in panel 1a of Figure 2-3.

Our next step would be to make observations relevant to testing our hypothesis. The shaded area in panel 1b of the figure represents perhaps hundreds of observations of different students, noting how many hours they studied and what grades they got. Finally, in panel 1c, we

compare the hypothesis and the observations. Because observations in the real world seldom, if ever, match our expectations perfectly, we must decide whether the match is close enough to confirm the hypothesis. In other words, can we conclude that the hypothesis describes the general pattern that exists, granting some variations in real life?

Now let's address the same research question by using the inductive method. We would begin—as in panel 2a of the figure—with a set of observations. Curious about the relationship between hours spent studying and grades earned, we might simply arrange to collect some relevant data. Then we'd look for a pattern that best

Church Involvement

Although many church leaders believe that the function of churches is to shape members' behavior in the community, the Glock study suggests that church involvement primarily reflects a need for comfort by those who are denied gratification in the secular society. How might churches apply these research results?

On the one hand, churches might adjust their programs to the needs that were drawing their members to participate. They might study members' needs for gratification and develop more programs to satisfy those needs.

On the other hand, churches could seek to remind members that the purpose of participation is to learn and practice proper behavior. Following that strategy would probably change participation patterns, attracting new participants to the church while driving away others.

represented or summarized our observations. In panel 2b of the figure, the pattern is shown as a curved line running through the center of the curving mass of points.

The pattern found among the points in this case suggests that with 1 to 15 hours of studying, each additional hour generally produces a higher grade on the exam. With 15 to about 25 hours, however, more study seems to lower the grade slightly. Studying more than 25 hours, on the other hand, results in a return to the initial pattern: More hours produce higher grades. Using the inductive method, then, we end up with a tentative conclusion about the pattern of the relationship between the two variables. The conclusion is tentative because the observations we've made cannot be taken as a test of the pattern—those observations are the source of the pattern we've created.

In actual practice, theory and research interact through a never-ending alternation of deduction and induction. Walter Wallace (1971) represents this process as a circle, which is presented in a modified form in Figure 2-4.

When Emile Durkheim ([1897] 1951) pored over table after table of official statistics on suicide rates in different areas, he was struck by the fact that Protestant countries consistently had higher suicide rates than did Catholic ones. Why would that be the case? His initial observations led him to create a theory of religion, social integration, anomie, and suicide. His theoretical explanations led to further hypotheses and further observations.

In summary, the scientific norm of logical reasoning provides a two-way bridge between theory and research. Scientific inquiry in practice typically involves an alternation between deduction and induction. During the deductive phase, we reason toward observations; during the inductive phase, we reason from observations. Both deduction and induction offer routes to the construction of social theories, and both logic and observation are essential.

Although both inductive and deductive methods are valid in scientific inquiry, individuals may feel more comfortable with one approach than the other. Consider this exchange in Sir Arthur Conan Doyle's "A Scandal in Bohemia," as Sherlock Holmes answers Dr. Watson's inquiry:

> "What do you imagine that it means?"
> "I have no data yet. It is a capital mistake to theorise before one has data. Insensibly one begins to twist facts to suit theories, instead of theories to suit facts."

> (Doyle [1891] 1892: 13)

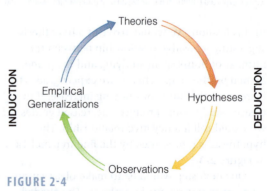

FIGURE 2-4

The Wheel of Science. The theory and research cycle can be compared to a relay race; although all participants do not necessarily start or stop at the same point, they share a common goal—to examine all levels of social life.

Source: Adapted from Walter Wallace, *The Logic of Science in Sociology* (Hawthorne, NY: Aldine, 1971).

Some social scientists would rally behind this inductive position (see especially the discussion of grounded theory in Chapter 10), whereas others would take a deductive stance. Most, however, concede the legitimacy of both. With this understanding of the deductive and inductive links between theory and research, let's delve a little more deeply into how theories are constructed using these two different approaches.

Deductive Theory Construction

To see what is involved in deductive theory construction and hypothesis testing, let's imagine that you're going to construct a deductive theory. How would you go about it?

Getting Started

The first step in deductive theory construction is to pick a topic that interests you. It can be broad, such as "What is the structure of society?" or narrower, as in "Why do people support or oppose a woman's right to an abortion?" Whatever the topic, it should be something you're interested in understanding and explaining.

Once you've picked your topic, you then undertake an inventory of what is known or thought about it. In part, this means writing down your own observations and ideas about it. Beyond that, you need to learn what other scholars have said about it. You can do this by talking to other people and by reading what others have written about it. Appendix A provides guidelines for using the library—you'll probably spend a lot of time there.

Your preliminary research will probably uncover consistent patterns discovered by prior scholars. For example, religious and political variables will stand out as important determinants of attitudes about abortion. Findings such as these will be quite useful to you in creating your own theory. We'll return to techniques of literature review as the book continues.

Throughout this process, introspection is helpful. If you can look at your own personal processes—including reactions, fears, and prejudices you aren't especially proud of—you may be able to gain important insights into human behavior in general.

Constructing Your Theory

Although theory construction is not a lockstep affair, the following list of elements in theory construction should organize the activity for you.

1. Specify the topic.
2. Specify the range of phenomena your theory addresses. Will your theory apply to all of human social life, will it apply only to U.S. citizens, only to young people, or what?
3. Identify and specify your major concepts and variables.
4. Find out what is known (or what propositions have been demonstrated) about the relationships among those variables.
5. Reason logically from those propositions to the specific topic you're examining.

We've already discussed items (1) through (3), so let's focus now on (4) and (5). As you identify the relevant concepts and discover what has already been learned about them, you can begin to create a propositional structure that explains the topic under study. For the most part, social scientists have not created formal, propositional theories. Still, looking at a well-reasoned example is useful. Let's look now at an example of how these building blocks fit together in actual deductive theory construction and empirical research.

An Example of Deductive Theory: Distributive Justice

A topic of central interest to scholars is the concept of distributive justice, people's perception of whether they're being treated fairly by life, whether they're getting "their share." Guillermina Jasso describes the theory of distributive justice more formally, as follows:

> The theory provides a mathematical description of the process whereby individuals, reflecting on their holdings of the goods they value (such as beauty, intelligence, or wealth), compare themselves to others, experiencing a fundamental instantaneous magnitude of the justice evaluation (J), which captures their sense of being fairly or unfairly treated in the distributions of natural and social goods.
>
> (1988: 11)

Notice that Jasso has assigned a letter to her key variable: J will stand for distributive justice. She does this to support her intention of stating her theory in mathematical formulas. Though theories are often expressed mathematically, we'll not delve too deeply into that practice here.

Jasso indicates that there are three kinds of postulates in her theory. "The first makes explicit the fundamental axiom which represents the substantive point of departure for the theory." She elaborates as follows:

> The theory begins with the received Axiom of Comparison, which formalizes the long-held view that a wide class of phenomena, including happiness, self-esteem, and the sense of distributive justice, may be understood as the product of a comparison process.

> *(1988: 11)*

Thus, our sense of whether we are receiving a "fair" share of the good things of life comes from comparing ourselves with others. If this seems obvious to you, that's good. Remember, axioms are the taken-for-granted beginnings of theory.

Jasso continues to do the groundwork for her theory. First, she indicates that our sense of distributive justice is a function of "Actual Holdings (*A*)" and "Comparison Holdings (*C*)" of some good. Let's consider money. My sense of justice in this regard is a function of how much I actually have, as compared with how much others have. By specifying the two components of the comparison, Jasso can use them as variables in her theory.

Jasso then offers a "measurement rule" that further specifies how the two variables, *A* and *C*, will be conceptualized. This step is needed because some of the goods to be examined are concrete and commonly measured (such as money), whereas others are less tangible (such as respect). The former kind, she says, will be measured conventionally, whereas the latter will be measured "by the individual's relative rank within a specially selected comparison group." The theory will provide a formula for making that measurement (Jasso 1988: 13).

Jasso continues in this fashion to introduce additional elements, weaving them into mathematical formulas for deriving predictions about the workings of distributive justice in a variety of social settings. Here is a sampling of where her theorizing takes her (1988: 14–15):

- Other things [being] the same, a person will prefer to steal from a fellow group member rather than from an outsider.

- The preference to steal from a fellow group member is more pronounced in poor groups than in rich groups.
- In the case of theft, informants arise only in cross-group theft, in which case they are members of the thief's group.
- Persons who arrive a week late at summer camp or for the freshman year of college are more likely to become friends of persons who play games of chance than of persons who play games of skill.
- A society becomes more vulnerable to deficit spending as its wealth increases.
- Societies in which population growth is welcomed must be societies in which the set of valued goods includes at least one quantity-good, such as wealth.

Jasso's theory leads to many other propositions, but this sampling should provide a good sense of where deductive theorizing can take you. To get a feeling for how she reasons her way to these propositions, let's look briefly at the logic involved in two of the propositions that relate to theft within and outside one's group.

- Other things [being] the same, a person will prefer to steal from a fellow group member rather than from an outsider.

Beginning with the assumption that thieves want to maximize their relative wealth, ask yourself whether that goal would be best served by stealing from those you compare yourself with or from outsiders. In each case, stealing will increase your Actual Holdings, but what about your Comparison Holdings?

A moment's thought should suggest that stealing from people in your comparison group will lower their holdings, further increasing your relative wealth. To simplify, imagine there are only two people in your comparison group: you and I. Suppose we each have $100. If you steal $50 from someone outside our group, you will have increased your relative wealth by 50 percent as compared with mine: $150 versus $100. But if you steal $50 from me, you will have increased your relative wealth 200 percent: $150 to my $50. Your goal is best served by stealing from within the comparison group.

- In the case of theft, informants arise only in cross-group theft, in which case they are members of the thief's group.

Can you see why it would make sense for informants: (1) to arise only in the case of cross-group theft and (2) to come from the thief's comparison group? This proposition again depends on the fundamental assumption that everyone wants to increase his or her relative standing. Suppose you and I are in the same comparison group, but this time the group contains additional people. If you steal from someone else within our comparison group, my relative standing in the group does not change. Although your wealth has increased, the average wealth in the group remains the same (because someone else's wealth has decreased by the same amount). I have no incentive to inform on you.

If you steal from someone outside our comparison group, your nefarious income increases the total wealth in our group, so my own wealth relative to that total is diminished. Because my relative wealth has suffered, I'm more likely to bring an end to your stealing.

This last deduction also begins to explain why informants are more likely to arrive from within the thief's comparison group. We've just seen how my relative standing was decreased by your theft. How about other members of the other group? Each of them would actually profit from the theft, because you would have reduced the total with which they compare themselves. Hence, the theory of distributive justice predicts that informants arise from the thief's comparison group.

This brief and selective peek into Jasso's derivations should give you some sense of the enterprise of deductive theory. Realize, of course, that the theory guarantees none of the given predictions. The role of research is to test each of them empirically to determine whether what makes sense (logic) occurs in practice (observation).

There are two important elements in science, then: logical integrity and empirical verification. Both are essential to scientific inquiry and discovery. Logic alone is not enough, but on the other hand, the mere observation and collection of empirical facts does not provide understanding—the telephone directory, for example, is not a scientific conclusion. Observation, however, can be the springboard for the construction of a social science theory, as we shall now see in the case of inductive theory.

Inductive Theory Construction

Quite often, social scientists begin constructing a theory through the inductive method by first observing aspects of social life and then seeking to discover patterns that may point to relatively universal principles. Barney Glaser and Anselm Strauss (1967) coined the term *grounded theory* in reference to this method.

Field research—the direct observation of events in progress—is frequently used to develop theories through observation (see Chapter 10). A long and rich anthropological tradition has seen this method used to good advantage.

Among social scientists of the twentieth century, no one was more adept at seeing the patterns of human behavior through observation than Erving Goffman:

> A game such as chess generates a habitable universe for those who can follow it, a plane of being, a cast of characters with a seemingly unlimited number of different situations and acts through which to realize their natures and destinies. Yet much of this is reducible to a small set of interdependent rules and practices. If the meaningfulness of everyday activity is similarly dependent on a closed, finite set of rules, then explication of them would give one a powerful means of analyzing social life.
>
> (1974: 5)

In a variety of research efforts, Goffman uncovered the rules of such diverse behaviors as living in a mental institution (1961) and managing the "spoiled identity" of disfiguration (1963). In each case, Goffman observed the phenomenon in depth and teased out the rules governing behavior. Goffman's research provides an excellent example of qualitative field research as a source of grounded theory.

Our earlier discussion of the Comfort hypothesis and church involvement shows that qualitative field research is not the only method of observation appropriate to the development of inductive theory. Here's another detailed example to illustrate further the construction of inductive theory using quantitative methods.

An Example of Inductive Theory: Why Do People Smoke Marijuana?

During the 1960s and 1970s, marijuana use on U.S. college campuses was a subject of considerable discussion in the popular press. Some people were troubled by marijuana's popularity; others welcomed it. What interests us here is why some students smoked marijuana and others didn't. A survey of students at the University of Hawaii (Takeuchi 1974) provided data to answer that question. While the reasons and practices regarding pot may have changed, the subtle redefinition of what needed explaining is still instructive.

At the time of the study, people were offering numerous explanations for drug use. Those who opposed drug use, for example, often suggested that marijuana smokers were academic failures trying to avoid the rigors of college life. Those in favor of marijuana, on the other hand, often spoke of the search for new values: Marijuana smokers, they said, were people who had seen through the hypocrisy of middle-class values.

David Takeuchi's 1974 analysis of the data gathered from University of Hawaii students, however, did not support any of the explanations being offered. Those who reported smoking marijuana had essentially the same academic records as those who didn't smoke it, and both groups were equally involved in traditional "school spirit" activities. Both groups seemed to feel equally well integrated into campus life.

There were differences, however:

1. Women were less likely than men to smoke marijuana.
2. Asian students (a large proportion of the student body) were less likely than non-Asians to smoke marijuana.
3. Students living at home were less likely to smoke marijuana than were those living in their own apartments.

As in the case of religiosity, the three variables independently affected the likelihood of a student's smoking marijuana. About 10 percent of the Asian women living at home had smoked marijuana, as compared with about 80 percent of the non-Asian men living in apartments. And, as in the religiosity study, the researchers discovered a powerful pattern of drug use before they had an explanation for that pattern.

In this instance, the explanation took a peculiar turn. Instead of explaining why some students smoked marijuana, the researchers explained why some didn't. Assuming that all students had some motivation for trying drugs, the researchers suggested that students differed in the degree of "social constraints" preventing them from following through on that motivation.

U.S. society is, on the whole, more permissive with men than with women when it comes to deviant behavior. Consider, for example, a group of men getting drunk and boisterous. We tend to dismiss such behavior with references to "camaraderie" and "having a good time," whereas a group of women behaving similarly would probably be regarded with at least some disapproval. We have an idiom, "Boys will be boys," but no comparable idiom for girls. The researchers reasoned, therefore, that women would have more to lose by smoking marijuana than would men. Being female, then, provided a constraint against smoking marijuana.

Students living at home had obvious constraints against smoking marijuana, as compared with students living on their own. Quite aside from differences in opportunity, those living at home were seen as being more dependent on their parents and hence more vulnerable to additional punishment for breaking the law.

Finally, the Asian subculture in Hawaii has traditionally placed a higher premium on obedience to the law than have other subcultures. As such, Asian students would have more to lose if they were caught violating the law by smoking marijuana.

Overall, then, a "social constraints" theory was offered as the explanation for observed differences in the likelihood of smoking marijuana. The more constraints a student had, the less likely he or she would be to smoke marijuana. It bears repeating that the researchers had no thoughts about such a theory when their research began. The theory came from an examination of the data.

The Links between Theory and Research

Throughout this chapter, we have seen various aspects of the links between theory and research in social science inquiry. In the deductive model, research is used to test theories. In the inductive model, theories are developed from the analysis of research data. This section looks more closely into the ways in which theory and research are related in actual social science inquiry.

Although we have discussed two idealized logical models for linking theory and research, social science inquiries have developed a great many variations on these themes. Sometimes theoretical issues are introduced merely as a background for empirical analyses. Other studies cite selected empirical data to bolster theoretical arguments. In neither case do theory and research really interact for the purpose of developing new explanations. Some studies make no use of theory at all, aiming specifically, for example, at an ethnographic description of a particular social situation, such as an anthropological account of food and dress in a particular society.

As you read social research reports, however, you'll often find that the authors are conscious of the implications of their research for social theories and vice versa. Here are a few examples to illustrate this point.

When W. Lawrence Neuman (1998) set out to examine the problem of monopolies (the "trust problem") in U.S. history, he saw the relevance of theories about how social movements transform society ("state transformation"). He became convinced, however, that existing theories were inadequate for the task before him:

> State transformation theory links social movements to state policy formation processes by focusing on the role of cultural meaning in organized political struggles. Despite a resemblance among concepts and concerns, constructionist ideas found in the social problems, social movements, and symbolic politics literatures have not been incorporated into the theory. In this paper, I draw on these three literatures to enhance state transformation theory.

(1998: 315)

Having thus modified state transformation theory, Neuman had a theoretical tool that could guide his inquiry and analysis into the political maneuverings related to monopolies beginning in the 1880s and continuing until World War I. Thus, theory served as a resource for research and at the same time was modified by it.

In a somewhat similar study, Alemseghed Kebede and J. David Knottnerus (1998) set out to investigate the rise of Rastafarianism in the Caribbean. However, they felt that recent theories on social movements had become too positivistic in focusing on the mobilization of resources. Resource mobilization theory, they felt,

> downplays the motivation, perceptions, and behavior of movement participants . . . and concentrates instead on the whys and hows of mobilization. Typically theoretical and research problems include: How do emerging movement organizations seek to mobilize and routinize the flow of resources and how does the existing political apparatus affect the organization of resources?

(1998: 500)

To study Rastafarianism more appropriately, the researchers felt the need to include several concepts from contemporary social psychology. In particular, they sought models to use in dealing with problems of meaning and collective thought.

Frederika Schmitt and Patricia Yancey Martin (1999) were particularly interested in discovering what produced successful rape crisis centers and how such centers dealt with the organizational and political environments within which they operated. The researchers found that theoretical constructs appropriate to their inquiry:

> This case study of unobtrusive mobilizing by [the] Southern California Rape Crisis Center uses archival, observational, and interview data to explore how a feminist organization worked to change police, schools, prosecutor[s], and some state and national organizations from 1974 to 1994. Mansbridge's concept of street theory and Katzenstein's concepts of unobtrusive mobilization and discursive politics guide the analysis.

(1999: 364)

In summary, there is no simple recipe for conducting social science research. It is far more open-ended than the traditional view of science suggests. Ultimately, science depends on two categories of activity: logic and observation. As you'll see throughout this book, they can be fit together in many patterns.

The Importance of Theory in the "Real World"

At this point you may be saying, "Sure, theory and research are OK, but what do they have to do with the real world?" As we'll see later in this book, there are many practical applications of social research, from psychology to social reform. Think, for instance, how someone could make use of David Takeuchi's research on marijuana use.

But how does theory work in such applications? In some minds, theoretical and practical matters are virtual opposites. Social scientists committed to the use of science know differently, however.

Lester Ward, the first president of the American Sociological Association, was committed to the application of social research in practice, or the use of that research toward specific ends. Ward distinguished pure and applied sociology as follows:

> Just as pure sociology aims to answer the questions What, Why, and How, so applied sociology aims to answer the question What for. The former deals with facts, causes, and principles, the latter with the object, end, or purpose.

(1906: 5)

No matter how practical and/or idealistic your aims, a theoretical understanding of the terrain may spell the difference between success and failure. As Ward saw it, "Reform may be defined as the desirable alteration of social structures. Any attempt to do this must be based on a full knowledge of the nature of such structures, otherwise its failure is certain" (1906: 4).

Suppose you were concerned about poverty in the United States. The sociologist Herbert Gans (1971) suggests that understanding the functions that poverty serves for people who are not poor is vital. For example, the persistence of poverty

means there will always be people willing to do the jobs no one else wants to do—and they'll work for very little money. The availability of cheap labor provides many affordable comforts for the nonpoor.

By the same token, poverty provides many job opportunities for social workers, unemployment office workers, police, and so forth. If poverty were to disappear, what would happen to social work colleges, for example?

I don't mean to suggest that people conspire to keep the poor in their place or that social workers secretly hope for poverty to persist. Nor do I want to suggest that the dark cloud of poverty has a silver lining. I merely want you to understand the point made by Ward, Gans, and many other sociologists: If you want to change society, you need to understand how it operates. As William White (1997) argues, "Theory helps create questions, shapes our research designs, helps us anticipate outcomes, helps us design interventions."

Research Ethics and Theory

In this chapter, we've seen how the paradigms and theories that guide research inevitably impact what is observed and how it is interpreted. Choosing a particular paradigm or theory does not guarantee a particular research conclusion, but it will affect what you look for and what you ignore. Whether you choose a functionalist or a conflict paradigm to organize your research on police–community relations will make a big difference.

This choice can produce certain ethical issues. Choosing a theoretical orientation for the purpose of encouraging a particular conclusion, for example, would generally be regarded as unethical. However, when researchers intend to bring about social change through their work, they usually choose a theoretical orientation appropriate to that intention. Let's say you're concerned about the treatment of homeless people by the police in your community. You might very well organize your research in terms of interactionist or conflict paradigms and theories that would reveal any instances of mistreatment that may occur. The danger lies in the bias this might cause in your research.

What do you think?...Revisited

As we've seen, many different paradigms have been suggested for the study of society. The opening *What Do You Think?* box asked which one was true. You should see by now that the answer is "None of the above." However, none of the paradigms is false, either.

By their nature, paradigms are neither true nor false. They are merely different ways of looking and of seeking explanations. Thus, they may be judged as useful or not useful in a particular situation, but not as true or false.

Imagine that you and some friends are in a totally darkened room. Each of you has a flashlight. When you turn on your own flashlight, you create a partial picture of what's in the room, whereby some things are revealed but others remain concealed. Now imagine your friends taking turns turning on their flashlights. Every person's flashlight presents a different picture of what's in the room, revealing part, but not all, of it.

Paradigms are like the flashlights in this gripping tale. Each offers a particular point of view that may or may not be useful in a given circumstance. None reveals the full picture, or the "truth."

Two factors counter this potential bias. First, as we'll see in the remainder of the book, social science research techniques—the various methods of observation and analysis—place a damper on our simply seeing what we expect. Even if you expect to find the police mistreating the homeless and use theories and methods that will reveal such mistreatment, you will not observe that which isn't there—if you apply those theories and methods appropriately.

Second, the collective nature of social research offers further protection. As we'll discuss more in Chapter 15, *peer review*, in which researchers evaluate each other's efforts, will point to instances of shoddy or biased research. Moreover, with several researchers studying the same phenomenon, perhaps using different paradigms, theories, and methods, the risk of biased research findings is further reduced.

MAIN POINTS

Introduction
- Theories seek to provide logical explanations.

Some Social Science Paradigms
- A paradigm is a fundamental model or scheme that organizes our view of something.
- Social scientists use a variety of paradigms to organize how they understand and inquire into social life.
- A distinction between types of theories that cut across various paradigms is macrotheory (theories about large-scale features of society) versus microtheory (theories about smaller units or features of society).
- The positivistic paradigm assumes that we can scientifically discover the rules governing social life.
- The conflict paradigm focuses on the attempt of one person or group to dominate others and to avoid being dominated.

- The symbolic interactionist paradigm examines how shared meanings and social patterns are developed in the course of social interactions.
- Ethnomethodology focuses on the ways people make sense out of life in the process of living it, as though each were a researcher engaged in an inquiry.
- The structural functionalist (or social systems) paradigm seeks to discover what functions the many elements of society perform for the whole system—for example, the functions of mothers, labor unions, and radio talk shows.
- Feminist paradigms, in addition to drawing attention to the oppression of women in most societies, highlight how previous images of social reality have often come from and reinforced the experiences of men.
- Like feminist paradigms, critical race theory both examines the disadvantaged position of a social group (African Americans) and offers a

different vantage point from which to view and understand society.

- Some contemporary theorists and researchers have challenged the long-standing belief in an objective reality that abides by rational rules. They point out that it is possible to agree on an "intersubjective" reality.

Elements of Social Theory

- The elements of social theory include observations, facts, and laws (which relate to the reality being observed), as well as concepts, variables, axioms or postulates, propositions, and hypotheses (which are logical building blocks of the theory itself).

Two Logical Systems Revisited

- In the traditional model of science, scientists proceed from theory to operationalization to observation. But this image is not an accurate picture of how scientific research is actually done.
- Social science theory and research are linked through two logical methods: Deduction involves the derivation of expectations or hypotheses from theories. Induction involves the development of generalizations from specific observations.
- Science is a process involving an alternation of deduction and induction.

Deductive Theory Construction

- Guillermina Jasso's theory of distributive justice illustrates how formal reasoning can lead to a variety of theoretical expectations that can be tested by observation.

Inductive Theory Construction

- David Takeuchi's study of factors influencing marijuana smoking among University of Hawaii students illustrates how collecting observations can lead to generalizations and an explanatory theory.

The Links between Theory and Research

- In practice, there are many possible links between theory and research and many ways of going about social inquiry.
- Using theories to understand how society works is key to offering practical solutions to society's problems.

The Importance of Theory in the "Real World"

- No matter what a researcher's aims are in conducting social research, a theoretical

understanding of his or her subject may spell the difference between success and failure.

- If one wants to change society, one needs to understand the logic of how it operates.

Research Ethics and Theory

- Researchers must guard against letting their choice of theory or paradigms bias their research results.
- The collective nature of social research offers protection against biased research findings.

KEY TERMS

critical realism	macrotheory
hypothesis	microtheory
interest convergence	null hypothesis
operational definition	paradigm
operationalization	

PROPOSING SOCIAL RESEARCH: THEORY

As this chapter has indicated, social research can be pursued within numerous theoretical paradigms—each suggesting a somewhat different way to approach the research question. In this portion of your proposal, you should identify the paradigm(s) that will shape the design of your research.

We've also seen that paradigms provide frameworks within which causal theories can be developed. Perhaps your research project will explore or test an existing theory. Or more ambitiously, you may propose your own theory or hypothesis for testing. This is the section of the proposal in which to describe this aspect of your project.

Not all research projects are formally organized around the creation and/or testing of theories and hypotheses. However, your research will involve theoretical concepts, which should be described in this section of the proposal. As we'll see more fully in Chapter 15, this portion of your proposal will reflect the literature on previous theories and research that have shaped your own thinking and research plans.

REVIEW QUESTIONS

1. Consider the possible relationship between education and prejudice (mentioned in Chapter 1). How might that relationship be examined through (a) deductive and (b) inductive methods?

2. Select a social problem that concerns you, such as war, pollution, overpopulation, prejudice, or poverty. Then, use one of the paradigms in this chapter to address that problem. What would be the main variables involved in the study of that problem, including variables that may cause it or hold the key to its solution?

3. What, in your own words, is the difference between a paradigm and a theory?

4. You have been hired to evaluate how well a particular health maintenance organization (HMO) serves the needs of its clients. How might you implement this study using each of the following: (a) the interactionist paradigm, (b) the social systems or functionalist paradigm, and (c) the conflict paradigm?

CHAPTER 3

The Ethics and Politics of Social Research

CHAPTER OVERVIEW

Social research takes place in a social context. Researchers must therefore take into account many ethical and political considerations alongside scientific ones in designing and executing their research. Often, however, clear-cut answers to thorny ethical and political issues are hard to come by.

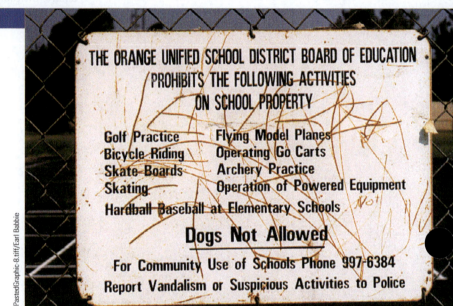

PastedGraphic-8.tiff/Earl Babbie

Introduction

To present a realistic and useful introduction to doing social research, this book must consider four main constraints on research projects: scientific, administrative, ethical, and political. Most of the book focuses on scientific and administrative constraints. We'll see that the logic of science suggests certain research procedures, but we'll also see that some scientifically "perfect" study designs are not administratively feasible because they would be too expensive or take too long to execute. Throughout the book, therefore, we'll deal with workable compromises.

Before we get to the scientific and administrative constraints on research, it's useful to explore the other important considerations in doing research in the real world: ethics and politics, which this chapter covers. Just as certain procedures are too impractical to use, others are either ethically prohibited or politically difficult or impossible. Here's a story to illustrate what I mean.

Several years ago, I was invited to sit in on a planning session to design a study of legal education in California. The joint project was to be conducted by a university research center and the state bar association. The purpose of the project was to improve legal education by learning which aspects of the law school experience were related to success on the bar exam. Essentially, the plan was to prepare a questionnaire that would get detailed information about the law school experiences of individuals. People would be required to answer the questionnaire when they took the bar exam. By analyzing how people with different kinds of law school experiences did on the bar exam, we could find out what sorts of things worked and what didn't. The findings of the research could be made available

to law schools, and ultimately legal education could be improved.

The exciting thing about collaborating with the bar association was that all the normally irritating logistical hassles would be handled. There would be no problem getting permission to administer questionnaires in conjunction with the exam, for example, and the problem of nonresponse could be eliminated altogether.

I left the meeting excited about the prospects for the study. When I told a colleague about it, I glowed about the absolute handling of the nonresponse problem. Her immediate comment turned everything around completely. "That's unethical. There's no law requiring the questionnaire, and participation in research has to be voluntary." The study wasn't done.

In retelling this story, it's obvious to me that requiring participation would have been inappropriate. You may have seen this even before I told you about my colleague's comment. I still feel a little embarrassed over the matter, but I have a specific purpose in telling this story about myself.

All of us consider ourselves to be ethical—not perfect perhaps, but as ethical as anyone else and perhaps more so than most. The problem in social research, as probably in life, is that ethical considerations are not always apparent to us. As a result, we often plunge into things without seeing ethical issues that may be apparent to others and may even be obvious to us when pointed out. When I reported back to the others in the planning group, for example, no one disagreed with the inappropriateness of requiring participation. Everyone was a bit embarrassed about not having seen it.

Any of us can immediately see that a study that requires small children to be tortured is

What do you think?

Whenever research ethics are discussed, the Nazi medical experiments of World War II often surface as the most hideous breach of ethical standards in history. Civilian and military prisoners in Nazi concentration camps were subjected to freezing; malaria; mustard gas; sulfanilamide; bone, muscle, and nerve transplants; jaundice; sterilization; spotted fever; various poisons; and phosphorus burns, among other tortures. Many died, and others were permanently maimed. All suffered tremendous physical and psychological pain. Some have argued, however, that the real breach of ethics did not lie in the suffering or the deaths per se. What could possibly be a worse ethical breach than that?

See the *What do you think?…Revisited* box toward the end of the chapter.

PastedGraphic-8.tiff/Earl Babbie

unethical. I know you'd speak out immediately if I suggested that we interview people about their sex lives and then publish what they said in the local newspaper. But, as ethical as you are, you'll totally miss the ethical issues in some other situations—we all do.

The first half of this chapter deals with the ethics of social research. In part, it presents some of the broadly agreed-on norms describing what's ethical in research and what's not. More important than simply knowing the guidelines, however, is becoming sensitized to the ethical component in research so that you'll look for it whenever you plan a study. Even when the ethical aspects of a situation are debatable, you should know that there's something to argue about. It's worth noting that many professions operate under ethical constraints and that these constraints differ from one profession to another. Thus, the ethics of priests, physicians, lawyers, reporters, and television producers differ. In this chapter, we'll look only at the ethical principles that govern social research.

Political considerations in research are subtle, ambiguous, and arguable. Notice that the law school example involves politics as well as ethics. Although social researchers have an ethical norm that participation in research should be voluntary, this norm clearly grows out of U.S. political norms protecting civil liberties. In some nations, the proposed study would not have been considered unethical at all.

In the second half of this chapter, we'll look at social research projects that were crushed or nearly crushed by political considerations. As with ethical concerns, there is often no "correct" take on a given situation. People of goodwill disagree. I won't try to give you a party line about what is and what is not politically acceptable. As with ethics, the point is to become sensitive to the political dimension of social research.

Ethical Issues in Social Research

In most dictionaries and in common usage, *ethics* is typically associated with *morality*, and both deal with matters of right and wrong. But what is right and what is wrong? What is the source of the distinction? For individuals the sources vary. They may include religion, political ideology, or the pragmatic observation of what seems to work and what doesn't.

Webster's New World Dictionary is typical among dictionaries in defining *ethical* as "conforming to the standards of conduct of a given profession or group." Although this definition may frustrate those in search of moral absolutes, what we regard as morality and ethics in day-to-day life is a matter of agreement among members of a group. And, not surprisingly, different groups agree on different codes of conduct. Part of living successfully in a particular society is knowing what that society considers ethical and unethical. The same holds true for the social research community.

Anyone involved in social science research, then, needs to be aware of the general agreements shared by researchers about what is proper and improper in the conduct of scientific inquiry. This section summarizes some of the most important ethical agreements that prevail in social research.

Voluntary Participation

Often, though not always, social research represents an intrusion into people's lives. The interviewer's knock on the door or the arrival of a questionnaire in the mail signals the beginning of an activity that the respondent has not requested and that may require significant time and energy. Participation in a social experiment disrupts the subject's regular activities.

Social research, moreover, often requires that people reveal personal information about themselves—information that may be unknown to their friends and associates. Further, social research often requires that such information be revealed to strangers. Other professionals, such as physicians and lawyers, also ask for such information. Their requests may be justified, however, by their aims: They need the information in order to serve the personal interests of the respondent. Social researchers can seldom make this claim. Like medical scientists, they can only argue that the research effort may ultimately help all humanity.

A major tenet of medical research ethics is that experimental participation must be voluntary. The same norm applies to social research. No one should be forced to participate. This norm is far easier to accept in theory than to apply in practice, however.

Again, medical research provides a useful parallel. Many experimental drugs used to be tested on prisoners. In the most rigorously ethical cases, the prisoners were told the nature and the possible dangers of the experiment, they were told that participation was completely voluntary, and they were further instructed that they could expect no special rewards—such as early parole—for participation. Even under these conditions, it was often clear that volunteers were motivated by the belief that they would personally benefit from their cooperation.

When the instructor in an introductory sociology class asks students to fill out a questionnaire that he or she hopes to analyze and publish, students should always be told that their participation in the survey is completely voluntary. Even so, most students will fear that nonparticipation will somehow affect their grade. The instructor should therefore be especially sensitive to such implications and make special provisions to eliminate them. For example, the instructor could ensure anonymity by leaving the room while the questionnaires are being completed. Or students could be asked to return the questionnaires by mail or to drop them in a box near the door just before the next course meeting.

As essential as it is, this norm of voluntary participation goes directly against several scientific concerns. In the most general terms, the scientific goal of *generalizability* is threatened if experimental subjects or survey respondents are all the kinds of people who willingly participate in such things. Because this orientation probably reflects other, more-general personality traits, the results of the research might not be generalizable to all kinds of people. Most clearly, in the case of a descriptive survey, a researcher cannot generalize the sample survey findings to

As in other aspects of life, there are limits to what is acceptable in social research.

an entire population unless a substantial majority of the scientifically selected sample actually participates—the willing respondents and the somewhat unwilling.

As you'll see in Chapter 10, field research has its own ethical dilemmas in this regard. Very often, the researcher cannot even reveal that a study is being done, for fear that that revelation might significantly affect the social processes being studied. Clearly, the subjects of study in such cases do not receive the opportunity to volunteer or refuse to participate. In cases in which you feel justified in violating the norm of voluntary participation, observing the other ethical norms of scientific research, such as bringing no harm to the people under study, becomes all the more important.

No Harm to the Participants

The need for norms against harming research subjects has been dramatized in part by horrendous violations by medical researchers. Perhaps at the top of list is the medical experimentation on prisoners of war by Nazi researchers in World War II. The subsequent war-crime trials at Nuremberg added the phrase "crimes against humanity" to the language of research and political ethics.

Less well-known were the Tuskegee Syphilis Experiments conducted by the U.S. Public Health Service between 1932 and 1972. The study followed the fate of nearly 400 impoverished, rural African American men suffering from syphilis. Even after penicillin had been accepted as an effective treatment for syphilis, the subjects were denied treatment—even kept from seeking treatment in the community—because the researchers wanted to observe the full progression of the disease. At times, diagnostic procedures such as spinal taps were falsely presented to subjects as cures for syphilis.

When the details of the Tuskegee Syphilis Experiments became widely known, the U.S. government was moved to take action, including a formal apology by then-president, Bill Clinton, and a program of financial reparations to the families of the subjects.

Perhaps the most concrete response to the Tuskegee scandal was the 1974 National Research Act that created the National Commission for the Protection of Human Subjects of Biomedical and Behavioral Research. The commission was charged with the task of determining the fundamental ethical principles that should guide research on human subjects. The commission subsequently published *The Belmont Report*, which elaborated on three key principles:

1. Respect for Persons—Participation must be completely voluntary and based on full understanding of what is involved. Moreover, special caution must be taken to protect minors and those lacking complete autonomy (e.g., prisoners).
2. Beneficence—Subjects must not be harmed by the research and, ideally, should benefit from it.
3. Justice—The burdens and benefits of research should be shared fairly within the society.

The National Research Act also established a requirement for institutional review boards (IRBs) through which universities would monitor compliance with ethical standards in research involving human subjects. We'll return to the role of IRBs later in this chapter.

Human research should never injure the people being studied, regardless of whether or not they volunteer for the study. Perhaps the clearest instance of this norm in social research practice concerns the revealing of information that would embarrass subjects or endanger their home life, friendships, jobs, and so forth. We'll discuss this aspect of the norm more fully in a moment.

Because subjects can be harmed psychologically in the course of a social research study, the researcher must look for the subtlest dangers and guard against them. Quite often, research subjects are asked to reveal deviant behavior, attitudes they feel are unpopular, or personal characteristics that may seem demeaning, such as low income, the receipt of welfare payments, and the like. Revealing such information usually makes subjects feel, at the very least, uncomfortable.

Social research projects may also force participants to face aspects of themselves that they don't normally consider. This can happen even when the information is not revealed directly to the researcher. In retrospect, a certain past behavior may appear unjust or immoral. The project, then, can cause continuing personal agony for the subject. If the study concerns codes of ethical conduct, for example, the subject may begin questioning his or her own morality, and that personal concern may last long after the research has been reported. For instance, probing questions can injure a fragile self-esteem.

When the psychologist Philip Zimbardo created his famous 1971 simulation of prison life—also known as the Stanford Prison Experiment—to study the dynamics of prisoner–guard interactions, Zimbardo employed Stanford students as subjects and assigned them roles as prisoners or guards at random. As you may be aware, the simulation became quickly and increasingly real for all the participants, including Zimbardo, who served as prison superintendent. It became evident that many of the student-prisoners were suffering psychological damage as a consequence of their mock incarceration, and some of the student-guards were soon exhibiting degrees of sadism that would later challenge their own self-images.

As these developments became apparent to Zimbardo, he terminated the experiment. Going beyond that, however, he created a debriefing program in which all the participants were counseled so as to avoid any lasting damage from the experience.

Clearly, just about any research you might conduct runs the risk of injuring other people in some way. Further, some study designs make injuries more likely than do others. If a particular research procedure seems likely to produce unpleasant effects for subjects—asking survey respondents to report deviant behavior, for example—the researcher should have the firmest of scientific grounds for doing it. If your research design is essential but is also likely to be unpleasant for subjects, you'll find yourself in an ethical netherworld and may face some personal agonizing. Although it has little value in itself, agonizing may be a healthy sign that you've become sensitive to the problem.

Increasingly, the ethical norms of voluntary participation and no harm to participants have become formalized in the concept of **informed consent**. This norm means that subjects must base their voluntary participation in research projects on a full understanding of the possible

risks involved. In a medical experiment, for example, prospective subjects will be presented with a discussion of the experiment and all the possible risks to themselves. They will be required to sign a statement indicating that they are aware of the risks and that they choose to participate anyway. Although the value of such a procedure is obvious when subjects will be injected with drugs designed to produce physical effects, for example, it's hardly appropriate when a participant-observer rushes to the scene of urban rioting to study deviant behavior. The researcher in this latter case is not excused from the norm of bringing no harm to those observed, but gaining informed consent is not the means to achieving that end.

On the other hand, consider the much-publicized Facebook experiments of 2012 that illustrate the two ethical principles just discussed. This experiment used the resource called "big data" (see Chapter 11) that are created by the millions and millions of Internet actions every day, hour, and minute. The researchers at Facebook thought it would be interesting to see how their subscribers were affected by the positive or negative input they received on their Facebook pages. So they fed some subscribers positive stories when they logged on and fed other subscribers negative stories. Then they monitored the positive or negative posts from the two groups of subscribers.

No one has suggested that any of the 700,000 guinea pigs were harmed by the experiment, but the researchers unquestionably violated the norm of informed consent. None of the experimental subjects had an opportunity to decide whether they wanted to participate in the research. Nor were they informed of any possible harm that might be done by the experiment—and the possibility for harm is not always obvious.

The controversy that arose over the Facebook study had a positive effect. As Jackman and Kanerva (2016) detail, Facebook conducted a study of research ethics and established a

---

informed consent A norm in which subjects base their voluntary participation in research projects on a full understanding of the possible risks involved.

How to Do It

The Basic Elements of Informed Consent

The federal regulations pertaining to what is expected in requests for human subjects to participate in research projects has been published by the Department of Health and Human Services, effective April 1, 2016.

1. A statement that the study involves research, an explanation of the purposes of the research and the expected duration of the subject's participation, a description of the procedures to be followed, and identification of any procedures which are experimental.
2. A description of any reasonably foreseeable risks or discomforts to the subject.
3. A description of any benefits to the subject or to others which may reasonably be expected from the research.
4. A disclosure of appropriate alternative procedures or courses of treatment, if any, that might be advantageous to the subject.
5. A statement describing the extent, if any, to which confidentiality of records identifying the subject will be maintained.
6. For research involving more than minimal risk, an explanation as to whether any compensation and an explanation as to whether any medical treatments are available if injury occurs and, if so, what they consist of, or where further information may be obtained.
7. An explanation of whom to contact for answers to pertinent questions about the research and research subject's rights, and whom to contact in the event of a research-related injury to the subject.
8. A statement that participation is voluntary, refusal to participate will involve no penalty or loss of benefits to which the subject is otherwise entitled, and the subject may discontinue participation at any time without penalty or loss of benefits to which the subject is otherwise entitled.

A web search will provide you with many samples of informed consent letters that you could use as models in your own research. It's worth noting that survey research and some other research techniques are exempted from the need to obtain informed consent.

Source: Code of Federal Regulations Title 21. http://www.accessdata.fda.gov/scripts/cdrh/cfdocs/cfcfr/cfrsearch.cfm?fr=50.25.

formal system for reviewing proposed research projects from an ethical standpoint. An internal review committee makes an initial evaluation of whether a research proposal deals with sensitive populations or subjects, requires additional expert consultation, etc. (2016: 9). If the answer is positive, the proposal is sent to an external IRB for their review.

One often unrecognized source of potential harm to subjects lies in the analysis and reporting of data. Every now and then, research subjects read the books published about the studies they participated in. Reasonably sophisticated subjects can locate themselves in the various indexes and tables. Having done so, they may find themselves characterized—though not identified by name—as bigoted, unpatriotic, irreligious, and so forth. At the very least, such characterizations will likely trouble them and threaten their self-images. Yet the whole purpose of the research project may be to explain why some people are prejudiced and others are not.

In one survey of churchwomen (Babbie 1967), ministers in a sample of churches were asked to distribute questionnaires to a specified sample of members, collect the questionnaires, and return them to the research office. One of these ministers read through the questionnaires from his sample before returning them. He then delivered a scathing sermon to his congregation, saying that many of them were atheists and were going to hell. Even though he could not identify the people who gave particular responses, it seems certain that the survey ended up harming many respondents.

Like voluntary participation, avoiding harm to people is easy in theory but often difficult in practice. Sensitivity to the issue and experience with its applications, however, should improve the researcher's tact in delicate areas of research.

In recent years, social researchers have gained support for abiding by this norm. Federal and other funding agencies typically require an independent evaluation of the treatment of human subjects for research proposals, and most universities now have human-subject committees to serve this evaluative function. Although sometimes troublesome and inappropriately applied, such requirements not only guard against unethical research but can also reveal ethical issues overlooked by even the most scrupulous researchers. See "How to Do It: The Basic Elements of Informed Consent," for guidelines from the U.S. Department of Health and Human Services.

Anonymity and Confidentiality

The clearest concern in guarding subjects' interests and well-being is the protection of their identity, especially in survey research. If revealing their survey responses would injure them in any way, adherence to this norm becomes all the more important. Two techniques—anonymity and confidentiality—assist researchers in this regard, although people often confuse the two.

Anonymity

A research project guarantees **anonymity** when the researcher—not just the people who read about the research—cannot link a given response with a given respondent. This implies that a typical interview survey respondent can never be considered anonymous, because an interviewer collects the information from an identifiable respondent. An example of anonymity is a mail survey in which nothing can identify the respondent when the questionnaire arrives at the research office.

As we'll see in Chapter 9, ensuring anonymity makes keeping track of who has or hasn't returned the questionnaires difficult. Despite this problem, you may be advised to pay the necessary price in some situations. In one study of drug use among university students, I decided that I specifically did not want to know the identity of respondents. I felt that honestly ensuring anonymity would increase the likelihood and accuracy of responses. Also, I did not want to be in the position of being asked by authorities for the names of drug offenders. In the few instances in which respondents volunteered their names, such information was immediately obliterated on the questionnaires.

Confidentiality

A research project guarantees **confidentiality** when the researcher can identify a given person's responses but essentially promises not to reveal them publicly. In an interview survey, for example, the researchers could make public the income reported by a given respondent, but they assure the respondent that this will not be done.

Whenever a research project is confidential rather than anonymous, it is the researcher's responsibility to make that fact clear to the respondent. Moreover, researchers should never use the term *anonymous* to mean *confidential*.

With few exceptions (such as surveys of public figures who agree to have their responses published), the information that respondents give must be kept at least confidential. This is not always an easy norm to follow, because, for example, the courts have not recognized social research data as the kind of "privileged communication" that is accepted in the case of priests and attorneys.

This unprotected guarantee of confidentiality produced a near disaster in 1991. Two years earlier, the Exxon *Valdez* supertanker had run aground in Prince William Sound in Alaska, spilling more than 10 million gallons of oil. The economic and environmental damage was widely reported.

Less attention was given to the psychological and sociological damage suffered by residents of the area. There were anecdotal reports of increased alcoholism, family violence, and other secondary consequences of the disruptions caused by the oil spill. Eventually, 22 communities in Prince William Sound and the Gulf of Alaska sued Exxon for the economic, social, and psychological damages suffered by their residents.

To determine the amount of damage done, the communities commissioned a San Diego research firm to undertake a household survey asking residents very personal questions about increased problems in their families. The sample of residents were asked to reveal painful and embarrassing information, under the guarantee of absolute confidentiality. Ultimately, the results of the survey confirmed that a variety of personal and family problems had increased substantially following the oil spill.

When Exxon learned that survey data would be presented to document the suffering, they

---

anonymity Anonymity is guaranteed in a research project when neither the researchers nor the readers of the findings can link a given response with a given respondent.

confidentiality A research project guarantees confidentiality when the researcher can identify a given person's responses but promises not to reveal them publicly.

took an unusual step: They asked the court to subpoena the survey questionnaires. The court granted the defendant's request and ordered the researchers to turn them over—with all identifying information. It appeared that Exxon's intention was to call survey respondents to the stand and cross-examine them regarding answers they had given to interviewers under the guarantee of confidentiality. Moreover, many of the respondents were Native Americans, whose cultural norms made such public revelations all the more painful.

Fortunately, the Exxon *Valdez* case was settled before the court decided whether it would force survey respondents to testify in open court. Unfortunately, the potential for disaster remains.

The seriousness of this issue is not limited to established research firms. Rik Scarce was a graduate student at Washington State University when he undertook participant observation among animal-rights activists. In 1990 he published a book based on his research, entitled *Ecowarriors: Understanding the Radical Environmental Movement*. In 1993, Scarce was called before a grand jury and asked to identify the activists he had studied. In keeping with the norm of confidentiality, the young researcher refused to answer the grand jury's questions and spent 159 days in the Spokane County jail. He reports,

> Although I answered many of the prosecutor's questions, on 32 occasions I refused to answer, saying, "Your question calls for information that I have only by virtue of a confidential disclosure given to me in the course of my research activities. I cannot answer the question without actually breaching a confidential communication. Consequently, I decline to answer the question under my ethical obligations as a member of the American Sociological Association and pursuant to any privilege that may extend to journalists, researchers, and writers under the First Amendment."

(Scarce 1999: 982)

At the time of his grand jury appearance and his incarceration, Scarce felt his ethical stand was strongly supported by the American Sociological Association (ASA) Code of Ethics, and the association filed a friend-of-the-court brief on his behalf. In 1997, the ASA revised its Code of Ethics and, while still upholding the norm of confidentiality, warned researchers to inform themselves regarding laws and rules that may limit their ability to promise confidentiality to research subjects.

You can use several techniques to guard against such dangers and ensure better performance on the guarantee of confidentiality. To begin, interviewers and others with access to respondent identifications should be trained in their ethical responsibilities. Beyond training, the most fundamental technique is to remove identifying information as soon as it's no longer necessary. In a survey, for example, all names and addresses should be removed from questionnaires and replaced by identification numbers. An identification file should be created that links numbers to names to permit the later correction of missing or contradictory information, but this file should not be available except for legitimate purposes.

Similarly, in an interview survey you may need to identify respondents initially so that you can recontact them to verify that the interview was conducted and perhaps to get information that was missing in the original interview. As soon as you've verified an interview and assured yourself that you don't need any further information from the respondent, however, you can safely remove all identifying information from the interview booklet. Often, interview booklets are printed so that the first page contains all the identifiers—it can be torn off once the respondent's identification is no longer needed. J. Steven Picou (1996a, 1996b) points out that even removing identifiers from data files does not always sufficiently protect respondent confidentiality, a lesson he learned during nearly a year in federal court. A careful examination of all the responses of a particular respondent sometimes allows others to deduce that person's identity. Imagine, for example, that someone said she was a former employee of a particular company. Knowing the person's gender, age, ethnicity, and other characteristics could make it possible for the company to identify that person.

Even if you intend to remove all identifying information, suppose you have not yet done so. What do you do when the police or a judge orders you to provide the responses given by your research subjects?

Upholding confidentiality is a real issue for practicing social researchers, even though they sometimes disagree about how to protect subjects. Harry O'Neill, the vice chair of the Roper Organization, for example, suggested that the best solution is to avoid altogether the ability to identify respondents with their responses:

> *So how is this accomplished? Quite simply by not having any respondent-identifiable information available for the court to request. In my initial contact with a lawyer-client, I make it unmistakably clear that, once the survey is completed and validated, all respondent-identifiable information will be removed and destroyed immediately. Everything else connected with the survey—completed questionnaires, data tapes, methodology, names of interviewers and supervisors—of course will be made available.*
>
> *(O'Neill 1992: 4)*

Board Chairman Burns Roper (1992: 5) disagreed, saying that such procedures might raise questions about the validity of the research methods. Instead, Roper said that he felt he must be prepared to go to jail if necessary. (He noted that Vice Chair O'Neill promised to visit him in that event.)

In 2002, the U.S. Department of Health and Human Services announced a program to issue a "Certificate of Confidentiality" to protect the confidentiality of research-subject data against forced disclosure by the police and other authorities. Not all research projects qualify for such protection, but it can provide an important support for research ethics in many cases.

> *Under section 301(d) of the Public Health Service Act (42 U.S.C. 241(d)) the Secretary of Health and Human Services may authorize persons engaged in biomedical, behavioral, clinical, or other research to protect the privacy of individuals who are the subjects of that research. This authority has been delegated to the National Institutes of Health (NIH).*
>
> *Persons authorized by the NIH to protect the privacy of research subjects may not be compelled in any Federal, State, or local civil, criminal, administrative, legislative, or other proceedings to identify them by name or other identifying characteristic.*
>
> *(U.S. Department of Health and Human Services 2002)*

The increased use of visual techniques in social research (e.g., video recording children at play, photographing hospice patients) has created a new problem for protecting subjects, as discussed by Rose Wiles and her colleagues (2012). The authors lay out some of the terrain for this issue: "concerns include the contexts in which images were produced and through which they may be consumed, the longevity of images in the public domain and the potential for future uses and secondary analysis of images" (2012: 41).

In all the aspects of research ethics discussed in this chapter, professional researchers avoid settling for mere rote compliance with established ethical rules. Rather, they continually ask what would be most appropriate in protecting the interests of those being studied. Here's the way Penny Edgell Becker addressed the issue of confidentiality in connection with a qualitative research project studying religious life in a community:

> *Following the lead of several recent studies, I identify the real name of the community, Oak Park, rather than reducing the complexity of the community's history to a few underlying dimensions or [creating] an "insider/outsider" dynamic where some small group of fellow researchers knows the community's real name and the rest of the world is kept in the dark.... In all cases individual identities are disguised, except for Jack Finney, the Lutheran pastor, who gave permission to be identified. "City Baptist" is a pseudonym used at the request of the church's leadership. The leaders of Good Shepherd Lutheran Church (GSLC) gave permission to use the church's real name.*
>
> *(1998: 452)*

Benjamin Saunders, Jenny Kitzinger, and Celia Kitzinger (2015) report on the problems that anonymity and confidentiality can present in qualitative studies of hospital patients and their families, and they call "anonymising data" a "balancing act." They can, of course, assign fictitious names to the subjects of their studies, but hospital staff can no doubt see through the deception if the patients' medical and family circumstances are detailed. And in an ironic twist, giving fictitious names to patients who died was sometimes seen as disrespectful by the deceased patients' survivors.

Deception

We've seen that the handling of subjects' identities is an important ethical consideration. Handling your own identity as a researcher can also be tricky. Sometimes it's useful and even necessary to identify yourself as a researcher to those you want to study. You'd have to be an experienced con artist to get people to participate in a laboratory experiment or complete a lengthy questionnaire without letting on that you were conducting research.

Even when you must conceal your research identity, you need to consider the following. Because deceiving people is unethical, deception within social research needs to be justified by compelling scientific or administrative concerns. Even then, the justification will be arguable.

Sometimes researchers admit that they're doing research but fudge about why they're doing it or for whom. Suppose a public welfare agency has asked you to conduct a study of living standards among aid recipients. Even if the agency is looking for ways of improving conditions, the recipient-subjects will likely fear a witch hunt for "cheaters." They might be tempted, therefore, to give answers that make them seem more destitute than they really are. Unless they provide truthful answers, however, the study will not produce accurate data that will contribute to an effective improvement of living conditions. What do you do?

One solution would be to tell subjects that you're conducting the study as part of a university research program—concealing your affiliation with the welfare agency. Doing that improves the scientific quality of the study, but it raises a serious ethical issue.

Lying about research purposes is common in laboratory experiments. Although it's difficult to conceal that you're conducting research, it's usually simple—and sometimes appropriate—to conceal your purpose. Many experiments in social psychology, for example, test the extent to which subjects will abandon the evidence of their own observations in favor of the views expressed by others. Figure 2-1 shows the stimulus from the classic Asch experiment—frequently replicated by psychology classes—in which subjects are shown three lines of differing lengths (A, B, and C) and asked to compare them with a fourth line (X). Subjects are then asked, "Which of the first three lines is the same length as the fourth?"

Recall from Chapter 2 that the purpose of the experiment is to see whether a person would give up his or her own judgment in favor of group agreement. I think you can see that conformity is a useful phenomenon to study and understand, and it couldn't be studied experimentally without deceiving the subjects. We'll examine a similar situation in the discussion of a famous experiment by Stanley Milgram later in this chapter. The question is, how do we get around the ethical issue that deception is necessary for an experiment to work?

One appropriate solution researchers have found is to debrief subjects following an experiment. **Debriefing** entails interviews to discover any problems generated by the research experience so that those problems can be corrected. Even though subjects can't be told the true purpose of the study prior to their participation in it, there's usually no reason they can't know afterward. Telling them the truth afterward may make up for having to lie to them at the outset. This must be done with care, however, making sure the subjects aren't left with bad feelings or doubts about themselves based on their performance in the experiment. If this seems complicated, it's simply the price we pay for using other people's lives as the subject matter for our research.

As a social researcher, then, you have many ethical obligations to the subjects in your studies. "Issues and Insights: Ethical Issues in Research on Human Sexuality" illustrates some of the ethical questions involved in a specific research area.

debriefing Interviewing subjects to learn about their experience of participation in the project and to inform them of any unrevealed purpose. This is especially important if there's a possibility that they have been damaged by that participation.

Issues and Insights

Ethical Issues in Research on Human Sexuality

by Kathleen McKinney
Department of Sociology, Illinois State University

When studying any form of human behavior, ethical concerns are paramount. This statement may be even truer for studies of human sexuality because of the topic's highly personal, salient, and perhaps threatening nature. Concern has been expressed by the public and by legislators about human sexuality research. Three commonly discussed ethical criteria have been related specifically to research in the area of human sexuality.

Informed Consent This criterion emphasizes the importance of both accurately informing your subject or respondent as to the nature of the research and obtaining his or her verbal or written consent to participate. Coercion is not to be used to force participation, and subjects may terminate their involvement in the research at any time. There are many possible violations of this standard. Misrepresentation or deception may be used when describing an embarrassing or personal topic of study, because the researchers fear high rates of refusal or false data. Covert research, such as some observational studies, also violates the informed consent standard, since subjects are unaware that they are being studied. Informed consent may create special problems with certain populations. For example, studies of the sexuality of children are limited by the concern that children may be cognitively and emotionally unable to give informed consent. Although there can be problems such as those discussed, most research is clearly voluntary, with informed consent from those participating.

Right to Privacy Given the highly personal nature of sexuality and society's tremendous concern with social control of sexuality, the right to privacy is a very important ethical concern for research in this area. Individuals may risk losing their jobs, having family difficulties, or being ostracized by peers if certain facets of their sexual lives are revealed. This is especially true for individuals involved in sexual behavior categorized as deviant. Violations of right to privacy occur when researchers identify members of certain groups they have studied, release or share an individual's data or responses, or covertly observe sexual behavior. In most cases, the right to privacy is easily maintained by the researchers. In survey research, self-administered questionnaires can be anonymous and interviews can be kept confidential. In case and observational studies, the identity of the person or group studied can be disguised in any publications. In most research methods, analysis and reporting of data should be at the group or aggregate level.

Protection from Harm Harm may include emotional or psychological distress, as well as physical harm. Potential for harm varies by research method; it is more likely in experimental studies in which the researcher manipulates or does something to the subject than in observational or survey research. Emotional distress, however, is a possibility in all studies of human sexuality. Respondents may be asked questions that elicit anxiety, dredge up unpleasant memories, or cause them to evaluate themselves critically. Researchers can reduce the potential for such distress during a study by using anonymous, self-administered questionnaires or well-trained interviewers and by wording sensitive questions carefully.

All three of these ethical criteria are quite subjective. Violations are sometimes justified by arguing that risks to subjects are outweighed by benefits to society. The issue here, of course, is who makes that critical decision. Usually, such decisions are made by the researcher and often a screening committee that deals with ethical concerns. Most creative researchers have been able to follow all three ethical guidelines and still do important research.

Analysis and Reporting

In addition to their ethical obligations to subjects, researchers have ethical obligations to their colleagues in the scientific community. These obligations concern the analysis of data and the way the results are reported.

In any rigorous study, the researcher should be more familiar than anyone else with the study's technical limitations and failures. Researchers have an obligation to make such shortcomings known to their readers—even if admitting qualifications and mistakes makes them feel foolish.

Negative findings, for example, should be reported if they are at all related to the analysis. There is an unfortunate myth in scientific reporting that only discoveries of strong, causal relationships among variables are worth reporting (journal editors are sometimes guilty of believing this as well). In science, however, it's often as important to know that two variables are *not* related as to know that they are.

Similarly, researchers must avoid the temptation to save face by describing their findings as the product of a carefully preplanned analytic strategy when that is not the case. Many findings arrive unexpectedly—even though they may seem obvious in retrospect. So an interesting relationship was uncovered by accident—so what? Embroidering such situations with descriptions of fictitious hypotheses is dishonest. It also does a disservice to less experienced researchers by leading them into thinking that all scientific inquiry is rigorously preplanned and organized.

Sadly, some "researchers" go several steps further toward dishonesty. Chapter 15 will deal with the problem of plagiarism—claiming someone else's work as your own—but every now and then you will read about cases in which claims to have conducted scientific studies are completely fraudulent and fictional. A recent example involved a Dutch psychology professor and dean who published a number of articles of popular interest—for example, one linked meat eating to selfishness; another claimed that public trash led to racist behavior—but it turned out that the studies he described were never conducted (Bhattacharjee 2013). While such misbehavior constitutes a small fraction of published research, it is common enough to warrant an online monitor of fraudulent research. (See an example of such an online monitor at http://retractionwatch.wordpress.com.)

In general, science progresses through honesty and openness; ego defenses and deception retard it. Researchers can best serve their peers—and scientific discovery as a whole—by telling the truth about all the pitfalls and problems they've experienced in a particular line of inquiry. Perhaps they'll save others from the same problems.

Finally, simple carelessness or sloppiness can be considered an ethical problem. If the research project uses up limited resources or imposes on subjects but produces no benefits, many in the research community would consider that an ethical violation. This is not to say that all research must produce positive results, but it should be conducted in a manner that promotes that possibility.

Institutional Review Boards

Research ethics in studies involving humans is now also governed by federal law. Any agency (such as a university or a hospital) wishing to receive federal research support must establish an institutional review board (IRB), a panel of faculty (and possibly others) who review all research proposals involving human subjects to guarantee that the subjects' rights and interests will be protected. The law applies specifically to federally funded research, but many universities apply the same standards and procedures to all research, including that funded by nonfederal sources and even research done at no cost, such as student projects.

The chief responsibility of an IRB is to ensure that the risks faced by human participants in research are minimal. In some cases, the IRB may refuse to approve a study or may ask the researcher to revise the study design. Where some minimal risks are deemed unavoidable, researchers must prepare an "informed consent" form that describes those risks clearly. Subjects may participate in the study only after they have read the statement and signed it.

Much of the original impetus for establishing IRBs had to do with medical experimentation on humans, and many social research study designs are generally regarded as exempt from IRB review. An example is an anonymous survey sent to a large sample of respondents. The guideline to be followed by IRBs, as contained in the Federal Exemption Categories (45 CFR 46.101 [b]), exempts a variety of research situations:

1. *Research conducted in established or commonly accepted educational settings, involving normal educational practices, such as (i) research on regular and special education instructional strategies, or (ii) research on the effectiveness of or the comparison among instructional techniques, curricula, or classroom management methods.*

2. *Research involving the use of educational tests (cognitive, diagnostic, aptitude, achievement), survey procedures, interview procedures or observation of public behavior, unless:*

   (i) information obtained is recorded in such a manner that human subjects can be identified, directly or through identifiers linked to the subjects; and (ii) any disclosure of the human subjects' responses outside the research could reasonably place the subjects at risk of criminal or civil liability or be damaging to the subjects' financial standing, employability, or reputation.

3. *Research involving the use of educational tests (cognitive, diagnostic, aptitude, achievement), survey procedures, interview procedures, or*

observation of public behavior that is not exempt under paragraph (b)(2) of this section, if:

> *(i) the human subjects are elected or appointed public officials or candidates for public office; or (ii) Federal statute(s) require(s) without exception that the confidentiality of the personally identifiable information will be maintained throughout the research and thereafter.*

4. *Research involving the collection or study of existing data, documents, records, pathological specimens, or diagnostic specimens, if these sources are publicly available or if the information is recorded by the investigator in such a manner that subjects cannot be identified, directly or through identifiers linked to the subjects.*

5. *Research and demonstration projects which are conducted by or subject to the approval of Department or Agency heads, and which are designed to study, evaluate, or otherwise examine:*

> *(i) Public benefit or service programs; (ii) procedures for obtaining benefits or services under those programs; (iii) possible changes in or alternatives to those programs or procedures; or (iv) possible changes in methods or levels of payment for benefits or services under those programs.*

6. *Taste and food quality evaluation and consumer acceptance studies, (i) if wholesome foods without additives are consumed or (ii) if a food is consumed that contains a food ingredient at or below the level and for a use found to be safe, or agricultural chemical or environmental contaminant at or below the level found to be safe, by the Food and Drug Administration or approved by the Environmental Protection Agency or the Food Safety and Inspection Service of the U.S. Department of Agriculture.*

Paragraph (2) of the excerpt exempts much of the social research described in this book. Nonetheless, universities sometimes apply the law's provisions inappropriately. As chair of a university IRB, for example, I was once asked to review the letter of informed consent that was to be sent to medical insurance companies, requesting their agreement to participate in a survey that would ask which medical treatments were covered under their programs. Clearly, the humans involved were not at risk in the sense anticipated by the law. In a case like that, the appropriate technique for gaining informed consent is to mail the questionnaire. If a company returns it, they've consented. If they don't, they haven't.

Other IRBs have suggested that researchers need to obtain permission before observing participants in public gatherings and events, before conducting surveys on the most mundane matters, and so forth. Christopher Shea (2000) has chronicled several such questionable applications of the law while supporting the ethical logic that originally prompted the law.

Don't think that these critiques of IRBs minimize the importance of protecting human subjects. Indeed, some universities exceed the federal requirements in reasonable and responsible ways: such as requiring IRB review of non–federally funded projects. Moreover, social researchers are particularly careful when dealing with vulnerable populations, such as young people and prisoners.

Research ethics is an ever-evolving subject, because new research techniques often require revisiting old concerns. For example, the increased use of public databases for secondary research has caused some IRBs to worry whether they would need to reexamine such projects as the General Social Survey every time a researcher proposed to use those data. Most IRBs have decided this is unnecessary. (See Skedsvold 2002 for more on this.)

Similarly, the prospects for research about and through the Internet has raised ethical concerns. For example, the American Association for the Advancement of Science held a workshop on this topic in November 1999. The overall conclusion of the report produced by the workshop summarizes some of the primary concerns already examined in this chapter:

> *The current ethical and legal framework for protecting human subjects rests on the principles of autonomy, beneficence, and justice. The first principle, autonomy, requires that subjects be treated with respect as autonomous agents and affirms that those persons with diminished autonomy are entitled to special protection. In practice, this principle is reflected in the process of informed consent, in which the risks and benefits of the research are disclosed to the subject. The second principle, beneficence, involves maximizing possible benefits and good for the subject, while minimizing the amount of possible harm and risks resulting from the research. Since the fruits of knowledge can come at a cost to those participating in research, the last principle, justice, seeks a fair distribution of the burdens and benefits associated with research, so that certain individuals or groups do not bear disproportionate risks while others reap the benefits.*

(Frankel and Siang 1999: 2–3)

The comments about research ethics and institutional review boards do not apply only to American research. Martyn Hammersley and Anna Traianou (2011) describe many of the same issues and problems in the case of British social researchers and the research ethics committees (RECs). Moreover, they report special problems faced by qualitative researchers, whose research designs may evolve over the course of a study. In some cases, the RECs have insisted on monitoring the ethical aspects of such research throughout the course of a study.

Professional Codes of Ethics

Ethical issues in social research are both important and ambiguous. For this reason, most of the professional associations of social researchers have created and published formal codes of conduct describing what is considered acceptable and unacceptable professional behavior. As one example, Figure 3-1 presents the code of conduct of the American Association for Public Opinion Research (AAPOR), an interdisciplinary research association in the social sciences.

AAPOR CODE OF PROFESSIONAL ETHICS AND PRACTICE

We—the members of the American Association for Public Opinion Research and its affiliated chapters—subscribe to the principles expressed in the following Code. Our goals are to support sound and ethical practice in the conduct of survey and public opinion research and in the use of such research for policy- and decision-making in the public and private sectors, as well as to improve public understanding of survey and public opinion research methods and the proper use of those research results.

We pledge ourselves to maintain high standards of scientific competence, integrity, and transparency in conducting, analyzing, and reporting our work; establishing and maintaining relations with survey respondents and our clients; and communicating with those who eventually use the research for decision-making purposes and the general public. We further pledge ourselves to reject all tasks or assignments that would require activities inconsistent with the principles of this Code.

The Code describes the obligations that we believe all research professionals have, regardless of their membership in this Association or any other, to uphold the credibility of survey and public opinion research.

It shall not be the purpose of this Code to pass judgment on the merits of specific research methods. From time to time, the AAPOR Executive Council may issue guidelines and recommendations on best practices with regard to the design, conduct, and reporting of surveys and other forms of public opinion research.

I. *Principles of Professional Responsibility in Our Dealings with People*

  A. Respondents and Prospective Respondents

    1. We shall avoid practices or methods that may harm, endanger, humiliate, or seriously mislead survey respondents or prospective respondents.

    2. We shall respect respondents' desires, when expressed, not to answer specific survey questions or provide other information to the researcher. We shall be responsive to their questions about how their contact information was secured.

    3. Participation in surveys and other forms of public opinion research is voluntary, except for the decennial census and a few other government surveys as specified by law. We shall provide all persons selected for inclusion with a description of the research study sufficient to permit them to make an informed and free decision about their participation. We shall make no false or misleading claims as to a study's sponsorship or purpose, and we shall provide truthful answers to direct questions about the research. If disclosure could substantially bias responses or endanger interviewers, it is sufficient to indicate that some information cannot be revealed or will not be revealed until the study is concluded.

    4. We shall not misrepresent our research or conduct other activities (such as sales, fundraising, or political campaigning) under the guise of conducting survey and public opinion research.

    5. Unless the respondent explicitly waives confidentiality for specified uses, we shall hold as privileged and confidential all information that could be used, alone or in combination with other reasonably available information, to identify a respondent with his or her responses. We also shall not disclose or use the names of respondents or any other personally-identifying information for non-research purposes unless the respondents grant us permission to do so.

FIGURE 3-1

Code of Ethics of the American Association for Public Opinion Research.

Source: Material taken from the AAPOR Code of Ethics and Practice. The 2015 revision of the code is accessible at http://www.aapor.org/Standards-Ethics/AAPOR-Code-of-Ethics.aspx.

6. We understand that the use of our research results in a legal proceeding does not relieve us of our ethical obligation to keep confidential all respondent-identifying information (unless waived explicitly by the respondent) or lessen the importance of respondent confidentiality.

B. Clients or Sponsors

1. When undertaking work for a private client, we shall hold confidential all proprietary information obtained about the client and about the conduct and findings of the research undertaken for the client, except when the dissemination of the information is expressly authorized by the client, or when disclosure becomes necessary under the terms of Section I-C or III-E of this Code. In the latter case, disclosures shall be limited to information directly bearing on the conduct and findings of the research.

2. We shall be mindful of the limitations of our techniques and capabilities and shall accept only those research assignments that we can reasonably expect to accomplish within these limitations.

C. The Public

1. We shall inform those for whom we conduct publicly released research studies that AAPOR Standards for Disclosure require the release of certain essential information about how the research was conducted, and we shall make all reasonable efforts to encourage clients to subscribe to our standards for such disclosure in their releases.

2. We shall correct any errors in our own work that come to our attention which could influence interpretation of the results, disseminating such corrections to all original recipients of our content.

3. We shall attempt, as practicable, to correct factual misrepresentations or distortions of our data or analysis, including those made by our research partners, co-investigators, sponsors, or clients. We recognize that differences of opinion in analysis are not necessarily factual misrepresentations or distortions. We shall issue corrective statements to all parties who were presented with the factual misrepresentations or distortions, and if such factual misrepresentations or distortions were made publicly, we shall correct them in as commensurate a public forum as is practicably possible.

D. The Profession

1. We recognize our responsibility to the science of survey and public opinion research to disseminate as freely as practicable the ideas and findings that emerge from our research.

2. We can point with pride to our membership in the Association and our adherence to this Code as evidence of our commitment to high standards of ethics in our relations with respondents, our clients or sponsors, the public, and the profession. However, we shall not cite our membership in the Association nor adherence to this Code as evidence of professional competence, because the Association does not so certify any persons or organizations.

II. *Principles of Professional Practice in the Conduct of Our Work*

A. We shall exercise due care in developing research designs and instruments, and in collecting, processing, and analyzing data, taking all reasonable steps to assure the reliability and validity of results.

1. We shall recommend and employ only those tools and methods of analysis that, in our professional judgment, are well suited to the research problem at hand.

2. We shall not knowingly select research tools and methods of analysis that yield misleading conclusions.

3. We shall not knowingly make interpretations of research results that are inconsistent with the data available, nor shall we tacitly permit such interpretations. We shall ensure that any findings we report, either privately or for public release, are a balanced and accurate portrayal of research results.

4. We shall not knowingly imply that interpretations should be accorded greater confidence than the data actually warrant. When we use samples to make statements about populations, we shall only make claims of precision that are warranted by the sampling frames and methods employed. For example, the reporting of a margin of sampling error based on an opt-in or self-selected volunteer sample is misleading.

5. We shall not knowingly engage in fabrication or falsification.

6. We shall accurately describe survey and public opinion research from other sources that we cite in our work, in terms of its methodology, content, and comparability.

B. We shall describe our methods and findings accurately and in appropriate detail in all research reports, adhering to the standards for disclosure specified in Section III.

III. *Standards for Disclosure*

Good professional practice imposes the obligation upon all survey and public opinion researchers to disclose certain essential information about how the research was conducted. When conducting publicly released research studies, full and complete disclosure to the public is best made at the time results are released, although some information may not be immediately available. When undertaking work for a private client, the same essential information should be made available to the client when the client is provided with the results.

FIGURE 3-1

(Continued)

A. We shall include the following items in any report of research results or make them available immediately upon release of that report.

   1. Who sponsored the research study, who conducted it, and who funded it, including, to the extent known, all original funding sources.

   2. The exact wording and presentation of questions and responses whose results are reported.

   3. A definition of the population under study, its geographic location, and a description of the sampling frame used to identify this population. If the sampling frame was provided by a third party, the supplier shall be named. If no frame or list was utilized, this shall be indicated.

   4. A description of the sample design, giving a clear indication of the method by which the respondents were selected (or self-selected) and recruited, along with any quotas or additional sample selection criteria applied within the survey instrument or post-fielding. The description of the sampling frame and sample design should include sufficient detail to determine whether the respondents were selected using probability or non-probability methods.

   5. Sample sizes and a discussion of the precision of the findings, including estimates of sampling error for probability samples and a description of the variables used in any weighting or estimating procedures. The discussion of the precision of the findings should state whether or not the reported margins of sampling error or statistical analyses have been adjusted for the design effect due to clustering and weighting, if any.

   6. Which results are based on parts of the sample, rather than on the total sample, and the size of such parts.

   7. Method and dates of data collection.

B. We shall make the following items available within 30 days of any request for such materials.

   1. Preceding interviewer or respondent instructions and any preceding questions or instructions that might reasonably be expected to influence responses to the reported results.

   2. Any relevant stimuli, such as visual or sensory exhibits or show cards.

   3. A description of the sampling frame's coverage of the target population.

   4. The methods used to recruit the panel, if the sample was drawn from a pre-recruited panel or pool of respondents.

   5. Details about the sample design, including eligibility for participation, screening procedures, the nature of any oversamples, and compensation/incentives offered (if any).

   6. Summaries of the disposition of study-specific sample records so that response rates for probability samples and participation rates for non-probability samples can be computed.

   7. Sources of weighting parameters and method by which weights are applied.

   8. Procedures undertaken to verify data. Where applicable, methods of interviewer training, supervision, and monitoring shall also be disclosed.

C. If response rates are reported, response rates should be computed according to AAPOR Standard Definitions.

D. If the results reported are based on multiple samples or multiple modes, the preceding items shall be disclosed for each.

E. If any of our work becomes the subject of a formal investigation of an alleged violation of this Code, undertaken with the approval of the AAPOR Executive Council, we shall provide additional information on the research study in such detail that a fellow researcher would be able to conduct a professional evaluation of the study.

FIGURE 3-1

(Continued)

Most professional associations have such codes of ethics. See, for example, the American Sociological Association (ASA), the American Psychological Association (APA), the American Political Science Association (APSA), and so forth. You can find many of these on the associations' websites. In addition, the Association of Internet Researchers (AoIR) has a code of ethics accessible online.

Two Ethical Controversies

As you may already have guessed, the adoption and publication of professional codes of conduct have not totally resolved the issue of research ethics. Social researchers still disagree on some general principles, and those who agree in principle often debate specifics.

This section briefly describes two research projects that have provoked ethical controversy and discussion. The first project studied homosexual behavior in public restrooms; the second examined obedience in a laboratory setting.

Trouble in the Tearoom

As a graduate student, Laud Humphreys became interested in the study of homosexual behavior. He developed a special interest in the casual and fleeting same-sex acts engaged in by some male nonhomosexuals. In particular, his research interest focused on homosexual acts between strangers meeting in the public restrooms in parks, called "tearooms" among homosexuals. The result was the publication in 1970 of the classic *Tearoom Trade*.

What particularly interested Humphreys about the tearoom activity was that the participants seemed otherwise to live conventional lives as "family men" and as accepted members of the community. They did nothing else that might qualify them as homosexuals. Thus, it was important to them that they remain anonymous in their tearoom visits. How would you study something like that?

Humphreys decided to take advantage of the social structure of the situation. Typically, the tearoom encounter involved three people: the two men actually engaging in the sexual act and a lookout, called the "watchqueen." Humphreys began showing up at public restrooms, offering to serve as watchqueen whenever it seemed appropriate. Because the watchqueen's payoff was the chance to watch the action, Humphreys was able to conduct field observations as he would in a study of political rallies or jaywalking behavior.

To round out his understanding of the tearoom trade, Humphreys needed to know something more about the people who participated. Because the men probably would not have been thrilled about being interviewed, Humphreys developed a different solution. Whenever possible, he noted the license numbers of participants' cars and tracked down their names and addresses through the police. Humphreys then visited the men at their homes, disguising himself enough to avoid recognition, and announced that he was conducting a survey. In that fashion, he collected the personal information he couldn't get in the restrooms.

As you can imagine, Humphreys' research provoked considerable controversy both inside and outside the social science community. Some critics charged Humphreys with a gross invasion of privacy in the name of science. What men did in public restrooms was their own business. Others were mostly concerned about the deceit involved—Humphreys had lied to the participants by leading them to believe he was only a voyeur-participant. Even people who felt that the tearoom participants were fair game for observation because they used a public facility protested the follow-up survey. They felt it was unethical for Humphreys to trace the participants to their homes and to interview them under false pretenses.

Still others justified Humphreys' research. The topic, they said, was worth study. It couldn't be studied any other way, and they regarded the deceit as essentially harmless, noting that Humphreys was careful not to harm his subjects in disclosing their tearoom activities. One result of Humphreys' research was to challenge some of the common stereotypes about the participants in anonymous sexual encounters in public places, showing them to be conventional in other aspects of their lives.

The *Tearoom Trade* controversy is still debated, and it probably always will be, because it stirs emotions and involves ethical issues people disagree about. What do you think? Was Humphreys ethical in doing what he did? Are there parts of the research that you believe were acceptable and other parts that were not?

Observing Human Obedience

The second illustration differs from the first in many ways. Whereas Humphreys' study involved participant observation, this study took place in the laboratory. Humphreys study was sociological and this one psychological. And whereas Humphreys examined behavior considered by many to be a form of deviance, the researcher in this study examined obedience and conformity.

One of the most unsettling clichés to come out of World War II was the German soldier's common excuse for atrocities: "I was only following orders." From the point of view that gave rise to this comment, any behavior—no matter how reprehensible—could be justified if someone else could be assigned responsibility for it. If a superior officer ordered a soldier to kill a baby, the fact of the order supposedly exempted the soldier from personal responsibility for the action.

Although the military tribunals that tried the war-crime cases did not accept this excuse, social researchers and others have recognized the extent to which this point of view pervades social life. People often seem willing to do things they know would be considered wrong by others if they can claim that some higher authority ordered them to do it. Such was the pattern of justification in the 1968 My Lai tragedy of the Vietnam War, when U.S. soldiers killed more than 300 unarmed civilians—some of them young children—simply because their village, My Lai, was believed to be a Vietcong stronghold. This sort of justification appears less dramatically in day-to-day civilian life. Few would disagree that this reliance on authority exists, yet Stanley Milgram's study (1963, 1965) of the topic provoked considerable controversy.

To observe people's willingness to harm others when following orders, Milgram brought 40 adult men from many different walks of life into a laboratory setting designed to create the phenomenon under study. If you had been a subject in the experiment, you would have had something like the following experience.

You've been informed that you and another subject are about to participate in a learning experiment. Through a draw of lots, you're assigned the job of "teacher" and your fellow subject the job of "pupil." The pupil is led into another room and strapped into a chair; an electrode is attached to his wrist. As the teacher, you're seated in front of an impressive electrical control panel covered with dials, gauges, and switches. You notice that each switch has a label giving a different number of volts, ranging from 15 to 315. The switches have other labels, too, some with the ominous phrases such as "Extreme Intensity Shock," "Danger—Severe Shock," and "XXX."

The experiment runs like this. You read a list of word pairs to the learner and then test his ability to match them up. Because you can't see him, a light on your control panel indicates his answer. Whenever the learner makes a mistake, you're instructed by the experimenter to throw one of the switches—beginning with the mildest—and administer a shock to your pupil. Through an open door between the two rooms, you hear your pupil's response to the shock. Then you read another list of word pairs and test him again.

As the experiment progresses, you administer ever more intense shocks, until your pupil screams for mercy and begs for the experiment to end. You're instructed to administer the next shock anyway. After a while, your pupil begins kicking the wall between the two rooms and continues to scream. The implacable experimenter tells you to give the next shock. Finally, you read a list and ask for the pupil's answer—but there is no reply whatever, only silence from the other room. The experimenter informs you that no answer is considered an error and instructs you to administer the next higher shock. This continues up to the "XXX" shock at the end of the series.

What do you suppose you really would have done when the pupil first began screaming? When he began kicking on the wall? Or when he became totally silent and gave no indication of life? You'd refuse to continue giving shocks, right? And surely the same would be true of most people.

So we might think—but Milgram found out otherwise. Of the first 40 adult men Milgram tested, nobody refused to continue administering the shocks until they heard the pupil begin kicking the wall between the two rooms. Of the 40, only 5 did so then. Two-thirds of the subjects, 26 of the 40, continued doing as they were told through the entire series—up to and including the administration of the highest shock.

As you've probably guessed, the shocks were phony, and the "pupil" was a confederate of the experimenter. Only the "teacher" was a real subject in the experiment. As a subject, you wouldn't actually have been hurting another person, but you would have been led to think you were. The experiment was designed to test your willingness to follow orders to the point of presumably killing someone.

Milgram's experiments have been criticized both methodologically and ethically. On the ethical side, critics have particularly cited the effects of the experiment on the subjects. Many seem to have experienced personally about as much pain as they thought they were administering to someone else. They pleaded with the experimenter to let them stop giving the shocks. They became extremely upset and nervous. Some had uncontrollable seizures.

How do you feel about this research? Do you think the topic was important enough to justify such measures? Would debriefing the subjects be sufficient to ameliorate any possible harm? Can you think of other ways the researcher might have examined obedience?

In recognition of the importance of ethical issues in social inquiry, the American Sociological Association has a website entitled, "Teaching Ethics throughout the Curriculum," which contains a wide variety of case studies as well as resources for dealing with them.

The National Institutes of Health has established an online course regarding the history, issues, and processes regarding human-subjects research. While it was specifically designed for researchers seeking federal funding for research, it is available to and useful for anyone with an interest in this topic.

The Politics of Social Research

As I indicated earlier, both ethics and politics hinge on ideological points of view. What is unacceptable from one point of view may be acceptable from another. Although political and ethical issues are often closely intertwined, I want to distinguish between them in two ways.

First, the ethics of social research deals mostly with the methods employed; political issues tend to center on the substance and use of research. Thus, for example, some critics raise ethical objections to the Milgram experiments, saying that the methods harmed the subjects. A political objection would be that obedience is not a suitable topic for study, either because (1) we should not tinker with people's willingness to follow orders from a higher authority or (2) because the results of the research could be misused to make people *more* obedient.

The second distinction between ethical and political aspects of social research is that there are no formal codes of accepted political conduct. Although some ethical norms have political aspects—for example, specific guidelines for not

harming subjects clearly relate to the U.S. protection of civil liberties—no one has developed a set of political norms that all social researchers accept.

The only partial exception to the lack of political norms is the generally accepted view that a researcher's personal political orientation should not interfere with or unduly influence his or her scientific research. It would be considered improper for a researcher to use shoddy techniques or to distort or lie about his or her research as a way of furthering the researcher's political views. As you can imagine, however, studies are often enough attacked for allegedly violating this norm.

Objectivity and Ideology

In Chapter 1, I suggested that social research can never be totally objective, because researchers are human and therefore necessarily subjective. Science, as a collective enterprise, achieves the equivalent of objectivity through intersubjectivity. That is, different scientists, having different subjective views, can and should arrive at the same results when they employ accepted research techniques. Essentially, this will happen to the extent that each can set personal values and views aside for the duration of the research.

The classic statement on objectivity and neutrality in social science is Max Weber's "Science as a Vocation" ([1925] 1946). In this lecture, Weber coined the phrase *value-free sociology* and urged that sociology, like other sciences, needed to remain unencumbered by personal values if it was to make a special contribution to society. Liberals and conservatives alike could recognize the "facts" of social science, regardless of how those facts accorded with their personal politics.

Most social researchers have agreed with this abstract ideal, but not all. Marxist and neo-Marxist scholars, for example, argue that social science and social action cannot and should not be separated. Explanations of the status quo in society, they contend, shade subtly into defenses of that same status quo. Simple explanations of the social functions of, say, discrimination can easily become justifications for its continuance. By the same token, merely studying society and its ills without a commitment to making society more humane has been called irresponsible.

In 2004 American Sociological Association President Michael Burawoy made *public sociology* the theme of the annual ASA meeting. Many scholars have espoused this term in recent years. Although these researchers may disagree on how sociology should affect society or which sectors of society should be affected, they generally agree that research should have an intentional impact. Recall our discussion of "applied" and "pure" research as a background for this movement in contemporary sociology (see Chapter 2). If you want to explore this further, you might examine a special symposium on the issue in *Contemporary Sociology* (2008).

In Chapter 10, we'll examine *participatory action research*, which is committed to using social research for the purposes designed and valued by the subjects of the research. Thus, for example, researchers who want to improve the conditions for workers at a factory would ask the workers to define the outcomes they would like to see and also ask them to have a hand in conducting the social research relevant to achieving them. The task of the researchers is to ensure that the workers have access to professional research methods.

Politics can intrude in research as in other aspects of life.

Quite aside from abstract disagreements about whether social science *should* be value-free, many have argued about whether particular research undertakings *are* value-free or whether they represent an intrusion of the researcher's own political values. Typically, researchers deny such intrusion, and others then challenge their denials. Let's look at some examples of the controversies surrounding this issue.

Social Research and Race

Nowhere have social research and politics been more controversially intertwined than in the area of race relations. Social researchers studied the topic for a long time. Often, the products of the social research found their way into practical politics. A few brief references should illustrate the point.

In 1896, when the U.S. Supreme Court established the principle of "separate but equal" as a means of reconciling the Fourteenth Amendment's guarantee of equality to African Americans with the norms of segregation, it neither asked for nor cited social research. Nonetheless, it is widely believed that the Court was influenced by the writings of William Graham Sumner, a leading social scientist of his era. Sumner was noted for his view that the mores and folkways of a society were relatively impervious to legislation and social planning. His view has often been paraphrased as "stateways do not make folkways." Thus, the Court ruled that it could not accept the assumption that "social prejudices may be overcome by legislation" and denied the wisdom of "laws which conflict with the general sentiment of the community" (Blaunstein and Zangrando 1970: 308). As many a politician has said, "You can't legislate morality."

When the doctrine of "separate but equal" was overturned in 1954 (*Brown v. Board of Education of Topeka*), the new Supreme Court decision was based in part on the conclusion that segregation had a detrimental effect on African American children. In drawing that conclusion, the Court cited several sociological and psychological research reports (Blaunstein and Zangrando 1970).

For the most part, social researchers in this century have supported the cause of African American equality in the United States, and their convictions often have provided the impetus for their research. Moreover, they've hoped that their research will lead to social change. There is no doubt, for example, that Gunnar Myrdal's classic two-volume study (1944) of race relations in the United States significantly influenced race relations themselves. Myrdal amassed a great deal of data to show that the position of African Americans directly contradicted U.S. values of social and political equality. Further, Myrdal did not attempt to hide his own point of view in the matter.

Many social researchers have become directly involved in the civil rights movement, some more radically than others. Given the broad support for ideals of equality, research conclusions supporting the cause of equality draw little or no criticism. To recognize how solid the general social science position is in this matter, we need to only examine a few research projects that have produced conclusions disagreeing with the predominant ideological position.

Most social researchers have—overtly, at least—supported the end of school segregation. Thus, an immediate and heated controversy was provoked in 1966 when James Coleman, a respected sociologist, published the results of a major national study of race and education. Contrary to general agreement, Coleman found little difference in academic performance between African American students attending integrated schools and those attending segregated ones. Indeed, such obvious things as libraries, laboratory facilities, and high expenditures per student made little difference. Instead, Coleman reported that family and neighborhood factors had the most influence on academic achievement.

Coleman's findings were not well received by many of the social researchers who had been active in the civil rights movement. Some scholars criticized Coleman's work on methodological grounds, but many others objected hotly on the ground that the findings would have segregationist political consequences. The controversy that raged around the Coleman report harkened back to the uproar provoked a year earlier by Daniel Moynihan (1965) in his critical analysis of the African American family in the United States. Whereas some felt Moynihan was blaming the victims, others objected to his tracing those problems to the legacy of slavery.

Another example of political controversy surrounding social research in connection with race concerns IQ scores. In 1969, Arthur Jensen, a Harvard psychologist, was asked to prepare an article for the *Harvard Educational Review* examining the data on racial differences in IQ test results (Jensen 1969). In the article, Jensen concluded that genetic differences between African Americans and whites accounted for the lower average IQ scores of African Americans. Jensen became so identified with that position that he appeared on college campuses across the country discussing it.

Jensen's research has been attacked on numerous methodological bases. Critics charged that much of the data on which Jensen's conclusion was based were inadequate and sloppy—there are many IQ tests, some worse than others. Similarly, it was argued that Jensen had not taken social–environmental factors sufficiently into account. Other social researchers raised still other methodological objections.

Beyond the scientific critique, however, many condemned Jensen as a racist. Hostile crowds drowned out his public presentations by booing. Ironically, Jensen's reception by several university audiences recalled the hostile reception received by abolitionists a century before, when the prevailing opinion favored leaving the institution of slavery intact.

Many social researchers limited their objections to the Moynihan, Coleman, and Jensen research to scientific, methodological grounds. The political firestorms ignited by these studies, however, demonstrate how ideology often shows up in matters of social research. Although the abstract model of science is divorced from ideology, the practice of science is not.

The Politics of Sexual Research

As I indicated earlier, the Laud Humphreys' study of "tearoom trade" raised ethical issues that researchers still discuss and debate. At the same time, much of the furor was related to the subject matter itself. As I have written elsewhere,

Laud Humphreys didn't just study S-E-X but observed and discussed homosexuality. *And it wasn't even the caring-and-committed-relationships-between-two-people-who-just-happen-to-be-of-the-same-sex homosexuality but tawdry encounters*

between strangers in public toilets. Only adding the sacrifice of Christian babies could have made this more inflammatory for the great majority of Americans in 1970.

(Babbie 2004: 12)

Although Humphreys' research topic was unusually provocative for many, tamer sexuality research has also engendered outcries of public horror. During the 1940s and 1950s, Alfred Kinsey, a biologist, published landmark studies of sexual practices of American men (1948) and women (1953). Kinsey's extensive interviewing allowed him to report on frequency of sexual activity, premarital and extramarital sex, homosexual behavior, and so forth. His studies produced public outrage and efforts to close his research institute at Indiana University.

Although today most people no longer get worked up about the Kinsey reports, Americans tend to remain touchy about research on sex. In 1987, the National Institutes of Health (NIH), charged with finding ways to combat the AIDS epidemic, found they needed hard data on contemporary sexual practices if they were to design effective anti-AIDS programs. Their request for research proposals resulted in a sophisticated study design by Edward O. Laumann and colleagues. The proposed study focused on the different patterns of sexual activity characterizing different periods of life, and it received rave reviews from the NIH and their consultants.

Enter Senator Jesse Helms (R-NC) and Congressman William Dannemeyer (R-CA). In 1989, having learned of the Laumann study, Helms and Dannemeyer began a campaign to block the study and shift the same amount of money to a teen abstinence-only program. Anne Fausto-Sterling, a biologist, sought to understand the opposition to the Laumann study.

> *The surveys, Helms argued, are not really intended "to stop the spread of AIDS. The real purpose is to compile supposedly scientific facts to support the left-wing liberal argument that homosexuality is a normal, acceptable life-style.... As long as I am able to stand on the floor of the U.S. Senate," he added, "I am never going to yield to that sort of thing, because it is not just another life-style; it is sodomy."*

(Fausto-Sterling 1992)

Helms was sufficiently persuasive as to win a 66–34 vote in favor of his amendment in the U.S. Senate. The House rejected the amendment, and it was dropped in conference committee, but government funding for the study was put on hold. Laumann and his colleagues then turned to the private sector and obtained funding for a smaller study, published in 1994 as *The Social Organization of Sexuality*.

Politics and the Census

There is probably a political dimension to every attempt to study human social behavior. Consider the decennial U.S. Census, mandated by the Constitution. The original purpose was to discover the population sizes of the various states to determine their proper representation in the House of Representatives. Whereas each state gets two senators, large states get more representatives than do small ones. So what could be simpler? Just count the number of people in each state.

From the beginning, there was nothing simple about counting heads in a dispersed, national population like the United States. Even the definition of a "person" was anything but straightforward. A slave, for example, counted as only three-fifths of a person for purposes of the census. This decreased the representation of the slaveholding Southern states, though counting slaves as whole people might have raised the dangerously radical idea that they should be allowed to vote.

Further, the logistical problems of counting people who reside in suburban tract houses, urban apartments, college dorms, military barracks, farms, cabins in the woods, and illegal housing units, as well as counting those who have no place to live, have always presented a daunting task. It's the sort of challenge social researchers tackle with relish. However, the difficulty of finding the hard-to-reach and the techniques created for doing so cannot escape the political net.

Kenneth Prewitt, who directed the Census Bureau from 1998 to 2001, describes some of the political aspects of counting heads:

> *Between 1910 and 1920, there was a massive wartime population movement from the rural, Southern states to industrial Northern cities.*

In 1920, for the first time in American history, the census included more city dwellers than rural residents. An urban America was something new and disturbing, especially to those who held to the Jeffersonian belief that independent farmers best protected democracy. Among those of this persuasion were rural, conservative congressmen in the South and West. They saw that reapportionment would shift power to factory-based unions and politically radical immigrants concentrated in Northeastern cities. Conservatives in Congress blocked reapportionment, complaining among other things that because January 1 was then census day, transient agricultural workers were "incorrectly" counted in cities rather than on the farms to which they would return in time for spring planting. (Census day was later shifted to April 1, where it has remained.) The arguments dragged out for a decade, and Congress was not reapportioned until after the next census.

(Prewitt 2003)

More recently, concern for undercounting the urban poor has become a political issue. The big cities, which have the most to lose from the undercounting, typically vote Democratic rather than Republican, so you can probably guess which party supports efforts to improve the counting and which party is less enthusiastic. By the same token, when social scientists have argued in favor of replacing the attempt at a total enumeration of the population with modern survey sampling methods (see Chapter 7 for more on sampling), they have enjoyed more support from Democrats, who would stand to gain from such a methodological shift, than from Republicans, who would stand to lose. Rather than suggesting that Democrats support science more than do Republicans, this situation offers another example of how the political context in which we live and conduct social research often affects that research.

Politics with a Little "p"

Political ideologies often confound social research, but the more personal "politics" of social research runs far deeper still. Social research in relation to contested social issues simply cannot remain antiseptically objective—particularly when differing ideologies are pitted against each other in a field of social science data.

The same is true when research is invoked in disputes between people with conflicting interests. For instance, social researchers who have served as "expert witnesses" in court would probably agree that the scientific ideal of a "search for truth" seems hopelessly naive in a trial or lawsuit. Although expert witnesses technically do not represent either side in court, they are, nonetheless, engaged by only one side to appear, and their testimony tends to support the side of the party who pays for their time. This doesn't necessarily mean that these witnesses will lie on behalf of their patrons, but the contenders in a lawsuit are understandably more likely to pay for expert testimony that supports their case than for testimony that attacks it.

Thus, as an expert witness, you appear in court only because your presumably scientific and honest judgment happens to coincide with the interests of the party paying you to testify. Once you arrive in court and swear to tell the truth, the whole truth, and nothing but the truth, however, you find yourself in a world foreign to the ideals of objective contemplation. Suddenly the norms are those of winning and losing. As an expert witness, of course, all you have to lose is your respectability (and perhaps the chance to earn fees as an expert witness in the future). Still, such stakes are high enough to create discomfort for most social researchers.

I recall one case in federal court when I was testifying on behalf of some civil service workers who had had their cost-of-living allowance cut on the basis of research I thought was rather shoddy. I was engaged to conduct more "scientific" research that would demonstrate the injustice worked against the civil servants (Babbie 1982: 232–43).

I took the stand, feeling pretty much like a respected professor and textbook author. In short order, however, I found I had moved from the academy to the hockey rink. Tests of statistical significance and sampling error were suddenly less relevant than a slap shot. At one point, an attorney from Washington lured me into casually agreeing that I was familiar with a certain professional journal. Unfortunately, the journal did not exist. I was mortified and suddenly found myself shifting domains. Without really thinking about it, I shifted from Mr. Rogers' Neighborhood to the dark alleys of ninja-professor. I would not be fully satisfied until I, in turn, could mortify the attorney, which I succeeded in doing.

Even though the civil servants got their cost-of-living allowance back, I have to admit I was also concerned with how I looked in front of the courtroom assemblage. I tell you this anecdote to illustrate the personal "politics" of human interactions involving presumably scientific and objective research. We need to realize that as human beings, social researchers are going to act like human beings, and we must take this into account in assessing their findings. This recognition neither invalidates their research nor provides an excuse for rejecting findings we happen to dislike, but it does need to be taken into account.

Similar questions are raised regularly outside the social sciences. For example, you've probably read reports about research demonstrating the safety of a new drug—research that was paid for by the very pharmaceutical company that developed the drug and was seeking FDA approval to sell it. Even if the research was of the highest quality, it is appropriate to question whether it was tainted by a conflict of interest. Similarly, when research sponsored by the coal or petroleum industries concludes that global climate change is not a human-caused problem, you shouldn't necessarily assume that the research was biased, but you should consider that possibility. At the very least, the sponsorship of such research should be made public.

Applying these kinds of concerns to survey research, the AAPOR, in 2009, established a "Transparency Initiative," requiring all association members, and urging all other survey researchers, to report openly and fully the details of their research methods. President of the AAPOR, Peter V. Miller, acknowledged that the program might be in for rough sledding:

> Recent events have taught us that disclosure itself can be manipulated. It is disturbingly easy to claim that polls have been conducted using particular methods, while, in truth, the work was not done or was done another way. While we must rely on the integrity of participants in the initiative, we cannot proceed on the basis of trust alone. We must develop ways to check the information we receive. The value of AAPOR's recognition depends on it.

> (2010: 606).

Politics in Perspective

Although the ethical and the political dimensions of research are in principle distinct, they do intersect. Whenever politicians or the public feel that social research is violating ethical or moral standards, they'll be quick to respond with remedies of their own. Moreover, the standards they defend may not be those of the research community. And even when researchers support the goals of measures directed at the way research is done, the means specified by regulations or legislation can hamstring research.

Today, the "politicization of science" is a particularly hot topic, with charges flung from both sides of the political spectrum. On the one hand, we can see renewed objections to the teaching of evolution while demands for the teaching of creationism have been replaced by support for intelligent design. In many of these regards, science is seen as a threat to religion-based views, and scientists are sometimes accused of having an antireligion agenda. On the other hand, a statement by the Union of Concerned Scientists (2005), cosigned by thousands of scientists, illustrates the concern that the concentration of political power in the hands of one party can threaten the independent functioning of scientific research:

> The United States has an impressive history of investing in scientific research and respecting the independence of scientists. As a result, we have enjoyed sustained economic progress and public health, as well as unequaled leadership within the global scientific community. Recent actions by political appointees, however, threaten to undermine this legacy by preventing the best available science from informing policy decisions that have serious consequences for our health, safety, and environment.

> Across a broad range of issues—from childhood lead poisoning and mercury emissions to climate change, reproductive health, and nuclear weapons—political appointees have distorted and censored scientific findings that contradict established policies. In some cases, they have manipulated the underlying science to align results with predetermined political decisions.

> (Union of Concerned Scientists 2005)

There are four main lessons that I hope you will take away from this discussion. First, science is not untouched by politics. The intrusion of

What do you think?...Revisited

The Nazi medical experiments were outrageous in many ways. Some of the experiments can only be described as ghoulish and sadistic. Often the scientific caliber of the experiments was shoddy. One could argue, however, that some people today suffer and even die from research. We often condone these risks by virtue of the benefits to humankind expected to follow from the research. Some of the Nazi doctors, no doubt, salved their own consciences with such justifications.

The Nazi medical experiments breached a fundamental ethical norm discussed in this chapter. This is reflected in the indictments

of the Nuremberg trials, which charged several medical personnel in the Nazi war machine with "plans and enterprises involving medical experiments *without the subjects' consent* on civilians and members of the armed forces of nations then at war with the German Reich [emphasis mine]" (*Trials of War Criminals* 1949–1953). Even if the most hideous experiments had not been conducted, and even accepting that there is always some risk when human research is undertaken, it is absolutely unacceptable to subject people to risks in research without their informed consent.

---

politics and related ideologies is not unique to social research; the natural sciences have experienced and continue to experience similar situations. However, social researchers study things that matter to people; things they have firm, personal feelings about; and things that affect their lives. Moreover, researchers are human beings, and their feelings often show through in their professional lives. To think otherwise would be naive.

Second, science proceeds in the midst of political controversy and hostility. Even when researchers get angry and call each other names, or when the research community comes under attack from the outside, scientific inquiry persists. Studies are done, reports are published, and new things are learned. In short, ideological disputes do not bring science to a halt, but they do make it more challenging—and more exciting.

Third, an awareness of ideological considerations enriches the study and practice of social research methods. Many of the established characteristics of science, such as intersubjectivity, function to cancel out or hold in check our human shortcomings, especially those we are unaware of. Otherwise, we

might look at the world and never see anything but a reflection of our personal biases and beliefs.

Finally, although researchers should not let their own values interfere with the quality and honesty of their research, this does not mean that researchers cannot or should not participate in public debates and express both their scientific expertise and personal values. You can do scientifically excellent research on racial prejudice, all the while being opposed to prejudice and saying so. Some would argue that social scientists, because of their scientific expertise in the workings of society, have an obligation to speak out, rather than leaving that role to politicians, journalists, and talk-show hosts. Herbert Gans writes of the need for "public sociologists":

A public sociologist is a public intellectual who applies sociological ideas and findings to social (defined broadly) issues about which sociology (also defined broadly) has something to say. Public intellectuals comment on whatever issues show up on the public agenda; public sociologists do so only on issues to which they can apply their sociological insights and findings.

MAIN POINTS

Introduction

- Ethical and political considerations, in addition to technical scientific and administrative concerns, shape social research problems.

Ethical Issues in Social Research

- What is considered ethical and unethical in research is ultimately a matter of what a community of people agree is right and wrong.

- Researchers agree that participation in research should normally be voluntary. This norm, however, can conflict with the scientific need for generalizability.

- Researchers agree that research should not harm those who participate in it, unless they willingly and knowingly accept the risks of harm by giving their informed consent.

- Whereas *anonymity* refers to the situation in which even the researcher cannot link specific information to the individuals it describes, *confidentiality* refers to the situation in which the researcher promises to keep information about subjects private. The most straightforward way to ensure confidentiality is to destroy identifying information as soon as it's no longer needed.

- Many research designs involve a degree of deception. Because deceiving people violates common standards of ethical behavior, deception in research requires a strong justification—and even then the justification may be challenged.

- Social researchers have ethical obligations to the community of researchers as well as to subjects. These obligations include reporting results fully and accurately as well as disclosing errors, limitations, and other shortcomings in the research.

- Institutional review boards review research proposals involving human subjects so that they can guarantee that the subjects' rights and interests will be protected.

- Professional associations in several disciplines publish codes of ethics to guide researchers. These codes are necessary and helpful, but they do not resolve all ethical questions.

Two Ethical Controversies

- Laud Humphreys' study of "tearoom" encounters and Stanley Milgram's study of obedience raised ethical issues that are debated to this day.

The Politics of Social Research

- Social research inevitably has a political and ideological dimension. Although science is neutral on political matters, scientists are not. Moreover, much social research inevitably involves the political beliefs of people outside the research community.

- Although most researchers agree that political orientation should not unduly influence research, in practice it can be very difficult to separate politics and ideology from the conduct of research. Some researchers maintain that research can and should be an instrument of social action and change. More subtly, a shared ideology can affect the way other researchers receive social research.

- Even though the norms of science cannot force individual researchers to give up their personal values, the intersubjective character of science provides a guard against "scientific" findings being the product of bias only.

KEY TERMS

anonymity	debriefing
confidentiality	informed consent

PROPOSING SOCIAL RESEARCH: ETHICAL ISSUES

If you are actually proposing a research project, you may be required to submit your proposal to your campus institutional review board (IRB). In that case, you'll need to find the proper forms and follow the procedures involved. The key concern here is the protection of research subjects: avoiding harm, safeguarding privacy, and so forth.

REVIEW QUESTIONS

1. Consider the following real and hypothetical research situations. What is the ethical component in each example? How do you feel about it? Do you think the procedures described are ultimately acceptable or unacceptable? You might find it useful to discuss some of these situations with classmates.

   a. A psychology instructor asks students in an introductory psychology class to complete questionnaires that the instructor will analyze and use in preparing a journal article for publication.

   b. After a field study of deviant behavior during a riot, law enforcement officials demand that the researcher identify the people who were observed looting. Rather than risk arrest as an accomplice after the fact, the researcher complies.

c. After completing the final draft of a book reporting a research project, the researcher-author discovers that 25 of the 2,000 survey interviews were falsified by interviewers. To protect the bulk of the research, the author leaves out this information and publishes the book.

d. Researchers obtain a list of right-wing radicals they wish to study. They contact the radicals with the explanation that each has been selected "at random" from among the general population to take a sampling of "public opinion."

e. A college instructor, who wants to test the effect of unfair berating, administers an hour exam to both sections of a specific course. The overall performance of the two sections is essentially the same. The grades of one section are artificially lowered, however, and the instructor berates the students for performing so badly. The instructor then administers the same final exam to both sections and discovers that the performance of the unfairly berated section is worse. The hypothesis is confirmed, and the research report is published.

f. In a study of sexual behavior, the investigator wants to overcome subjects' reluctance to report what they might regard as shameful behavior. To get past their reluctance, subjects are asked, "Everyone masturbates now and then; about how much do you masturbate?"

g. A researcher studying dorm life on campus discovers that 60 percent of the residents regularly violate restrictions on alcohol consumption. Publication of this finding would probably create a furor in the campus community. Because no extensive analysis of alcohol use is planned, the researcher decides to keep this finding quiet.

h. To test the extent to which people may try to save face by expressing attitudes on matters they are wholly uninformed about, the researcher asks for subjects' attitudes regarding a fictitious issue.

i. A research questionnaire is circulated among students as part of their university registration packet. Although students are not told they must complete the questionnaire, the hope is that they will believe they must—thus ensuring a higher completion rate.

j. A researcher pretends to join a radical political group in order to study it and is successfully accepted as a member of the inner planning circle. What should the researcher do if the group makes plans for the following?

   1. A peaceful, though illegal, demonstration

   2. The bombing of a public building during a time when it is sure to be unoccupied

   3. The assassination of a public official

2. Review the discussion of the Milgram experiment on obedience. How would you design a study to accomplish the same purpose while avoiding the ethical criticisms leveled at Milgram? Would your design be equally valid? Would it have the same effect?

3. Suppose a researcher who is personally in favor of small families (as a response to the problem of overpopulation) wants to conduct a survey to determine why some people want many children and others don't. What personal-involvement problems would the researcher face and how could she or he avoid them?

4. What ethical issues should the researcher in item 3 take into account in designing the survey?

CHAPTER 4

Research Design

CHAPTER OVERVIEW

Here you'll see the wide variety of research designs available to social researchers as well as how to design a study—that is, specifying exactly who or what is to be studied when, how, and for what purpose.

© ESTUDI M6/Shutterstock.com

Learning Objectives

After studying this chapter, you will be able to . . .

- Discuss the role of exploration, description, and explanation in social research.
- Discuss the logic and procedures of idiographic explanation.
- Name and discuss the legitimate and false criteria for nomothetic explanation.
- Distinguish between necessary and sufficient causes, giving examples.
- Identify some of the common units of analysis in social research and explain why that concept is important.
- Identify and describe some common study designs based on the time dimension.
- Explain and illustrate some advantages of mixed-mode designs.
- Identify and discuss the key elements in the design of a research project.
- Describe the elements and structure of a research proposal.
- Give examples of how research design can have ethical implications.

Introduction

Science is dedicated to "finding out." No matter what you want to find out, though, you'll likely discover many ways to go about finding it. That's true in life in general. Suppose, for example, that you want to find out whether a particular automobile—say, the new Burpo-Blasto—would be a good car for you. You could, of course, buy one and find out that way. Or you could talk to a lot of B-B owners or to people who considered buying one and didn't. You might check the classified ads to see if there are a lot of B-Bs being sold cheap. You could read consumer-magazine or online evaluations of Burpo-Blastos. A similar situation occurs in scientific inquiry.

Ultimately, scientific inquiry comes down to making observations and interpreting what you've observed, the subjects of Parts 3 and 4 of this book. Before you can observe and analyze, however, you need a plan. You need to determine what you're going to observe and analyze: why and how. That's what research design is all about.

Although the details vary according to what you wish to study, you face two major tasks in any research design. First, you must specify as clearly as possible what it is you want to find out. Second, you must determine the best way to do it. Interestingly, if you can handle the first consideration fully, you'll probably have addressed the second already. As mathematicians say, a properly framed question contains the answer.

Let's say you're interested in conducting social research on terrorism. When Jeffrey Ross (2004) addressed this issue, he found the existing studies used a variety of qualitative and quantitative approaches. Qualitative researchers, for example, generated original data through

- Autobiographies
- Incident reports and accounts
- Hostages' experiences with terrorists
- Firsthand accounts of implementing policies

Ross goes on to discuss some of the secondary materials used by qualitative researchers: "biographies of terrorists, case studies of terrorist organizations, case studies on types of terrorism, case studies on particular terrorist incidents, and case studies of terrorism in selected regions and countries" (2004: 27). Quantitative researchers, on the other hand, addressed terrorism in a variety of ways, including analyses of media coverage, statistical modeling of terrorist events, and the use of various databases relevant to the topic. As you'll see in this chapter, any research topic can be approached from many different directions. Each of the topics we'll examine is relevant to both qualitative and quantitative studies, though some topics may be more relevant to one than to the other approach.

This chapter provides a general introduction to research design; the other chapters in Part 2 elaborate on specific aspects of it. In practice, all aspects of research design are interrelated. As

What do you think?

In the following letter published in a college newspaper, the Provost objects to data that had been previously reported.

Provost says percentage was wrong

I am writing to clarify a misstatement in an editorial in the April 19 The Panther. *As recently as last fall, the concept behind this statement was presented to your staff.*

This current use of erroneous numbers demands correction. The figure used in the statement, "With about 52 percent of the faculty being part-time…" is absolutely incorrect.

Since the thrust of the editorial is Chapman's ability to live up to its desire to "nurture and help develop students," a proper measure of the difference between full-time faculty presence and that of part-time faculty is how many credits or courses are taught.

For the past four years, full-time faculty have taught about 70 percent of the credits in which students enroll each semester.

Thus, a large majority of our faculty are here full-time: teaching classes, advising students, attending meetings, inter- acting with students in the hallways and dining rooms.

Once again, I welcome the opportunity to present the truth.

Might I suggest that a future edition of The Panther *be devoted to the contributions of part-time faculty.*

Harry L. Hamilton, Provost

Sometimes, data seem as though they dropped out of the sky, making no sense. Which side is correct in this case: the original newspaper report or the Provost's account? Or are both sides correct? If so, why?

See the *What do you think? … Revisited* box toward the end of the chapter.

Earl Babbie

you read through Part 2, the interrelationships among parts will become clearer.

We'll start by briefly examining the main purposes of social research, learning about both idiographic and nomothetic approaches. Then, we'll consider units of analysis—the "what or whom" you want to study. Next we'll consider alternative ways of handling time in social research, or how to study a moving target that changes over time.

With these ideas in hand, we'll turn to how to design a research project. This overview of the research process serves two purposes: In addition to describing how you might go about designing a study, it provides a map of the remainder of this book.

Next, we'll look at the elements of a research proposal. Often you'll need to detail your intentions before you actually conduct your research in order to obtain funding for a major project or perhaps to get your instruc- tor's approval for a class project. You'll see that

the research proposal provides an excellent op- portunity for you to consider all aspects of your research in advance. Also, this section should help you with the continuing, end-of-chapter exercise concerning research proposals, in the event that you are doing that.

Finally, we'll consider the ethical implications of this research design. As you read through this chapter, think about how the practice of social research in this regard can raise larger issues.

Three Purposes of Research

Social research can serve many purposes. Three of the most common and useful purposes are exploration, description, and explanation. Although most studies have more than one of these purposes, examining them separately is useful because each has different implications for other aspects of research design.

Exploration

Much of social research is conducted to explore a topic, that is, to start to familiarize a researcher with that topic. This approach typically occurs when a researcher examines a new interest or when the subject of study itself is relatively new.

As an example, let's suppose that widespread taxpayer dissatisfaction with the government erupts into a taxpayers' revolt. People begin refusing to pay their taxes, and they organize themselves around that issue. You might like to learn more about the movement: How widespread is it? What levels and degrees of support exist within the community? How is the movement organized? What kinds of people are active in it? An exploratory study could help you find at least approximate answers to some of these questions. You might check figures with tax-collecting officials, gather and study the literature of the movement, attend meetings, and interview leaders.

Exploratory studies are also appropriate for more-persistent phenomena. Suppose you're unhappy with your college's graduation requirements and want to help change them. You might study the history of such requirements at the college and meet with college officials to learn the reasons for the current standards. You could talk to several students to get a rough idea of their sentiments on the subject. Although this last activity would not necessarily yield an accurate picture of student opinion, it could suggest what the results of a more extensive study might be.

Sometimes exploratory research is pursued through the use of focus groups, or guided small-group discussions. This technique is frequently used in market research, which we'll examine further in Chapter 10.

Exploratory studies are most typically done for three purposes: (1) to satisfy the researcher's curiosity and desire for better understanding, (2) to test the feasibility of undertaking a more extensive study, and (3) to develop the methods to be employed in any subsequent study.

Exploratory studies are quite valuable in social science research. They're essential whenever a researcher is breaking new ground, and they almost always yield new insights into a topic for research. Exploratory studies are also a source of grounded theory, as discussed in Chapter 10.

The chief shortcoming of exploratory studies is that they seldom provide satisfactory answers to research questions, although they can hint at the answers and can suggest which research methods could provide definitive answers. The reason exploratory studies are seldom definitive in themselves has to do with representativeness; that is, the people you study in your exploratory research may not be typical of the larger population that interests you. Once you understand representativeness, you'll be able to know whether a given exploratory study actually answered its research problem or only pointed the way toward an answer. (Representativeness is discussed at length in Chapter 7.)

Description

Many social science studies aim at describing situations and events. The researcher observes and then describes what was observed. Because scientific observation is careful and deliberate, scientific descriptions are typically more accurate and precise than are casual ones.

For example, the goal of the U.S. Census is to describe accurately and precisely a wide variety of the population characteristics of the United States, as well as areas such as states and counties. Other examples of descriptive studies include the creation of age–gender profiles of populations by demographers, the computation of crime rates for different cities, and a product-marketing survey that describes the people who use, or would use, a particular product.

Research design involves the creation and integration of many diverse elements.

Many qualitative studies aim primarily at description. An anthropological ethnography, for example, may try to detail the particular culture of some preliterate society. At the same time, such studies are seldom limited to a merely descriptive purpose. Researchers usually go on to examine why the observed patterns exist and what they imply.

Explanation

The third general purpose of social science research is to explain things. Descriptive studies answer questions of what, where, when, and how; explanatory studies address questions of why. So when William Sanders (1994) set about describing the varieties of gang violence, he also wanted to reconstruct the process that brought about violent episodes among the gangs of different ethnic groups.

Reporting the voting intentions of an electorate is descriptive, but reporting why some people plan to vote for Candidate A and others for Candidate B is explanatory. Reporting why some cities have higher crime rates than others involves explanation, as does identifying *variables* that explain why some cities have higher crime rates than others.

Let's look at a specific case. Recent years have seen a radical shift in American attitudes toward marijuana. Support for the use of medical marijuana has increased in many states and, at this writing, recreational use of marijuana is legal in Colorado, Washington, Alaska, Oregon, California, Nevada, Michigan, Massachusetts, Vermont, and Maine. What factors do you suppose might shape people's attitudes toward the legalization of marijuana? To answer this, you might first consider whether men and women differ in their opinions. An explanatory analysis of the 2016 General Social Survey (GSS) data indicates that 65 percent of men and 56 percent of women said that marijuana should be legalized.

What about political orientation? The GSS data show that 74 percent of liberals said marijuana should be legalized, as compared with 54 percent of moderates and 42 percent of conservatives. Further, 62 percent of Democrats, as compared with 57 percent of Independents and 42 percent of Republicans, supported such legalization.

Given these statistics, you might begin to develop an explanation for attitudes toward marijuana legalization. Further study of gender and political orientation might then lead to a deeper explanation of these attitudes.

In Chapter 1, we noted that there were two different approaches to explanation in social research (and in everyday life). Let's return to those now.

In the remainder of this chapter, we'll examine some general approaches to research and the elements of research design from which you can choose. As you do this, keep in mind that the advanced planning for research may not fit perfectly the situations you will confront in the field. The "How to Do It: Putting Social Research to Work" reports on a graduate student's experience in the field.

Idiographic Explanation

As you will recall from Chapter 1, idiographic explanation seeks an exhaustive understanding of the causes producing events and situations in a single or limited number of cases. If you wished to understand why a student protest broke out on a particular college campus, you would seek to root out everything that contributed to that result. You would consider the history of the college, its organizational structure, the nature of the student body, the actions of influential individuals (administrators, faculty, students, others), the context of student activities nationally, triggering events (e.g., shutting down a student organization, arresting a student), and so forth. You'll know your analysis is complete when the explanatory factors you have assembled made the protest inevitable, and when the absence of any of those factors might have kept it from happening.

There is no statistical test that can tell you when you have achieved this analytical success, however. This conclusion rests on the "art" of social research, which is achieved primarily through experience: by reading the analyses of others and by conducting your own. Here are a few techniques to consider.

● *Pay attention to the explanations offered by the people living the social processes you are studying.* It is important that you not believe everything

How to Do It

Putting Social Research to Work

Jacob Perry is a graduate student at the Clinton School of Public Service in Little Rock, Arkansas, and he chose to do his semester of fieldwork abroad in North Africa. He quickly discovered that the research techniques he had mastered in his studies did not necessarily fit into the research situation. Here's how he described it:

> We Americans have our quite fixed ideas about what research is: intense preparation, thorough literature review, ample discussions to outline details, timeline of events to take place, schedule of responsibilities and activities to perform, etc. These simply must happen in order to perform reputable research. And everything must be agreed upon before the research begins. That is the American way (granted, it is also somewhat of an internationally accepted way as well) and it is in many ways a reflection of our organized, prompt, accountable, obligation and time-oriented culture. However, I am in Morocco performing a research project, and Moroccan culture has quite different views on time, responsibilities, and planning. Time here is neither rigid nor fixed. It is not linear but rather cyclical, meaning time is not lost—it is simply recovered later. Life is now; it is present; it is mostly unplanned.

Jacob found ways to develop rapport with his fellow researchers in order to put his research training to good use.

> I spent the first two weeks building trust, familiarity, comfort—relations! This has been vital to the project's progress, as our team is able to discuss openly, honestly, and sometimes aggressively to arrive at an agreement. I also speak the local language, which has allowed my Moroccan partners to remain in their linguistic comfort zone, remain in control of conversations concerning project work, and not feel they are being neo-colonized by a foreigner here to "save" their country.

Ultimately, Jacob felt the project had been successful, and it is obvious that it was a powerful learning experience for the young social researcher.

> I am so pleased with this project in Morocco because it is a partnership between myself and Moroccans who have already established the goals, needs, and approach to developing their society. This project depends entirely on local expertise and initiative. I am here because of a mutually expressed interest by myself and locals who want to improve their city. And the project has progressed because of continuous discussions in which all team members offer their perspectives. In short, learning, considering, and respecting local culture is necessary for international development work to be successful, and I am seeing a great example of effective development work on the ground in Morocco.

Source: Private communication June 24, 2013.

you are told, of course, but don't make the opposite mistake of thinking you understand the situation better than those living there. (Social researchers have sometimes been accused of a certain degree of arrogance in this respect.) If there is wide agreement as to the importance of a certain factor, that should increase your confidence that it was a cause of the event under study. This would be more so if participants with very different points of view agree on that point. In the case of the student protest, administrators and students are likely to have very different opinions about what happened, but if they all agree that the arrest of a student activist was a triggering event, then it probably was an important cause.

- *Comparisons with similar situations, either in different places or at different times in the same place, can be insightful.* Perhaps the campus in question has had previous protests, or perhaps there was a time when a protest almost occurred but didn't. Knowledge of such instances can provide useful comparisons and contrasts to the case under study. Similarly, protests or nonprotests at other campuses can offer useful comparisons.

The Logic of Nomothetic Explanation

The preceding examination of what factors might cause attitudes about legalizing marijuana illustrates nomothetic explanation, as discussed in Chapter 1. Recall that in this model, we try to find a few factors (independent variables) that can account for much of the variation in a given phenomenon. This explanatory model stands in contrast to the idiographic model, in which we seek a complete, in-depth understanding of a single case.

In contrast to the idiographic approach, a nomothetic approach might suggest that overall political orientations account for much of the difference of opinion about legalizing marijuana. Because this model is inherently probabilistic, it is more open than the idiographic model to misunderstanding and misinterpretation. Let's examine what social researchers mean when they say one variable (nomothetically) causes another. Then, we'll look at what they don't mean.

Criteria for Nomothetic Causality

There are three main criteria for nomothetic causal relationships in social research: (1) the variables must be correlated, (2) the cause takes place before the effect, and (3) the variables are nonspurious.

Correlation

Unless some actual relationship—a statistical **correlation**—is found between two variables, we can't say that a causal relationship exists. Our analysis of GSS data suggested that political orientation was a cause of attitudes about legalizing marijuana. Had the same percentage of liberals and conservatives supported legalization, we could hardly say that political orientations caused the attitude. Though this criterion is obvious, it emphasizes the need to base social research assertions on actual observations rather than on assumptions.

Time Order

Next, we can't say a causal relationship exists unless the cause precedes the effect in time.

---

correlation An empirical relationship between two variables such that (1) changes in one are associated with changes in the other, or (2) particular attributes of one variable are associated with particular attributes of the other. Thus, for example, we say that *education* and *income* are correlated in that higher levels of education are associated with higher levels of income. Correlation in and of itself does not constitute a causal relationship between the two variables, but it is one criterion of causality.

spurious relationship A coincidental statistical correlation between two variables, shown to be caused by some third variable.

Notice that it makes more sense to say that most children's religious affiliations are caused by those of their parents than to say that parents' affiliations are caused by those of their children—even though it would be possible for you to change your religion and for your parents to follow suit. Remember, nomothetic explanation deals with general patterns but not all cases.

In our marijuana example, it would make sense to say that gender causes, to some extent, attitudes toward legalization, whereas it would make no sense to say that opinions about marijuana determine a person's gender. Notice, however, that the time order connecting political orientations and attitudes about legalization is less clear, although we sometimes reason that general orientations cause specific opinions. And sometimes our analyses involve two or more independent variables that were established at the same time: looking at the effects of gender and race on voting behavior, for example. As we'll see in the next chapter, the issue of time order can be a complex matter.

Nonspuriousness

The third requirement for a causal relationship is that the effect cannot be explained in terms of some third variable. For example, there is a correlation between ice-cream sales and deaths due to drowning: the more ice cream sold, the more drownings, and vice versa. There is, however, no direct link between ice cream and drowning. The third variable at work here is *season* or *temperature*. Most drowning deaths occur during summer—the peak period for ice-cream sales.

Here are two more examples of **spurious relationships**, or ones that aren't genuine. There is a negative relationship between the number of mules and the number of PhD's in towns and cities: the more mules, the fewer PhD's and vice versa. Perhaps you can think of another variable that would explain this apparent relationship. The answer is *rural versus urban settings*. There are more mules (and fewer PhD's) in rural areas, whereas the opposite is true in cities.

Or, consider the positive correlation between shoe size and math ability among schoolchildren. Here, the third variable that explains the puzzling relationship is *age*. Older children have bigger feet and more highly developed math skills, on average, than younger children do.

Observed Correlation

Positive (direct) correlation

Bigger shoe size is associated with greater math skill, and vice versa.

Spurious causal relationships

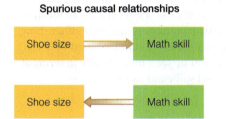

Neither shoe size nor math skill is a cause of the other.

Actual causal relationships

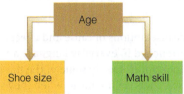

The underlying variable of age causes both bigger shoe size and greater math skill, thus explaining the observed correlation.

FIGURE 4-1

An Example of a Spurious Causal Relationship. Finding an empirical correlation between two variables does not necessarily establish a causal relationship. Sometimes the observed correlation is the incidental result of other causal relationships, involving other variables.

See Figure 4-1 for an illustration of this spurious relationship. Notice that observed associations go in both directions. That is, as one variable occurs or changes, so does the other.

At the top of the diagram, we see the observed correlation between shoe size and math ability. The double-headed arrow indicates that we don't know which variable might cause the other. To the bottom left of the diagram, we suggest that thinking math ability causes shoe size or that shoe size causes math ability are spurious, not real causal relationships. The bottom right of the diagram indicates what is actually at work. A third variable, *age*, (1) "causes" shoe size, in that older kids have bigger feet and (2) older kids know more math.

The list goes on. Areas with many storks have high birth rates. Those with few storks have low birth rates. Do storks really deliver babies? Birth rates are higher in the country than in the city; more storks live in the country than the city. The third variable here is *urban/rural areas*.

Finally, the more fire trucks that put out a fire, the more damage to the structure. Can you guess what the third variable is? In this case, it is the *size* of the fire.

Thus, when social researchers say there is a causal relationship between, say, education and racial tolerance, they mean (1) there is a statistical correlation between the two variables, (2) a person's educational level occurred before their current level of tolerance or prejudice, and (3) there is no third variable that can explain away the observed correlation as spurious.

Nomothetic Causal Analysis and Hypothesis Testing

The nomothetic model of causal analysis lends itself to hypothesis testing (see Chapter 2). To do this, you would carefully specify the variables you think are causally related, as well as specifying the manner in which you will measure them. (These steps will be discussed in detail in Chapter 5 under the terms *conceptualization* and *operationalization*.)

In addition to hypothesizing that two variables will be correlated, you could specify the level of *statistical significance* you will require in order to conclude that a genuine relationship exists. As you will learn in Chapter 7 (The Logic of Sampling), it is possible for factors such as sampling error to create the illusion of correlations where none exists in the larger population.

Finally, you could specify the tests for spuriousness that any observed relationship must survive. Not only would you hypothesize, for example, that increased education will reduce levels of prejudice, but you would specify further that the hypothesized relationship will not be the product of, say, political orientations.

False Criteria for Nomothetic Causality

Because notions of cause and effect are well entrenched in everyday language and logic, it's important to specify some of the things social researchers do not mean when they speak of causal relationships. When they say that one variable causes another, they do not necessarily mean to suggest complete causation, to account for exceptional cases, or to claim that the causation exists in a majority of cases.

Complete Causation

Whereas an idiographic explanation of causation is relatively complete, a nomothetic explanation is probabilistic and usually incomplete. As we've seen, social researchers may say that political orientations cause attitudes toward legalizing marijuana even though not all liberals approve or all conservatives disapprove. Thus, we say that political orientation is one of the causes of the attitude, but not the only one.

Exceptional Cases

In nomothetic explanations, exceptions do not disprove a causal relationship. For example, it is consistently found that women are more religious than men in the United States. Thus, gender may be a cause of religiosity, even if your uncle is a religious zealot or you know a woman who is an avowed atheist. Those exceptional cases do not disprove the overall, causal pattern.

Majority of Cases

Causal relationships can be true even if they do not apply in a majority of cases. For example, we say that children who are not supervised after school are more likely to become delinquent than are those who are supervised; hence, lack of supervision is a cause of delinquency. This causal relationship holds true even if only a small percentage of those not supervised become delinquent. As long as they are more likely to be delinquent than those who are supervised, we say there is a causal relationship.

The social science view of causation may vary from what you are accustomed to, because people commonly use the term *cause* to mean something that completely causes another thing. The somewhat different standard used by social researchers can be seen more clearly in terms of necessary and sufficient causes.

Necessary and Sufficient Causes

A necessary cause represents a condition that must be present for the effect to follow. For example, it is necessary for you to take college courses in order to get a degree. Take away the courses, and the degree never happens. However, simply taking the courses is not a sufficient cause of getting a degree. You need to take the right ones and pass them. Similarly, being female is a necessary condition of becoming pregnant, but it is not a sufficient cause. Otherwise, all women would get pregnant.

Figure 4-2 illustrates this relationship between the variables of *sex* and *pregnancy* as a matrix showing the possible outcomes of combining these variables.

A sufficient cause, on the other hand, represents a condition that, if it is present, guarantees the effect in question. This is not to say that a sufficient cause is the only possible cause of a particular effect. For example, skipping an exam in this course would be a sufficient cause for failing it, though students could fail it other ways as well. Thus, a cause can be sufficient, but not necessary. Figure 4-3 illustrates the relationship between taking or not taking the exam and either passing or failing it.

The discovery of a cause that is both necessary and sufficient is, of course, the most satisfying outcome in research. If juvenile delinquency were the effect under examination, it would be nice to discover a single condition that (1) must be present for delinquency to

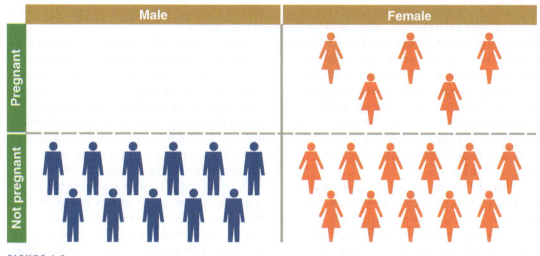

FIGURE 4-2

Necessary Cause. Being female is a necessary cause of pregnancy, that is, you can't get pregnant unless you're female.

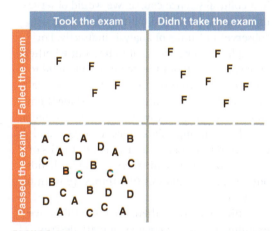

FIGURE 4-3

Sufficient Cause. Not taking the exam is a sufficient cause of failing it, even though there are other ways of failing (such as answering randomly).

develop and (2) always results in delinquency. In such a case, you would surely feel that you knew precisely what caused juvenile delinquency.

Unfortunately, when analyzing the nomothetic relationships among variables, we never discover single causes that are absolutely necessary and absolutely sufficient. It is not uncommon, however, to find causal factors that are either 100 percent necessary (you must be female to become pregnant) or 100 percent sufficient (skipping an exam will cause you to fail it).

In the idiographic analysis of single cases, you may reach a depth of explanation from which it is reasonable to assume that things could not have turned out differently, suggesting that you have determined the sufficient causes for a particular result. (Anyone with all the same details of your genetic inheritance, upbringing, and subsequent experiences would have ended up going to college.) At the same time, there could always be other causal paths to the same result. Thus, the idiographic causes are sufficient but not necessary.

Units of Analysis

In social research, there is virtually no limit to what or whom can be studied, or the **units of analysis**. This topic is relevant to all forms of social research, although its implications are clearest in the case of nomothetic, quantitative studies.

The idea for units of analysis may seem slippery at first, because research—especially nomothetic research—often studies large collections of people or things, or aggregates. It's important to distinguish between the unit of analysis and the aggregates that we generalize about.

> **units of analysis** The what or whom being studied. In social science research the most typical units of analysis are individual people.

For instance, a researcher may study a class of people, such as Democrats, college undergraduates, or African American women under age 30. But if the researcher is interested in exploring, describing, or explaining how different groups of individuals behave as *individuals*, the unit of analysis is the individual, not the group. This is so even though the researcher then proceeds to generalize about aggregates of individuals, as in saying that more Democrats than Republicans favor legalizing marijuana.

Think of it this way: Having an attitude about marijuana is something that can be an attribute only of an individual, not a group; that is, there is no one group "mind" that can have an attitude. So even when we generalize about Democrats, we're generalizing about an attribute they possess as individuals.

In contrast, we may sometimes want to study groups, considered as individual "actors" or entities that have attributes as groups. For instance, we might want to compare the characteristics of different types of street gangs. In that case our unit of analysis would be gangs (not members of gangs), and we might proceed to make generalizations about different types of gangs. For example, we might conclude that male gangs are more violent than female gangs. Each gang (unit of analysis) would be described in terms of two variables: (1) What *gender* are the members? and (2) How *violent* are its activities? So we might study 52 gangs, reporting that 40 were male and 12 were female, and so forth. The "gang" would be the unit of analysis, even though some of the characteristics were drawn from the components (members) of the gangs.

Social researchers perhaps most typically choose individual people as their units of analysis. You might note the characteristics of individual people—gender, age, region of birth, attitudes, and so forth. You could then combine these descriptions to provide a composite picture of the group the individuals represent, whether a street-corner gang or a whole society.

For example, you might note the age and gender of each student enrolled in Political Science 110 and then characterize the group of students as being 53 percent men and 47 percent women and as having a mean age of 18.6 years. Although the final description would be of the class as a whole, the description would be based on characteristics that members of the class have as individuals.

The same distinction between units of analysis and aggregations occurs in explanatory studies. Suppose you wished to discover whether students with good study habits received better grades in Political Science 110 than did students with poor study habits. You would operationalize the variable *study habits* and measure this variable, perhaps in terms of hours of study per week. You might then aggregate students with good study habits and those with poor study habits and see which group received the best grades in the course. The purpose of the study would be to explain why some groups of students do better in the course than do others, but the unit of analysis would still be individual students.

Units of analysis in a study are usually also the units of observation. Thus, to study success in a political science course, we would observe individual students. Sometimes, however, we "observe" our units of analysis indirectly. For example, suppose we want to find out whether disagreements about the death penalty tend to cause divorce. In this case, we might "observe" individual husbands and wives by asking them about their attitudes toward capital punishment, in order to distinguish couples who agree and disagree on this issue. In this case, our units of observation are individual wives and husbands, but our units of analysis (the things we want to study) are couples.

Units of analysis, then, are those things we examine in order to create summary descriptions of all such units and to explain differences among them. In most research projects, the unit of analysis will probably be clear to you. When the unit of analysis is not clear, however, it's essential to determine what it is; otherwise, you cannot determine what observations are to be made about whom or what.

Some studies try to describe or explain more than one unit of analysis. In these cases, the researcher must anticipate what conclusions she or he wishes to draw with regard to which units of analysis. For example, we may want to discover what kinds of college students (individuals) are most successful in their careers after graduation; we may also want to learn what kinds of colleges (organizations) produce the most successful graduates.

Here's an example that illustrates the complexity of units of analysis. Murder is a fairly personal matter: One individual kills another individual. However, when Charis Kubrin and Ronald Weitzer (2003: 157) ask, "Why do these neighborhoods generate high homicide rates?" the unit of analysis in that question is "neighborhood." You can probably imagine some kinds of neighborhood (such as poor, urban) that would have high homicide rates and some (such as wealthy, suburban) that would have low homicide rates. In this particular conversation, the unit of analysis (neighborhood) would be categorized in terms of variables such as *economic level*, *locale*, and *homicide rate*.

In their analysis, however, Kubrin and Weitzer were also interested in different types of homicide: in particular, those that occurred in retaliation for some earlier event, such as an assault or insult. Can you identify the unit of analysis common to all of the following excerpts?

1. The sample of killings...
2. The coding instrument includes over 80 items related to the homicide.
3. Of the 2,161 homicides that occurred from 1985 [to] 1995...
4. Of those with an identified motive, 19.5 percent (n = 337) are retaliatory (Kubrin and Weitzer 2003: 163).

In each of these excerpts, the unit of analysis is *homicide* (also called *killing* or *murder*). Sometimes you can identify the unit of analysis in the description of the sampling methods, as in the first excerpt. A discussion of classification methods might also identify the unit of analysis, as in the second excerpt (80 ways to code the homicides). Often, numerical summaries point the way: 2,161 homicides; 19.5 percent (of the homicides). With a little practice you'll be able to identify the units of analysis in most social research reports, even when more than one is used in a given analysis.

To explore this topic in more depth, let's consider several common units of analysis in social research.

Individuals

As mentioned earlier, individual human beings are perhaps the most typical units of analysis for social research. We tend to describe and explain

social groups and interactions by aggregating and manipulating the descriptions of individuals.

Any type of individual can be the unit of analysis for social research. This point is more important than it may seem at first. The norm of generalized understanding in social research should suggest that scientific findings are most valuable when they apply to all kinds of people. In practice, however, social researchers seldom study all kinds of people. At the very least, their studies are typically limited to the people living in a single country, though some comparative studies stretch across national boundaries. Often, however, studies are quite circumscribed.

Examples of classes of individuals that might be chosen for study include students, gays and lesbians, autoworkers, voters, single parents, and faculty members. Note that each of these terms implies some population of individuals.

Groups

Social groups can also be units of analysis in social research. That is, we may be interested in characteristics that belong to one group, considered as a single entity. If you were to study the members of a criminal gang to learn about criminals, the individual (criminal) would be the unit of analysis; but if you studied all the gangs in a city to learn the differences, say, between big gangs and small ones, between "uptown" and "downtown" gangs, and so forth, you would be interested in gangs rather than their individual members. In this case, the unit of analysis would be the gang, a social group.

Here's another example. Suppose you were interested in the question of access to computers in different segments of society. You might describe families in terms of total annual income and according to whether or not they had computers. You could then aggregate families and describe the mean income of families and the percentage with computers. You would then be in a position to determine whether families with higher incomes were more likely to have computers than were those with lower incomes. In this case, the unit of analysis would be families.

As with other units of analysis, we can derive the characteristics of social groups from those of their individual members. Thus, we might describe a family in terms of the age, race, or education of its head. In a descriptive study, we might find the percentage of all families that have a college-educated head of family. In an explanatory study, we might determine whether such families have, on average, more or fewer children than do families headed by people who have not graduated from college. In each of these examples, the family is the unit of analysis. In contrast, had we asked whether college-educated individuals have more or fewer children than do their less-educated counterparts, then the individual would have been the unit of analysis.

Organizations

Formal social organizations can also be the units of analysis in social research. For example, a researcher might study corporations, by which he or she implies a population of all corporations. Individual corporations might be characterized in terms of their number of employees, net annual profits, gross assets, number of defense contracts, percentage of employees from racial or ethnic minority groups, and so forth. We might determine whether large corporations hire a larger or smaller percentage of minority-group employees than do small corporations. Other examples of formal social organizations suitable as units of analysis include church congregations, colleges, army divisions, academic departments, and supermarkets.

Figure 4-4 provides a graphic illustration of some different units of analysis and the statements that might be made about them.

Social Interactions

Sometimes social interactions are the relevant units of analysis. Instead of studying individual humans, you can study what goes on between them: telephone calls, kisses, dancing, arguments, fistfights, e-mail exchanges, chat-room discussions, and so forth. As you saw in Chapter 2, social interaction is the basis for one of the primary theoretical paradigms in the social sciences, and the number of units of analysis that social interactions provide is nearly infinite.

social artifact Any product of social beings or their behavior. It can be a unit of analysis.

Even though individuals are usually the actors in social interactions, there is a difference between (1) comparing the kinds of people who subscribe to different Internet service providers (individuals being the unit of analysis) and (2) comparing the length of chat-room discussions on those same providers (the discussion being the unit of analysis).

Social Artifacts

Another unit of analysis is the **social artifact**, or any product of social beings or their behavior. One class of artifacts includes concrete objects such as books, poems, paintings, automobiles, buildings, songs, pottery, jokes, student excuses for missing exams, and scientific discoveries.

As these examples suggest, just as people or social groups imply populations, each social object implies a set of all objects of the same class: all books, all novels, all biographies, all introductory sociology textbooks, all cookbooks, all press conferences. In a study using books as the units of analysis, an individual book might be characterized by size, weight, length, price, content, number of pictures, number sold, or description of its author. Then the population of all books or of a particular kind of book could be analyzed for the purpose of description or explanation: what kinds of books sell best and why, for example.

Social interactions form another class of social artifacts suitable for social research. For example, we might characterize weddings as racially or religiously mixed or not, as religious or secular in ceremony, as resulting in divorce or not, or by descriptions of one or both of the marriage partners (such as "previously married," "Oakland Raider fan," "wanted by the FBI"). When a researcher reports that weddings between partners of different religions are more likely to be performed by secular authorities than are those between partners of the same religion, the weddings are the units of analysis, not the individuals involved.

Other social interactions that might be units of analysis include friendship choices, court cases, traffic accidents, divorces, fistfights, ship launchings, airline hijackings, race riots, final exams, student demonstrations, and congressional hearings. Congressional hearings, for instance, could be characterized by whether or not they occurred during an election campaign,

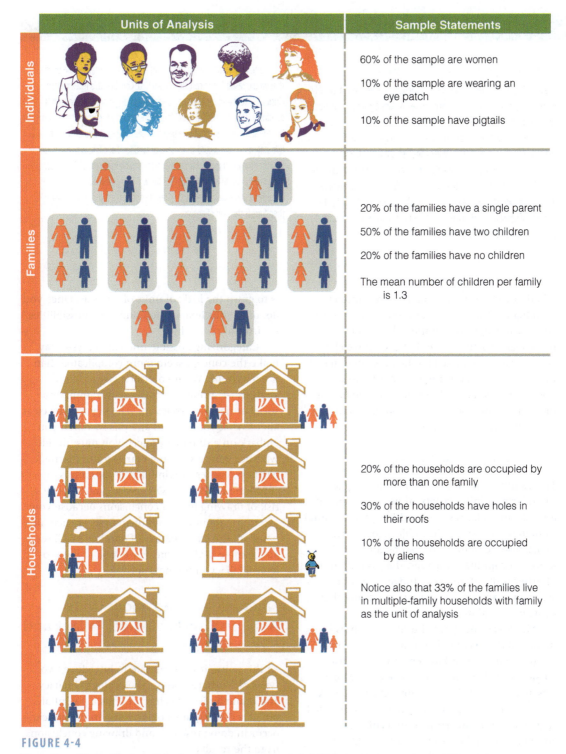

Units of Analysis	Sample Statements
Individuals	60% of the sample are women 10% of the sample are wearing an eye patch 10% of the sample have pigtails
Families	20% of the families have a single parent 50% of the families have two children 20% of the families have no children The mean number of children per family is 1.3
Households	20% of the households are occupied by more than one family 30% of the households have holes in their roofs 10% of the households are occupied by aliens Notice also that 33% of the families live in multiple-family households with family as the unit of analysis

FIGURE 4-4

Illustrations of Units of Analysis. Units of analysis in social research can be individuals, groups, or even nonhuman entities.

How to Do It

Identifying the Unit of Analysis

The unit of analysis is an important element in research design and in data analysis. However, students sometimes find it elusive. The easiest way to identify the unit of analysis is to examine a statement regarding the variables under study.

Consider the following statement: "The average household income was $40,000." *Income* is the variable of interest, but who or what *has* income? Households. We would arrive at the given statement by examining the incomes of several households. To calculate the mean (average) income, we would add up all the household incomes and divide by the number of households. Household is the unit of analysis. It is the unit being analyzed in terms of the variable, *income*.

One way of identifying the unit of analysis is to imagine the process that would result in the conclusion reached. Consider this research conclusion: "Twenty-four percent of the families have more than one adult earning at least $30,000 a year." To be sure, adults are earning the income, but the statement is about whether families have such adults. To make this statement, we would study several families. For each, we would ask whether they had more than two adults earning in excess of $30,000; each family would be scored as "yes" or "no" in that respect. Finally, we would calculate the percentage of families scored as "yes." The family, therefore, is the unit of analysis.

whether the committee chairs were running for a higher office, whether these chairs had received campaign contributions from interested parties, and so on. Notice that even if we characterized and compared the hearings in terms of the committee chairs, the hearings themselves—not the individual chairpersons—would be our units of analysis. See "How to Do It: Identifying the Unit of Analysis" for more.

Units of Analysis in Review

The examples in this section should suggest the nearly infinite variety of possible units of analysis in social research. Although individual human beings are typical objects of study, many research questions can be answered more appropriately through the examination of other units of analysis. Indeed, social researchers can study just about anything that bears on social life.

Moreover, the types of units of analysis named in this section do not begin to exhaust the possibilities. This has been a topic of discussion and elaboration for some time. Morris Rosenberg (1968: 234–48), for example, speaks of individual, group, organizational, institutional, spatial, cultural, and societal units of analysis. John Lofland and colleagues (2006: 122–32) speak of practices, episodes, encounters, roles and social types, social and personal relationships, groups and cliques, organizations, settlements and habitats, subcultures, and lifestyles as suitable units of study. The important thing here

is to grasp the logic of units of analysis. Once you do, only your imagination limits the possibilities for fruitful research.

Categorizing possible units of analysis may make the concept seem more complicated than it needs to be. What you call a given unit of analysis—a group, a formal organization, or a social artifact—is irrelevant. The key is to be clear about what your unit of analysis is. When you embark on a research project, you must decide whether you're studying marriages or marriage partners, crimes or criminals, corporations or corporate executives. Otherwise, you run the risk of drawing invalid conclusions because your assertions about one unit of analysis are actually based on the examination of another. We'll see an example of this issue as we look at the ecological fallacy in the next section.

Faulty Reasoning about Units of Analysis: The Ecological Fallacy and Reductionism

At this point, it's appropriate to introduce two types of faulty reasoning: the ecological fallacy and reductionism. Each represents a potential pitfall regarding units of analysis, and either can occur in doing research and drawing conclusions from the results.

The Ecological Fallacy

In this context, *ecological* refers to groups or sets or systems: something larger than individuals.

The **ecological fallacy** is the assumption that something learned about an ecological unit says something about the individuals making up that unit. Let's consider a hypothetical illustration of this fallacy.

Suppose we're interested in learning something about the nature of electoral support received by a female political candidate in a recent citywide election. Let's assume we have the vote tally for each precinct so we can tell which precincts gave her the greatest support and which the least. Assume also that we have census data describing some characteristics of these precincts. Our analysis of such data might show that precincts with relatively young voters gave the female candidate a greater proportion of their votes than did precincts with older voters. We might be tempted to conclude from these findings that young voters are more likely to vote for female candidates than are older voters—in other words, that age affects support for the woman. In reaching such a conclusion, we run the risk of committing the ecological fallacy because it may have been the older voters in those "young" precincts who voted for the woman. Our problem is that we've examined *precincts* as our units of analysis but wish to draw conclusions about *voters*.

The same problem would arise if we discovered that crime rates were higher in cities having large African American populations than in those with few African Americans. We would not know whether the crimes were actually committed by African Americans. Or if we found suicide rates higher in Protestant countries than in Catholic ones, we still could not know for sure that more Protestants than Catholics committed suicide.

In spite of these hazards, social researchers very often have little choice but to address a particular research question through an ecological analysis. Perhaps the most appropriate data are simply not available. For example, the precinct vote tallies and the precinct characteristics mentioned in our initial example might be easy to obtain, but we may not have the resources to conduct a post-election survey of individual voters. In such cases, we may reach a tentative conclusion, recognizing and noting the risk of an ecological fallacy.

Although you should be careful not to commit the ecological fallacy, don't let these warnings lead you into committing what we might call the "individualistic fallacy." Some people who approach social research for the first time have trouble reconciling general patterns of attitudes and actions with individual exceptions. But generalizations and probabilistic statements are not invalidated by such exceptions. Your knowing a rich Democrat, for example, doesn't deny the fact that most rich people vote Republican—as a general pattern. Similarly, if you know someone who has gotten rich without any formal education, that doesn't deny the general pattern of higher education relating to higher income.

The ecological fallacy deals with something else altogether—confusing units of analysis in such a way that we base conclusions about individuals solely on the observation of groups. Although the patterns observed among variables at the level of groups may be genuine, the danger lies in reasoning from the observed attributes of groups to the attributes of the individuals who made up those groups, when we have not actually observed individuals. "Applying Concepts in Everyday Life: Red Families and Blue Families" illustrates some of the complexities presented by different units of analysis.

Reductionism

A second type of potentially faulty reasoning related to units of analysis is reductionism. **Reductionism** involves attempts to explain a particular phenomenon in terms of limited and/or lower-order concepts. The reductionist explanation is not altogether wrong; it is simply too limited. Thus, you might attempt to predict this year's winners and losers in the National Basketball Association by focusing on the abilities of the individual players on each team. This is certainly neither stupid nor irrelevant, but the success or failure of teams involves more than just the individuals on them; it involves coaching, teamwork, strategies, finances, facilities, fan loyalty, and so forth. To understand why some teams do better

---

ecological fallacy Erroneously basing conclusions about individuals solely on the observation of groups.

reductionism A fault of some researchers: a strict limitation (reduction) of the kinds of concepts to be considered relevant to the phenomenon under study.

Applying Concepts in Everyday Life

Red Families and Blue Families

During recent American political campaigns, concern for "family values" has often been featured as a hot-button issue. Typically, conservatives and Republicans have warned of the decline of such traditional values, citing divorce rates, teen pregnancies, same-sex marriage, and such. This is, however, a more complex matter than would fit on a bumper sticker.

In their analysis of conservative "red families" and liberal "blue families," Naomi Cahn and June Carbone report:

> Red family champions correctly point out that growing numbers of single-parent families threaten the well-being of the next generation, and they accurately observe that greater male fidelity and female "virtue" strengthen relationships. Yet red regions of the country have higher teen pregnancy rates, more shotgun marriages, and lower average ages at marriage and first birth.

(2010: 2)

Reviewing the Cahn–Carbone study, Jonathan Rauch headlines the question, "Do 'Family Values' Weaken Families?" and summarizes the data thusly:

> Six of the seven states with the lowest divorce rates in 2007, and all seven with the lowest teen birthrates in 2006, voted blue in both elections. Six of the seven states with the highest divorce rates in 2007, and five of the seven with the highest teen birthrates, voted red. It's as if family strictures undermine family structures.

(Rauch 2010)

Assuming that young people are going to have sex, Cahn and Carbone argue that the "traditional family values" that oppose sex education, contraception, and abortion will result in unplanned births that will typically be dealt with by forcing the young parents to marry. This, in turn, may interrupt their educations, limit their employment opportunities, lead to poverty, and result in unstable marriages that may not survive. This interpretation of the data may be completely valid, but can you recognize a methodological issue that might be raised? Think about the ecological fallacy.

The units of analysis used in these analyses are the 50 states of the union. The variables correlated are (1) overall voting patterns of the states and (2) family-problem rates in the states. States voting Republican overall have more problems than those voting Democratic overall. However, the data do not guarantee that Republican families or teenagers in Republican families have more problems than their Democratic counterparts. The ecological data suggest that that's the case, but it is possible that Democrats in Republican states have the most family problems and Republicans in Democratic states have the least. It is unlikely but it is possible.

To be more confident about the conclusions drawn here, we would need to do a study in which the family or the individual was the unit of analysis.

Sources: Naomi Cahn and June Carbone, *Red Families v. Blue Families: Legal Polarization and the Creation of Culture* (New York: Oxford University Press, 2010); Jonathan Rauch, "Do 'Family Values' Weaken Families?" *Dallas Morning News*, May 2010, http://www.dallasnews.com/opinion/commentary/2010/05/28/Jonathan-Rauch-Do-family-9089.

than others, you would make *team* the unit of analysis, and the *quality of players* would be one variable you would probably want to use in describing and classifying the teams.

Thus, different academic disciplines approach the same phenomenon quite differently. Sociologists tend to consider sociological variables (such as *values, norms*, and *roles*), economists ponder economic variables (such as *supply and demand* and *marginal value*), and psychologists examine psychological variables (such as *personality types* and *traumas*). Explaining all or most

human behavior in terms of economic factors is called *economic reductionism;* explaining all or most human behavior in terms of psychological factors is called *psychological reductionism;* and so forth. Notice how this issue relates to the discussion of theoretical paradigms in Chapter 2.

For many social scientists, the field of **sociobiology** is a prime example of reductionism, suggesting that all social phenomena can be explained in terms of biological factors. Thus, for example, Edward O. Wilson, sometimes referred to as the father of sociobiology, sought to explain altruistic behavior in human beings in terms of our genetic makeup (1975). In his neo-Darwinian view, Wilson suggests that humans have evolved in such a way that individuals sometimes need to sacrifice themselves for the benefit of the whole species. Some people might

sociobiology A paradigm based on the view that social behavior can be explained solely in terms of genetic characteristics and behavior.

explain such sacrifice in terms of ideals or warm feelings between humans. However, genes are the essential unit in Wilson's paradigm, producing his famous dictum that human beings are "only DNA's way of making more DNA."

Reductionism of any type tends to suggest that particular units of analysis or variables are more relevant than others. Suppose we ask what caused the American Revolution. Was it a shared commitment to the value of individual liberty? The economic plight of the colonies in relation to Britain? The megalomania of the founders? As soon as we inquire about *the* single cause, we run the risk of reductionism. If we were to regard shared values as the cause of the American Revolution, our unit of analysis would be the individual colonist. An economist, though, might choose the thirteen colonies as units of analysis and examine the economic organizations and conditions of each. A psychologist might choose individual leaders as the units of analysis for purposes of examining their personalities.

Like the ecological fallacy, reductionism can occur when we use inappropriate units of analysis. The appropriate unit of analysis for a given research question, however, is not always clear. Social researchers, especially across disciplinary boundaries, often debate this issue.

The Time Dimension

So far in this chapter, we've regarded research design as a process for deciding what aspects we'll observe, of whom, and for what purpose. Now we must consider a set of time-related options that cuts across each of these earlier considerations. We can choose to make observations more or less at one time or over a long period.

Time plays many roles in the design and execution of research, quite aside from the time it takes to do research. Earlier we noted that the time sequence of events and situations is critical to determining causation (a point we'll return to in Part 4). Time also affects the generalizability of research findings. Do the descriptions and explanations resulting from a particular study accurately represent the situation of ten years ago, ten years from now, or only the present? Researchers have two principal options for dealing with the issue of time in the design

of their research: cross-sectional studies and longitudinal studies.

Cross-Sectional Studies

A **cross-sectional study** involves observations of a sample, or cross section, of a population or phenomenon that are made at one point in time. Exploratory and descriptive studies are often cross-sectional. A single U.S. Census, for instance, is a study aimed at describing the U.S. population at a given time.

Many explanatory studies are also cross-sectional. A researcher conducting a large-scale national survey to examine the sources of racial and religious prejudice would, in all likelihood, be dealing with a single time frame—taking a snapshot, so to speak, of the sources of prejudice at a particular point in history.

Explanatory cross-sectional studies have an inherent problem. Although their conclusions are based on observations made at only one time, typically they aim at understanding causal processes that occur over time. This is akin to determining the speed of a moving object from a high-speed, still photograph.

Yanjie Bian, for example, conducted a survey of workers in Tianjin, China, to study stratification in contemporary urban Chinese society. In undertaking the survey in 1988, however, he was conscious of the important changes brought about by a series of national campaigns, such as the Great Proletarian Cultural Revolution, dating from the Chinese Revolution in 1949 (which brought the Chinese Communists into power) and continuing into the present.

> These campaigns altered political atmospheres and affected people's work and nonwork activities. Because of these campaigns, it is difficult to draw conclusions from a cross-sectional social survey, such as the one presented in this book, about general patterns of Chinese workplaces and their effects on workers. Such conclusions may be limited to one period of time and are subject to further tests based on data collected at other times.
>
> *(1994: 19)*

cross-sectional study A study based on observations representing a single point in time.

As you'll see, this textbook repeatedly addresses the problem of using a "snapshot" to make generalizations about social life. One solution is suggested by Bian's final comment—about data collected "at other times": Social research often involves revisiting phenomena and building on the results of earlier research.

Longitudinal Studies

In contrast to cross-sectional studies, a **longitudinal study** is designed to permit observations of the same phenomenon over an extended period. For example, a researcher can participate in and observe the activities of a UFO cult from its inception to its demise. Other longitudinal studies use records or artifacts to study changes over time. In analyses of newspaper editorials or Supreme Court decisions over time, for example, the studies are longitudinal whether the researcher's actual observations and analyses were made at one time or over the course of the actual events under study.

Many field research projects, involving direct observation and perhaps in-depth interviews, are naturally longitudinal. Thus, for example, when Ramona Asher and Gary Fine (1991) studied the life experiences of the wives of alcoholic men, these researchers were in a position to examine the evolution of the wives' troubled marital relationships over time, sometimes even including the reactions of the subjects to the research itself.

In the classic study *When Prophecy Fails* (1956), Leon Festinger, Henry Reicker, and Stanley Schachter set out to learn what happened to a flying saucer cult when its predictions of an alien encounter failed to come true. Would the cult members close down the group, or would they become all the more committed to their beliefs? A longitudinal study was required to provide an answer. (The cult redoubled their efforts to get new members.)

Longitudinal studies can be more difficult for quantitative studies such as large-scale surveys.

---

longitudinal study A study design involving data collected at different points in time.

trend study A type of longitudinal study in which a given characteristic of some population is monitored over time.

Nonetheless, they are often the best way to study changes over time. There are three special types of longitudinal studies that you should know about: trend studies, cohort studies, and panel studies.

Trend Studies

A **trend study** is a type of longitudinal study that examines changes within a population over time. A simple example is a comparison of U.S. Censuses over a period of decades, showing shifts in the makeup of the national population. A similar use of archival data was made by Michael Carpini and Scott Keeter (1991), who wanted to know whether contemporary U.S. citizens were better or more poorly informed about politics than were citizens of an earlier generation. To find out, they compared the results of several Gallup polls conducted during the 1940s and 1950s with a 1989 survey that asked several of the same questions tapping political knowledge.

Overall, the analysis suggested that contemporary citizens were slightly better informed than were earlier generations. In 1989, 74 percent of the sample could name the vice president of the United States, compared with 67 percent in 1952. Substantially higher percentages could explain presidential vetoes and congressional overrides of vetoes than could people in 1947. On the other hand, more of the 1947 sample could identify their U.S. representative (38 percent) than could the 1989 sample (29 percent).

An in-depth analysis, however, indicates that the slight increase in political knowledge resulted from the fact that the people in the 1989 sample were more highly educated than were those from earlier samples. When educational levels were taken into account, the researchers concluded that political knowledge has actually declined within specific educational groups.

Every trend study must begin with a first stage, and the 2014 Chapman survey of American Fears is a good example of a trend study in the making. Day, Bader, and Gordon (2014) have set out to trace trends in what Americans are afraid of, timing the annual surveys to coincide with Halloween. In 2014, the top personal fears and concerns were:

- Safety in different spaces
- Anxiety about one's future (illness, job stability, etc.)

- Internet-related fears (identity theft, government surveillance, etc.)
- Criminal victimization (murder, mugging, mass shootings, etc.)
- Phobias (clowns, tight spaces, blood, etc.)

In 2018 (Day, Bader, and Gordon 2018) the leading fears had shifted substantially:

- Corruption of government officials
- Pollution of oceans, rivers, and lakes
- Pollution of drinking water
- Not having enough money for the future
- People I love becoming seriously ill

Cohort Studies

In a **cohort study**, a researcher examines specific subpopulations, or *cohorts*, as they change over time. Typically, a cohort is an age group, such as people born during the 1950s, but it can also be some other time grouping, such as people born during the Vietnam War, people who got married in 2012, and so forth. An example of a cohort study would be a series of national surveys, conducted perhaps every 20 years, to study the attitudes of the cohort born during World War II toward U.S. involvement in global affairs. A sample of people 15 to 20 years of age might be surveyed in 1960, another sample of those 35 to 40 years of age in 1980, and another sample of those 55 to 60 years of age in 2000. Although the specific set of people studied in each survey would differ, each sample would represent the cohort born between 1940 and 1945.

Figure 4-5 offers a graphic illustration of a cohort design. In the example, three studies are being compared: one was conducted in 1990, another in 2000, and the third in 2010. Those who were 20 years old in the 1990 study are compared with those who were 30 in the 2000 study and those who were 40 in the 2010 study. Although the subjects being described in each of the three groups are different, each set of subjects represents the same cohort: those who were born in 1970.

In one study, Eric Plutzer and Michael Berkman (2005) used a cohort design to reverse a prior conclusion regarding aging and support for education. Logically, as people grow well beyond the child-rearing years, we might expect them to reduce their commitment to educational funding. Moreover, cross-sectional data support that expectation. The researchers present several data sets showing those over 65 voicing less support for education funding than did those under 65.

Such simplistic analyses, however, leave out an important variable: increasing support for educational funding in U.S. society over time in general. The researchers add to this the concept

> **cohort study** A study in which some specific subpopulation, or cohort, is studied over time, although data may be collected from different members in each set of observations.

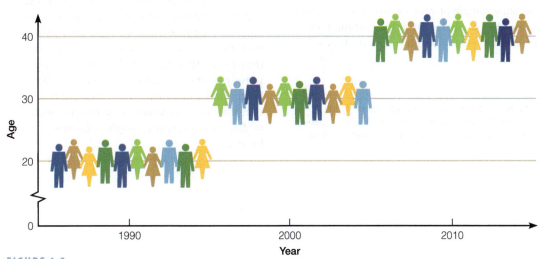

FIGURE 4-5

A Cohort Study Design. Each of the three groups shown here is a sample representing people who were born in 1970.

of "generational replacement," meaning that the older respondents in a survey grew up during a time when there was less support for education in general, whereas the younger respondents grew up during a time of greater overall support.

A cohort analysis allowed the researchers to determine what happened to the attitudes of specific cohorts over time. Here, for example, are the percentages of Americans born during the 1940s who felt that educational spending was too low, when members of that cohort were interviewed over time (Plutzer and Berkman 2005: 76):

Year Interviewed	Percent Who Say Educational Funding Is Too Low
1970s	58
1980s	66
1990s	74
2000s	79

As these data indicate, those who were born during the 1940s have steadily increased their support for educational funding as they have passed through and beyond the child-rearing years.

Panel Studies

Though similar to trend and cohort studies, a **panel study** examines the same set of people each time. For example, we could interview the same sample of voters every month during an election campaign, asking for whom they intended to vote. Though such a study would

---

panel study A type of longitudinal study in which data are collected from the same set of people (the sample or panel) at several points in time.

---

allow us to analyze overall trends in voter preferences for different candidates, it would also show the precise patterns of persistence and change in intentions. For example, a trend study that showed that Candidates A and B each had exactly half of the voters on September 1 and on October 1 as well could indicate that none of the electorate had changed voting plans, that all of the voters had changed their intentions, or something in between. A panel study would eliminate this confusion by showing what kinds of voters switched from A to B and what kinds switched from B to A, as well as other facts.

Joseph Veroff, Shirley Hatchett, and Elizabeth Douvan (1992) wanted to learn about marital adjustment among newlyweds and focused on differences between white and African American couples. To get subjects for study, they selected a sample of couples who applied for marriage licenses in Wayne County, Michigan, in April through June 1986.

Concerned about the possible impact their research might have on the couples' marital adjustment, the researchers divided their sample in half at random: an experimental group and a control group (concepts we'll explore further in Chapter 8). Couples in the former group were intensively interviewed over a four-year period, whereas the latter group was contacted only briefly each year.

By studying the same couples over time, the researchers could follow the specific problems that arose and the way the couples dealt with them. As a by-product of their research, they found that those studied the most intensely seemed to achieve a somewhat better marital adjustment. The researchers felt that the interviews may have forced couples to discuss matters they might otherwise have buried.

Panel mortality is a fundamental problem in panel studies: subjects dropping out of the study. Over the years, researchers have developed many techniques for tracking down missing subjects. Bryan Rhodes and Ellen Marks (2011) used Facebook as a vehicle for tracking down members of a longitudinal study who had been unreachable by telephone or mail. They were successful in locating a third of the subjects.

Roger Tourangeau and Cong Ye (2009) were also intent on decreasing panel mortality. Specifically, they considered positive and negative inducements for subjects to continue. To find out, they randomly divided their panel survey sample in half and gave the two groups different pleas to continue. In one subsample, they stressed the benefits to be gained if everyone continued with the study. In the other subsample, they stressed how the study would be hurt by people dropping out. The latter, negative, message increased continued participation by ten percentage points.

As Steven Farrall et al. (2016) point out, panel mortality can be especially problematic for qualitative longitudinal studies, since the relatively small sample sizes rule out the possibility of weighting as a solution. One effective technique was to maintain contact sheets with names, addresses, and phone numbers of family and friends likely to know how to reach a missing study participant. They found family members more reliable than friends in that regard and found sisters were the best bet. Other researchers found it useful to stay in touch with participants with birthday cards, holiday greetings, and the like.

Comparing the Three Types of Longitudinal Studies

To reinforce the distinctions among trend, cohort, and panel studies, let's contrast the three study designs in terms of the same variable: *religious affiliation*. A trend study might look at shifts in U.S. religious affiliations over time, as the Gallup poll does on a regular basis. A cohort study might follow shifts in religious affiliations among "the 9/11 generation," specifically, say, people who were 10 to 20 years of age on September 11, 2001. We could study a sample of people 20 to 30 years old in 2011, a new sample

of people who were 30 to 40 in 2021, and so forth throughout their life span. A panel study could start with a sample of the whole population or of some special subset and study those specific individuals over time. Notice that only the panel study would give a full picture of the shifts among the various categories of affiliations, including "none." Cohort and trend studies would uncover only net changes.

Longitudinal studies have an obvious advantage over cross-sectional ones in providing information describing processes over time. But this advantage often comes at a heavy cost in both time and money, especially in a large-scale survey. Observations may have to be made at the time events are occurring, and the method of observation may require many research workers.

Panel studies, which offer the most comprehensive data on changes over time, face a special problem: panel mortality, as discussed above. Some of the respondents studied in the first wave of the survey may not participate in later waves. (This is comparable to the problem of experimental mortality discussed in Chapter 8.) The danger is that those who drop out of the study may not be typical, thereby distorting the results of the study.

Figure 4-6 provides a schematic comparison of the several study types we have been discussing.

Approximating Longitudinal Studies

Longitudinal studies do not always provide a feasible or practical means of studying processes that take place over time. Fortunately, researchers often can draw approximate conclusions about such processes even when only cross-sectional data are available. Here are some ways to do that.

Sometimes cross-sectional data imply processes over time on the basis of simple logic. For example, in the study of student drug use conducted at the University of Hawaii that I mentioned in Chapter 2, students were asked to report whether they had ever tried each of

panel mortality The failure of some panel subjects to continue participating in the study.

	Cross-Sectional	Longitudinal		
		Trend	Cohort	Panel
Snapshot in time	X			
Measurements across time		X	X	X
Follow age group across time			X	
Study same people over time				X

FIGURE 4-6

Comparing Types of Study Design.

several illegal drugs. The study found that some students had tried both marijuana and LSD, some had tried only one, and others had tried neither. Because these data were collected at one time, and because some students presumably would experiment with drugs later on, it would appear that such a study could not tell whether students were more likely to try marijuana or LSD first.

A closer examination of the data showed, however, that although some students reported having tried marijuana but not LSD, there were no students in the study who had tried only LSD. From this finding it was inferred—as common sense suggested—that marijuana use preceded LSD use. If the process of drug experimentation occurred in the opposite time order, then a study at a given time should have found some students who had tried LSD but not marijuana, and it should have found no students who had tried only marijuana.

Researchers can also make logical inferences whenever the time order of variables is clear. If we discovered in a cross-sectional study of college students that those educated in private high schools received better college grades than did those educated in public high schools, we would conclude that the type of high school attended affected college grades, not the other way around. Thus, even though our observations were made at only one time, we would feel justified in drawing conclusions about processes taking place across time.

Very often, age differences discovered in a cross-sectional study form the basis for inferring processes across time. Suppose you're interested

in the pattern of worsening health over the course of the typical life cycle. You might study the results of annual checkups in a large hospital. You could group health records according to the ages of those examined and rate each age group in terms of several health conditions—sight, hearing, blood pressure, and so forth. By reading across the age-group ratings for each health condition, you would have something approximating the health history of individuals. Thus, you might conclude that the average person develops vision problems before hearing problems. You would need to be cautious in this assumption, however, because the differences might reflect society-wide trends. For instance, improved hearing examinations instituted in the schools might have affected only the young people in your study.

Asking people to recall their pasts is another common way of approximating observations over time. Researchers use that method when they ask people where they were born or when they graduated from high school or whom they voted for in 1988. Qualitative researchers often conduct in-depth "life history" interviews. For example, C. Lynn Carr (1998) used this technique in a study of "tomboyism." Her respondents, ages 25 to 40, were asked to reconstruct aspects of their lives from childhood on, including experiences of identifying themselves as tomboys.

The danger in this technique is evident. Sometimes people have faulty memories; sometimes they lie. When people are asked in post-election polls whom they voted for, the results inevitably show more people voting for the winner than actually did so on election day. As part of a series of in-depth interviews, such a report can be validated in the context of other reported details; however, we should regard with caution results based on a single question in a survey.

Cohorts can also be used to infer processes over time from cross-sectional data. For example, when Prem Saxena and colleagues (2004) wanted to examine whether wartime conditions would affect the age at which people married, they used cross-sectional data from a survey of Lebanese women. During the Lebanese Civil War, from 1975 to 1990, many young men migrated to other countries. By noting the year in which the survey respondents first married, they could determine that the average age at first marriage increased with the onset of the war.

For a more in-depth and comprehensive analysis of longitudinal methodologies, you might consider the book edited by Peter Lynn (2009). The several authors cover more aspects of this subject than would be feasible in this introductory textbook.

This discussion of the way time figures into social research suggests several questions you should confront in your own research projects. In designing any study, be sure to look at both the explicit and the implicit assumptions you're making about time. Are you interested in describing some process that occurs over time, or are you simply going to describe what exists now? If you want to describe a process occurring over time, will you be able to make observations at different points in the process, or will you have to approximate such observations by drawing logical inferences from what you can observe now? If you opt for a longitudinal design, which method best serves your research purposes?

Examples of Research Strategies

As the preceding discussions have implied, social research follows many paths. The following short excerpts from a variety of completed studies further illustrate this point. As you read them, note both the content of each study and the method used to study the chosen topic. Does the study seem to be exploring, describing, or explaining (or some combination of these)? What are the sources of data in each study? Can you identify

the unit of analysis? Is the dimension of time relevant? If so, how will it be handled?

- This case study of unobtrusive mobilizing by Southern California Rape Crisis Center uses archival, observational, and interview data to explore how a feminist organization worked to change police, schools, prosecutors, and some state and national organizations from 1974 to 1994. (Schmitt and Martin 1999: 364)
- Using life-history narratives, the present study investigates processes of agency and consciousness among 14 women who identified themselves as tomboys. (Carr 1998: 528)
- By drawing on interviews with activists in the former Estonian Soviet Socialist Republic, we specify the conditions by which accommodative and oppositional subcultures exist and are successfully transformed into social movements. (Johnston and Snow 1998: 473)
- This paper presents the results of an ethnographic study of an AIDS service organization located in a small city. It is based on a combination of participant observation, interviews with participants, and review of organizational records. (Kilburn 1998: 89)
- Using interviews obtained during fieldwork in Palestine in 1992, 1993, and 1994, and employing historical and archival records, I argue that Palestinian feminist discourses were shaped and influenced by the sociopolitical context in which Palestinian women acted and with which they interacted. (Abdulhadi 1998: 649)
- This article reports on women's experiences of breastfeeding in public, as revealed through in-depth interviews with 51 women. (Stearns 1999: 308)
- Using interview and observational field data, I demonstrate how a system of temporary employment in a participative workplace both exploited and shaped entry-level workers' aspirations and occupational goals. (V. Smith 1998: 411)
- I collected data [on White Separatist rhetoric] from several media of public discourse, including periodicals, books, pamphlets, transcripts from radio and television talk shows, and newspaper and magazine accounts. (Berbrier 1998: 435)
- In the analysis that follows, racial and gender inequality in employment and retirement will be analyzed, using a national sample of persons who began receiving Social Security Old Age benefits in 1980–81. (Hogan and Perrucci 1998: 528)

• Drawing from interviews with female crack dealers, this paper explores the techniques they use to avoid arrest. (Jacobs and Miller 1998: 550)

Mixed Modes

In this chapter, I have mentioned a number of ways to conduct social research: experiments, survey research (telephone, in person, online), field research, and so forth. In my observation over time, researchers have often spoken of the value of using more than one approach to understanding a social phenomenon. But, as researchers find techniques they are comfortable with and adept at, the support for multiple techniques has been talked about more than practiced. However, this may be changing. Partly in response to the growing problems faced by survey researchers, perhaps, a review of the literature will produce increasing numbers of studies actually using mixed modes.

The U.S. Energy Information Administration's Residential Energy Consumption Survey (RECS), for example, seeks to get detailed information on America's energy use. Household interviews provide much of the needed information, but the researchers discovered that household subjects usually could not provide special details of their energy consumption. So the agency collects billing data from the energy suppliers and matches those data to the households interviewed (Worthy and Mayclin 2013).

On the other side of the globe, Peggy Koopman-Boyden and Margaret Richardson used diaries and focus groups to study the activities of New Zealand seniors in a three-year study that aimed "to analyse the experiences and perceptions of elders and organizational representatives with whom they interact in everyday encounters" (2013: 392). Participating seniors were asked to maintain logs of their interactions, and they were invited to participate in periodic discussions of their experiences. In addition to providing researchers with a greater depth, breadth, and richness of data, they report that the participants often indicated that they had benefited from the combination of methods, as in the focus-group discussions of the experience of keeping a research diary.

Moving north from New Zealand to India, Prem Saxena and Dhirendra Kumar examined the risk of mortality among seniors after retirement, focusing on the importance of work for defining social position. In the context of a number of social psychology studies of problems regarding retirement, the researchers found a data source in the Office of the Accountant General to add to the previous studies, allowing them to examine overall patterns of mortality after retirement. Ultimately, they concluded that "In developing countries few people look forward to retirement, while the majority dread it. However, the way pensioners react depends mainly upon their social liability and their unmet needs at the time of retirement" (1997: 122).

While social researchers have been using mixed modes of inquiry for a long time, this approach has begun attracting more attention and, more important, more actual use in recent years. I think you can expect to see more mixed modes in the future and may utilize this approach yourself.

How to Design a Research Project

You have now seen some of the options available to social researchers in designing projects. I know there are a lot of pieces, and the relationships among them may not be totally clear, so here's a way of pulling the parts together. Let's assume you were to undertake research. Where would you start? Then, where would you go?

Although research design occurs at the beginning of a research project, it involves all the steps of the subsequent project. This discussion, then, provides guidance on how to start a research project and gives an overview of the topics that follow in later chapters of this book.

Figure 4-7 presents a schematic view of the traditional image of research. I present this view reluctantly, because it may suggest more of a step-by-step order to research than actual practice bears out. Nonetheless, this idealized overview of the process provides a context for the specific details of particular components of social research. Essentially, it is another and more-detailed picture of the scientific process presented in Chapter 2.

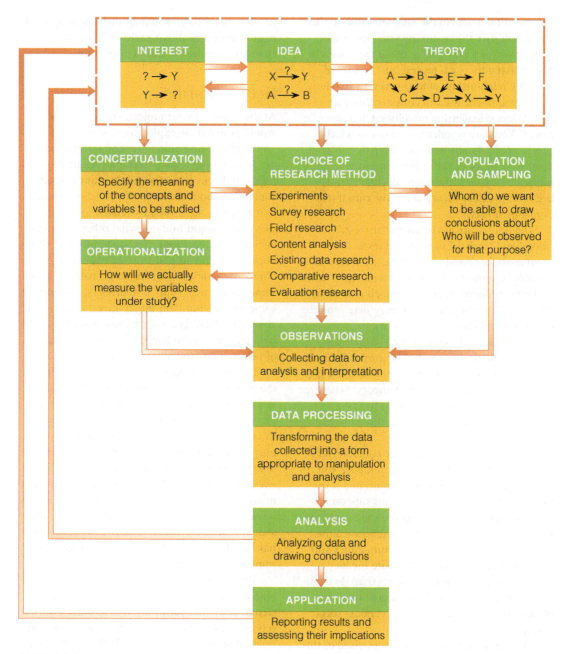

FIGURE 4-7

The Research Process. Here are some of the key elements that we'll be examining throughout this book: the pieces that make up the whole of social research.

At the top of the diagram are interests, ideas, and theories, the possible beginning points for a line of research. The letters (A, B, X, Y, and so forth) represent variables or concepts such as *prejudice* or *alienation*. Thus, you might have a general interest in finding out what causes some

people to be more prejudiced than others, or you might want to know some of the consequences of alienation. Alternatively, your inquiry might begin with a specific idea about the way things are. For example, you might have the idea that working on an assembly line causes alienation.

The question marks in the diagram indicate that you aren't sure things are the way you suspect they are—that's why you're doing the research. Notice that a theory is represented as a set of complex relationships among several variables.

Or you might want to consider this question: "How is leadership established in a juvenile gang?" You may wonder the extent to which age, strength, family and friendship ties, intelligence, or other variables figure into the determination of who runs things. We don't always begin with a clear theory about the causal relationships at play.

The double arrows between "interest," "idea," and "theory" suggest that researchers often move back and forth across these several possible beginnings. An initial interest may lead to the formulation of an idea, which may be fit into a larger theory, and the theory may produce new ideas and create new interests.

Any or all of these three may suggest the need for empirical research. The purpose of such research can be to explore an interest, test a specific idea, or validate a complex theory. Whatever the purpose, the researcher needs to make a variety of decisions, as indicated in the remainder of the diagram.

To make this discussion more concrete, let's take a specific research example. Suppose you're concerned with the issue of abortion and want to learn why some college students support abortion rights and others oppose them. Let's say you've gone a step further and formed the impression that students in the humanities and social sciences seem generally more inclined to support the idea of abortion rights than do those in the natural sciences. (That kind of thinking often leads people to design and conduct social research.)

In terms of the options we've discussed in this chapter, you probably have both descriptive and explanatory interests: What percentage of the student body supports a woman's right to an abortion (description), and what causes some to support it and others to oppose it (explanation)? The units of analysis in this case would be individuals: college students. Let's assume you would be satisfied to learn something about the way things are now. You might then decide that a cross-sectional study would suit your purposes. Although this would provide you with no direct evidence of processes over time, you might be able to approximate some longitudinal analyses if you pursued changes in students' attitudes over time.

Getting Started

At the outset of your project, your interests would probably be exploratory. At this point, you might choose among several possible activities in exploring student attitudes about abortion rights. To begin with, you might want to read something about the issue. If you have a hunch that attitudes are somehow related to college major, you might find out what other researchers have written about that. Appendix A of this book will help you make use of your college library. In addition, you would probably talk to some people who support abortion rights and some who do not. You might attend meetings of abortion-related groups. All these activities could help prepare you to handle the various decisions of research design we're about to examine.

Before designing your study, you must define the purpose of your project. What kind of study will you undertake—exploratory, descriptive, explanatory? Do you plan to write a research paper to satisfy a course or thesis requirement? Is your purpose to gain information that will support you in arguing for or against abortion rights? Do you want to write an article for the campus newspaper or an academic journal? In reviewing the previous research literature regarding abortion rights, you should note the design decisions other researchers have made, always asking whether the same decisions would satisfy your purpose.

Usually, your purpose for undertaking research can be expressed as a report. A good first step in designing your project is to outline such a report (see Chapter 15 for more). Although your final report may not look much like your initial image of it, this exercise will help you figure out which research designs are most appropriate. During this step, clearly describe the kinds of statements you want to make when the research is complete. Here are some examples of such statements: "Students frequently mentioned abortion rights in the context of discussing social issues that concerned them personally." "X percent of State U. students favor a woman's right to choose an abortion." "Engineers are (more/less) likely than sociologists to favor abortion rights."

Conceptualization

Once you have a well-defined purpose and a clear description of the kinds of outcomes you want to achieve, you can proceed to the next step in the design of your study—conceptualization. We often talk quite casually about social science concepts such as prejudice, alienation, religiosity, and liberalism, but we need to clarify what we mean by these concepts in order to draw meaningful conclusions about them. Chapter 5 examines this process of conceptualization in depth. For now, let's see what it might involve in the case of our hypothetical example.

If you're going to study how college students feel about abortion and why, the first thing you'll have to specify is what you mean by "the right to an abortion." Because support for abortion probably varies according to the circumstances, you'll want to pay attention to the different conditions under which people might approve or disapprove of abortion: for example, when the woman's life is in danger, in the case of rape or incest, or simply as a matter of personal choice.

Similarly, you'll need to specify exact meanings for all the other concepts you plan to study. If you want to study the relationship of opinion about abortion to college major, you'll have to decide whether you want to consider only officially declared majors or to include students' intentions as well. What will you do with those who have no major?

In surveys and experiments, such concepts must be specified in advance. In less tightly structured research, such as open-ended interviews, an important part of the research may involve the discovery of different dimensions, aspects, or nuances of concepts. In such cases, the research itself may uncover and report aspects of social life that were not evident at the outset of the project.

Choice of Research Method

As we'll discuss in Part 3, each research method has its strengths and weaknesses, and certain concepts are more appropriately studied by some methods than by others. In our study of attitudes toward abortion rights, a survey might be the most appropriate method: either interviewing students or asking them to fill out a questionnaire. Surveys are particularly well suited to the study of public opinion. Of course, you could also make good use of the other methods presented in Part 3. For example, you might use the method of content analysis to examine letters to the editor and analyze the different opinions letter writers have of abortion. Field research would provide an avenue to understanding how people interact with one another regarding the issue of abortion, how they discuss it, and how they change their minds. Usually the best study design uses more than one research method, taking advantage of their different strengths. If you look back at the brief examples of actual studies at the end of the preceding section, you'll see several instances in which the researchers used many methods in a single study.

Operationalization

Once you've specified the concepts to be studied and chosen a research method, the next step is operationalization, or deciding on your measurement techniques (discussed further in Chapters 5 and 6). The meaning of variables in a study is determined in part by how they are measured. Part of the task here is deciding how the desired data will be collected: direct observation, review of official documents, a questionnaire, or some other technique.

If you decide to use a survey to study attitudes toward abortion rights, part of operationalization is determining the wording of questionnaire items. For example, you might operationalize your main variable by asking respondents whether they would approve a woman's right to have an abortion under each of the conditions you've conceptualized: in the case of rape or incest, if her life were threatened by the pregnancy, and so forth. You would have designed the questionnaire so that it asked respondents to express approval or disapproval for each situation. Similarly, you would have specified exactly how respondents would indicate their college major and what choices to provide those who have not declared a major.

Population and Sampling

In addition to refining concepts and measurements, you must decide whom or what to study. The population for a study is that group (usually of people) about whom we want to draw conclusions. We're almost never able to study all the members of the population that interests us, however, and we can never make every possible observation of them. In every case, then, we select a sample from among the data that might be collected and studied. The sampling of information, of course, occurs in everyday life and often produces biased observations. (Recall the discussion of "selective observation" in Chapter 1.) Social researchers are more deliberate in their sampling of what will be observed.

Chapter 7 describes methods for selecting samples that adequately reflect the whole population that interests us. Notice in Figure 4-7 that decisions about population and sampling are related to decisions about the research method to be used. Whereas probability-sampling techniques would be relevant to a large-scale survey or a content analysis, a field researcher might need to select only those informants who will yield a balanced picture of the situation under study, and an experimenter might assign subjects to experimental and control groups in a manner that creates comparability.

In your hypothetical study of abortion attitudes, the relevant population would be the student population of your college. As you'll discover in Chapter 7, however, selecting a sample will require you to get more specific than that. Will you include part-time as well as full-time students? Only degree candidates or everyone? International students as well as U.S. citizens? Undergraduates, graduate students, or both? There are many such questions—each of which must be answered in terms of your research purpose. If your purpose is to predict how students would vote in a local referendum on abortion, you might want to limit your population to those eligible and likely to vote.

Observations

Having decided what to study among whom by what method, you're now ready to make observations—to collect empirical data. The chapters of Part 3, which describe the various research methods, give the different observation techniques appropriate to each.

To conduct a survey on abortion, you might want to print questionnaires and mail them to a sample selected from the student body.

Alternatively, you could arrange to have a team of interviewers conduct the survey over the telephone. The relative advantages and disadvantages of these and other possibilities are discussed in Chapter 9.

Data Processing

Depending on the research method chosen, you'll have amassed a volume of observations in a form that probably isn't immediately interpretable. If you've spent a month observing a street-corner gang firsthand, you'll now have enough field notes to fill a book. In a historical study of ethnic diversity at your school, you may have amassed volumes of official documents, interviews with administrators and others, and so forth. Chapters 13 and 14 describe some of the ways social science data are processed for quantitative or qualitative analysis.

In the case of a survey, the "raw" observations are typically in the form of questionnaires with boxes checked, answers written in spaces, and the like. The data-processing phase for a survey typically involves the classification (coding) of written-in answers and the transfer of all information to a computer.

Analysis

Once the collected data are in a suitable form, you're ready to interpret them for the purpose of drawing conclusions that reflect the interests, ideas, and theories that initiated the inquiry. Chapters 13 and 14 describe a few of the many options available to you in analyzing data. In Figure 4-7, notice that the results of your analyses feed back into your initial interests, ideas, and theories. Often this feedback represents the beginning of another cycle of inquiry.

In the survey of student attitudes about abortion rights, the analysis phase would pursue both descriptive and explanatory aims. You might begin by calculating the percentages of students who favored or opposed each of the several different versions of abortion rights. Taken together, these several percentages would provide a good picture of student opinion on the issue.

Moving beyond simple description, you might describe the opinions of subsets of the student body, such as different college majors. Provided that your design called for tracking other information about respondents, you could also look at men versus women; freshmen, sophomores, juniors, seniors, and graduate students; or other categories that you've included. The description of subgroups could then lead you into an explanatory analysis.

Application

The final stage of the research process involves the uses made of the research you've conducted and the conclusions you've reached. To start, you'll probably want to communicate your findings so that others will know what you've learned. You may want to prepare—and even publish—a written report. Perhaps you'll make oral presentations, such as papers delivered to professional and scientific meetings. Other students would also be interested in hearing what you've learned about them.

You may want to go beyond simply reporting what you've learned to discussing the implications of your findings. Do your findings say anything about actions that might be taken in support of policy goals? Both the proponents and the opponents of abortion rights would be interested.

Karen Akerlof and Chris Kennedy (2013) have provided an omnibus analysis of ways in which social research can be used to design and evaluate programs to combat environmental degradation. They identify five major areas:

1. Promote favorable attitudes.
2. Increase personal agency.
3. Facilitate emotional motivation.
4. Communicate supportive social norms.
5. Alter the environmental context; design the choice.

Finally, be sure to consider what your work suggests in regard to further research on your subject. What mistakes should be corrected in future studies? What avenues that were opened up only slightly in your study should be pursued further?

Research Design in Review

As this overview shows, research design involves a set of decisions regarding what topic is to be studied, among what population, with what research methods, for what purpose.

Although you'll want to consider many ways of studying a subject—and use your imagination as well as your knowledge of a variety of methods—research design is the process of focusing your perspective for the purposes of a particular study.

If you're doing a research project for one of your courses, many aspects of research design may be specified for you in advance, including the method (such as an experiment) or the topic (as in a course on a particular subject). The following summary assumes that you're free to choose both your topic and your research strategy.

In designing a research project, you'll find it useful to begin by assessing three things: your interests, your abilities, and the available resources. Each of these considerations will suggest a large number of possible studies.

Simulate the beginning of a somewhat conventional research project: Ask yourself what you're interested in understanding. Surely you have several questions about social behavior and attitudes. Why are some people politically liberal and others politically conservative? Why are some people more religious than others? Why do people join white supremacist groups? Do colleges and universities still discriminate against minority faculty members? Why would a woman stay in an abusive relationship? Spend some time thinking about the kinds of questions that interest and concern you.

Once you have a few questions you'd be interested in answering for yourself, think about the kind of information needed to answer them. What research units of analysis would provide the most relevant information: college students, corporations, voters, cities, or unions? This question will probably be inseparable from the question of research topics. Then ask which aspects of the units of analysis would provide the information you need in order to answer your research question.

Once you have some ideas about the kind of information relevant to your purpose, ask yourself how you might go about getting that information. Are the relevant data likely to be available somewhere already (say, in a government publication), or would you have to collect them yourself? If you think you would have to collect them, how would you go about doing it?

Would you need to survey a large number of people, or interview a few people in depth? Could you learn what you need to know by attending meetings of certain groups? Could you glean the data you need from books in the library?

As you answer these questions, you'll find yourself well into the process of research design. Keep in mind your own research abilities and the resources available to you. There is little point in designing a perfect study that you can't actually carry out. You may want to try a research method you have not used before so you can learn from it, but be careful not to put yourself at too great a disadvantage.

Once you have a general idea of what you want to study and how, carefully review previous research in journals and books to see how other researchers have addressed the topic and what they have learned about it. Your review of the literature may lead you to revise your research design: Perhaps you'll decide to use a previous researcher's method or even replicate an earlier study. The independent replication of research projects is a standard procedure in the physical sciences, and it is just as important in the social sciences, although social researchers tend to overlook that. Or, you might want to go beyond replication and study some aspect of the topic that you feel previous researchers overlooked.

Here's another approach you might take. Suppose a topic has been studied previously using field research methods. Can you design an experiment that would test the findings those earlier researchers produced? Or can you think of existing statistics that could be used to test their conclusions? Did a mass survey yield results that you would like to explore in greater detail through on-the-spot observations and in-depth interviews? The use of several different research methods to test the same finding is sometimes called *triangulation*, and you should always keep it in mind as a valuable research strategy. Because each research method has particular strengths and weaknesses, there is always a danger that research findings will reflect, at least in part, the method of inquiry. In the best of all worlds, your own research design should bring more than one research method to bear on the topic.

The Research Proposal

Quite often, in the design of a research project, you'll have to lay out the details of your plan for someone else's review or approval. For a course project, for example, your instructor might very well want to see a "proposal" before you set off to work. Later in your career, if you wanted to undertake a major project, you might need to obtain funding from a foundation or government agency, who would definitely want a detailed proposal that describes how you would spend their money. You might respond to a request for proposals (RFP), which both public and private agencies often circulate in search of someone to do research for them.

We now turn to a brief discussion of how you might prepare a research proposal. This will give you one more overview of the whole research process that the rest of this book details.

Elements of a Research Proposal

Although some funding agencies (or your instructor, for that matter) may have specific requirements for the elements or structure of a research proposal, here are some basic components you should include. I've posed some questions that should help you establish some key elements of your proposal.

Problem or Objective

What exactly do you want to study? Why is it worth studying? Does the proposed study have practical significance? Does it contribute to the construction of social theories?

Literature Review

What have others said about this topic? What theories address it, and what do they say? What previous research exists? Are there consistent findings, or do past studies disagree? Does the body of existing research have flaws that you think you can remedy?

You'll find that reading social science research reports requires special skills. If you need to undertake a review of the literature at this point in your course, you may want to skip ahead to Chapter 15. It will familiarize you with the different types of research literature, how to find what you want, and how to read it. There is a special discussion of how to use electronic resources online and how to avoid being misled by information on the Internet.

In part, the data-collection method(s) you intend to use in your study will shape your review of the literature. Reviewing the designs of previous studies using that same technique can give you a head start in planning your own study. At the same time, you should focus your search on your research topic, regardless of the methods other researchers have used. So, if you're planning field research on, say, interracial marriages, you might gain some useful insights from the findings of surveys on the topic; further, past field research on interracial marriages could be invaluable while you design a survey on the topic.

Because the literature review will appear early in your research proposal, you should write it with an eye toward introducing the reader to the topic you'll address, laying out in a logical manner what has already been learned on the topic by past researchers, then leading up to the holes or loose ends in our knowledge of the topic, which you propose to remedy. Or, a little differently, your review of the literature may point to inconsistencies or disagreements among existing findings. In that case, your proposed research will aim to resolve the ambiguities that plague us. I don't know about you, but I'm already excited about the research you're proposing to undertake.

Subjects for Study

Whom or what will you study in order to collect data? Identify the subjects in general, theoretical terms, and in specific, more concrete terms, identify who is available for study and how you'll reach them. Will it be appropriate to select a sample? If so, how will you do that? If there is any possibility that your research will affect those you study, how will you ensure that the research does not harm them?

Beyond these general questions, the specific research method you'll use will further specify the matter. If you're planning to undertake an experiment, a survey, or field research, for example, the techniques for subject selection will vary quite a bit. Lucky for you, Chapter 7 of this book discusses sampling techniques for both qualitative and quantitative studies.

Measurement

What are the key variables in your study? How will you define and measure them? Do your definitions and measurement methods duplicate or differ from those of previous research on this topic? If you have already developed your measurement device (a questionnaire, for example) or will be using something previously developed by others, it might be appropriate to include a copy of it in an appendix to your proposal.

Data-Collection Methods

How will you actually collect the data for your study? Will you conduct an experiment or a survey? Will you undertake field research or will you focus on the reanalysis of statistics already created by others? Perhaps you'll use more than one method.

Analysis

Indicate the kind of analysis you plan to conduct. Spell out the purpose and logic of your analysis. Are you interested in precise description? Do you intend to explain why things are the way they are? Do you plan to account for variations in some quality, such as why some students are more liberal than others? What possible explanatory variables will your analysis consider, and how will you know if you've explained variations adequately?

Schedule

Providing a schedule for the various stages of research is often appropriate. Even if you don't do this for the proposal, do it for yourself. Without a timeline for accomplishing the several stages of research and keeping track of how you're doing, you may end up in trouble.

Budget

When you ask someone to cover the costs of your research, you need to provide a budget that specifies where the money will go. Large, expensive projects include budgetary categories such as personnel, computers, supplies, telephones, and postage. Even if you'll be paying

What do you think?...Revisited

When the Provost and the student newspaper seemed to disagree over the extent of part-time faculty teaching, they used different units of analysis. The newspaper said 52 percent of the faculty were part-time; the Provost said about 70 percent of the credits were taught by full-time faculty. The table here demonstrates how they could both be right, given that the typical full-time faculty member teaches three courses, or nine credits, whereas the typical part-time faculty member teaches one course, or three credits. For simplicity, I've assumed that there are 100 faculty members.

In this hypothetical illustration, full-time faculty taught 432 of the 588 credits, or 73 percent. As you can see, being clear about what the unit of analysis matters a great deal.

Faculty Status	Number	Credits Taught by Each	Total Credits Taught
Full-time	48	9	432
Part-time	52	3	156
Total			588

for your project yourself, you should spend some time anticipating expenses: office supplies, photocopying, computer software, transportation, and so on.

Institutional Review Board

Depending on the nature of your research design, you may need to submit your proposal to the campus institutional review board for approval to ensure the protection of human subjects. Your instructor can advise you on this.

As you can see, if you're interested in conducting a social research project, it's a good idea to prepare a research proposal for your own purposes, even if you aren't required to do so by your instructor or a funding agency. If you're going to invest your time and energy in such a project, you should do what you can to ensure a return on that investment.

Now that you've had a broad overview of social research, you can move on to the remaining chapters in this book and learn

exactly how to design and execute each specific step. If you've found a research topic that really interests you, you'll want to keep it in mind as you see how you might go about studying it.

The Ethics of Research Design

Designing a research project needs to include serious considerations of the ethical dimension. To begin, if your study requires the participation of human subjects, you must determine that the likely benefits of the research will do justice to the time and effort you'll ask them to contribute.

You'll also want to design the study in concurrence with the ethical guidelines discussed in Chapter 3. For example, you should ensure that the subjects' privacy and well-being are protected. As I indicated earlier, having your research design reviewed by an institutional review board may be appropriate.

MAIN POINTS

Introduction

- Any research design requires researchers to specify as clearly as possible what they want to find out and then determine the best way to do it.

Three Purposes of Research

- The principal purposes of social research include exploration, description, and explanation. Research studies often combine more than one purpose.

- Exploration is the attempt to develop an initial, rough understanding of some phenomenon.

- Description is the precise measurement and reporting of the characteristics of some population or phenomenon under study.

- Explanation is the discovery and reporting of relationships among different aspects of the phenomenon under study. Descriptive studies answer the question "What's so?"; explanatory ones tend to answer the question "Why?"

Idiographic Explanation

- Idiographic explanation seeks an exhaustive understanding of the causes producing events and situations in a single or limited number of cases.

- Pay attention to the explanations offered by the people living the social processes you are studying.

- Comparisons with similar situations, either in different places or at different times in the same place, can be insightful.

The Logic of Nomothetic Explanation

- Both idiographic and nomothetic models of explanation rest on the idea of causation. The idiographic model aims at a complete understanding of a particular phenomenon, using all relevant causal factors. The nomothetic model aims at a general understanding—not necessarily complete—of a class of phenomena, using a small number of relevant causal factors.

- There are three basic criteria for establishing causation in nomothetic analyses: (1) The variables must be empirically associated, or correlated; (2) the causal variable must occur earlier in time than the variable it is said to affect; and (3) the observed effect cannot be explained as the effect of a different variable.

Necessary and Sufficient Causes

- Mere association, or correlation, does not in itself establish causation. A spurious causal relationship is an association that in reality is caused by one or more other variables. We will examine this at length in Chapter 15 on the logic of multivariate analysis.

- A necessary cause is one that must be present for the effect to occur.

- A sufficient cause is one will always produce the effect in question.

Units of Analysis

- Units of analysis are the people or things whose characteristics social researchers observe, describe, and explain. Typically, the unit of analysis in social research is the individual person, but it may also be a social group, a formal organization, a social interaction, a social artifact, or another phenomenon such as lifestyles.

- The ecological fallacy involves applying conclusions drawn from the analysis of groups (election precincts) to individuals (voters).

- Reductionism is the attempt to understand a complex phenomenon in terms of a narrow set of concepts, such as attempting to explain the American Revolution solely in terms of economics (or political idealism or psychology) when there were many causes.

The Time Dimension

- The research of social processes that occur over time presents challenges that can be addressed through cross-sectional studies or longitudinal studies.

- Cross-sectional studies are based on observations made at one time. Although conclusions drawn from such studies are limited by this characteristic, researchers can sometimes use such studies to make inferences about processes that occur over time.

- In longitudinal studies, observations are made at many times. Such observations may be made of samples drawn from general populations (trend studies), samples drawn from more-specific subpopulations (cohort studies), or the same sample of people each time (panel studies).

Mixed Modes

- Most studies use a single method for collecting data (e.g., survey, experiment, field research), but using more than one method in a given study can yield a more comprehensive understanding.

How to Design a Research Project

- Research design starts with an initial interest, idea, or theoretical expectation and proceeds through a series of interrelated steps that narrow the focus of the study so that concepts, methods, and procedures are well defined. A good research plan accounts for all these steps in advance.

- At the outset, a researcher specifies the meaning of the concepts or variables to be studied (conceptualization), chooses a research method or methods (such as experiments versus surveys), and specifies the population to be studied and, if applicable, how it will be sampled.

- The researcher operationalizes the proposed concepts by stating precisely how the variables in the study will be measured. Research then proceeds through observation, processing the data, analysis, and application, such as reporting the results and assessing their implications.

The Research Proposal

- A research proposal provides a preview of why a study will be undertaken and how it will be conducted. Researchers must often get permission or necessary resources in order to proceed with a project. Even when not required, a proposal is a useful device for planning.

The Ethics of Research Design

- Your research design should indicate how your study will abide by the ethical strictures of social research.

- It may be appropriate for an institutional review board to review your research proposal.

KEY TERMS

cohort study	reductionism
correlation	social artifact
cross-sectional study	sociobiology
ecological fallacy	spurious relationship
longitudinal study	trend study
panel mortality	units of analysis
panel study	

PROPOSING SOCIAL RESEARCH: DESIGN

This chapter has laid out many different ways social research can be structured. In designing your research project, you'll need to specify which of these ways you'll use. Is your purpose that of exploring a topic, providing a detailed description, or explaining the social differences and processes you may observe? If you're planning a causal analysis, for example, you should say something about how you'll organize and pursue that goal.

Will your project collect data at one point in time or compare data across time? What data-collection technique(s) will you employ?

REVIEW QUESTIONS

1. One example in this chapter suggested that political orientations cause attitudes toward legalizing marijuana. Can you make an argument that the time order is just the opposite of what was assumed?

2. Here are some examples of real research topics. For each excerpt, can you name the unit of analysis? (The answers are at the end of this chapter.)

   a. Women watch TV more than men because they are likely to work fewer hours outside the home than men....Black people watch an average of approximately three-quarters of an hour more television per day than white people. (Hughes 1980: 290)

   b. Of the 130 incorporated U.S. cities with more than 100,000 inhabitants in 1960, there were 126 that had at least two short-term nonproprietary general hospitals accredited by the American Hospital Association. (Turk 1980: 317)

   c. The early TM [transcendental meditation] organizations were small and informal. The Los Angeles group, begun in June 1959, met at a member's house where, incidentally, Maharishi was living. (Johnston 1980: 337)

   d. However, it appears that the nursing staffs exercise strong influence over . . . a decision to change the nursing care system. . . . Conversely, among those decisions dominated by the administration and the medical staffs....(Comstock 1980: 77)

   e. Though 667,000 out of 2 million farmers in the United States are women, women historically have not been viewed as farmers, but rather, as the farmer's wife. (Votaw 1979: 8)

   f. The analysis of community opposition to group homes for the mentally handicapped...indicates that deteriorating neighborhoods are most likely to organize in opposition, but that upper-middle class neighborhoods are most likely to enjoy private access to local officials. (Graham and Hogan 1990: 513)

   g. Some analysts during the 1960s predicted that the rise of economic ambition and political militancy among blacks would foster discontent with the "otherworldly" black mainline churches. (Ellison and Sherkat 1990: 551)

   h. This analysis explores whether propositions and empirical findings of contemporary theories of organizations directly apply to both private product producing organizations (PPOs) and public human service organizations (PSOs). (Schiflett and Zey 1990: 569)

i. This paper examines variations in job title structures across work roles. Analyzing 3,173 job titles in the California civil service system in 1985, we investigate how and why lines of work vary in the proliferation of job categories that differentiate ranks, functions, or particular organizational locations. (Strang and Baron 1990: 479)

3. Review the logic of spuriousness. Can you think up an example in which an observed relationship between two variables could actually be explained away by a third variable?

4. Make up a research example—different from those discussed in the text—that illustrates a researcher committing the ecological fallacy. How would you modify the example to avoid this trap?

ANSWERS TO UNITS OF ANALYSIS QUIZ, REVIEW QUESTION 2

a. Men and women, black and white people (individuals)

b. Incorporated U.S. cities (groups)

c. Transcendental meditation organizations (groups)

d. Nursing staffs (groups)

e. Farmers (individuals)

f. Neighborhoods (groups)

g. Blacks (individuals)

h. Service and production organizations (formal organizations)

i. Job titles (artifacts)

CHAPTER 5

Conceptualization, Operationalization, and Measurement

Dragon Images/Shutterstock.com

CHAPTER OVERVIEW

The interrelated steps of conceptualization, operationalization, and measurement allow researchers to turn a general idea for a research topic into useful and valid measurements in the real world. An essential part of this process involves transforming the relatively vague terms of ordinary language into precise objects of study with well-defined and measurable meanings.

Introduction

This chapter and the next one deal with how researchers move from a general idea about what to study to effective and well-defined measurements in the real world. This chapter discusses the interrelated processes of conceptualization, operationalization, and measurement. Chapter 6 builds on this foundation to discuss types of measurements that are more complex.

Consider a notion such as "satisfaction with college." I'm sure you know some people who are very satisfied, some who are very dissatisfied, and many who fall between those extremes. Moreover, you can probably place yourself somewhere along that satisfaction spectrum. Although this probably makes sense to you as a general matter, how would you go about measuring these differences among students, so you could place them along that spectrum?

Some of the comments students make in conversations (such as "This place sucks") would tip you off as to where they stood. In a more active effort, you could think of questions you might ask students (as in "How satisfied are you...?"), such that their answers would indicate their satisfaction. Perhaps certain behaviors—such as class attendance, use of campus facilities, or setting the dean's office on fire—would suggest different levels of satisfaction. As you think about ways of measuring satisfaction with college, you're engaging in the subject matter of this chapter.

We begin by confronting the hidden concern people sometimes have about whether it's truly possible to measure the stuff of life: love, hate, prejudice, religiosity, radicalism, alienation. Over the next few pages, we'll see that researchers can measure anything that exists. Once that point has been established, we'll turn to the steps involved in actual measurement.

Measuring Anything That Exists

Earlier in this book, I said that one of the two pillars of science is observation. Because this word can suggest a casual, passive activity, scientists often use the term *measurement* instead, meaning careful, deliberate observations of the real world for the purpose of describing objects and events in terms of the attributes composing a variable.

Like the people in the opening "What Do You Think?" box, you may have some reservations about the ability of science to measure the really important aspects of human social existence. If you've read research reports dealing with something like liberalism or religion or prejudice, you may have been dissatisfied with the way the researchers measured whatever they were studying. You may have felt that they were too superficial, missing the aspects that really matter most. Maybe they measured religiosity as the number of times a person went to religious services, or liberalism by how people voted in a single election. Your dissatisfaction would surely have increased if you had found yourself being misclassified by the measurement system.

What do you think?

People sometimes doubt the social researcher's ability to measure and study things that matter. For many people, for example, religious faith is a deep and important part of life. Yet both the religious and the nonreligious might question the social researcher's ability to measure how religious a given person or group is and—even more difficult—why some people are religious and others are not.

   This chapter will show that social researchers can't say definitively who is religious and who is not, nor what percentage of people in a particular population are religious, but they can to an extent determine the causes of religiosity. How can that be?

   See the *What do you think? ... Revisited* box toward the end of the chapter.

Steve Broer/Shutterstock.com

Your dissatisfaction reflects an important fact about social research: Most of the variables we want to study do not actually exist in the physical way that rocks exist. Indeed, they are made up. Moreover, they seldom have a single, unambiguous meaning.

To see what I mean, suppose we want to study *political party affiliation*. To measure this variable, we might consult the list of registered voters to note whether the people we were studying were registered as Democrats or Republicans and take that as a measure of their party affiliation. But we could also simply ask someone what party they identify with and take their response as our measure. Notice that these two different measurement possibilities reflect somewhat different definitions of *political party affiliation*. They might even produce different results: Someone may have registered as a Democrat years ago but gravitated more and more toward a Republican philosophy over time. Or someone who is registered with neither political party may, when asked, say she is affiliated with the one she feels the most kinship with.

Similar points apply to *religious affiliation*. Sometimes this variable refers to official membership in a particular church, temple, mosque, or other; other times it simply means whatever religion, if any, one identifies oneself with. Perhaps to you it means something else, such as attendance at religious services.

The truth is that neither *party affiliation* nor *religious affiliation* has any real meaning, if by

"real" we mean corresponding to some objective aspect of reality. These variables do not exist in nature. They are merely terms we have made up and assigned specific meanings to for some purpose, such as doing social research.

Let's take a closer look by considering a variable of interest to many social researchers (and many other people as well)—*prejudice*.

In the early 1970s, Elijah Anderson spent three years observing life in a black, working-class neighborhood in South Chicago, focusing on Jelly's, a combination bar and liquor store. Although some people had the idea that impoverished neighborhoods in the inner city were socially chaotic and disorganized, Anderson's study, like those of other social scientists, clearly demonstrated there was a definite social structure that guided the behavior of its participants. Much of his interest centered on systems of social status and how the 55 or so regulars at Jelly's worked those systems to establish themselves among their peers.

In the second edition of his classic study of urban life, *A Place on the Corner: A Study of Black Street Corner Men* (University of Chicago Press, 2004), Elijah Anderson returned to Jelly's and the surrounding neighborhood to discover several changes. These were largely due to the outsourcing of manufacturing jobs overseas, which had brought economic and mental depression to many of the residents. The nature of social organization had also consequently changed in many ways.

Sociologist Elijah Anderson

PETER TOBIA/KRT/Newscom

For a research-methods student, the book offers many insights into the process of establishing rapport with people being observed in their natural surroundings. Anderson also offers excellent examples of how concepts are established in qualitative research.

You might do an Internet search for "Elijah Anderson 'A Place on the Corner'" to learn more about how this seasoned researcher goes about discovering the dynamics of street-corner life. What you learn will be applicable to many other research settings.

The definition of terms and variables is critical in social research, and in life, for that matter. During the 2012 congressional elections, Missouri's then-congressman, Todd Akin, stirred up a controversy when he proclaimed that "legitimate rape" rarely resulted in pregnancy; this, in his view, would rule out rape as a justification for abortion. Akin's unusual terminology would suggest, by taking the concept a step further, the possibility of "legitimate murder" or "legitimate bank robbery." In analysis, the congressman probably meant the phrase to be something more like "genuine rape," which still has a fuzzy meaning, and still leaves each person to define it for him or herself. Whatever terminology he used, it leaves the definition at a "you know what I mean" level. But what *do* we know? Obviously, the term means different things to different people. As college and university administrations have begun to acknowledge and deal with the alarming prevalence of rape on campus, for example, they have sometimes been stymied by ambiguity as to what constitutes "rape."

Surely we would agree that if a man knocks a woman unconscious or holds a knife to her throat in order to have sex with her against her will, that would qualify as rape. Even the threat of violence or the use of the date-rape drug, Rohypnol (roofies), would surely qualify. But beyond that, it gets muddy. We should probably distinguish rape from other forms of sexual assault, such as groping. But when does sexual assault become rape? Is penetration required for it to be classified as rape?

The issue of "consent" can be complicated as well. In particular, there are debates over the conditions that might render a woman incapable of giving consent. Certainly unconsciousness qualifies, but how about intoxication? And how intoxicated and under what conditions? For those men who have trouble understanding that "no" means "no," the question of when "yes" really means "yes" is even more elusive.

Rape is a very real problem and we need to find real solutions. Notice how critical definition is in dealing with something seemingly obvious at first glance. And moving beyond rape, what is your definition of *terrorist*, *bigot*, *liberal*, *conservative*, *Christian*, or *Muslim*? These are terms we use all the time, but how would you define them?

Conceptions, Concepts, and Reality

As you and I wander down the road of life, we've already observed a lot of things and known they were real through our observations, and we've heard reports from other people that have seemed real. For example:

- We personally heard people say nasty things about minority groups.
- We heard people say women are inferior to men.
- We read about African Americans being lynched.
- We read that women and minorities earn less than white men do for the same work.
- We learned about "ethnic cleansing" and wars in which one ethnic group tried to eradicate another.

With additional experience, we noticed something more. People who participated in lynching were also quite likely to call African Americans ugly names. A lot of them, moreover, seemed to want women to "stay in their place." Eventually it dawned on us that these several tendencies often appeared together in the same people and also had something in common. At some point, someone had a bright idea: "Let's use the word *prejudiced* for people like that. We can use the term even if they don't do all those things—as long as they're pretty much like that."

Being basically agreeable and interested in efficiency, we agreed to go along with the system. That's where the word *prejudice* came from. We never observed it. We just agreed to use it as a shortcut, a name that represents a collection of apparently related phenomena that we've each observed in the course of life. In short, we made it up.

Here's another clue that prejudice isn't something that exists apart from our rough agreement to use the term in a certain way. Each of us develops our own mental image of what the set of real phenomena we've observed represents in general and what these phenomena have in common. When I say the word *prejudice*, it evokes a mental image in your mind, just as it evokes one in mine. It's as though file drawers in our minds contained thousands of sheets of paper, with each sheet of paper labeled in the upper right-hand corner. A sheet of paper in each of our minds has the label "prejudice" on it. On your sheet are all the things you've been told about prejudice and everything you've observed that seems to be an example of it. My sheet has what I've been told about it plus all the things I've observed that seem examples of it—and mine isn't the same as yours.

The technical term for those mental images, those sheets of paper in our mental file drawers, is *conception*. That is, I have a conception of prejudice, and so do you. We can't communicate these mental images directly, so we use the terms written in the upper right-hand corner of our own mental sheets of paper as a way of communicating about our conceptions and the things we observe that are related to those conceptions. These terms make it possible for us to

Conventionalization and operationalization involve the search for ways to measure the variations that interest us.

communicate and eventually agree on what we specifically mean by those terms. In social research, the process of coming to an agreement about what terms mean is *conceptualization*, and the result is called a **concept**.

In the big picture, language and communication work only to the extent that our mental file-sheet entries overlap considerably. The similarities on those sheets represent the agreements existing in our society. As we grow up, we're all taught approximately the same thing when we're first introduced to a particular term, though our nationality, gender, race, ethnicity, region, language, and other cultural factors can shade our understanding of concepts.

Dictionaries formalize the agreements our society has about such terms. Each person, then, shapes his or her mental images to correspond with such agreements. But because all of us have different experiences and observations, no two people end up with exactly the same set of entries on any sheet in their file systems. If we want to measure "prejudice" or "compassion," we must first stipulate what, exactly, counts as prejudice or compassion for our purposes.

Concepts as Constructs

Abraham Kaplan distinguishes three classes of things that scientists measure. The first class is *direct observables*: the things we can observe rather simply and directly, like the color of an apple or what a person in front of us is wearing. The second class, *indirect observables*, requires "relatively more subtle, complex, or indirect observations"

concept A family of conceptions, such as "chair," representing the whole class of actual chairs.

TABLE 5-1
Three Things Social Scientists Measure

	Examples
Direct observables	Physical characteristics (sex, height, skin color) of a person being observed and/or interviewed
Indirect observables	Characteristics of a person as indicated by answers given in a self-administered questionnaire
Constructs	Level of alienation, as measured by a scale that is created by combining several direct and/or indirect observables

(1964: 55). When we note a person's check mark beside "female" in a questionnaire, we've indirectly observed that person's gender. History books or minutes of corporate board meetings provide indirect observations of past social actions. Finally, the third class of observables consists of *constructs*—theoretical creations that are based on observations but that cannot be observed directly or indirectly. A good example is intelligence quotient, or IQ. It is constructed mathematically from observations of the answers given to a large number of questions on an IQ test. No one can directly or indirectly observe IQ. It is no more a "real" characteristic of people than is compassion or prejudice. (See Table 5-1 for a summary.)

To summarize, concepts are constructs derived by mutual agreement from mental images (conceptions). Our conceptions summarize collections of seemingly related observations and experiences. The observations and experiences are real, at least subjectively, but conceptions, and the concepts derived from them, are only mental creations. The terms associated with concepts are merely devices created for the purposes of filing and communication. A term like *prejudice* is, objectively speaking, only a collection of letters. It has no intrinsic reality beyond that. It has only the meaning we agree to give it.

Usually, however, we fall into the trap of believing that terms for constructs do have intrinsic meaning, that they name real entities in the world. That danger seems to grow stronger when we begin to take terms seriously and attempt to use them precisely. Further, the danger increases in the presence of experts who appear to know more than we do about what the terms really mean: Yielding to authority is easy in such a situation.

Once we assume that terms like *prejudice* and *compassion* have real meanings, we begin the tortured task of discovering what those real meanings are and what constitutes a genuine measurement of them. Regarding constructs as real is called *reification*. Does this discussion imply that compassion, prejudice, and similar constructs cannot be measured? Interestingly, the answer is no. (And a good thing, too, or a lot of us social researcher types would be out of work.) I've said that we can measure anything that's real. Constructs aren't real in the way that trees are real, but they do have another important virtue: They are useful. That is, they help us organize, communicate about, and understand things that are real. They help us make predictions about real things. Some of those predictions even turn out to be true. Constructs can work this way because, although not real or observable in themselves, they have a definite relationship to things that are real and observable. The bridge from direct and indirect observables to useful constructs is the process called conceptualization.

Conceptualization

The process through which we specify what we mean when we use particular terms in research is called **conceptualization**. Suppose we want to find out, for example, whether women are more compassionate than men. I suspect many people assume this is the case, but it might be interesting to find out if it's really so. We can't meaningfully study the question, let alone agree on the answer, without some working agreements about the meaning of compassion. They are "working" agreements in the sense that they allow us to work on the question. We don't need to agree or even pretend to agree that a particular specification is ultimately the best one.

---

conceptualization The mental process whereby fuzzy and imprecise notions (concepts) are made more specific and precise. So you want to study prejudice. What do you mean by "prejudice"? Are there different kinds of prejudice? What are they?

Conceptualization, then, produces a specific, agreed-on meaning for a concept, for the purposes of research. This process of specifying exact meaning involves describing the indicators we'll be using to measure our concept and the different aspects of the concept, called dimensions.

Indicators and Dimensions

Conceptualization gives definite meaning to a concept by specifying one or more indicators of what we have in mind. An **indicator** is a sign of the presence or absence of the concept we're studying. Here's an example.

We might agree that visiting children's hospitals during Christmas and Hanukkah is an indicator of compassion. Putting little birds back in their nests might be agreed on as another indicator, and so forth. If the unit of analysis for our study is the individual person, we can then observe the presence or absence of each indicator for each person under study. Going beyond that, we can add up the number of indicators of compassion observed for each individual. We might agree on ten specific indicators, for example, and find six present in our study of Pat, three for John, nine for Mary, and so forth.

Returning to our question about whether men or women are more compassionate, we might calculate that the women we studied displayed an average of 6.5 indicators of compassion and the men an average of 3.2. On the basis of our quantitative analysis of group difference, we might therefore conclude that women are, on the whole, more compassionate than men.

Usually, though, it's not that simple. Imagine you're interested in understanding a small religious cult, particularly their views on various groups: gays, nonbelievers, feminists, and others. In fact, they suggest that anyone who refuses to join their group and abide by its teachings will "burn in hell." In the context of your interest in compassion, they

don't seem to have much. And yet, the group's literature often speaks of their compassion for others. You want to explore this seeming paradox.

To pursue this research interest, you might arrange to interact with cult members, getting to know them and learning more about their views. You could tell them you were a social researcher interested in learning about their group, or perhaps you would just express an interest in learning more without saying why.

In the course of your conversations with group members and perhaps your attendance at religious services, you would put yourself in situations where you could come to understand what the cult members mean by compassion. You might learn, for example, that members of the group were so deeply concerned about sinners burning in hell that they were willing to be aggressive, even violent, to make people change their sinful ways. Within their own paradigm, then, cult members would see beating up gays, prostitutes, and abortion doctors as acts of compassion.

Social researchers focus their attention on the meanings that people under study give to words and actions. Doing so can often clarify the behaviors observed: At least now you understand how the cult can see violent acts as compassionate. On the other hand, paying attention to what words and actions mean to the people under study almost always complicates the concepts researchers are interested in.

Whenever we take our concepts seriously and set about specifying what we mean by them, we discover disagreements and inconsistencies. Not only do you and I disagree, but each of us is likely to find a good deal of muddiness within our own mental images. If you take a moment to look at what you mean by compassion, you'll probably find that your image contains several kinds of compassion. That is, the entries on your mental file sheet can be combined into groups and subgroups, say, compassion toward friends, coreligionists, humans, and birds. You can also find several different strategies for making combinations. For example, you might group the entries into feelings and actions.

The technical term for such groupings is **dimension**: a specifiable aspect of a concept. For instance, we might speak of the "feeling dimension" of compassion and its "action dimension." In a different grouping scheme, we might distinguish "compassion for humans" from

---

indicator An observation that we choose to consider as a reflection of a variable we wish to study. Thus, for example, attending religious services might be considered an indicator of *religiosity*.

dimension A specifiable aspect of a concept. "Religiosity," for example, might be specified in terms of a belief dimension, a ritual dimension, a devotional dimension, a knowledge dimension, and so forth.

"compassion for animals." Or we might see compassion as helping people have what we want for them versus what they want for themselves. Still differently, we might distinguish "compassion as forgiveness" from "compassion as pity."

Thus, we could subdivide compassion into several clearly defined dimensions. A complete conceptualization involves both specifying dimensions and identifying the various indicators for each.

When Jonathan Jackson set out to measure "fear of crime," he considered eleven different dimensions, such as:

- *the frequency of worry about becoming a victim of three personal crimes and two property crimes in the immediate neighbourhood . . . ;*
- *estimates of likelihood of falling victim to each crime locally;*
- *perceptions of control over the possibility of becoming a victim of each crime locally;*
- *perceptions of the seriousness of the consequences of each crime;*
- *beliefs about the incidence of each crime locally;*
- *perceptions of the extent of social physical incivilities in the neighbourhood;*
- *perceptions of community cohesion, including informal social control and trust/social capital.*

(Jackson 2005: 301)

Sometimes conceptualization aimed at identifying different dimensions of a variable leads to a different kind of distinction. We may conclude that we've been using the same word for meaningfully distinguishable concepts. In the following example, the researchers find, first, that "violence" is not a sufficient description of "genocide" and, second, that the concept "genocide" itself comprises several distinct phenomena. Let's look at the process they went through to come to this conclusion.

When Daniel Chirot and Jennifer Edwards attempted to define the concept of "genocide," they found that existing assumptions were not precise enough for their purposes:

The United Nations originally defined it as an attempt to destroy "in whole or in part, a national, ethnic, racial, or religious group." If genocide is distinct from other types of violence, it requires its own unique explanation.

(2003: 14)

Notice the final comment in this excerpt, as it provides an important insight into why researchers are so careful in specifying the concepts they study. If genocide, such as the Holocaust, were simply another example of violence, like assaults and homicides, then what we know about violence in general might explain genocide. If it differs from other forms of violence, then we may need a different explanation for it. So the researchers began by suggesting that, for their purposes, "genocide" was a concept distinct from "violence."

Then, as Chirot and Edwards examined historical instances of genocide, they began concluding that the motivations for launching genocidal mayhem differed sufficiently to represent four distinct phenomena that were all called "genocide":

1. *Convenience*: Sometimes the attempt to eradicate a group of people serves a function for the eradicators, such as Julius Caesar's attempt to eradicate tribes defeated in battle, fearing they would be difficult to rule. Or when gold was discovered on Cherokee land in the southeastern United States in the early nineteenth century, the Cherokee were forcibly relocated to Oklahoma in an event known as the "Trail of Tears," which ultimately killed as many as half of those forced to leave.
2. *Revenge*: When the Chinese of Nanking bravely resisted the Japanese invaders in the early years of World War II, the conquerors felt they had been insulted by those they regarded as inferior beings. Tens of thousands were slaughtered in the "Rape of Nanking" in 1937–1938.
3. *Fear*: The ethnic cleansing that occurred during the 1990s in the former Yugoslavia was at least partly motivated by economic competition and worries that the growing Albanian population of Kosovo was gaining political strength through numbers. Similarly, the Hutu attempt to eradicate the Tutsis of Rwanda grew out of a fear that returning Tutsi refugees would seize control of the country. Often, intergroup fears such as these grow out of long histories of atrocities, often inflicted in both directions.
4. *Purification*: The Nazi Holocaust, probably the most publicized case of genocide, was intended as a purification of the "Aryan race." While Jews were the main target, gypsies,

homosexuals, and many other groups were also included. Other examples include the witch hunt against Communists in 1965–1966 and the attempt to eradicate all non-Khmer Cambodians under Pol Pot in the 1970s. (2003: 15–18)

No single theory of genocide could explain these various forms of mayhem. Indeed, this act of conceptualization suggests four distinct phenomena, each needing a different set of explanations.

Specifying the different dimensions of a concept, then, often paves the way for a more sophisticated understanding of what we're studying. To take another example, we might observe that women are more compassionate in terms of feelings, and men more so in terms of actions—or vice versa. Whichever turned out to be the case, we would not be able to say whether men or women are really more compassionate. Our research would have shown that there is no single answer to the question. That alone represents an advance in our understanding of reality.

The Interchangeability of Indicators

There is another way that the notion of indicators can help us in our attempts to understand reality by means of "unreal" constructs. Suppose, for the moment, that you and I have compiled a list of 100 indicators of compassion and its various dimensions. Suppose further that we disagree widely on which indicators give the clearest evidence of compassion or its absence. If we pretty much agree on some indicators, we could focus our attention on those, and we would probably agree on the answer they provided. We would then be able to say that some people are more compassionate than others in some dimension. But suppose we don't really agree on any of the possible indicators. Surprisingly, we can still reach an agreement on whether men or women are the more compassionate. How we do that has to do with the interchangeability of indicators.

The logic works like this. If we disagree totally on the value of the indicators, one solution would be to study all of them. Suppose that women turn out to be more compassionate than men on all 100 indicators—on all the indicators you favor and on all of mine. Then

we would be able to agree that women are more compassionate than men even though we still disagree on exactly what compassion means in general.

The interchangeability of indicators means that if several different indicators all represent, to some degree, the same concept, then all of them will behave the same way that the concept would behave if it were real and could be observed. Thus, given a basic agreement about what "compassion" is, if women are generally more compassionate than men, we should be able to observe that difference by using any reasonable measure of compassion. If, on the other hand, women are more compassionate than men on some indicators but not on others, we should see whether the two sets of indicators represent different dimensions of compassion.

You've now seen the fundamental logic of conceptualization and measurement. The discussions that follow are mainly refinements and extensions of what you've just read. Before turning to a technical elaboration of measurement, however, we need to fill out the picture of conceptualization by looking at some of the ways social researchers provide the meanings of terms that have standards, consistency, and commonality.

Real, Nominal, and Operational Definitions

As we've seen, the design and execution of social research requires us to clear away the confusion over concepts and reality. To this end, logicians and scientists have distinguished three kinds of definitions: real, nominal, and operational.

The first of these reflects the reification of terms. As Carl Hempel cautions,

> A "real" definition, according to traditional logic, is not a stipulation determining the meaning of some expression but a statement of the "essential nature" or the "essential attributes" of some entity. The notion of essential nature, however, is so vague as to render this characterization useless for the purposes of rigorous inquiry.
>
> (1952: 6)

In other words, trying to specify the "real" meaning of concepts only leads to a quagmire: It mistakes a construct for a real entity.

How to Do It

Conceptualization

By this point in the chapter, you should be gaining an appreciation for the ambiguity of language. This creates special challenges for social researchers, but it is no less significant in daily life. George Lakoff, a professor of linguistics and cognitive science at the University of California–Berkeley, has written widely about the ways in which language choices have shaped political debate in the United States (Lakoff 2002). Specifically, he suggests that conservatives have been generally more adept than liberals in this regard. Thus, for example, paying taxes could reasonably be seen as a patriotic act, paying your fair share of the cost of an orderly society. Avoiding taxes, within this construction, would be unpatriotic. Instead of "tax avoidance" or "tax evasion," however, we frequently hear calls for "tax relief," which creates an image of citizens being unfairly burdened by government and revolting against that injustice. The intellectual act of conceptualization has real consequences in your daily life.

The specification of concepts in scientific inquiry depends instead on nominal and operational definitions. A *nominal definition* is one that is simply assigned to a term without any claim that the definition represents a "real" entity. Nominal definitions are arbitrary—I could define compassion as "plucking feathers off helpless birds" if I wanted to—but as definitions they can be more, or less, useful. For most purposes, especially communication, that last definition of compassion would be useless. Most nominal definitions represent some consensus, or convention, about how a particular term is to be used.

An *operational definition*, as you may recall from Chapter 2, specifies precisely how a concept will be measured—that is, the operations we choose to perform. An operational definition is nominal rather than real, but it achieves maximum clarity about what a concept means in the context of a given study. In the midst of disagreement and confusion over what a term "really" means, we can specify a working definition for the purposes of an inquiry. Wishing to examine socioeconomic status (SES) in a study, for example, we may simply specify that we are going to treat SES as a combination of income and educational attainment. In this decision, we rule out other possible aspects of SES: occupational status, money in the bank, property, lineage, lifestyle, and so forth. Our findings will then be interesting to the extent that our definition of SES is useful for our purpose. See "How to Do It: Conceptualization," for more on this.

Clearly, the wording of questions can affect the answers we get, but how can we know which words have what effect? **Cognitive interviewing** is a useful tool for this purpose. Once you have drafted potential questions, you can have people answer and comment on the questions as they do so.

Camilla Priede and Stephen Farrall (2011) describe some different ways to conduct cognitive interviewing in evaluating the quality of research questions. For example, the "thinking aloud" method encourages respondents to talk about the questions as they answer them: for example, "I assume here that you wanted to know if I *knew* I was using marijuana." Alternatively, researchers can use "verbal probing" to elicit feedback from respondents: either preplanned questions or those that occur to the researcher during the course of the cognitive interview: "How did you feel when you realized the brownies contained marijuana?"

Ralph Fevre and his colleagues (2010) were interested in the phenomenon of bullying in the workplace in different European countries. Ultimately, they would do quantitative surveys, but they wanted to be sure they used the most appropriate measures of bullying. To do this, a major portion of their research design involved qualitative interviews in which subjects were asked potential survey questions that were followed by further questioning to fully understand the answers given. This process indicated, for example, that respondents often thought

---

cognitive interviewing Testing potential questions in an interview setting, probing to learn how respondents understand or interpret the questions.

of bullying in a school context, which might confuse their understanding of bullying in the workplace. The extensive, in-depth interviews helped them frame survey questions that elicited responses appropriate to the topic.

The impact of wording choices is a never-ending concern for social researchers, and the testing of different words will never end. For example, Eleanor Singer and Mick Couper (2014) suspected that opinions on the hotly controversial topic of abortion might be affected by whether the object of an abortion was referred to as a baby or a fetus. To find out, they conducted a test in which some survey respondents were asked for their opinions about abortion using the term *fetus* and other respondents were asked the same question using the term *baby*. To their surprise, the wording difference in this instance made no significant difference.

Hyunjoo Park, M. Mandy Sha, and Yulin Pan (2014) found that cognitive interviewing needed special care in some non–English-speaking cultures. In Korean culture, for example, criticizing question wording might be regarded as impolite, thus challenging the effectiveness of cognitive interviewing. Nonetheless, with proper attention to cultural norms, they found that it could be done effectively.

Creating Conceptual Order

The clarification of concepts is a continuing process in social research. Catherine Marshall and Gretchen Rossman (1995: 18) speak of a "conceptual funnel" through which a researcher's interest becomes increasingly focused. Thus, a general interest in social activism could narrow to "individuals who are committed to empowerment and social change," and further focus on discovering "what experiences shaped the development of fully committed social activists." This focusing process is inescapably linked to the language we use.

In some forms of qualitative research, the clarification of concepts is a key element in the collection of data. Suppose you were conducting interviews and observations of a radical political group devoted to combating oppression in U.S. society. Imagine how the meaning of oppression would shift as you delved more and more deeply into the members' experiences

and worldviews. For example, you might start out thinking of oppression in physical and perhaps economic terms. The more you learned about the group, however, the more you might appreciate the possibility of psychological oppression.

The same point applies even to contexts in which meanings might seem more fixed. In the analysis of textual materials, for example, social researchers sometimes speak of the "hermeneutic circle," a cyclical process of ever-deeper understanding.

> The understanding of a text takes place through a process in which the meaning of the separate parts is determined by the global meaning of the text as it is anticipated. The closer determination of the meaning of the separate parts may eventually change the originally anticipated meaning of the totality, which again influences the meaning of the separate parts, and so on.
>
> *(Kvale 1996: 47)*

Consider the concept of "prejudice." Suppose you needed to write a definition of the term. You might start out thinking about racial/ethnic prejudice. At some point you would realize you should probably allow for gender prejudice, religious prejudice, antigay prejudice, and the like in your definition. Examining each of these specific types of prejudice would affect your overall understanding of the general concept. As your general understanding changed, however, you would likely see each of the individual forms somewhat differently.

The continual refinement of concepts occurs in all social research methods. Often you'll find yourself refining the meaning of important concepts even as you write up your final report.

Although conceptualization is a continuing process, it's vital to address it specifically at the beginning of any study design, especially rigorously structured research designs such as surveys and experiments. In a survey, for example, operationalization results in a commitment to a specific set of questionnaire items that will represent the concepts under study. Without that commitment, the study could not proceed.

Even in less-structured research methods, however, we need to begin with an initial set of anticipated meanings that can be refined during data collection and interpretation. No one seriously believes we can observe life with no preconceptions; for this reason, scientific observers must be conscious of and explicit about these conceptual starting points.

Let's explore initial conceptualization as it applies to structured inquiries such as surveys and experiments. Though specifying nominal definitions focuses our observational strategy, it does not allow us to observe. As a next step we must specify exactly what we're going to observe, how we'll observe it, and how we'll interpret various possible observations. All these further specifications make up the operational definition of the concept.

In the example of socioeconomic status, we might decide to measure SES in terms of income and educational attainment. We might then ask survey respondents two questions:

1. What was your total family income during the past 12 months?
2. What is the highest level of school you completed?

To organize our data, we would probably want to specify a system for categorizing the answers people give us. For income, we might use categories such as "under $25,000," "$25,000 to $49,999," and so on. Educational attainment might be similarly grouped in categories: less than high school, high school, college, graduate degree. Finally, we would specify the way a person's responses to these two questions would be combined to create a measure of SES.

In this way, we would create a working and workable definition of SES. Although others might disagree with our conceptualization and operationalization, the definition would have one essential scientific virtue: It would be absolutely specific and unambiguous. Even if someone disagreed with our definition, that person would have a good idea of how to interpret our research results, because what we meant by SES—reflected in our analyses and conclusions—would be precise and clear.

Next is a diagram showing the progression of measurement steps from our vague sense of what a term means to specific measurements in a fully structured scientific study.

Measurement Step	Example: "Social Class"
Conceptualization	What are the different meanings and dimensions of the concept "social class"?
Nominal definition	For our study, we define "social class" as representing economic differences: specifically, income.
Operational definition	We measure economic differences by responses to the survey question "What was your annual income, before taxes, last year?"
Measurements in the real world	The interviewer asks, "What was your annual income, before taxes, last year?"

An Example of Conceptualization: The Concept of Anomie

To look at the overall process of conceptualization in research, let's look briefly at the history of a specific social science concept. Researchers studying urban riots often focus on the part played by feelings of powerlessness. Social researchers sometimes use the word *anomie* in this context. This term was first introduced into social science by Emile Durkheim, the great French sociologist, in his classic 1897 book, *Suicide.*

Using only government publications on suicide rates in different regions and countries, Durkheim produced a work of analytic genius. To determine the effects of religion on suicide, he compared the suicide rates of predominantly Protestant countries with those of predominantly Catholic ones, Protestant regions of Catholic countries with Catholic regions of Protestant countries, and so forth. To determine the possible effects of the weather, he compared suicide rates in northern and southern countries and regions, and he examined the different suicide rates across the months and seasons of the year. Thus, he could draw conclusions about a supremely individualistic and personal act without having any data about the individuals engaging in it.

At a more general level, Durkheim suggested that suicide also reflects the extent to which a society's agreements are clear and stable. Noting that times of social upheaval and change often present individuals with grave uncertainties about what is expected of them, Durkheim suggested that such uncertainties cause confusion, anxiety, and even self-destruction. To describe this societal condition of normlessness,

Durkheim chose the term *anomie*. Durkheim did not make this word up. Used in both German and French, it literally meant "without law." The English term *anomy* had been used for at least three centuries before Durkheim to mean disregard for divine law. However, Durkheim created the social science concept of anomie.

Since Durkheim's time, social scientists have found anomie a useful concept, and many have expanded on Durkheim's use. Robert Merton, in a classic article entitled "Social Structure and Anomie" (1938), concluded that anomie results from a disparity between the goals and means prescribed by a society. Monetary success, for example, is a widely shared goal in our society, yet not all individuals have the resources to achieve it through acceptable means. An emphasis on the goal itself, Merton suggested, produces normlessness, because those denied the traditional avenues to wealth go about getting it through illegitimate means. Merton's discussion, then, could be considered a further conceptualization of the concept of anomie.

Although Durkheim originally used the concept of anomie as a characteristic of societies, as did Merton after him, other social scientists have used it to describe individuals. To clarify this distinction, some scholars have chosen to use *anomie* in reference to its original, societal meaning and to use the term *anomia* in reference to the individual characteristic. In a given society, then, some individuals experience anomia, and others do not. Elwin Powell, writing 20 years after Merton, provided the following conceptualization of anomia (though using the term *anomie*) as a characteristic of individuals:

> When the ends of action become contradictory, inaccessible or insignificant, a condition of anomie arises. Characterized by a general loss of orientation and accompanied by feelings of "emptiness" and apathy, anomie can be simply conceived as meaninglessness.

(1958: 132)

Powell went on to suggest that there were two distinct kinds of anomia and to examine how the two rose out of different occupational experiences to result, at times, in suicide. In his study, however, Powell did not measure anomia per se; he studied the relationship between suicide and occupation, making inferences about the two

kinds of anomia. Thus, the study did not provide an operational definition of anomia, only a further conceptualization.

Although many researchers have offered operational definitions of anomia, one name stands out over all. Two years before Powell's article appeared, Leo Srole (1956) published a set of questionnaire items that he said provided a good measure of anomia. It consists of five statements that subjects were asked to agree or disagree with:

1. *In spite of what some people say, the lot of the average man is getting worse.*
2. *It's hardly fair to bring children into the world with the way things look for the future.*
3. *Nowadays a person has to live pretty much for today and let tomorrow take care of itself.*
4. *These days a person doesn't really know who he can count on.*
5. *There's little use writing to public officials because they aren't really interested in the problems of the average man.*

(1956: 713)

In the decades following its publication, the Srole scale has become a research staple for social scientists. You'll likely find this particular operationalization of anomia used in many of the research projects reported in academic journals.

This abbreviated history of anomie and anomia as social science concepts illustrates several points. First, it is a good example of the process through which general concepts become operationalized measurements. This is not to say that the issue of how to operationalize anomie/anomia has been resolved once and for all. Scholars will surely continue to reconceptualize and reoperationalize these concepts for years to come, continually seeking ever more useful measures.

The Srole scale illustrates another important point. Letting conceptualization and operationalization be open-ended does not necessarily produce anarchy and chaos, as you might expect. Order often emerges. For one thing, although we could define *anomia* any way we choose—in terms of, say, shoe size—we would likely define it in ways not too different from other people's mental images. If you were to use a really offbeat definition, people would probably ignore you.

A second source of order is that, as researchers discover the utility of a particular conceptualization and operationalization of a concept, they're likely to adopt it, which leads to standardized definitions of concepts. Besides the Srole scale, examples include IQ tests and a host of demographic and economic measures developed by the U.S. Census Bureau. Using such established measures has two advantages: They have been extensively pretested and debugged, and studies using the same scales can be compared. If you and I do separate studies of two different groups and use the Srole scale, we can compare our two groups on the basis of anomia.

Social scientists, then, can measure anything that's real; through conceptualization and operationalization, they can even do a pretty good job of measuring things that aren't. Granting that such concepts as socioeconomic status, prejudice, compassion, and anomia aren't ultimately real, social scientists can create order in handling them.

Definitions in Descriptive and Explanatory Studies

As described in Chapter 4, two general purposes of research are description and explanation. The distinction between them has important implications for definition and measurement. If it seems that description is simpler than explanation, you may be surprised to learn that definitions are more problematic for descriptive research than for explanatory research. Before we turn to other aspects of measurement, you'll need a basic understanding of why this is so (we'll discuss this point more fully in Part 4).

It's easy to see the importance of clear and precise definitions for descriptive research. If we want to describe and report the unemployment rate in a city, our definition of "being unemployed" is obviously critical. That definition will depend on our definition of another term: the labor force. If it seems patently absurd to regard a three-year-old child as being unemployed, it is because such a child is not considered a member of the labor force. Thus, we might follow the U.S. Census Bureau's convention and exclude all people under age 14 from the labor force.

This convention alone, however, would not give us a satisfactory definition, because it would count as unemployed such people as high school students, the retired, the disabled, and homemakers who don't want to work outside the home. We might follow the Census convention further by defining the labor force as "all persons 14 years of age and over who are employed, looking for work, or waiting to be called back to a job from which they have been laid off or furloughed." If a student, homemaker, or retired person is not looking for work, such a person would not be included in the labor force. Unemployed people, then, would be those members of the labor force, as defined, who are not employed.

But what does "looking for work" mean? Must a person register with the state employment service or go from door to door asking for employment? Or would it be sufficient to want a job or be open to an offer of employment? Conventionally, "looking for work" is defined operationally as saying yes in response to an interviewer's asking, "Have you been looking for a job during the past seven days?" (Seven days is the period most often specified, but for some research purposes it might make more sense to shorten or lengthen it.)

As you can see, the conclusion of a descriptive study about the unemployment rate depends directly on how each issue with regard to definition is resolved. Increasing the period during which people are counted as looking for work would add more unemployed people to the labor force as defined, thereby increasing the reported unemployment rate. If we follow another convention and speak of the civilian labor force and the civilian unemployment rate, we're excluding military personnel; that, too, increases the reported unemployment rate, because military personnel would be employed—by definition. Thus, the descriptive statement that the unemployment rate in a city is 3 percent, or 9 percent, or whatever it might be, depends directly on the operational definitions used.

This example is relatively clear because there are several accepted conventions relating to the labor force and unemployment. Now, consider how difficult it would be to get agreement about the definitions you would need in order to say, "Forty-five percent of the students at this institution are politically conservative." Like the unemployment rate, this percentage would depend directly on the definition of what is being measured—in this case, political conservatism. A different definition might result in the conclusion "Five percent of the student body are politically conservative."

What percentage of the population do you suppose is "disabled"? That's the question Lars Gronvik (2009) asked in Sweden. He analyzed several databases that encompassed four different definitions or measures of disability in Swedish society. One study asked people if they had hearing, seeing, walking, or other functional problems. Two other measures were based on whether people received one of two forms of government disability support. Another study asked people whether they believed they were disabled.

The four measures indicated different population totals for those citizens defined as "disabled," and each measure produced different demographic profiles that included variables such as gender, age, education, living arrangement, education, and labor-force participation. As you can see, it is impossible to answer a descriptive question such as this without specifying the meaning of terms.

Ironically, definitions are less problematic in the case of explanatory research. Let's suppose we're interested in explaining political conservatism. Why are some people conservative and others not? More specifically, let's suppose we're interested in whether conservatism increases with age. What if you and I have twenty-five different operational definitions of conservative and we can't agree on which definition is best? As we saw in the discussion of indicators, this is not necessarily an insurmountable obstacle to our research. Suppose we found old people to be more conservative than young people in terms of all twenty-five definitions. Clearly, the exact definition wouldn't matter much. We would conclude that old people are generally more conservative than young people—even though we couldn't agree about exactly what *conservative* means.

In practice, explanatory research seldom results in findings quite as unambiguous as this example suggests; nonetheless, the general pattern is quite common in actual research. There are consistent patterns of relationships in human social life that result in consistent research findings. However, such consistency does not appear in a descriptive situation. Changing definitions almost inevitably result in different descriptive conclusions.

Operationalization Choices

In discussing conceptualization, I frequently have referred to operationalization, for the two are intimately linked. To recap: Conceptualization is the refinement and specification of abstract concepts, and operationalization is the development of specific research procedures (operations) that will result in empirical observations representing those concepts in the real world.

As with the methods of data collection, social researchers have a variety of choices when operationalizing a concept. Although the several choices are intimately interconnected, I've separated them for the sake of discussion. Realize, however, that operationalization does not proceed through a systematic checklist.

Range of Variation

In operationalizing any concept, researchers must be clear about the range of variation that interests them. The question is, to what extent are we willing to combine attributes in fairly gross categories?

Let's suppose you want to measure people's incomes in a study by collecting the information from either records or interviews. The highest annual incomes people receive run into the millions of dollars, but not many people earn that much. Unless you're studying the very rich, keeping track of extremely high categories probably won't add much to your study. Depending on whom you study, you'll probably want to establish a highest income category with a much lower floor—maybe $250,000 or more. Although this decision will lead you to throw together people who earn a trillion dollars a year with paupers earning a mere $250,000, they'll survive it, and that mixing probably won't hurt your research any, either. The same decision faces you at the other end of the income spectrum. In studies of the general U.S. population, a bottom category of $25,000 or less usually works fine.

In studies of attitudes and orientations, the question of range of variation has another dimension. Unless you're careful, you may end up measuring only half an attitude without really meaning to. Here's an example of what I mean.

Suppose you're interested in people's attitudes toward expanding the use of nuclear-power generators. If you reasonably guess or have experienced that some people consider nuclear power the greatest thing since the wheel, whereas other people have absolutely no interest in it, it makes sense to ask people how much they favor expanding the use of nuclear energy and to give them answer categories ranging from "Favor it very much" to "Don't favor it at all."

This operationalization, however, conceals half the attitudinal spectrum regarding nuclear energy. Many people have feelings that go beyond simply not favoring it: They are, with greater or lesser degrees of intensity, actively opposed to it. In this instance, there is considerable variation on the left side of zero. Some oppose it a little, some quite a bit, and others a great deal. To measure the full range of variation, then, you'd want to operationalize attitudes toward nuclear energy with a range from favoring it very much, through no feelings one way or the other, to opposing it very much.

This consideration applies to many of the variables that social researchers study. Virtually any public issue involves both support and opposition, each in varying degrees. Political orientations range from very liberal to very conservative, and depending on the people you're studying, you may want to allow for radicals on one or both ends. Similarly, people are not just more or less religious; some are antireligious.

The point is not that you must measure the full range of variation in every case. You should, however, consider whether you need to do so, given your particular research purpose. If the difference between not religious and antireligious isn't relevant to your research, forget it. Someone has defined pragmatism as "knowing that any difference that makes no difference is no difference." Be pragmatic.

Finally, decisions about the range of variation should be governed by the expected distribution of attributes among the subjects of the study. In a study of college professors' attitudes toward the value of higher education, you could probably stop at no value and not worry about those who might consider higher education dangerous to students' health. (If you were studying students, however....)

Variations between the Extremes

Degree of precision is a second consideration in operationalizing variables. What it boils down to is how fine you will make the distinctions among the various possible attributes composing a given variable. Does it matter for your purposes whether a person is 17 or 18 years old, or could you conduct your inquiry by throwing them together in a group labeled 10 to 19 years old? Don't answer too quickly. If you wanted to study rates of voter registration and participation, you'd definitely want to know whether the people you studied were old enough to vote. In general, if you're going to measure age, you must look at the purpose and procedures of your study and decide whether fine or gross differences in age are important to you. In a survey, you'll need to make these decisions in order to design an appropriate questionnaire. In the case of in-depth interviews, these decisions will condition the extent to which you probe for details.

The same thing applies to other variables. If you measure political affiliation, will it matter to your inquiry whether a person is a conservative Democrat rather than a liberal Democrat, or will knowing just the party suffice? In measuring religious affiliation, is it enough to know that a person is a Protestant, or do you need to know the denomination? Do you simply need to know whether or not a person is married, or will it make a difference to know if he or she has never married or is separated, widowed, or divorced?

There is, of course, no general answer to such questions. The answers come out of the purpose of a given study, or why we are making a particular measurement. I can give you a useful guideline, though. Whenever you're not sure how much detail to pursue in a measurement, go after too much rather than too little. When a subject in an in-depth interview volunteers that she is 37 years old, record "37" in your notes, not "in her thirties." When you're analyzing the data, you can always combine precise attributes into more-general categories, but you can never separate any variations you lumped together during observation and measurement.

A Note on Dimensions

We've already discussed dimensions as a characteristic of concepts. When researchers get down to the business of creating operational measures of variables, they often discover—or worse, never notice—that they haven't been exactly clear about which dimensions of a variable they're really interested in. Here's an example.

Let's suppose you're studying people's attitudes toward government, and you want to include an examination of how people feel about government corruption, as they understand that term. Here are just a few of the dimensions you might examine:

- Do people think there is corruption in government?
- How much corruption do they think there is?
- How certain are they in their judgment of how much corruption there is?
- How do they feel about corruption in government as a problem in society?
- What do they think causes corruption?
- Do they think that corruption is inevitable?
- What do they feel should be done about corruption?
- What are they willing to do personally to eliminate corruption in government?
- How certain are they that they would be willing to do what they say they would do?

The list could go on and on—how people feel about corruption in government has many dimensions. It's essential to be clear about which ones are important in our inquiry; otherwise, you may measure how people feel about corruption when you really wanted to know how much they think there is, or vice versa.

Once you've determined how you're going to collect your data (for example, survey, field research) and decided on the relevant range of variation, the degree of precision needed between the extremes of variation, and the specific dimensions of the variables that interest you, you may have another choice: a mathematical–logical one. That is, you may need to decide what level of measurement to use. To discuss this point, we need to take another look at attributes and their relationship to variables.

Defining Variables and Attributes

An attribute, you'll recall, is a characteristic or quality of something. "Female" is an example. So is "old" or "student." Variables, on the other hand, are logical sets of attributes. Thus, *gender* is a variable composed of the attributes "female" and "male." What could be simpler?

The conceptualization and operationalization processes can be seen as the specification of variables and the attributes composing them. Thus, in the context of a study of unemployment, *employment status* is a variable having the attributes "employed" and "unemployed"; the list of attributes could also be expanded to include the other possibilities discussed earlier, such as "homemaker."

Every variable must have two important qualities. First, the attributes composing it should be exhaustive. For the variable to have any utility in research, we must be able to classify every observation in terms of one of the attributes composing the variable. We'll run into trouble if we conceptualize the variable *political party affiliation* in terms of the attributes "Republican" and "Democrat," because some of the people we set out to study will identify with the Green Party, the Libertarian Party, or some other organization, and some (often a large percentage) will tell us they have no party affiliation. We could make the list of attributes exhaustive by adding "other" and "no affiliation." Whatever we do, we must be able to classify every observation.

At the same time, attributes composing a variable must be mutually exclusive. That is, we must be able to classify every observation in terms of one and only one attribute. For example, we need to define "employed" and "unemployed" in such a way that nobody can be both at the same time. That means being able to classify the person who is working at a job but is also looking for work. (We might run across a fully employed mud wrestler who is looking for the glamour and excitement of being a social researcher.) In this case, we might define the attributes so that "employed" takes precedence over "unemployed," and anyone working at a job is employed regardless of whether he or she is looking for something better.

The process of conceptualizing variables is situation-dependent. What works in one situation won't necessarily work elsewhere. Malcom Williams and Kerryn Husk (2013) have examined in detail the many problems involved in measuring ethnicity. To begin, there are no absolute definitions of various ethnic groups; they are a matter of social conventions, which are understood differently by different people and which change over time. Although they focused on Cornwall County in Britain, the authors' analysis applies to the measurement of ethnicity more broadly and, indeed, applies to conceptualization in general.

Two general conclusions can be drawn. First, the conceptualization of variables depends, obviously perhaps, on the population being studied. A survey conducted in Cornwall might include the ethnic category of "Cornish," whereas you wouldn't do that in a survey of Arkansas. Second, conceptualization should be tailored to the purposes of the study. In the case of ethnicity, four or five broad ethnic categories might surface in one study, while the intentions of another might require much finer distinctions.

Levels of Measurement

All variables are composed of attributes, but as we are about to see, the attributes of a given variable can have a variety of different relationships to one another. In this section, we'll examine four levels of measurement: nominal, ordinal, interval, and ratio.

Nominal Measures

Variables whose attributes are simply different from one another are called nominal measures. Examples include *gender, religious affiliation, political party affiliation, birthplace, college major,* and *hair color.* Although the attributes composing each of these variables—as male and female compose the variable gender—are distinct from one another, they have no additional structures. **Nominal measures** merely offer names or labels for characteristics.

Imagine a group of people characterized in terms of one such variable, and physically grouped by the applicable attributes. For example, say we've asked a large gathering of people to stand together in groups according to the states in which they were born: all those born in Vermont in one group, those born in California in another, and so forth. The variable

is *birthplace*; the attributes are born in California, born in Vermont, and so on. All the people standing in a given group have at least one thing in common and differ from the people in all other groups in that same regard. Where the individual groups form, how close they are to one another, or how the groups are arranged in the room is irrelevant. All that matters is that all the members of a given group share the same state of birth and that each group has a different shared state of birth. All we can say about two people in terms of a nominal variable is that they are either the same or different.

Ordinal Measures

Variables with attributes we can logically rank-order are **ordinal measures**. The different attributes of ordinal variables represent relatively more or less of the variable. Variables of this type include *social class, conservatism, alienation, prejudice,* and *intellectual sophistication.* In addition to saying whether two people are the same or different in terms of an ordinal variable, we can also say one is "more" than the other—that is, more conservative, more religious, older, and so forth.

In the physical sciences, hardness is the most frequently cited example of an ordinal measure. We can say that one material (for example, diamond) is harder than another (say, glass) if the former can scratch the latter and not vice versa. By attempting to scratch various materials with other materials, we might eventually be able to arrange several materials in a row, ranging from the softest to the hardest. We could never say how hard a given material was in absolute terms; we could only say how hard in relative terms—which materials it is harder than and which softer than.

---

nominal measure A variable whose attributes have only the characteristics of exhaustiveness and mutual exclusiveness—in other words, a level of measurement describing a variable that has attributes that are merely different, as distinguished from ordinal, interval, or ratio measures. *Gender* is an example of a nominal measure.

ordinal measure A level of measurement describing a variable with attributes we can rank-order along some dimension. An example is *socioeconomic status*, composed of the attributes *high, medium,* and *low.*

Let's pursue the earlier example of grouping the people at a social gathering. This time, imagine that we ask all the people who have graduated from college to stand in one group, all those with only a high school diploma to stand in another group, and all those who have not graduated from high school to stand in a third group. This manner of grouping people satisfies the requirements for exhaustiveness and mutual exclusiveness discussed earlier. In addition, however, we might logically arrange the three groups in terms of the relative amount of formal education (the shared attribute) each had. We might arrange the three groups in a row, ranging from most to least formal education. This arrangement would provide a physical representation of an ordinal measure. If we knew which groups two individuals were in, we could determine that one had more, less, or the same formal education as the other.

Notice in this example that it doesn't matter how close to or far apart from one another the educational groups are. The college and high school groups might be 5 feet apart, and the less-than-high-school group 500 feet farther down the line. These actual distances don't have any meaning. The high school group, however, should be between the less-than-high-school group and the college group, or else the rank order will be incorrect.

Interval Measures

For the attributes composing some variables, the actual distance separating those attributes does have meaning. Such variables are **interval measures**. For these, the logical distance between attributes can be expressed in meaningful standard intervals.

For example, in the Fahrenheit temperature scale, the difference, or distance, between 80 degrees and 90 degrees is the same as that between 40 degrees and 50 degrees. However, 80 degrees Fahrenheit is not twice as hot as 40 degrees, because in both the Fahrenheit and Celsius scales, "zero" is arbitrary; that is, zero degrees does not really mean total lack of heat. Similarly, minus 30 degrees on either scale doesn't represent 30 degrees less than no heat. (In contrast, the Kelvin scale is based on an absolute zero, which does mean a complete lack of heat.)

About the only interval measures commonly used in social science research are constructed measures such as standardized intelligence tests that have been more or less accepted. The interval separating IQ scores of 100 and 110 may be regarded as the same as the interval between 110 and 120, by virtue of the distribution of observed scores obtained by many thousands of people who have taken the tests over the years. But it would be incorrect to infer that someone with an IQ of 150 is 50 percent more intelligent than someone with an IQ of 100. (A person who received a score of 0 on a standard IQ test could not be regarded, strictly speaking, as having no intelligence, although we might feel he or she was unsuited to be a college professor or even a college student. But perhaps a dean…?)

When comparing two people in terms of an interval variable, we can say they are different from one another (nominal), and that one is more than another (ordinal). In addition, we can say "how much" more in terms of the scores themselves.

Ratio Measures

Most of the social science variables meeting the minimum requirements for interval measures also meet the requirements for ratio measures. In **ratio measures**, the attributes composing a variable, besides having all the structural characteristics mentioned previously, are based on a true zero point. The Kelvin temperature scale is one such measure. Examples from social research include age, length of residence in a given place, number of organizations belonged to, number of times attending religious services during a particular period, number of times married, and number of Arab friends.

Returning to the illustration of methodological party games, we might ask a gathering of people to group themselves by age. All the one-year-olds would stand (or sit or lie) together, the

---

interval measure A level of measurement describing a variable whose attributes are rank-ordered and have equal distances between adjacent attributes. The *Fahrenheit temperature scale* is an example of this, because the distance between 17 and 18 is the same as that between 89 and 90.

ratio measure A level of measurement describing a variable with attributes that have all the qualities of nominal, ordinal, and interval measures and in addition are based on a "true zero" point. *Age* is an example of a ratio measure.

two-year-olds together, the three-year-olds, and so forth. The fact that members of a single group share the same age and that each different group has a different shared age satisfies the minimum requirements for a nominal measure. Arranging the several groups in a line from youngest to oldest meets the additional requirements of an ordinal measure and lets us determine whether one person is older than, younger than, or the same age as another. If we space the groups equally far apart, we satisfy the additional requirements of an interval measure and will be able to say how much older one person is than another. Finally, because one of the attributes included in age represents a true zero (babies carried by women

about to give birth), the phalanx of hapless partygoers also meets the requirements of a ratio measure, permitting us to say that one person is twice as old as another. (Remember this in case you're asked about it in a workbook assignment.) Another example of a ratio measure is income, which extends from an absolute zero to approximately infinity, if you happen to be the founder of Microsoft.

Comparing two people in terms of a ratio variable, then, allows us to determine (1) that they are different (or the same), (2) that one is more than the other, (3) how much they differ, and (4) the ratio of one to another. Figure 5-1 illustrates the four levels of measurement we've just discussed.

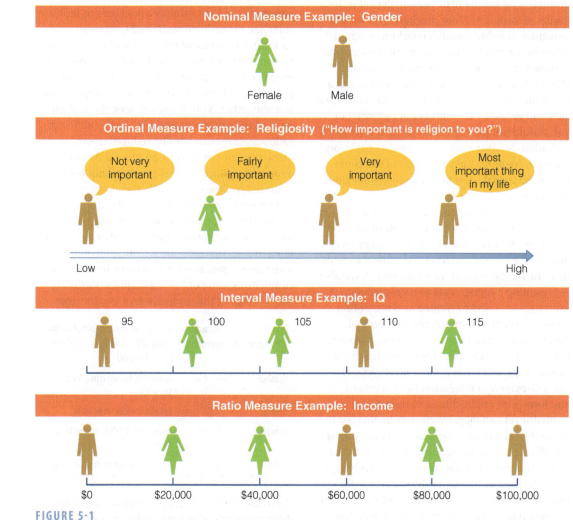

FIGURE 5-1

Levels of Measurement. Often you can choose among different levels of measurement—nominal, ordinal, interval, or ratio—carrying progressively more amounts of information.

Implications of Levels of Measurement

Because it's unlikely that you'll undertake the physical grouping of people just described (try it once, and you won't be invited to many more parties), I should draw your attention to some of the practical implications of the differences that have been distinguished. These implications appear primarily in the analysis of data (discussed in Part 4), but you need to anticipate such implications when you're structuring any research project.

Certain quantitative analysis techniques require variables that meet certain minimum levels of measurement. To the extent that the variables to be examined in a research project are limited to a particular level of measurement—say, ordinal—you should plan your analytical techniques accordingly. More precisely, you should anticipate drawing research conclusions appropriate to the levels of measurement used in your variables. For example, you might reasonably plan to determine and report the mean age of a population under study (add up all the individual ages and divide by the number of people), but you should not plan to report the mean religious affiliation, because that's a nominal variable, and the mean requires ratio-level data. (You could report the modal—the most common—religious affiliation.)

At the same time, you can treat some variables as representing different levels of measurement. Ratio measures are the highest level, descending through interval and ordinal to nominal, the lowest level of measurement. A variable representing a higher level of measurement—say, ratio—can also be treated as representing a lower level of measurement—say, ordinal. Recall, for example, that age is a ratio measure. If you wished to examine only the relationship between age and some ordinal-level variable—say, self-perceived religiosity: high, medium, and low—you might choose to treat age as an ordinal-level variable as well. You might characterize the subjects of your study as being young, middle-aged, and old, specifying what age range determines each of these groupings. Finally, age might be used as a nominal-level variable for certain research purposes. People might be grouped as being born during the Iraq War or not. Another nominal measurement, based on birth date

rather than just age, would be the grouping of people by astrological signs.

The level of measurement you'll seek, then, is determined by the analytic uses you've planned for a given variable, as you keep in mind that some variables are inherently limited to a certain level. If a variable is to be used in a variety of ways, requiring different levels of measurement, the study should be designed to achieve the highest level required. For example, if the subjects in a study are asked their exact ages, they can later be organized into ordinal or nominal groupings.

You don't necessarily need to measure variables at their highest level of measurement, however. If you're sure to have no need for ages of people at higher than the ordinal level of measurement, you may simply ask people to indicate their age range, such as 20 to 29, 30 to 39, and so forth. In a study of the wealth of corporations, rather than seek more-precise information, you may use Dun & Bradstreet ratings to rank corporations. Whenever your research purposes are not altogether clear, however, seek the highest level of measurement possible. Again, although ratio measures can later be reduced to ordinal ones, you cannot convert an ordinal measure to a ratio one. More generally, you cannot convert a lower-level measure to a higher-level one. That is a one-way street worth remembering.

The level of measurement is significant in terms of the arithmetic operations that can be applied to a variable and the statistical techniques using those operations. The accompanying table summarizes some of the implications, including ways of stating the comparison of two incomes.

Level of Measurement	Arithmetic Operations	How to Express the Fact That Jan Earns $80,000 a Year and Andy Earns $40,000
Nominal	= ≠	Jan and Andy earn *different* amounts.
Ordinal	> <	Jan earns *more* than Andy.
Interval	+ −	Jan earns *$40,000 more* than Andy.
Ratio	% ×	Jan earns *twice* as much as Andy.

Typically, a research project will tap variables at different levels of measurement. For example, William and Denise Bielby (1999) set out to

Applying Concepts in Everyday Life

On to Hollywood

So, you want to be a Hollywood screenwriter. How might you use the results of the Bielby and Bielby (1999) study to enhance your career? Say you didn't do so well and instead started a school for screenwriters.

How could the results of the study be used to plan courses? Finally, how might the results be useful to you if you were a social activist committed to fighting discrimination in the "culture industry"?

examine the world of film and television, using a nomothetic, longitudinal approach (take a moment to remind yourself what that means). In what they referred to as the "culture industry," the authors found that reputation (an ordinal variable) is the best predictor of screenwriters' future productivity. More interestingly, they found that screenwriters who were represented by "core" (or elite) agencies were far more likely not only to find jobs (a nominal variable) but also to find jobs that paid more (a ratio variable). In other words, the researchers found that an agency's reputation (ordinal) was a key independent variable for predicting a screenwriter's success. The researchers also found that being older (ratio), being female (nominal), belonging to an ethnic minority (nominal), and having more years of experience (ratio) were disadvantageous for a screenwriter. On the other hand, higher earnings from previous years (measured in ordinal categories) led to more success in the future. In the researchers' terms, "success breeds success" (Bielby and Bielby 1999: 80). See "Applying Concepts in Everyday Life: On to Hollywood," for more on the Bielby study.

Single or Multiple Indicators

With so many alternatives for operationalizing social research variables, you may find yourself worrying about making the right choices. To counter this feeling, let me add a dash of certainty and stability.

Many social research variables have fairly obvious, straightforward measures. No matter how you cut it, gender usually turns out to be a matter of male or female: a nominal-level variable that can be measured by a single observation—either through looking (well, not always) or through asking a question (usually).

In a study involving the size of families, you'll want to think about adopted and foster children as well as blended families, but it's usually pretty easy to find out how many children a family has. For most research purposes, the resident population of a country is the resident population of that country—you can find the number on the Web. A great many variables, then, have obvious single indicators. If you can get one piece of information, you have what you need.

Sometimes, however, there is no single indicator that will give you the measure of a chosen variable. As discussed earlier in this chapter, many concepts are subject to varying interpretations— each with several possible indicators. In these cases, you'll want to make several observations for a given variable. You can then combine the several pieces of information you've collected to create a composite measurement of the variable in question. Chapter 6 is devoted to ways of doing that, so here let's look at just one simple illustration.

Consider the concept "college performance." All of us have noticed that some students perform well in college courses and others do not. In studying these differences, we might ask what characteristics and experiences are related to high levels of performance (many researchers have done just that). How should we measure overall performance? Each grade in any single course is a potential indicator of college performance, but it also may not typify the student's general performance. The solution to this problem is so firmly established that it is, of course, obvious: the grade point average (GPA). To obtain a composite measure, we assign numerical scores to each letter grade, total the points earned by a given student, and divide by the number of courses taken. (If the courses vary in number of credits, we adjust the point values accordingly.)

It's often appropriate to create such composite measures in social research. We can create composite measures to describe individuals, as in the GPA example, or to describe groups such as colleges, churches, or nations.

Some Illustrations of Operationalization Choices

To bring together all the operationalization choices available to the social researcher and to show the potential in those possibilities, let's look at some of the distinct ways you might address various research problems. The alternative ways of operationalizing the variables in each case should demonstrate the opportunities that social research can present to our ingenuity and imaginations. To simplify matters, I have not attempted to describe all the research conditions that would make one alternative superior to the others, though in a given situation they would not all be equally appropriate.

1. Are women more compassionate than men?

   a. Select a group of subjects for study, with equal numbers of men and women. Present them with hypothetical situations that involve someone being in trouble. Ask them what they would do if they were confronted with that situation. What would they do, for example, if they came across a small child who was lost and crying for his or her parents? Consider any answer that involves helping or comforting the child as an indicator of compassion. See whether men or women are more likely to indicate they would be compassionate.

   b. Set up an experiment in which you pay a small child to pretend that he or she is lost. Put the child to work on a busy sidewalk and observe whether men or women are more likely to offer assistance. Also be sure to count the total number of men and women who walk by, because there may be more of one than the other. If that's the case, simply calculate the percentage of men and the percentage of women who help.

   c. Select a sample of people and do a survey in which you ask them what organizations they belong to. Calculate whether women or men are more likely to belong to those that seem to reflect compassionate feelings. To take account of men who belong to more organizations than do women in general—or vice versa—do this: For each person you study, calculate the percentage of his or her organizational memberships that reflect compassion. See if men or women have a higher average percentage.

2. Are sociology students or accounting students better informed about world affairs?

   a. Prepare a short quiz on world affairs and arrange to administer it to the students in a sociology class and in an accounting class at a comparable level. If you want to compare sociology and accounting majors, be sure to ask students what they are majoring in.

   b. Get the instructor of a course in world affairs to give you the average grades of sociology and accounting students in the course.

   c. Take a petition to sociology and accounting classes that urges that "the United Nations headquarters be moved to New York City." Keep a count of how many in each class sign the petition and how many inform you that the UN headquarters is already located in New York City.

3. Who are the most popular instructors on your campus—those in the social sciences, the natural sciences, or the humanities?

   a. If your school has formal student evaluations of instructors, review some recent results and compute the average ratings of each group.

   b. Begin visiting the introductory courses given in each group of disciplines and measure the attendance rate of each class.

   c. In December, select a group of faculty in each of the three disciplines and ask them to keep a record of the numbers of holiday greeting cards and presents they receive from admiring students. See who wins.

The point of these examples is not necessarily to suggest respectable research projects but to illustrate the many ways variables can be operationalized. "How to Do It: Measuring College Satisfaction" gives a brief overview of the preceding steps in terms of a concept mentioned at the outset of this chapter.

How to Do It

Measuring College Satisfaction

Early in this chapter, we considered "satisfaction with college" as an example of a concept we may often talk about casually. To study such a concept, however, we need to engage in the processes of conceptualization and operationalization. After I've sketched out the process, you might try your hand at expanding on my comments.

What are some of the dimensions of college satisfaction? Here are a few to get you started:

- Academic quality: faculty, courses, majors
- Physical facilities: classrooms, dorms, cafeteria, grounds
- Athletics and extracurricular activities
- Costs and availability of financial aid
- Sociability of students, faculty, staff
- Security, crime on campus

What other dimensions might be relevant to students' satisfaction or dissatisfaction with their school?

How would you measure each of these dimensions? One method would be to ask a sample of students, "How would you rate your level of satisfaction with each of the following?" giving them a list of items similar to those listed here and providing a set of categories for them to use (such as very satisfied, satisfied, dissatisfied, very dissatisfied). But suppose you didn't have the time or money to conduct a survey and were interested in comparing overall levels of satisfaction at several schools. What data about schools (the unit of analysis) might give you the answer you were interested in? Retention rates might be one general indicator of satisfaction. Can you think of others?

Notice that you can measure college quality both positively and negatively. Modern classrooms with WiFi access would count positively, whereas the number of crimes on campus would count negatively. But the latter could be used as a measure of college quality: with low crime rates counting as high quality.

Operationalization Goes On and On

Although I've discussed conceptualization and operationalization as activities that precede data collection and analysis—for example, you must design questionnaire items before you send out a questionnaire—these two processes continue throughout any research project, even if the data have been collected in a structured mass survey. As we've seen, in less-structured methods such as field research, the identification and specification of relevant concepts is inseparable from the ongoing process of observation.

Imagine, for example, that you're doing a qualitative, observational study of members of a new religious cult, and, in part, you want to identify those members who are more religious and those who are less religious. You may begin with a focus on certain kinds of ritual behavior, only to eventually discover that the members of the group place a higher premium on religious experience or steadfast beliefs.

The open-endedness of conceptualization and operationalization is perhaps more obvious in qualitative than in quantitative research, since changes can be made at any point during data collection and analysis. In quantitative methods such as survey research or experiments, you'll be required to commit yourself to particular measurement structures. Once a questionnaire has been printed and administered, for example, there is no longer an easy opportunity for changing it, even when the unfolding of the research suggests improvements. Even in the case of a survey questionnaire, however, you may have some flexibility in how you measure variables during the analysis phase, as we'll see in the following chapter.

As I mentioned, however, the qualitative researcher has greater flexibility in this regard. Things you notice during in-depth interviews, for example, may suggest a set of questions different from those you initially planned, allowing you to pursue unanticipated avenues. Then later, as you review and organize your notes for analysis, you may again see unanticipated patterns and redirect your analysis.

Regardless of whether you're using qualitative or quantitative methods, you should always be open to reexamining your concepts and definitions. The ultimate purpose of social research is to clarify the nature of social life. The validity and utility of what you learn in this regard doesn't depend on when you first figured out how to look at things any more than it matters whether you got the idea from a learned textbook, a dream, or your brother-in-law.

Criteria of Measurement Quality

This chapter has come some distance. It began with the bold assertion that social scientists can measure anything that exists. Then we discovered that most of the things we might want to measure and study don't really exist. Next we learned that it's possible to measure them anyway. Now we'll discuss some of the yardsticks against which we judge our relative success or failure in measuring things—even things that don't exist.

Precision and Accuracy

To begin, measurements can be made with varying degrees of precision. As we saw in the discussion of operationalization, precision concerns the fineness of distinctions made between the attributes that compose a variable. The description of a woman as "43 years old" is more precise than "in her forties." Saying a street-corner gang was formed "in the summer of 1996" is more precise than saying "during the 1990s."

As a general rule, precise measurements are superior to imprecise ones, as common sense suggests. There are no conditions under which imprecise measurements are intrinsically superior to precise ones. Even so, exact precision is not always necessary or desirable. If knowing that a woman is in her forties satisfies your research requirements, then any additional effort invested in learning her precise age is wasted.

> **reliability** That quality of measurement methods that suggests that the same data would have been collected each time in repeated observations of the same phenomenon. In the context of a survey, we would expect that the question "Did you attend religious services last week?" would have higher reliability than the question "About how many times have you attended religious services in your life?" This is not to be confused with *validity*.

The operationalization of concepts, then, must be guided partly by an understanding of the degree of precision required. If your needs are not clear, *be more precise rather than less.*

Don't confuse precision with accuracy, however. Describing someone as "born in New England" is less precise than "born in Stowe, Vermont"—but suppose the person in question was actually born in Boston. The less precise description, in this instance, is more accurate, a better reflection of the real world.

Precision and accuracy are obviously important qualities in research measurement, and they probably need no further explanation. When social researchers construct and evaluate measurements, however, they pay special attention to two technical considerations: reliability and validity.

Reliability

In the abstract, **reliability** is a matter of whether a particular technique, applied repeatedly to the same object, yields the same result each time. Let's say you want to know how much I weigh. (No, I don't know why.) As one technique, say you ask two different people to estimate my weight. If the first person estimates 150 pounds and the other estimates 300, we have to conclude that the technique of having people estimate my weight isn't very reliable.

Suppose, as an alternative, that you use a bathroom scale as your measurement technique. I step on the scale twice, and you note the same result each time. The scale has presumably reported the same weight both times, indicating that the scale provides a more reliable technique for measuring a person's weight than does asking people to estimate it.

Reliability, however, does not ensure accuracy any more than does precision. Suppose I've set my bathroom scale to shave five pounds off my weight just to make me feel better. Although you would (reliably) report the same weight for me each time, you would always be wrong. This new element, called bias, is discussed in Chapter 7. For now, just be warned that reliability does not ensure accuracy.

Let's suppose we're interested in studying morale among factory workers in two different kinds of factories. In one set of factories, workers have specialized jobs, reflecting an extreme division of labor. Each worker contributes a tiny part to the overall process performed on a long assembly line. In the other set of factories, each worker performs many tasks, and small teams of workers complete the whole process.

How should we measure morale? Following one strategy, we could observe the workers in each factory, noticing such things as whether they joke with one another, whether they smile and laugh a lot, and so forth. We could ask them how they like their work and even ask them whether they think they would prefer their current arrangement or the other one being studied. By comparing what we observed in the different factories, we might reach a conclusion about which assembly process produces the higher morale. Notice that I've just described a qualitative measurement procedure.

Now let's look at some reliability problems inherent in this method. First, how you and I are feeling when we do the observing will likely color what we see. We may misinterpret what we observe. We may see workers kidding each other but think they're having an argument. We may catch them on an off day. If we were to observe the same group of workers several days in a row, we might arrive at different evaluations on each day. If several observers evaluated the same behavior, on the other hand, they similarly might arrive at different conclusions about the workers' morale.

Here's another, quantitative approach to assessing morale. Suppose we check the company records to see how many grievances have been filed with the union during some fixed period. Presumably, this would be an indicator of morale: the more grievances, the lower the morale. This measurement strategy would appear to be more reliable: Counting up the grievances over and over, we should keep arriving at the same number.

If you're thinking that the number of grievances doesn't necessarily measure morale, you're worrying about validity, not reliability. We'll discuss validity in a moment. The point for now is that the last method is more like my bathroom scale—it gives consistent results.

In social research, reliability problems crop up in many forms. Reliability is a concern every time a single observer is the source of data, because we have no certain guard against the impact of that observer's subjectivity. We can't tell for sure how much of what's reported originated in the situation observed and how much came from the observer.

Subjectivity is a problem not only with single observers, however. Survey researchers have known for a long time that different interviewers, because of their own attitudes and demeanors, get different answers from respondents. Or, if we were to conduct a study of newspapers' editorial positions on some public issue, we might create a team of coders to take on the job of reading hundreds of editorials and classifying them in terms of their position on the issue. Unfortunately, different coders will code the same editorial differently. Or we might want to classify a few hundred specific occupations in terms of some standard coding scheme, say a set of categories created by the Department of Labor or by the Census Bureau. You and I would not place all those occupations in the same categories.

Each of these examples illustrates problems of reliability. Similar problems arise whenever we ask people to give us information about themselves. Sometimes we ask questions that people don't know the answers to: How many times, if any, have you been to religious services this year? Sometimes we ask people about things they consider totally irrelevant: Are you satisfied with China's current relationship with Albania? In such cases, people will answer differently at different times because they're making up answers as they go. Sometimes we explore issues so complicated that a person who had a clear opinion in the matter might arrive at a different interpretation of the question when asked a second time.

So how do you create reliable measures? If your research design calls for asking people for information, you can be careful to ask only about things the respondents are likely to know the answer to. Ask about things relevant to them, and be clear in what you're asking. Of course, these techniques don't solve every possible reliability problem. Fortunately, social researchers have developed several techniques for cross-checking the reliability of the measures they devise.

Test–Retest Method

Sometimes it's appropriate to make the same measurement more than once, a technique called the *test–retest method*. If you don't expect the information being sought to change, then you should expect the same response each time. If answers vary, the measurement method may, to the extent of that variation, be unreliable. Here's an illustration.

In their classic research on Health Hazard Appraisal (HHA), a part of preventive medicine, Jeffrey Sacks, W. Mark Krushat, and Jeffrey Newman (1980) wanted to determine the risks associated with various background and life-style factors, making it possible for physicians to counsel their patients appropriately. By knowing patients' life situations, physicians could advise them on their potential for survival and on how to improve that potential. This purpose, of course, depended heavily on the accuracy of the information gathered about each subject in the study.

To test the reliability of their information, Sacks and his colleagues had all 207 subjects complete a baseline questionnaire that asked about their characteristics and behavior. Three months later, a follow-up questionnaire asked the same subjects for the same information, and the results of the two surveys were compared. Overall, only 15 percent of the subjects reported the same information in both studies.

Sacks and his colleagues reported the following:

> Almost 10 percent of subjects reported a different height at follow-up examination. Parental age was changed by over one in three subjects. One parent reportedly aged 20 chronologic years in three months. One in five ex-smokers and ex-drinkers have apparent difficulty in reliably recalling their previous consumption pattern.
>
> (1980: 730)

Some subjects had erased all traces of previously reported heart murmurs, diabetes, emphysema, arrest records, and thoughts of suicide. One subject's mother, deceased in the first questionnaire, was apparently alive and well in time for the second. One subject had one ovary missing in the first study but present in the second. In another case, an ovary present in the first study was missing (and had been for ten years!) in the second study. One subject was reportedly 55 years old in the first study and 50 years old three months later. (You have to wonder whether the physician-counselors could ever have the impact on their patients that their patients' memories had.) Thus, test–retest revealed that this data-collection method was not especially reliable.

Split-Half Method

As a general rule, it's always good to make more than one measurement of any subtle or complex social concept, such as prejudice, alienation, or social class. This procedure lays the groundwork for another check on reliability. Let's say you've created a questionnaire that contains ten items you believe measure prejudice against women. Using the split-half technique, you would randomly assign those ten items to two sets of five. As we saw in the discussion of interchangeability of indicators, each set should provide a good measure of prejudice against women, and the two sets should classify respondents the same way. If the two sets of items classify people differently, you most likely have a problem of reliability in your measure of the variable.

Using Established Measures

Another way to help ensure reliability in getting information from people is to use measures that have proved their reliability in previous research. If you want to measure anomia, for example, you might want to follow Leo Srole's lead.

The heavy use of measures, though, does not guarantee their reliability. For example, the Scholastic Assessment Test (SAT) has, for decades, been an established standard for judging whether students are prepared for college study. However, the test has been revised repeatedly over the years. The most recent, 2016, revision aimed at better reflecting what students learn in high school.

Reliability of Research Workers

As we've seen, measurement unreliability can also be generated by research workers: interviewers and coders, for example. There are several ways to check on reliability in such cases. To guard against interviewer unreliability, it's common practice in surveys to have a supervisor call

Applying Concepts in Everyday Life

Pregnant Chads and Voter Intentions

Replication in measurement is exactly the issue that was raised in Florida during the 2000 presidential election in which George W. Bush was declared the winner over Al Gore. Specifically, given the thousands of ballots rejected by the vote-counting machines, many people questioned the reliability of this method of measuring votes. In many cases, holes in ballots were not punched completely, leaving small pieces of card stock—"chads"—still attached to and/or partially covering the holes. Had the election been a survey of voting intentions, the researchers would have reviewed the ballots rejected by the machine and sought to make judgments regarding those intentions. Notice that decisions

about hanging and pregnant "chads" would have concerned measurement procedures. Much of the debate hinged on how much each type of chad reflected voter intent: What does a bump in a chad "really" mean? Without agreement on what really constituted a vote, researchers would have simply scored ballots in terms of which candidate was apparently selected and why—that is, the basis on which the researcher's decision was made (for example, "pregnant chad"). Of course, there would first have to be clear operational definitions of what "hanging," "pregnant," and other sorts of chads were. It would certainly have been possible to determine the results in terms of the standards used—that is, how many of each type of vote was counted for each candidate.

a subsample of the respondents on the telephone to verify selected pieces of information.

Replication works in other situations also. If you're worried that newspaper editorials or occupations may not be classified reliably, you could have each independently coded by several coders. Those cases that are classified inconsistently can then be evaluated more carefully and resolved.

Finally, clarity, specificity, training, and practice can prevent a great deal of unreliability and grief. If you and I spent some time reaching a clear agreement on how to evaluate editorial positions on an issue—discussing various positions and reading through several together—we could probably do a good job of classifying them in the same way independently. See "Applying Concepts in Everyday Life: Pregnant Chads and Voter Intentions," for more on reliability.

The reliability of measurements is a fundamental issue in social research, and we'll return to it more than once in the chapters ahead. For now, however, let's recall that even total reliability doesn't ensure that our procedures measure what we think they measure. Now let's plunge into the question of validity.

Validity

In conventional usage, **validity** refers to the extent to which an empirical measure adequately reflects the real meaning of the concept under consideration. A measure of social class should measure social class, not political orientations. A measure of political orientations should measure

political orientations, not sexual permissiveness. *Validity* means that we are actually measuring what we say we are measuring.

Whoops! I've already committed us to the view that concepts don't have real meanings. How can we ever say whether a particular measure adequately reflects the concept's meaning, then? Ultimately, of course, we can't. At the same time, as we've already seen, all of social life, including social research, operates on agreements about the terms we use and the concepts they represent. There are several criteria of success in making measurements that are appropriate to these agreed-on meanings of concepts.

First, there's something called **face validity**. Particular empirical measures may

---

validity A term describing a measure that accurately reflects the concept it is intended to measure. For example, your IQ would seem to be a more valid measure of your intelligence than would the number of hours you spend in the library. Though the ultimate validity of a measure can never be proved, we may agree to its relative validity on the basis of face validity, criterion-related validity, content validity, construct validity, internal validation, and external validation. This must not be confused with *reliability*.

face validity The quality of an indicator that makes it seem to be a reasonable measure of some variable. That the frequency of attendance at religious services is some indication of a person's religiosity seems to make sense without a lot of explanation. It has face validity.

or may not jibe with our common agreements and our individual mental images concerning a particular concept. For example, you and I might quarrel about whether counting the number of grievances filed with the union is an adequate measure of worker morale. Still, we'd surely agree that the number of grievances has something to do with morale. That is, the measure is valid "on its face," whether or not it's adequate. If I were to suggest that we measure morale by finding out how many books the workers took out of the library during their off-duty hours, you'd undoubtedly raise a more serious objection: That measure wouldn't have much face validity.

Second, I've already pointed to many of the more formally established agreements that define some concepts. The Census Bureau, for example, has created operational definitions of such concepts as family, household, and employment status that seem to have a workable validity in most studies using these concepts.

Three additional types of validity also specify particular ways of testing the validity of measures. The first, **criterion-related validity**, sometimes called *predictive validity*, is based on some external criterion. For example, the validity of College Board exams is shown in their ability to predict students' success in college. The validity of a written drivers' test is determined, in this sense, by the relationship between the scores people get on the test and their subsequent driving records. In these examples, college success and driving ability are the criteria.

To test your understanding of criterion-related validity, see whether you can think of behaviors that might be used to validate each of the following attitudes:

- Is very religious
- Supports equality of men and women

- Supports far-right militia groups
- Is concerned about the environment

Some possible validators would be, respectively, attends religious services, votes for women candidates, belongs to the NRA, and belongs to the Sierra Club.

Sometimes it's difficult to find behavioral criteria that can be taken to validate measures as directly as in such examples. In those instances, however, we can often approximate such criteria by applying a different test. We can consider how the variable in question ought, theoretically, to relate to other variables. **Construct validity** is based on the logical relationships among variables.

Suppose, for example, that you want to study the sources and consequences of marital satisfaction. As part of your research, you develop a measure of marital satisfaction, and you want to assess its validity.

In addition to developing your measure, you'll have developed certain theoretical expectations about the way the variable *marital satisfaction* relates to other variables. For example, you might reasonably conclude that satisfied husbands and wives will be less likely than dissatisfied ones to cheat on their spouses. If your measure relates to marital fidelity in the expected fashion, that constitutes evidence of your measure's construct validity. If satisfied marriage partners are as likely to cheat on their spouses as are the dissatisfied ones, however, that would challenge the validity of your measure.

Tests of construct validity, then, can offer a weight of evidence that your measure either does or does not reflect the quality you want it to measure, but this evidence is not definitive proof. Although I've suggested that tests of construct validity are less compelling than tests of criterion validity, there is room for disagreement about which kind of test a particular comparison variable (driving record, marital fidelity) represents in a given situation. It's less important to distinguish the two types of validity tests than to understand the logic of validation that they have in common: If we've succeeded in measuring some variable, then our measures should relate in some logical way to other measures.

criterion-related validity The degree to which a measure relates to some external criterion. Also called *predictive validity.*

construct validity The degree to which a measure relates to other variables as expected within a system of theoretical relationships.

Issues and Insights

Validity and Social Desirability

A particular challenge in measurement occurs when the attitudes or behaviors asked about are generally considered socially undesirable, whether it involves an actual crime or something like not voting. (Post-election surveys show a higher percentage reporting that they voted than actually turned up at the polls.)

One technique has been to downplay the misreporting involved, but this has unexpected consequences. For example, the format of asking "Do you feel that the president should do X or do you feel that the president should do Y?" may seem a little too blunt, and researchers have sought to soften it by prefacing the question with: "Some people feel that the president should do X while others feel the president should do Y." Experiments by Yeager and Krosnick (2011, 2012) utilizing both face-to-face and Internet surveys suggest this is not a useful variation.

It may affect respondents' assumptions about how others feel, but it does not seem to improve reports of respondents' own opinions. In fact, where independent checks on attitudes and behaviors were possible, the some/other format *reduced* the validity of reports. Adolescents, for example, tended to report more deviant behavior than they had actually done. As a bottom line, the "softened" format requires more words (and time) and makes the questions more complicated without adding to the validity of responses.

Sources: David Scott Yeager and Jon A. Krosnick. 2012. "Does Mentioning 'Some People' and 'Other People' in an Opinion Question Improve Measurement Quality?" *Public Opinion Quarterly* 76 (1): 131–41; Yeager and Krosnick. (2011). "Does Mentioning 'Some People' and 'Other People' in a Survey Question Increase the Accuracy of Adolescents' Self-Reports?" *Developmental Psychology* 47 (6): 1674–9. doi:10.1037/a0025440.

Finally, **content validity** refers to how much a measure covers the range of meanings included within a concept. For example, a test of mathematical ability cannot be limited to addition but also needs to cover subtraction, multiplication, division, and so forth. Or, if we're measuring prejudice, do our measurements reflect all types of prejudice, including prejudice against racial and ethnic groups, religious minorities, women, the elderly, and so on?

"Issues and Insights: Validity and Social Desirability" examines the special challenges involved in asking people to report deviant attitudes or behaviors.

Figure 5-2 illustrates the difference between validity and reliability. If you think of measurement as analogous to shooting repeatedly at the bull's-eye on a target, you'll see that reliability looks like a "tight pattern," regardless of where the shots hit, because reliability is a function of consistency. Validity, on the other hand, is a function of shots being arranged around the bull's-eye. The failure of reliability in the figure is randomly distributed around the target; the failure of validity is systematically off the mark. Notice that neither an unreliable nor an invalid measure is likely to be useful.

Who Decides What's Valid?

Our discussion of validity began with a reminder that we depend on agreements to determine what's real, and we've just seen some of the ways social scientists can agree among themselves that they have made valid measurements. There is yet another way of looking at validity.

Social researchers sometimes criticize themselves and one another for implicitly assuming that they are somewhat superior to those they study. For example, researchers often seek to uncover motivations that the social actors themselves are unaware of. You think you bought that new Burpo-Blasto because of its high performance and good looks, but we know you're really trying to achieve a higher social status.

This implicit sense of superiority would fit comfortably with a totally positivistic approach (the biologist feels superior to the frog on the lab table), but it clashes with the more humanistic and typically qualitative approach taken by many social scientists. We'll explore this issue more deeply in Chapter 10.

In seeking to understand the way ordinary people make sense of their worlds, ethnomethodologists have urged all social scientists to pay

---

content validity The degree to which a measure covers the range of meanings included within a concept.

Reliable but not valid

Valid but not reliable

Valid *and* reliable

FIGURE 5-2

An Analogy to Validity and Reliability. A good measurement technique should be both valid (measuring what it is intended to measure) and reliable (yielding a given measurement dependably).

more respect to these natural social processes of conceptualization and shared meaning. At the very least, behavior that may seem irrational from the scientist's paradigm may make logical sense when viewed from the actor's point of view.

Clifford Geertz (1973) appropriates the term *thick description* in reference to the goal of understanding, as deeply as possible, the meanings that elements of a culture have for those who live within that culture. He recognizes that the outside observer will never grasp those meanings fully, however, and warns that "cultural analysis is intrinsically incomplete." He then elaborates:

> There are a number of ways to escape this—turning culture into folklore and collecting it, turning it into traits and counting it, turning it into institutions and classifying it, turning it into structures and toying with it. But they are escapes. The fact is that to commit oneself to a semiotic concept of culture and an interpretive approach to the study of it is to commit oneself to a view of ethnographic assertion as, to borrow W. B. Gallie's by now famous phrase, "essentially contestable." Anthropology, or at least interpretive anthropology, is a science whose progress is marked less by a perfection of consensus than by a refinement of debate. What gets better is the precision with which we vex each other.

> (1973: 29)

Ultimately, social researchers should look to both colleagues and subjects as sources of agreement on the most useful meanings and measurements of the concepts they study. Sometimes one source will be more useful, sometimes the other. But neither should be dismissed.

Tension between Reliability and Validity

Clearly, we want our measures to be both reliable and valid. Often, however, a tension arises between the criteria of reliability and validity, forcing a trade-off between the two.

Recall the example of measuring morale in different factories. The strategy of immersing yourself in the day-to-day routine of the assembly line, observing what goes on, and talking to the workers would seem to provide a more valid measure of morale than would counting grievances. It just seems obvious that we'd get a clearer sense of whether the morale was high or low using this first method.

As I pointed out earlier, however, the counting strategy would be more reliable. This situation reflects a more general strain in research measurement. Most of the really interesting concepts we want to study have many subtle nuances, and it's hard to specify precisely what we mean by them. Researchers sometimes speak of such concepts as having a "richness of meaning." Although scores of books and articles have been written on anomie/anomia, for example, they still haven't exhausted its meaning.

Very often, then, specifying reliable operational definitions and measurements seems to rob concepts of their richness of meaning. Positive morale is much more than a lack of grievances filed with the union; anomia is much more than what is measured by the five items created by Leo Srole. Yet, the more variation and richness we allow for a concept, the more potential

What do you think?...Revisited

Can social scientists measure religiosity and determine its causes? As you've seen, making descriptive statements about religiosity is harder than making explanatory ones. Any assertion about who is or is not religious depends directly on the definitions used. By one definition, nearly all of the population would be deemed religious; by another definition, few would be so designated.

As the discussion of the interchangeability of indicators suggested, however, we can be more confident and definitive in speaking about the causes of religiosity. For example, it's often reported that U.S. women are more religious than U.S. men. This assertion is based on the observation that women are more religious than men on virtually all indicators of religiosity: attendance at religious services, prayer, beliefs, and so forth. So, even if we disagree on which of these indicators is the best or truest measure of what we mean by the term *religiosity*, women would be more religious than men regardless of the indicator chosen.

we create for disagreement on how it applies to a particular situation, thus reducing reliability.

To some extent, this dilemma explains the persistence of two quite different approaches to social research: quantitative, nomothetic, structured techniques such as surveys and experiments on the one hand and qualitative, idiographic methods such as field research and historical studies on the other. In the simplest generalization, the former methods tend to be more reliable, the latter more valid.

By being forewarned, you'll be effectively forearmed against this persistent and inevitable dilemma. If there is no clear agreement on how to measure a concept, measure it several different ways. If the concept has several dimensions, measure each. Above all, know that the concept does not have any meaning other than what you and I give it. The only justification for giving any concept a particular meaning is utility. Measure concepts in ways that help us understand the world around us.

The Ethics of Measurement

Measurement decisions can sometimes be judged by ethical standards. We've seen that most of the concepts of interest to social researchers are open to varied meanings. Suppose, for example, that you're interested in sampling public opinion on the abortion issue in the United States. Notice the difference it would make if you conceptualized one side of the debate as "pro-choice" or as "pro-abortion." If your personal bias made you want to minimize support for the former position, you might be tempted to frame the concept and the measurements based on it in terms of people being "pro-abortion," thereby eliminating all those who were not especially fond of abortion per se but felt a woman should have the right to make that choice for herself. To pursue this strategy, however, would violate accepted research ethics.

Or as another example, consider these questions asked in a mailed survey from the National Rifle Association. (2017 National Gun Owner's Action Survey)

> *6. Do you oppose any United Nations treaty that strips the U.S. of its sovereignty and gives U.N. diplomats the power to regulate every rifle, pistol and shotgun you own?*
>
> ❑*Yes* ❑*No* ❑*No opinion*

> *8. Radical billionaires like Michael Bloomberg are spending tens of millions of dollars to push gun bans, ammo bans, and gun owner registration at the state and local levels. Do you support NRA's efforts to fight back and defend your rights in your state?*
>
> ❑*Yes* ❑*No* ❑*No opinion*

MAIN POINTS

Introduction

- The interrelated processes of conceptualization, operationalization, and measurement allow researchers to move from a general idea about what they want to study to effective and well-defined measurements in the real world.

Measuring Anything That Exists

- Conceptions are mental images we use as summary devices for bringing together observations and experiences that seem to have something in common. We use terms or labels to reference these conceptions.

- Concepts are constructs; they represent the agreed-on meanings we assign to terms. Our concepts don't exist in the real world, so they can't be measured directly, but we can measure the things that our concepts summarize.

Conceptualization

- Conceptualization is the process of specifying observations and measurements that give concepts definite meaning for the purposes of a research study.

- Conceptualization includes specifying the indicators of a concept and describing its dimensions. Operational definitions specify how variables relevant to a concept will be measured.

Definitions in Descriptive and Explanatory Studies

- Precise definitions are even more important in descriptive than in explanatory studies. The degree of precision needed varies with the type and purpose of a study.

Operationalization Choices

- Operationalization is an extension of conceptualization that specifies the exact procedures that will be used to measure the attributes of variables.

- Operationalization involves a series of interrelated choices: specifying the range of variation that is appropriate for the purposes of a study, determining how precisely to measure variables, accounting for relevant dimensions of variables, clearly defining the attributes of variables and their relationships, and deciding on an appropriate level of measurement.

- Researchers must choose from four types of measures that capture increasing amounts of information: nominal, ordinal, interval, and ratio. The most appropriate level depends on the purpose of the measurement.

- Sometimes a variable can be measured by a single indicator (e.g., gender) while others (e.g., prejudice) may need multiple indicators.

- A given variable can sometimes be measured at different levels. When in doubt, researchers should use the highest level of measurement appropriate to that variable so that they can capture the greatest amount of information.

- Operationalization begins in the design phase of a study and continues through all phases of the research project, including the analysis of data.

Criteria of Measurement Quality

- Criteria of the quality of measures include precision, accuracy, reliability, and validity.

- Whereas *reliability* means getting consistent results from the same measure, *validity* refers to getting results that accurately reflect the concept being measured.

- Researchers can test or improve the reliability of measures through the test–retest method, the split-half method, the use of established measures, and the examination of work performed by research workers.

- The yardsticks for assessing a measure's validity include face validity, criterion-related validity, construct validity, and content validity.

- Creating specific, reliable measures often seems to diminish the richness of meaning our general concepts have. This problem is inevitable. The best solution is to use several different measures, tapping the various aspects of a concept.

The Ethics of Measurement

- Conceptualization and measurement must not be guided by bias or preferences for particular research outcomes.

KEY TERMS

cognitive interviewing	indicator
concept	interval measure
conceptualization	nominal measure
construct validity	ordinal measure
content validity	ratio measure
criterion-related validity	reliability
dimension	validity
face validity	

PROPOSING SOCIAL RESEARCH: MEASUREMENT

This chapter has taken us deeper into the matter of measurement. In previous exercises, you've identified the concepts and variables you want to address in your research project. Now you'll need to get more specific in terms of conceptualization and operationalization. You'll want to conclude this portion of the proposal process with a description of the manner in which you'll make distinctions in relation to your variables. If you want to compare liberals and conservatives, for example, precisely how will you identify subjects' political orientations?

The ease or difficulty of this exercise will vary with the type of data collection you are planning. Typically, quantitative studies, such as surveys, are easier in this regard, because you can report the questionnaire items you'll use for measurements. Your task may be harder in qualitative research such as participation—you may need to work "in the trenches" before you can really understand what are the most helpful measurement methods for your study. In qualitative research, there is more opportunity for you to modify the ways variables are measured as the study unfolds, taking advantage of insights gained as you go. However, the basic purpose here is just as important as it is in quantitative research. You need to present clear ideas about how you'll at least begin to measure what it is you want to measure.

Quality criteria such as precision, accuracy, reliability, and validity are important in all kinds of social research projects. So, keep these criteria in mind as you write this part of your proposal.

REVIEW QUESTIONS

1. Pick a social science concept such as liberalism or alienation, then specify that concept so that it could be studied in a research project. How would you specify the indicators you would use, as well as the dimensions you wish to include in and exclude from your conceptualization?

2. If you wanted to compare two societies on anomie, or normlessness, what indicators might you look at? For example, which statistical indicators might you examine? Which nonstatistical, qualitative indicators?

3. What level of measurement—nominal, ordinal, interval, or ratio—describes each of the following variables?

   a. Race (white, African American, Asian, and so on)

   b. Order of finish in a race (first, second, third, and so on)

   c. Number of children in families

   d. Populations of nations

   e. Attitudes toward nuclear energy (strongly approve, approve, disapprove, strongly disapprove)

   f. Region of birth (Northeast, Midwest, and so on)

   g. Political orientation (very liberal, somewhat liberal, somewhat conservative, very conservative)

4. Suppose you've been contracted by Cengage to interview a group of students to evaluate their level of satisfaction with this textbook. How would you measure satisfaction in this case?

CHAPTER 6

Index, Scales, and Typologies

CHAPTER OVERVIEW

Researchers often need to employ multiple indicators to measure a variable adequately and validly. Indexes, scales, and typologies are useful composite measures made up of several indicators of variables.

Picture Partners/Alamy Stock Photo

Learning Objectives

After studying this chapter, you will be able to...

- Distinguish between indexes and scales as they are used in social research.
- Learn the steps involved in constructing and validating an index.

- Name and describe several of the scales commonly used in social research.
- Differentiate typologies from scales and indexes, giving examples of typologies.

Introduction

As we saw in Chapter 5, many social science concepts have complex and varied meanings. Making measurements that capture such concepts can be a challenge. Recall our discussion of content validity, which concerns whether we've captured all the different dimensions of a concept.

To achieve broad coverage of the various dimensions of a concept, we usually need to make multiple observations pertaining to it. Thus, for example, Bruce Berg (1989: 21) advises in-depth interviewers to prepare "essential questions," which are "geared toward eliciting specific, desired information." In addition, the researcher should prepare extra questions: "questions roughly equivalent to certain essential ones, but worded slightly differently."

Multiple indicators are used with quantitative data as well. Though you can sometimes construct a single questionnaire item that captures the variable of interest—"Gender: ❑—Male ❑ Female" is a simple example—other variables are less straightforward and may require you to use several questionnaire items to measure them adequately.

Quantitative data analysts have developed specific techniques for combining indicators into a single measure. This chapter discusses the construction of two types of composite measures of variables—indexes and scales. Although scales and indexes can be used in any form of social research, they are most common in survey research and other quantitative methods. A short section at the end of the chapter considers typologies, which are relevant to both qualitative and quantitative research.

Composite measures are frequently used in quantitative research, for several reasons. First,

social scientists often wish to study variables that have no clear and unambiguous single indicators. Single indicators do suffice for some variables, such as *age*. We can determine a survey respondent's age simply by asking, "How old are you?" Similarly, we can determine a newspaper's circulation merely by looking at the figure the newspaper reports. In the case of complex concepts, however, researchers can seldom develop single indicators before they actually do the research. This is especially true with regard to attitudes and orientations. Rarely can a survey researcher, for example, devise single questionnaire items that adequately reflect respondents' degrees of prejudice, religiosity, political orientations, alienation, and the like. More likely, the researcher will devise several items, each of which provides some indication of the variables. Taken individually, each of these items is likely to prove invalid or unreliable for many respondents. A composite measure, however, can overcome this problem.

Second, researchers may wish to employ a rather refined ordinal measure of a particular variable—*alienation*, say—arranging cases in several ordinal categories from very low to very high, for example. In this case, a single data item might not have enough categories to provide the desired range of variation. However, an index or a scale formed from several items can provide the needed range.

Finally, indexes and scales are efficient devices for data analysis. If considering a single data item gives us only a rough indication of a given variable, considering several data items can give us a more comprehensive and a more accurate indication. For example, a single newspaper editorial may provide some indication of the political orientation of that newspaper.

What do you think?

Earl Babbie

Often, data analysis aims at reducing a mass of observations to a more manageable form. Our use of concepts to stand for many similar observations is one example. The trick is to have the reduction represent the original observations well enough to be accurate and useful.

Sometimes this sort of reduction can be accomplished in the analysis of quantitative data. You could, for example, ask people to answer five different questions, reduce each person's answers to a single number, and then use that number to reproduce that person's answers. So, if you told me that you had assigned someone a score of 3, I would be able to tell you how he or she answered each of the original five questions.

How in the world could such a little bit of information communicate so much?

See the *What do you think? … Revisited* box toward the end of the chapter.

Examining several editorials would probably give us a better assessment, but the manipulation of several data items simultaneously could be very complicated. Indexes and scales (especially scales) are efficient data-reduction devices: They allow us to summarize several indicators in a single numerical score, while sometimes nearly maintaining the specific details of all the individual indicators.

Indexes versus Scales

The terms *index* and *scale* are commonly used imprecisely and interchangeably in social research literature. These two types of measures do have some characteristics in common, but in this book we'll distinguish between them. However, you should be warned of a growing tendency in the literature to use the term *scale* to refer to both indexes and scales.

First, let's consider what they have in common. Both scales and indexes are ordinal measures of variables. Both rank-order the units of analysis in terms of specific variables such as *religiosity, alienation, socioeconomic status, prejudice,* or *intellectual sophistication.* A person's score on either a scale or an index of religiosity, for

> **index** A type of composite measure that summarizes and rank-orders several specific observations and represents some more-general dimension.
>
> **scale** A type of composite measure composed of several items that have a logical or empirical structure among them. Examples of scales include the Bogardus social distance, Guttman, Likert, and Thurstone scales.

example, indicates his or her relative religiosity vis-à-vis other people.

Further, both scales and indexes are composite measures of variables: measurements based on more than one data item. Thus, a survey respondent's score on an index or scale of religiosity is determined by the responses given to several questionnaire items, each of which provides some indication of religiosity. Similarly, a person's IQ score is based on answers to a large number of test questions. The political orientation of a newspaper might be represented by an index or scale score reflecting the newspaper's editorial policy on various political issues.

Despite these shared characteristics, distinguishing between indexes and scales is useful. In this book we'll do so through the manner in which scores are assigned. We construct an **index** simply by accumulating scores assigned to individual indicators. We might measure prejudice, for example, by counting the number of prejudiced statements each respondent agreed with. We construct a **scale,** however, by assigning scores to patterns of responses, recognizing that some items reflect a relatively weak degree of the variable, whereas others reflect something stronger. For example, agreeing that "Women are different from men" is, at best, weak evidence of sexism as compared with agreeing that "Women should not be allowed to vote." A scale takes advantage of differences in intensity among the attributes of the same variable to identify distinct patterns of response.

Let's consider this simple example of sexism a bit further. Imagine asking people to agree or disagree with the two statements just presented.

Index-Construction Logic

Here are several types of political actions people may have taken. By and large, the different actions represent similar *degrees* of political activism.

To create an *index* of overall political activism, we might give people 1 point for each of the actions they've taken.

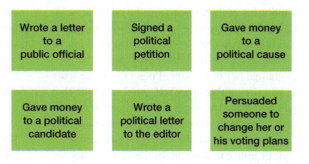

Scale-Construction Logic

Here are some political actions that represent very different degrees of activism: for example, running for office represents a higher degree of activism than simply voting does. It seems likely, moreover, that anyone who has taken one of the more demanding actions would have taken all the easier ones as well.

To construct a *scale* of political activism, we might score people according to which of the following "ideal" patterns comes closest to describing them.

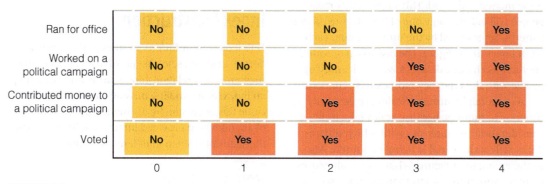

FIGURE 6-1

Indexes versus Scales. Both indexes and scales seek to measure variables such as *political activism*. Whereas indexes count the number of indicators of the variable, scales take account of the differing intensities of those indicators.

Some might agree with both, some might disagree with both. But suppose I told you someone agreed with one and disagreed with the other: Could you guess which statement they agreed with and which they did not? I would guess the person in question agreed that women were different but disagreed that they should be prohibited from voting. On the other hand, I doubt that anyone would want to prohibit women from voting and assert that there is no difference between men and women. That would make no sense.

Now consider this. The two responses we wanted from each person would technically yield four response patterns: agree/agree, agree/disagree, disagree/agree, and disagree/disagree. We've just seen, however, that only three of the four patterns make any sense or are likely to occur. Where indexes score people on the basis

of their responses, scales score people on the basis of response patterns: We determine what the logical response patterns are and score people in terms of the pattern their responses most closely resemble.

Figure 6-1 illustrates the difference between indexes and scales. Let's assume we want to develop a measure of political activism, distinguishing those people who are very active in political affairs, those who don't participate much at all, and those who are somewhere in between.

The first part of Figure 6-1 illustrates the logic of indexes. The figure shows six different political actions. Although you and I might disagree on some specifics, I think we could agree that the six actions represent roughly the same degree of political activism.

Using these six items, we could construct an index of political activism by giving each person

1 point for each of the actions he or she has taken. If you wrote to a public official and signed a petition, you'd get a total of 2 points. If I gave money to a candidate and persuaded someone to change her or his vote, I'd get the same score as you. Using this approach, we'd conclude that you and I had the same degree of political activism, even though we had taken different actions.

The second part of Figure 6-1 describes the logic of scale construction. In this case, the actions clearly represent different degrees of political activism—ranging from simply voting to running for office. Moreover, it seems safe to assume a pattern of actions in this case. For example, all those who contributed money probably also voted. Those who worked on a campaign probably also gave some money and voted. This suggests that most people will fall into only one of five idealized action patterns, represented by the number under each set of boxes in the figure. The discussion of scales, later in this chapter, describes ways of identifying people with the type they most closely represent.

As you might surmise, scales are generally superior to indexes, because scales take into consideration the intensity with which different items reflect the variable being measured. Also, as the example in Figure 6-1 shows, scale scores convey more information than do index scores. Again, be aware that the term *scale* is commonly misused to refer to measures that are only indexes. Merely calling a given measure a scale instead of an index doesn't make it better.

There are two other misconceptions about scaling that you should know. First, whether the combination of several data items results in a scale almost always depends on the particular sample of observations under study. Certain items may form a scale within one sample but not within another. For this reason, do not assume that a given set of items is a scale simply because it has turned out that way in an earlier study.

Second, the use of specific scaling techniques— such as Guttman scaling, which will be discussed— does not ensure the creation of a scale. Rather, such techniques let us determine whether or not a set of items constitutes a scale.

An examination of actual social science research reports will show that researchers use indexes much more often than they use scales. Ironically, however, the methodological literature contains little if any discussion of index construction, whereas discussions of scale construction abound. There appear to be two reasons for this disparity. First, indexes are more frequently used because scales are often difficult or impossible to construct from the data at hand. Second, methods of index construction seem so obvious and straightforward that they aren't discussed much.

Constructing indexes is not simple, however. The general failure to develop index-construction techniques has resulted in many bad indexes in social research. With this in mind, I've devoted over half of this chapter to the methods of index construction. With a solid understanding of the logic of this activity, you'll be better equipped to try constructing scales.

Index Construction

Let's look now at four main steps in the construction of an index: selecting possible items, examining their empirical relationships, scoring and validating an index, and handling missing data. We'll conclude this discussion by examining the construction of an index that provided interesting findings about the status of women in different countries.

Item Selection

The first step in creating an index is selecting items for a composite index, which is created to measure some variable.

Face Validity

The first criterion for selecting items to be included in an index is face validity (or logical validity).

Composite measures involve the combination of elements to create something new. Sometimes it works, sometimes it doesn't.

Earl Babbie

If you want to measure political conservatism, for example, each of your items should appear on its face to indicate conservatism (or its opposite, liberalism). Political party affiliation would be one such item. Another would be an item asking people to approve or disapprove of the views of a well-known conservative public figure. In constructing an index of religiosity, you might consider items such as attendance at religious services, acceptance of certain religious beliefs, and frequency of prayer; each of these appears to offer some indication of religiosity.

Unidimensionality

The methodological literature on conceptualization and measurement stresses the need for unidimensionality in scale and index construction; that is, a composite measure should represent only one dimension of a concept. Thus, items reflecting religiosity should not be included in a measure of political conservatism, even though the two variables might be empirically related.

General or Specific

Although measures should tap the same dimension, the general dimension you're attempting to measure may have many nuances. In the example of religiosity, the indicators mentioned previously—ritual participation, belief, and so on—represent different types of religiosity. If you want to focus on ritual participation in religion, you should choose items specifically indicating this type of religiosity: attendance at religious services and other rituals such as confession, bar mitzvah, bowing toward Mecca, and the like. If you want to measure religiosity in a more general way, you should include a balanced set of items, representing each of the different types of religiosity. Ultimately, the nature of the items included will determine how specifically or generally the variable is measured.

Variance

In selecting items for an index, you must also be concerned with the amount of variance they provide. If an item is intended to indicate political conservatism, for example, you should note what proportion of respondents would be identified as conservatives by the item. If a given item identified no one as a conservative or everyone as a conservative—for example, if nobody indicated approval of a radical-right political figure—that item would not be very useful in the construction of an index.

To guarantee variance, you have two options. First, you can select several items that generate responses that divide people about equally in terms of the variable; for example, about half conservative and half liberal. Although no single response would justify characterizing a person as very conservative, a person who responded as a conservative on all items might be so characterized.

The second option is to select items that differ in variance. One item might identify about half the subjects as conservative, while another might identify few of the respondents as conservative. Note that this second option is necessary for scaling, and it's reasonable for index construction as well.

Examination of Empirical Relationships

The second step in index construction is to examine the empirical relationships among the items being considered for inclusion. (See Chapter 14 for more.) An empirical relationship is established when respondents' answers to one question—in a questionnaire, for example—help us predict how they will answer other questions. If two items are empirically related to each other, we can reasonably argue that each reflects the same variable, and we can include them both in the same index. There are two types of possible relationships among items: bivariate and multivariate.

Bivariate Relationships among Items

A bivariate relationship is, simply put, a relationship between two variables. Suppose we want to measure respondents' support for U.S. participation in the United Nations. One indicator of different levels of support might be the question "Do you feel the U.S. financial support of the UN is ❏ Too high ❏ About right ❏ Too low?"

A second indicator of support for the United Nations might be the question "Should the United States contribute military personnel to UN peacekeeping actions? ❏ Strongly approve ❏ Mostly approve ❏ Mostly disapprove ❏ Strongly disapprove."

Both of these questions, on their face, seem to reflect different degrees of support for the

United Nations. Nonetheless, some people might feel that the United States should give more money but not provide troops. Others might favor sending troops but cutting back on financial support.

If the two items both reflect degrees of the same thing, however, we should expect responses to the two items to generally correspond with one another. Specifically, those who approve of military support should be more likely to favor financial support than would those who disapprove of military support. Conversely, those who favor financial support should be more likely to favor military support than would those who disapprove of financial support. If these expectations are met, we say there is a bivariate relationship between the two items.

Here's another example. Suppose we want to determine the degree to which respondents feel women have the right to an abortion. We might ask (1) "Do you feel a woman should have the right to an abortion when her pregnancy was the result of incest?" and (2) "Do you feel a woman should have the right to an abortion if continuing her pregnancy would seriously threaten her life?"

Some respondents might agree with item (1) and disagree with item (2); others will do just the reverse. If both items tap into some general opinion people have about the issue of abortion, then the responses to these two items should be related to one another. Those who support the right to an abortion in the case of incest should be more likely to support it if the woman's life is threatened than would those who disapproved of abortion in the case of incest. This would be another example of a bivariate relationship between the two items.

To determine the relative strengths of relationships among the several pairs of items, you should examine all the possible bivariate relationships among the several items being considered for inclusion in an index. Percentage tables or more-advanced statistical techniques may be used for this purpose. How we evaluate the strength of the relationships, however, can be rather subtle. "How to Do It: 'Cause' and 'Effect' Indicators" examines some of these subtleties.

Be wary of items that are not related to one another empirically: It's unlikely that they measure the same variable. You should probably drop any item that's not related to several other items.

At the same time, a very strong relationship between two items presents a different problem. If two items are perfectly related to each other, then only one needs to be included in the index; because it completely conveys the indications provided by the other, nothing would be added by including the other item. (This problem will become even clearer in a later section.)

Here's an example to illustrate the testing of bivariate relationships in index construction. I once conducted a survey of medical school faculty members to find out about the consequences of a "scientific perspective" on the quality of patient care provided by physicians. The primary intent was to determine whether scientifically inclined doctors treated patients more impersonally than did other doctors.

The survey questionnaire offered several possible indicators of respondents' scientific perspectives. Of those, three items appeared to provide especially clear indications of whether the doctors were scientifically oriented:

1. *As a medical school faculty member, in what capacity do you feel you can make your greatest teaching contribution: as a practicing physician or as a medical researcher?*
2. *As you continue to advance your own medical knowledge, would you say your ultimate medical interests lie primarily in the direction of total patient management or the understanding of basic mechanisms? [The purpose of this item was to distinguish those who were mostly interested in overall patient care from those mostly interested in biological processes.]*
3. *In the field of therapeutic research, are you generally more interested in articles reporting evaluations of the effectiveness of various treatments or articles exploring the basic rationale underlying the treatments? [Similarly, I wanted to distinguish those more interested in articles dealing with patient care from those more interested in biological processes.]*

(Babbie 1970: 27–31)

For each of these items, we might conclude that respondents who chose the second answer are more scientifically oriented than respondents

How to Do It

"Cause" and "Effect" Indicators

by Kenneth Bollen

Department of Sociology, University of North Carolina, Chapel Hill

While it often makes sense to expect indicators of the same variable to be positively related to one another, as discussed in the text, this is not always the case.

Indicators should be related to one another if they are essentially "effects" of a variable. For example, to measure self-esteem, we might ask a person to indicate whether he or she agrees or disagrees with the statements (1) "I am a good person" and (2) "I am happy with myself." A person with high self-esteem should agree with both statements, whereas one with low self-esteem would probably disagree with both. Since each indicator depends on or "reflects" self-esteem, we expect each to be positively correlated with the others. More generally, indicators that depend on the same variable should be associated with one another if they are valid measures.

But, this is not the case when the indicators are the "cause" rather than the "effect" of a variable. In this situation, the indicators may correlate positively, negatively, or not at all. For example, we could use gender and race as indicators of the variable *exposure to discrimination*. Being nonwhite or female increases the likelihood of experiencing discrimination, so both are good indicators of the variable. But we would not expect the race and gender of individuals to be strongly associated.

Or, we may measure *social interaction* with three indicators: time spent with friends, time spent with family, and time spent with coworkers. Though each indicator is valid, it need not be positively correlated with the others. Time spent with friends, for instance, may be inversely related to time spent with family. Here, the three indicators "cause" the degree of social interaction.

As a final example, exposure to stress may be measured by whether a person recently experienced divorce, death of a spouse, or loss of a job. Though any of these events may indicate stress, they need not correlate with one another.

In short, we expect an association between indicators that depend on or "reflect" a variable, that is, if they are the "effects" of the variable. But if the variable depends on the indicators—if the indicators are the "causes"—those indicators may be either positively or negatively correlated, or even unrelated. Therefore, we should decide whether indicators are causes or effects of a variable before using their intercorrelations to assess their validity.

---

who chose the first answer. Though this comparative conclusion is reasonable, we should not be misled into thinking that respondents who chose the second answer to a given item are scientists in any absolute sense. They are simply more scientifically oriented than those who chose the first answer to the item.

To see this point more clearly, let's examine the distribution of responses to each item. From the first item—greatest teaching contribution—only about one-third of the respondents appeared scientifically oriented. That is, a little over one-third said they could make their greatest teaching contribution as medical researchers. In response to the second item—ultimate medical interests—approximately two-thirds chose the scientific answer, saying they were more interested in learning about basic mechanisms than learning about total patient management. In response to the third item—reading preferences—about 80 percent chose the scientific answer.

These three questionnaire items can't tell us how many "scientists" there are in the sample, because none of them is related to a set of criteria for what constitutes being a scientist in any absolute sense. Using the items for this purpose would present us with the problem of three quite different estimates of how many scientists there were in the sample.

However, these items do provide us with three independent indicators of respondents' relative inclinations toward science in medicine. Each item separates respondents into the more scientific and the less scientific. But each grouping of more or less scientific respondents will have a somewhat different membership from the others. Respondents who seem scientific in terms of one item will not seem scientific in terms of another. Nevertheless, to the extent that each item measures the same general dimension, we should find some correspondence among the several groupings. Respondents who appear scientific in terms of one item should be more likely to appear scientific in their response to another item than do those who appear nonscientific in their response to the first. In other words, we should find an association or correlation between the responses given to two items.

Figure 6-2 shows the associations among the responses to the three items. Three bivariate

a.

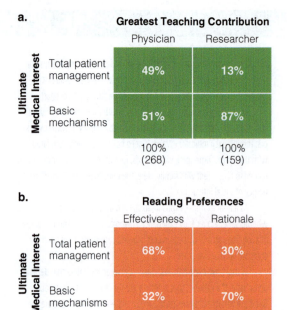

	Greatest Teaching Contribution	
	Physician	Researcher
Ultimate Medical Interest Total patient management	49%	13%
Basic mechanisms	51%	87%
	100% (268)	100% (159)

b.

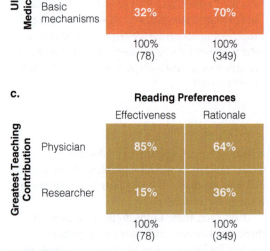

	Reading Preferences	
	Effectiveness	Rationale
Ultimate Medical Interest Total patient management	68%	30%
Basic mechanisms	32%	70%
	100% (78)	100% (349)

c.

	Reading Preferences	
	Effectiveness	Rationale
Greatest Teaching Contribution Physician	85%	64%
Researcher	15%	36%
	100% (78)	100% (349)

FIGURE 6-2

Bivariate Relationships among Scientific Orientation Items. If several indicators are measures of the same variable, then they should be empirically correlated with one another, as you can observe in this case. Those who choose the scientific orientation on one item are more likely to choose the scientific orientation on another item.

tables are presented, showing the distribution of responses for each pair of items. An examination of the three bivariate relationships presented in the figure supports the suggestion that the three items all measure the same variable: *scientific orientation.* To see why this is so, let's begin by looking at the first bivariate relationship in the figure. Panel a shows that faculty who responded that "researcher" was the role in which they could make their greatest teaching contribution were more likely to identify their ultimate medical interests as "basic mechanisms" (87 percent)

than were those who answered "physician" (51 percent). The fact that the "physicians" are about evenly split in their ultimate medical interests is irrelevant for our purposes. It's only relevant that they are less scientific in their medical interests than are the "researchers." The strength of this relationship can be summarized as a 36-percentage-point difference.

The same general conclusion applies to the other bivariate relationships. The strength of the relationship between reading preferences and ultimate medical interests (panel b) may be summarized as a 38-percentage-point difference, and the strength of the relationship between reading preferences and the two teaching contributions (panel c) as a 21-percentage-point difference. In summary, then, each single item produces a different grouping of "scientific" and "nonscientific" respondents. However, the responses given to each of the items correspond, to a greater or lesser degree, to the responses given to each of the other items.

Initially, the three items were selected on the basis of face validity—each appeared to give some indication of faculty members' orientations to science. By examining the bivariate relationship between the pairs of items, we have found support for the expectation that they all measure basically the same thing. However, that support does not sufficiently justify including the items in a composite index. Before combining them in a single index, we need to examine the multivariate relationships among the several variables.

Multivariate Relationships among Items

Whereas a bivariate relationship deals with two variables at a time, a *multivariate relationship* uses more than two variables. To present the trivariate relationships among the three variables in our example, we would first categorize the sample of medical school respondents into four groups according to (1) their greatest teaching contribution and (2) their reading preferences. Figure 6-3 does just that. The numbers in parentheses indicate the number of respondents in each group. Thus, 66 of the faculty members who said they could best teach as physicians also said they preferred articles dealing with the effectiveness of treatments. Then, for each of the four groups, we would determine the percentage of those who say they are ultimately more interested in basic mechanisms. So, for

Percent Interested in Basic Mechanisms

		Greatest Teaching Contribution	
		Physician	Researcher
Reading Preferences	Effectiveness of treatments	27% (66)	58% (12)
	Rationale behind treatments	58% (219)	89% (130)

FIGURE 6-3

Trivariate Relationships among Scientific Orientation Items. Indicators of the same variable should be correlated in a multivariate analysis as well as in bivariate analyses. Those who chose the scientific responses on greatest teaching contribution and reading preferences are the most likely to choose the scientific response on the third item.

example, of the 66 faculty mentioned, 27 percent are primarily interested in basic mechanisms, as the figure shows.

The arrangement of the four groups is based on a previously drawn conclusion regarding scientific orientations. The group in the upper left corner of the table is presumably the least scientifically oriented, based on greatest teaching contribution and reading preference. The group in the lower right corner is presumably the most scientifically oriented in terms of those items.

Recall that expressing a primary interest in basic mechanisms was also taken as an indication of scientific orientation. As we should expect, then, those in the lower right corner are the most likely to give this response (89 percent), and those in the upper left corner are the least likely (27 percent). The respondents who gave mixed responses in terms of teaching contributions and reading preferences have an intermediate rank in their concern for basic mechanisms (58 percent in both cases).

Figure 6-3 tells us many things. First, we may note that the original relationships between pairs of items are not significantly affected by the presence of a third item. Recall, for example, that the relationship between teaching contribution and ultimate medical interest was summarized as a 36-percentage-point difference. Looking at the figure, we see that among only those respondents who are most interested in articles dealing with the effectiveness of treatments, the relationship between teaching contribution and ultimate medical interest is 31 percentage points (58 percent minus 27 percent: first row). The same is

true among those most interested in articles dealing with the rationale for treatments (89 percent minus 58 percent: second row). The original relationship between teaching contribution and ultimate medical interest is essentially the same as in Figure 6-2, even among respondents who were judged to be scientific or nonscientific in terms of reading preferences.

We can draw the same conclusion from the columns in Figure 6-3. Recall that the original relationship between reading preferences and ultimate medical interests was summarized as a 38-percentage-point difference. Looking only at the "physicians" in Figure 6-3, we see that the relationship between the other two items is now 31 percentage points. The same relationship is found among the "researchers" in the second column.

The importance of these observations becomes clearer when we consider what might have happened. In Figure 6-4, hypothetical data tell a much different story than do the actual data in Figure 6-3. As you can see, Figure 6-4 shows that the original relationship between teaching contribution and ultimate medical interest persists, even when reading preferences are introduced into the picture. In each row of the table, the "researchers" are more likely to express an interest in basic mechanisms than are the "physicians." Looking down the columns, however, we note that there is no relationship between reading preferences and ultimate medical interest. If we know whether a respondent feels that he or she can best teach as a physician or as a researcher, knowing the respondent's reading preference adds nothing to our evaluation of his or her scientific orientation. If something like Figure 6-4 resulted from the actual data, we

Percent Interested in Basic Mechanisms

		Greatest Teaching Contribution	
		Physician	Researcher
Reading Preferences	Effectiveness of treatments	51% (66)	87% (12)
	Rationale behind treatments	51% (219)	87% (130)

FIGURE 6-4

Hypothetical Trivariate Relationship among Scientific Orientation Items. This hypothetical relationship would suggest that not all three indicators would contribute effectively to a composite index.

would conclude that reading preference should not be included in the same index as teaching contribution, because it contributed nothing to the composite index.

This example used only three questionnaire items. If more were being considered, then more-complex multivariate tables would be in order, constructed of four, five, or more variables. The purpose of this step in index construction, again, is to discover the simultaneous interaction of the items in order to determine which should be included in the same index.

Index Scoring

When you've chosen the best items for the index, you next assign scores for particular responses, thereby creating a single composite index out of the several items. There are two basic decisions to be made in this step.

First, you must decide the desirable range of the index scores. Certainly a primary advantage of an index over a single item is the range of gradations it offers in the measurement of a variable. As noted earlier, *political conservatism* might be measured from "very conservative" to "not at all conservative" (or "very liberal"). How far to the extremes, then, should the index extend?

In this decision, the question of variance enters once more. Almost always, as the possible extremes of an index are extended, fewer cases are found at each end. The researcher who wishes to measure political conservatism to its greatest extreme may find there is almost no one in that category.

The first decision, then, concerns the conflicting desire for (1) a range of measurement in the index and (2) an adequate number of cases at each point in the index. You'll be forced to reach some kind of compromise between these conflicting desires.

The second decision concerns the actual assignment of scores for each particular response. Basically you must decide whether to give each item in the index equal weight or different weights. Although there are no firm rules, I suggest—and practice tends to support this method—that items be weighted equally unless there are compelling reasons for differential weighting. That is, the burden of proof should be on differential weighting; equal weighting should be the norm.

Of course, this decision must be related to the earlier issue regarding the balance of items chosen. If the index is to represent the composite of slightly different aspects of a given variable, then you should give each aspect the same weight. In some instances, however, you may feel that, say, two items reflect essentially the same aspect and the third reflects a different aspect. If you wished to have both aspects equally represented by the index, you might decide to give the different item a weight equal to the combination of the two similar ones. In such a situation, you might want to assign a maximum score of 2 to the different item and a maximum score of 1 to each of the similar ones.

Although the rationale for scoring responses should take such concerns into account, you'll typically experiment with different scoring methods, examining the relative weights given to different aspects but at the same time worrying about the range and distribution of cases provided. Ultimately, the scoring method chosen will represent a compromise among these several demands. Of course, as in most research activities, such a decision is open to revision on the basis of later examinations. Validation of the index, to be discussed shortly, may lead you to recycle your efforts toward constructing a completely different index.

In the example taken from the medical school faculty survey, I decided to weight the items equally, since I'd chosen them, in part, because they represented slightly different aspects of the overall variable *scientific orientation*. On each of the items, the respondents were given a score of 1 for choosing the "scientific" response to the item and a score of 0 for choosing the "nonscientific" response. Each respondent, then, could receive a score of 0, 1, 2, or 3. This scoring method provided what was considered a useful range of variation—four index categories—and also provided enough cases for analysis in each category.

Here's a similar example of index scoring, from a study of work satisfaction. One of the key variables was *job-related depression*, measured by an index composed of the following four items, which asked workers how they felt when thinking about themselves and their jobs:

- "I feel downhearted and blue."
- "I get tired for no reason."
- "I find myself restless and can't keep still."
- "I am more irritable than usual."

Applying Concepts in Everyday Life

What Is the Best College in the United States?

Each year *U.S. News & World Report* publishes a special issue ranking the nation's colleges and universities. Their rankings reflect an index created from several items: educational expenditures per student, graduation rates, selectivity (percent accepted of those applying), average SAT scores of first-year students, and similar indicators of quality.

Typically, Harvard is ranked the number one school in the nation, followed by Yale and Princeton. However, the 1999 "America's Best Colleges" issue shocked educators, prospective college students, and their parents. The California Institute of Technology had leaped from ninth place in 1998 to first place a year later. Although Harvard, Yale, and Princeton still did well, they had been supplanted. What had happened at Caltech to produce such a remarkable surge in quality?

The answer was to be found at *U.S. News & World Report*, not at Caltech. The newsmagazine changed the structure of the ranking index in 1999, which made a big difference in how schools fared.

Bruce Gottlieb (1999) gives this example of how the altered scoring made a difference.

> So, how did Caltech come out on top? Well, one variable in a school's ranking has long been educational expenditures per student, and Caltech has traditionally been tops in this category. But until this year, U.S. News *considered only a school's ranking in this category—first, second, etc.—rather than how much it spent relative to other schools. It didn't matter whether Caltech beat Harvard by $1 or by $100,000. Two other schools that rose in their rankings this year were MIT (from fourth to third) and Johns Hopkins (from 14th to seventh). All three have high per-student expenditures and all three are especially strong in the hard sciences. Universities are allowed to count their research budgets in their per-student expenditures, though students get no direct benefit from costly research their professors are doing outside of class.*

> In its "best colleges" issue two years ago, U.S. News *made precisely this point, saying it considered only the rank ordering of per-student expenditures, rather than the actual amounts, on the grounds that expenditures at institutions with large research programs and medical schools are substantially higher than those at the rest of the schools in the category. In other words, just two years ago, the magazine felt it unfair to give Caltech, MIT, and Johns Hopkins credit for having lots of fancy laboratories that don't actually improve undergraduate education.*

Gottlieb reviewed each of the changes in the index and then asked how 1998's ninth-ranked Caltech would have done had the revised indexing formula been in place a year earlier. His conclusion: Caltech would have been first in 1998 as well. In other words, the apparent improvement was solely a function of how the index was scored.

For a very different ranking of colleges and universities, you might be interested in the "Webometrics Ranking," which focuses on schools' presence on the Web. This website details the items included in the index, as well as how they are combined to produce an overall ranking of the world's institutions of higher education. As of January 2008, MIT was the top-ranked American university, but you'll have to examine the methodological description to know what that means.

Composite measures such as scales and indexes are valuable tools for understanding society. However, it's important that we know how those measures are constructed and what that construction implies.

So, what's really the best college in the United States? It depends on how you define "best." There is no "really best," only the various social constructions we can create.

Sources: "America's Best Colleges," *U.S. News & World Report,* August 30, 1999; Bruce Gottlieb, Cooking the School Books: How U.S. News Cheats in Picking Its 'Best American Colleges.' *Slate,* August 31, 1999, http://www.slate.com/articles/news_and_politics /crapshoot/1999/09/cooking_the_school_books.html.

The researchers, Amy Wharton and James Baron, report, "Each of these items was coded: 4–5 = often, 3–5 = sometimes, 2–5 = rarely, 1–5 = never" (1987: 578). They go on to explain how they measured other variables examined in the study:

> Job-related self-esteem was based on four items asking respondents how they saw themselves in their work: happy/sad; successful/not successful; important/not important; doing their best/ not doing their best. Each item ranged from 1 to 7, where 1 indicates a self-perception of not being happy, successful, important, or doing one's best.
>
> (1987: 578)

As you look through the social research literature, you'll find numerous similar examples of cumulative indexes being used to measure variables. Sometimes the indexing procedures are controversial, as evidenced in "Applying Concepts in Everyday Life: What Is the Best College in the United States?"

Although it's often appropriate to examine the relationships among indicators of a variable being measured by an index or scale, you should realize that the indicators are sometimes independent of one another. For example, Stacy De Coster notes that the indicators of family stress may be independent of one another, though they contribute to the same variable:

> Family Stress is a scale of stressful events within the family. The experience of any one of these events—parent job loss, parent separation, parent illness—is independent of the other events. Indeed, prior research on events utilized in stress scales has demonstrated that the events in these scales typically are independent of one another and reliabilities on the scales [are] low.
>
> (2005: 176)

If the indicators of a variable are logically related to one another, on the other hand, that relationship should be used as a criterion for determining which are the better indicators.

Handling Missing Data

Regardless of your data-collection method, you'll frequently face the problem of missing data. In a content analysis of the political orientations of newspapers, for example, you may discover that a particular newspaper has never taken an editorial position on one of the issues being studied. In an experimental design involving several retests of subjects over time, some subjects may be unable to participate in some of the sessions. In virtually every survey, some respondents fail to answer some questions (or choose a "don't know" response). Although missing data present problems at all stages of analysis, they're especially troublesome in index construction. However, several methods for dealing with these problems exist.

First, if there are relatively few cases with missing data, you may decide to exclude them from the construction of the index and the analysis. (I did this in the medical school faculty example.) The primary concerns in this instance are whether the numbers available for analysis will remain sufficient and whether the exclusion will result in a biased sample whenever the index is used in the analysis. The latter possibility can be examined through a comparison—on other relevant variables—of those who would be included in or excluded from the index.

Second, you may sometimes have grounds for treating missing data as one of the available

responses. For example, if a questionnaire has asked respondents to indicate their participation in various activities by checking "yes" or "no" for each, many respondents may have checked some of the activities "yes" and left the remainder blank. In such a case, you might decide that a failure to answer meant "no," and score missing data in this case as though the respondents had checked the "no" space.

Third, a careful analysis of missing data may yield an interpretation of their meaning. In constructing a measure of political conservatism, for example, you may discover that respondents who failed to answer a given question were generally as conservative on other items as were those who gave the conservative answer. Whenever the analysis of missing data yields such interpretations, then, you may decide to score such cases accordingly.

There are many other ways of handling the problem of missing data. If an item has several possible values, you might assign the middle value to cases with missing data; for example, you could assign a 2 if the values are 0, 1, 2, 3, and 4. For a continuous variable such as *age*, you could similarly assign the mean to cases with missing data (more on this in Chapter 14). Or you can supply missing data by assigning values at random. All of these are conservative solutions, because any such changes weaken the "purity" of your index and reduce the likelihood that it will relate to other variables in ways you may have hypothesized.

If you're creating an index out of several items, you can sometimes handle missing data by using proportions based on what is observed. Suppose your index is composed of six indicators and you have only four observations for a particular subject. If the subject has earned 4 points out of a possible 4, you might assign an index score of 6; if the subject has 2 points (half the possible score on four items), you could assign a score of 3 (half the possible score on six observations).

The choice of a particular method to be used depends so much on the research situation that I can't reasonably suggest a single "best" method or rank the several I've described. Excluding all cases with missing data can bias the representativeness of the findings, but including such cases by assigning scores to missing data can influence the nature of the findings. The safest and best

Issues and Insights

How Healthy Is Your State?

Since 1990, United Health Foundation, the American Public Health Association, and Partnership for Prevention have collaborated on an annual evaluation of the health status of each of the 50 states. The following table displays the findings for overall rankings from the 2016 report. The scores indicate where each state stands in comparison to the nation as a whole.

The scores are shown as standard deviations from the national average. While you may not have studied this statistical technique, you can still determine whether your state is above or below the national average, indicated by a plus or minus sign. The healthiest state in 2016 was Hawaii and Mississippi the least healthy.

You may be interested in seeing how your state ranks.

2016 Ranking

Rank	State	Overall Score*
1	Hawaii	0.905
2	Massachusetts	0.760
3	Connecticut	0.747
4	Minnesota	0.727
5	Vermont	0.709
6	New Hampshire	0.696
7	Washington	0.582
8	Utah	0.578
9	New Jersey	0.571
10	Colorado	0.559
11	North Dakota	0.473
12	Nebraska	0.432
13	New York	0.430
14	Rhode Island	0.422
15	Idaho	0.356
16	California	0.346
17	Iowa	0.343
18	Maryland	0.322
19	Virginia	0.264
20	Wisconsin	0.220
21	Oregon	0.211
22	Maine	0.192
23	Montana	0.178
24	South Dakota	0.169
25	Wyoming	0.116
26	Illinois	0.084
27	Kansas	−0.012
28	Pennsylvania	−0.016
29	Arizona	−0.020
30	Alaska	−0.031
31	Delaware	−0.077
32	North Carolina	−0.194
33	Texas	−0.208

2016 Alphabetical Ranking

Rank	State	Overall Score*
47	Alabama	−0.793
30	Alaska	−0.031
29	Arizona	−0.020
48	Arkansas	−0.834
16	California	0.346
10	Colorado	0.559
3	Connecticut	0.747
31	Delaware	−0.077
36	Florida	−0.307
41	Georgia	−0.464
1	Hawaii	0.905
15	Idaho	0.356
26	Illinois	0.084
39	Indiana	−0.372
17	Iowa	0.343
27	Kansas	−0.012
45	Kentucky	−0.651
49	Louisiana	−1.043
22	Maine	0.192
18	Maryland	0.322
2	Massachusetts	0.760
34	Michigan	−0.251
4	Minnesota	0.727
50	Mississippi	−1.123
37	Missouri	−0.338
23	Montana	0.178
12	Nebraska	0.432
35	Nevada	−0.304
6	New Hampshire	0.696
9	New Jersey	0.571
38	New Mexico	−0.363
13	New York	0.430
32	N orth Carolina	−0.194

Issues and Insights (*Continued*)

2016 Ranking (*Continued*)

Rank	State	Overall Score*
34	Michigan	−0.251
35	Nevada	−0.304
36	Florida	−0.307
37	Missouri	−0.338
38	New Mexico	−0.363
39	Indiana	−0.372
40	Ohio	−0.391
41	Georgia	−0.464
42	South Carolina	−0.532
43	West Virginia	−0.595
44	Tennessee	−0.626
45	Kentucky	−0.651
46	Oklahoma	−0.691
47	Alabama	−0.793
48	Arkansas	−0.834
49	Louisiana	−1.043
50	Mississippi	−1.123

2016 Alphabetical Ranking (*Continued*)

Rank	State	Overall Score*
11	North Dakota	0.473
40	Ohio	−0.391
46	Oklahoma	−0.691
21	Oregon	0.211
28	Pennsylvania	−0.016
14	Rhode Island	0.422
42	South Carolina	−0.532
24	South Dakota	0.169
44	Tennessee	−0.626
33	Texas	−0.208
8	Utah	0.578
5	Vermont	0.709
19	Virginia	0.264
7	Washington	0.582
43	West Virginia	−0.595
20	Wisconsin	0.220
25	Wyoming	0.116

* Scores presented in this table indicate the weighted number of standard deviations a state is above or below the national norm.

Source: http://assets.americashealthrankings.org/app/uploads/ahr16-complete-v2.pdf, p. 6.

Since you are, by now, a critical consumer of social research, I can hear you demanding, "Wait a minute, how did they measure healthy?" Good question. The table, "Weight of Individual Measures," is from the 2015 report and provides a summary of the components included in the report's definition of what constitutes good or bad health. You'll see that the indicators encompass a number of categories. Some represent positive indications (e.g., high school graduation rates) and some are negative indicators (e.g., smoking and binge drinking). Moreover, the table shows the weight assigned to each indicator in the construction of a state's overall score.

Weight of Individual Measures

Determinants (Name of Measure)	Percentage of Total	Effect on Score
Behaviors		
Prevalence of smoking	7.5	Negative
Prevalence of binge drinking	5.0	Negative
Prevalence of obesity	7.5	Negative
High school graduation	5.0	Positive
Community and Environment		
Violent crime	5.0	Negative
Occupational fatalities	2.5	Negative
Infectious disease	5.0	Negative
Children living in poverty	5.0	Negative
Air pollution	5.0	Negative

Weight of Individual Measures (*Continued*)

Determinants (Name of Measure)	Percentage of Total	Effect on Score
Public and Health Policies		
Lack of health insurance	5.0	Negative
Public health funding	2.5	Positive
Immunization coverage	5.0	Positive
Clinical care		
Early prenatal care	5.0	Positive
Primary care physicians	5.0	Positive
Preventable hospitalizations	5.0	Negative
Outcomes		
Poor mental health days	2.5	Negative
Poor physical health days	2.5	Negative
Geographic disparity	5.0	Negative
Infant mortality	5.0	Negative
Cardiovascular deaths	2.5	Negative
Cancer deaths	2.5	Negative
Premature death	5.0	Negative
Overall health ranking	100.0	—

It would be a good idea for you to review each indicator to see whether you agree that it reflects on how healthy states are. Perhaps you can think of other indicators that might have been used.

The full report provides a wealth of thoughtful discussion on why each of these indicators was chosen, and I'd encourage you to download the data from http://www.americashealthrankings.org/reports/annual.

Source: United Health Foundation, Public Health Association, and Partnership for Prevention, "America's Health Rankings: A Call to Action for Individuals and Their Communities." 2016 United Health Foundation. Table "2016 Overall Rankings" taken from Tables 1 and 2, page 8.

method is to construct the index using alternative methods and see whether the same findings follow from each. Understanding your data is the final goal of analysis anyway.

Now that we've covered several aspects of index construction, see "Issues and Insights: How Healthy Is Your State?" for more on choosing indicators and scoring items.

Index Validation

Up to this point, we've discussed all the steps in the selection and scoring of items that result in a composite index purporting to measure some variable. If each of the preceding steps is carried out carefully, the likelihood of the index actually measuring the variable is enhanced. To demonstrate success, however, we need to validate the index. Following the basic logic of validation, we assume that the index provides a measure of some variable; that is, the scores on the index arrange cases in a rank order in terms of that variable. An index of political conservatism rank-orders people in terms of their relative conservatism. If the index does that successfully, then people scored as relatively conservative on the index should appear relatively conservative on all other indications of political orientation, such as their responses to other questionnaire items. There are several methods of validating an index.

Item Analysis

The first step in index validation is an internal validation called item analysis. In **item analysis**, you examine the extent to which the composite index is related to (or predicts responses to) the individual items it comprises. Here's an illustration of this step.

In the index of scientific orientations among medical school faculty, for example, index scores ranged from 0 (most interested in patient care) to 3 (most interested in research). Now let's consider one of the items in the index: whether respondents wanted to advance their own knowledge more with regard to total patient management or more in the area of basic mechanisms. The latter were treated as being more scientifically oriented than the former. The following empty table shows how we would examine the relationship between the index and the individual item.

	Index of Scientific Orientations			
	0	1	2	3
Percent who said they were more interested in basic mechanisms	??	??	??	??

If you take a minute to reflect on the table, you may see that we already know the numbers that go in two of the cells. To get a score of 3 on the index, respondents had to say "basic mechanisms" in response to this question and give the "scientific" answers to the other two items as well. Thus, 100 percent of the 3s on the index said "basic mechanisms." By the same token, all the 0s had to answer this item with "total patient management." Thus, 0 percent of those respondents said "basic mechanisms." Here's how the table looks with the information we already know.

	Index of Scientific Orientations			
	0	1	2	3
Percent who said they were more interested in basic mechanisms	0	??	??	100

If the individual item is a good reflection of the overall index, we should expect the 1s and 2s to fill in a progression between 0 percent and 100 percent. More of the 2s should choose "basic

item analysis An assessment of whether each of the items included in a composite measure makes an independent contribution or merely duplicates the contribution of other items in the measure.

mechanisms" than 1s. This is not guaranteed by the way the index was constructed, however; it is an empirical question—one we answer in an item analysis. Here's how this particular item analysis turned out.

	Index of Scientific Orientations			
	0	1	2	3
Percent who said they were more interested in basic mechanisms	0	16	91	100

As you can see, in accord with our assumption that the 2s are more scientifically oriented than the 1s, we find that a higher percentage of the 2s (91 percent) than the 1s (16 percent) say "basic mechanisms."

	Index of Scientific Orientations			
	0	1	2	3
Percent who said they were more interested in basic mechanisms	0	4	14	100
Percent who said they preferred reading about rationales	0	80	97	100

An item analysis of the other two components of the index yields similar results, as follows.

Each of the items, then, seems an appropriate component in the index. Each seems to reflect the same quality measured by the index as a whole.

In a complex index containing many items, this step provides a convenient test of the independent contribution of each item to the index. If a given item is found to be poorly related to the index, it may be assumed that other items in the index cancel out the contribution of that item, and it should be excluded from the index. In other words, if the item in question contributes nothing to the index's power, it should be excluded.

Although item analysis is an important first test of the index's validity, it is scarcely sufficient. If the index adequately measures a given variable, it should successfully predict other indications of that variable. To test this, we must turn to items not included in the index.

External Validation

People scored as politically conservative on an index should appear conservative by other measures as well, such as their responses to other items in a questionnaire. Of course, we're talking about relative conservatism, because we can't make an absolute definition of what constitutes conservatism. However, the respondents scored as the most conservative on the index should be the most conservative in answering other questions. Those scored as the least conservative on the index should be the least conservative on other items. Indeed, the ranking of groups of respondents on the index should predict the ranking of those groups in answering other questions dealing with political orientations.

In our example of the scientific orientation index, several questions in the questionnaire offered the possibility of such **external validation**. Table 6-1 presents some of these items, which provide several lessons regarding index validation. First, we note that the index strongly predicts the responses to the validating items in the sense that the rank order of scientific responses among the four groups is the same as the rank order provided by the index itself. That is, the percentages reflect greater scientific orientation as you read across the rows of the table. At the same time, each item gives a different description of scientific orientations overall. For

TABLE 6-1

Validating the Index of Scientific Orientations

	Index of Scientific Orientations			
	Low 0	1	2	High 3
Percent interested in attending scientific lectures at the medical school	34	42	46	65
Percent who say faculty members should have experience as medical researchers	43	60	65	89
Percent who would prefer faculty duties involving research activities only	0	8	32	66
Percent who engaged in research during the preceding academic year	61	76	94	99

example, the last validating item indicates that the great majority of all faculty were engaged in research during the preceding year. If this were the only indicator of scientific orientation, we would conclude that nearly all faculty were scientific. Nevertheless, those who were scored

as more scientific on the index are more likely to have engaged in research than are those who were scored as relatively less scientific. The third validating item provides a different descriptive picture: Only a minority of the faculty overall say they would prefer duties limited exclusively to research. (Only among those who scored 3 on the index do a majority agree with that statement.) Nevertheless, the percentages giving this answer correspond to the scores assigned on the index.

Bad Index versus Bad Validators

Nearly every index constructor at some time must face the apparent failure of external items to validate the index. If the internal item analysis shows inconsistent relationships between the items included in the index and the index itself, something is wrong with the index. But if the index fails to predict strongly the external validation items, the conclusion to be drawn is more ambiguous. You must choose between two possibilities: (1) the index does not adequately measure the variable in question, or (2) the validation items do not adequately measure the variable and therefore do not provide a sufficient test of the index.

Having worked long and conscientiously on the construction of an index, you'll likely find the second conclusion compelling. Typically, you'll feel that you have included the best indicators of the variable in the index; the validating items are, therefore, second-rate indicators. Nevertheless, you should recognize that the index is purportedly a very powerful measure of the variable; thus, it should be somewhat related to any item that taps the variable even poorly.

When external validation fails, you should reexamine the index before deciding that the validating items are insufficient. One way is to examine the relationships between the validating items and the individual items included in the index. If you discover that some of the index

---

external validation The process of testing the validity of a measure, such as an index or scale, by examining its relationship to other presumed indicators of the same variable. If the index really measures prejudice, for example, it should correlate with other indicators of prejudice.

items relate to the validators and others do not, you'll have your understanding of the index as it was initially constituted.

There is no cookbook solution to this dilemma; it is an agony serious researchers must learn to survive. Ultimately, the wisdom of your decision to accept an index will be determined by the usefulness of that index in your later analyses. Perhaps you'll initially decide that the index is a good one and that the validators are defective, but you'll later find that the variable in question (as measured by the index) is not related to other variables in the ways you expected. You may then have to compose a new index.

The Status of Women: An Illustration of Index Construction

For the most part, I've talked about index construction in the context of survey research, but other types of research also lend themselves to this kind of composite measure. For example, when the United Nations (2015) set about examining the status of women in the world, they chose to create two indexes, reflecting two different dimensions.

The Human Development Index (HDI) compared women with men in terms of three indicators: life expectancy, education, and income. These indicators are commonly used in monitoring the status of women in the world. In 2014, Norway, Denmark, and the Netherlands ranked highest on this measure.

The second index, the Gender Inequality Index (GII), aimed more at power issues and comprised three different indicators:

- The proportion of parliamentary seats held by women
- The proportion of administrative, managerial, professional, and technical positions held by women
- A measure of access to jobs and wages

The leaders in gender equality were Slovenia, Switzerland, Germany, Denmark, and Austria. Having two different measures of gender equality allowed the researchers to make more-sophisticated distinctions. For example, in several countries—the United States and Australia, for example—women fared relatively well on the HDI but quite poorly on the GII; thus, although

they were doing fairly well in terms of income, education, and life expectancy, they were still denied access to power. And whereas the HDI scores were higher in the wealthier nations than in the poorer ones, GII scores showed that women's empowerment did not seem to depend on national wealth, with many poor, developing countries outpacing some rich, industrial ones in regard to such empowerment.

By examining several different dimensions of the variables involved in their study, the UN researchers also uncovered an aspect of women's earnings that generally goes unnoticed. Population Communications International (1996) summarizes the finding nicely:

> *Every year, women make an invisible contribution of eleven trillion U.S. dollars to the global economy, the UNDP [United Nations Development Programme] report says, counting both unpaid work and the underpayment of women's work at prevailing market prices. This "underevaluation" of women's work not only undermines their purchasing power, says the 1995 HDR [Human Development Report], but also reduces their already low social status and affects their ability to own property and use credit. Mahbub ul Haq, the principal author of the report, says that "if women's work were accurately reflected in national statistics, it would shatter the myth that men are the main breadwinners of the world." The UNDP report finds that women work longer hours than men in almost every country, including both paid and unpaid duties. In developing countries, women do approximately 53% of all work and spend two-thirds of their work time on unremunerated activities. In industrialized countries, women do an average of 51% of the total work, and—like their counterparts in the developing world—perform about two-thirds of their total labor without pay. Men in industrialized countries are compensated for two-thirds of their work.*

(Population Communications International 1996: 1)

"Issues and Insights: Indexing the World" gives some other examples of indexes that have been created to monitor the state of the world.

As you can see, indexes can be constructed from many different kinds of data for a variety of purposes. (See "Applying Concepts in Everyday Life: Assessing Women's Status" for more on this topic.) Now we'll turn our attention from the construction of indexes to an examination of scaling techniques.

Issues and Insights

Indexing the World

If you search the Web for indexes, you'll be handsomely rewarded. Here are just a few examples of the ways in which people have used the logic of social indexes to monitor the state of the world or large portions of it.

The well-being of nations is commonly measured in economic terms, such as the gross domestic product (GDP) per capita, average income, or stock market averages. In 1972, however, the mountainous kingdom of Bhutan drew global attention by proposing an index of "gross national happiness," augmenting economic factors with measures of physical and mental health, freedom, environment, marital stability, and other indicators of noneconomic well-being.

- Columbia University's "Environmental Sustainability Index" is one of several measures that seek to monitor nations' environmental impact on the planet.
- Vision of Humanity offers an index of peace in the world, indexing the peacefulness of 165 countries in the world.

- The well-being of America's young people is the focus of the "Child and Youth Well-Being Index," housed at Duke University.
- The World Economic Forum offers a "Global Competitiveness Index" to rate 142 economies around the world.
- *Money Magazine* has indexed the 100 best places to live in America, using factors such as economics, housing, schools, health, crime, weather, and public facilities.
- The Heritage Foundation offers an "Index of Economic Freedom" for those planning business ventures around the world.
- For Christians who believe in prophecies of the end of times, "The Rapture Index" uses 45 indicators—including inflation, famine, floods, liberalism, and Satanism—to gauge of how close or far away the end is.

See if you can find some other similar indexes.

Applying Concepts in Everyday Life

Assessing Women's Status

In our discussion of the Gender Inequality Index (GII), we analyzed the status of women in countries around the world. How might you use the logic of this analysis to examine and assess the status of women in a particular organization, such as the college you attend or a corporation you're familiar with?

Scale Construction

Good indexes provide an ordinal ranking of cases on a given variable. All indexes are based on this kind of assumption: A senator who voted for seven out of ten conservative bills is considered to be more conservative than one who voted for only four of them. What an index may fail to take into account, however, is that not all indicators of a variable are equally important or equally strong. The first senator might have voted for the seven least-conservative bills, whereas the second senator might have voted for the four most-conservative bills. (The second senator might have considered the other six bills too liberal and voted against them.)

Scales offer more assurance of ordinality by reflecting the intensity structures among the indicators. The several items going into a composite measure may have different intensities in terms of the variable. Many methods of scaling are available. To illustrate the variety of techniques at hand, we'll look at four scaling procedures, along with a technique called the semantic differential. Although these examples focus on questionnaires, the logic of scaling, like that of indexing, applies to other research methods as well.

Bogardus Social Distance Scale

Let's suppose you're interested in the extent to which U.S. citizens are willing to associate with, say, sex offenders. You might ask the following questions:

1. Are you willing to let sex offenders live in your country?
2. Are you willing to let sex offenders live in your community?
3. Are you willing to let sex offenders live in your neighborhood?
4. Would you be willing to let a sex offender live next door to you?
5. Would you let your child marry a sex offender?

These questions increase in terms of how closely the respondents want to associate with sex offenders. Beginning with the original concern to measure willingness to associate with sex offenders, you have thus developed several questions indicating differing degrees of intensity on this variable. The kinds of items presented constitute a **Bogardus social distance scale** (created by Emory Bogardus). This scale is a measurement technique for determining the willingness of people to participate in social relations—of varying degrees of closeness—with other kinds of people.

The clear differences of intensity suggest a structure among the items. Presumably, if a person is willing to accept a given kind of association, he or she would be willing to accept all those preceding it in the list—those with lesser intensities. For example, the person who is willing to permit sex offenders to live in the neighborhood will surely accept them in the community and the nation but may or may not be willing to accept them as next-door neighbors or relatives. This, then, is the logical structure of intensity inherent among the items.

Empirically, one would expect to find the largest number of people accepting co-citizenship and the fewest accepting intermarriage. In this sense, we speak of "easy items" (for example, residence in the United States) and "hard items" (for example, intermarriage). More people agree to the easy items

than to the hard ones. With some inevitable exceptions, logic demands that once a person has refused a relationship presented in the scale, he or she will also refuse all the harder ones that follow it.

The Bogardus social distance scale illustrates the important economy of scaling as a data-reduction device. By knowing how many relationships with sex offenders a given respondent will accept, we know which relationships were accepted. Thus, a single number can accurately summarize five or six data items without a loss of information.

Motoko Lee, Stephen Sapp, and Melvin Ray (1996) noticed an implicit element in the Bogardus social distance scale: It looks at social distance from the point of view of the majority group in a society. These researchers decided to turn the tables and create a "reverse social distance" scale: looking at social distance from the perspective of the minority group. Here's how they framed their questions (1996: 19):

> *Considering typical Caucasian Americans you have known, not any specific person nor the worst or the best, circle Y or N to express your opinion.*
>
> Y N 5. *Do they mind your being a citizen in this country?*
>
> Y N 4. *Do they mind your living in the same neighborhood?*
>
> Y N 3. *Would they mind your living next to them?*
>
> Y N 2. *Would they mind your becoming a close friend to them?*
>
> Y N 1. *Would they mind your becoming their kin by marriage?*

As with the original scale, the researchers found that knowing the number of items minority respondents agreed with also told the researchers which ones were agreed with—99 percent of the time in this case.

Thurstone Scales

Often, the inherent structure of the Bogardus social distance scale is not appropriate to the variable being measured. Indeed, such a logical structure among several indicators is seldom apparent. A **Thurstone scale** (created by Louis Thurstone) is an attempt to develop a format for generating groups of indicators of a variable that have at least an empirical structure among them.

One of the basic formats is that of "equal-appearing intervals." A group of judges is given perhaps a hundred items felt to be indicators of a

Bogardus social distance scale A measurement technique for determining the willingness of people to participate in social relations—of varying degrees of closeness—with other kinds of people. It is an especially efficient technique in that one can summarize several discrete answers without losing any of the original details of the data.

Thurstone scale A type of composite measure, constructed in accordance with the weights assigned by "judges" to various indicators of some variables.

given variable. Each judge is then asked to estimate how strong an indicator of a variable each
item is by assigning scores of perhaps 1 to 13.
If the variable were *prejudice,* for example, the
judges would be asked to assign the score of
1 to the very weakest indicators of prejudice, the
score of 13 to the strongest indicators, and intermediate scores to those in between.

Once the judges have completed this task,
the researcher examines the scores assigned to
each item to determine which items produced
the greatest agreement among the judges. The
items on which the judges disagreed broadly
would be rejected as ambiguous. Among those
items producing general agreement in scoring,
one or more would be selected to represent each
scale score from 1 to 13.

The items selected in this manner might then
be included in a survey questionnaire. Respondents
who appeared prejudiced on items representing
a strength of 5 would then be expected to appear
prejudiced on those having lesser strengths, and if
some of those respondents did not appear prejudiced on the items with a strength of 6, it would be
expected that they would also not appear prejudiced on those with greater strengths.

If the Thurstone scale items were adequately
developed and scored, the economy and effectiveness of data reduction inherent in the Bogardus social distance scale would appear. A single
score might be assigned to each respondent (the
strength of the hardest item accepted), and that
score would adequately represent the responses
to several questionnaire items. And, as is true of
the Bogardus scale, a respondent who scored 6
might be regarded as more prejudiced than one
who scored 5 or less.

Thurstone scaling is not often used in research today, primarily because of the tremendous expenditure of energy and time required
to have 10 to 15 judges score the items. Because
the quality of their judgments would depend
on their experience with the variable under
consideration, professional researchers might
be needed. Moreover, the meanings conveyed
by the several items indicating a given variable tend to change over time. Thus, an item
might have a given weight at one time and
quite a different weight later on. To be effective,
a Thurstone scale would have to be updated
periodically.

Likert Scaling

You may sometimes hear people refer to a questionnaire item containing response categories
such as "strongly agree," "agree," "disagree,"
and "strongly disagree" as a Likert scale. This is
technically a misnomer, although Rensis Likert
(pronounced "LICK-ert") did create this commonly used question format. Likert also created a
technique for combining the items into a scale, but
while Likert's scaling technique is rarely used, his
answer format is one of the most frequently used
formats in survey research.

The particular value of this format is the
unambiguous ordinality of response categories.
If respondents were permitted to volunteer or
select such answers as "sort of agree," "pretty
much agree," "really agree," and so forth, you
would find it impossible to judge the relative
strength of agreement intended by the various respondents. The Likert format solves this
problem.

Though seldom used, Likert's scaling method
is fairly easy to understand, based on the relative
intensity of different items. As a simple example,
suppose we wish to measure prejudice against
women. To do this, we create a set of 20 statements, each of which reflects that prejudice. One
of the items might be "Women can't drive as well
as men." Another might be "Women shouldn't
be allowed to vote." Likert's scaling technique
would demonstrate the difference in intensity
between these items as well as pegging the intensity of the other 18 statements.

Let's suppose we ask a sample of people
to agree or disagree with each of the 20 statements. Simply giving one point for each of the
indicators of prejudice against women would
yield the possibility of index scores ranging
from 0 to 20. A true **Likert scale** goes one step

Likert scale A type of composite measure developed by Rensis Likert in an attempt to improve the
levels of measurement in social research through
the use of standardized response categories in survey
questionnaires to determine the relative intensity of
different items. Likert items are those using such
response categories as "strongly agree," "agree,"
"disagree," and "strongly disagree." Such items may
be used in the construction of true Likert scales as
well as other types of composite measures.

beyond that and calculates the average index score for those agreeing with each of the individual statements. Let's say that all those who agreed that women are poorer drivers than are men had an average index score of 1.5 (out of a possible 20). Those who agreed that women should be denied the right to vote might have an average index score of, say, 19.5—indicating the greater degree of prejudice reflected in that response.

As a result of this item analysis, respondents could be rescored to form a scale: 1.5 points for agreeing that women are poorer drivers, 19.5 points for saying women shouldn't vote, and points for other responses reflecting how those items related to the initial, simple index. If those who disagreed with the statement "I might vote for a woman for president" had an average index score of 15, then the scale would give 15 points to people disagreeing with that statement.

As I've said earlier, Likert scaling is seldom used today. The item format devised by Likert, however, is one of the most commonly used formats in contemporary questionnaire design. Typically, it's now used in the creation of simple indexes. With, say, five response categories, scores of 0 to 4 or 1 to 5 might be assigned, taking the direction of the items into account (for example, assign a score of 5 to "strongly agree" for positive items and to "strongly disagree" for negative items). Each respondent would then be assigned an overall score representing the summation of the scores he or she received for responses to the individual items.

The Likert format has become a popular one in online surveys, which led Hotaka Maeda (2015) to explore the possible impact of different display options. Will the answers be arranged from positive to negative (e.g., Strongly Agree . . . Strongly Disagree)? Also, does it make a difference whether the answers are offered horizontally or vertically? Overall, Maeda found that the various options made little difference, though there was a slight bias toward the leftmost answer when presented horizontally. This would make little difference if the purpose were to compare respondents, since that slight bias should apply to all. At the same time, Maeda found that respondents answered horizontal arrays somewhat more quickly than vertical ones.

Semantic Differential

Like the Likert format, the **semantic differential** asks respondents to choose between two opposite positions. Here's how it works.

Suppose you're evaluating the effectiveness of a new music-appreciation lecture on subjects' appreciation of music. As a part of your study, you want to play some musical selections and have the subjects report their feelings about them. A good way to capture those feelings would be to use a semantic differential format.

To begin, you must determine the dimensions along which subjects should judge each selection. Then you need to find two opposite terms representing the polar extremes along each dimension. Let's suppose one dimension that interests you is simply whether subjects enjoyed the piece or not. Two opposite terms in this case could be "enjoyable" and "unenjoyable." Similarly, you might want to know whether they regarded the individual selections as "complex" or "simple," "harmonic" or "discordant," and so forth.

Once you have determined the relevant dimensions and have found terms to represent the extremes of each, you might prepare a rating sheet to be completed by each subject for each piece of music. Figure 6-5 shows what it might look like.

On each line of the rating sheet, the subject would indicate how he or she felt about the piece of music: whether it was enjoyable or unenjoyable, for example, and whether it was "somewhat" or "very much" that way. To avoid creating a biased pattern of responses to such items, it's a good idea to vary the placement of terms that are likely to be related to each other. Notice, for example, that "discordant" and "traditional" are on the left side of the sheet, with "harmonic" and "modern" on the right. Most likely, those selections scored as "discordant" would also be scored as "modern" as opposed to "traditional."

semantic differential A questionnaire format in which the respondent is asked to rate something in terms of two opposite adjectives (e.g., rate textbooks as "boring" or "exciting"), using qualifiers such as "very," "somewhat," "neither," "somewhat," and "very" to bridge the distance between the two opposites.

	Very Much	Somewhat	Neither	Somewhat	Very Much	
Enjoyable	☐	☐	☐	☐	☐	Unenjoyable
Simple	☐	☐	☐	☐	☐	Complex
Discordant	☐	☐	☐	☐	☐	Harmonic
Traditional	☐	☐	☐	☐	☐	Modern

FIGURE 6-5

Semantic Differential: Feelings about Musical Selections. The semantic differential asks respondents to describe something or someone in terms of opposing adjectives.

Both the Likert and semantic differential formats have a greater rigor and structure than do other question formats. As I've indicated earlier, these formats produce data suitable to both indexing and scaling.

Guttman Scaling

Researchers today often use the scale developed by Louis Guttman. Like Bogardus, Thurstone, and Likert scaling, Guttman scaling is based on the fact that some items under consideration may prove to be more extreme indicators of the variable than others. One example should suffice to illustrate this pattern.

The construction of a **Guttman scale** would begin with some of the same steps that initiate index construction. You would begin by examining the face validity of items available for analysis. Then, you would examine the bivariate and perhaps multivariate relations among those items. In scale construction, however, you would also look for relatively "hard" and "easy" indicators of the variable being examined.

Earlier, when we talked about attitudes regarding a woman's right to have an abortion, we discussed several conditions that can affect people's opinions: whether the woman is married, whether her health is endangered, and so forth. These differing conditions provide an excellent illustration of Guttman scaling.

Here are the percentages of the people in the 2014 GSS sample who supported a woman's right to an abortion, under three different conditions:

Woman's health is seriously endangered	88%
Pregnant as a result of rape	78%
Woman is not married	43%

The different percentages supporting abortion under the three conditions suggest something about the different levels of support that each item indicates. For example, if someone would support abortion when the mother's life is seriously endangered, that's not a very strong indicator of general support for abortion, because almost everyone agreed with that. Supporting abortion for unmarried women seems a much stronger indicator of support for abortion in general—fewer than half the sample took that position.

Guttman scaling is based on the notion that anyone who gives a strong indicator of some variable will also give the weaker indicators. In this case, we would assume that anyone who supported abortion for unmarried women would also support it in the case of rape or of the woman's health being threatened. Table 6-2 tests this assumption by presenting the number of respondents who gave each of the possible response patterns.

The first four response patterns in the table compose what we would call the *scale types*: those patterns that form a scalar structure. Following those respondents who supported abortion under all three conditions (line 1), we see that those with only two pro-choice responses (line 2) have chosen the two easier ones; those with only one such response (line 3) chose the easiest of the three (the woman's health being endangered). And finally, there are some respondents who opposed abortion in all three circumstances (line 4).

Guttman scale A type of composite measure used to summarize several discrete observations and to represent some more-general variable.

TABLE 6-2

Scaling Support for Choice of Abortion, 2014

	Women's Health Endangered	Result of Rape	Woman Unmarried	Number of Cases
Scale types	+	+	+	699
	+	+	−	508
	+	−	−	176
	−	−	−	156
			Total =	1,539
Mixed types	−	+	−	35
	+	−	+	7
	−	−	+	2
	−	+	+	1
			Total =	45

The second part of the table presents *mixed types*, or those response patterns that violate the scalar structure of the items. The most radical departures from the scalar structure are the last two response patterns: those who accepted only the hardest item and those who rejected only the easiest one.

The final column in the table indicates the number of survey respondents who gave each of the response patterns. The great majority (1,539, or 97 percent) fit into one of the scale types. The presence of mixed types, however, indicates that the items do not form a perfect Guttman scale. (It would be extremely rare for such data to form a Guttman scale perfectly.)

Recall at this point that one of the chief functions of scaling is efficient data reduction. Scales provide a technique for presenting data in a summary form while maintaining as much of the original information as possible. When the scientific orientation items were formed into an index in our earlier discussion, respondents were given one point for each scientific response they gave. If these same three items were scored as a Guttman scale, some respondents would be assigned scale scores that would permit the most accurate reproduction of their original responses to all three items.

In the present example of attitudes regarding abortion, respondents fitting into the scale types would receive the same scores as were assigned in the index construction. Persons selecting all three pro-choice responses would still be scored 3,

those who selected pro-choice responses to the two easier items and were opposed on the hardest item would be scored 2, and so on. For each of the four scale types, we could predict accurately from their scores all the actual responses given by all the respondents.

The mixed types in the table present a problem, however. The first mixed type (− + −) was scored 1 on the index to indicate only one pro-choice response. But, if 1 were assigned as a scale score, we would predict that the 35 respondents in this group had chosen only the easiest item (approving abortion when the woman's life was endangered), and we would be making two errors for each such respondent: thinking their response pattern was (+ − −) instead of (− + −). Scale scores are assigned, therefore, with the aim of minimizing the errors that would be made in reconstructing the original responses.

Table 6-3 illustrates the index and scale scores that would be assigned to each of the response patterns in our example. Note that one error is made for each respondent in the mixed types. This is the minimum we can hope for in a mixed-type pattern. In the first mixed type, for example, we would erroneously predict

TABLE 6-3

Index and Scale Scores

	Response Pattern	Number of Cases	Index Scores	Scale Scores	Total Scale Errors
Scale types	+ + +	699	3	3	0
	+ + −	508	2	2	0
	+ − −	176	1	1	0
	− − −	156	0	0	0
Mixed types	− + −	35	1	2	35
	+ − +	7	2	3	7
	− − +	2	1	0	2
	− + +	1	2	3	1

$$\text{Coefficient of reproducibility} = 1 - \frac{\text{number of errors}}{\text{number of guesses}}$$

$$= 1 - \frac{45}{1,584 \times 3} = 1 - \frac{45}{4,752}$$

$$= 0.9905 = 99\%$$

Note: This table presents one common method for scoring mixed types, but you should be advised that other methods can also be used.

a pro-choice response to the easiest item for each of the 35 respondents in this group, making a total of 35 errors.

The extent to which a set of empirical responses form a Guttman scale is determined by the accuracy with which the original responses can be reconstructed from the scale scores. For each of the 1,584 respondents in this example, we'll predict three questionnaire responses, for a total of 4,752 predictions. Table 6-3 indicates that we'll make 45 errors using the scale scores assigned. The percentage of correct predictions is called the *coefficient of reproducibility*: the percentage of original responses that could be reproduced by knowing the scale scores used to summarize them. In the present example, the coefficient of reproducibility is 99 percent.

Except in the case of perfect (100 percent) reproducibility, there is no way of saying that a set of items does or does not form a Guttman scale in any absolute sense. Virtually all sets of such items approximate a scale. As a general guideline, however, coefficients of 90 or 95 percent are the commonly used standards in this regard. If the observed reproducibility exceeds the coefficient you've specified, you'll probably decide to score and use the items as a scale.

The decision concerning criteria in this regard is, of course, arbitrary. Moreover, a high degree of reproducibility does not ensure that the scale constructed in fact measures the concept under consideration, although it increases confidence that all the component items measure the same thing. Also, you should realize that a high coefficient of reproducibility is most likely when few items are involved.

One concluding remark with regard to Guttman scaling: It's based on the structure observed among the actual data under examination. This important point is often misunderstood. It does not make sense to say that a set of questionnaire items (perhaps developed and used by a previous researcher) constitutes a Guttman scale. Rather, we can say only that they form a scale within a given body of data being analyzed. Scalability, then, is a sample-dependent, empirical matter. Although a set of items may form a Guttman scale among one sample of survey respondents, for example, there is no guarantee that this set will form such a scale among another sample. In this sense, then,

a set of questionnaire items in and of themselves never forms a scale, but a set of empirical observations may.

This concludes our discussion of indexing and scaling. Like indexes, scales are composite measures of a variable, typically broadening the meaning of the variable beyond what might be captured by a single indicator. Both scales and indexes seek to measure variables at the ordinal level of measurement. Unlike indexes, however, scales take advantage of any intensity structure that may be present among the individual indicators. To the extent that such an intensity structure is found and the data from the people or other units of analysis comply with the logic of that intensity structure, we can have confidence that we've created an ordinal measure. At present there are a number of data analysis programs, such as the Statistical Package for Social Sciences (SPSS), which can calculate various scale scores for you, but it is important that you understand the logic of scaling before using those analytic aids.

Typologies

This chapter now ends with a short discussion of typology construction and analysis. Recall that indexes and scales are constructed to provide ordinal measures of given variables. We attempt to assign index or scale scores to cases in such a way as to indicate a rising degree of prejudice, religiosity, conservatism, and so forth. In such cases, we're dealing with single dimensions.

Often, however, the researcher wishes to summarize the intersection of two or more variables, thereby creating a set of categories or types, which we call a **typology**. You may, for example, wish to examine the political orientations of newspapers separately in terms of domestic issues and foreign policy. The fourfold presentation in Table 6-4 describes such a typology.

Newspapers in cell A of the table are conservative on both foreign policy and domestic

---

typology The classification (typically nominal) of observations in terms of their attributes on two or more variables. The classification of newspapers as liberal-urban, liberal-rural, conservative-urban, or conservative-rural would be an example.

TABLE 6-4

A Political Typology of Newspapers

		Foreign Policy	
		Conservative	Liberal
Domestic Policy	Conservative	A	B
	Liberal	C	D

policy; those in cell D are liberal on both. Those in cells B and C are conservative on one and liberal on the other.

As another example, Rodney Coates (2006) created a typology of "racial hegemony" from two dimensions:

1. Political Ideology
   a. Democratic
   b. Non-Democratic
2. Military and Industrial Sophistication
   a. Low
   b. High

He then used the typology to examine modern examples of colonial rule, with specific reference to race relations. The cases he looked at allowed him to illustrate and refine the typology. He points out that such a device represents Weber's "ideal type":

> As stipulated by Weber, ideal types represent a type of abstraction from reality. These abstractions, constructed from the logical extraction of elements derived from specific examples, provide a theoretical model by which and from which we may examine reality.

(Coates 2006: 87)

Frequently, you arrive at a typology in the course of an attempt to construct an index or scale. The items that you felt represented a single variable appear to represent two. You might have been attempting to construct a single index of political orientations for newspapers but discovered—empirically—that foreign and domestic politics had to be kept separate.

In any event, you should be warned against a difficulty inherent in typological analysis. Whenever the typology is used as the independent variable, there will likely be no problem. In the preceding example, you might compute the percentages of newspapers in each cell that normally endorse Democratic candidates; you could then easily examine the effects of both foreign and domestic policies on political endorsements.

It's extremely difficult, however, to analyze a typology as a dependent variable. If you want to discover why newspapers fall into the different cells of typology, you're in trouble. That becomes apparent when we consider the ways you might construct and read your tables. Assume, for example, that you want to examine the effects of community size on political policies. With a single dimension, you could easily determine the percentages of rural and urban newspapers that were scored conservative and liberal on your index or scale.

With a typology, however, you would have to present the distribution of the urban newspapers in your sample among types A, B, C, and D. Then you would repeat the procedure for the rural ones in the sample and compare the two distributions. Let's suppose that 80 percent of the rural newspapers are scored as type A (conservative on both dimensions) as compared with 30 percent of the urban ones. Moreover, suppose that only 5 percent of the rural newspapers are scored as type B (conservative only on domestic issues) as compared with 40 percent of the urban ones. It would be incorrect to conclude from an examination of type B that urban newspapers are more conservative on domestic issues than are rural ones because 85 percent of the rural newspapers, compared with 70 percent of the urban ones, have this characteristic. The relative scarcity of rural newspapers in type B is due to their concentration in type A. It should be apparent that an interpretation of such data would be very difficult for anything other than description.

In reality, you'd probably examine two such dimensions separately, especially if the dependent variable has more categories of responses than does the example given.

Don't think that typologies should always be avoided in social research; often they provide the most appropriate device for understanding the data. To examine the pro-choice orientation in depth, you might create a typology involving both abortion and capital punishment. Libertarianism could be seen in terms of both economic and social permissiveness. You have been warned, however, against the special difficulties involved in using typologies as dependent variables.

What do you think?...Revisited

If I were to tell you that we had given each respondent one point for every association they were willing to have with sex offenders, and I told you further that a particular respondent had been given a score of 3, would you be able to reproduce each of these five answers?

Although this logic is very clear in the case of the Bogardus social distance scale, we've also seen how social researchers approximate that structure in creating other types of scales, such as the Thurstone and Guttman scales, which also take account of differing intensities among the indicators of a variable.

1.	Are you willing to let sex offenders live in your country?	YES
2.	Are you willing to let sex offenders live in your community?	YES
3.	Are you willing to let sex offenders live in your neighborhood?	YES
4.	Would you be willing to let a sex offender live next door to you?	NO
5.	Would you let your child marry a sex offender?	NO

MAIN POINTS

Introduction

- Single indicators of variables seldom capture all the dimensions of a concept, have sufficient validity to warrant their use, or permit the desired range of variation to allow ordinal rankings. Composite measures, such as scales and indexes, solve these problems by including several indicators of a variable in one summary measure.

Indexes versus Scales

- Although both indexes and scales are intended as ordinal measures of variables, scales typically satisfy this intention better than do indexes.

- Whereas indexes are based on the simple cumulation of indicators of a variable, scales take advantage of any logical or empirical intensity structures that exist among a variable's indicators.

Index Construction

- The principal steps in constructing an index include selecting possible items, examining their empirical relationships, scoring the index, handling missing data, and validating it.

- Criteria of item selection include face validity, unidimensionality, the degree of specificity with which a dimension is to be measured, and the amount of variance provided by the items.

- If different items are indeed indicators of the same variable, then they should be related empirically to one another. In constructing an index, the researcher needs to examine bivariate and multivariate relationships among the items.

- Index scoring involves deciding the desirable range of scores and determining whether items will have equal or different weights.

- Various techniques allow items to be used in an index in spite of missing data.

- Item analysis is a type of internal validation based on the relationship between individual items in the composite measure and the measure itself. External validation refers to the relationships between the composite measure and other indicators of the variable—indicators not included in the measure.

Scale Construction

- Four types of scaling techniques were discussed: the Bogardus social distance scale, a device for measuring the varying degrees to which a person would be willing to associate with a given class of people; Thurstone scaling, a technique that uses judges to determine the intensities of different indicators; Likert scaling, a measurement technique based on the use of standardized response categories; and Guttman scaling, a method of discovering and using the empirical intensity structure among several indicators of a given variable. Guttman scaling is probably the most popular scaling technique in social research today.

- The semantic differential is a question format that asks respondents to make ratings that fall between two extremes, such as "very positive" and "very negative."

Typologies

- A typology is a nominal composite measure often used in social research. Typologies can be used effectively as independent variables, but interpretation is difficult when they are used as dependent variables.

KEY TERMS

Bogardus social distance scale	Likert scale
external validation	scale
Guttman scale	semantic differential
index	Thurstone scale
item analysis	Typology

PROPOSING SOCIAL RESEARCH: COMPOSITE MEASURES

This chapter has extended the issue of measurement to include those in which variables are measured by more than one indicator. What you've learned here may extend the discussion of measurement in your proposal. As in the case of operationalization, you may find this easier to formulate in the case of quantitative studies, but the logic of multiple indicators may be applied to all research methods.

If your study will involve the use of composite measures, you should identify the type(s), the indicators to be used in their construction, and the methods you'll use to create and validate them. If the study you're planning in this series of exercises will not include composite measures, you can test your understanding of the chapter by exploring ways in which they could be used, even if you need to temporarily vary the data-collection method and/or variables you have in mind.

REVIEW QUESTIONS

1. In your own words, what is the difference between an index and a scale?

2. Suppose you wanted to create an index for rating the quality of colleges and universities. What are three data items that might be included in such an index?

3. Why do you suppose Thurstone scales have not been used more widely in the social sciences?

4. What would be some questionnaire items that could measure attitudes toward nuclear power and that would probably form a Guttman scale?

CHAPTER 7

The Logic of Sampling

iStock.com/Gary Blakeley

Learning Objectives

After studying this chapter, you will be able to . . .

- Highlight some of the key events in the development of sampling in social research.
- Describe what is meant by "nonprobability sampling" and identify several techniques.
- Identify and explain the key elements in probability sampling.
- Explain the relationship between populations and sampling frames in social research.
- Identify and describe several types of probability sampling designs.
- Describe the steps involved in selecting a multistage cluster sample.
- Discuss the key advantages of probability sampling.
- Explain how the sampling design of a study could have ethical implications.

Introduction

One of the most visible uses of survey sampling lies in the political polling that is subsequently tested by election results. Whereas some people doubt the accuracy of sample surveys, others complain that political polls take all the suspense out of campaigns by foretelling the result.

Going into the 2008 presidential elections, pollsters were in agreement as to who would win, in contrast to their experiences in 2000 and 2004, which were closely contested races. Table 7-1 reports polls conducted during the few days preceding the election. Despite some variations, the overall picture they present is amazingly consistent and pretty well matches the election results.

Now, how many interviews do you suppose it took each of these pollsters to come within a couple of percentage points in estimating the behavior of more than 131 million voters? Often fewer than 2,000! In this chapter, we're going to find out how social researchers can achieve such wizardry.

In the 2016 presidential election, the pre-election polls again clustered closely around the actual popular votes for Hillary Clinton and Donald Trump. Most correctly predicted that Secretary Clinton would win the popular vote by 2 or 3 percentage points.

Of course, the president is not elected by the nation's overall popular vote, but by the electoral college, determined by how the votes go in the individual states. Relatively small victories totaling 107,000 votes in three swing states—Michigan, Pennsylvania, Wisconsin—gave Trump all those states' electoral votes, and the presidency, while Clinton won the popular vote by 2.8 million (Washington Post 2016).

FiveThirtyEight.com offers a useful analysis and rating of the many polling companies active in forecasting political outcomes.

TABLE 7-1
Election-Eve Polls Reporting Presidential Voting Plans, 2008

Poll	Date Ended	Obama	McCain
Fox	Nov 2	54	46
NBC/WSJ	Nov 2	54	46
Marist College	Nov 2	55	45
Harris Interactive	Nov 3	54	46
Reuters/C-SPAN/Zogby	Nov 3	56	44
ARG	Nov 3	54	46
Rasmussen	Nov 3	53	47
IBD/TIPP	Nov 3	54	46
DailyKos.com/Research 2000	Nov 3	53	47
GWU	Nov 3	53	47
Marist College	Nov 3	55	45
Actual vote	Nov 4	54	46

Note: For simplicity, since there were no "undecideds" in the official results and each of the third-party candidates received less than one percentage of the vote, I've apportioned the undecided and other votes according to the percentages saying they were voting for Obama or McCain.

Source: Poll data are adapted from http://www.pollster.com/polls/us/08-us-pres-ge-mvo.php. The official election results are from the Federal Election Commission, http://www.fec.gov/pubrec/fe2008/2008presgeresults.pdf.

What do you think?

In 1936, the *Literary Digest* collected the voting intentions of 2 million voters in order to predict whether Franklin D. Roosevelt or Alf Landon would be elected president of the United States. During more-recent election campaigns, with many more voters going to the polls, national polling firms have typically sampled around 2,000 voters across the country.

Which technique do you think is the most effective? Why? See the *What do you think? ... Revisited* box toward the end of the chapter.

For another powerful illustration of the potency of sampling, look at Figure 7-1 for a graph of then-president George W. Bush's approval ratings prior to and following the September 11, 2001, terrorist attacks on the United States. The data reported by several different polling agencies describe the same pattern.

Political polling, like other forms of social research, rests on observations. But neither pollsters nor other social researchers can observe everything that might be relevant to their interests. A critical part of social research, then, is deciding what to observe and what not to observe. If you want to study voters, for example, which voters should you study?

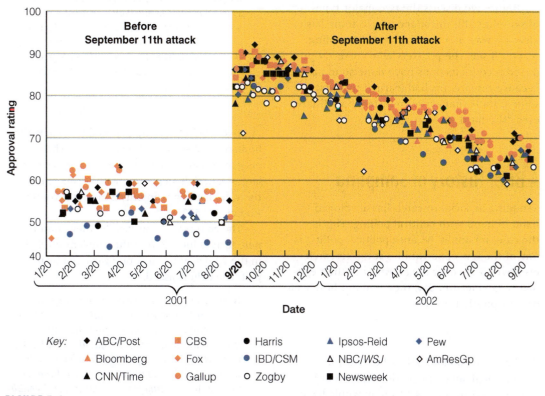

FIGURE 7-1

Bush Approval: Raw Poll Data. This graph demonstrates how independent polls produce the same picture of reality. It also shows the impact of a national crisis on the president's popularity: in this case, the 9/11 terrorist attack and then-president George W. Bush's popularity.

Source: drlimerick.com

The process of selecting observations is called *sampling*. Although sampling can mean any procedure for selecting units of observation—for example, interviewing every tenth passerby on a busy street—the key to generalizing from a sample to a larger population is probability sampling, which involves the important idea of random selection.

Much of this chapter is devoted to the logic and skills of probability sampling. This topic is more rigorous and precise than some of the other topics in this book. Whereas social research as a whole is both art and science, sampling leans toward science. Although this subject is somewhat technical, the basic logic of sampling is not difficult to understand. In fact, the logical neatness of this topic can make it easier to comprehend than, say, conceptualization.

Although probability sampling is central to social research today, we'll also examine a variety of nonprobability methods. These methods have their own logic and can provide useful samples for social inquiry.

Before we discuss the two major types of sampling, I'll introduce you to some basic ideas by way of a brief history of sampling. As you'll see, the pollsters who correctly predicted recent elections have done so in part because researchers had learned to avoid some pitfalls that earlier pollsters had discovered "the hard way."

A Brief History of Sampling

Sampling in social research has developed hand in hand with political polling. This is the case, no doubt, because political polling is one of the few opportunities social researchers have to discover the accuracy of their estimates. On election day, they find out how well or how poorly they did.

President Alf Landon

President Alf Landon? Who's he? Did you sleep through an entire presidency in your U.S. history class? No—but Alf Landon would have been president if a famous poll conducted by the *Literary Digest* had proved to be accurate. The *Literary Digest* was a popular newsmagazine

published between 1890 and 1938. In 1916, *Digest* editors mailed postcards to people in six states, asking them whom they were planning to vote for in the presidential campaign between Woodrow Wilson and Charles Evans Hughes. Names were selected for the poll from telephone directories and automobile registration lists. Based on the postcards sent back, the *Digest* correctly predicted that Wilson would be elected. In the elections that followed, the *Literary Digest* expanded the size of its poll and made correct predictions in 1920, 1924, 1928, and 1932.

In 1936 the *Digest* conducted its most ambitious poll: 10 million ballots were sent to people listed in telephone directories and on lists of automobile owners. Over 2 million people responded, giving the Republican contender, Alf Landon, a stunning 57 to 43 percent landslide over the incumbent, President Franklin Roosevelt. The editors modestly cautioned,

> We make no claim to infallibility. We did not coin the phrase "uncanny accuracy" which has been so freely applied to our Polls. We know only too well the limitations of every straw vote, however enormous the sample gathered, however scientific the method. It would be a miracle if every State of the forty-eight behaved on Election Day exactly as forecast by the Poll.

(Literary Digest *1936a: 6*)

Two weeks later, the *Digest* editors knew the limitations of straw polls even better: The voters gave Roosevelt a second term in office by the largest landslide in history, with 61 percent of the vote. Landon won only 8 electoral votes to Roosevelt's 523.

The editors were puzzled by their unfortunate turn of luck. Part of the problem surely lay in the 22 percent return rate garnered by the poll. The editors asked,

> Why did only one in five voters in Chicago to whom the Digest *sent ballots take the trouble to reply? And why was there a preponderance of Republicans in the one-fifth that did reply?... We were getting better cooperation in what we have always regarded as a public service from Republicans than we were getting from Democrats. Do Republicans live nearer to mailboxes? Do Democrats generally disapprove of straw polls?*

(Literary Digest *1936b: 7*)

Actually, there was a better explanation—and it lay in what is technically called the *sampling frame* used by the *Digest*. In this case the sampling frame consisted of telephone subscribers and automobile owners. In the context of 1936, this design selected a disproportionately wealthy sample of the voting population, especially coming on the tail end of the worst economic depression in the nation's history. The sample effectively excluded poor people, and the poor voted predominantly for Roosevelt's New Deal recovery program. The *Digest*'s poll may or may not have correctly represented the voting intentions of telephone subscribers and automobile owners. Unfortunately for the editors, it decidedly did not represent the voting intentions of the population as a whole.

Basing its decision on early political polls that showed Dewey leading Truman, the *Chicago Tribune* sought to scoop the competition with this unfortunate headline.

President Thomas E. Dewey

The 1936 election also saw the emergence of a young pollster whose name would become synonymous with public opinion. In contrast to the *Literary Digest*, George Gallup correctly predicted that Roosevelt would beat Landon. Gallup's success in 1936 hinged on his use of something called *quota sampling*, which we'll examine later in the chapter. For now, it's enough to know that quota sampling is based on a knowledge of the characteristics of the population being sampled: the proportion of men, the proportion of women, that proportions of various incomes, ages, and so on. Quota sampling selects people to match a set of these characteristics: the right number of poor, white, rural men; the right number of rich, African American, urban women; and so on. The quotas are based on those variables most relevant to the study. In the case of Gallup's poll, the sample selection was based on levels of income; the selection procedure ensured the right proportion of respondents at each income level.

Gallup and his American Institute of Public Opinion used quota sampling to good effect in 1936, 1940, and 1944—correctly picking the presidential winner each time. Then, in 1948, Gallup and most political pollsters suffered the embarrassment of picking Governor Thomas Dewey of New York

over the incumbent, President Harry Truman. The pollsters' miscue continued right up to election night. A famous photograph shows a jubilant Truman—whose followers' battle cry was "Give 'em hell, Harry!"—holding aloft a newspaper with the banner headline "Dewey Defeats Truman."

Several factors accounted for the pollsters' failure in 1948. First, most pollsters stopped polling in early October, despite a steady trend toward Truman toward the end of the campaign. In addition, many voters were undecided throughout the campaign, and they went disproportionately for Truman when they stepped into the voting booth.

More important, Gallup's failure rested on the unrepresentativeness of his samples. Quota sampling—which had been effective in earlier years—was Gallup's undoing in 1948. This technique requires that the researcher know something about the total population (of voters, in this instance). For national political polls, such information came primarily from census data. By 1948, however, World War II had produced a massive movement from the country to cities, radically changing the character of the U.S. population from what the 1940 census showed, and Gallup relied on 1940 census data. City dwellers, moreover, tended to vote Democratic; hence, the overrepresentation of rural voters in his poll had the effect of underestimating the number of Democratic votes.

Two Types of Sampling Methods

By 1948 some academic researchers had already been experimenting with a form of sampling based on probability theory. This technique involves the selection of a "random sample" from a list containing the names of everyone in the population being sampled. By and large, the probability-sampling methods used in 1948 were far more accurate than quota-sampling techniques.

Today, probability sampling remains the primary method of selecting large, representative samples for social research, including national political polls. At the same time, probability sampling can be impossible or inappropriate in many research situations. Accordingly, before turning to the logic and techniques of probability sampling, we'll first take a look at techniques for nonprobability sampling and how they're used in social research.

Nonprobability Sampling

Social research is often conducted in situations that do not permit the kinds of probability samples used in large-scale social surveys. Suppose you wanted to study homelessness: There is no list of all homeless individuals, nor are you likely to create such a list. Moreover, as you'll see, there are times when probability sampling would not be appropriate even if it were possible. Many such situations call for **nonprobability sampling**.

In this section, we'll examine four types of nonprobability sampling: reliance on available subjects, purposive or judgmental sampling, snowball sampling, and quota sampling. We'll conclude with a brief discussion of techniques for obtaining information about social groups through the use of informants.

> **nonprobability sampling** Any technique in which samples are selected in some way not suggested by probability theory. Examples include reliance on available subjects as well as purposive (judgmental), snowball, and quota sampling.

Reliance on Available Subjects

Relying on available subjects, such as stopping people at a street corner or some other location, is sometimes called "convenience" or "haphazard" sampling. This is a common method for journalists in their "person-on-the-street" interviews, but it is an extremely risky sampling method for social research. Clearly, this method does not permit any control over the representativeness of a sample. It's justified only if the researcher wants to study the characteristics of people passing the sampling point at specified times or if less risky sampling methods are not feasible. Even when this method is justified on grounds of feasibility, researchers must exercise great caution in generalizing from their data. Also, they should alert readers to the risks associated with this method.

University researchers frequently conduct surveys among the students enrolled in large lecture classes. The ease and frugality of this method explains its popularity, but it seldom produces data of any general value. It may be useful for pretesting a questionnaire, but such a sampling method should not be used for a study purportedly describing the student body as a whole.

Consider this report on the sampling design in an examination of knowledge and opinions about nutrition and cancer among medical students and family physicians:

> *The fourth-year medical students of the University of Minnesota Medical School in Minneapolis comprised the student population in this study. The physician population consisted of all physicians attending a "Family Practice Review and Update" course sponsored by the University of Minnesota Department of Continuing Medical Education.*
>
> *(Cooper-Stephenson and Theologides 1981: 472)*

After all is said and done, what will the results of this study represent? They do not provide a meaningful comparison of medical students and family physicians in the United States or even in Minnesota. Who were the physicians who attended the course? We can guess that they were probably more concerned about their continuing education than were other physicians, but we can't say for sure.

Although such studies can provide useful insights, we must take care not to overgeneralize from them.

Purposive or Judgmental Sampling

Sometimes it's appropriate to select a sample on the basis of knowledge of a population, its elements, and the purpose of the study. This type of sampling is called **purposive sampling** (or *judgmental sampling*). In the initial design of a questionnaire, for example, you might wish to select the widest variety of respondents to test the broad applicability of questions. Although the study findings would not represent any meaningful population, the test run might effectively uncover any peculiar defects in your questionnaire. This situation would be considered a pretest, however, rather than a final study.

In some instances, you may wish to study a small subset of a larger population in which many members of the subset are easily identified, but the enumeration of them all would be nearly impossible. For example, you might want to study the leadership of a student-protest movement; many of the leaders are visible, but it would not be feasible to define and sample all leaders. In studying all or a sample of the most visible leaders, you may collect data sufficient for your purposes.

Or let's say you want to compare left-wing and right-wing students. Because you may not be able to enumerate and sample from all such students, you might decide to sample the memberships of left- and right-leaning groups, such as the Green Party and the Young Americans for Freedom. Although such a sample design would not provide a good description of either left-wing or right-wing students as a whole, it might suffice for general comparative purposes.

Field researchers are often particularly interested in studying deviant cases—those do not fit into patterns of mainstream attitudes and behaviors—in order to improve their understanding of the more usual pattern. For example, you might gain important insights into the nature of school spirit, as exhibited at a pep rally, by interviewing people who did not appear to be caught up in the emotions of the crowd or by interviewing students who did not attend the

rally at all. Selecting deviant cases for study is another example of purposive study.

In qualitative research projects, the sampling of subjects may evolve as the structure of the situation being studied becomes clearer and certain types of subjects seem more central to understanding than others. Let's say you're conducting an interview study among the members of a radical political group on campus. You may initially focus on friendship networks as a vehicle for the spread of group membership and participation. In the course of your analysis of the earlier interviews, you may find several references to interactions with faculty members in one of the social science departments. As a consequence, you may expand your sample to include faculty in that department and other students that they interact with. This is called "theoretical sampling," since the evolving theoretical understanding of the subject steers the sampling in certain directions.

Snowball Sampling

Another nonprobability-sampling technique, which some consider to be a form of accidental sampling, is called **snowball sampling**. This procedure is appropriate when the members of a special population are difficult to locate, such as homeless individuals, migrant workers, or undocumented immigrants. In snowball sampling, the researcher collects data on the few members of the target population he or she can locate, then asks those individuals to provide the information needed to locate other members of that population whom they happen to know. "Snowball" refers to the process of accumulation as each located subject suggests other subjects. Because this procedure also results in samples with questionable representativeness, it's used primarily for exploratory purposes. Sometimes, the term *chain referral* is used in reference to

---

purposive sampling A type of nonprobability sampling in which the units to be observed are selected on the basis of the researcher's judgment about which ones will be the most useful or representative. Also called *judgmental sampling*.

snowball sampling A nonprobability-sampling method, often employed in field research, whereby each person interviewed may be asked to suggest additional people for interviewing.

snowball sampling and other, similar techniques in which the sample unfolds and grows from an initial selection.

Suppose you wish to learn a community organization's pattern of recruitment over time. You might begin by interviewing fairly recent recruits, then asking them who introduced them to the group. You might then interview the people named, asking *them* who introduced them to the group. You might then interview the next round of people named, and so forth. Or, in studying a loosely structured political group, you might ask one of the participants who he or she believes to be the most influential members of the group. You might interview those people and, in the course of the interviews, ask who they believe to be the most influential. In each of these examples, your sample would "snowball" as each of your interviewees suggested other people to interview.

In another example, Karen Farquharson (2005) provides a detailed discussion of how she used snowball sampling to discover a network of tobacco policy makers in Australia: both those at the core of the network and those on the periphery.

Kath Browne (2005) used snowballing through social networks to develop a sample of nonheterosexual women in a small town in the United Kingdom. She reports that her own membership in such networks greatly facilitated this type of sampling and that potential subjects in the study were more likely to trust her than to trust heterosexual researchers.

In more general, theoretical terms, Chaim Noy argues that the process of selecting a snowball sample reveals important aspects of the populations being sampled: "the dynamics of natural and organic social networks" (2008: 329). Do the people you interview know others like themselves? Are they willing to identify those people to researchers? In this way, snowball sampling can be more than a simple technique for finding people to study. It, in itself, can be a revealing part of the inquiry.

> **quota sampling** A type of nonprobability sampling in which units are selected for a sample on the basis of prespecified characteristics, so that the total sample will have the same distribution of characteristics assumed to exist in the population being studied.

Jaime Waters (2015) discovered some of the limitations of snowball sampling. In an attempt to study adult (over 40) users of illegal drugs, he discovered that his initial subjects were reluctant or unable to identify other users. Partly, this seemed to reflect a feeling that they had more to lose if their drug use were discovered. Also, he found that his adult users were not as involved in drug-using networks as younger users. Still, snowball sampling is sometimes an effective way to reach hard-to-find subjects.

Quota Sampling

Quota sampling is the method that helped George Gallup avoid disaster in 1936—and set up the disaster of 1948. Like probability sampling, quota sampling addresses the issue of representativeness, although the two methods approach the issue quite differently.

Quota sampling begins with a matrix, or table, describing the characteristics of the target population. Depending on your research purposes, you may need to know what proportion of the population is male and what proportion female, as well as what proportions of each sex fall into various categories of age, educational level, ethnic group, and so forth. In establishing a national quota sample, you might need to know what proportion of the national population is urban, Eastern, male, under 25, white, working class, and the like, and all the possible combinations of these attributes.

Once you've created such a matrix and assigned a relative proportion to each cell in the matrix, you proceed to collect data from people having all the characteristics of a given cell. You then assign to all the people in a given cell a weight appropriate to their portion of the total population. When all the sample elements are so weighted, the overall data should provide a reasonable representation of the total population.

Although quota sampling resembles probability sampling, it has several inherent problems. First, the quota frame (the proportions that different cells represent) must be accurate, and it's often difficult to get up-to-date information for this purpose. The Gallup failure to predict Truman as the presidential victor in 1948 stemmed partly from this problem. Second, the selection of sample elements within a given cell may be biased even if its proportion of the population is accurately estimated. Instructed to interview five people who meet a given, complex set of characteristics, an interviewer may still avoid people living at the top of seven-story walk-ups, having particularly run-down homes, or owning vicious dogs.

In recent years, some researchers have attempted to combine probability and quota-sampling methods, but the effectiveness of this effort remains to be seen. At present, you should treat quota sampling warily if your purpose is statistical description.

At the same time, the logic of quota sampling can sometimes be applied usefully to a field research project. In the study of a formal group, for example, you might wish to interview both leaders and nonleaders. In studying a student political organization, you might want to interview radical, moderate, and conservative members of that group. You may be able to achieve sufficient representativeness in such cases by using quota sampling to ensure that you interview both men and women, both younger and older people, and so forth.

J. Michael Brick (2011), in pondering the future of survey sampling, suggested the possibility of a rebirth for quota sampling. Perhaps it is a workable solution to the problem of representativeness that bedevils falling response rates and online surveys. We'll return to this issue in Chapter 9 on survey research.

Selecting Informants

When field research involves the researcher's attempt to understand some social setting—a juvenile gang or local neighborhood, for example—much of that understanding will come from a collaboration with some members of the group being studied. Whereas social researchers speak of *respondents* as people who provide information about themselves, allowing the researcher to construct a composite picture of the group those respondents represent, an **informant** is a member of the group who can talk directly about the group per se.

Anthropologists in particular depend on informants, but other social researchers rely on them as well. If you wanted to learn about informal social networks in a local public-housing project, for example, you would do well to locate individuals who understand what you are looking for and help you find it.

When Jeffrey Johnson (1990) set out to study a salmon-fishing community in North Carolina, he used several criteria to evaluate potential informants. Did their positions allow them to interact regularly with other members of the camp, for example, or were they isolated? (He found that the carpenter had a wider range of interactions than did the boat captain.) Was their information about the camp limited to their specific jobs, or did it cover many aspects of the operation? These and other criteria helped determine how useful the potential informants might be to his study. We'll return to this example in a bit.

Usually, you'll want to select informants who are somewhat typical of the groups you're studying. Otherwise, their observations and opinions may be misleading. Interviewing only physicians will not give you a well-rounded view of how a community medical clinic is working, for example. Along the same lines, an anthropologist who interviews only men in a society where women are sheltered from outsiders will get a biased view. Similarly, although informants fluent in English are convenient for English-speaking researchers from the United States, they do not typify the members of many societies or even many subgroups within English-speaking countries.

Simply because they're the ones willing to work with outside investigators, informants will almost always be somewhat "marginal" or atypical within their group. Sometimes this is obvious. Other times, however, you'll learn about their marginality only in the course of your research.

In Johnson's study, a county agent identified one fisherman who seemed squarely in the mainstream of the community. Moreover, he was cooperative and helpful to Johnson's research. The more

informant Someone who is well versed in the social phenomenon that you wish to study and who is willing to tell you what he or she knows about it. Not to be confused with a respondent.

Earl Babbie

With so many possible informants, how can the researcher begin to choose?

Johnson worked with the fisherman, however, the more he found the man to be a marginal member of the fishing community.

> First, he was a Yankee in a southern town. Second, he had a pension from the Navy [so he was not seen as a "serious fisherman" by others in the community].... Third, he was a major Republican activist in a mostly Democratic village. Finally, he kept his boat in an isolated anchorage, far from the community harbor.
>
> (Johnson 1990: 56)

Informants' marginality may not only bias the view you get but also limit their access (and hence yours) to the different sectors of the community you wish to study.

These comments should give you some sense of the concerns involved in nonprobability sampling, typically used in qualitative research projects. I conclude with the following injunction from John Lofland, a particularly thoughtful and experienced qualitative researcher:

> Your overall goal is to collect the richest possible data. By rich data, we mean a wide and diverse range of information collected over a relatively prolonged period of time in a persistent and systematic manner. Ideally, such data enable you to grasp the meanings associated with the actions of those you are studying and to understand the contexts in which those actions are embedded.
>
> (Lofland et al. 2006: 15)

probability sampling The general term for samples selected in accordance with probability theory, typically involving some random-selection mechanism. Specific types of probability sampling include EPSEM, PPS, simple random sampling, and systematic sampling.

In other words, nonprobability sampling does have its uses, particularly in qualitative research projects. But researchers must take care to acknowledge the limitations of nonprobability sampling, especially regarding accurate and precise representations of populations. This point will become clearer as we discuss the logic and techniques of probability sampling.

The Logic and Techniques of Probability Sampling

Although appropriate to some research purposes, nonprobability-sampling methods cannot guarantee that the sample we observed is representative of the whole population. When researchers want precise, statistical descriptions of large populations—for example, the percentage of the population that is unemployed, that plans to vote for Candidate X, or that feels a rape victim should have the right to an abortion—they turn to **probability sampling**. All large-scale surveys use probability-sampling methods.

Although the application of probability sampling involves a somewhat sophisticated use of statistics, the basic logic of probability sampling is not difficult to understand. If all members of a population were identical in all respects—all demographic characteristics, attitudes, experiences, behaviors, and so on—there would be no need for careful sampling procedures. In this extreme case of perfect homogeneity, in fact, any single case would suffice as a sample to study characteristics of the whole population.

In fact, of course, the human beings who compose any real population are quite heterogeneous, varying in many ways. Figure 7-2 offers a simplified illustration of a heterogeneous population: The 100 members of this small population differ by gender and race. We'll use this hypothetical micropopulation to illustrate various aspects of probability sampling.

The fundamental idea behind probability sampling is this: In order to provide useful descriptions of the total population, a sample of individuals from a population must contain

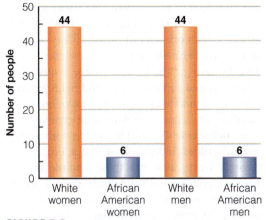

FIGURE 7-2

A Population of 100 Folks. Typically, sampling aims at reflecting the characteristics and dynamics of large populations. For the purpose of some simple illustrations, let's assume our total population has only 100 members.

essentially the same variations that exist in the population. This isn't as simple as it might seem, however. Let's take a minute to look at some of the ways researchers might go astray. Then, we'll see how probability sampling provides an efficient method for selecting a sample that should adequately reflect variations that exist in the population.

Conscious and Subconscious Sampling Bias

At first glance, it may look as though sampling is pretty straightforward. To select a sample of 100 university students, you might simply interview the first 100 students you find walking around campus. Although untrained researchers often use this kind of sampling method, it runs a high risk of introducing biases into the samples.

In connection with sampling, *bias* simply means that those selected are not typical or representative of the larger populations they've been chosen from. This kind of bias does not have to be intentional. In fact, it's virtually inevitable when you pick people by the seat of your pants.

Figure 7-3 illustrates what can happen when researchers simply select people who are convenient for study. Although women make up 50 percent of our micropopulation, the people closest to the researcher (in the lower right corner) happen to be 70 percent women, and although the population is 12 percent African American, none were selected for the sample.

Beyond the risks inherent in simply studying people who are convenient, other problems can arise. To begin, the researcher's personal leanings

FIGURE 7-3

A Sample of Convenience: Easy, but Not Representative. Selecting and observing those people who are most readily at hand is the simplest method, perhaps, but it's unlikely to provide a sample that accurately reflects the total population.

may affect the sample to the point where it does not truly represent the student population. Suppose you're a little intimidated by students who look particularly "cool," feeling that they might ridicule your research effort. You might consciously or subconsciously avoid interviewing such people. Or, you might feel that the attitudes of "super-straight-looking" students would be irrelevant to your research purposes, and so you avoid interviewing them.

Even if you sought to interview a "balanced" group of students, you wouldn't know the exact proportions of different types of students making up such a balance, and you wouldn't always be able to identify the different types just by watching them walk by.

Further, even if you made a conscientious effort to interview, say, every tenth student entering the university library, you could not be sure of a representative sample, because different types of students visit the library with different frequencies. Your sample would overrepresent students who visit the library more often than do others.

Similarly, the "public opinion" call-in polls—in which radio stations or newspapers ask people to call specified telephone numbers, text, or tweet to register their opinions—cannot be trusted to represent general populations. At the very least, not everyone in the population will even be aware of the poll. This problem also invalidates polls by magazines and newspapers who publish questionnaires for readers to complete and mail in. Even among those who are aware of such polls, not all will express an opinion, especially if doing so will cost them a stamp, an envelope, and their time. Similar considerations apply to polls taken over the Internet.

Ironically, the failure of such polls to represent all opinions equally was inadvertently acknowledged by Phillip Perinelli (1986), a staff manager of AT&T Communications' DIAL-IT 900 Service, which offers a call-in poll facility to organizations. Perinelli attempted to counter criticisms by saying, "The 50-cent charge assures that only interested parties respond and helps assure also that no individual 'stuffs' the ballot box." Social researchers cannot determine general public opinion while considering "only interested parties." This excludes those who don't care 50-cents' worth, as well as those who recognize that such polls are not valid. Both types of people may have opinions and may even vote on election day. Perinelli's assertion that the 50-cent charge will prevent ballot stuffing actually means that only those who can afford it will engage in ballot stuffing.

The possibilities for inadvertent sampling bias are endless and not always obvious. Fortunately, several techniques can help us avoid bias.

Representativeness and Probability of Selection

Although the term **representativeness** has no precise, scientific meaning, it carries a commonsense meaning that makes it useful here. For our purpose, a sample is representative of the population from which it is selected if the aggregate characteristics of the sample closely approximate those same aggregate characteristics in the population. If, for example, the population contains 50 percent women, then a sample must contain "close to" 50 percent women to be representative. Later, we'll discuss "how close" in detail. See "Applying Concepts in Everyday Life: Representative Sampling" for more on this.

Note that samples need not be representative in all respects; representativeness concerns only those characteristics that are relevant to the substantive interests of the study. However, you may not know in advance which characteristics are relevant.

A basic principle of probability sampling is that a sample will be representative of the population from which it is selected if all members of the population have an equal chance of being selected for the sample. (We'll see shortly that the size of the sample selected also affects the degree of

---

representativeness That quality of a sample of having the same distribution of characteristics as the population from which it was selected. By implication, descriptions and explanations derived from an analysis of the sample may be assumed to represent similar ones in the population. Representativeness is enhanced by probability sampling and provides for generalizability and the use of inferential statistics.

Applying Concepts in Everyday Life

Representative Sampling

Representativeness applies to many areas of life, not just survey sampling. Consider quality control, for example. Imagine running a company that makes light bulbs. You want to be sure that they actually light up, but you can't test them all. You could, however, devise a method of selecting a sample of bulbs drawn from different times in the production day, on different machines, in different factories, and so forth.

Sometimes the concept of representative sampling serves as a protection against overgeneralization, discussed in Chapter 1.

Suppose you go to a particular restaurant and don't like the food or service. You're ready to cross it off your list of dining possibilities, but then you think about it—perhaps you hit them on a bad night. Perhaps the chef had just discovered her boyfriend in bed with that "witch" from the Saturday wait staff and her mind wasn't on her cooking. Or perhaps the "witch" was serving your table and kept looking over her shoulder to see if anyone with a meat cleaver was bursting out of the kitchen. In short, your first experience might not have been representative.

---

representativeness.) Samples that have this quality are often labeled **EPSEM** samples (EPSEM stands for "equal probability of selection method"). Later we'll discuss variations of this principle, which forms the basis of probability sampling.

Moving beyond this basic principle, we must realize that samples—even carefully selected EPSEM samples—seldom, if ever, perfectly represent the populations from which they are drawn. Nevertheless, probability sampling offers two special advantages.

First, probability samples, although never perfectly representative, are typically more representative than other types of samples, because the biases previously discussed are avoided. In practice, a probability sample is more likely than a nonprobability sample to be representative of the population from which it is drawn.

Second, and more important, probability theory permits us to estimate the accuracy or representativeness of the sample. Conceivably, an uninformed researcher might, through wholly haphazard means, select a sample that nearly perfectly represents the larger population. The odds are against doing so, however, and we would be unable to estimate the likelihood that he or she has achieved representativeness. The probability sample, on the other hand, can provide an accurate estimate of success or failure. Shortly we'll see exactly how this estimate can be achieved.

I've said that probability sampling ensures that samples are representative of the population we wish to study. As we'll see in a moment, probability sampling rests on the use of a random-selection procedure. To develop this idea, though,

we need to give more-precise meaning to two important terms: *element* and *population*.

An **element** is that unit about which information is collected and that provides the basis of analysis. Typically, in survey research, elements are people or certain types of people. However, other kinds of units can constitute the elements of social research: Families, social clubs, or corporations might be the elements of a study. In a given study, elements are often the same as units of analysis, though the former are used in sample selection and the latter in data analysis.

Up to now we've used the term *population* to mean the group or collection that we're interested in generalizing about. More formally, a **population** is the theoretically specified aggregation of study elements. Whereas the vague term *Americans* might be the target for a study, the delineation of the population would include the definition of the element "Americans" (for example, citizenship, residence) and the time referent for the study (Americans as of when?). Translating the abstract "adult New Yorkers" into a workable population would require a specification of the age defining *adult*

EPSEM (equal probability of selection method) A sample design in which each member of a population has the same chance of being selected for the sample.

element That unit of which a population is composed and that is selected for a sample. Elements are distinguished from units of analysis, which are used in data analysis.

population The theoretically specified aggregation of the elements in a study.

and the boundaries of New York. Specifying "college student" would include a consideration of full- and part-time students, degree candidates and non–degree candidates, undergraduate and graduate students, and so forth.

A **study population** is that aggregation of elements from which the sample is actually selected. As a practical matter, researchers are seldom in a position to guarantee that every element meeting the theoretical definitions laid down actually has a chance of being selected in the sample. Even where lists of elements exist for sampling purposes, the lists are usually somewhat incomplete. Some students are always inadvertently omitted from student rosters. Some telephone subscribers have unlisted numbers.

Often, researchers decide to limit their study populations more severely than indicated in the preceding examples. National polling firms may limit their national samples to the 48 adjacent states, omitting Alaska and Hawaii for practical reasons. A researcher wishing to sample psychology professors may limit the study population to those in psychology departments, omitting those in other departments. Whenever the population under examination is altered in such fashion, you must make the revisions clear to your readers.

Random Selection

With these definitions in hand, we can define the ultimate purpose of sampling: to select a set of elements from a population in such a way that descriptions of those elements accurately portray the total population from which the elements are selected. Probability sampling enhances the likelihood of accomplishing this aim and also provides methods for estimating the degree of probable success.

Random selection is the key to this process. In **random selection**, each element has an equal chance of being selected independently of any other event in the selection process. Flipping a coin is the most frequently cited example: Provided that the coin is perfect (that is, not biased in terms of coming up heads or tails), the "selection" of a head or a tail is independent of previous selections of heads or tails. No matter how many heads turn up in a row, the chance that the next flip will produce "heads" is exactly 50–50. Rolling a perfect set of dice is another example.

Such images of random selection, though useful, seldom apply directly to sampling methods in social research. More typically, social researchers use tables of random numbers or computer programs that provide a random selection of sampling units. A **sampling unit** is that element or set of elements considered for selection at some stage of sampling. A little later, we'll see how computers are used to select random telephone numbers for interviewing, a technique called random-digit dialing.

There are two reasons for using random-selection methods. First, this procedure serves as a check on conscious or subconscious bias on the part of the researcher. The researcher who selects cases on an intuitive basis might very well select cases that would support his or her research expectations or hypotheses. Random selection erases this danger. Second, and more important, random selection offers access to the body of probability theory, which provides the basis for estimating the characteristics of the population as well as estimates of the precision of sample results. Now let's examine probability theory in greater detail.

Probability Theory, Sampling Distributions, and Estimates of Sampling Error

Probability theory is a branch of mathematics that provides the tools researchers need (1) to devise sampling techniques that produce representative samples and (2) to statistically analyze the results of their sampling. More formally, probability theory provides the basis for estimating the parameters of a population. A **parameter** is the summary description of a given variable in a population. The mean income of all families in a city is a

---

study population That aggregation of elements from which a sample is actually selected.

random selection A sampling method in which each element has an equal chance of being selected independently of any other event in the selection process.

sampling unit That element or set of elements considered for selection in some stage of sampling.

parameter The summary description of a given variable in a population.

How would researchers conduct a random sample of this neighborhood? What are the pitfalls they would need to avoid?

Earl Babbie

parameter; so is the age distribution of the city's population. When researchers generalize from a sample, they're using sample observations to estimate population parameters. Probability theory enables them to make these estimates and also to arrive at a judgment of how likely it is that the estimates will accurately represent the actual parameters in the population. So, for example, probability theory allows pollsters to infer from a sample of 2,000 voters how a population of 100 million voters is likely to vote—and to specify exactly what the probable margin of error in the estimates is.

Probability theory accomplishes these seemingly magical feats by way of the concept of sampling distributions. A single sample selected from a population will give an estimate of the population parameter. Other samples would give the same or slightly different estimates. Probability theory tells us about the distribution of estimates that would be produced by a large number of such samples.

The logic of sampling error can be applied to different kinds of measurements: mean income or mean age, for example. Measurements expressed as percentages, however, provide the simplest introduction to this general concept.

To see how this works, we'll look at two examples of sampling distributions, beginning with a simple example in which our population consists of just ten cases.

The Sampling Distribution of Ten Cases

Suppose that there are ten people in a group and that each has a certain amount of money in his or her pocket. To simplify, let's assume that one person has no money, another has one dollar, another has two dollars, and so forth up to the person with nine dollars. Figure 7-4 presents the population of ten people.

FIGURE 7-4

A Population of Ten People with $0 to $9. Let's imagine a population of only ten people with differing amounts of money in their pockets—ranging from $0 to $9.

Our task is to determine the average amount of money one person has: specifically, the mean number of dollars. If you simply add up the money shown in Figure 7-4, you'll find that the total is $45, so the mean is $4.50. Our purpose in the rest of this exercise is to estimate that mean without actually observing all ten individuals. We'll do that by selecting random samples from the population and using the means of those samples to estimate the mean of the whole population.

To start, suppose we were to select—at random—a sample of only one person from the ten. Our ten possible samples thus consist of the ten cases shown in Figure 7-4.

The ten dots shown on the graph in Figure 7-5 represent these ten samples. Because we're taking samples of only one, they also represent the "means" we would get as estimates of the population. The distribution of the dots on the graph is called the sampling distribution. Obviously, it wouldn't be a very good idea to select a sample of only one, because we'll very likely miss the true mean of $4.50 by quite a bit.

Now suppose we take a sample of two. As shown in Figure 7-6, increasing the sample size improves our estimations. There are now forty-five possible samples: [$0, $1], [$0, $2], . . . [$7, $8], [$8, $9]. Moreover, some of those samples produce the same means.

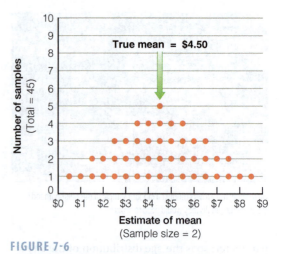

FIGURE 7-6

The Sampling Distribution of Samples of 2. After merely increasing our sample size to two, the possible samples provide somewhat better estimates of the mean. We couldn't get either $0 or $9, and the estimates are beginning to cluster around the true value of the mean: $4.50.

For example, [$0, $6], [$1, $5], and [$2, $4] all produce means of $3. In Figure 7-6, the three dots shown above the $3 mean represent those three samples.

Moreover, the forty-five samples are not evenly distributed, as they were when the sample size was only one. Rather, they cluster somewhat around the true value of $4.50. Only two possible samples deviate by as much as $4 from the true value ([$0, $1] and [$8, $9]), whereas five of the samples give the true estimate of $4.50; another eight samples miss the mark by only 50 cents (plus or minus).

Now suppose we select even larger samples. What do you suppose that will do to our estimates of the mean? Figure 7-7 presents the sampling distributions of samples of 3, 4, 5, and 6.

The progression of sampling distributions is clear. Every increase in sample size improves the distribution of estimates of the mean. The limiting case in this procedure, of course, is to select a sample of ten. There would be only one possible sample (everyone) and it would give us the true mean of $4.50. As we'll see shortly, this principle applies to actual sampling of meaningful populations. The larger the sample selected, the more accurate it is as an estimation of the population from which it was drawn.

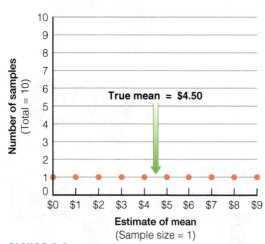

FIGURE 7-5

The Sampling Distribution of Samples of 1. In this simple example, the mean amount of money these people have is $4.50 ($45/10). If we picked ten different samples of one person each, our "estimates" of the mean would range across the board.

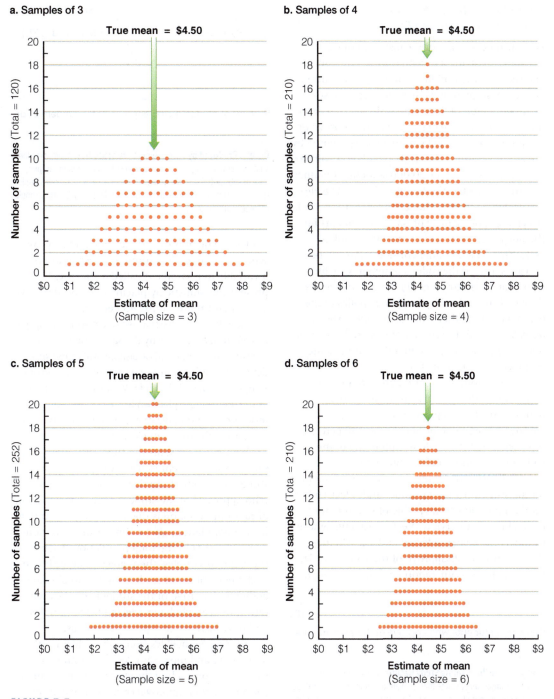

FIGURE 7-7

The Sampling Distributions of Samples of 3, 4, 5, and 6. As we increase the sample size, the possible samples cluster ever more tightly around the true value of the mean. The chance of extremely inaccurate estimates is reduced at the two ends of the distribution, and the percentage of the samples near the true value keeps increasing.

Sampling Distribution and Estimates of Sampling Error

Let's turn now to a more realistic sampling situation involving a much larger population and see how the notion of sampling distribution applies. Assume that we wish to study the student population of State University (SU) to determine the percentage of students who approve or disapprove of a student conduct code proposed by the administration. The study population will be the aggregation of, say, 20,000 students contained in a student roster: the sampling frame. The elements will be the individual students at SU. We'll select a random sample of, say, 100 students for the purposes of estimating the entire student body. The variable under consideration will be attitudes toward the code, a binomial variable comprising the attributes *approve* and *disapprove*. (The logic of probability sampling applies to the examination of other types of variables, such as *mean income*, but the computations are somewhat more complicated. Consequently, this introduction focuses on binomials.)

The horizontal axis of Figure 7-8 presents all possible values of this parameter in the population—from 0 percent to 100 percent approval. The midpoint of the axis—50 percent—represents half the students approving of the code and the other half disapproving.

To choose our sample, we give each student on the student roster a number and select 100 random numbers from a table of random numbers. Then we interview the 100 students whose numbers have been selected and ask whether they approve or disapprove of the student code. Suppose that this operation gives us forty-eight

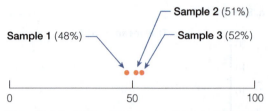

Percent of students who approve of the student code

FIGURE 7-9

Results Produced by Three Hypothetical Studies. Assuming a large student body, let's suppose that we selected three different samples, each of substantial size. We would not necessarily expect those samples to perfectly reflect attitudes in the whole student body, but they should come reasonably close.

students who approve of the code and fifty-two who disapprove. This summary description of a variable in a sample is called a **statistic**. We present this statistic by placing a dot on the *x* axis at the point representing 48 percent.

Now let's suppose that we select another sample of 100 students in exactly the same fashion and measure their approval or disapproval of the student code. Perhaps fifty-one students in the second sample approve of the code. We place another dot in the appropriate place on the *x* axis. Repeating this process once more, we may discover that fifty-two students in the third sample approve of the code.

Figure 7-9 presents the three different sample statistics representing the percentages of students in each of the three random samples who approved of the student code. The basic rule of random sampling is that such samples, drawn from a population, give estimates of the parameter that exists in the total population. Each of the random samples, then, gives us an estimate of the percentage of students in the total student body who approve of the student code. Unhappily, however, we have selected three samples and now have three separate estimates.

To rescue ourselves from this problem, let's draw more and more samples of 100 students each, question each of the students in the samples concerning their approval or disapproval of the code, and plot the new sample statistics on our summary graph. In drawing many such samples, we discover that some of the new samples provide duplicate estimates, as in the illustration of ten cases. Figure 7-10 shows the sampling distribution of, say, hundreds of samples. This is often referred to as a *normal curve*.

Percent of students who approve of the student code

FIGURE 7-8

Range of Possible Sample Study Results. Shifting to a more realistic example, let's assume that we want to sample student attitudes concerning a proposed conduct code. Let's assume that 50 percent of the whole student body approves and 50 percent disapproves—though the researcher doesn't know that.

statistic The summary description of a variable in a sample, used to estimate a population parameter.

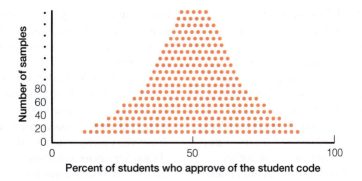

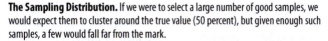

Percent of students who approve of the student code

FIGURE 7-10

The Sampling Distribution. If we were to select a large number of good samples, we would expect them to cluster around the true value (50 percent), but given enough such samples, a few would fall far from the mark.

Note that by increasing the number of samples of students selected and interviewed, we've also increased the range of estimates provided by the sampling operation. In one sense we've increased our dilemma in attempting to guess the parameter in the population. Probability theory, however, provides certain important rules regarding the sampling distribution presented in Figure 7-10.

First, if many independent random samples are selected from a population, the sample statistics provided by those samples will be distributed around the population parameter in a known way. Thus, although Figure 7-10 shows a wide range of estimates, more of them fall near 50 percent than elsewhere in the graph. Probability theory tells us, then, that the true value is in the vicinity of 50 percent.

Second, probability theory gives us a formula for estimating how closely the sample statistics are clustered around the true value. To put it another way, probability theory enables us to estimate the **sampling error**—the degree of error to be expected for a given sample design. This formula contains three factors: the parameter (*P*), the sample size, and the standard error (a measure of sampling error):

$$s = \sqrt{\frac{P \times Q}{n}}$$

The *P* and *Q* in the formula equal the population parameters for the binomial: If 60 percent of the student body approves of the code and 40 percent disapproves, *P* and *Q* are 60 percent and 40 percent, respectively, or 0.6 and 0.4.

The symbol *n* equals the number of cases in each sample, and *s* is the standard error.

Let's assume that the population parameter in the student example is 50 percent approving of the code and 50 percent disapproving. Recall that we've been selecting samples of 100 cases each. When these numbers are put into the formula, we find that the standard error equals 0.05, or 5 percent.

In probability theory, the *standard error* is a valuable piece of information because it indicates the extent to which the sample estimates will be distributed around the population parameter. (If you're familiar with the standard deviation in statistics, you may recognize that the standard error, in this case, is the standard deviation of the sampling distribution.) Specifically, probability theory indicates that certain proportions of the sample estimates will fall within specified increments—each equal to one standard error—from the population parameter. Approximately 34 percent (0.3413) of the sample estimates will fall within one standard error increment above the population parameter, and another 34 percent will fall within one standard error below the parameter. In our example, the standard error increment is 5 percent, so we know that 34 percent of our samples will give estimates of student approval between 50 percent (the parameter) and 55 percent (one standard error above);

sampling error The degree of error to be expected in probability sampling. The formula for determining sampling error contains three factors: the parameter, the sample size, and the standard error.

another 34 percent of the samples will give estimates between 50 percent and 45 percent (one standard error below the parameter). Taken together, then, we know that roughly two-thirds (68 percent) of the samples will give estimates within 5 percentage points of the parameter.

Moreover, probability theory dictates that roughly 95 percent of the samples will fall within plus or minus two standard errors of the true value, and 99.9 percent of the samples will fall within plus or minus three standard errors. In our present example, then, we know that only one sample out of a thousand would give an estimate lower than 35 percent approval or higher than 65 percent.

The proportion of samples falling within one, two, or three standard errors of the parameter is constant for any random-sampling procedure such as the one just described, providing that a large number of samples are selected. The size of the standard error in any given case, however, is a function of the population parameter and the sample size. If we return to the formula for a moment, we note that the standard error will increase as a function of an increase in the quantity P times Q (PQ). Note further that this quantity reaches its maximum in the situation of an even split in the population. If $P = 0.5$, $PQ = 0.25$; if $P = 0.6$, $PQ = 0.24$; if $P = 0.8$, $PQ = 0.16$; if $P = 0.99$, $PQ = 0.0099$. By extension, if P is either 0.0 or 1.0 (either 0 percent or 100 percent approve of the student code), the standard error will be 0. If everyone in the population has the same attitude (no variation), then every sample will give exactly that estimate.

The standard error is also a function of the sample size—an inverse function. As the sample size increases, the standard error decreases. As the sample size increases, more and more samples will be clustered nearer to the true value. Another general guideline is evident in the formula: Because of the square-root formula, the standard error is reduced by half if the sample size is quadrupled.

confidence level The estimated probability that a population parameter lies within a given confidence interval. Thus, we might be 95 percent confident that between 35 and 45 percent of all voters favor Candidate A.

confidence interval The range of values within which a population parameter is estimated to lie.

In our present example, samples of 100 produce a standard error of 5 percent; to reduce the standard error to 2.5 percent, we must increase the sample size to 400.

All of this information is provided by established probability theory in reference to the selection of large numbers of random samples. (If you've taken a statistics course, you may know this as the central tendency theorem.) If the population parameter is known and many random samples are selected, we can predict how many of the sampling estimates will fall within specified intervals from the parameter.

Recognize that this discussion illustrates only the logic of probability sampling; it does not describe the way research is actually conducted. Usually, we don't know the parameter: The very reason we conduct a sample survey is to estimate that value. Moreover, we don't actually select large numbers of samples: We select only one sample. Nevertheless, the preceding discussion of probability theory provides the basis for inferences about the typical social research situation. Knowing what it would be like to select thousands of samples allows us to make assumptions about the one sample we do select and study.

Confidence Levels and Confidence Intervals

Whereas probability theory specifies that 68 percent of that fictitious large number of samples would produce estimates falling within one standard error of the parameter, we can turn the logic around and infer that any single random sample has a 68 percent chance of falling within that range. This observation leads us to the two key components of sampling-error estimates: confidence level and confidence interval. We express the precision of our sample statistics in terms of a level of confidence that the statistics fall within a specified interval from the parameter. For example, we may say we are 95 percent confident that our sample statistics (for example, 50 percent favor the new student code) are within plus or minus 10 percentage points of the population parameter. As the confidence interval is expanded for a given statistic, our confidence increases. For example, we may say that we are 99.9 percent confident that our statistic falls within three standard errors of the true value. (Now perhaps you can appreciate the humorous quip of unknown origin: Statistics means never having to say you are certain.)

Although we may be confident (at some level) of being within a certain range of the parameter, we've already noted that we seldom know what the parameter is. To resolve this problem, we substitute our sample estimate for the parameter in the formula; that is, lacking the true value, we substitute the best available guess.

The result of these inferences and estimations is that we can estimate a population parameter and also the expected degree of error on the basis of one sample drawn from a population. Beginning with the question "What percentage of the student body approves of the student code?" you could select a random sample of 100 students and interview them. You might then report that your best estimate is that 50 percent of the student body approves of the code and that you are 95 percent confident that between 40 and 60 percent (plus or minus two standard errors) approve. The range from 40 to 60 percent is the confidence interval. (At the 68 percent confidence level, the confidence interval would be 45 to 55 percent.)

The logic of confidence levels and confidence intervals also provides the basis for determining the appropriate sample size for a study. Once you've decided on the degree of sampling error you can tolerate, you'll be able to calculate the number of cases needed in your sample. Thus, for example, if you want to be 95 percent confident that your study findings are accurate within plus or minus 5 percentage points of the population parameters, you should select a sample of at least 400. (Appendix E is a convenient guide in this regard.)

This, then, is the basic logic of probability sampling. Random selection permits the researcher to link findings from a sample to the body of probability theory so as to estimate the accuracy of those findings. All statements of accuracy in sampling must specify both a confidence level and a confidence interval. The researcher must report that he or she is x percent confident that the population parameter lies between two specific values. In this example, I've demonstrated the logic of sampling error using a variable analyzed in percentages. Although different statistical procedures would be required to calculate the standard error for a mean, for example, the overall logic is the same.

Notice that nowhere in this discussion of sample size and accuracy of estimates did we consider the size of the population being studied. This is because the population size is almost always irrelevant. A sample of 2,000 respondents drawn properly to represent Vermont voters will be no more accurate than a sample of 2,000 drawn properly to represent all voters in the United States, even though the Vermont sample would be a substantially larger proportion of that small state's voters than would the same number chosen to represent the nation's voters. The reason for this counterintuitive fact is that the equations for calculating sampling error all assume that the populations being sampled are infinitely large, so every sample would equal 0 percent of the whole.

Of course, this is not literally true in practice. A sample of 2,000 represents 0.63 percent of the Vermonters who voted for president in the 2016 election, and a sample of 2,000 U.S. voters represents 0.0014 percent of the national electorate. Nonetheless, both of these proportions are small enough to approach the ideal of a sample taken from an infinitely large population. Further, that proportion remains irrelevant unless a sample represents, say, 5 percent or more of the population it's drawn from. In those rare cases of large proportions being selected, a "finite population correction" can be calculated to adjust the confidence intervals.

The following formula calculates the proportion to be multiplied against the calculated error:

$$\text{finite population correction} = \sqrt{\frac{N - n}{N - 1}}$$

In the formula, N is the population size and n is the size of the sample. Notice that in the extreme case where you studied the whole population (hence, $N - n$), the formula would yield zero as the finite population correction. Multiplying zero times the sampling error calculated by the earlier formula would give a final sampling error of zero, which would, of course, be precisely the case because you wouldn't have sampled at all.

Lest you weary of the statistical nature of this discussion, it's useful to realize what an amazing thing we've been examining. There is remarkable order within what might seem random and chaotic. One of the researchers to whom we owe this observation is Sir Francis Galton (1822–1911):

Order in Apparent Chaos. I know of scarcely anything so apt to impress the imagination as the wonderful form of cosmic order expressed by the "Law of Frequency of Error." The law would have been personified by the Greeks and deified, if they had known of it. It reigns with serenity and in complete self-effacement amidst the wildest confusion. The huger the mob, and the greater the apparent anarchy, the more perfect is its sway. It is the supreme law of Unreason.

(Galton 1889: 66)

Two cautions are in order before we conclude this discussion of the basic logic of probability sampling. First, the survey uses of probability theory as discussed here are technically not wholly justified. The theory of sampling distribution makes assumptions that almost never apply in survey conditions. The exact proportion of samples contained within specified increments of standard errors, for example, mathematically assumes an infinitely large population, an infinite number of samples, and sampling with replacement—that is, every sampling unit selected is "thrown back into the pot" and could be selected again. Second, our discussion has greatly oversimplified the inferential jump from the distribution of several samples to the probable characteristics of one sample.

I offer these cautions to provide perspective on the uses of probability theory in sampling. Social researchers often appear to overestimate the precision of estimates produced by the use of probability theory. As I'll mention elsewhere in this chapter and throughout the book, variations in sampling techniques and nonsampling factors may further reduce the legitimacy of such estimates. For example, those selected in a sample who fail or refuse to participate further detract from the representativeness of the sample.

Over the years, the public has grown accustomed to having the media reports ranges of sampling error to accompany political poll results, for example. However, there is still room for improvement in this regard. During the 2015–2016 presidential campaigns, there were as many as seventeen Republican candidates at one point. This was a problem for the logistics of televised debates. The general solution was to limit the debates to the most popular candidates, leaving out those who had little or no public support. But how was the popularity decision to be made?

Each of the debate lineups was selected somewhat differently, but the basic logic was to determine the target number, say ten, and decide on a set of national polls that would be conducted prior to the debate. Each candidate would receive a score equal to the average popularity that candidate received in the set of specified polls. As reasonable as this might seem, the difference in popularity scores of candidate 10 and candidate 11 was always well within the margin of sampling error, sometimes a fraction of a percentage point. It was a meaningless statistical difference but a profound political difference.

Lest this seem a partisan criticism of the Republican Party, the Democrats used precisely the same hair-splitting technique in the run-up to their 2020 Presidential primary.

Though they can be misused, the calculations discussed in this section can be extremely valuable to you in understanding and evaluating your data. Although the calculations do not provide estimates as precise as some researchers might assume, they can be quite valid for practical purposes. They are unquestionably more valid than less rigorously derived estimates based on less rigorous sampling methods. Most important, being familiar with the basic logic underlying the calculations can help you react sensibly both to your own data and to those reported by others.

Populations and Sampling Frames

The preceding section introduced the theoretical model for social research sampling. Although as students, research consumers, and researchers we need to understand that theory, appreciating the less-than-perfect conditions that exist in the field is no less important. In this section we'll look at one aspect of field conditions that requires a compromise with idealized theoretical conditions and assumptions: the congruence of or disparity between populations of sampling frames.

Simply put, a **sampling frame** is the list or quasi-list of elements from which a probability

sampling frame The list or quasi-list of units composing a population from which a sample is selected. If the sample is to be representative of the population, it is essential that the sampling frame include all (or nearly all) members of the population.

sample is selected. If a sample of students is selected from a student roster, the roster is the sampling frame. If the primary sampling unit for a complex population sample is the census block, the list of census blocks composes the sampling frame—in the form of a printed booklet or, better, some digital format permitting computer manipulation. Here are some reports of sampling frames appearing in research journals. In each example I've italicized the actual sampling frames:

> *The data for this research were obtained from a random sample of* parents of children in the third grade in public and parochial schools in Yakima County, Washington.
>
> *(Petersen and Maynard 1981: 92)*

> *The sample at Time 1 consisted of 160 names drawn randomly from the* telephone directory of Lubbock, Texas.
>
> *(Tan 1980: 242)*

> *The data reported in this paper…were gathered from a probability sample of* adults aged 18 and over residing in households in the 48 contiguous United States. *Personal interviews with 1,914 respondents were conducted by the Survey Research Center of the University of Michigan during the fall of 1975.*
>
> *(Jackman and Senter 1980: 345; emphasis mine)*

Properly drawn samples provide information appropriate for describing the population of elements composing the sampling frame—nothing more. I emphasize this point in view of the all-too-common tendency for researchers to select samples from a given sampling frame and then make assertions about a population similar to, but not identical to, the population defined by the sampling frame.

For example, take a look at this report, which discusses the drugs most frequently prescribed by U.S. physicians:

> *Information on prescription drug sales is not easy to obtain. But Rinaldo V. DeNuzzo, a professor of pharmacy at the Albany College of Pharmacy, Union University, Albany, NY, has been tracking prescription drug sales for 25 years by polling nearby drugstores. He publishes the results in an industry trade magazine, MM&M.*
>
> *DeNuzzo's latest survey, covering 1980, is based on reports from 66 pharmacies in 48 communities*

> *in New York and New Jersey. Unless there is something peculiar about that part of the country, his findings can be taken as representative of what happens across the country.*
>
> *(Moskowitz 1981: 33)*

What is striking in the excerpt is the casual comment about whether there is anything peculiar about New York and New Jersey. There is. The lifestyle in these two states hardly typifies the lifestyles in the other 48. We cannot assume that residents in these large, urbanized, Eastern Seaboard states necessarily have the same prescription-drug–use patterns that residents of Mississippi, Nebraska, or Vermont have.

Does the survey even represent prescription patterns in New York and New Jersey? To determine that, we would have to know something about the way the 48 communities and the 66 pharmacies were selected. We should be wary in this regard, in view of the reference to "polling nearby drugstores." As we'll see, there are several methods for selecting samples that ensure representativeness, and unless they're used, we shouldn't generalize from the study findings.

A sampling frame, then, must be consonant with the population we wish to study. In the simplest sample design, the sampling frame is a list of the elements composing the study population. In practice, though, existing sampling frames often define the study population rather than the other way around. That is, we often begin with a population in mind for our study; then we search for possible sampling frames. Having examined and evaluated the frames available for our use, we decide which frame presents a study population most appropriate to our needs.

Studies of organizations are often the simplest from a sampling standpoint because organizations typically have membership lists. In such cases, the list of members constitutes an excellent sampling frame. If a random sample is selected from a membership list, the data collected from that sample may be taken as representative of all members—if all members are included in the list.

Populations that can be sampled from good organizational lists include elementary school, high school, and university students and faculty; church members; factory workers; fraternity or sorority members; members of social, service, or political clubs; and members of professional associations.

The preceding comments apply primarily to local organizations. Often, statewide or national organizations do not have a single membership list. There is, for example, no single list of Episcopalian church members. However, a slightly more complex sample design could take advantage of local church membership lists by first sampling churches and then subsampling the membership lists of the churches selected. (More about that later.)

Other lists that may be available contain the names of automobile owners, welfare recipients, taxpayers, business-permit holders, licensed professionals, and so forth. Although it may be difficult to gain access to some of these lists, they provide excellent sampling frames for specialized research purposes.

Certain lists of individuals may be especially relevant to the research needs of a particular study. For example, government agencies maintain lists of registered voters, and some political pollsters use those lists to do registration-based sampling. (In some cases, however, such files are not up to date; further, a person who is registered to vote may not actually vote in a given election.)

The sampling elements in a study need not be individual persons. Lists of other types of elements also exist: universities, corporations, cities, academic journals, newspapers, unions, political clubs, professional associations, and so forth.

Telephone directories were once used for "quick-and-dirty" public opinion polls. It's undeniable that they're easy and inexpensive to use—no doubt the reason for their popularity. And, if you want to make assertions about telephone subscribers, a directory is a fairly good sampling frame. (Realize, of course, that a given directory will include neither new subscribers nor those who have requested unlisted numbers, or cell phone numbers. Sampling is further complicated by the directories' inclusion of nonresidential listings.)

The earliest telephone surveys had a rather bad reputation among professional researchers. Telephone surveys are limited by definition to people who have telephones. Years ago, this method produced a substantial social-class bias by excluding poor people from the surveys. This was vividly demonstrated by the *Literary Digest* fiasco of 1936. Recall that, even though voters were contacted by mail, the sample was partially selected from telephone subscribers, who were hardly typical in a nation struggling with the Great

Depression. By 2009, however, 95.7 percent of all households had telephones, so the earlier form of class bias has virtually disappeared (U.S. Bureau of the Census 2012: 712, Table 1132).

A related sampling problem involved unlisted numbers. A survey sample selected from the pages of a local telephone directory would totally omit all those people—typically richer—who requested that their numbers not be published. This potential bias was erased through a technique that has advanced telephone sampling substantially: random-digit dialing (RDD).

Imagine selecting a set of seven-digit telephone numbers at random. Even people whose numbers were unlisted would have the same chance of selection as would those in the directory. However, if you simply dialed randomly selected numbers, a high proportion of those would turn out to be "not in service," government offices, commercial enterprises, and so forth. Fortunately, it's possible to obtain ranges of numbers that are mostly active residential numbers. Selecting a set of those numbers at random will provide a representative sample of residential households. As a consequence, RDD has become a standard procedure in telephone surveys.

The growth in popularity of cell phones has created a new source of concern for survey researchers. The Telephone Consumer Protection Act of 1991 put limitations on telephone solicitations and, because calls to a cell phone may incur an expense to the target of the call (depending on their service plan), the Act made it illegal for automatic dialing systems (e.g., the robocalls alerting you to a special sale on widgets) to call cell phones (FCC 2015). But where does this leave survey researchers, who aren't selling anything? While efforts are underway to officially exempt research projects from that ruling, AAPOR (2010) advises members that:

> To ensure compliance with this federal law, in the absence of express prior consent from a sampled cell phone respondent, telephone research call centers should have their interviewers manually dial cell phone numbers (i.e., where a human being physically touches the numerals on the telephone to dial the number).

Those who use cell phones exclusively, moreover, tend to be younger than the general population. In 2004, they were more likely to vote for John Kerry than older voters were.

In 2008, they were more likely than the average voter to support Barack Obama. In a study of this matter, Scott Keeter and Kylie McGeeney (2015) found that a distinct bias by age and the variables closely related to it (such as marital status) distinguished those who were reachable only by cell phone and those reachable by landline.

> For example, young adults, Hispanics, renters and the poor (as defined by the U.S. Census Bureau's poverty thresholds) are all far more likely to be cell-only. To the extent that cell-only households are underrepresented in our samples, these groups are also underrepresented.
>
> *(Keeter and McGeeney, 2015)*

At the 2008 meeting of the American Association for Public Opinion Research (AAPOR), several research papers examined the implications of cell-phone popularity. Overall, most of the researchers found that ignoring people who use only cell phones did not seriously bias survey results, in most cases; this is because these people represented a relatively small portion of all telephone customers. That situation has changed quickly and substantially, as reported by the National Center for Health Statistics (2017):

> The second 6 months of 2016 was the first time that a majority of American homes had only wireless telephones. Preliminary results from the July–December 2016 National Health Interview Survey (NHIS) indicate that 50.8% of American homes did not have a landline telephone but did have at least one wireless telephone. . .

In part, researchers have sought to address the dramatic increase in cell phones by augmenting RDD sampling with address-based sampling (ABS), based on U.S. Postal Service lists of residential addresses. It will be possible for researchers

Cell phones have complicated survey sampling.

to match some of the residential addresses with phone numbers by using directories that contain both pieces of information. The nonmatched addresses will be mailed a request for a phone number that researchers can then call. The mailing will also include a 1-800 number that potential survey respondents can call if they have no telephone number of their own. If two sampling frames are employed, however, it is important to either (1) rule out duplicate residences before sampling or (2) identify respondents who have both cell phones and landlines so that their responses can be weighted half as much as those with only one chance of being selected into the sample. The preferred method is still under study and debate (Boyle, Lewis, and Tefft 2010).

In an updated report from the Pew Center (Christian et al. 2010), special attention was paid to differences in the estimation of opinions and behaviors using landline-only samples and dual-frame samples, including cell-phone-only households.

> The items selected include nearly all of the key indicators regularly tracked by our two centers (e.g., presidential approval, party affiliation, internet use, broadband adoption, sending and receiving text messages on a cell phone), as well as a sampling of other important measures that were timely or are asked intermittently (e.g., agreement with the Tea Party, approval of health care legislation, use of cell phones to play music).
>
> *(2010: 2)*

Overall, the Pew researchers found relatively little bias due to sample design.

> Despite the growth in cell-only households, the magnitude of possible non-coverage bias remains relatively small for the majority of measures tested. Of 72 questions examined, 43 of them show differences of 0, 1 or 2 percentage points between the landline and dual frame weighted samples.
>
> *(2010: 2)*

This is an issue that will be followed closely by survey researchers in the years to come, as cell phones, presumably, will become ever-more dominant.

If cell phones have created a problem for sampling frames and sampling, that challenge pales in comparison with the new interest in online surveys. As we shall see in Chapter 9, there are substantial advantages to conducting surveys via the Web, but obtaining responses from a sample representing the population of interest (e.g., adult population, voters) is especially tricky. At the most basic, not everyone participates on the Internet, comparable to the problem early in telephone polls, when not everyone had a telephone. Beyond that basic problem, those who are active on the Web participate to differing degrees and visit different websites.

Though a variety of methods are being tested at present, there is no clear solution to the problem of representativeness. As mentioned earlier in this chapter, we may see a rebirth of quota-sampling techniques as a way of making online samples represent the larger populations we are interested in studying. Another way of framing this issue is in terms of sampling error, as discussed earlier.

The calculation of sampling error, often called the *margin of error* (MOE), has a solid statistical grounding, providing it satisfies two criteria: (1) a probability sample is selected from a complete listing of the population of interest, and (2) everyone selected participates in the study. Now the bad news: rarely is either of these criteria fully satisfied. In our discussion of sampling frames, we saw that it is often difficult to obtain a complete listing or quasi-listing of the entire population under study. In the case of telephone surveys, we have seen the difficulty in accommodating the proliferation of cell phones. And the second criterion has been steadily undercut by falling response rates. Thus, survey researchers struggle to weigh their results to arrive at representative estimates of the population.

This problem is even greater in the case of nonprobability, "opt-in" polls on the Web. Potential respondents are invited to visit a website to participate in a poll. Some do, some don't, but there is no way of calculating conventional probability statistics regarding sampling error, because no probability sample was selected. In response to this dilemma, you may find reports of "credibility intervals," which resemble margins of sampling error in form (e.g., 95 percent confident of a 3-percentage-point error). Such estimates are based on the researchers choosing models that they believe distinguish opt-in respondents and the larger population of interest. This is a relatively new technique and somewhat controversial among researchers. The American Association for Public Opinion Research suggests caution in using or accepting credibility intervals.

Review of Populations and Sampling Frames

Because social research literature gives surprisingly little attention to the issues of populations and sampling frames, I've devoted special consideration to them here by providing a summary of the main guidelines to remember:

1. Findings based on a sample can be taken as representing only the aggregation of elements that compose the sampling frame.
2. Often, sampling frames do not truly include all the elements their names might imply. Omissions are almost inevitable. Thus, a first concern of the researcher must be to assess the extent of the omissions and to correct them if possible. (Of course, the researcher may feel that he or she can safely ignore a small number of omissions that cannot easily be corrected.)
3. To be generalized even to the population composing the sampling frame, all elements must have equal representation in the frame. Typically, each element should appear only once. Elements that appear more than once will have a greater probability of selection, and the sample will, overall, overrepresent those elements.

Other, more practical matters relating to populations and sampling frames will be treated elsewhere in this book. For example, the form of the sampling frame—such as a list in a publication, 3-by-5 card file, hard-drive file, or other digital storage method—can affect how easy it is to use. And ease of use may often take priority over scientific considerations: An "easier" list may be chosen over a "harder" one, even though the latter is more appropriate to the target population. Every researcher should carefully weigh the relative advantages and disadvantages of such alternatives.

Types of Sampling Designs

Up to this point, we've focused on simple random sampling. Indeed, the body of statistics typically used by social researchers assumes such a sample. As you'll see shortly, however, you have several options in choosing your sampling method, and you'll seldom if ever choose simple random sampling. There are two reasons for this. First, with all but the simplest sampling frame, simple random sampling is not feasible. Second, and

How to Do It

Using a Table of Random Numbers

In social research, it's often appropriate to select a set of random numbers from a table such as the one in Appendix B. Here's how to do that.

Suppose you want to select a simple random sample of 400 people (or other units) out of a population totaling 9,300.

1. To begin, number the members of the population: in this case, from 1 to 9,300. Now the problem is to select 400 random numbers. Once you've done that, your sample will consist of the people having the numbers you've selected. (*Note*: It's not essential to actually number them, as long as you're sure of the total. If you have them in a list, for example, you can always count through the list after you've selected the numbers.)

2. The next step is to determine the number of digits you'll need in the random numbers you select. In our example, there are 9,300 members of the population, so you'll need four-digit numbers to give everyone a chance of selection. (If there were 11,825 members of the population, you'd need to select five-digit numbers.) Thus, we want to select 400 random numbers in the range from 0001 to 9300.

3. On the first page of Appendix B, note that there are several rows and columns of five-digit numbers and that there are two pages. The table represents a series of random numbers in the range from 00001 to 99999. To use the table for your hypothetical sample, you have to answer these questions:

   a. How will you create four-digit numbers out of five-digit numbers?
   b. What pattern will you follow while moving through the table to select your numbers?
   c. Where will you start?

   Each of these questions has several satisfactory answers. The key is to create a plan and follow it. Here's an example.

4. To create four-digit numbers from five-digit numbers, let's agree to select five-digit numbers from the table but consider only the leftmost four digits in each case. If we picked the first number on the first page—51426—we would only consider the 5142. (We could agree to take the digits farthest to the right, 1426.) The key is to make a plan and stick with it. For convenience, let's use the leftmost four digits.

5. We can also choose to progress through the tables in any way we want: down the columns, up them, left to right or right to left, or diagonally. Again, any of these plans will work fine as long as we stick to it. For convenience, let's agree to move down the columns. When we get to the bottom of one column, we'll go to the top of the next.

6. Now, where do we start? You can close your eyes and stick a pencil into the table and start wherever the pencil point lands. (I know it doesn't sound scientific, but it works.) Or, if you're afraid you'll hurt the book or miss it altogether, close your eyes and make up a column number and a row number. ("I'll pick the number in the sixth row of column 2.") Start with that number.

7. Let's suppose we decide to start with the sixth number in column 2. On the first page of Appendix B, you'll see that the starting number is 09599. We have selected 0959 as our first random number, and we have 399 more to go. Moving down the second column, we select 3333, 1342, 3695, 4484, 7074, 3158, 9435—that's a problem, as there are only 9,300 people in the population. The solution is simple: ignore that number. If you happen to get the same number twice, ignore it the second time. Technically, this is called "random selection without replacement." After skipping 9435, we proceed to 6661, 4592, and so forth. When we get to the bottom of column 2, move to the top of column 3.

8. That's it. You keep up the procedure until you've selected 400 random numbers. Returning to your list, your sample consists of person number 0959, person number 3333, person number 1342, and so forth.

---

probably surprisingly, simple random sampling may not be the most accurate method available. Let's turn now to a discussion of simple random sampling and the other available options.

Simple Random Sampling

As noted, **simple random sampling** is the basic sampling method assumed in the statistical computations of social research. Because the mathematics of random sampling are especially complex, we'll detour around them in favor of describing the ways of employing this method in the field.

Once a sampling frame has been properly established, to use simple random sampling the researcher assigns a single number to each element in the list, not skipping any number in the process. A table of random numbers (Appendix B) is then used to select elements for the sample. "How to Do It: Using a Table of Random Numbers" explains its use.

> **simple random sampling** A type of probability sampling in which the units composing a population are assigned numbers. A set of random numbers is then generated, and the units having those numbers are included in the sample.

If your sampling frame is in a machine-readable form, such as a hard-drive file or flash drive, a simple random sample can be selected automatically by computer. (In effect, the computer program numbers the elements in the sampling frame, generates its own series of random numbers, and prints out the list of elements selected.)

Figure 7-11 offers an illustration of simple random sampling. Note that the members of our hypothetical micropopulation have been numbered from 1 to 100. Moving to the small table of random numbers provided, we decide to use the last two digits of the third column and to begin with the third number from the top. This yields person number 12 as the first one selected for the sample. Number 97 is next, and so forth. (Person 100 would have been selected if "00" had come up in the list.)

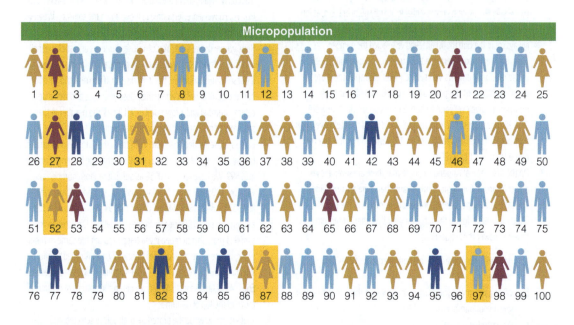

FIGURE 7-11

A Simple Random Sample. Having numbered everyone in the population, we can use a table of random numbers to select a representative sample from the overall population. Anyone whose number is chosen from the table is in the sample.

Systematic Sampling

Simple random sampling is seldom used in practice. As you'll see, it's not usually the most efficient method, and it can be laborious if done manually. Typically, simple random sampling requires a list of elements. When such a list is available, researchers usually employ systematic sampling instead.

In **systematic sampling**, every kth element in the total list is chosen (systematically) for inclusion in the sample. If the list contained 10,000 elements and you wanted a sample of 1,000, you would select every tenth element for your sample. To ensure against any possible human bias in using this method, you should select the first element at random. Thus, in the preceding example, you would begin by selecting a random number between one and ten. The element having that number is included in the sample, as is every tenth element following it. This method is technically referred to as a *systematic sample with a random start*. Two terms are frequently used in connection with systematic sampling. The **sampling interval** is the standard distance between elements selected in the sample: ten in the preceding sample. The **sampling ratio** is the proportion of elements in the population that are selected: 1/10 in the example.

$$\text{sampling interval} = \frac{\text{population size}}{\text{sample size}}$$

$$\text{sampling ratio} = \frac{\text{sample size}}{\text{population size}}$$

In practice, systematic sampling is virtually identical to simple random sampling. If the list of elements is indeed randomized before sampling, one might argue that a systematic sample drawn from that list is in fact a simple random sample. By now, debates over the relative merits of simple random sampling and systematic sampling have been resolved largely in favor of the latter, simpler method. Empirically, the results are virtually identical. And, as you'll see in a later section, systematic sampling, in some instances, is slightly more accurate than simple random sampling.

There is one danger involved in systematic sampling. The arrangement of elements in the list can make systematic sampling unwise. Such an arrangement is usually called *periodicity*. If the list of elements is arranged in a cyclical pattern that coincides with the sampling interval, a grossly biased sample can be drawn. Here are two examples that illustrate this danger.

In a classic study of soldiers during World War II, the researchers selected a systematic sample from unit rosters. Every tenth soldier on the roster was selected for the study. The rosters, however, were arranged in a table of organizations: sergeants first, then corporals and privates, squad by squad. Each squad had ten members. As a result, every tenth person on the roster was a squad sergeant. The systematic sample selected contained only sergeants. It could, of course, have been the case that no sergeants were selected for much the same reason.

As another example, suppose we select a sample of apartments in an apartment building. If the sample is drawn from a list of apartments arranged in numerical order (for example, 101, 102, 103, 104, 201, 202, and so on), there is a danger of the sampling interval coinciding with the number of apartments on a floor or some multiple thereof. Then the samples might include only northwest-corner apartments or only apartments near the elevator. If these types of apartments have some other particular characteristic in common (for example, higher rent), the sample will be biased. The same danger would appear in a systematic sample of houses in a subdivision arranged with the same number of houses on a block.

In considering a systematic sample from a list, then, you should carefully examine the nature of that list. If the elements are arranged in any particular order, you should figure out whether that order will bias the sample to be selected, then you should take steps to counteract any possible bias (for example, take a simple random sample from cyclical portions).

---

systematic sampling A type of probability sampling in which every kth unit in a list is selected in the sample—for example, every 25th student in the college directory of students.

sampling interval The standard distance (k) between elements selected from a population for a sample.

sampling ratio The proportion of elements in the population that are selected to be in a sample.

Usually, however, systematic sampling is superior to simple random sampling, in convenience if nothing else. Problems in the ordering of elements in the sampling frame can usually be remedied quite easily.

Stratified Sampling

So far we've discussed two methods of sample selection from a list: random and systematic. **Stratification** is not an alternative to these methods; rather, it represents a possible modification of their use.

Simple random sampling and systematic sampling both ensure a degree of representativeness and permit an estimate of the error present. Stratified sampling is a method for obtaining a greater degree of representativeness by decreasing the probable sampling error. To understand this method, we must return briefly to the basic theory of sampling distribution.

Recall that sampling error is reduced by two factors in the sample design. First, a large sample produces a smaller sampling error than does a small sample. Second, a homogeneous population produces samples with smaller sampling errors than does a heterogeneous population. If 99 percent of the population agrees with a certain statement, it's extremely unlikely that any probability sample will greatly misrepresent the extent of agreement. If the population is split 50–50 on the statement, then the sampling error will be much greater.

Stratified sampling is based on this second factor in sampling theory. Rather than selecting a sample from the total population at large, the researcher ensures that appropriate numbers of elements are drawn from homogeneous subsets of that population. To get a stratified sample of university students, for example, you would first organize your population by college class and then draw appropriate numbers of freshmen, sophomores, juniors, and seniors. In a nonstratified

---

stratification The grouping of the units composing a population into homogeneous groups (or strata) before sampling. This procedure, which may be used in conjunction with simple random, systematic, or cluster sampling, improves the representativeness of a sample, at least in terms of the variables used for stratification.

---

sample, representation by class would be subject to the same sampling error as would other variables. In a sample stratified by class, the sampling error on this variable is reduced to zero.

More-complex stratification methods are also possible. In addition to stratifying by class, you might also stratify by gender, by GPA, and so forth. In this fashion, you might be able to ensure that your sample would contain the proper numbers of male sophomores with a 3.5 GPA, of female sophomores with a 4.0 GPA, and so forth.

The ultimate function of stratification, then, is to organize the population into homogeneous subsets (with heterogeneity between subsets) and to select the appropriate number of elements from each. To the extent that the subsets are homogeneous on the stratification variables, they may be homogeneous on other variables as well. Because age is related to college class, a sample stratified by class will also be more representative in terms of age than an unstratified sample. Because occupational aspirations still seem to be related to gender, a sample stratified by gender will be more representative in terms of occupational aspirations.

The choice of stratification variables typically depends on what variables are available. Gender can often be determined in a list of names. University lists are typically arranged by class. Lists of faculty members may indicate their departmental affiliation. Government agency files may be arranged by geographic region. Voter registration lists are arranged according to precinct.

In selecting stratification variables from among those available, however, you should be concerned primarily with those that are presumably related to variables you want to represent accurately. Because gender is related to many variables and is often available for stratification, it is often used. Education is related to many variables, but it is often not available for stratification. Geographic location within a city, state, or nation is related to many things. Within a city, stratification by geographic location usually increases representativeness in social class, ethnic group, and so forth. Within a nation, it increases representativeness in a broad range of attitudes as well as in social class and ethnicity.

When you're working with a simple list of all elements in the population, two methods of stratification predominate. In one method, you sort the population elements into discrete groups based on whatever stratification variables are being used. On the basis of the relative proportion of the population represented by a given group, you select—randomly or systematically—several elements from that group constituting the same proportion of your desired sample size. For example, if sophomore men with a 4.0 GPA compose 1 percent of the student population and you desire a sample of 1,000 students, you would select 10 sophomore men with a 4.0 average.

The other method is to group students as described and then put those groups together in a continuous list, beginning with all male freshmen with a 4.0 average and ending with all female seniors with a 1.0 or below. You would then select a systematic sample, with a random start, from the entire list. Given the arrangement of the list, a systematic sample would select

proper numbers (within an error range of 1 or 2) from each subgroup. (*Note:* A simple random sample drawn from such a composite list would cancel out the stratification.)

Figure 7-12 offers an illustration of stratified, systematic sampling. As you can see, we lined up our micropopulation according to gender and race. Then, beginning with a random start of 3, we've taken every tenth person thereafter, resulting in a list of 3, 13, 23, . . . 93.

Stratified sampling ensures the proper representation of the stratification variables; this, in turn, enhances the representation of other variables related to them. Taken as a whole, then, a stratified sample is more likely than a simple random sample to be more representative on several variables. Although the simple random sample is still regarded as somewhat sacred, it should now be clear that you can often do better.

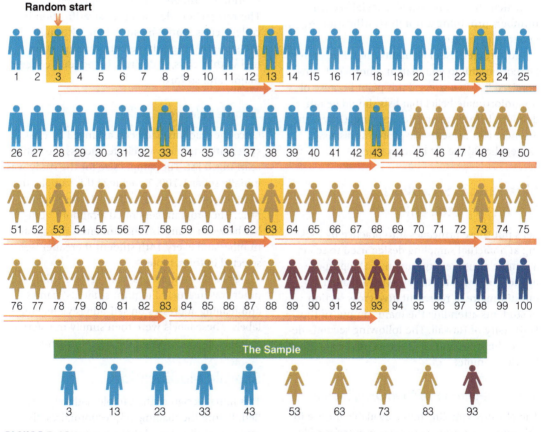

FIGURE 7-12

A Stratified, Systematic Sample with a Random Start. A stratified, systematic sample involves two stages. First the members of the population are gathered into homogeneous strata; this simple example merely uses gender and race as stratification variables, but more could be used. Then every *k*th (in this case, every tenth) person in the stratified arrangement is selected for the sample.

Implicit Stratification in Systematic Sampling

I mentioned that systematic sampling can, under certain conditions, be more accurate than simple random sampling. This is the case whenever the arrangement of the list creates an implicit stratification. As already noted, if a list of university students is arranged by class, then a systematic sample provides a stratification by class, whereas a simple random sample would not.

In a study of students at the University of Hawaii, after stratification by school class, the students were arranged by their student identification numbers. At that time, student social security numbers served as their student IDs. The first three digits of the social security number indicate the state in which the number was issued. As a result, within a class, students were arranged by the state in which they were issued a social security number, providing a rough stratification by geographic origin.

An ordered list of elements, therefore, may be more useful to you than an unordered, randomized list. I've stressed this point because of the unfortunate belief that lists should be randomized before systematic sampling. Only if the arrangement presents the problems discussed earlier should the list be rearranged.

Illustration: Sampling University Students

Let's put these principles into practice by looking at an actual sampling design used to select a sample of university students. The purpose of the study was to survey, with a mailed questionnaire, a representative cross section of students attending the main campus of the University of Hawaii. The following sections describe the steps and decisions involved in selecting that sample.

Study Population and Sampling Frame

The obvious sampling frame available for use in this sample selection was the computerized file maintained by the university administration. The file contained students' names, local and permanent addresses, social security numbers, and a variety of other information such as field of study, class, age, and gender.

The computer database, however, contained information on all people who could, by any conceivable definition, be called students, many of whom seemed inappropriate to the purposes of the study. As a result, researchers needed to define the study population in a somewhat more restricted fashion. The final definition included those 15,225 day-program degree candidates registered for the fall semester on the Manoa campus of the university, including all colleges and departments, both undergraduate and graduate students, and both U.S. and foreign students. The computer program used for sampling, therefore, limited consideration to students fitting this definition.

Stratification

The sampling program also permitted stratification of students before sample selection. The researchers decided that stratification by college class would be sufficient, although the students might have been further stratified within class, if desired, by gender, college, major, and so forth.

Sample Selection

Once the students had been arranged by class, a systematic sample was selected across the entire rearranged list. The sample size for the study was initially set at 1,100. To achieve this sample, the sampling program was set for a 1/14 sampling ratio. The program generated a random number between 1 and 14; the student having that number and every 14th student thereafter was selected in the sample.

Once the sample had been selected, the computer was instructed to print students' names and mailing addresses on self-adhesive mailing labels. These labels were then simply transferred to envelopes for mailing the questionnaires.

Sample Modification

This initial design of the sample had to be modified. Before the mailing of questionnaires, the researchers discovered that unexpected expenses in the production of the questionnaires made it impossible to cover the costs of mailing to all 1,100 students. As a result, one-third of the mailing labels were systematically selected (with a random start) for exclusion from the sample. The final sample for the study was thereby reduced to 733 students.

I mention this modification in order to illustrate the frequent need to alter a study plan in midstream. Because the excluded students were systematically omitted from the initial systematic sample, the remaining 733 students could still be taken to reasonably represent the study population. The reduction in sample size did, of course, increase the range of sampling error.

Multistage Cluster Sampling

The preceding sections have dealt with reasonably simple procedures for sampling from lists of elements. Such a situation is ideal. Unfortunately, however, much interesting social research requires the selection of samples from populations that cannot easily be listed for sampling purposes: the population of a city, state, or nation; all university students in the United States; and so forth. In such cases, the sample design must be much more complex. Such a design typically involves the initial sampling of *clusters* (groups of elements), followed by the selection of elements within each of the selected clusters.

Cluster sampling may be used when it's either impossible or impractical to compile an exhaustive list of the elements composing the target population, such as all church members in the United States. Often, however, the population elements are already grouped into subpopulations, and a list of those subpopulations either exists or can be created practically. For example, church members in the United States belong to discrete churches, which are either listed or could be. Following a cluster-sample format, then, researchers would sample the list of churches in some manner (for example, a stratified, systematic sample). Next, they would obtain lists of members from each of the selected churches. Each of the lists would then be sampled, to provide samples of church members for study.

Another typical situation concerns sampling among population areas such as a city. Although there is no single list of a city's population, citizens reside on discrete city blocks or census blocks. Researchers can therefore select a sample of blocks initially, create a list of people living on each of the selected blocks, and take a subsample of the people on each block.

In a more complex design, researchers might (1) sample blocks, (2) list the households on each selected block, (3) sample the households, (4) list the people residing in each household, and (5) sample the people within each selected household. This multistage sample design leads ultimately to a selection of a sample of individuals but does not require the initial listing of all individuals in the city's population.

Multistage cluster sampling, then, involves the repetition of two basic steps: listing and sampling. The list of primary sampling units (churches, blocks) is compiled and, perhaps, stratified for sampling. Then a sample of those units is selected. The selected primary sampling units are then listed and perhaps stratified. The list of secondary sampling units is then sampled, and so forth.

The listing of households on even the selected blocks is, of course, a labor-intensive and costly activity—one of the elements making face-to-face household surveys quite expensive. Vincent Iannacchione, Jennifer Staab, and David Redden (2003) reported some initial success using postal mailing lists for this purpose, and more researchers are considering them, as we've seen. Although the lists are not perfect, they may be close enough to warrant the significant savings in cost.

Multistage cluster sampling makes possible those studies that would otherwise be impossible. Specific research circumstances often call for special designs, as demonstrated in "Issues and Insights: Sampling Iran."

---

cluster sampling A multistage sampling in which natural groups (clusters) are sampled initially, with the members of each selected group being subsampled afterward. For example, you might select a sample of U.S. colleges and universities from a directory, get lists of the students at all the selected schools, then draw samples of students from each.

Issues and Insights

Sampling Iran

Whereas most of the examples given in this textbook are taken from their country of origin (the United States), the basic methods of sampling would apply in other national settings as well. At the same time, researchers may need to make modifications appropriate to local conditions. In selecting a national sample of Iran, for example, Hamid Abdollahyan and Taghi Azadarmaki (2000: 21) from the University of Tehran began by stratifying the nation on the basis of cultural differences, dividing the country into nine cultural zones as follows:

1. Tehran
2. Central region including Isfahan, Arak, Qum, Yazd, and Kerman
3. The southern provinces including Hormozgan, Khuzistan, Bushehr, and Fars
4. The marginal western region including Lorestan, Charmahal and Bakhtiari, and Kogiluyeh and Eelam
5. The western provinces including western and eastern Azarbaijan, Zanjan, Ghazvin, and Ardebil
6. The eastern provinces including Khorasan and Semnan

7. The northern provinces including Gilan, Mazandran, and Golestan
8. Systan
9. Kurdistan

Within each of these cultural areas, the researchers selected samples of census blocks and, on each selected block, a sample of households. Their sample design made provisions for getting the proper numbers of men and women as respondents within households and provisions for replacing the households where no one was at home.

Though the United States and Iran are politically and culturally quite different, the sampling methods appropriate for selecting a representative sample of populations are the same. Later in this chapter, when you review a detailed description of sampling the household population of an American city, you will find it strikingly similar to the methods used in Iran by Abdollahyan and Azadarmaki.

Source: Hamid Abdollahyan and Taghi Azadarmaki. 2000. "Sampling Design in a Survey Research: The Sampling Practice in Iran." Paper presented at the meeting of the American Sociological Association, Washington, DC, August 12–16.

Multistage Designs and Sampling Error

Although cluster sampling is highly efficient, the price of that efficiency is a less-accurate sample. A simple random sample drawn from a population list is subject to a single sampling error, but a two-stage cluster sample is subject to two sampling errors. First, the initial sample of clusters will represent the population of clusters only within a range of sampling error. Second, the sample of elements selected within a given cluster will represent all the elements in that cluster only within a range of sampling error. Thus, for example, a researcher runs a certain risk of selecting a sample of disproportionately wealthy city blocks, plus a sample of disproportionately wealthy households within those blocks. The best solution to this problem lies in the number of clusters selected initially and the number of elements within each cluster.

Typically, researchers are restricted to a total sample size; for example, you may be limited to conducting 2,000 interviews in a city. Given this broad limitation, however, you have several options in designing your cluster sample. At the extremes, you could choose one cluster and select 2,000 elements within that cluster, or you could select 2,000 clusters with one element selected within each. Of course, neither approach is advisable, but a broad range of choices lies between them. Fortunately, the logic of sampling distributions provides a general guideline for this task.

Recall that sampling error is reduced by two factors: an increase in the sample size and increased homogeneity of the elements being sampled. These factors operate at each level of a multistage sample design. A sample of clusters will best represent all clusters if a large number are selected and if all clusters are very much alike. A sample of elements will best represent all elements in a given cluster if a large number are selected from the cluster and if all the elements in the cluster are very much alike.

With a given total sample size, however, if the number of clusters is increased, the number of elements within a cluster must be decreased, and vice versa. In the first case, the representativeness of the clusters is increased at the expense of more poorly representing the elements composing each cluster. Fortunately, homogeneity can be used to ease this dilemma.

Typically, the elements composing a given natural cluster within a population are more homogeneous than are all elements composing the total population. The members of a given church are more alike than are members of all churches; the residents of a given city block are more alike than are the residents of a whole city. As a result, relatively few elements may be needed to represent a given natural cluster adequately, although a larger number of clusters may be needed to adequately represent the diversity found among the clusters. This fact is most clearly seen in the extreme case of very different clusters composed of identical elements within each. In such a situation, a large number of clusters would adequately represent all its members. Although this extreme situation never exists in reality, it's closer to the truth in most cases than its opposite: identical clusters composed of grossly divergent elements.

The general guideline for cluster design, then, is to maximize the number of clusters selected while decreasing the number of elements within each cluster. However, this scientific guideline must be balanced against an administrative constraint. The efficiency of cluster sampling is based on the ability to minimize the listing of population elements. By initially selecting clusters, you need only list the elements composing the selected clusters, not all elements in the entire population. Increasing the number of clusters, however, goes directly against this efficiency factor. A small number of clusters may be listed more quickly and more cheaply than a large number. (Remember that all the elements in a selected cluster must be listed even if only a few are to be chosen in the sample.)

The final sample design will reflect these two constraints. In effect, you'll probably select as many clusters as you can afford. Lest this issue be left too open-ended at this point, here is one general guideline. Population researchers conventionally aim at the selection of 5 households per census block. If a total of 2,000 households are to be interviewed, you would aim at 400 blocks with 5 household interviews in each. Figure 7-13 presents an overview of this process.

Before turning to other, more-detailed procedures available to cluster sampling, let me reiterate that this method almost inevitably involves a loss of accuracy. The manner in which this appears, however, is somewhat complex. First, as noted earlier, a multistage sample design is subject to a sampling error at each stage. Because the sample size is necessarily smaller at each stage than the total sample size, the sampling error at each stage will be greater than would be the case for a single-stage random sample of elements. Second, sampling error is estimated on the basis of observed variance among the sample elements. When those elements are drawn from among relatively homogeneous clusters, the estimated sampling error will be too optimistic and must be corrected in light of the cluster sample design.

Stratification in Multistage Cluster Sampling

Thus far, we've looked at cluster sampling as though a simple random sample were selected at each stage of the design. In fact, stratification techniques can be used to refine and improve the sample being selected.

The basic options here are essentially the same as those in single-stage sampling from a list. In selecting a national sample of churches, for example, you might initially stratify your list of churches by denomination, geographic region, size, rural or urban location, and perhaps some measure of social class.

Once the primary sampling units (churches, blocks) have been grouped according to the relevant, available stratification variables, either simple random or systematic sampling can be used to select the sample. You might select a specified number of units from each group, or stratum, or you might arrange the stratified clusters in a continuous list and systematically sample that list.

To the extent that clusters are combined into homogeneous strata, the sampling error at this stage will be reduced. The primary goal of stratification, as before, is homogeneity.

Stratification could, of course, take place at each level of sampling. The elements listed within a selected cluster might be stratified before the next stage of sampling. Typically, however, this is not done. (Recall the assumption of relative homogeneity within clusters.)

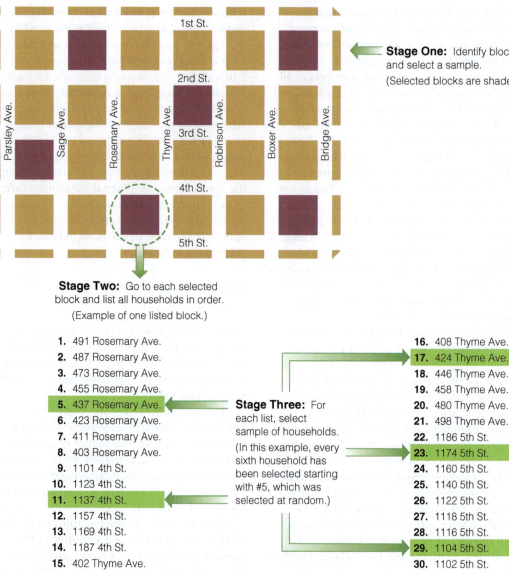

FIGURE 7-13

Multistage Cluster Sampling. In multistage cluster sampling, we begin by selecting a sample of the clusters (in this case, city blocks). Then, we make a list of the elements (households, in this case) and select a sample of elements from each of the selected clusters.

Probability Proportionate to Size (PPS) Sampling

This section introduces you to a more sophisticated form of cluster sampling, one that is used in many large-scale survey-sampling projects. In the preceding discussion, I talked about selecting a random or systematic sample of clusters and then a random or systematic sample of elements within each cluster selected. Notice that this produces an overall sampling scheme in which every element in the whole population has the same probability of selection.

Let's say we're selecting households within a city. If there are 1,000 city blocks and we initially select a sample of 100, that means that each block has a 100/1,000 or 0.1 chance of being selected. If we next select 1 household in 10 from those residing on the selected blocks, each household has a 0.1 chance of selection within its block. To calculate the overall probability of a household being selected, we simply multiply the probabilities at the individual steps in sampling. That is, each household has a 1/10 chance of its block being selected and a 1/10 chance of that specific household being selected if the block is one of those chosen. Each household, in this case, has a $1/10 \times 1/10 = 1/100$ chance of selection overall. Because each household would have the same chance of selection, the sample so selected should be representative of all households in the city.

There are dangers in this procedure, however. In particular, the variation in the size of blocks (measured in numbers of households) presents a problem. Let's suppose that half the city's population resides in 10 densely packed blocks filled with high-rise apartment buildings, and suppose that the rest of the population lives in single-family dwellings spread out over the remaining 990 blocks. When we first select our sample of 1/10 of the blocks, it's quite possible that we'll miss all of the 10 densely packed high-rise blocks. No matter what happens in the second stage of sampling, our final sample of households will be grossly unrepresentative of the city, comprising only single-family dwellings.

Whenever the clusters sampled are of greatly differing sizes, it's appropriate to use a modified sampling design called **PPS (probability proportionate to size)**. This design guards against the problem I've just described and still produces a final sample in which each element has the same chance of selection.

As the name suggests, each cluster is given a chance of selection proportionate to its size. Thus, a city block with 200 households has twice the chance of selection as one with only 100 households. Within each cluster, however, a fixed number of elements is selected, say, 5 households per block. Notice how this procedure results in each household having the same probability of selection overall.

Let's look at households of two different city blocks. Block A has 100 households, whereas Block B has only 10. In PPS sampling, we would give Block A 10 times as good a chance of being selected as Block B. So if, in the overall sample design, Block A has a 1/20 chance of being selected, that means Block B would only have a 1/200 chance. Notice that this means that all the

households on Block A would have a 1/20 chance of having their block selected, whereas Block B households would have only a 1/200 chance.

If Block A is selected and we're taking 5 households from each selected block, then the households on Block A have a 5/100 chance of being selected for the block's sample. Because we can multiply probabilities in a case like this, we see that every household on Block A had an overall chance of selection equal to $1/20 \times 5/100 = 5/2,000 = 1/400$.

If Block B happens to be selected, on the other hand, its households stand a much better chance of being among the 5 chosen there: 5/10. When this is combined with their relatively poorer chance of having their block selected in the first place, however, they end up with the same chance of selection as those on Block A: $1/200 \times 5/10 = 5/2,000 = 1/400$.

Further refinements to this design make it a very efficient and effective method for selecting large cluster samples. For now, however, it's enough to understand the basic logic involved.

Disproportionate Sampling and Weighting

Ultimately, a probability sample is representative of a population if all elements in the population have an equal chance of selection for that sample. Thus, in each of the preceding discussions, we've noted that the various sampling procedures result in an equal chance of selection—even though the ultimate selection probability is the product of several partial probabilities.

More generally, however, a probability sample is one in which each population element has a known nonzero probability of selection—even though different elements may have different probabilities. If controlled probability-sampling procedures have been used, any such sample may be representative of the population from which it is drawn if each sample element is

---

PPS (probability proportionate to size) This refers to a type of multistage cluster sample in which clusters are selected, not with equal probabilities (see *EPSEM*) but with probabilities proportionate to their sizes—as measured by the number of units to be subsampled.

assigned a weight equal to the inverse of its probability of selection. Thus, where all sample elements have had the same chance of selection, each is given the same weight: 1. This is called a *self-weighting sample*.

Sometimes it's appropriate to give some cases more weight than others, a process called **weighting**. Disproportionate sampling and weighting come into play in two basic ways. First, you may sample subpopulations disproportionately to ensure sufficient numbers of cases from each for analysis. For example, a given city may have a suburban area containing one-fourth of its total population. Yet you might be especially interested in a detailed analysis of households in that area and may feel that one-fourth of this total sample size would be too few. As a result, you might decide to select the same number of households from the suburban area as from the remainder of the city. Households in the suburban area, then, are given a disproportionately better chance of selection than those located elsewhere in the city.

As long as you analyze the two area samples separately or comparatively, you need not worry about the differential sampling. If you want to combine the two samples to create a composite picture of the entire city, however, you must take the disproportionate sampling into account. If n is the number of households selected from each area, then the households in the suburban area had a chance of selection equal to n divided by one-fourth of the total city population. Because the total city population and the sample size are the same for both areas, the suburban-area households should be given a weight of $1/4n$, and the remaining households should be given a weight of $3/4n$. This weighting procedure could be simplified by merely giving a weight of 3 to each of the households selected outside the suburban area. (This procedure gives a proportionate representation to each sample element. The population figure would have to be included in the weighting if population estimates were desired.)

weighting Assigning different weights to cases that were selected into a sample with different probabilities of selection. In the simplest scenario, each case is given a weight equal to the inverse of its probability of selection. When all cases have the same chance of selection, no weighting is necessary.

Here's an example of the problems that can be created when disproportionate sampling is not accompanied by a weighting scheme. When the *Harvard Business Review* decided to survey its subscribers on the issue of sexual harassment at work, it seemed appropriate to oversample women because female subscribers were vastly outnumbered by male subscribers. Here's how G. C. Collins and Timothy Blodgett explained the matter:

> We also skewed the sample another way: to ensure a representative response from women, we mailed a questionnaire to virtually every female subscriber, for a male/female ratio of 68% to 32%. This bias resulted in a response of 52% male and 44% female (and 4% who gave no indication of gender)—compared to HBR's U.S. subscriber proportion of 93% male and 7% female.
>
> *(1981: 78)*

Notice a couple of things in this quotation. First, it would be nice to know a little more about what "virtually every female" means. Evidently, the authors of the study didn't send questionnaires to all female subscribers, but there's no indication of who was omitted and why. Second, they didn't use the term *representative* in its normal social science usage. What they mean, of course, is that they wanted to get a substantial or "large enough" response from women, and oversampling is a perfectly acceptable way of accomplishing that.

By sampling more women than a straightforward probability sample would have produced, the authors were able to "select" enough women (812) to compare with the men (960). Thus, when they report, for example, that 32 percent of the women and 66 percent of the men agree that "the amount of sexual harassment at work is greatly exaggerated," we know that the female response is based on a substantial number of cases. That's good. There are problems, however.

To begin with, subscriber surveys are always problematic. In this case, the best the researchers can hope to talk about is "what subscribers to *Harvard Business Review* think." In a loose way, it might make sense to think of that population as representing the more sophisticated portion of corporate management. Unfortunately, the overall response rate was 25 percent. Although that's quite good for subscriber surveys, it's a low response rate in terms of generalizing from probability samples.

Beyond that, however, the disproportionate sample design creates a further problem. When the authors state that 73 percent of respondents favor company policies against harassment (Collins and Blodgett 1981: 78), that figure is undoubtedly too high, because the sample contains a disproportionately high percentage of women, who are more likely to favor such policies. Further, when the researchers report that top managers are more likely to feel that claims of sexual harassment are exaggerated than are middle- and lower-level managers (1981: 81), that finding is also suspect. As the researchers report, women are disproportionately represented in lower management. That alone might account for the apparent differences among levels of management regarding harassment. In short, the failure to take account of the oversampling of women confounds all survey results that do not separate the findings by gender. The solution to this problem would have been to weight the responses by gender, as described earlier in this section.

In recent election-campaign polls, survey weighting has become a central topic, as some polling agencies weight their results on the basis of party affiliation and other variables, whereas others do not. Weighting in this instance involves assumptions regarding the differential participation of Republicans and Democrats in opinion polls and on Election Day—plus a determination of how many Republicans and Democrats there are. This will likely remain a topic of debate among pollsters and politicians in the years to come.

Probability Sampling in Review

Much of this chapter has been devoted to the key sampling method used in controlled survey research: probability sampling. In each of the variations examined, we've seen that elements are chosen for study from a population on a basis of random selection with known nonzero probabilities.

Depending on the field situation, probability sampling can be either very simple or extremely difficult, time consuming, and expensive. Whatever the situation, however, it remains the most effective method for the selection of study elements. There are two reasons for that.

First, probability sampling avoids researchers' conscious or subconscious biases in element selection. If all elements in the population have an equal (or unequal and subsequently weighted) chance of selection, there is an excellent chance that the sample so selected will closely represent the population of all elements.

Second, probability sampling permits estimates of sampling error. Although no probability sample will be perfectly representative in all respects, controlled selection methods permit the researcher to estimate the degree of expected error.

In this lengthy chapter, we've taken on a basic issue in much social research: selecting observations that will tell us something more general than the specifics we've actually observed. This issue confronts field researchers, who face more action and more actors than they can observe and record fully, as well as political pollsters who want to predict an election but can't interview all voters. As we proceed through the book, we'll see in greater detail how social researchers have found ways to deal with this issue.

The Ethics of Sampling

The key purpose of the sampling techniques discussed in this chapter is to allow researchers to make relatively few observations but gain an accurate picture of a large population. Quantitative studies using probability sampling should result in a statistical profile, based on the sample, that closely mirrors the profile that would have been gained from observing the whole population. In addition to using legitimate sampling techniques, researchers should be careful to point out the possibility of errors: sampling error, flaws in the sampling frame, nonresponse error, or anything else that might make the results misleading.

Sometimes, more typically in qualitative studies, the purpose of sampling may be to capture the breadth of variation within a population rather than to focus on the "average" or "typical" member of that population. Although this is a legitimate and valuable approach, readers may mistake the display of differences to reflect the distribution of characteristics in the population. As such, the researcher should ensure that the reader is not misled.

What do you think?...Revisited

Contrary to common sense, we've seen that the number of people selected in a sample, while important, is less important than how people are selected. The *Literary Digest* mailed ballots to 10 million people and received 2 million back from voters around the country. However, the people they selected for their enormous sample—auto owners and telephone subscribers—were not representative of the population in 1936, during the Great Depression. Overall, the sample population was wealthier than was the voting population at large. Because rich people

are more likely than the general public to vote Republican, the *Literary Digest* tallied the voting intentions of a disproportionate number of Republicans.

The probability-sampling techniques used today allow researchers to select smaller, more representative samples. Even a couple of thousand respondents, properly selected, can accurately predict the behavior of 100 million voters.

MAIN POINTS

Introduction

- Social researchers must select observations that will allow them to generalize to people and events not observed. Often this involves sampling—selecting people to observe.

- Understanding the logic of sampling is essential to doing social research.

A Brief History of Sampling

- Sometimes you can and should select probability samples using precise statistical techniques, but at other times nonprobability techniques are more appropriate.

Nonprobability Sampling

- Nonprobability-sampling techniques include reliance on available subjects, purposive (judgmental) sampling, snowball sampling, and quota sampling. In addition, researchers studying a social group may make use of informants. Each of these techniques has its uses, but none of them ensures that the resulting sample will be representative of the population being sampled.

The Logic and Techniques of Probability Sampling

- Probability-sampling methods provide an excellent way of selecting representative samples from large, known populations. These methods counter the problems of conscious and subconscious sampling bias by giving each element in the population a known (nonzero) probability of selection.

- Random selection is often a key element in probability sampling.

- The most carefully selected sample will never provide a perfect representation of the population from which it was selected. There will always be some degree of sampling error.

- By predicting the distribution of samples with respect to the target parameter, probability-sampling methods make it possible to estimate the amount of sampling error expected in a given sample.

- The expected error in a sample is expressed in terms of confidence levels and confidence intervals.

Populations and Sampling Frames

- A sampling frame is a list or quasi-list of the members of a population. It is the resource used in the selection of a sample. A sample's representativeness depends directly on the extent to which a sampling frame contains all the members of the total population that the sample is intended to represent.

- We've seen that the proliferation of cell phones is complicating the task of listing the desired population and selecting a sample.

Types of Sampling Designs

- Several sampling designs are available to researchers.

- Simple random sampling is logically the most fundamental technique in probability sampling, but it is seldom used in practice.

- Systematic sampling involves the selection of every kth member from a sampling frame. This method is more practical than simple random sampling and, with a few exceptions, is functionally equivalent.

- Stratification, the process of grouping the members of a population into relatively homogeneous strata before sampling, improves the representativeness of a sample by reducing the degree of sampling error.

Multistage Cluster Sampling

- Multistage cluster sampling is a relatively complex sampling technique that is frequently used when a list of all the members of a population does not exist. Typically, researchers must balance the number of clusters and the size of each cluster to achieve a given sample size.

Stratification can be used to reduce the sampling error involved in multistage cluster sampling.

- Probability proportionate to size (PPS) is a special, efficient method for multistage cluster sampling.

- If the members of a population have unequal probabilities of selection for the sample, researchers must assign weights to the different observations made in order to provide a representative picture of the total population. Basically, the weight assigned to a particular sample member should be the inverse of its probability of selection.

Probability Sampling in Review

- Probability sampling remains the most effective method for the selection of study elements because (1) it allows researchers to avoid biases in element selection and (2) it permits estimates of error.

The Ethics of Sampling

- Probability sampling always carries a risk of error; researchers must inform readers of any errors that might make results misleading.

- When nonprobability-sampling methods are used to obtain the breadth of variations in a population, researchers must take care not to mislead readers into confusing variations with what's typical in the population.

KEY TERMS

cluster sampling	random selection
confidence interval	representativeness
confidence level	sampling error
element	sampling frame
EPSEM (equal probability of selection method)	sampling interval
	sampling ratio
informant	sampling unit
nonprobability sampling	simple random sampling
parameter	snowball sampling
population	statistic
PPS (probability proportionate to size)	stratification
	study population
probability sampling	systematic sampling
purposive sampling	weighting
quota sampling	

PROPOSING SOCIAL RESEARCH: SAMPLING

In this portion of the proposal, you'll describe how you'll select from among all the possible observations you might make. Depending on the data-collection method you plan to employ, either probability or nonprobability sampling may be appropriate for your study. Similarly, this aspect of your proposal may involve the sampling of subjects or informants, or it could involve the sampling of corporations, cities, books, and so forth.

Your proposal, then, must specify what units you'll be sampling among, the data you'll use for purposes of your sample selection (your sampling frame, for instance), and the actual sampling methods you plan to use.

REVIEW QUESTIONS

1. Review the discussion of the 1948 Gallup poll that predicted that Thomas Dewey would defeat Harry Truman for president. What are some ways Gallup could have modified his quota-sampling design to avoid the error?

2. Using Appendix B of this book, select a simple random sample of 10 numbers in the range from 1 to 9,876. What is each step in the process?

3. What are the steps involved in selecting a multi-stage cluster sample of students taking first-year English in U.S. colleges and universities?

4. In Chapter 9, we'll discuss surveys conducted on the Internet. Can you anticipate possible problems concerning sampling frames, representativeness, and the like? Do you see any solutions?

CHAPTER 8

Experiments

CHAPTER OVERVIEW

An experiment is a mode of observation that enables researchers to probe causal relationships. Many experiments in social research are conducted under the controlled conditions of a laboratory, but experimenters can also take advantage of natural occurrences to study the effects of events in the social world.

Jacobs Stock Photography/Getty Images

Learning Objectives

After studying this chapter, you will be able to . . .

- Give examples of topics well suited to experimental studies and topics that would not be appropriate.
- Diagram and explain the key elements in the classical experiment.
- Discuss three methods for selecting and assigning subjects in an experiment.
- Understand several types of experimental designs.
- Provide examples illustrating the use of the experimental model in social research.

- Discuss the advantages and disadvantages of Web-based experiments.
- Describe what is meant by "natural" experiments, giving examples to illustrate.
- Identify and discuss both the strengths and weaknesses of experiments in social research.
- Explain some of the ethical issues involved in the use of the experimental model.

Introduction

This chapter addresses the *controlled experiment*: a research method commonly associated with the natural sciences. Although this is not the approach most commonly used in the social sciences, Part 3 begins with this method because it illustrates fundamental elements in the logic of explanatory research. If you can grasp the logic of the controlled experiment, you'll find it a useful backdrop for understanding techniques that are more commonly used. Of course, this chapter will also present some of the inventive ways social scientists have conducted experiments, and it will demonstrate some basic experimental techniques.

At the most basic level, experiments involve (1) taking action and (2) observing the consequences of that action. Social researchers typically select a group of subjects, do something to them, and observe the effect of what was done.

It's worth noting that experiments are often used in nonscientific human inquiry. In preparing a stew, for example, we add salt, taste, add more salt, and taste again. In defusing a bomb, we clip the red wire, observe whether the bomb explodes, clip the blue wire, and. . . .

We also experiment copiously in our attempt to develop an overall understanding of the world we live in. All skills are learned through experimentation: eating, walking, riding a bicycle, and so forth. This chapter will discuss some ways social researchers use experiments to develop generalized understandings. We'll see that, like other methods available to the social researcher, experimenting has its special strengths and weaknesses.

Topics Appropriate for Experiments

Experiments are more appropriate for some topics and research purposes than for others. Experiments are especially well suited to research projects involving relatively limited and well-defined concepts and propositions. In terms of the traditional image of science, discussed earlier in this book, the experimental model is especially appropriate for hypothesis testing. Because experiments focus on determining causation, they're also better suited to explanatory than to descriptive purposes.

Let's assume, for example, that we want to discover ways of reducing prejudice against Muslims. We hypothesize that learning about the contribution of Muslims to U.S. history will reduce prejudice, and we decide to test this hypothesis experimentally. To begin, we might test a group of experimental subjects to determine their levels of prejudice against Muslims. Next, we might show them a documentary film depicting the many important ways Muslims have contributed to the scientific, literary, political, and social development of the nation. Finally, we would measure our subjects' levels of prejudice against Muslims to determine whether the film has actually reduced prejudice.

What do you think?

The impact of the observer raises many serious questions regarding the usefulness of experiments in social research. How can the manipulation of people in a controlled, experimental environment tell us anything about "natural" human behavior? After all is said and done, doesn't an experiment simply tell us how people behave when they participate in an experiment?

See the *What do you think? ... Revisited* box toward the end of the chapter.

Earl Babbie

Experimentation has also been successful in the study of small-group interaction. Thus, we might bring together a small group of experimental subjects and assign them a task, such as making recommendations for popularizing car pools. Then we would observe how the group organizes itself and deals with the problem. Over the course of several such experiments, we might systematically vary the nature of the task or the rewards for handling the task successfully. By observing differences in the way groups organize themselves and operate under these varying conditions, we could learn a great deal about the nature of small-group interaction and the factors that influence it. For example, attorneys sometimes present evidence in different ways to different mock juries, to see which method is the most effective.

Earl Babbie

Experiments often involve putting people in unusual, controlled situations to see how they will respond.

We typically think of experiments as being conducted in laboratories. Indeed, most of the examples in this chapter involve such a setting. This need not be the case, however. Increasingly, social researchers are using the World Wide Web as a vehicle for conducting experiments. Further, sometimes we can construct what are called *natural experiments*: "experiments" that occur in the regular course of social events. The latter portion of this chapter deals with such research.

The Classical Experiment

In both the natural and the social sciences, the most conventional type of experiment involves three major pairs of components: (1) independent and dependent variables, (2) pretesting and posttesting, and (3) experimental and control groups. This section looks at each of these components and the way they're put together in the execution of an experiment.

Independent and Dependent Variables

Essentially, an experiment examines the effect of an independent variable on a dependent variable. Typically, the independent variable takes the form of an experimental stimulus, which is either present or absent. That is, the stimulus is a *dichotomous* variable, having two attributes—present or not present. In this typical model, the experimenter compares what happens when the stimulus is present to what happens when it is not.

In the example concerning prejudice against Muslims, *prejudice* is the dependent variable and *exposure to Muslim history* is the independent variable. The researcher's hypothesis suggests that prejudice depends, in part, on a lack of knowledge of Muslim history. The purpose of the experiment is to test the validity of this hypothesis by presenting some subjects with an appropriate stimulus, such as a documentary film. In other words, the independent variable is the cause and the dependent variable is the effect. Thus, we might say that watching the film caused a change in prejudice or that reduced prejudice was an effect of watching the film.

The independent and dependent variables appropriate for experimentation are nearly limitless. Moreover, a given variable might serve as an independent variable in one experiment and as a dependent variable in another. For example, *prejudice* is the dependent variable in the previous example, but it might be the independent variable in an experiment examining the effect of prejudice on voting behavior.

To be used in an experiment, both independent and dependent variables must be operationally defined. Such operational definitions might involve a variety of observational methods. Responses to a questionnaire, for example, might be the basis for defining prejudice. Speaking to or ignoring Muslims, or agreeing or disagreeing with them, might be elements in the operational definition of interaction with Muslims in a small-group setting.

Conventionally, in the experimental model, dependent and independent variables must be operationally defined before the experiment begins. However, as you'll see in connection with survey research and other methods, it's sometimes appropriate to make a wide variety of observations during data collection and then determine the most useful operational definitions of variables during later analyses. Ultimately, however, experimentation, like other quantitative methods, requires specific and standardized measurements and observations.

Pretesting and Posttesting

In the simplest experimental design, **pretesting** occurs first, whereby subjects are measured in terms of a dependent variable. Then the subjects are exposed to a stimulus representing an independent variable. Finally, in **posttesting**, they are remeasured in terms of the dependent variable. Any differences between the first and last measurements on the dependent variable are then attributed to the independent variable.

In the example of prejudice and exposure to Muslim history, we would begin by pretesting the extent of prejudice among our experimental subjects. Using a questionnaire asking about attitudes toward Muslims, for example, we could measure the extent of prejudice exhibited by each individual subject and the average prejudice level of the whole group. After exposing the subjects to the Muslim history film, we could administer the same questionnaire again. Responses given in this posttest would permit us to measure the later extent of prejudice for each subject and the average prejudice level of the group as a whole. If we discovered a lower level of prejudice during the second administration of the questionnaire, we might conclude that the film had indeed reduced prejudice.

In the experimental examination of attitudes such as prejudice, we face a special practical problem relating to validity. As you may already have imagined, the subjects might respond differently to the questionnaires the second time even if their attitudes remain unchanged. During the first administration of the questionnaire, the subjects may be unaware of its purpose. By the second measurement, they may have figured out that the researchers are interested in measuring their prejudice. Because no one wishes to seem prejudiced, the subjects may "clean up" their answers the second time around. Thus, the film will *seem* to have reduced prejudice although, in fact, it has not.

This is an example of a more general problem that plagues many forms of social research: The very act of studying something may change it. The techniques for dealing with this problem in the context of experimentation will be discussed in various places throughout the chapter. The first technique involves the use of control groups.

pretesting The measurement of a dependent variable among subjects before they are exposed to a stimulus representing an independent variable.

posttesting The remeasurement of a dependent variable among subjects after they've been exposed to a stimulus representing an independent variable.

Experimental and Control Groups

Laboratory experiments seldom, if ever, involve only the observation of an **experimental group** to which a stimulus has been administered. In addition, the researchers observe a **control group**, which does not receive the experimental stimulus.

In the example of prejudice and Muslim history, we might examine two groups of subjects. To begin, we give each group a questionnaire designed to measure their prejudice against Muslims. Then we show the film only to the experimental group. Finally, we administer a posttest of prejudice to both groups. Figure 8-1 illustrates this basic experimental design.

Using a control group allows the researcher to detect any effects of the experiment itself. If the posttest shows that the overall level of prejudice exhibited by the control group has dropped as much as that of the experimental group, then the apparent reduction in prejudice must be a function of the experiment or of some external factor rather than a function of the film. If, on the other hand, prejudice is reduced only in the experimental group, this reduction would seem to be a consequence of exposure to the film, because that's the only difference between the two groups. Alternatively, if prejudice is reduced in both groups but to a greater degree in the experimental group than in the control group, that, too, would be grounds for assuming that the film reduced prejudice.

The need for control groups in social research became clear in connection with a series of studies of employee satisfaction, conducted by F. J. Roethlisberger and W. J. Dickson (1939) in the late 1920s and early 1930s. These researchers were interested in discovering what kinds of changes in working conditions would improve employee satisfaction and productivity. To pursue

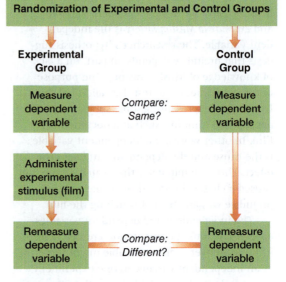

FIGURE 8-1

Diagram of Basic Experimental Design. The fundamental purpose of an experiment is to isolate the possible effect of an independent variable (called the *stimulus* in experiments) on a dependent variable. Members of the experimental group(s) are exposed to the stimulus and those in the control group(s) are not.

this objective, they studied working conditions in the telephone "bank wiring room" of the Western Electric Works in the Chicago suburb of Hawthorne, Illinois.

To the researchers' great satisfaction, they discovered that improving the working conditions increased satisfaction and productivity consistently. As the workroom was brightened up through better lighting, for example, productivity went up. When lighting was further improved, productivity went up again.

To further substantiate their scientific conclusion, the researchers then dimmed the lights. Whoops—productivity again improved!

At this point it became evident then that the wiring-room workers were responding more to the attention given them by the researchers than to the improved working conditions. As a result of this phenomenon, often called the *Hawthorne effect*, social researchers have become more sensitive to and cautious about the possible effects of experiments themselves. In the wiring-room study, the use of a proper control group— one that was studied intensively without any other changes in the working conditions—would have pointed to the existence of this effect.

---

experimental group In experimentation, a group of subjects to whom an experimental stimulus is administered.

control group In experimentation, a group of subjects to whom no experimental stimulus is administered and who resemble the experimental group in all other respects. The comparison of the control group and the experimental group at the end of the experiment points to the effect of the experimental stimulus.

The need for control groups in experimentation has been nowhere more evident than in medical research. Time and again, patients who participate in medical experiments have appeared to improve, but it has been unclear how much of the improvement has come from the experimental treatment and how much from the experiment. In testing the effects of new drugs, then, medical researchers frequently administer a *placebo*—a "drug" with no relevant effect, such as sugar pills—to a control group. Thus, the control-group patients believe they, like the experimental group, are receiving an experimental drug. Often, they improve. If the new drug is effective, however, those receiving the actual drug will improve more than those receiving the placebo.

In social science experiments, control groups provide an important guard against not only the effects of the experiments themselves but also against the effects of any events outside the laboratory during the experiments. In the example of the study of prejudice, suppose that a popular Muslim leader is assassinated in the middle of, say, a weeklong experiment. Such an event might horrify the experimental subjects, requiring them to examine their own attitudes toward Muslims, resulting in reduced prejudice. Because such an effect should happen about equally for members of the control group and the experimental group, a greater reduction of prejudice among the experimental group would, again, point to the impact of the experimental stimulus: the documentary film.

Sometimes an experimental design will require more than one experimental or control group. In the case of the documentary film, for example, we might also want to examine the impact of reading a book on Muslim history. In that case, we might have one group see the film and read the book, another group only see the movie, still another group only read the book, and the control group do neither. With this kind of design, we could determine the impact of each stimulus separately, as well as their combined effect.

The Double-Blind Experiment

Like patients who improve when they merely think they're receiving a new drug, sometimes experimenters tend to prejudge results. In medical research, the experimenters may be more likely to "observe" improvements among patients receiving the experimental drug than among those receiving the placebo. (This would be most likely, perhaps, for the researcher who developed the drug.) A **double-blind experiment** eliminates this possibility, because neither the subjects nor the experimenters know which is the experimental group and which is the control group. In the medical case, those researchers responsible for administering the drug and for noting improvements would not be told which subjects were receiving the drug and which the placebo. Conversely, the researcher who knew which subjects were in which group would not administer the experiment.

In social science experiments, as in medical ones, the danger of experimenter bias is further reduced to the extent that the operational definitions of the dependent variables are clear and precise. For example, medical researchers would be less likely to unconsciously bias their reading of a patient's temperature than they would be to bias their assessment of how lethargic the patient was. Similarly, the small-group researcher would be less likely to misperceive which subject spoke, or to whom he or she spoke, than whether the subject's comments sounded cooperative or competitive, a more subjective judgment that's difficult to define in precise behavioral terms.

The role of the placebo may be more complex than you think, according to a 2010 medical experiment on irritable bowel syndrome. One group of sufferers was given pills in a bottle marked "Placebo" and it was explained that a placebo, sometimes called a sugar pill, contained no active ingredients. Subjects were told that people sometimes seemed to benefit from the placebos. A control group was given no treatment at all. After 21 days the placebo group had improved significantly, while the control group had not.

This study was further complicated, however, by the fact that those receiving the placebo pills also received examinations and counseling sessions, while the control group received no

---

double-blind experiment An experimental design in which neither the subjects nor the experimenters know which is the experimental group and which is the control.

attention at all. Perhaps, as the researchers acknowledge, the positive results were produced by the comprehensive treatment package, not by the placebo pills alone. Also, they note, the measures of improvement were self-assessments. It is possible that physiological measurements might have shown no improvement. But, to complicate matters further, isn't "feeling better" the goal of such treatments?

As I've indicated several times, seldom can we devise operational definitions and measurements that are wholly precise and unambiguous. This is another reason why employing a double-blind design in social research experiments might be appropriate.

Selecting Subjects

In Chapter 7 we discussed the logic of sampling, which involves selecting a sample that is representative of some populations. Similar considerations apply to experiments. Because most social researchers work in colleges and universities, it seems likely that most social research laboratory experiments are conducted with college undergraduates as subjects. Typically, the experimenter asks students enrolled in his or her classes to participate in experiments or advertises for subjects in a college newspaper. Subjects may or may not be paid for participating in such experiments. (See Chapter 3 for more on the ethical issues involved in this situation.)

In relation to the norm of generalizability in science, this tendency clearly represents a potential defect in social research. Simply put, college undergraduates do not typify the public at large. There is a danger, therefore, that we may learn much about the attitudes and actions of college undergraduates but not about social attitudes and actions in general.

However, this potential defect is less significant in explanatory research than in descriptive research. Although it is true that having noted the level of prejudice among a group of college undergraduates in our pretesting, we would have little confidence that the same level existed among the public at large. On the other hand, if we found that a documentary film reduced whatever level of prejudice existed among those undergraduates, we would have

more confidence—without being certain—that it would have a similar effect in the community at large. Social processes and patterns of causal relationships appear to be more generalizable and more stable than specific characteristics such as an individual's level of prejudice.

This problem of generalizing from students isn't always seen as problematic, as Jerome Taylor reports in a commentary on research concerning the common cold, a disease he traces back to ancient Egypt. This elusive illness attacks only humans and chimpanzees, so you can probably guess which subjects medical researchers have tended to select. However, you might be wrong:

> Chimpanzees were too expensive to import en masse, so during the first half of the 20th century British scientists began looking into how the common cold worked by conducting experiments on medical students at St. Bartholomew's Hospital in London.
>
> (Taylor 2008)

Aside from the question of generalizability, the cardinal rule of subject selection and experimentation concerns the comparability of experimental and control groups. Ideally, the control group represents what the experimental group would have been like if it had *not* been exposed to the experimental stimulus. The logic of experiments requires, therefore, that experimental and control groups be as similar as possible. There are several ways to accomplish this, as will be discussed next.

Probability Sampling

The discussions of the logic and techniques of probability sampling in Chapter 7 outline one method for selecting two groups that are similar to each other. Beginning with a sampling frame composed of all the people in the population under study, the researcher might select two probability samples. If these samples each resemble the total population from which they're selected, they'll also resemble each other.

Recall also, however, that the degree of resemblance (representativeness) achieved by probability sampling is largely a function of the sample size. As a general guideline, probability samples of less than 100 are not likely to be representative, and social science experiments seldom involve that many subjects in either experimental or control groups. As a result, then, probability sampling is seldom used in experiments to select subjects from a larger population. Researchers do, however, use the logic of random selection when they assign subjects to groups.

Randomization

Having recruited, by whatever means, a total group of subjects, the experimenter may randomly assign those subjects to either the experimental or the control group. Such **randomization** might be accomplished by numbering all of the subjects serially and selecting numbers by means of a random-number table, or the experimenter might assign the odd-numbered subjects to the experimental group and the even-numbered subjects to the control group.

Let's return again to the basic concept of probability sampling. If we recruit 40 subjects (in response to a newspaper advertisement, for example), there's no reason to believe that the 40 subjects represent the entire population from which they've been drawn. Nor can we assume that the 20 subjects randomly assigned to the experimental group represent that larger population. We can have greater confidence, however, that the 20 subjects randomly assigned to the experimental group will be reasonably similar to the 20 assigned to the control group.

Following the logic of our earlier discussions of sampling, we can see our 40 subjects as a population from which we select two probability samples—each consisting of half the population. Because each sample reflects the characteristics of the total population, the two samples will mirror each other.

As we saw in Chapter 7, our assumption of similarity in the two groups depends in part on the number of subjects involved. In the extreme case, if we recruited only two subjects and assigned, by the flip of a coin, one as the experimental subject and one as the control, there would be no reason to assume that the two subjects are similar to each other. With larger numbers of subjects, however, randomization makes good sense.

Matching

Another way to achieve comparability between the experimental and control groups is through **matching**. This process is similar to the quota-sampling methods discussed in Chapter 7. If 12 of our subjects are young white men, we might assign 6 of those at random to the experimental group and the other 6 to the control group. If 14 are middle-aged African American women, we might assign 7 to each group.

The overall matching process could be most efficiently achieved through the creation of a quota matrix constructed of all the most relevant characteristics. Figure 8-2 provides a simplified illustration of such a matrix. In this example, the experimenter has decided that the relevant characteristics are race, age, and gender. Ideally, the quota matrix is constructed to result in an even number of subjects in each cell of the matrix. Then, half the subjects in each cell go into the experimental group and half into the control group.

Alternatively, we might recruit more subjects than our experimental design requires. We might then examine many characteristics of the large initial group of subjects. Whenever we discover a pair of quite similar subjects, we might assign one at random to the experimental group and the other to the control group. Potential subjects who are unlike anyone else in the initial group might be left out of the experiment altogether.

Whatever method we employ, the desired result is the same. The overall average description of the experimental group should be the same as that of the control group. For example, they should have about the same average age, the same proportions of males and females, the same racial composition, and so forth. This test of comparability should be used whether

---

randomization A technique for assigning experimental subjects to experimental and control groups randomly.

matching In connection with experiments, the procedure whereby pairs of subjects are matched on the basis of their similarities on one or more variables, and one member of the pair is assigned to the experimental group and the other to the control group.

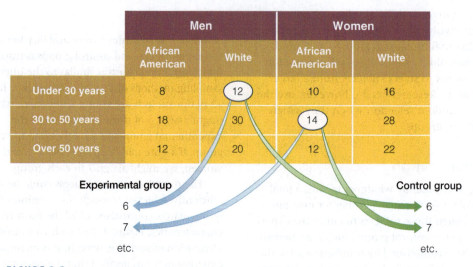

	Men		Women	
	African American	White	African American	White
Under 30 years	8	12	10	16
30 to 50 years	18	30	14	28
Over 50 years	12	20	12	22

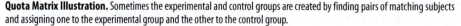

Experimental group   Control group

6   6

7   7

etc.   etc.

FIGURE 8-2

Quota Matrix Illustration. Sometimes the experimental and control groups are created by finding pairs of matching subjects and assigning one to the experimental group and the other to the control group.

the two groups are created through probability sampling or through randomization.

Thus far, I've referred to the "relevant" variables without saying clearly what those variables are. Of course, I can't give a definite answer to this question, any more than I could specify in Chapter 7 which variables should be used in stratified sampling. Which variables are relevant ultimately depends on the nature and purpose of the experiment. As a general rule, however, the control and experimental groups should be comparable in terms of those variables most likely to be related to the dependent variable under study. In a study of prejudice, for example, the two groups should be alike in terms of education, ethnicity, and age, among other characteristics. In some cases, moreover, we may delay assigning subjects to experimental and control groups until we've initially measured the dependent variable. Thus, for example, we might administer a questionnaire measuring subjects' prejudice and then match the experimental and control groups to assure ourselves that the two groups exhibit the same overall level of prejudice.

Matching or Randomization?

When assigning subjects to the experimental and control groups, you should be aware of two arguments in favor of randomization over matching. First, you may not be in a position to

know in advance which variables will be relevant for the matching process. Second, most of the statistics used to analyze the results of experiments assume randomization. Failure to design your experiment that way, then, makes your later use of those statistics less meaningful.

On the other hand, randomization makes sense only if you have a fairly large pool of subjects so that the laws of probability sampling apply. With only a few subjects, matching would be a better procedure.

Sometimes researchers can combine matching and randomization. When conducting an experiment in the educational enrichment of young adolescents, for example, Milton Yinger and his colleagues (1977) needed to assign a large number of students, ages 13 and 14, to several different experimental and control groups to ensure the comparability of students composing each of the groups. They achieved this goal using the following method.

Beginning with a pool of subjects, the researchers first created strata of students nearly identical to one another in terms of some 15 variables. From each of the strata, students were randomly assigned to the different experimental and control groups. In this fashion, the researchers actually improved on conventional randomization. Essentially, they used a stratified sampling procedure (recall the discussion in Chapter 7), except that they employed far more stratification variables than are typically used in, say, survey sampling.

Thus far, I've described the classical experiment—the experimental design that best represents the logic of causal analysis in the laboratory. In practice, however, social researchers use a great variety of experimental designs. In the next section, we'll look at some of these approaches.

Variations on Experimental Design

In their classic book on research design, *Experimental and Quasi-Experimental Designs for Research,* Donald Campbell and Julian Stanley (1963) describe sixteen different experimental and quasi-experimental designs. This section summarizes a few of these variations to help show the potential for experimentation in social research.

Preexperimental Research Designs

To begin, Campbell and Stanley discuss three *preexperimental* designs, not to recommend them but because they're frequently used in less-than-professional research. These designs are called *preexperimental* to indicate that they do not meet the scientific standards of experimental designs, and sometimes they may be used because the conditions for full-fledged experiments are impossible to meet. In the first such design—the *one-shot case study*—a single group of subjects is measured on a dependent variable following the administration of some experimental stimulus. Suppose, for example, that we show the previously mentioned Muslim history film to a group of people and then administer a questionnaire that seems to measure prejudice against Muslims. Suppose further that the answers given to the questionnaire seem to represent a low level of prejudice. We might be tempted to conclude that the film reduced prejudice. Lacking a pretest, however, we can't be sure. Perhaps the questionnaire doesn't really represent a very sensitive measure of prejudice, or perhaps the group we're studying was low in prejudice to begin with. In either case, the film might have made no difference, though our experimental results might mislead us into thinking it did.

The second preexperimental design discussed by Campbell and Stanley adds a pretest for the experimental group but lacks a control group. This design—which the authors call the *one-group pretest–posttest design*—suffers from the possibility that some factor other than the independent variable might cause a change between the pretest and posttest results, such as the scenario described earlier concerning the assassination of a respected Muslim leader. Thus, although we can see that prejudice has been reduced, we can't be sure the film caused that reduction.

To round out the possibilities for preexperimental designs, Campbell and Stanley point out that some research is based on experimental and control groups but has no pretests. They call this design the *static-group comparison.* For example, we might show the Muslim history film to one group but not to another and then measure prejudice in both groups. If the experimental group had less prejudice at the conclusion of the experiment, we might assume the film was responsible. But unless we had randomized our subjects, we would have no way of knowing that the two groups had the same degree of prejudice initially; perhaps the experimental group started out with less prejudice.

Figure 8-3 illustrates these three preexperimental research designs, using a different research question: "Does exercise cause weight reduction?" To make the several designs clearer, the figure shows individuals rather than groups, but the same logic pertains to group comparisons. Let's review the three preexperimental designs in this new example.

The one-shot case study design represents a common form of logical reasoning in everyday life. Asked whether exercise causes weight reduction, we may bring to mind an example that would seem to support the proposition: someone who exercises and is thin. There are problems with this reasoning, however. Perhaps the person was thin long before beginning to exercise. Or perhaps he became thin for some other reason, such as eating less or suffering from an illness. The observations shown in the diagram do not guard against these other possibilities. Moreover, the observation that the man in the diagram is in trim shape depends on our intuitive idea of what constitutes trim and overweight body shapes. All told, this is very weak evidence for testing the relationship between exercise and weight loss.

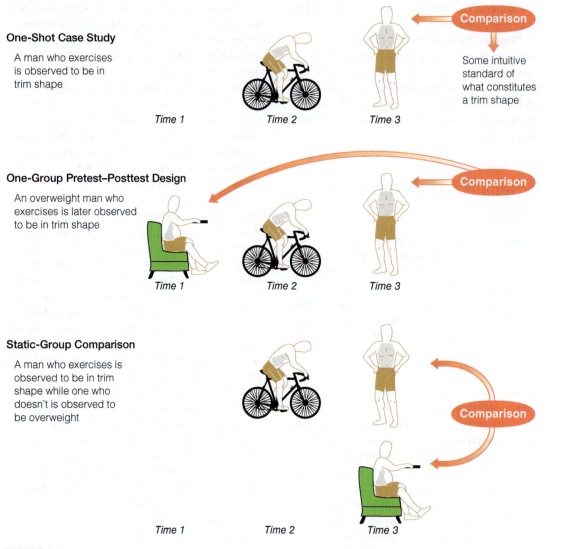

One-Shot Case Study

A man who exercises is observed to be in trim shape

Time 1   *Time 2*   *Time 3*

Comparison

Some intuitive standard of what constitutes a trim shape

One-Group Pretest–Posttest Design

An overweight man who exercises is later observed to be in trim shape

Time 1   *Time 2*   *Time 3*

Comparison

Static-Group Comparison

A man who exercises is observed to be in trim shape while one who doesn't is observed to be overweight

Time 1   *Time 2*   *Time 3*

Comparison

FIGURE 8-3

Three Preexperimental Research Designs. These preexperimental designs anticipate the logic of true experiments but remain open to errors of interpretation. Can you see the errors that might be made in each of these designs? The various risks are solved by the addition of control groups, pretesting, and posttesting.

The one-group pretest–posttest design offers somewhat better evidence that exercise produces weight loss. Specifically, we've ruled out the possibility that the man was thin before beginning to exercise. However, we still have no assurance that it was his exercising that caused him to lose weight.

Finally, the static-group comparison eliminates the problem of our questionable definition of what constitutes trim or overweight body shapes. In this case, we can compare the shapes of the man who exercises and the one who does not.

This design, however, reopens the possibility that the man who exercises was thin to begin with.

Validity Issues in Experimental Research

At this point, I want to present in a more systematic way the factors that affect experimental research—those I've already discussed as well as additional factors. First we'll look at what Campbell and Stanley call the sources of *internal invalidity*, reviewed and expanded in a follow-up book by Thomas Cook and Donald Campbell (1979). Then we'll consider the problem of generalizing experimental results to the "real" world, referred to as *external invalidity*. Having examined these, we'll be in a position to appreciate the advantages of some of the more sophisticated experimental and quasi-experimental designs that social science researchers sometimes use.

Sources of Internal Invalidity

The problem of **internal invalidity** refers to the possibility that the conclusions drawn from experimental results may not accurately reflect what has gone on in the experiment itself. The threat of internal invalidity is present whenever anything other than the experimental stimulus can affect the dependent variable.

Donald Campbell and Julian Stanley (1963: 5–6) and Thomas Cook and Donald Campbell (1979: 51–55) point to several sources of internal invalidity. I will touch on eight of them here to illustrate this concern:

1. *History.* During the course of the experiment, historical events may occur that confound the experimental results. The assassination of a Muslim leader during the course of an experiment on reducing anti–Muslim prejudice is one example.
2. *Maturation.* People are continually growing and changing, and such changes affect the results of the experiment. In a long-term experiment, the fact that the subjects grow older (and wiser?) can have an effect. In shorter experiments, they can grow tired, sleepy, bored, or hungry—or change in other ways that affect their behavior in the experiment.
3. *Testing.* Often the process of testing and retesting influences people's behavior, thereby confounding the experimental results. Suppose we administer a questionnaire to a group as a way of measuring their prejudice. Then we administer an experimental stimulus and remeasure their prejudice. As we saw earlier, by the time we conduct the posttest, the subjects will probably have become more sensitive to the issue of prejudice and will be more thoughtful in their answers. In fact, they may have figured out that we're trying to find out how prejudiced they are, and, because few people want to appear prejudiced, they may give answers that they think the researchers are seeking or that will make themselves "look good."

4. *Instrumentation.* The process of measurement in pretesting and posttesting brings in some of the issues of conceptualization and operationalization discussed earlier in the book. For example, if we use different measures of the dependent variable (say, different questionnaires about prejudice), how can we be sure they're comparable? Perhaps prejudice will seem to decrease simply because the pretest measure was more sensitive than the posttest measure. Or if the measurements are being made by the experimenters, their standards or abilities may change over the course of the experiment.
5. *Statistical regression.* Sometimes it's appropriate to conduct experiments on subjects who start out with extreme scores on the dependent variable. If you were testing a new method for teaching math to hard-core failures in math, you would want to conduct your experiment on people who previously have done extremely poorly in math. But consider for a minute what's likely to happen to the math achievement of such people over time without any experimental interference. They're starting out so low that they can only stay at the bottom or improve: They can't get worse. Even without any experimental stimulus, then, the group as a whole is likely to show some improvement over time. Referring to a *regression to the mean,* statisticians often point out that extremely tall people as a group are likely to have children shorter than themselves, and extremely short people as a group are likely to have children taller than themselves. There is a danger, then, that changes occurring by virtue of subjects starting out in extreme positions will be attributed erroneously to the effects of the experimental stimulus.
6. *Selection biases.* We discussed selection bias earlier when we examined different ways of selecting subjects for experiments and assigning them to experimental and control groups. Comparisons have no meaning unless the groups are comparable at the start of an experiment.
7. *Experimental mortality.* Although some social experiments could, I suppose, kill subjects, *experimental mortality* refers to a more general and less extreme problem. Often, experimental subjects will drop out of the experiment before

---

internal invalidity Refers to the possibility that the conclusions drawn from experimental results may not accurately reflect what went on in the experiment itself.

it's completed, and this can affect statistical comparisons and conclusions. In the classical experiment involving an experimental and a control group, each with a pretest and posttest, suppose that the bigots in the experimental group are so offended by the Muslim history film that they leave before it's over. Those subjects sticking around for the posttest will have been less prejudiced to start with, so the group results will reflect a substantial "decrease" in prejudice.

8. *Demoralization.* On the other hand, feelings of deprivation within the control group may result in some giving up. In educational experiments, control-group subjects may feel the experimental group is being treated better and they may become demoralized, stop studying, act up, or get angry.

These, then, are some of the sources of internal invalidity in experiments, as cited by Campbell, Stanley, and Cook. Aware of these pitfalls, experimenters have devised designs aimed at managing them. The classical experiment, coupled with proper subject selection and assignment, addresses each of these problems. Let's look again at that study design, presented in Figure 8-4, as it applies to our hypothetical study of prejudice.

If we use the experimental design shown in Figure 8-4, we should expect two findings from our Muslim history film experiment. For the experimental group, the level of prejudice measured in their posttest should be less than that found in their pretest. In addition, when the two posttests are compared, less prejudice should be found in the experimental group than in the control group.

This design also guards against the problem of history, in that anything occurring outside the experiment that might affect the experimental group should also affect the control group. Consequently, the two posttest results should still differ. The same comparison guards against problems of maturation as long as the subjects have been randomly assigned to the two groups. Testing and instrumentation can't be problems, because both the experimental and control groups are subject to the same tests and experimenter effects. If the subjects have been assigned to the two groups randomly, statistical regression should affect both equally, even if people with extreme scores on prejudice (or whatever the dependent variable is) are being studied. Selection bias is ruled out by the random assignment of subjects. Experimental mortality is more complicated to handle, but the data provided in this study design offer several ways to deal with it. Pretest measurements would let us discover any differences in the dropouts of the experimental

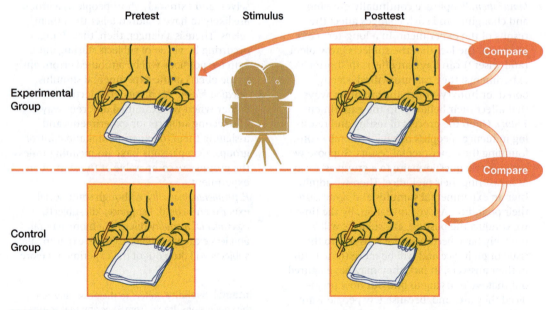

FIGURE 8-4

The Classical Experiment: Using a Muslim History Film to Reduce Prejudice. This diagram illustrates the basic structure of the classical experiment as a vehicle for testing the impact of a film on prejudice. Notice how the control group, the pretesting, and the posttesting function.

and control groups. Slight modifications to the design—administering a placebo (such as a film having nothing to do with Muslims) to the control group, for example—can make the problem even easier to manage. Finally, demoralization can be watched for and taken into account in evaluating the results of the experiment.

Sources of External Invalidity

Internal invalidity accounts for only some of the complications faced by experimenters. In addition, there are problems of what Campbell and Stanley call **external invalidity**, which relates to the generalizability of experimental findings to the "real" world. Even if the results of an experiment provide an accurate gauge of what happened during that experiment, do they really tell us anything about life in the wilds of society?

Campbell and Stanley describe four forms of this problem; I'll present one of them to you as an illustration. The generalizability of experimental findings is jeopardized, as the authors point out, if there's an interaction between the testing situation and the experimental stimulus (1963: 18). Here's an example of what they mean.

Staying with the study of prejudice and the Muslim history film, let's suppose that our experimental group—in the classical experiment—has less prejudice in its posttest than in its pretest, and that its posttest shows less prejudice than that of the control group. We can be confident that the film actually reduced prejudice among our experimental subjects. But would it have the same effect on the public if the film were shown in theaters or on television? We can't be sure, because the film might be effective only when people have been sensitized to the issue of prejudice, as the subjects may have been while taking the pretest. This is an example of interaction between the testing and the stimulus. The classical experimental design cannot control for that possibility. Fortunately, experimenters have devised other designs that can.

The *Solomon four-group design* (Campbell and Stanley 1963: 24–25) addresses the problem of testing interaction with the stimulus. As the name suggests, it involves four groups of subjects, assigned randomly from a pool. Figure 8-5 presents this design.

Notice that Groups 1 and 2 in Figure 8-5 compose the classical experiment. Group 3 is administered the experimental stimulus without

a pretest, and Group 4 is only posttested. This latest experimental design permits four meaningful comparisons. If the Muslim history film really reduces prejudice—unaccounted for by the problem of internal invalidity and unaccounted for by an interaction between the testing and the stimulus—we should expect four findings:

1. In Group 1, posttest prejudice should be less than pretest prejudice.
2. In Group 2, prejudice should be the same in the pretest and the posttest.
3. The Group 1 posttest should show less prejudice than the Group 2 posttest.
4. The Group 3 posttest should show less prejudice than the Group 4 posttest.

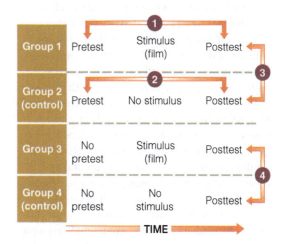

Expected Findings

1. In Group 1, posttest prejudice should be less than pretest prejudice.
2. In Group 2, prejudice should be the same in the pretest and the posttest.
3. The Group 1 posttest should show less prejudice than the Group 2 posttest does.
4. The Group 3 posttest should show less prejudice than the Group 4 posttest does.

FIGURE 8-5

The Solomon Four-Group Design. The classical experiment runs the risk that pretesting will have an effect on subjects, so the Solomon four-group design adds experimental and control groups that skip the pretest. Thus, it combines the classical experiment and the after-only design or "static-group comparison."

---

external invalidity Refers to the possibility that conclusions drawn from experimental results may not be generalizable to the "real" world.

Notice that finding number 4 rules out any interaction between the testing and the stimulus. Remember that these comparisons are meaningful only if subjects have been assigned randomly to the different groups, thereby providing groups of equal prejudice initially, even though their preexperimental prejudice is measured only in Groups 1 and 2.

There is a side benefit to this research design, as the authors point out. Not only does the Solomon four-group design rule out interactions between testing and the stimulus, it also provides data for comparisons that will reveal the amount of such interaction that occurs in the classical experimental design. This knowledge would allow a researcher to review and evaluate the value of any prior research that used the simpler design.

The last experimental design I'll mention here is what Campbell and Stanley (1963: 25–26) call the *posttest-only control-group design*; it consists of the second half—Groups 3 and 4—of the Solomon design (refer again to Figure 8-5). As the authors argue persuasively, with proper randomization, only Groups 3 and 4 are needed for a true experiment that controls for the problems of internal invalidity as well as for the interaction between testing and stimulus. With randomized assignment to experimental and control groups (which distinguishes this design from the static-group comparison discussed earlier), the subjects will be initially comparable on the dependent variable—comparable enough to satisfy the conventional statistical tests used to evaluate the results—so it's not necessary to measure them. Indeed, Campbell and Stanley suggest that the only justification for pretesting in this situation is tradition. Experimenters have simply grown accustomed to pretesting and feel more secure with research designs that include it. Be clear, however, that this point applies only to experiments in which subjects have been assigned to experimental and control groups randomly, because that's what justifies the assumption that the groups are equivalent—without actually measuring them to find out.

This discussion has introduced the intricacies of experimental design, its problems, and some solutions. Of course, researchers use a great many other possible experimental designs as well. Some involve more than one stimulus and combinations of stimuli. Others involve several tests of the dependent variable over time and the administration of the stimulus at different times for different groups. If you're interested in pursuing this topic, you might want to look at the Campbell and Stanley book.

Examples of Experimentation

Experiments have been used to study a wide variety of topics in the social sciences. Some experiments take place within laboratory situations; others occur out in the "real world"—these are referred to as **field experiments**. The following discussion will give you a glimpse of both. We'll begin with an example of a field experiment.

In George Bernard Shaw's well-loved play, *Pygmalion*—the basis for the musical *My Fair Lady*—Eliza Doolittle speaks of the powers others have in determining our social identity. Here's how she distinguishes between the ways she's treated by her tutor, Professor Higgins, and by Higgins's friend, Colonel Pickering:

> You see, really and truly, apart from the things anyone can pick up (the dressing and the proper way of speaking, and so on), the difference between a lady and a flower girl is not how she behaves, but how she's treated. I shall always be a flower girl to Professor Higgins, because he always treats me as a flower girl, and always will, but I know I can be a lady to you, because you always treat me as a lady, and always will.

(Act 5)

The sentiment Eliza expresses here is basic social science, addressed more formally by sociologists such as Charles Horton Cooley ("looking-glass self") and George Herbert Mead ("the generalized other"). The basic point is that who we think we are—our self-concept—and how we behave is largely a function of how others see and treat us. Further, the way others perceive us is largely conditioned by their expectations. If they've been told we're stupid, for example, they're likely to see us that way—and we may come to see ourselves that way and, in fact, act stupidly. "Labeling theory" addresses the phenomenon of people acting in accord with the ways they are perceived and labeled by others. These theories

---

field experiment A formal experiment conducted outside the laboratory, in a natural setting.

have served as the premise for numerous movies, such as the 1983 film *Trading Places*, in which Eddie Murphy and Dan Ackroyd play a derelict converted into a stockbroker and vice versa.

The tendency to see in others what we've been led to expect takes its name from Shaw's play and is called the *Pygmalion effect*. This effect is nicely suited to controlled experiments. In one of the best-known experiments on this topic, Robert Rosenthal and Lenore Jacobson (1968) administered what they called a "Harvard Test of Inflected Acquisition" to students in a West Coast school. Subsequently, they met with the students' teachers to present the results of the test. In particular, Rosenthal and Jacobson identified certain students as very likely to exhibit a sudden spurt in academic abilities during the coming year, based on the results of the test.

When IQ test scores were compared later, the researchers' predictions proved accurate. The students identified as "spurters" far exceeded their classmates during the following year, suggesting that the predictive test was a powerful one. In fact, the test was a hoax! The researchers had made their predictions randomly among both good and poor students. What they told the teachers did not really reflect students' test scores at all. The progress made by the spurters was simply a result of the teachers' expecting the improvement and paying more attention to those students, encouraging them, and rewarding them for achievements. (Notice the similarity between this situation and the Hawthorne effect, discussed earlier in this chapter.)

The Rosenthal–Jacobson study attracted a lot of popular, as well as scientific, attention. Subsequent experiments have focused on specific aspects of what has become known as the *attribution process*, or the *expectations communication model*. This research, largely conducted by psychologists, parallels research primarily by sociologists, which takes a slightly different focus and is often gathered under the label *expectations-states theory*. The psychological studies focus on situations in which the expectations of a dominant individual affect the performance of subordinates—as in the case of a teacher and students or a boss and employees. The sociological research has tended to focus more on the role of expectations among equals in small, task-oriented groups. In a jury, for example, how do jurors initially evaluate each other, and how do those initial assessments affect their later interactions?

Here's a different kind of social science experiment. Shelley Correll, Stephen Benard, and In Paik (2007) were interested in learning whether race, gender, and/or parenthood might produce discrimination in hiring. Specifically, they wanted to find out if there was a "motherhood penalty." They decided to explore this topic with an experiment using college undergraduates.

The student-subjects chosen for the study were told that a new communications company was looking for someone to manage the marketing department of their East Coast office.

> *They heard that the communications company was interested in receiving feedback from younger adults since young people are heavy consumers of communications technology. To further increase their task orientation, participants were told that their input would be incorporated with the other information the company collects on applicants and would impact actual hiring decisions.*

(Correll, Benard, and Paik 2007: 1311)

The researchers had created several résumés describing fictitious candidates for the manager's position. Initially, the résumés had no indication of race, gender, or parenthood. A group of subjects was asked to evaluate the quality of the candidates. The group decided that the résumés reflected equivalent quality.

In the next part of the experiment, the résumés were augmented with additional information. Gender became apparent when names were added to the résumés. Moreover, the use of typically African American names (such as Latoya and Ebony for women, Tyrone and Jamal for men) or typically white names (such as Allison and Sarah for women, Brad and Matthew for men) allowed subjects to guess the candidates' races. Finally, including participation in a parent–teacher association (PTA) or listing names of children identified some candidates as parents. Over the course of the experiment, these different status indicators were added to the same résumés used in the initial trial. Thus, a particular résumé might appear as an African American mother, a white non-mother, a white father, and so forth. Of course, no student-subject would evaluate the same résumé with different status indicators.

Finally, the experimental subjects were given sets of résumés to evaluate in several ways. For example, they were asked how competent they felt the candidates were and how committed they seemed. They were asked to suggest a salary that might be offered a given candidate and to predict how likely it was that the candidate would eventually be promoted within the organization. They were even asked to indicate how many days the candidate should be allowed to miss work or come late before being fired.

Since each of the résumés was evaluated with different status indicators attached, the experimenters could determine whether those statuses made a difference. Specifically, they could test for the existence of a motherhood penalty. And they found it. Among other things,

- Mothers were judged to be less competent and less committed than non-mothers.
- Students offered mothers lower salaries than they did non-mothers and would allow mothers fewer missed or late days on the job.
- They felt that mothers were less likely to be promoted than non-mothers.
- They recommended hiring non-mothers almost twice as often as they did mothers.

Rounding out the analysis of gender and parenthood, the researchers found that, while the differences were smaller for men than for women, fathers were rated *higher* than non-fathers—just the opposite of the pattern found among women candidates.

The motherhood penalty was found among both white and African American candidates. Moreover, it did not matter what the gender of the subject evaluators were. Both women and men rated mothers lower than non-mothers.

Web-Based Experiments

Increasingly, researchers are using the World Wide Web as a vehicle for conducting social science experiments. Because representative samples are not essential in most experiments, researchers can often use volunteers who respond to invitations online. To get a better idea of this form of experimentation, go to www.socialpsychology.org/expts.htm. This website offers links to numerous professional and student research projects on such topics as "interpersonal relations," "beliefs and attitudes," and "personality and individual differences." In addition, the site offers resources for conducting Web experiments. Participating as a subject will provide data

for other researchers, may offer you some insights into your own thinking patterns, and can suggest experiments you might want to conduct.

Web experiments have raised new ethical questions as we saw in Chapter 3. Here's the abstract of an online experiment:

> We show, via a massive (N = 689,003) experiment on Facebook, that emotional states can be transferred to others via emotional contagion, leading people to experience the same emotions without their awareness. We provide experimental evidence that emotional contagion occurs without direct interaction between people (exposure to a friend expressing an emotion is sufficient), and in the complete absence of nonverbal cues.
>
> (Kramer et al. 2014)

What is most interesting about this online research report lies in the numerous comments. Most raise ethical concerns, chiefly concerning informed consent. More generally, every new research technique will raise new issues of research ethics as well as scientific validity. Social research is unlikely to ever become dull.

"Natural" Experiments

Although people tend to equate the terms *experiment* and *laboratory experiment*, we've seen that experiments are sometimes conducted outside the lab (field experiments) and can be conducted on the Web. Other important social science experiments occur outside controlled settings altogether, often in the course of normal social events. Sometimes nature designs and executes experiments that we can observe and analyze; sometimes social and political decision makers serve this natural function.

Imagine, for example, that a hurricane has struck a particular town. Some residents of the town suffer severe financial damages, whereas others escape relatively lightly. What, we might ask, are the behavioral consequences of suffering a natural disaster? Are those who suffer the most more likely to take precautions against future disasters than are those who suffer the least? To answer these questions, we might interview residents of the town some time after the hurricane. We might question them regarding the precautions they had taken before the hurricane and those they're currently taking. We could then compare the precautionary actions of the people who suffered a great deal from the hurricane with those taken by citizens who suffered relatively little. In this fashion, we might take advantage of a natural experiment, which we could not have arranged even if we'd been perversely willing to do so.

Because in natural experiments the researcher must take things pretty much as they occur, such experiments raise many of the validity problems discussed earlier. Thus, when Stanislav Kasl, Rupert Chisolm, and Brenda Eskenazi (1981) chose to study the impact that the Three Mile Island (TMI) nuclear accident in Pennsylvania had on plant workers, they had to be especially careful when devising the study design:

> Disaster research is necessarily opportunistic, quasi-experimental, and after-the-fact. In the terminology of Campbell and Stanley's classical analysis of research designs, our study falls into the "static-group comparison" category, considered one of the weak research designs. However, the weaknesses are potential and their actual presence depends on the unique circumstances of each study.
>
> (1981: 474)

The foundation of this study was a survey of the people who had been working at Three Mile Island on March 28, 1979, when the cooling system failed in the number 2 reactor and began melting the uranium core. The survey was conducted 5 to 6 months after the accident. Among other things, the survey questionnaire measured workers' attitudes toward working at nuclear power plants. If they had measured only the TMI workers' attitudes after the accident, the researchers would have had no idea whether attitudes had changed as a consequence of the accident. But they improved their study design by selecting another, nearby—seemingly comparable—nuclear power plant (abbreviated as PB) and surveyed workers there as a control group: hence their reference to a static-group comparison.

Even with an experimental and a control group, the authors were wary of potential problems in their design. In particular, their design was based on the idea that the two sets of workers were equivalent to each other, except for the single fact of the accident. The researchers could have assumed this if they had been able to assign workers to the two plants randomly, but

of course they couldn't. Instead, they compared characteristics of the two groups to see whether they were equivalent. Ultimately, the researchers concluded that the two sets of workers were very much alike, and the plant the employees worked at was merely a function of where they lived.

Even granting that the two sets of workers were equivalent, the researchers faced another problem of comparability. They could not contact all the workers who had been employed at TMI at the time of the accident. The researchers discuss the problem as follows:

> One special attrition problem in this study was the possibility that some of the no-contact nonrespondents among the TMI subjects, but not PB subjects, had permanently left the area because of the accident. This biased attrition would, most likely, attenuate the estimated extent of the impact. Using the evidence of disconnected or "not in service" telephone numbers, we estimate this bias to be negligible (1 percent).
>
> (Kasl, Chisolm, and Eskenazi 1981: 475)

The TMI example points both to the special problems involved in natural experiments and to the possibility of taking those problems into account. Social research generally requires ingenuity and insight, and natural experiments are certainly no exception.

Earlier in this chapter, we used a hypothetical example of studying whether an ethnic history film reduced prejudice. Sandra Ball-Rokeach, Joel Grube, and Milton Rokeach (1981) were able to address that topic in real life through a natural experiment. In 1977, the television dramatization of Alex Haley's *Roots,* a historical saga about African Americans, was presented by ABC on eight consecutive nights. It garnered the largest audiences in television history at that time. Ball-Rokeach and her colleagues wanted to know whether *Roots* changed white Americans' attitudes toward African Americans. Their opportunity arose in 1979, when a sequel—*Roots: The Next Generation*—was televised. Although it would have been nice (from a researcher's point of view) to assign random samples of Americans either to watch or not watch the show, this wasn't possible. Instead, the researchers selected four samples in Washington State and mailed questionnaires (before the broadcast) that measured attitudes toward African Americans. Following the last episode of the show, respondents were called and asked how many, if any, episodes they had watched. Subsequently, questionnaires were sent to respondents, remeasuring their attitudes toward African Americans.

By comparing attitudes before and after for both those who watched the show and those who didn't, the researchers reached several conclusions. For example, they found that people with already egalitarian attitudes were much more likely to watch the show than were those who were more prejudiced toward African Americans: a self-selection phenomenon. Comparing the before and after attitudes of those who watched the show, moreover, suggested that the show itself had little or no effect. Those who watched it were no more egalitarian afterward than they had been before.

This example anticipates the subject of Chapter 12, evaluation research, which can be seen as a special type of natural experiment. As you'll see, evaluation research involves taking the logic of experimentation into the field to observe and evaluate the effects of stimuli in real life. Because this is an increasingly important form of social research, an entire chapter is devoted to it.

Strengths and Weaknesses of the Experimental Method

Experiments are the primary tool for studying causal relationships. However, like all research methods, experiments have both strengths and weaknesses.

The chief advantage of a controlled experiment lies in the isolation of the experimental variable's impact over time. This is seen most clearly in terms of the basic experimental model. A group of experimental subjects are found, at the outset of the experiment, to have a certain characteristic; following the administration of an experimental stimulus, they are found to have a different characteristic. To the extent that subjects have experienced no other stimuli, we may conclude that the change of characteristics is caused by the experimental stimulus.

Further, because individual experiments are often rather limited in scope, requiring relatively little time and money and relatively few subjects, we often can replicate a given experiment several times using many different groups of subjects.

(This isn't always the case, of course, but it's usually easier to repeat experiments than, say, surveys.) As in all other forms of scientific research, replication of research findings strengthens our confidence in the validity and generalizability of those findings.

The greatest weakness of laboratory experiments lies in their artificiality. Social processes that occur in a laboratory setting might not necessarily occur in natural social settings. For example, a Muslim history film might genuinely reduce prejudice among a group of experimental subjects. This would not necessarily mean, however, that the same film shown in neighborhood movie theaters throughout the country would reduce prejudice among the general public. Artificiality is not as much of a problem, of course, for natural experiments as for those conducted in the laboratory.

In discussing several of the sources of internal and external invalidity mentioned by Campbell, Stanley, and Cook, we saw that we can create experimental designs that logically control such problems. This possibility points to one of the great advantages of experiments: They lend themselves to a logical rigor that is often much more difficult to achieve in other modes of observation.

Ethics and Experiments

As you've seen, many important ethical issues come up in the conduct of social science experiments. I'll mention only two here.

First, experiments almost always involve deception. In most cases, explaining the purpose of the experiment to subjects would probably cause them to behave differently—trying to look good, for example. It's important, therefore, to determine (1) whether a particular deception is essential to the experiment and (2) whether the value of what may be learned from the experiment justifies the ethical violation.

Second, experiments typically intrude on the lives of the subjects. Experimental researchers commonly put subjects in unusual situations and ask them to undergo unusual experiences. Rarely, if ever, do they physically injure the subjects (don't do that, by the way); however, psychological damage to subjects may occur, as some of the examples in this chapter illustrate. As with the matter of deception, then, researchers must balance the potential value of the research against the potential damage to subjects.

What do you think?...Revisited

As we've seen, the impact of the experiment itself on subjects' responses is a major concern in social research. Several elements of experimental designs address this concern. First, the use of control groups allows researchers to account for any effects of the experiment that are not related to the stimulus. Second, the Solomon four-group design tests for the possible impact of pretests on the dependent variable. And, finally, so-called natural experiments are done in real-life situations, imposing an experimental template over naturally occurring events.

Thus, although the impact of the observer can affect experimental results negatively, researchers have developed methods for addressing it.

MAIN POINTS

Introduction

- In experiments, social researchers typically select a group of subjects, do something to them, and observe the effect of what was done.

Topics Appropriate for Experiments

- Experiments provide an excellent vehicle for the controlled testing of causal processes.

The Classical Experiment

- The classical experiment tests the effect of an experimental stimulus (the independent variable) on a dependent variable through the pretesting and posttesting of experimental and control groups.

- A double-blind experiment guards against experimenter bias because neither the experimenter nor the subject knows which subjects are in the control and experimental groups.

Selecting Subjects

- It's generally less important that a group of experimental subjects be representative of some larger population than that experimental and control groups be similar to each other.

- Probability sampling, randomization, and matching are all methods of achieving comparability in the experimental and control groups. Randomization is the generally preferred method. In some designs, it can be combined with matching.

Variations on Experimental Design

- Campbell and Stanley describe three forms of preexperiments: the one-shot case study, the one-group pretest–posttest design, and the static-group comparison.

- Campbell and Stanley list, among others, eight sources of internal invalidity in experimental design: history, maturation, testing, instrumentation, statistical regression, selection biases, experimental mortality, and demoralization. The classical experiment with random assignment of subjects guards against each of these.

- Experiments also face problems of external invalidity, in that experimental findings might not reflect real life.

- The interaction of testing with the stimulus is an example of external invalidity that the classical experiment does not guard against.

- The Solomon four-group design and other variations on the classical experiment can safeguard against external invalidity.

- Campbell and Stanley suggest that, given proper randomization in the assignment of subjects to the experimental and control groups, there is no need for pretesting in experiments.

Examples of Experimentation

- In a controlled field experiment, researchers exposed the Pygmalion effect as one phenomenon that researchers must account for in experimental design.

- One recent experiment in a laboratory setting showed that a "motherhood penalty" exists in the work world.

Web-Based Experiments

- The World Wide Web has become an increasingly common vehicle for performing social science experiments.

"Natural" Experiments

- Natural experiments often occur in the course of social life in the real world, and social researchers can implement them in somewhat the same way they would design and conduct laboratory experiments.

Strengths and Weaknesses of the Experimental Method

- Like all research methods, experiments have strengths and weaknesses.

- The primary weakness of experiments is artificiality: What happens in an experiment may not reflect what happens in the outside world.

- The strengths of experimentation include the isolation of the independent variable, which permits causal inferences; the relative ease of replication; and scientific rigor.

Ethics and Experiments

- Experiments typically involve deceiving subjects.
- By their intrusive nature, experiments open the possibility of inadvertently causing damage to subjects.

KEY TERMS

control group	internal invalidity
double-blind experiment	matching
experimental group	posttesting
external invalidity	pretesting
field experiment	randomization

PROPOSING SOCIAL RESEARCH: EXPERIMENTS

In the next series of exercises, we focus on specific data-collection techniques, beginning here with experiments. If you're doing these exercises as part of an assignment in the course, your instructor will tell you whether you should skip those chapters dealing with methods you won't be using. If you're doing them on your own, to improve your understanding of the topics in the book, I suggest that you do all of these exercises. You can temporarily modify your proposed data-collection method and explore how you would research your topic using the method at hand.

In the case of experimentation, your proposal should make clear why you chose the experimental model over other forms of research and how it best serves your goals. More specifically, you'll want to describe the experimental stimulus and how it will be administered, as well as detailing the experimental and control groups you'll use. You'll need to describe the pretesting and posttesting that will be involved in your experiment. Where will you conduct your experiments—in a laboratory setting or under natural circumstances?

If you plan to conduct a double-blind experiment, you should describe how you'll accomplish it. You may also want to explore some of the internal and external problems of validity that might complicate the analysis of your results.

Finally, the experimental model is typically used to test specific hypotheses, so you should specify how you'll accomplish that in your study. What standard will determine whether hypotheses are accepted or rejected?

REVIEW QUESTIONS

1. What are some examples of internal invalidity? Pick four of the eight sources discussed in the book and make up your own examples to illustrate each.

2. Think of a recent natural disaster you've witnessed or read about. What research question might be studied by treating that disaster as a natural experiment? In two or three paragraphs, outline how the study might be done.

3. Say you want to evaluate a new operating system or other software. How might you set up an experiment to see what people really think of it? Keep in mind the use of control groups and the placebo effect.

4. Think of a recent, highly publicized trial. How might the attorneys have used mock juries to evaluate different strategies for presenting evidence?

CHAPTER 9

Survey Research

CHAPTER OVERVIEW

Researchers have many methods for collecting data through surveys—from mail questionaires to personal interviews to online surveys conducted over the Internet. Social researchers should know how to select an appropriate method and how to implement it effectively.

Chris Ryan/OJO Images/Getty Images

Introduction

Survey research is a frequently used mode of observation in the social sciences. In a typical survey, the researcher selects a sample of respondents and administers a standardized questionnaire to each person in the sample. Chapter 7 discussed sampling techniques in detail. This chapter discusses how to prepare a questionnaire and describes the various options for administering it so that respondents answer questions adequately.

This chapter also briefly discusses *secondary analysis*, the analysis of survey data collected by someone else. This use of survey results has become an important aspect of survey research in recent years, and it's especially useful for students and others with scarce research funds. The chapter closes with a look at the ethical implications of survey research.

Let's begin by looking at the kinds of topics that researchers can appropriately study through survey research.

Topics Appropriate for Survey Research

Surveys may be used for descriptive, explanatory, and exploratory purposes. They are chiefly used in studies that have individual people as the units of analysis. Although this method can be used for other units of analysis, such as groups or interactions, some individual persons must serve as **respondents** or informants. Thus, we could undertake a survey in which divorces were the unit of analysis, but we would need to administer the survey questionnaire to the participants in the divorces (or to some other respondents).

Survey research is probably the best method available to the social researcher who is interested in collecting original data for describing a population too large to observe directly. Careful probability sampling provides a group of respondents whose characteristics may be taken to reflect those of the larger population, and carefully constructed standardized questionnaires provide data in the same form from all respondents.

Surveys are also excellent vehicles for measuring attitudes and orientations in a large population. Public opinion polls—for example, Pew, Gallup, Harris, Roper, and a number of university survey centers—are well-known examples of this use. Indeed, polls have become so prevalent that at times the public seems

respondent A person who provides data for analysis by responding to a survey questionnaire.

What do you think?

All of us get telephone calls asking us to participate in a "survey," but some are actually telemarketing calls.

> Caller: The first question is "Who was the first president of the United States?"
> You: Crazy Horse.
> Caller: Close enough. You have just won …

This is obviously not legitimate. But subtler techniques can entangle you in a phony sales pitch before you know what's happening. When you get a call announcing that you've been selected for a "survey," how can you tell whether it's genuine?

See the *What do you think?…Revisited* box toward the end of the chapter.

iStock.com/killerb10

unsure what to think of them. Pollsters are criticized by those who don't think (or want to believe) that polls are accurate (candidates who are "losing" in polls often tell voters not to trust the polls). But polls are also criticized for being too accurate—for example, when exit polls on Election Day are used to predict a winner before the actual voting is complete.

The general attitude toward public opinion research is further complicated by scientifically unsound "surveys" that nonetheless capture people's attention because of the topics they cover and/or their "findings." Sometimes, people use the pretense of survey research for quite different purposes, as suggested in the "What Do You Think?" box. For example, you may have received a telephone call indicating that you've been selected for a survey, only to find that the first question was "How would you like to make thousands of dollars a week right there in your own home?" Unfortunately, a few unscrupulous telemarketers prey on the general cooperation people have given to survey researchers.

By the same token, political parties and charitable organizations have begun conducting phony "surveys." Often under the guise of collecting public opinion about some issue, callers ultimately ask respondents for a monetary contribution.

Recent political campaigns have produced another form of bogus survey, called the *push poll*. Here's what the American Association for Public Opinion Research had to say in condemning this practice:

A "push poll" is a telemarketing technique in which telephone calls are used to canvass potential voters, feeding them false or misleading "information" about a candidate under the pretense of taking a poll to see how this "information" affects voter preferences. In fact, the intent is not to measure public opinion but to manipulate it—to "push" voters away from one candidate and toward the opposing candidate. Such polls defame selected candidates by spreading false or misleading information about them. The intent is to disseminate campaign propaganda under the guise of conducting a legitimate public opinion poll.

(Bednarz 1996)

In short, the labels "survey" and "poll" are sometimes misused. Done properly, however, survey research can be a useful tool for social inquiry. Designing useful (and trustworthy) survey research begins with formulating good questions. Let's turn to that topic now.

Guidelines for Asking Questions

In social research, variables are often operationalized when researchers ask people questions as a way of getting data for analysis and interpretation. Sometimes the questions are asked by an interviewer; sometimes they are written down and given to respondents for completion. In other cases, several general guidelines can help researchers frame and ask questions that serve as excellent operationalizations of variables while avoiding pitfalls that can result in useless or even misleading information.

Surveys include the use of a **questionnaire**—an instrument designed to elicit information that will be useful for analysis. Although some of the specific points to follow are more appropriate for structured questionnaires than for the more open-ended questionnaires used in qualitative, in-depth interviewing, the underlying logic is valuable whenever we ask people questions in order to gather data.

Choose Appropriate Question Forms

Let's begin with some of the options available to you in creating questionnaires. These options include whether to use questions or statements and to choose open-ended or closed-ended questions.

Questions and Statements

Although the term *questionnaire* suggests a collection of questions, a typical questionnaire often presents as many statements as questions. Often, the researcher wants to determine the extent to which respondents hold a particular attitude or perspective. If you can summarize the attitude in a fairly brief statement, you can present that statement and ask respondents whether they agree or disagree with it. As you may remember, Rensis Likert greatly formalized this procedure through the creation of the Likert scale, a format in which respondents are asked to strongly agree, agree, disagree, or strongly disagree, or perhaps strongly approve, approve, and so forth.

You can use both questions and statements profitably. Using both in a given questionnaire gives you more flexibility in the design of items and can make the questionnaire more interesting as well.

Open-Ended and Closed-Ended Questions

In asking questions, researchers have two options. They can ask **open-ended questions**, in which case the respondent is asked to provide his or her own answer to the question. For example, the respondent may be asked, "What do you feel is the most important issue facing the United States today?" and be provided with a space to write in the answer (or be asked to report it verbally to an interviewer). As we'll see in Chapter 10, in-depth, qualitative interviewing relies almost exclusively on open-ended questions. However, they are also used in survey research.

In the case of **closed-ended questions**, the respondent is asked to select an answer from among a list provided by the researcher. Closed-ended questions are quite popular in survey research because they provide a greater uniformity of responses and are more easily processed than open-ended ones.

Open-ended responses must be coded before they can be processed for computer analysis (see Chapter 14). This coding process often requires the researcher to interpret the meaning of responses, opening the possibility of misunderstanding and researcher bias. Further, some respondents might give answers that are essentially irrelevant to the researcher's intent. Closed-ended responses, on the other hand, can often be transferred directly into a computer format.

The chief shortcoming of closed-ended questions lies in the researcher's structuring of responses. When the relevant answers to a given question are quite clear, there should be no problem. In other cases, however, the researcher's structuring of responses might overlook some important responses. In asking about "the most important issue facing the United States," for example, his or her checklist of issues might omit certain issues that respondents would have said were important.

The construction of closed-ended questions should be guided by two structural requirements. First, the response categories should

---

questionnaire A document containing questions and other types of items designed to solicit information appropriate for analysis. Questionnaires are used primarily in survey research but also in experiments, field research, and other modes of observation.

open-ended questions Questions for which the respondent is asked to provide his or her own answers. In-depth, qualitative interviewing relies almost exclusively on open-ended questions.

closed-ended questions Survey questions in which the respondent is asked to select an answer from among a list provided by the researcher. These are popular in survey research because they provide a greater uniformity of responses and are more easily processed than open-ended questions.

be exhaustive: They should include all the possible responses that might be expected. Often, researchers ensure this by adding a category such as "Other (Please specify: _____)." Second, the answer categories must be mutually exclusive: The respondent should not feel compelled to select more than one. (In some cases, you may wish to solicit multiple answers, but these can create difficulties in data processing and analysis later on.) To ensure that your categories are mutually exclusive, carefully consider each combination of categories, asking yourself whether a person could reasonably choose more than one answer. In addition, it's useful to add an instruction for the respondent to select the one best answer, but this technique cannot serve as a substitute for a carefully constructed set of responses.

Make Items Clear

It should go without saying that questionnaire items should be clear and unambiguous, but the broad proliferation of unclear and ambiguous questions in surveys makes the point worth emphasizing. Often we can become so deeply involved in the topic under examination that opinions and perspectives are clear to us but not to our respondents, many of whom have paid little or no attention to the topic. Or, if we have only a superficial understanding of the topic, we may fail to specify the intent of a question sufficiently. The question "What do you think about the proposed peace plan?" may evoke in the respondent a counterquestion: "Which proposed peace plan?" Questionnaire items should be precise so that the respondent knows exactly what the researcher is asking.

Similarly, the use of the term *Native American* to mean American Indian often produces an overrepresentation of that ethnic group in surveys. Clearly, many respondents understand the term to mean *born in the United States*.

Avoid Double-Barreled Questions

Frequently, researchers ask respondents for a single answer to a question that actually has multiple parts. These types of queries are often termed *double-barreled questions* and seem to happen most often when the researcher has personally identified with a complex question. For example, you might ask respondents to agree or disagree with the statement "The United States should abandon its space program and spend the money on domestic programs." Although many people would unequivocally agree with the statement and others would unequivocally disagree, still others would be unable to answer. Some would want to abandon the space program and give the money back to the taxpayers. Others would want to continue the space program but also put more money into domestic programs. These latter respondents could neither agree nor disagree without misleading you.

As a general rule, whenever the word *and* appears in a question or questionnaire statement, check whether you're asking a double-barreled question.

Respondents Must Be Competent to Answer

In asking respondents to provide information, you should continually ask yourself whether they can do so reliably. In a study of child-rearing, you might ask respondents to report the age at which they first talked back to their parents. Quite aside from the problem of defining *talking back to parents*, it's doubtful that most respondents would remember with any degree of accuracy.

As another example, student-government leaders occasionally ask their constituents to indicate how students' fees ought to be spent. Typically, respondents are asked to indicate the percentage of available funds that should be devoted to a long list of activities. Without a fairly good knowledge of the nature of those activities and the costs involved in them, the respondents cannot provide meaningful answers. Administrative costs, for example, will receive little support although they may be essential to the programs as a whole.

One group of researchers examining the driving experience of teenagers insisted on asking an open-ended question concerning the number of miles driven since receiving a license, even though consultants argued that few drivers could estimate such information with any accuracy. In response, some teenagers reported driving hundreds of thousands of miles.

Respondents Must Be Willing to Answer

Often, we would like to learn things from people that they are unwilling to share with us. For example, Yanjie Bian indicates that getting candid answers from people in China has often been difficult.

> [Here] people are generally careful about what they say on nonprivate occasions in order to survive under authoritarianism. During the Cultural Revolution between 1966 and 1976, for example, because of the radical political agenda and political intensity throughout the country, it was almost impossible to use survey techniques to collect valid and reliable data inside China about the Chinese people's life experiences, characteristics, and attitudes towards the Communist regime.
>
> (1994: 19–20)

Sometimes, U.S. respondents may say they're undecided when, in fact, they have an opinion but think they're in the minority. Under that condition, they may be reluctant to tell a stranger (the interviewer) what that opinion is. Given this problem, the Gallup Organization, for example, has used a "secret ballot" format, which simulates actual election conditions, in that the "voter" enjoys complete anonymity. In an analysis of the Gallup poll election data from 1944 to 1988, Andrew Smith and G. F. Bishop (1992) found that this technique substantially reduced the percentage of respondents who said they were undecided about how they would vote.

This problem is not limited to survey research, however. Richard Mitchell faced a similar problem in his field research among U.S. survivalists:

> Survivalists, for example, are ambivalent about concealing their identities and inclinations. They realize that secrecy protects them from the ridicule of a disbelieving majority, but enforced separatism diminishes opportunities for recruitment and information exchange....
>
> "Secretive" survivalists eschew telephones, launder their mail through letter exchanges, use nicknames and aliases, and carefully conceal their addresses from strangers. Yet once I was invited to group meetings, I found them cooperative respondents.
>
> (1991: 100)

Questions Should Be Relevant

Similarly, questions asked in a questionnaire should be relevant to most respondents. When attitudes are requested on a topic that few respondents have thought about or really care about, the results are not likely to be useful. Of course, because the respondents might express attitudes even though they have never given any thought to the issue, researchers run the risk of being misled.

This point is illustrated occasionally when researchers ask for responses relating to fictitious people and issues. In one political poll I conducted, I asked respondents whether they were familiar with each of 15 political figures in the community. As a methodological exercise, I made up a name: Tom Sakumoto. In response, 9 percent of the respondents said they were familiar with him. Of those respondents familiar with him, about half reported seeing him on television and reading about him in the newspapers.

When you obtain responses to fictitious issues, you can disregard those responses. But when the issue is real, you may have no way of telling which responses genuinely reflect attitudes and which reflect meaningless answers to an irrelevant question.

Ideally, we would like respondents simply to report that they don't know, have no opinion, or are undecided in instances where that is the case. Unfortunately, however, they often make up answers.

Short Items Are Best

In the interests of being unambiguous and precise and of pointing to the relevance of an issue, researchers tend to create long and complicated items. That should be avoided. Respondents are often unwilling to study an item in order to understand it. The respondent should be able to read an item quickly, understand its intent, and select or provide an answer without difficulty. In general, assume that respondents will read items quickly and give quick answers. Accordingly, provide clear, short items that respondents will not misinterpret under those conditions.

Avoid Negative Items

The appearance of a negation in a questionnaire item paves the way for easy misinterpretation. Asked to agree or disagree with the statement "The United States should not recognize Cuba," a sizable portion of the respondents will read over the word *not* and answer on that basis. Thus, some will agree with the statement when they're in favor of recognition, and others will agree when they oppose it. And you may never know which are which.

Similar considerations apply to other "negative" words. In a study of support for civil liberties, for example, respondents were asked whether they felt "the following kinds of people should be prohibited from teaching in public schools" and were presented with a list including such items as a Communist, a Ku Klux Klansman, and so forth. The response categories "yes" and "no" were given beside each entry. A comparison of the responses to this item with other items reflecting support for civil liberties strongly suggested that many respondents gave the answer "yes" to indicate willingness for such a person to teach, rather than to indicate that such a person should be prohibited from teaching. (A later study in the series giving as answer categories "permit" and "prohibit" produced much clearer results.)

Avoid Biased Items and Terms

Recall from our discussion of conceptualization and operationalization in Chapter 5 that there are no ultimately true meanings for any of the concepts we typically study in social science. *Prejudice* has no ultimately correct definition; whether a given person is prejudiced depends on our definition of that term. The same general principle applies to the responses we get from people completing a questionnaire.

The meaning of someone's response to a question depends in large part on its wording. This is true of every question and answer. Some questions seem to encourage particular responses more than other questions do. In the context of questionnaires, **bias** refers to any property of questions that encourages respondents to answer in a particular way.

Most researchers recognize the likely effect of a leading question that begins, "Don't you agree with the president of the United States that…," and no reputable researcher would use such an item. Unhappily, the biasing effect of items and terms is far subtler than this example suggests.

The mere identification of an attitude or position with a prestigious person or agency can bias responses. The item "Do you agree or disagree with the recent Supreme Court

decision that . . ." would have a similar effect. Such wording may not produce consensus or even a majority in support of the position identified with the prestigious person or agency, but it will likely increase the level of support over what would have been obtained without such identification.

Sometimes the impact of different forms of question wording is relatively subtle. For example, when Kenneth Rasinski (1989) analyzed the results of several General Social Survey studies of attitudes toward government spending, he found that the way programs were identified affected the amount of public support they received. Here are some comparisons:

More Support	Less Support
"Assistance to the poor"	"Welfare"
"Halting rising crime rate"	"Law enforcement"
"Dealing with drug addiction"	"Drug rehabilitation"
"Solving problems of big cities"	"Assistance to big cities"
"Improving conditions of blacks"	"Assistance to blacks"
"Protecting social security"	"Social security"

In 1986, for example, 62.8 percent of the respondents said too little money was being spent on "assistance to the poor," while in a matched survey that year, only 23.1 percent said we were spending too little on "welfare."

In this context, be wary of what researchers call the *social desirability* of questions and answers. Whenever we ask people for information, they answer through a filter of what will make them look good. This is especially true if they're interviewed face to face. Thus, for example, during the 2008 Democratic primary, many voters who might have been reluctant to vote for an African American (Barack Obama) or a woman (Hillary Clinton) might have also been reluctant to admit their racial or gender prejudice to a survey interviewer. (Others, to be sure, were not reluctant to say how they felt.)

Frauke Kreuter, Stanley Presser, and Roger Tourangeau (2008) conducted an experiment on the impact of three data-collection techniques

---

bias That quality of a measurement device that tends to result in a misrepresentation, in a particular direction, of what is being measured.

on respondents' willingness to provide sensitive information that might not reflect positively on themselves, such as failing a class or being put on probation in college. Respondents were least likely to volunteer such information when interviewed in a conventional telephone interview. They were somewhat more willing when interviewed by an interactive recording, and they were most likely to provide such information when questioned in a Web survey.

The best way to guard against this problem is to imagine how you would feel giving each of the answers you intend to offer to respondents. If you would feel embarrassed, perverted, inhumane, stupid, irresponsible, or otherwise socially disadvantaged by any particular response, give serious thought to how willing others would be to provide those answers.

The Centers for Disease Control and Prevention (Choi and Pak 2005) have provided an excellent analysis of both obvious and not-so-obvious ways in which your choice of terms can bias and otherwise confuse responses to questionnaires. Among other things, they warn against using ambiguous, technical, uncommon, or vague words. Their thorough analysis provides many concrete illustrations.

These, then, are some general guidelines for writing questions to elicit data for analysis and interpretation. Next we look at how to construct questionnaires.

Questionnaire Construction

Questionnaires are used in connection with many modes of observation in social research. Although structured questionnaires are essential to and most-directly associated with survey research, they are also widely used in experiments, field research, and other data-collection activities. For this reason, questionnaire construction can be an important practical skill for researchers. As we discuss the established techniques for constructing questionnaires, let's begin with some issues concerning questionnaire format.

General Questionnaire Format

The format of a questionnaire is just as important as the nature and wording of the questions asked. An improperly laid out questionnaire can lead respondents to miss questions, confuse them

about the nature of the data desired, and even lead them to throw the questionnaire away.

As a general rule, a questionnaire should be adequately spaced and have an uncluttered layout. If a self-administered questionnaire is being designed, inexperienced researchers tend to fear that their questionnaire will look too long; as a result, they squeeze several questions onto a single line, abbreviate questions, and use as few pages as possible. These efforts are ill-advised and even dangerous. Putting more than one question on a line will cause some respondents to miss the second question altogether. Some respondents will misinterpret abbreviated questions. More generally, respondents who find they have spent considerable time on the first page of what seemed a short questionnaire will be more demoralized than respondents who quickly complete the first several pages of what initially seemed a rather long form. Moreover, the latter will have made fewer errors and will not have been forced to reread confusing, abbreviated questions. Nor will they have been forced to write a long answer in a tiny space.

Similar problems can arise for interviewers in a face-to-face or telephone interview. Like respondents with a self-administered questionnaire, interviewers may miss questions, lose their place, and generally become frustrated and flustered. Interview questionnaires need to be laid out in a way that supports the interviewer's work, including special instructions and guidelines.

The desirability of spreading out questions in the questionnaire cannot be overemphasized. Squeezed-together questionnaires are disastrous, whether they are to be completed by the respondents themselves or administered by trained interviewers. And processing such questionnaires is a nightmare.

Formats for Respondents

In one of the most common types of questionnaire items, the respondent is expected to check one response from a series. In my experience, boxes adequately spaced apart provide the best format for this purpose. Modern word processing makes the use of boxes a practical technique; setting boxes in type can also be accomplished easily and neatly. You can approximate boxes by using brackets.

Rather than providing boxes to be checked, you might print a code number beside each response and ask the respondent to circle the appropriate number (see Figure 9-1).

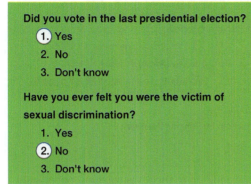

FIGURE 9-1
Circling the Answer.

This method has the added advantage of specifying the code number to be entered later during the processing stage (see Chapter 14). If numbers are to be circled, however, you should provide clear and prominent instructions to the respondent, because many will be tempted to cross out the appropriate number, which makes data processing even more difficult. (Note that the circled-response technique can be used more safely when interviewers administer the questionnaires, because the interviewers themselves record the responses.)

Contingency Questions

Quite often in questionnaires, certain questions will be relevant to some of the respondents and irrelevant to others. In a study of birth control methods, for instance, you would probably not want to ask men if they take birth control pills.

This sort of situation often arises when researchers wish to ask a series of questions about a certain topic. You may want to ask whether your respondents belong to a particular organization and, if so, how often they attend meetings, whether they have held office in the organization, and so forth. Or, you might want to ask whether respondents have heard anything about a certain political issue and then learn the attitudes of those who have heard of it.

Each subsequent question in series such as these is called a **contingency question**: Whether it is to be asked and answered is contingent on responses to the first question in the series. The proper use of contingency questions can facilitate

the respondents' task in completing the questionnaire, because they are not faced with trying to answer questions irrelevant to them.

There are several formats for contingency questions. The one shown in Figure 9-2 is probably the clearest and most effective. Note two key elements in this format. First, the contingency question is isolated from the other questions by being set off to the side and enclosed in a box. Second, an arrow connects the contingency question to the answer on which it is contingent. In the illustration, only respondents who answer "yes" are expected to answer the contingency question. The rest of the respondents should simply skip it.

Note that the questions shown in Figure 9-2 could have been dealt with in a single question. The question might have read, "How many times, if any, have you smoked marijuana?" The response categories, then, might have read: "Never," "Once," "2 to 5 times," and so forth. This single question would apply to all respondents, and each would find an appropriate

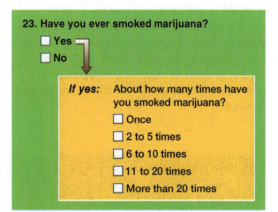

FIGURE 9-2
Contingency Question Format. Contingency questions offer a structure for exploring subject areas logically in some depth.

contingency question A survey question intended for only some respondents, determined by their responses to some other question. For example, all respondents might be asked whether they belong to the Cosa Nostra, and only those who said "yes" would be asked how often they go to company meetings and picnics. The latter would be a contingency question.

answer category. Such a question, however, might put some pressure on respondents to report having smoked marijuana, because the main question asks how many times they have smoked it, even though it allows for those *exceptional cases who have never smoked marijuana even once.* (The emphasis used in the previous sentence gives a fair indication of how respondents might read the question.) The contingency question format illustrated in Figure 9-2 should reduce the subtle pressure on respondents to report having smoked marijuana.

Used properly, even rather complex sets of contingency questions can be constructed without confusing the respondent. Figure 9-3 illustrates a more complicated example.

Sometimes a set of contingency questions is long enough to extend over several pages. Suppose you're studying the political activities of college students, and you wish to ask a large number of questions of students who have voted in a national, state, or local election. You could separate out the relevant respondents with an initial question such as "Have you ever voted in a national, state, or local election?" but it would be confusing to place the contingency questions in a box stretching over several pages. It would make more sense to enter instructions in parentheses after each answer telling respondents to answer or skip the contingency questions. Figure 9-4 provides an illustration of this method.

13. Have you ever voted in a national, state, or local election?
☐ Yes (Please answer questions 14–25.)
☐ No (Please skip questions 14–25. Go directly to question 26 on page 8.)

FIGURE 9-4
Instructions for Skipping Questions.

In addition to these instructions, you should place an instruction at the top of each page containing only the contingency questions. For example, you might say, "This page is only for respondents who have voted in a national, state, or local election." Clear instructions such as these spare respondents the frustration of reading and puzzling over questions that are irrelevant to them and increase the likelihood that *only* those for whom the questions are relevant will respond.

Matrix Questions

Quite often you'll want to ask several questions that have the same set of answer categories. This is typically the case whenever the Likert response categories are used. In such cases, constructing a matrix of items and answers is often possible, as illustrated in Figure 9-5.

This format offers several advantages over other configurations. First, it uses space efficiently. Second, respondents will probably complete such a set of questions more quickly than other formats would allow. In addition, using a matrix may increase the comparability of responses given to different questions—for the respondent as well as for the researcher. Because respondents can quickly review their answers to earlier items in the set, they might choose between, say, "strongly agree" and "agree" on a given statement by comparing the strength of their agreement with their earlier responses in the set.

This approach, however, holds some inherent dangers. Its advantages may encourage you to structure an item so that the responses fit into the matrix format, when a different, more idiosyncratic set of responses might be more appropriate. Also, the matrix-question format can foster a "response-set" among some respondents:

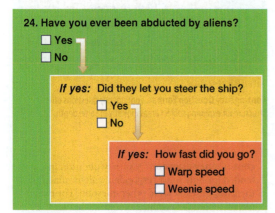

FIGURE 9-3

Contingency Table. Sometimes it will be appropriate for certain kinds of respondents to skip over inapplicable questions. To avoid confusion, you should provide clear instructions to that end.

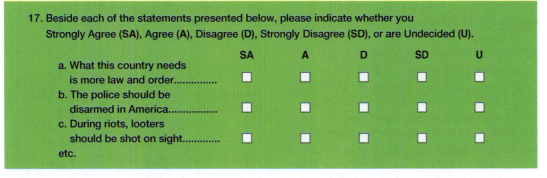

17. Beside each of the statements presented below, please indicate whether you
Strongly Agree (SA), Agree (A), Disagree (D), Strongly Disagree (SD), or are Undecided (U).

	SA	A	D	SD	U
a. What this country needs is more law and order...............	☐	☐	☐	☐	☐
b. The police should be disarmed in America.................	☐	☐	☐	☐	☐
c. During riots, looters should be shot on sight..............	☐	☐	☐	☐	☐
etc.					

FIGURE 9-5

Matrix Question Format. Matrix questions offer an efficient format for presenting closed-ended questionnaire items that have the same response categories.

They may develop a pattern of, say, agreeing with all the statements. This would be especially likely if the set of statements began with several that indicated a particular orientation (for example, a liberal political perspective) with only a few later ones representing the opposite orientation. Respondents might assume that all the statements represent the same orientation and, reading quickly, misread some of them, thereby giving the wrong answers. This problem can be reduced somewhat by alternating statements representing different orientations and by making all statements short and clear.

Ordering Items in a Questionnaire

The order in which questionnaire items are presented can also affect responses. First, the appearance of one question can affect the answers given to later ones. For example, if several questions have been asked about the dangers of terrorism to the United States and then a question asks respondents to list (open-ended) things that they believe represent dangers to the United States, terrorism will receive more citations than would otherwise be the case. In this situation, asking the open-ended question first is best.

Some researchers attempt to overcome this effect by randomizing the order of items. This effort is usually futile. In the first place, a randomized set of items will probably strike respondents as chaotic and worthless. The random order also makes answering more difficult because respondents must continually switch their attention from one topic to another. Finally,

even a randomized ordering of items will have the effect discussed previously—except that researchers will have no control over the effect.

The safest solution is sensitivity to the problem. Although you can't avoid the effect of item order, try to estimate what that effect will be so that you can interpret results meaningfully. If the order of items seems especially important in a given study, you might construct more than one version of the questionnaire with different orderings of the items in each. You'll then be able to determine the effects by comparing responses to the various versions. At the very least, you should pretest your questionnaire in the different forms. (We'll discuss pretesting in a moment.)

Questionnaire Instructions

Every questionnaire, whether it is to be completed by respondents or administered by interviewers, should contain clear instructions and introductory comments where appropriate.

It's useful to begin every self-administered questionnaire with basic instructions for completing it. Although many people these days have experience with forms and questionnaires, you should begin by telling them exactly what you want: that they are to indicate their answers to certain questions by placing a check mark or an X in the box beside the appropriate answer or by writing in their answer when asked to do so. If many open-ended questions are used, respondents should receive some guidelines about whether brief or lengthy answers are expected. If you wish to encourage your respondents to elaborate on their responses to closed-ended questions, that should be made clear to them.

If a questionnaire has subsections—political attitudes, religious attitudes, background data—introduce each with a short statement concerning its content and purpose. For example, "In this section, we would like to know what people consider the most important problems in the community." Demographic items at the end of a self-administered questionnaire might be introduced thus: "Finally, we would like to know just a little about you so we can see how different types of people feel about the issues we have been examining."

Short introductions such as these help the respondent make sense of the questionnaire. They make the questionnaire seem less chaotic, especially when it captures a variety of data. And they help put the respondent in the proper frame of mind for answering the questions.

Some questions may require special instructions to facilitate proper answering. This is especially true if a given question varies from the general instructions pertaining to the whole questionnaire. The following three examples illustrate this situation.

Despite attempts to provide mutually exclusive answers in closed-ended questions, often more than one answer will apply for respondents. If you want a single answer, you should make this perfectly clear in the question. An example would be "From the list below, please check the primary reason for your decision to attend college." Often the main question can be followed by a parenthetical note: "Please check the one best answer." If, on the other hand, you want the respondent to check as many answers as apply, you should make this clear.

When the respondent needs to rank-order a set of answer categories, the instructions should indicate this, and a different type of answer format should be used (for example, blanks instead of boxes). These instructions should indicate how many answers are to be ranked (for example, all, only the first and second, only the first and last, the most important and least important). These instructions should also spell out the order of ranking (for example, "Place a 1 beside the most important item, a 2 beside the next most important, and so forth"). Rank-ordering of responses is often difficult for respondents, however, because they may have to read and reread the list

several times, so this technique should only be used in situations where no other method will produce the desired result.

Pretesting the Questionnaire

No matter how carefully researchers design a data-collection instrument such as a questionnaire, there is always the possibility—indeed the certainty—of error. They will always make some mistake: an ambiguous question, question that people cannot answer, or some other violation of the rules just discussed.

The surest protection against such errors is to pretest the questionnaire in full or in part. Give the questionnaire to the ten people in your bowling league, for example. It's not usually essential that the pretest subjects compose a representative sample, although you should use people for whom the questionnaire is at least relevant.

By and large, it's better to ask people to complete the questionnaire than to read through it looking for errors. All too often, a question seems to make sense on a first reading but proves impossible to answer.

Stanley Presser and Johnny Blair (1994) describe several different pretesting strategies and report on the effectiveness of each. They also provide data on the cost of the various methods. Paul Beatty and Gordon Willis (2007) offer a useful review of "cognitive interviewing." In this case, the pretesting of a questionnaire also has respondents comment on the questionnaire itself in a way that helps the researchers learn whether the questions are communicating effectively and collecting the information sought.

There are many more tips and guidelines for questionnaire construction, but covering them all would take a book in itself. Now I'll complete this discussion with an excerpt taken from a real questionnaire, showing how some of these comments find substance in practice.

A Sample Questionnaire

Figure 9-6 is part of a questionnaire used by the University of Chicago's National Opinion Research Center in its General Social Survey (GSS). The questionnaire deals with people's attitudes toward the government and is designed to be self-administered, though most of the GSS is conducted in face-to-face interviews.

10. Here are some things the government might do for the economy. Circle one number for each action to show whether you are in favor of it or against it.

> 1. **Strongly in favor of**
> 2. **In favor of**
> 3. **Neither in favor of nor against**
> 4. **Against**
> 5. **Strongly against**

PLEASE CIRCLE A NUMBER

a. Control of wages by legislation	1	2 3 4 5	28/
b. Control of prices by legislation	1	2 3 4 5	29/
c. Cuts in government spending	1	2 3 4 5	30/
d. Government financing of projects to create new jobs	1	2 3 4 5	31/
e. Less government regulation of business	1	2 3 4 5	32/
f. Support for industry to develop new products and technology	1	2 3 4 5	33/
g. Supporting declining industries to protect jobs	1	2 3 4 5	34/
h. Reducing the work week to create more jobs	1	2 3 4 5	35/

11. Listed below are various areas of government spending. Please indicate whether you would like to see more or less government spending in each area. Remember that if you say "much more," it might require a tax increase to pay for it.

> 1. **Spend much more**
> 2. **Spend more**
> 3. **Spend the same as now**
> 4. **Spend less**
> 5. **Spend much less**
> 8. **Can't choose**

PLEASE CIRCLE A NUMBER

a. The environment	1	2 3 4 5 8	36/
b. Health	1	2 3 4 5 8	37/
c. The police and law enforcement	1	2 3 4 5 8	38/
d. Education	1	2 3 4 5 8	39/
e. The military and defense	1	2 3 4 5 8	40/
f. Retirement benefits	1	2 3 4 5 8	41/
g. Unemployment benefits	1	2 3 4 5 8	42/
h. Culture and the arts	1	2 3 4 5 8	43/

12. If the government *had* to choose between keeping down inflation or keeping down unemployment, to which do you think it should give highest priority?

Keeping down inflation 1   44/
Keeping down unemployment 2
Can't choose 8

13. Do you think that labor unions in this country have too much power or too little power?

Far too much power 1   45/
Too much power 2
About the right amount of power 3
Too little power 4
Far too little power 5
Can't choose 8

FIGURE 9-6

A Sample Questionnaire. This questionnaire excerpt is from the General Social Survey, a major source of data for analysis by social researchers around the world.

14. How about business and industry, do they have too much power or too little power?

Far too much power	1
Too much power	2
About the right amount of power	3
Too little power	4
Far too little power	5
Can't choose	8

46/

15. And what about the federal government, does it have too much power or too little power?

Far too much power	1
Too much power	2
About the right amount of power	3
Too little power	4
Far too little power	5
Can't choose	8

47/

16. In general, how good would you say labor unions are for the country as a whole?

Excellent	1
Very good	2
Fairly good	3
Not very good	4
Not good at all	5
Can't choose	8

48/

17. What do you think the government's role in each of these industries should be?

> 1. Own it
> 2. Control prices and profits but not own it
> 3. Neither own it nor control its prices and profits
> 8. Can't choose

PLEASE CIRCLE A NUMBER

a. Electric power	1	2	3	8	49/
b. The steel industry	1	2	3	8	50/
c. Banking and insurance	1	2	3	8	51/

18. On the whole, do you think it should or should not be the government's responsibility to . . .

> 1. Definitely should be
> 2. Probably should be
> 3. Probably should not be
> 4. Definitely should not be
> 8. Can't choose

PLEASE CIRCLE A NUMBER

a. Provide a job for everyone who wants one	1	2	3	4	8	52/
b. Keep prices under control	1	2	3	4	8	53/
c. Provide health care for the sick	1	2	3	4	8	54/
d. Provide a decent standard of living for the old	1	2	3	4	8	55/

FIGURE 9-6 (Continued)

You may notice several mysterious-looking numbers in the right-hand margins of this sample survey. These reflect a critical aspect of questionnaire design: precoding. Because the information collected by questionnaires is typically transformed into some type of computer format, it's usually appropriate to include data-processing instructions on the questionnaire itself. These instructions indicate where specific pieces of information will be stored in the machine-readable data files.

Self-Administered Questionnaires

So far we've discussed how to formulate questions and how to design effective questionnaires. As important as these tasks are, the labor will be wasted unless the questionnaire produces useful data—which means that respondents have actually completed the questionnaire. We turn now to the major methods of getting responses to questionnaires.

I've referred several times in this chapter to interviews versus self-administered questionnaires. Along these lines, there are four main methods of administering survey questionnaires to a sample of respondents: traditional self-administered questionnaires, in which respondents are asked to complete a hard copy of the questionnaire themselves; surveys administered by interviewers in face-to-face encounters; surveys conducted by telephone; and surveys presented online. This section and the next three discuss each of these methods in turn.

The most common form of self-administered questionnaire is the mail survey. However, several other techniques are used as well. At times, it may be appropriate to administer a questionnaire to a group of respondents gathered at the same place at the same time. A survey of students taking introductory psychology might be conducted in this manner during class. High school students might be surveyed during homeroom period.

Some recent experimentation has been conducted with regard to the home delivery of questionnaires. A research worker delivers the questionnaire to the home of sample respondents and explains the study. Then the questionnaire is left for the respondent to complete, and the researcher picks it up later.

Home delivery and the mail can also be used in combination. Questionnaires are mailed to families, and then research workers visit homes to pick up the questionnaires and check them for completeness. Or research workers can hand deliver questionnaires with a request that the respondents mail the completed questionnaires to the research office.

On the whole, when a research worker either delivers the questionnaire, picks it up, or both, the completion rate seems higher than it is in straightforward mail surveys. Additional experimentation with this technique will likely point to other ways to improve completion rates while reducing costs. The remainder of this section, however, is devoted specifically to the mail survey, which remains the typical form of the self-administered questionnaire.

Mail Distribution and Return

The basic method for collecting data through the mail has been to send a questionnaire accompanied by a letter of explanation and a self-addressed, stamped envelope. The respondent is expected to complete the questionnaire, put it in the envelope, and return it. If, by any chance, you've received such a questionnaire and failed to return it, it would be valuable to recall the reasons you had for not returning it and keep them in mind any time you plan to send questionnaires to others.

A common reason for not returning questionnaires is that it's too much trouble. To overcome this problem, researchers have developed several ways to make returning them easier. For instance, a self-mailing questionnaire requires no return envelope: When the questionnaire is folded a particular way, the return address appears on the outside. The respondent therefore doesn't have to worry about losing the envelope.

More-elaborate designs are available also. The university student questionnaire to be described later in this chapter was bound in a booklet with a special, two-panel back cover. Once the questionnaire was completed, the respondent needed only to fold out the extra panel, wrap it around the booklet, and seal the whole thing with the adhesive strip running along the edge of the panel. The foldout panel contained the researcher's return address and postage. When the study was repeated a couple of years later, the design was improved. Both the front and back covers had foldout panels: one for sending the questionnaire out and the other for getting it back—thus avoiding the use of envelopes altogether.

The point here is that anything you can do to make the job of completing and returning the questionnaire easier will improve your study. Imagine receiving a questionnaire that made no provisions for its return to the researcher. Suppose you had to (1) find an envelope, (2) write the address on it, (3) figure out how much postage it required, and (4) put the stamps on it. How likely is it that you would return the questionnaire?

A few brief comments on postal options are in order. On outgoing mail, your main choices are first-class postage and bulk rate. First class is more certain, but bulk rate is far cheaper. (Check your local post office for rates and procedures.) On return mail, your choices are postage stamps and business-reply permits. Here, the cost differential is more complicated. If you use stamps, you pay for them whether people return their questionnaires or not. With the business-reply permit, you pay for only those that are used, but you pay an additional surcharge of about a nickel. This means that stamps are cheaper if a lot of questionnaires are returned, but business-reply permits are cheaper if fewer are returned (and there is no way for you to know in advance how many will be returned).

There are many other considerations involved in choosing among the several postal options. Some researchers, for example, feel that the use of postage stamps communicates more "humanness" and sincerity than do bulk rate and business-reply permits. Others worry that respondents will remove the stamps and use them for some purpose other than returning the questionnaires. Because both bulk rate and business-reply permits require establishing accounts at the post office, you'll probably find stamps much easier to use for small-scale surveys.

Monitoring Returns

The mailing of questionnaires sets up a new research question that may prove valuable to a study: How many and which respondents will complete and return the questionnaires?

Researchers shouldn't sit back idly as questionnaires are returned; instead, they should undertake a careful recording of the varying rates of return among respondents.

An invaluable tool in this activity is a return-rate graph. The day on which questionnaires were mailed is labeled Day 1 on the graph, and on every day thereafter the number of returned questionnaires is logged on the graph. It's usually best to compile two graphs. One shows the number returned each day—rising, then dropping. The second reports the cumulative number or percentage. In part, this activity provides the researchers with gratification, as they get to draw a picture of their successful data collection. More important, however, it is their guide to how the data collection is going. If follow-up mailings are planned, the graph provides a clue about when such mailings should be launched. (The dates of subsequent mailings should be noted on the graph.)

As completed questionnaires are returned, each should be opened, scanned, and assigned an identification (ID) number. These numbers should be assigned serially as the questionnaires are returned, even if other ID numbers have already been assigned. Two examples should illustrate the important advantages of this procedure.

Let's assume you're studying attitudes toward a political figure. In the middle of the data collection, the media break the story that the politician is having extramarital affairs. By knowing the date of that public disclosure and the dates when questionnaires were received, you'll be in a position to determine the effects of the disclosure. (Recall the discussion in Chapter 8 of history in connection with experiments.)

In a less sensational way, serialized ID numbers can be valuable in estimating nonresponse biases in the survey. Barring more-direct tests of bias, you may wish to assume that those who failed to answer the questionnaire will be more like respondents who delayed answering than like those who answered right away. An analysis of questionnaires received at different points in the data collection might then be used for estimates of sampling bias. For example, if the GPAs reported by student respondents decrease steadily through the data collection, with those replying right away having higher GPAs and those replying later having lower GPAs, you might tentatively conclude that those who failed to answer at all have yet lower GPAs. Although it would not be advisable to make statistical estimates of bias in this fashion, you could take advantage of approximate estimates based on the patterns you've observed.

If respondents have been identified for purposes of follow-up mailing, then preparations for those mailings should be made as the questionnaires are returned. The case study later in this section discusses this process in greater detail.

Follow-up Mailings

Follow-up mailings can be administered in several ways. In the simplest, nonrespondents are sent a letter of additional encouragement to participate. A better method is to send a new copy of the survey questionnaire with the follow-up letter. If potential respondents have not returned their questionnaires after 2 or 3 weeks, the questionnaires have probably been lost or misplaced. Receiving a follow-up letter might encourage them to look for the original questionnaire, but if they can't find it easily, the letter may go for naught.

The methodological literature strongly suggests that follow-up mailings provide an effective method for increasing return rates in mail surveys. In general, the longer a potential respondent delays replying, the less likely it is that he or she will do so at all. Properly timed follow-up mailings, then, provide additional stimuli to respond.

The effects of follow-up mailings will be seen in the response-rate curves recorded during data collection. The initial mailings will be followed by a rise and subsequent fall of returns; the follow-up mailings will spur a resurgence of returns; and more follow-ups will do the same. In practice, using three mailings (an original and two follow-ups) seems to be the most efficient.

The timing of follow-up mailings matters as well. Here, the methodological literature offers less-precise guides, but it has been my experience that 2 or 3 weeks is a reasonable space between mailings. (This period might be increased by a few days if the mailing time—out and in—is more than 2 or 3 days.)

If the individuals in the survey sample are not identified on the questionnaires, it might not be possible to remail only to nonrespondents.

In such a case, send your follow-up mailing to all members of the sample, thanking those who have already participated and encouraging those who have not participated to do so. (The case study reported later describes another method you can use in an anonymous mail survey.)

Response Rates

A question that new survey researchers frequently ask concerns the percentage return rate, or the **response rate**, that a mail survey should achieve. The body of inferential statistics used in connection with survey analysis assumes that all members of the initial sample complete the survey. Because this almost never happens, nonresponse bias becomes a concern, with the researcher testing (and hoping) for the possibility that the respondents look essentially like a random sample of the initial sample, and thus a somewhat smaller random sample of the total population.

Nevertheless, overall response rate is one guide to the representativeness of the sample respondents. A high response rate means less chance of significant nonresponse bias than does a low rate. Conversely, a low response rate is a danger signal, because the nonrespondents are likely to differ from the respondents in ways other than just their willingness to participate in your survey. Richard Bolstein (1991), for example, found that those who did not respond to a pre-election political poll were less likely to vote than were those who did participate. Estimating the turnout rate from the survey respondents, then, would have overestimated the number who would show up at the polls. Ironically, of course, since the nonrespondents were unlikely to vote, the preferences of the survey participants might offer a good estimate of the election results.

In the book *Standard Definitions*, the American Association for Public Opinion Research (AAPOR)

---

response rate The number of people participating in a survey divided by the number selected in the sample, in the form of a percentage. This is also called the *completion rate* or, in self-administered surveys, the *return rate*: the percentage of questionnaires sent out that are returned.

(2008) defines the response rate, and further distinguishes contact rates, refusal rates, and cooperation rates.

- Response rates: *The number of complete interviews with reporting units divided by the number of eligible reporting units in the sample. The report provides six definitions of response rates, ranging from the definition that yields the lowest rate to the definition that yields the highest rate, depending on how partial interviews are considered and how cases of unknown eligibility are handled.*
- Cooperation rates: *The proportion of all cases interviewed of all eligible units ever contacted. The report provides four definitions of cooperation rates, ranging from a minimum or lowest rate, to a maximum or highest rate.*
- Refusal rates: *The proportion of all cases in which a housing unit or the respondent refuses to be interviewed, or breaks-off an interview, of all potentially eligible cases. The report provides three definitions of refusal rates, which differ in the way they treat dispositions of cases of unknown eligibility.*
- Contact rates: *The proportion of all cases in which some responsible housing unit member was reached. The report provides three definitions of contact rates.*

(Aapor 2008: 4–5)

While response rates logically affect the quality of survey data, this is not always, in fact, the case, as Robert Groves (2006) points out. With recent declines in response rates, this is a topic under careful study by survey researchers. At the same time, higher response rates are a goal.

Response rates are important, since the nonrespondents may differ significantly from respondents, resulting in the survey results giving an inaccurate picture of the population under study. Let's say men are more likely to respond to a particular survey than women. Descriptive analyses of the survey results would overestimate the percentage of men in the population sampled.

As Ashley and Presser (2017) found, however, the impact of nonresponse bias is a complex one. In an extensive study involving several large surveys, they found that while univariate analyses (e.g., percentage of men) were highly susceptible to nonresponse bias, analyses of relationships among variables were less susceptible. If you found, for example, that women were more likely to vote Democratic than men, you could be confident that bivariate relationship was also true in the population, even if your sample had too many men and too few women.

Don Dillman (2009) has spent decades painstakingly assessing the various techniques that survey researchers have used to increase return rates on mail surveys, and he evaluates the impact of each. More important, Dillman stresses the necessity of paying attention to all aspects of the study—what he calls the "Tailored Design Method"—rather than one or two special gimmicks.

Having said all this, there is no absolutely acceptable level of response to a mail survey—except for 100 percent. While it is possible to achieve response rates of 70 percent or more, most mail surveys probably fall below that level. Thus, it's important to test for nonresponse bias wherever possible.

Compensation for Respondents

It is fairly common practice to pay experimental and focus group subjects for their participation, though it has been rare in other research methods. Whether to pay survey respondents is sometimes discussed—and often controversial.

In addition to cash payments, researchers have sometimes employed gift certificates, contributions to charities, lotteries, and other prize drawings. In a survey of New Zealanders, Mike Brennan and Jan Charbonneau (2009) sent chocolates as an incentive for participation.

Some researchers have provided incentives to all those selected in the sample during the first contact. In the case of cash incentives in mail surveys, this means respondents get the incentive whether they participate or not. In other cases, the researchers have provided or offered incentives in follow-up contacts with nonrespondents, though this creates a problem of inequity, with the most cooperative people getting no compensation.

In a 1999 review of studies of this topic, Singer, Groves, and Corning found that with very few exceptions, response rates are increased by the use of incentives in mail surveys, face-to-face interviews, and telephone polls. Also, the authors found no evidence of negative effects on the quality of responses collected. A decade later, Petrolia and Bhattacharee (2009) reviewed past experience with incentives and conducted their own study. They confirmed that incentives increase response rates, and they found that pre-paid incentives had a greater effect than those introduced later in the process.

J. Michael Brick and his colleagues (2012) report high response rates with a two-stage mail survey. They begin with an address-based sampling of households and send a short demographic questionnaire to learn relevant characteristics about the members of the household. Next, they select a subsample among those identified as appropriate for the particular survey focus and send a follow-up questionnaire. Both mailings were accompanied by a $1 cash incentive, and additional phone calls and postcard reminders were used with nonrespondents.

A Case Study

The steps involved in the administration of a mail survey are many and can best be appreciated in a walk-through of an actual study. Accordingly, this section concludes with a detailed description of how the student survey we discussed in Chapter 7 (as an illustration of systematic sampling) was administered. This study does not represent the theoretical ideal for such studies, but in that regard it serves present purposes all the better. The study was conducted by the students in my graduate seminar in survey research methods.

As you may recall, 1,100 students were selected from the university registration records through a stratified, systematic sampling procedure. For each student selected, the computer produced six self-adhesive mailing labels.

By the time we were ready to distribute the questionnaires, it became apparent that our meager research funds wouldn't cover several mailings to the entire sample of 1,100 students (questionnaire printing costs were higher than anticipated). As a result, we chose a systematic two-thirds sample of the mailing labels, yielding a subsample of 733 students.

Earlier, we had decided to keep the survey anonymous in the hope of encouraging candid responses to some sensitive questions. (Later surveys of the same issues among the same population indicated that this anonymity was unnecessary.) Thus, the questionnaires would carry no

identification of students on them. At the same time, we hoped to reduce the follow-up mailing costs by mailing only to nonrespondents.

To achieve both of these aims, a special postcard method was devised. Each student was mailed a questionnaire that carried no identifying marks, plus a postcard addressed to the research office—with one of the student's mailing labels affixed to the reverse side of the card. The introductory letter asked the student to complete and return the questionnaire and to return the postcard simultaneously. Receiving the postcard would tell us—without indicating which questionnaire it was—that the student had returned his or her questionnaire. This procedure would then facilitate follow-up mailings.

The 32-page questionnaire was printed in booklet form. The three-panel cover described earlier in this chapter permitted the questionnaire to be returned without an additional envelope.

A letter introducing the study and its purposes was printed on the front cover of the booklet. It explained why the study was being conducted (to learn how students feel about a variety of issues), how students had been selected for the study, the importance of each student's responding, and the mechanics of returning the questionnaire.

Students were assured that their responses to the survey were anonymous, and the postcard method was explained. A statement followed about the auspices under which the study was being conducted, and a telephone number was provided for those who might want more information about the study. (Five students called for information.)

By printing the introductory letter on the questionnaire, we avoided the necessity of enclosing a separate letter in the outgoing envelope, thereby simplifying the task of assembling mailing pieces.

The materials for the initial mailing were assembled as follows. (1) One mailing label for each student was stuck on a postcard. (2) Another label was stuck on an outgoing manila envelope. (3) One postcard and one questionnaire were placed in each envelope—with a glance to ensure that the name on the postcard and on the envelope were the same in each case.

The distribution of the survey questionnaires had been set up for a bulk-rate mailing. Once the questionnaires had been stuffed into envelopes, they were grouped by ZIP Code, tied in bundles, and delivered to the post office.

Shortly after the initial mailing, questionnaires and postcards began arriving at the research office. Questionnaires were opened, scanned, and assigned identification numbers, as described earlier in this chapter. For every postcard received, a search was made for that student's remaining labels, and they were destroyed.

After 2 or 3 weeks, the remaining mailing labels were used to organize a follow-up mailing. This time a special, separate letter of appeal was included in the mailing piece. The new letter indicated that many students had returned their questionnaires already, and it was very important for all others to do so as well.

The follow-up mailing stimulated a resurgence of returns, as expected, and the same logging procedures were continued. The returned postcards told us which additional mailing labels to destroy. Unfortunately, time and financial pressures made it impossible to undertake a third mailing, as had been initially planned, but the two mailings resulted in an overall return rate of 62 percent.

This illustration should give you a fairly good sense of what's involved in the execution of mailed self-administered questionnaires. Let's turn now to the second principal method of conducting surveys: in-person interviews.

Interview Surveys

The **interview** is an alternative method of collecting survey data. Rather than asking respondents to read questionnaires and enter their own answers, researchers send interviewers to ask the questions orally and to record respondents' answers. Interviewing is typically done in a face-to-face encounter, but telephone interviewing, discussed in the next section, follows most of the same guidelines.

---

interview A data-collection encounter in which one person (an interviewer) asks questions of another (a respondent). Interviews may be conducted face to face or by telephone.

Most interview surveys require more than one interviewer, although you might undertake a small-scale interview survey yourself. Portions of this section will discuss methods for training and supervising a staff of interviewers who are assisting you with a survey.

This section deals specifically with survey interviewing. Chapter 10 discusses the less-structured, in-depth interviews often conducted in qualitative field research.

The Role of the Survey Interviewer

There are several advantages to having a questionnaire administered by an interviewer rather than a respondent. To begin with, interview surveys typically attain higher response rates than do mail surveys. A properly designed and executed interview survey ought to achieve a completion rate of at least 80 to 85 percent. (Federally funded surveys often require one of these response rates.) Respondents seem more reluctant to turn down an interviewer standing on their doorstep than to throw away a mailed questionnaire.

The presence of an interviewer also generally decreases the number of "don't knows" and "no answers." If minimizing such responses is important to the study, the interviewer can be instructed to probe for answers ("If you had to pick one of the answers, which do you think would come closest to your feelings?").

Interviewers can also serve as a guard against questionnaire items that are confusing. If the respondent clearly misunderstands the intent of a question, the interviewer can clarify matters, thereby obtaining relevant responses. (As we'll discuss shortly, such clarifications must be strictly controlled through formal specifications.)

Finally, the interviewer can observe respondents as well as ask questions. For example, the interviewer can note the respondent's gender without having to ask. Similar observations can be made regarding the quality of the dwelling, the presence of various possessions, the respondent's ability to speak English, the respondent's general reactions to the study, and so forth. In one survey of students, respondents were given a short, self-administered questionnaire to complete—concerning sexual attitudes and behavior—during the course of the interview.

While a student completed the questionnaire, the interviewer made detailed notes regarding the dress and grooming of the respondent.

This procedure raises an ethical issue. Some researchers have objected that such practices violate the spirit of the agreement by which the respondent has allowed the interview. Although ethical issues are seldom clear-cut in social research, it's important to be sensitive to them (see Chapter 3).

Survey research is of necessity based on an unrealistic stimulus–response theory of cognition and behavior. Researchers must assume that a questionnaire item will mean the same thing to every respondent, and every given response will mean the same thing when given by different respondents. Although this is an impossible goal, survey questions are drafted to achieve the ideal as closely as possible.

The interviewer must also fit into this ideal situation. The interviewer's presence should not affect a respondent's perception of a question or the answer given. In other words, the interviewer should be a neutral medium through which questions and answers are transmitted.

As such, different interviewers should obtain exactly the same responses from a given respondent. (Recall our earlier discussions of reliability.) This neutrality has a special importance in area samples. To save time and money, a given interviewer is typically assigned to complete all the interviews in a particular geographic area—a city block or a group of nearby blocks. If the interviewer does anything to affect the responses obtained, the bias thus interjected might be interpreted as a characteristic of that area.

Let's suppose that a survey is being done to determine attitudes toward low-cost housing in order to help in the selection of a site for a new government-sponsored development. An interviewer assigned to a given neighborhood might—through word or gesture—communicate his or her own distaste for low-cost housing developments. Respondents might therefore tend to give responses in general agreement with the interviewer's own position. The results of the survey would indicate that the neighborhood in question strongly resists construction of the development in their area when in fact their apparent resistance simply reflects the interviewer's attitudes.

General Guidelines for Survey Interviewing

The manner in which interviews ought to be conducted will vary somewhat by the survey population and, to some degree, by the nature of the survey content. Nevertheless, some general guidelines apply to most interviewing situations.

Appearance and Demeanor

As a rule, interviewers should dress in a fashion similar to that of the people they'll be interviewing. A richly dressed interviewer will probably have difficulty getting good cooperation and responses from poorer respondents; a poorly dressed interviewer will have similar difficulties with richer respondents. To the extent that the interviewer's dress and grooming differ from those of the respondents, it should be in the direction of cleanliness and neatness in modest apparel. If cleanliness is not next to godliness, it appears to at least be next to neutrality. Although middle-class neatness and cleanliness may not be accepted by all sectors of U.S. society, they remain the primary norm and are the most likely to be acceptable to the largest number of respondents.

Dress and grooming are typically regarded as signs of a person's attitudes and orientations. At the time this is being written, torn jeans, green hair, and razor-blade earrings may communicate—correctly or incorrectly—that the interviewer is politically radical, sexually permissive, in favor of drug use, and so forth. Any of these impressions could bias responses or affect the willingness of people to be interviewed.

In demeanor, interviewers should be pleasant if nothing else. Because they'll be prying into a respondent's personal life and attitudes, they must communicate a genuine interest in getting to know the respondent without appearing to pry. They must be relaxed and friendly without being too casual or clingy. Good interviewers also have the ability to determine very quickly the kind of person the respondent will feel most comfortable with, the kind of person the respondent would most enjoy talking to. Clearly, the interview will be more successful if the interviewer can become the kind of person the respondent is comfortable with. Further, because respondents are asked to volunteer a portion of their time and to divulge personal information, they deserve the most enjoyable experience the researcher and interviewer can provide.

Familiarity with the Questionnaire

If an interviewer is unfamiliar with the question-naire, the study suffers and the respondent bears an unfair burden. The interview is likely to take more time than necessary and be unpleasant. Moreover, the interviewer cannot acquire familiarity by skim-ming through the questionnaire two or three times. He or she must study it carefully, question by ques-tion, and must practice reading it aloud.

Ultimately, the interviewer must be able to read the questionnaire items to respondents without error and without stumbling over words and phrases. A good model is the actor reading lines in a play or movie. The lines must be read as though they constituted a natural conversa-tion, but that conversation must follow exactly the language set down in the questionnaire.

By the same token, the interviewer must be familiar with the specifications prepared in con-junction with the questionnaire. Inevitably some questions will not exactly fit a given respondent's situation, and the interviewer must determine how the question should be interpreted in that situation. The specifications provided to the in-terviewer should give adequate guidance in such cases, but the interviewer must know the orga-nization and contents of the specifications well enough to refer to them efficiently. It would be better for the interviewer to leave a given ques-tion unanswered than to spend 5 minutes search-ing through the specifications for clarification or trying to interpret the relevant instructions.

Following Question Wording Exactly

The first part of this chapter discussed the sig-nificance of question wording for the responses obtained. A slight change in the wording of a given question may lead a respondent to an-swer "yes" rather than "no." It follows that interviewers must be instructed to follow the wording of questions exactly. Otherwise, all the effort that the developers have put into care-fully phrasing the questionnaire items to obtain the information they need and to ensure that respondents interpret items precisely as in-tended will be wasted.

While I hope the logic of this injunction is clear, it's not necessarily a closed discussion. For example, Giampietro Gobo (2006) argues that we might consider giving interviewers more latitude, especially when respondents make errors that are apparent to the interviewer on the spot. Allowing the interviewer to intervene, as he notes, does increase the possibility that the interviewer's understanding and opinions may influence the data collected.

Recording Responses Exactly

Whenever the questionnaire contains open-ended questions, which solicit the respondent's own answer, the interviewer must record that answer exactly as given. No attempt should be made to summarize, paraphrase, or correct bad grammar.

This exactness is especially important because the interviewer will not know how the responses are to be coded. Indeed, the researchers them-selves may not know the coding until they've read a hundred or so responses. For example, the questionnaire might ask respondents how they feel about the traffic situation in their com-munity. One respondent might answer that there are too many cars on the roads and that some-thing should be done to limit their numbers. Another might say that more roads are needed. If the interviewer recorded these two responses with the same summary—"congested traffic"— the researchers would not be able to take advan-tage of the important differences in the original responses.

Sometimes, verbal responses are too inarticu-late or ambiguous to permit interpretation. How-ever, the interviewer may be able to understand the intent of the response through the respon-dent's gestures or tone. In such a situation, the interviewer should still record the exact verbal response but should also add marginal comments giving both the interpretation and the reasons for arriving at it.

More generally, researchers can use any marginal comments explaining aspects of the response not conveyed in the verbal recording, such as the respondent's apparent anger, embar-rassment, uncertainty in answering, and so forth. In each case, however, the exact verbal response should also be recorded.

Probing for Responses

Sometimes respondents in an interview will give an inappropriate or incomplete answer. In such cases, a **probe**, or request for an elaboration, can be useful. For example, a closed-ended question may present an attitudinal statement and ask the respondent to strongly agree, agree somewhat, disagree somewhat, or strongly disagree. The respondent, however, may reply: "I think that's true." The interviewer should follow this reply with "Would you say you strongly agree or agree somewhat?" If necessary, interviewers can explain that they must check one or the other of the categories provided. If the respondent adamantly refuses to choose, the interviewer should write in the exact response given by the respondent.

Probes are more frequently required in eliciting responses to open-ended questions. For example, in response to a question about traffic conditions, the respondent might simply reply, "Pretty bad." The interviewer could obtain an elaboration on this response through a variety of probes. Sometimes the best probe is silence; if the interviewer sits quietly with pencil poised, the respondent will probably fill the pause with additional comments. (Newspaper reporters use this technique effectively.) Appropriate verbal probes might be "How is that?" or "In what ways?" Perhaps the most generally useful probe is "Anything else?"

Often, interviewers need to probe for answers that will be sufficiently informative for analytic purposes. In every case, however, such probes must be completely neutral; they must not in any way affect the nature of the subsequent response. Whenever you anticipate that a given question may require probing for appropriate responses, you should provide one or more useful probes next to the question in the questionnaire. This practice has two important advantages. First, you'll have more time to devise the best, most neutral probes. Second, all interviewers will use the same probes whenever they're needed. Thus, even if the probe isn't perfectly neutral, all respondents will be presented with the same stimulus. This is the same logical guideline that we discussed for question wording. Although a question should not be loaded or biased, every respondent must be presented with the same question, even if it is biased.

Coordination and Control

Most interview surveys require the assistance of several interviewers. In large-scale surveys, interviewers are hired and paid for their work. Student researchers might find themselves recruiting friends to help them interview. Whenever more than one interviewer is involved in a survey, their efforts must be carefully controlled. This control has two aspects: training interviewers and supervising them after they begin work.

The interviewers' training session should begin with the description of what the study is all about. Even though the interviewers may be involved only in the data-collection phase of the project, it will be useful for them to understand what will be done with the interviews they conduct and what purpose will be served. Morale and motivation are usually lower when interviewers don't know what's going on.

The training on how to interview should begin with a discussion of general guidelines and procedures, such as those discussed earlier in this section. Then the whole group should go through the questionnaire together—question by question. Don't simply ask if anyone has any questions about the first page of the questionnaire. Read the first question aloud, explain the purpose of the question, and then entertain any questions or comments the interviewers may have. Once all their questions and comments have been handled, go on to the next question in the questionnaire.

It's always a good idea to prepare specifications to accompany an interview questionnaire. *Specifications* are explanatory and clarifying comments about handling difficult or confusing situations that may occur with regard to particular questions in the questionnaire. When drafting the questionnaire, try to think of all the problem cases that might arise—the bizarre (or not so bizarre) circumstances that might make a

---

probe A technique employed in interviewing to solicit a more complete answer to a question. It is a nondirective phrase or question used to encourage a respondent to elaborate on an answer. Examples include "Anything more?" and "How is that?"

question difficult to answer. The survey specifications should provide detailed guidelines on how to handle such situations. For example, even as simple a matter as age might present problems. Suppose a respondent says he or she will be 25 next week. The interviewer might not be sure whether to take the respondent's current age or the nearest one. The specifications for that question should explain what should be done. (Probably, you would specify that the age as of last birthday should be recorded in all cases.)

If you've prepared a set of specifications, review them with the interviewers when you go over the individual questions in the questionnaire. Make sure your interviewers fully understand the specifications and the reasons for them as well as the questions themselves.

This portion of the interviewer training is likely to generate many troublesome questions from your interviewers. They'll ask, "What should I do if … ?" In such cases, avoid giving a quick, offhand answer. If you have specifications, show how the solution to the problem could be determined from the specifications. If you do not have specifications, show how the preferred handling of the situation fits within the general logic of the question and the purpose of the study. Giving unexplained answers to such questions will only confuse the interviewers and cause them to take their work less seriously. If you don't know the answer to such a question when it is asked, admit it and ask for some time to decide on the best answer. Then think out the situation carefully and be sure to give all the interviewers your answer, explaining your reasons.

Once you've gone through the whole questionnaire, conduct one or two demonstration interviews in front of everyone. Preferably, you should interview someone other than one of the interviewers. Realize that your interview will be a model for those you're training, so make it good. It would be best, moreover, if the demonstration interview were done as realistically as possible. Do not pause during the demonstration to point out how you've handled a complicated situation: Handle it, and then explain later. It is irrelevant if the person you're interviewing gives real answers or takes on some hypothetical identity for the purpose, as long as the answers consistently represent the identity being presented.

After the demonstration interviews, pair off your interviewers and have them practice on each other, going through the entire process twice, reversing roles for the second round. Interviewing is the best training for interviewing. As your interviewers practice on each other, wander around and listen in on the practice so you'll know how well they're doing. Once the practice is completed, the whole group should discuss their experiences and ask any other questions they may have.

The final stage of the training for interviewers should involve some "real" interviews. Have them conduct some interviews under the actual conditions that will pertain to the final survey. You may want to assign them people to interview, or you may let them pick people themselves. Do not have them practice on people you've selected in your sample, however. After each interviewer has completed three to five interviews, have him or her check back with you. Look over the completed questionnaires for any evidence of misunderstanding. Again, answer any questions that the interviewers may have. Once you're convinced that a given interviewer knows what to do, assign some actual interviews, using the sample you've selected for the study.

It's essential to continue supervising the work of interviewers over the course of the study. You should check in with them after they conduct no more than twenty or thirty interviews. You might assign twenty interviews, have the interviewer bring back those questionnaires when they're completed, look them over, and assign another twenty or so. Although this may seem overly cautious, you must continually protect yourself against any misunderstandings not evident early in the study. Moreover, Kristen Olson and Andy Peytchev (2007) have discovered that an interviewer's behavior changes over the course of a survey project. For example, as time goes on, interviewers tend to speed up, presenting the interview more quickly, and they are more likely to judge respondents as uninterested in it.

If you're the only interviewer in your study, these comments may not seem relevant. However, it would be wise, for example, to prepare specifications for potentially troublesome questions in your questionnaire. Otherwise, you run the risk of making ad hoc decisions during the course of the study that you'll later regret or forget. Also, the emphasis on practice applies to the one-person project as much as to the complex, funded survey with a large interviewing staff.

Telephone Surveys

For years telephone surveys had a rather bad reputation among professional researchers. By definition, telephone surveys are limited to people who have telephones. Years ago, this method produced a substantial social-class bias by excluding poor people from the surveys. This was vividly demonstrated by the *Literary Digest* fiasco of 1936. Recall that, even though voters were contacted by mail, the sample was partially selected from telephone subscribers, who were hardly typical in a nation just recovering from the Great Depression. As we saw in Chapter 7, virtually all American households now have telephones, so the earlier form of class bias has substantially diminished.

Positive and Negative Factors

Telephone surveys have many advantages, which underlie the popularity of this method. Probably the greatest advantages are money and time, in that order. In a face-to-face household interview, you may drive several miles to a respondent's home, find no one there, return to the research office, and drive back the next day—possibly finding no one there again. It's cheaper and quicker to let your fingers make the trips.

When interviewing by telephone, you can dress any way you please without affecting the answers respondents give. And sometimes respondents will be more honest in giving socially disapproved answers if they don't have to look you in the eye. Similarly, it may be possible to probe into more-sensitive areas, though this isn't necessarily the case. People are, to some extent, more suspicious when they can't see the person who is asking them questions—which is perhaps a consequence of "surveys" aimed at selling magazine subscriptions and time-share condominiums.

Interviewers can communicate a lot about themselves over the phone, however, even though they can't be seen. For example,

researchers worry about the impact of an interviewer's name (particularly if ethnicity is relevant to the study) and debate the ethics of having all interviewers use bland "stage names" such as Smith or Jones. (Female interviewers sometimes ask permission to do this, to avoid subsequent harassment from men they interview.)

Telephone surveys can allow greater control over data collection if several interviewers are engaged in the project. If all the interviewers are calling from the research office, they can get clarification from the person in charge whenever problems occur, as they inevitably do. Alone in the boondocks, an in-person interviewer may have to wing it between weekly visits with the interviewing supervisor.

Finally, another important factor involved in the growing use of telephone surveys has to do with personal safety. Don Dillman describes the situation this way:

> Interviewers must be able to operate comfortably in a climate in which strangers are viewed with distrust and must successfully counter respondents' objections to being interviewed. Increasingly, interviewers must be willing to work at night to contact residents in many households. In some cases, this necessitates providing protection for interviewers working in potentially dangerous locations.
>
> (1978: 4)

Concerns for safety thus work two ways to hamper face-to-face interviews. Potential respondents may refuse to be interviewed, fearing the stranger-interviewer. And the interviewers themselves may incur some risks. All this is made even worse by the possibility of the researchers being sued for huge sums if anything goes wrong.

There are problems involved in telephone interviewing, however. As I've already mentioned, the method is hampered by the proliferation of bogus "surveys" that are actually sales campaigns disguised as research. If you have any questions about any such call you receive, by the way, ask the interviewer directly whether you've been selected for a survey only or if a sales "opportunity" is involved. It's also a good idea, if you have any doubts, to get the interviewer's name, phone number, and company. Hang up if the caller refuses to provide any of these.

For the researcher, the ease with which people can hang up is another shortcoming of telephone surveys. Once you've been let inside someone's home for an interview, the respondent is unlikely to order you out of the house in the middle of the interview. It's much easier to terminate a telephone interview abruptly, saying something like, "Whoops! Someone's at the door. I gotta go," or "OMIGOD! The pigs are eating my Volvo!" (That sort of evasion is much harder to fake when the interviewer is sitting in your living room.)

Research has shown that several factors, including voice mail and answering machines, have reduced response rates in telephone surveys (Tuckel and O'Neill 2002).

Computer-Assisted Telephone Interviewing (CATI)

In Chapter 14, we'll be looking at some of the ways computers have influenced the conduct of social research—particularly data processing and analysis. Computers are also changing the nature of telephone interviewing. One innovation is computer-assisted telephone interviewing (CATI). This method is increasingly used by academic, government, and commercial survey researchers. Here's a general example of what using CATI can look like.

Imagine an interviewer wearing a telephone headset, sitting in front of a computer monitor. The central computer has been programmed to select a telephone number at random and dials it. On the video screen is an introduction ("Hello, my name is . . .") and the first question to be asked ("Could you tell me how many people live at this address?").

When the respondent answers the phone, the interviewer says hello, introduces the study, and asks the first question displayed on the screen. When the respondent answers the question, the interviewer types that answer into the computer terminal—either the verbatim response to an open-ended question or the code category for the appropriate answer to a closed-ended question. The answer is immediately stored in the computer. The second question appears on the video screen and is asked, and the answer is entered into the computer. Thus, the interview continues.

In addition to the obvious advantages in terms of data collection, CATI automatically prepares the data for analysis; in fact, the researcher can begin analyzing the data before the interviewing is complete, thereby gaining an advanced view of how the analysis will turn out.

It is also possible to go a step further than computer-*assisted* interviews. With the innovation of so-called robopolls, the entire interview is conducted by a programmed recording that can interpret the spoken answers of respondents. This discussion may remind you of the robocalls in which a recorded voice presents a political or commercial message once you answer your phone. Robopolls go a step further through the use of *Interactive Voice Recognition (IVR)*. The computer is programmed to interpret the respondent's answers, record them, and determine how to continue the interview appropriately.

Clearly this method is cost-effective by cutting out the labor cost of hiring human beings as interviewers. It has been viewed with suspicion and/or derision by some survey researchers, but in its evaluation of the 2008 primary polling, the American Association for Public Opinion Research (AAPOR) reported no difference in the accuracy of results produced by CATI or IVR (AAPOR 2009).

During the 2010 midterm election campaigns, survey-watcher Nate Silver (2010b) found that robopolls tended to produce results slightly more favorable to Republicans than did conventional methods. Silver also found that robopolls might produce different answers to sensitive questions. He looked at California's Proposition 19, which would have legalized and taxed the personal use of marijuana. Silver found that:

> The methodologies split in the support they show for the initiative. The three automated surveys all have Prop 19 passing by a double-digit margin. The human-operator polls, meanwhile, each show it trailing narrowly.
>
> (Silver 2010a)

Ultimately, Proposition 19 failed by a two-to-one margin. The next edition of this textbook may revise the discussion of robopolls, though it is not clear now what the fate of this technique will be.

Response Rates in Interview Surveys

Earlier in this chapter we looked at the issue of response rates in mail surveys, and this is an equally important issue for interview surveys. In Chapter 7, when we discussed formulas for calculating sampling error to determine the accuracy of survey estimates, the implicit assumption was that everyone selected in a sample would participate—which is almost never the case. Lacking perfection, researchers must maximize participation by those selected. Although interview surveys tend to produce higher response rates than do mail surveys, interview success in telephone surveys has declined substantially.

By analyzing response-rate trends in the University of Michigan's Survey of Consumer Attitudes, Richard Curtin, Stanley Presser, and Eleanor Singer (2005) sketched a pattern of general decline over the years. Between 1979 and 1996, the response rate in this telephone survey dropped from 72 to 60 percent, representing an average annual decline of three-quarters of a percent. Since 1996, the rate of decline has doubled. The increased nonresponses reflected both refusals and those they were unable to contact.

By contrast, the General Social Survey (GSS), using personal interviews, experienced response rates between 73.5 and 82.4 percent in the years from 1975 to 1998. In the 2014 survey, however, the GSS completion rate was 69 percent. Their decline came primarily from refusals, because household interviews produce higher rates of contact than do telephone surveys.

In general, both household and telephone surveys have experienced a decline in response rates. A special issue of *Public Opinion Quarterly* (2006) was devoted to an analysis of the many dimensions of the decline in response rates in household surveys. As the analyses show, lower response rates do not necessarily produce inaccurate estimates of the population being studied. Nonetheless, this complex issue defies a simple summary.

Former director of the U.S. Census Robert Groves detailed some of the factors complicating modern survey research:

> Walled subdivisions, locked apartment buildings, telephone answering machines, telephone caller ID, and a host of other access impediments for survey researchers grew in this era. Response rates continued to deteriorate. Those household surveys devoted to high response rates experienced continuous inflation of costs due to increased effort to contact and interview the public. Face-to-face interviews continued to decline in volume, often limited to the first wave of longitudinal surveys.

> *(2011: 866)*

Many researchers believe that the widespread growth of telemarketing has played a large role in the problems experienced by legitimate telephone surveys, and they hope that the state and national "do not call" lists may ease that problem. Further, we saw that other factors such as answering machines and voice mail also contribute to these problems (Tuckel and O'Neill 2002). Response rate will likely remain an important issue in survey research.

Again, as a consumer of social research, you should be wary of "surveys" whose apparent purpose is to raise money for the sponsor. This practice has been common in mail surveys, and soon expanded to the realm of "fax surveys," evidenced by a fax entitled "Should Hand Guns Be Outlawed?" Two fax numbers were provided for expressing either a "Yes" or "No" opinion. The smaller print noted, "Calls to these numbers cost $2.95 per minute, a small price for greater democracy. Calls take approx. 1 or 2 minutes." You can imagine where the $2.95 went. I imagine you can give your own examples of similar e-mail "surveys."

These are challenging times for legitimate surveys, particularly for political pollsters. It is a time of experimentation and some have been generally more successful than others. You might be interested in Nate Silver's FiveThirtyEight rating of pollsters at https://projects.fivethirtyeight.com/pollster-ratings/.

Online Surveys

An increasingly popular method of survey research involves the use of the Internet and the World Wide Web—two of the most far-reaching developments of the late twentieth century. Mick Couper and Peter Miller give an excellent introduction to the timeline of this new face of social research:

> Despite their relatively short history, Web surveys have already had a profound effect on survey research. The first graphic browser (NCSA Mosaic) was released in 1992, with Netscape Navigator following in 1994 and Internet Explorer in 1995. The first published papers on Web surveys appeared in 1996. Since then, there has been a virtual explosion of interest in the Internet generally, and World Wide Web specifically, as a tool for survey data collection.
>
> *(2008: 831)*

Three years later, Mick Couper (2011) reflected on the probable role of online surveys in the future of social research:

> The newer modes have tended to supplement rather than replace existing modes, in part because even though they address some problems (e.g., improvements in measurement, reductions in cost), they may not solve others (e.g., coverage, nonresponse). In other words, there is no one mode that can be all things to all research questions. Multiple modes, and mixes of mode, will continue to be a fact of life for survey research for the foreseeable future.
>
> *(2011: 901)*

While this section will examine various aspects of online survey research, you should be forewarned that this technique is developing so quickly that new innovations will surely have arisen by the time this book reaches your hands. To stay abreast of these developments, your best single source is the American Association for Public Opinion Research (AAPOR) and two key publications: *Public Opinion Quarterly* (POQ) and the online journal *Survey Practice*. While neither of these is dedicated solely to online research, an increasing percentage of their articles address that topic. University survey research offices such as those at the University of California (Berkeley), University of Michigan, NORC at the University of Chicago, and many other institutions around the globe are very active in developing this new technique. Similarly, commercial research firms such as Pew, Harris, Nielsen, and others are equally involved.

As we saw in Chapter 7 on sampling, one immediate objection that many social researchers make to online surveys concerns representativeness: Will the people who can be surveyed online be representative of meaningful populations, such as all U.S. adults, all voters, and so on? This was the criticism that researchers raised earlier with regard to surveys via fax and telephone surveys.

Early in the development of online surveys, Camilo Wilson (1999), founder of Cogix, pointed out that some respondent populations are ideally suited to this technique: specifically, those who visit a particular website. For example, Wilson indicates that market research for online companies should be conducted online, and his firm has developed software called ViewsFlash for precisely that purpose. Although website surveys could easily collect data from all who visit a particular site, Wilson suggests that survey sampling techniques can provide sufficient consumer data without irritating thousands or millions of potential customers. As we saw in Chapter 7, much methodological research is being devoted to ways of achieving representative sampling of general populations with online surveys.

Let's turn now to some of the other methodological aspects of online surveys that are currently being examined and experimented with.*

Online Devices

At the outset, online surveys were aimed at users of personal computers, most typically desktop models. As the distinction between desktop and laptop computer capabilities narrowed, both devices were considered proper ways of participating in online surveys. Notice, however, that the growing use of laptop computers for this purpose broadened the variety of environments in which respondents might participate. This was only the beginning, however.

* In beginning this section of the chapter, I want to acknowledge Michael Link of the Nielsen Company, for his excellent, online seminar, "Leveraging New Technologies," conducted as part of AAPOR's Webinar Series on December 5, 2012. While I have not quoted directly from the seminar, I have benefited greatly from the overview and detailing of variations it provided.

When I attended the first meeting of the Chinese Survey Research Association in Shanghai in 2010, I was struck by the vitality of the researchers reporting on their studies in a country where sociology had been removed from universities from 1949 to 1979. Most of the articles I looked at were in Chinese, which was a problem for me. However, many articles included photographs to illustrate some of the new techniques being used, and I was struck by the number of smartphones and other mobile devices pictured. This interest is hardly limited to Chinese research.

Tablets and smartphones have been rapidly gaining in computing power, and they are increasingly being used as vehicles for online surveys. Respondents, probably indirectly, propelled researchers to develop survey formats compatible with mobile devices: As respondents attempted, sometimes unsuccessfully, to use smartphones and digital tablets to complete questionnaires designed only for desktop computers, survey researchers caught on to the need—and the potential—for adapting their questionnaires to the range of devices that respondents might want to use. Screen size, of course, is a major concern, but so are the various navigation systems used by different devices.

Research supports that survey formats must accommodate respondents' device preferences. For example, Morgan M. Millar and Don A. Dillman (2012) conducted an experiment in which they attempted to encourage respondents to participate in a survey using their smartphones while allowing the use of other devices such as tablets or laptops. The researchers reported only a slight increase in smartphone use by respondents who were urged to use the device, as compared with those who were given no encouragement.

This line of methodological research will continue, but consider this: We will surely see the development of new devices, some we can't currently imagine, which will have to be accommodated in the future.

Many of the cautions urged in relation to online surveys today are similar to those urged in relation to *telephone surveys* when I first began writing research methods textbooks. Mick Couper makes a similar observation:

Several years ago, I predicted that the rapid spread of electronic data collection methods such as the Internet would produce a bifurcation in the survey industry between high-quality surveys based on probability samples and using traditional data collection methods, on the one hand, and surveys focused more on low cost and rapid turnaround than on representativeness and accuracy on the other. In hindsight, I was wrong, and I underestimated the impact of the Web on the survey industry. It has become much more of a fragmentation than a bifurcation (in terms of Web surveys at least), with vendors trying to find or create a niche for their particular approach or product. No longer is it just "quick and dirty" in one corner and "expensive but high quality" in the other; rather, there is a wide array of approaches representing varying levels of quality and cost.

(2001: 466)

The early discussions of online surveys mostly assumed that respondents were sitting at their desk computers or perhaps with laptops. That image is changing rapidly, however. More recently, Tom Wells, Justin Bailey, and Michael Link (2013) have offered an excellent "state of the art" review of tablet-based surveys at present. To be clear, however, this aspect of survey research is evolving faster than anyone imagined.

Instrument Design

Over the years, members of industrialized nations have become familiar with the format and process of self-administered questionnaires, but, as just mentioned, the Web presents a new challenge for many. Leah Christian, Don Dillman, and Jolene Smyth provide a wealth of guidance on formatting Web surveys. Their aim is, as their article title states, "helping respondents get it right the first time" (2007).

The initial temptation, of course, is to simply import the digital file for the mail questionnaire to a Web survey. However, there are two problems with doing this. First, the mail format doesn't necessarily fit on a computer screen, let alone the screen of a tablet or smartphone. On the other hand, the e-devices offer possibilities unattainable with words on paper. I am unable to list those possibilities for you now, because they are still being developed, but I can connect you with some of the possibilities and challenges currently underway or on the radar.

For example, researchers like Roger Tourangeau, Mick P. Couper, and Frederick G. Conrad (2013) were concerned about whether the placement of answers in a list would affect respondents' choices. Their conclusion, based on the review of several studies, is that "up means good." When several opinion choices are arranged vertically, respondents are more likely to select the topmost choice.

Jason Husser and Kenneth Fernandez (2013) examined whether it was better to have an online respondent enter numerical answers by clicking the answer, typing it, or dragging it along a scale to indicate the answer. With a limited number of responses, clicking radio buttons was fastest, but a long list of possible answers makes dragging the sliding scale more practical.

Those who use the Internet regularly are familiar with emoticons such as the "smiley face." While these graphics could be printed in a mail questionnaire, they seem more at home online. Matthias Emde and Marek Fuchs (2012) undertook an experiment to determine the possibility of using a range of faces (sad to happy) in place of radio buttons labeled from bad to good. They concluded that this format change did not affect responses. Thus, these types of formatting options can be chosen on purely aesthetic grounds. There is no reason not to make surveys appealing.

Lawrence Malakhoff and Matt Jans (2011) explore some of the more advanced possibilities for online survey research. While the survey interview involves a person showing up on your doorstep or a voice coming over your phone, they suggest that an animated avatar might be used to conduct an online interview, and they have begun experimenting with gender and other differences for the animated interviewer. The avatar interviewer can be programmed to change facial expressions based on the respondent's answers. Going one step (or several) further, it would be possible to use the respondents' webcams to monitor their facial expressions and log that data along with the answers provided verbally.

Improving Response Rates

Online surveys appear to have response rates approximately comparable to those of mail surveys, according to a large-scale study of Michigan State University students (Kaplowitz, Hadlock,

and Levine 2004), especially when the online survey is accompanied by a postcard reminder encouraging respondents to participate. While producing a comparable response rate, the cost of an online survey is substantially less than that of a conventional mail survey. The cost of paper, printing, and postage alone can constitute a large expense in the latter.

In another study of ways to improve response rates in online surveys, Stephen Porter and Michael Whitcomb (2003) found that some of the techniques effective in mail surveys, such as personalizing the appeal or varying the apparent status of the researcher, had little or no impact in the new medium. At the same time, specifying that the respondents had been specially selected for the survey and setting a deadline for participation did increase response rates.

Kylie McGeeney (2015) at the Pew Center offers a useful checklist for improving Web survey completion rates:

1. Software should be mobile optimized.
2. Shorter is better.
3. Avoid fancy features.
4. No grids.
5. Ask multiple questions on the same screen.
6. Maximize use of the smartphone screen.
7. Use a unique URL in the survey invitation.
8. Invite respondents through a text message.

The relative youth of online surveys makes them a fertile ground for innovation and experimentation. For example, survey researchers have often worried that respondents to self-administered questionnaires may spend more of their attention on the first responses in a list, skipping quickly over those farther down. To test this possibility, Mirta Galesic and colleagues (2008) employed a special eye-tracking computer monitor that unobtrusively followed respondents' eye movements as they completed an online survey. The result: Respondents did, in fact, spend more time on the early choices, sometimes failing to read the whole list before clicking their choice on the screen.

The years ahead will see many more experiments aimed at improving the effectiveness of online surveys.

You are reading this discussion at an exciting time, when online survey methodology is evolving. For example, in an effort to increase response rates for Web surveys, Morgan Millar and Don Dillman (2012) achieved modest increases by sending respondents an e-mail reminder to participate in the survey. Because a large percentage of cell-phone owners have smartphones, they were offered the opportunity to complete the survey on those devices instead of going to a computer. As the authors point out, further experimentation with e-mail reminders will require tailoring survey formats to accommodate smartphones, as discussed earlier. The Pew Center (Smith 2013) estimates that more than three in five American adults own a smartphone, and as many as nine in ten own some kind of cellular phone.

For now, Mick P. Couper's *Designing Effective Web Surveys* (2008) offers a comprehensive guide to this new technique, based on what we have learned about it to date. If you are interested in experimenting with web surveys on your own, see "How to Do It: Conducting an Online Survey."

Mixed-Mode Surveys

In Chapter 4, I introduced the idea of mixed modes, indicating that different research techniques could be combined in a given study, such as a survey, combined with a review of existing data and in-depth field observations and interviews. Although researchers have sometimes combined face-to-face, mail, and telephone surveys, the advent of online surveys has

increased attention to the potential of combining survey techniques.

As Don Dillman (2012) points out, the logistical advantages of online surveys are somewhat offset by the difficulty of getting representative samples. Thus, researchers sometimes use an address-based sampling as the basis for a mail survey, which invites recipients to respond online if that's convenient for them or by mail if it is not.

As Edith de Leeuw (2010) points out, this is not a new idea.

> Already in 1788, Sir John Sinclair used a mixed-mode approach. Lacking funds for a full statistical census, Sinclair used a cost-effective mail survey among ministers of all parishes in the Church of Scotland. To achieve a high response Sinclair also used follow-up letters and finally "statistical missionaries," who personally visited the late responders to hurry ministerial replies.

This combination of survey techniques evidently produced a 100 percent completion rate.

The special advantages of Internet surveys (mass scale and cost) have added new impetus for combining survey modes. In addition to sampling issues, survey researchers are also attentive to response effects that may be caused by the different modes. That is, whether people would answer a given question the same online as in a mail questionnaire or a telephone interview. Initial studies suggest relatively small effects (de Leeuw and Hox 2011), but this will be a subject of methodological research for years to come.

Comparison of the Different Survey Methods

Now that we've seen several ways to collect survey data, let's take a moment to compare them directly.

Self-administered questionnaires are generally cheaper and quicker than face-to-face interview surveys. These considerations are likely to be important for an unfunded student who wishes to undertake a survey for a term paper or thesis. Moreover, if you use the self-administered mail format, it costs no more to conduct a national survey than a local one of the same sample size. In contrast, a national interview survey utilizing face-to-face contacts would cost far more than a local one. Also, mail surveys typically require a small staff: One person can conduct a reasonable mail survey alone, although you shouldn't underestimate the work involved. Further, respondents are sometimes reluctant to report controversial or deviant attitudes or behaviors in interviews but are willing to respond to an anonymous self-administered questionnaire.

Interview surveys also offer many advantages. For example, they generally produce fewer incomplete questionnaires. Although respondents may skip questions in a self-administered questionnaire, interviewers are trained not to do so. In CATI surveys, the computer offers a further check on this. Interview surveys, moreover, typically achieve higher completion rates than do self-administered questionnaires.

Although self-administered questionnaires may be more effective for sensitive issues, interview surveys are definitely more effective for complicated ones. Prime examples include the enumeration of household members and the determination of whether a given address corresponds to more than one housing unit. Although the concept of "housing unit" has been refined and standardized by the Bureau of the Census and interviewers can be trained to deal with the concept, it's extremely difficult to communicate in a self-administered questionnaire. This advantage of interview surveys pertains generally to all complicated contingency questions.

With interviews, you can conduct a survey based on a sample of addresses or phone numbers rather than on names. An interviewer can arrive at an assigned address or call the assigned number, introduce the survey, and even—following instructions—choose the appropriate person at that address to respond to the survey. In contrast, self-administered questionnaires addressed to "occupant" receive a notoriously low response.

Finally, as we've seen, interviewers questioning respondents face to face can make important observations aside from responses to questions asked in the interview. In a household interview, they may note the characteristics of the neighborhood, the dwelling unit, and so forth. They may also note characteristics of the respondents or the quality of their interaction with the respondents—whether the respondent had difficulty communicating, was hostile, seemed to be lying, and so on. A student using this textbook recently pointed out another advantage of face-to-face interviews. In his country, where literacy rates are relatively low in some areas, people would not be able to read a self-administered questionnaire and record their answers—but they could be interviewed.

The chief advantages of telephone surveys over those conducted face to face center primarily on time and money. Telephone interviews are much cheaper and can be mounted and executed quickly. Also, interviewers are safer when interviewing people in high-crime areas. Moreover, the impact of the interviewers on responses is somewhat lessened when they can't be seen by the respondents. As only one indicator of the popularity of telephone interviewing, when Johnny Blair and his colleagues (1995) compiled a bibliography on sample designs for telephone interviews, they listed over 200 items.

Online surveys have many of the strengths and weaknesses of mail surveys. Once the available software has been further developed, they are likely to be substantially cheaper. An important weakness, however, lies in the difficulty of ensuring that respondents to an online survey will be representative of some more-general population.

Martyn Denscombe (2009) used matched samples of students to test the nonresponse rates produced by conventional, paper questionnaires with those administered online. (Students did not get to choose the method; they were randomly assigned.) Overall, the online surveys produced somewhat lower nonresponse rates, and this difference was more pronounced for open-ended questions.

Online surveys are particularly appropriate for certain targeted groups and for research specifically based on Web participation. An online survey would be perfect for studying the feelings of people who have purchased items from Seller #12345 on eBay, for example. This advantage may become more significant if and when our lives become increasingly organized around our Web participation.

As respondents become more accustomed to online surveys, they may be able to ease some of the problems that have plagued telephone surveys: allowing for longer and more complex surveys, for example. Online respondents, like those completing mail questionnaires, will have more time to reflect on their responses. And online surveys may lend themselves to experimental designs more easily than other methods. Surely, online survey methodology will continue to evolve, as previously experienced with other survey techniques.

With the growth of online surveys, we have seen an increased interest in and use of paradata—a wealth of data generated by computer in the course of a survey. How long did a respondent take before answering each question? Did men or women take longer to answer a particular question? Did conservative or liberal responses come more quickly? Already such data are being used for studies of survey methodology, but they also can provide data useful to understanding human behavior, as social scientists are wont to do.

Clearly, each survey method has its place in social research. Ultimately, you must balance the advantages and disadvantages of the different methods in relation to your research needs and your resources. As we have just seen, researchers sometimes employ mixed-mode surveys in the same study, combining more than one of the techniques we've examined, such as mail and interview. While this option has been employed for some time, Edith D. de Leeuw (2010) updated the discussion by bringing online surveys into the mix. All in all, these are exciting times for the measurement of public opinion.

Michael W. Link and his colleagues (2014), in a task force report for the American Association for Public Opinion Research (AAPOR), suggest that important changes lie ahead for public opinion measurement.

> *Public opinion research is entering a new era, one in which traditional survey research may play a less dominant role. The proliferation of new technologies, such as mobile devices and social media platforms, are changing the societal landscape across which public opinion researchers operate. The ways in which people both access and share information about opinions, attitudes, and behaviors have gone through perhaps a greater transformation in the last decade than in any previous point in history and this trend appears likely to continue.*

(2014: 3)

Strengths and Weaknesses of Survey Research

Regardless of the specific method used, surveys—like other modes of observation in social research—have special strengths and weaknesses. You should keep these in mind when determining whether a survey is appropriate for your research goals.

Surveys are particularly useful in describing the characteristics of a large population. A carefully selected probability sample in combination with a standardized questionnaire offers the possibility of making refined descriptive assertions about a student body, a city, a nation, or any other large population. Surveys determine unemployment rates, voting intentions, and the like with uncanny accuracy. Although the examination of official documents—such as marriage, birth, or death records—can provide equal accuracy for a few topics, no other method of observation can provide this general capability.

Surveys—especially self-administered ones—make large samples feasible. Surveys of 2,000 respondents are not unusual. A large number of cases is very important for both descriptive and explanatory analyses, especially wherever several variables are to be analyzed simultaneously.

In one sense, surveys are flexible. They allow you to ask many questions on a given topic, giving you considerable flexibility in your analyses. Whereas an experimental design may require you to commit yourself in advance to a particular operational definition of a concept, surveys let you develop operational definitions from actual observations.

Finally, standardized questionnaires have an important strength with regard to measurement generally. Earlier chapters have discussed the ambiguous nature of most concepts: Ultimately, they have no real meanings. One person's religiosity is quite different from another's. Although you must be able to define concepts in the ways most relevant to your research goals, you may not find it easy to apply the same definitions uniformly to all subjects. The survey researcher is bound to this requirement, however, by having to ask exactly the same questions of all subjects and having to impute the same intent to all respondents giving a particular response.

Survey research also has several weaknesses. First, the requirement of standardization often seems to result in the fitting of round pegs into square holes. Standardized questionnaire items often represent the least common denominator in assessing people's attitudes, orientations, circumstances, and experiences. By designing questions that will be at least minimally appropriate for all respondents, you may miss what is most appropriate for many of them. In this sense, surveys often appear superficial in their coverage of complex topics. Although sophisticated analyses can partly offset this problem, it's inherent in survey research.

Similarly, survey research can seldom deal with the context of social life. Although questionnaires can provide information in this area, the survey researcher rarely develops a feel for the total life situation in which respondents are thinking and acting that, say, the participant-observer can (see Chapter 10).

In many ways, surveys are inflexible. Studies involving direct observation can be modified as field conditions warrant, but surveys typically require that an initial study design remain unchanged throughout. As a field researcher, for example, you can become aware of an important new variable operating in the phenomenon you're studying and begin making careful observations of it. The survey researcher would probably be unaware of the new variable's importance and could do nothing about it in any event.

Finally, surveys are subject to the artificiality mentioned in connection with experiments. Finding out that a person gives conservative answers in a questionnaire does not necessarily mean the person is conservative; finding out that a person gives prejudiced answers in a questionnaire does not necessarily mean the person is prejudiced. This shortcoming is especially salient in the realm of action. Surveys cannot measure social action; they can only collect self-reports of recalled past action or of prospective or hypothetical action.

The problem of artificiality has two aspects. First, the topic of study may not be amenable to measurement through questionnaires. Second, the act of studying that topic—an attitude, for example—may affect it. A survey respondent may have given no thought to whether the governor should be impeached until asked for his or her opinion by an interviewer. He or she may form an opinion on the spot.

Survey research is generally weak on validity and strong on reliability. In comparison with field research, for example, the artificiality of the survey format puts a strain on validity. As an illustration, people's opinions on issues seldom take the form of strongly agreeing, agreeing, disagreeing, or strongly disagreeing with a specific statement. Their survey responses in such cases must be regarded as approximate indicators of what the researchers had in mind when they framed the questions. This comment, however, needs to be held in the context of earlier discussions of the ambiguity of validity itself. To say something is a valid or an invalid measure assumes the existence of a "real" definition of what's being measured, and many scholars now reject that assumption.

Reliability is a clearer matter. By presenting all subjects with a standardized stimulus, survey research goes a long way toward eliminating unreliability in observations made by the researcher. Moreover, careful wording of the questions can also significantly reduce the subject's own unreliability.

As with all methods of observation, a full awareness of the inherent or probable weaknesses of survey research can partially resolve them in some cases. Ultimately, though, researchers find the safest ground when they employ several research methods in studying a given topic.

Secondary Analysis

As a mode of observation, survey research involves the following steps: (1) questionnaire construction, (2) sample selection, and (3) data collection, through either interviewing or self-administered questionnaires. As you've gathered, surveys are usually major undertakings. It's not unusual for a large-scale survey to take several months or even more than a year to progress from conceptualization to data in hand. (Smaller-scale surveys can, of course, be done more quickly.) Through a method called *secondary analysis,* however, researchers can pursue their particular social research interests—analyzing survey data from, say, a national sample of 2,000 respondents—while avoiding the enormous expenditure of time and money such a survey entails.

Secondary analysis is a form of research in which the data collected and processed by one researcher are reanalyzed—often for a different purpose—by another. Beginning in the 1960s, survey researchers became aware of the potential value in archiving survey data for analysis by scholars who had nothing to do with the survey design and data collection. Even when one researcher had conducted a survey and analyzed the data, those same data could be further analyzed by others who had slightly different interests. Thus, if you were interested in the relationship between political views and attitudes toward gender equality, you could examine that research question through the analysis of any data set that happened to contain questions relating to those two variables.

The initial data archives were very much like book libraries, with a couple of differences. First, instead of books, the data archives contained data sets: first as punched cards, then as magnetic tapes. Today they're typically contained on computer drives, portable electronic storage devices, or online servers. Second, whereas you're expected to return books to a conventional library, you can keep the data obtained from a data archive.

The best-known current example of secondary analysis is the General Social Survey (GSS). The National Opinion Research Center (NORC) at the University of Chicago conducts this major national survey, currently every other year, to collect data on a large number of social science variables. These surveys are conducted precisely for the purpose of making data available to scholars at little or no cost and are supported by a combination of private and government funding. Recall that the GSS was created by James A. Davis in 1972; it's currently directed by Davis, Tom W. Smith, and Peter V. Marsden. Their considerable ongoing efforts make an unusual contribution to social science research and to education in social science.

Numerous other resources are available for identifying and acquiring survey data for secondary analysis. The Roper Center for Public Opinion Research at the University of Connecticut is one excellent resource. The center also publishes the journal *Public Perspective,* which is focused on public opinion polling. Polling the Nations is an online repository for thousands of polls conducted in the United States and seventy other nations. A paid subscription allows users to obtain specific data results from studies they specify, rather than obtaining whole studies.

Whereas secondary analysis typically involves obtaining a data set and undertaking an extensive investigation, I would like you to consider another approach as well. Often you can do limited analyses for just a little investment of time. Let's say you're writing a term paper about the impact of religion in contemporary

---

secondary analysis A form of research in which the data collected and processed by one researcher are reanalyzed—often for a different purpose—by another. This is especially appropriate in the case of survey data. Data archives are repositories or libraries for the storage and distribution of data for secondary analysis.

American life. You want to comment on the role of the Roman Catholic Church in the debate over abortion. Although you might get away with an offhand, unsubstantiated assertion, imagine how much more powerful your paper would be if you supported your position with additional information. Refer to Figure 9-7 and follow the steps listed there to learn how to access data relevant to this research topic.

1. Go to the SDA analysis site at http://sda .berkeley.edu/sdaweb/analysis/?dataset=gss14, which was introduced in Chapter 1.
2. In the codebook listing on the left of the figure, locate the survey items dealing with abortion—by selecting the appropriate entry under the "Controversial Social Issues" listing.
3. For purposes of this illustration, let's see how members of the different religious groups responded in regard to women being allowed to choose an abortion "for any reason."
4. Type the name of this item—ABANY—where I've entered it in Figure 9-7.
5. Scroll through the variables in the column to the left and locate the variable label for Religious Affiliation. That will give you the variable label, RELIG, and you can enter it where I have in Figure 9-7. To see current opinions on this topic, specify the year 2014 as I've done in the figure.
6. Click the button labeled "Run the Table." You should be rewarded with the table shown in Figure 9-8.

The results of your analysis, shown in Figure 9-8, may surprise you. Whereas Catholics are less supportive of abortion (39.8 percent) than are Jews (82.5 percent) and

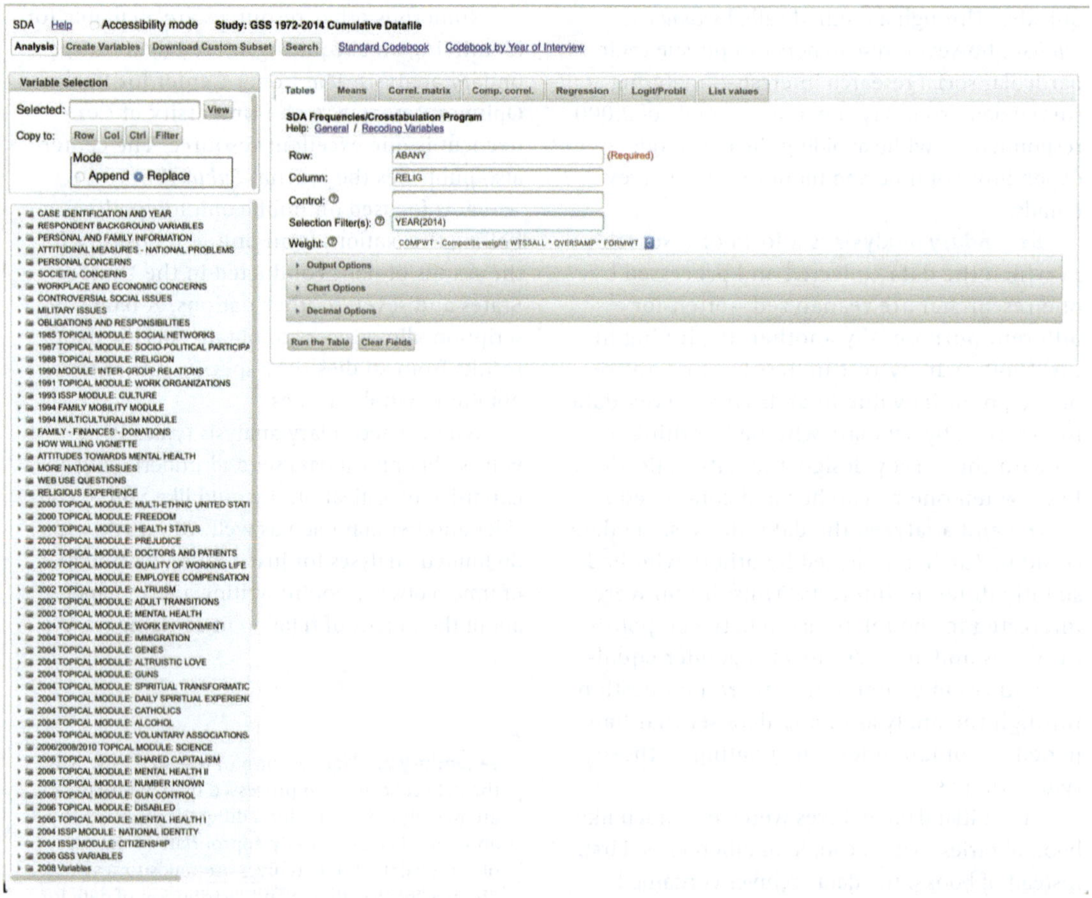

FIGURE 9-7

Requesting an Analysis of GSS Data.

SDA 4.1: Tables

General Social Survey Cumulative Datafile 1972-2014

Oct 18, 2019 (Fri 06:55 AM PDT)

		Variables			
Role	**Name**	**Label**	**Range**	**MD**	**Dataset**
Row	ABANY	ABORTION IF WOMAN WANTS FOR ANY REASON	1-2	0,8,9	1
Column	RELIG	RS RELIGIOUS PREFERENCE	1-13	0,98,99	1
Weight	COMPWT	Composite weight = WTSSALL * OVERSAMP * FORMWT	.1913-11.1261		1
Filter	YEAR(2014)	GSS YEAR FOR THIS RESPONDENT	1972-2014		1

Frequency Distribution

RELIG

Cells contain: -Column percent -Weighted N	1 PROTESTANT	2 CATHOLIC	3 JEWISH	4 NONE	5 OTHER	6 BUDDHISM	7 HINDUISM	9 MOSLEM/ISLAM	10 ORTHODOX-CHRISTIAN	11 CHRISTIAN	12 NATIVE AMERICAN	13 INTER-NONDENOMINATIONAL	*ROW TOTAL*
ABANY 1: YES	35.1 / 250.0	39.8 / 162.4	82.5 / 20.1	66.0 / 231.3	56.2 / 11.8	63.8 / 14.9	100.0 / 8.7	19.8 / 1.3	71.7 / 7.5	39.6 / 34.9	100.0 / .9	33.3 / .4	45.0 / 744.4
2: NO	64.9 / 461.7	60.2 / 246.1	17.5 / 4.3	34.0 / 118.9	43.8 / 9.2	36.2 / 8.4	.0 / .0	80.2 / 5.4	28.3 / 3.0	60.4 / 53.2	.0 / .0	66.7 / .9	55.0 / 911.2
COL TOTAL	100.0 / 711.7	100.0 / 408.6	100.0 / 24.4	100.0 / 350.2	100.0 / 21.0	100.0 / 23.3	100.0 / 8.7	100.0 / 6.8	100.0 / 10.5	100.0 / 88.2	100.0 / .9	100.0 / 1.3	100.0 / 1,655.5

Color coding: <-2.0 <-1.0 <0.0 >0.0 >1.0 >2.0 Z

N in each cell: Smaller than expected   Larger than expected

FIGURE 9-8

Impact of Religion on Attitude toward Abortion.

those with no religion (66 percent), they are slightly more supportive than are Protestants (35.1 percent).

Imagine a term paper that says, "Whereas the Roman Catholic Church has taken a strong, official position on abortion, many Catholics do not necessarily agree, as shown in Table . . . ." Moreover, this might be just the beginning of an analysis that looks a bit more deeply into the matter, as described in Chapter 14, on quantitative analysis.

The key advantage of secondary analysis is that it's cheaper and faster than doing original surveys, and, depending on who did the original survey, you may benefit from the work of topflight professionals. The ease of secondary analysis has also enhanced the possibility of meta-analysis, in which a researcher brings together a body of past research on a particular topic. To gain confidence in your understanding of the relationship between religion and abortion, for example, you could go beyond the GSS to analyze similar data collected in dozens or even hundreds of other studies.

There are disadvantages inherent in secondary analysis, however. The key problem involves the recurrent question of validity. When one researcher collects data for one particular purpose, you have no assurance that those data will be appropriate for your research interests. Typically, you'll find that the original researcher asked a question that "comes close" to measuring what you're interested in, but you'll wish the question had been asked just a little differently—or that another, related question had also been asked. For example, you may want to study the degree of people's religiosity, but the survey data available to you only asked about attendance at worship services. Your quandary, then, is whether the question that was asked provides a valid measure of the variable you want to analyze.

Nevertheless, secondary analysis can be immensely useful. Moreover, it illustrates once again the range of possibilities available in finding the answers to questions about social life. Although no single method unlocks all puzzles, there is no limit to the ways in which you can find out about things. And when you zero in on an issue from several independent directions, you gain that much more expertise.

I've discussed secondary analysis in this chapter on survey research because it's the type of analysis most associated with the technique. However, the reanalysis of social research data is not limited to data collected in surveys. For example, when Dana Berkowitz and Maura Ryan (2011) set out to study how lesbian and gay parents deal with gender socialization for the adoptive children, they were able to find the qualitative data they needed in

What do you think?...Revisited

Professional survey research has been damaged in recent years by the actions of telemarketers who pretend they're conducting surveys. Potential respondents sometimes refuse to participate in a legitimate survey because they suspect that it's really a sales call.

Here are a few ways to determine the legitimacy of a survey.

1. Ask who is conducting the survey. The caller probably said something about it quickly at the outset, but ask him or her to repeat the information so you can write it down.

2. Ask for the telephone number of a supervisor or manager so you can call the people running the survey.

3. Ask whether the call involves a sales solicitation.

If the caller is reluctant to answer any of these questions, assume that the call is not a professional survey. You may respond as you deem appropriate. One possibility: Let them recite their entire sales pitch, and then ask, "Could you repeat that?"

---

the qualitative interview records of two earlier studies of lesbian and gay parents. In taking a step beyond utilizing secondary studies, Nigel Fielding (2004) has examined the possibilities for the archiving and reanalysis of qualitative data as well.

Ethics and Survey Research

Survey research almost always involves a request that people provide us with information about themselves that is not readily available. Sometimes, we ask for information (such as attitudes and behaviors) that would be embarrassing to the respondents if that information became publicly known. In some cases, such revelations could result in the loss of a job or a marriage. Hence, maintaining the norm of confidentiality,

mentioned earlier in the book, is particularly important in survey research.

Another ethical concern relates to the possibility of psychological injury that can be done to respondents. Even if the information they provide is kept confidential, simply forcing them to think about some matters can be upsetting. Imagine asking people for their attitudes toward suicide when one of them has recently experienced the suicide of a family member or close friend. Or asking people to report on their attitudes about different racial groups, which may cause them to reflect on whether they may be racist or at least appear as such to the interviewers. The possibilities for harming survey respondents are endless. Although this fact should not prevent you from doing surveys, it should increase your efforts to avoid the problem wherever possible.

MAIN POINTS

Introduction

- Survey research, a popular social research method, is the administration of questionnaires to a sample of respondents selected from some population.

Topics Appropriate for Survey Research

- Survey research is especially appropriate for making descriptive studies of large populations; survey data may be used for explanatory purposes as well.

Guidelines for Asking Questions

- Questionnaires provide a method of collecting data by (1) asking people questions or (2) asking them to agree or disagree with statements

representing different points of view. Questions can be open-ended (wherein respondents supply their own answers) or closed-ended (wherein they select from a list of provided answers).

- Items in a questionnaire should observe several guidelines: (1) The items must be clear and precise; (2) the items should ask only about one thing (double-barreled questions should be avoided); (3) respondents must be competent to answer the item; (4) respondents must be willing to answer the item; (5) questions should be relevant to the respondent; (6) items should ordinarily be short; (7) negative items should be avoided so as not to confuse respondents; (8) the items should be worded so as to avoid biasing responses.

Questionnaire Construction

- The format of a questionnaire can influence the quality of data collected.
- A clear format for contingency questions is necessary to ensure that the respondents answer all the questions intended for them.
- The matrix question is an efficient format for presenting several items sharing the same response categories.
- The order of items in a questionnaire can influence the responses given.
- Clear instructions are important for getting appropriate responses in a questionnaire.
- Questionnaires should be pretested before being administered to the study sample.
- Questionnaires can be administered in four basic ways: as self-administered paper questionnaires, face-to-face interviews, telephone surveys, or online surveys.

Self-Administered Questionnaires

- It's generally advisable to plan follow-up mailings in the case of self-administered questionnaires, and send new questionnaires to respondents who fail to respond to the initial appeal. Properly monitoring questionnaire returns provides a good guide for determining when a follow-up mailing is appropriate.
- It is important to monitor the rate of responses to a mail survey and do follow-up mailings if necessary to achieve a high response rate.

Interview Surveys

- The essential characteristic of interviewers is that they be neutral; their presence in the data-collection process must not have any effect on the responses given to questionnaire items.
- Interviewers must be carefully trained to be familiar with the questionnaire, to follow the question wording and question order exactly, and to record responses exactly as they are given.
- Interviewers can use probes to elicit an elaboration on an incomplete or ambiguous response. Probes should be neutral. Ideally, all interviewers should use the same probes.
- Interview surveys often require an additional level of personnel: supervisors to train and manage the interviewers.

Telephone Surveys

- Telephone surveys can be cheaper and more efficient than face-to-face interviews, and they can permit greater control over data collection. The development of computer-assisted telephone interviewing (CATI) is especially promising.
- Robopolls are computer-executed phone surveys that involve no human interviewers.

Online Surveys

- Online surveys have a similar response rate to that for mailed surveys and are less costly.
- The main problem with online surveys is representativeness.
- It is important to tailor questionnaires to accommodate whatever devices (e.g., laptop, tablet, smartphone) that respondents will use.

Mixed-Mode Surveys

- Increasingly, researchers are combining different modes (e.g., interview, online) in a single survey, benefiting from the advantages of each.
- The best combination of modes will depend on the respondents and the topic and format of the survey.

Comparison of the Different Survey Methods

- The advantages of a self-administered questionnaire over an interview survey are economy, speed, lack of interviewer bias, and the possibility of anonymity and privacy to encourage candid responses on sensitive issues.
- The advantages of an interview survey over a self-administered questionnaire are fewer incomplete questionnaires and fewer misunderstood questions, generally higher completion rates, and greater flexibility in terms of sampling and special observations.
- The principal advantages of telephone surveys over face-to-face interviews are the savings in cost and time. There is also a safety factor: In-person interviewers might be required to conduct surveys in high-crime areas, which could pose a safety issue; telephone interviews, by design, eliminate such risks.
- Online surveys have many of the strengths and weaknesses of mail surveys. Although they are cheaper to conduct, it can be difficult to ensure that the respondents represent a more-general population.

Strengths and Weaknesses of Survey Research

- Survey research in general offers advantages in terms of economy, the amount of data that can be collected, and the chance to sample a large population. The standardization of the data collected represents another special strength of survey research.
- Survey research has the weaknesses of being somewhat artificial, potentially superficial, and relatively inflexible. It's difficult to use surveys to gain a full sense of social processes in their natural settings. In general, survey research is comparatively weak on validity and strong on reliability.

Secondary Analysis

- Secondary analysis provides social researchers with an important option for "collecting" data cheaply and easily but at a potential cost in validity.

Ethics and Survey Research

- Surveys often ask for private information, which researchers must keep confidential.

- Because asking questions can cause psychological discomfort or harm to respondents, the researcher should minimize this risk.

KEY TERMS

bias	probe
closed-ended questions	questionnaire
contingency question	respondent
interview	response rate
open-ended questions	secondary analysis

PROPOSING SOCIAL RESEARCH: SURVEY RESEARCH

If you're planning a survey, you'll already have described the sampling technique you plan to employ, and your discussion of measurement will have included at least portions of your questionnaire. At this point you need to describe the type of survey you plan to conduct: self-administered, telephone, face-to-face, or Internet. Whichever you choose, there will be numerous logistical details to spell out in the proposal. How will you deal with nonrespondents, for example? Will you have a follow-up mailing in a self-administered questionnaire, follow-up calls in a telephone survey, and so forth? Will you have a target completion rate?

In the case of interview surveys, it will be appropriate to say something about the way you'll select and train the interviewers. You should also say something about the time frame within which the survey will be conducted.

REVIEW QUESTIONS

1. What closed-ended questions could you construct from each of the following open-ended questions?

   a. What was your family's total income last year?

   b. How do you feel about the space program?

   c. How important is religion in your life?

   d. What was your main reason for attending college?

   e. What do you feel is the biggest problem facing your community?

2. What are the main advantages and disadvantages of conducting surveys over the Internet?

3. A newspaper headline proclaims, "Most Americans Oppose Abortion, According to New Survey." What methodological details do you want to know about the survey to help you interpret the results?

4. Look at your appearance right now. What aspects of your attire, hairstyle, or hygiene might create a problem if you were interviewing a general cross section of the public?

CHAPTER 10

Qualitative Field Research

Qualitative field research enables researchers to observe social life in its natural habitat: to go where the action is and watch. This type of research can produce a richer understanding of many social phenomena than can be achieved through other observational methods, provided that the researcher observes in a deliberate, well-planned, and active way.

iStock.com/PaulaConnelly

Introduction

Several chapters ago, I suggested that you've been doing social research all your life. This idea should become even clearer as we turn to what probably seems like the most obvious method of making observations: qualitative field research. In a sense, we do field research whenever we observe or participate in social behavior and try to understand it, whether in a college classroom, in a doctor's waiting room, or on an airplane. Whenever we report our observations to others, we're reporting our field research efforts.

Such research is at once very old and very new in social science, stretching at least from the nineteenth century studies of preliterate societies, through firsthand examinations of urban community life in the "Chicago School" of the 1930s and 1940s, to contemporary observations of Web chat-room interactions. Many of the techniques discussed in this chapter have been used by social researchers for centuries. Within the social sciences, anthropologists are especially associated with this method and have contributed greatly to its development as a scientific technique. Moreover, many people who might not, strictly speaking, be regarded as social science researchers employ something similar to field research. Welfare department caseworkers are one example; newspaper reporters are another.

To take this last example further, consider that interviewing is a technique common to both journalism and sociology. A journalist uses the data to report a subject's attitude, belief, or experience—that's usually it. Sociologists, on the other hand, treat an interview as data that need to be analyzed in depth; their ultimate goal is to understand social life in the context of theory, using established analytic techniques. Although sociology and journalism use similar techniques, the two disciplines view and use data differently.

Although many of the techniques involved in field research are "natural" activities, they are also skills to be learned and honed. This chapter discusses these skills in some detail, examining some of the major paradigms of field research and describing some of specific techniques that make scientific field research more useful than the casual observation that we all engage in.

I use the term *qualitative field research* to distinguish this type of observation method from methods designed to produce data appropriate for quantitative (statistical) analysis. Thus, surveys provide data from which to calculate the percentage unemployed in a population, mean incomes, and so forth. Field research more typically yields qualitative data: observations not easily reduced to numbers. Thus, for example, a field researcher may note the "paternalistic demeanor" of leaders at a political rally or the "defensive evasions" of a city official at a public hearing without trying to express either the paternalism or the defensiveness as numerical quantities or degrees. Although field research can be used to collect quantitative data—for example, noting the number of interactions of various specified types within a field setting— typically, field research is qualitative.

Field observation also differs from some other models of observation in that it's not just a data-collecting activity. Frequently, perhaps typically,

What do you think?

The impact of the observer is a fundamental issue in social research. If you participate in the events you're studying, observing them directly, up close and personal, won't your presence change things? How can you observe something as though you aren't actually there observing it? In other words, how close is too close?

See the *What do you think? ... Revisited* box toward the end of the chapter.

Earl Babbie

it's a theory-generating activity as well. As a field researcher, you'll seldom approach your task with precisely defined hypotheses to be tested. More typically, you'll attempt to make sense out of an ongoing process that cannot be predicted in advance—making initial observations, developing tentative general conclusions that suggest particular types of further observations, making those observations and thereby revising your conclusions, and so forth. In short, the alternation of induction and deduction discussed in Part 1 of this book is perhaps nowhere more evident and essential than in good field research. For expository purposes, however, this chapter focuses primarily on some of the theoretical foundations of field research and on techniques of data collection. Chapter 13 discusses how to analyze qualitative data. Keep in mind that the types of methods researchers use depend in part on the specific research questions they want to answer.

Topics Appropriate for Field Research

One of the key strengths of field research is how comprehensive a perspective it can give to researchers. By going directly to the social phenomenon under study and observing it as completely as possible, researchers can develop a deeper and fuller understanding of it. As such, this mode of observation is especially, though not exclusively, appropriate for research topics and social studies that appear to defy simple quantification. Field researchers may recognize several nuances of attitude and behavior that might escape researchers using other methods.

Field research is well suited to the study of social processes over time. Thus, the field researcher might be in a position to examine the rumblings and final explosion of a riot as the events actually occur rather than afterward in a reconstruction of the events.

Finally, field research is especially appropriate for the study of attitudes and behaviors that are best understood within their natural setting, as opposed to the somewhat artificial settings of experiments and surveys. For example, field research provides a superior method for studying the dynamics of religious conversion at a revival meeting, just as a statistical analysis of membership rolls would be a better way of discovering whether men or women were more likely to convert.

Or consider the insightful study of high school culture by Murray Milner Jr., appropriately entitled, *Freaks, Geeks, and Cool Kids* (2004). Milner was interested in exploring two questions: (1) Why do teenagers behave in the ways they do? and (2) How do their behaviors fit into the structure of the larger society?

Perhaps you can relate personally to one of the key starting points in Milner's study of teenage life: the feeling that they are largely powerless in many aspects of their lives.

> They must attend school for most of the day and they have only very limited influence on what happens there. They are pressured to learn complex and esoteric knowledge like algebra, chemistry, and European history, which rarely has immediate relevance to their day-to-day lives.
>
> (2004: 4)

Milner goes on to identify one area where teenagers exercise a special kind of power:

They do, however, have one crucial kind of power: the power to create an informal social world in which they evaluate one another. That is, they can and do create their own status systems—usually based on criteria that are quite different from those promoted by parents or teachers.

(2004: 4)

Status systems constitute a central concept for social scientists. Milner's expertise on the Indian caste system figured into his examination and understanding of high school youth culture.

Other good places to apply field research methods include campus demonstrations, courtroom proceedings, labor negotiations, public hearings, or similar events taking place within a relatively limited area and time. Researchers can combine several such observations in a more comprehensive examination over time and space.

Racism is a common topic for social research, ranging from mild prejudices to extreme and often violent race hatred. Peter Simi and Robert Futrell (2015) sought to understand the latter: white Aryan supremacists and neo-Nazis. They chose to do so with direct observation, interviews, and content analysis between 1996 and 2014.

The authors point out that only a few decades ago, white supremacists felt free to express their views openly in many parts of the country—especially in the Deep South. As American sensibilities regarding race have shifted, white supremacist advocacy has required the creation of "free spaces." This was a key concept, essential to finding relevant subjects for study.

Aryan free spaces may take the form of ordinary and benign settings and activities, but the content of the talk, rituals, and symbolism is anchored in white power ideology. For instance, most Aryan homes do not stand out as dens of hatred to neighbors or casual passersby. Outwardly, they tend to blend into their neighborhoods, apartment buildings, and communities. Inside, however, swastikas decorate the walls, white power literature lines the bookshelves, family pictures are full of Aryan symbolism, and mealtime prayers stress white power visions. Aryan homes are refuges from the mainstream world where members escape into a context defined by their white power beliefs.

(2015: 2)

The researchers conducted 222 interviews by phone and in person with 128 current and past leaders and followers in the white power movement. They describe their sample as purposive and snowball, methods you will recall from Chapter 7 of this book. They describe the data-collection process as anything but simple.

Gaining access to Aryan free spaces was not easy. Our approach was time consuming, labor intensive, and emotionally draining as we tried to overcome our gut feelings of shock, revulsion, rage, and sadness at the things we saw and heard. Our research goal was to understand Aryans on their own terms in their natural settings. This required listening to them with the discipline to temper our reactions to what our subjects said. It meant repeatedly reading over Aryans' views about the world and taking those views seriously. It was necessary to exclude our own moral and ethical values and assumptions in order to understand and interpret the meaning of Aryans' point of view.

(2015: 7)

Simi and Futrell's experience was perhaps an extreme case of a common aspect of this research method. You must be willing to set aside your own opinions, values, and beliefs in order to elicit and understand the views of those you are studying.

In *Analyzing Social Settings*, John Lofland and colleagues (2006: 123–32) discuss several elements of social life appropriate for field research.

A. *Practices:* Various kinds of behavior, such as talking or reading a book
B. *Episodes:* A variety of events such as divorce, crime, and illness
C. *Encounters:* Two or more people meeting and interacting
D. *Roles and Social Types:* The analysis of the positions people occupy and the behavior associated with those positions—occupations, family roles, ethnic groups
E. *Social and Personal Relationships:* Behavior appropriate to pairs or sets of roles—mother–son relationships, friendships, and the like
F. *Groups and Cliques:* Small groups, such as friendship cliques, athletic teams, and work groups
G. *Organizations:* Formal organizations, such as hospitals or schools

H. *Settlements and Habitats:* Small-scale "societies" such as villages, ghettos, and neighborhoods, as opposed to large societies such as nations, which are difficult to study

I. *Subcultures and Lifestyles:* How large numbers of people adjust to life in groups such as a "ruling class" or an "urban underclass"

In all these social settings, field research can reveal things that would not otherwise be apparent. Here's a concrete example.

One issue I'm particularly interested in (Babbie 1985) is the nature of responsibility for public matters: Who's responsible for maintaining the things that we share? Who's responsible for keeping public spaces—parks, malls, buildings, and so on—clean? Who's responsible for seeing that broken street signs get fixed? Or if a strong wind knocks over garbage cans and rolls them around the street, who's responsible for getting them out of the street?

On the surface, the answers to these questions are pretty clear. We have formal and informal agreements in our society that assign responsibility for these activities. Government custodians are the ones who keep public places clean. Transportation department employees take care of the street signs, and perhaps the police deal with the garbage cans rolling around on a windy day. And when these responsibilities are not fulfilled, we tend to look for someone to blame.

What fascinates me is the extent to which the assignment of responsibility for public things to specific individuals not only relieves others of the responsibility but actually prohibits them from taking it on. It's my notion that it has become unacceptable for someone like you or me to take personal responsibility for public matters that haven't been assigned to us.

Let me illustrate what I mean. If you were walking through a public park and you threw down a bunch of trash, you'd discover that your action was unacceptable to those around you. People would glare at you, grumble to each other; perhaps someone would say something to you about it. Whatever the form, you'd be subjected to definite, negative sanctions for littering. Now, here's the irony. If you were walking through that same park, came across a bunch of trash that someone else had dropped, and cleaned it up, it's likely that your action would

also produce negative sanctions from those around you.

When I first began discussing this pattern with students, most felt the notion was absurd. Although littering would bring negative sanctions, cleaning up a public place would obviously bring positive ones: People would be pleased with us for doing it. Certainly, all my students said they would be pleased if someone cleaned up a public place. It seemed likely that everyone else would be pleased, too, if we asked them how they would react to someone's cleaning up litter in a public place or otherwise taking personal responsibility for fixing some social problem.

To settle the issue, I suggested that my students start fixing the public problems they came across in the course of their everyday activities. As they did so, I asked them to note the answers to two questions:

1. How did they feel while they were fixing a public problem they had not been assigned responsibility for?
2. How did others around them react?

My students picked up litter, fixed street signs, put knocked-over traffic cones back in place, cleaned and decorated communal lounges in their dorms, trimmed trees that blocked visibility at intersections, repaired public playground equipment, cleaned public restrooms, and took care of a hundred other public problems that weren't "their responsibility."

Most reported feeling very uncomfortable doing whatever they did. They felt foolish, goody-goody, conspicuous, and all the other feelings that usually keep us from performing these activities routinely. In almost every case, the reactions of those around them increased their discomfort. One student was removing a damaged and long-unused newspaper box from the bus stop, where it had been a problem for months, when the police arrived, having been summoned by a neighbor. Another student decided to clean out a clogged storm drain on his street and found himself being yelled at by a neighbor who insisted that the mess should be left for the street cleaners. Everyone who picked up litter was sneered at, laughed at, and generally put down. One young man was picking up litter scattered around a trash can when a passerby sneered, "Clumsy!" It became clear to us that there are only three acceptable explanations for picking up litter in a public place:

1. You did it and got caught—somebody forced you to clean up your mess.
2. You did it and felt guilty.
3. You're stealing litter.

In the normal course of life in the United States, it's simply not acceptable for people to take responsibility for public things.

Clearly, we could not have discovered the nature and strength of agreements about taking personal responsibility for public things except through field research. Social norms suggest that taking responsibility is a good thing, sometimes referred to as good citizenship. Asking people what they thought about taking responsibility would have produced a solid consensus that it was good. Only going out into life, doing it, and watching what happened gave us an accurate picture.

As an interesting footnote to this story, my students and I found that whenever people could get past their initial reactions and discover that the students were simply taking responsibility for fixing things for the sake of having them work, the passersby tended to assist. Although there are some very strong agreements making it "unsafe" to take responsibility for public things, the willingness of one person to rise above those agreements seemed to make it safe for others to do so, and they did.

Field research is not to be confused with journalism. Social scientists and journalists may use similar techniques, but they have quite a different relationship to data. For instance, individual interviewing is a common technique in journalism and sociology; nevertheless, sociologists are not simply concerned with reporting about a subject's attitude, belief, or experience. A sociologist's goal is to treat an interview as data that need to be analyzed to understand social life more generally.

Anne Byrne, John Canavan, and Michelle Millar (2009) suggest that this distinction can go even deeper. The voice-centered relational communications method focuses on who is speaking in and who is listening, taking account of the difference between the two actors and the impact of those differences. Often, the listener is the researcher. This approach shows up during interviews and during the analysis of transcripts.

The authors say about their study that dealt with Irish teenagers:

> *One of the challenging dimensions of the work was that it brought us face to face with a reality that demanded that we act with or on behalf of the teenagers. The work of relationship building is time consuming and energy sapping—many research approaches do not require the formation of "caring relationships" with the researched. Building relationships between old and young, from different class backgrounds and diverse life experiences require a sustained and shared commitment from all.*

(2009: 75)

Two important aspects of qualitative research need to be stressed. First, a wide range of studies fall under the umbrella "qualitative field research." As we'll see in this chapter, various epistemologies within different paradigms have quite different approaches to basic questions such as "What are data?" "How should we collect data?" and "How should we analyze data?" Second, we should remember that the questions we want to answer in our research determine the types of methods we need to use. A question such as "How do women construct their everyday lives in order to perform their roles as mothers, partners, and breadwinners?" could be addressed by in-depth interviews and direct observations. The assessment of advertising campaigns might profit from focus-group discussions. In most cases, we'll find that researchers have alternative methods to choose from.

In summary, field research offers the advantage of probing social life in its natural habitat. Although some things can be studied adequately in questionnaires or in the laboratory, others cannot. And direct observation in the field lets researchers observe subtle communications and other events that might not be anticipated or measured otherwise.

Special Considerations in Qualitative Field Research

Every research method presents specific issues and concerns, and qualitative field research is no exception. When you use field research methods, you're confronted with decisions about the role you'll play as an observer and your relations with the people you're observing. Let's examine some of the issues involved in these decisions.

The Various Roles of the Observer

In field research, observers can play any of several roles, including participating in what they want to observe (this was the situation of the students who fixed public things). In this chapter, I've used the term *field research* rather than the frequently used term *participant-observation*, because field researchers need not always participate in what they're studying, though they usually will study it directly at the scene of the action. As Catherine Marshall and Gretchen Rossman point out:

> The researcher may plan a role that entails vary-ing degrees of "participantness"—that is, the degree of actual participation in daily life. At one extreme is the full participant, who goes about ordinary life in a role or set of roles constructed in the setting. At the other extreme is the complete observer, who engages not at all in social interac-tion and may even shun involvement in the world being studied. And, of course, all possible comple-mentary mixes along the continuum are available to the researcher.

(1995: 60)

The complete participant, in this sense, may be a genuine participant in what he or she is studying (for example, a participant in a campus demonstration) or may pretend to be a genuine participant. In any event, if you're acting as a complete participant, you would let people see you only as a participant, not as a researcher. For instance, if you're studying a group made up of uneducated and inarticulate people, it would not be appropriate for you to talk and act like a university professor or student.

This type of research introduces an ethical issue, one on which social researchers them-selves are divided. Is it ethical to deceive the people you're studying, in the hope that they will confide in you in ways that they would not if you identified yourself as a researcher? Do the potential benefits to be gained from the research offset such considerations? Although many professional associations have addressed this issue, the norms to be followed remain somewhat ambiguous when applied to specific situations.

Related to this ethical consideration is a scientific one. No researcher deceives his or her subjects solely for the purpose of deception. Rather, it's done in the belief that the data will be more valid and reliable—that the subjects will be more natural and honest if they do not know the researcher is doing a research project. If the people being studied know they're being studied, they might modify their behavior in a variety of ways. This problem is known as **reactivity**.

First, they might expel the researcher. Second, they might modify their speech and behavior to appear more respectable than would otherwise be the case. Third, the social process itself might be radically changed. Students making plans to burn down the uni-versity administration building, for example, might give up the plan altogether once they learn that one of their group is a social scientist conducting a research project.

On the other side of the coin, if you're a complete participant, you may affect what you're studying. To play the role of participant, you must participate. Yet, your participation may significantly affect the social process you're studying. Suppose, for example, that you're asked for your ideas about what the group should do next. No matter what you say, you will affect the process in some fashion. If the group follows your suggestion, your influence on the process is obvious. If the group decides not to follow your suggestion, the process whereby the suggestion is rejected may affect what happens next. Finally, if you indicate that you just don't know what should be done next, you may be adding to a general feeling of uncertainty and indecisiveness in the group.

Ultimately, anything the participant-observer does or does not do will have some effect on what's being observed; it's simply inevitable. More seriously, the research effort may have an important effect on what happens. There is no complete protection against this effect, though

---

reactivity The problem that the subjects of social research may react to the fact of being studied, thus altering their behavior from what it would have been normally.

sensitivity to the issue may provide a partial protection. (This influence, called the Hawthorne effect, was discussed more fully in Chapter 8.)

Because of these ethical and scientific considerations, the field researcher frequently chooses a different role from that of complete participant. You could participate fully with the group under study but make it clear that you were also undertaking research. As a member of the volleyball team, for example, you might use your position to launch a study in the sociology of sports, letting your teammates know what you're doing. There are dangers in this role also, however. The people being studied may shift much of their attention to the research project rather than focus on the natural social process, so that the process being observed is no longer typical. Or, conversely, you yourself may come to identify too much with the interests and viewpoints of the participants. You may begin to "go native" and lose much of your scientific detachment.

At the other extreme, the complete observer studies a social process without becoming a part of it in any way. Quite possibly, because of the researcher's unobtrusiveness, the subjects of study might not realize they're being studied. Sitting at a bus stop to observe people jaywalking nearby is one example. Although the complete observer is less likely to affect what's being studied and less likely to "go native" than the complete participant, she or he is also less likely to develop a full appreciation of what's being studied. Observations may be more sketchy and transitory.

Fred Davis (1973) characterizes the extreme roles that observers might play as "the Martian" and "the Convert." The latter involves delving deeper and deeper into the phenomenon under study, running the risk of "going native." We'll examine this risk further in the next section.

To appreciate the "Martian" approach, imagine that you were sent to observe some newfound life on Mars. Probably you would feel yourself inescapably separate from the Martians. Some social scientists adopt this degree of separation when observing cultures or social classes different from their own.

Marshall and Rossman (1995: 60–61) also note that the researcher can vary the amount of time spent in the setting being observed; that is, researchers can be a full-time presence on the

scene or just show up now and then. Moreover, they can focus their attention on a limited aspect of the social setting or seek to observe all of it—framing an appropriate role to match their aims.

When Jeffrey Kidder set out to study the culture of bike messengers in New York City, he found it appropriate to identify his research role to some of those he observed but not others:

> While I did have an academic motivation in working as a messenger, it should be made clear that my participation within the messenger world was neither forced nor faked. To the contrary, my lifelong interest in bicycles and alternative transportation melded seamlessly with the messenger lifestyle.
>
> During the course of my fieldwork, most of the messengers with whom I came in contact were unaware of my research; this was a matter of necessity. In New York City, a messenger crosses paths with hundreds of messengers a day. The numerous individuals that helped form my understandings of messenger style could not all be approached to sign consent forms. Messengers with whom I had reoccurring contact were informed of my sociological interest.
>
> (2005: 349)

Different situations ultimately require different roles for the researcher. Unfortunately, there are no clear guidelines for making this choice—you must rely on your understanding of the situation and your own good judgment. In making your decision, however, you must be guided by both methodological and ethical considerations. Because these often conflict, your decision will frequently be difficult, and you may sometimes find that your role limits your study.

Relations to Subjects

Having introduced the different roles field researchers might play in connection with their observations, we now focus more specifically on how researchers may relate to the subjects of their study and to the subjects' points of view.

We've already noted the possibility of pretending to occupy social statuses we don't really occupy. Consider now how you would think and feel in such a situation.

Suppose you've decided to study a religious cult that has enrolled many people in your neighborhood. You might study the group by joining it or pretending to join it. Take a moment

Field research is a hands-on process, which involves going to the scene of the action and checking it out.

to ask yourself what the difference is between "really" joining and "pretending" to join. The main difference is whether or not you actually take on the beliefs, attitudes, and other points of view shared by the "real" members. If the cult members believe that Jesus will come next Thursday night to destroy the world and save the members of the cult, do you believe it or do you simply pretend to believe it?

Traditionally, social scientists have tended to emphasize the importance of *objectivity* in such matters. In this example, that injunction would be to avoid getting swept up in the beliefs of the group. Without denying the advantages associated with such objectivity, social scientists today also recognize the benefits gained by immersing themselves in the points of view they're studying, what Lofland and associates (2006: 70) refer to as "selective competence" or "insider knowledge, skill, or understanding." Ultimately, you will not be able to fully understand the thoughts and actions of the cult members unless you can adopt their points of view as true—at least temporarily. To fully appreciate the phenomenon you've set out to study, you need to believe that Jesus is coming Thursday night. In some settings, this can also help you gain rapport with your subjects, a topic we'll return to later in this chapter.

Adopting an alien point of view is an uncomfortable prospect for most people. It can be hard enough merely to learn about views that seem strange to you; you may sometimes find it hard just to tolerate certain views, but to take them on as your own is ten times worse. Robert Bellah

(1970, 1974) has offered the term *symbolic realism* to indicate the need for social researchers to treat the beliefs they study as worthy of respect rather than as objects of ridicule. The difficulty of this task led William Shaffir and Robert Stebbins to conclude that "fieldwork must certainly rank with the more disagreeable activities that humanity has fashioned for itself" (1991: 1).

There is, of course, a danger in adopting the points of view of the people you're studying. When you abandon your objectivity in favor of adopting such views, you lose the possibility of seeing and understanding the phenomenon within frames of reference unavailable to your subjects. On the one hand, accepting the belief that the world will end Thursday night allows you to appreciate aspects of that belief available only to believers; stepping outside that view, however, makes it possible for you to consider some reasons why people might adopt such a view. You may discover that some did so as a consequence of personal trauma (such as unemployment or divorce), whereas others were brought into the fold through their participation in particular social networks (for example, their whole bowling team joined the cult). Notice that the cult members might disagree with those "objective" explanations, and you might not come up with them because of the extent that you operated legitimately within the group's views.

Anthropologists sometimes use the term *emic perspective* in reference to taking on the point of view of those being studied. In contrast, the *etic perspective* maintains a distance from the native point of view in the interest of achieving more objectivity.

The apparent dilemma here is that both of these postures offer important advantages but also seem mutually exclusive. In fact, you can assume both postures. Sometimes you can simply shift viewpoints at will. When appropriate, you can fully assume the beliefs of the cult; later, you can step outside those beliefs (more accurately, you can step inside the viewpoints associated with social science). As you become more adept at this kind of research, you may come to hold contradictory viewpoints simultaneously, rather than switching back and forth.

During my study of trance channeling— people who allow spirits to occupy their bodies and speak through them—I found that I could participate fully in channeling sessions without becoming alienated from conventional social science. Rather than "believing" in the reality of channeling, I found it possible to suspend disbeliefs in that realm: neither believing it to be genuine (like most of the other participants) nor disbelieving it (like most scientists). Put differently, I was open to either possibility. Notice how this differs from our normal need to "know" whether such things are legitimate or not.

Social researchers often refer to the concerns just discussed as a matter of *reflexivity*, in the sense of things acting on themselves. Thus, your own characteristics can affect what you see and how you interpret it. The issue is broader than that, however, and applies to the subjects as well as to the researcher. Imagine yourself interviewing a homeless person (1) on the street, (2) in a homeless shelter, or (3) in a social welfare office. The research setting could affect the person's responses. In other words, you might get different results because of where you conducted the interview. Moreover, you might act differently as a researcher in those different settings. If you reflect on this issue, you'll be able to identify other aspects of the research encounter that complicate the task of "simply observing what's so."

The problem we've just been discussing could be seen as psychological, occurring mostly inside the researchers' or subjects' heads. There is a corresponding problem at a social level, however. When you become deeply involved in the lives of the people you're studying, you're likely to be moved by their personal problems and crises. Imagine, for example, that one of the cult members becomes ill and needs a ride to the hospital. Should you provide transportation? Sure. Suppose someone wants to borrow money to buy a stereo. Should you loan it? Probably not. Suppose they need the money for food?

There are no black-and-white rules for resolving situations such as these, but you should realize that you'll need to deal with them regardless of whether or not you reveal that you're a researcher. Such problems do not tend to arise in other types of research—surveys and experiments, for example—but they are part and parcel of field research.

Caroline Knowles (2006) raises a somewhat different issue with regard to your relationship with subjects in the field. In her interview study of British expatriates living in Hong Kong, she noticed that some were particularly difficult for her to deal with because of the attitudes they expressed, their rude interaction styles, and/or the nature of the relationship she was establishing with them. When she found herself writing research notes explaining why the project would not profit from her interviewing them further, she forced herself to look more deeply into the interactional dynamics in question—with an emphasis on her side of the relationships. She examined *why* certain informants made her uncomfortable and then pressed through the discomfort to continue interviewing. In the end, she gained a much deeper understanding of her subjects than would have been possible if she had limited herself to those who were cooperative and nice.

Similarly, Alex Broom, Kelly Hand, and Philip Tovey (2009) examined the impact of gender when conducting in-depth interviews with cancer patients. Did it matter whether patients were interviewed by someone of the same or of the opposite gender? As you've probably guessed, it did. Prostate cancer patients were more graphic in describing their experiences to a male interviewer than to a female one. Similarly, a breast cancer patient's feelings of disfigurement, for example, were expressed differently to male and female interviewers. Before you decide that gender matching is the best policy, notice that a cancer patient's overall experience includes both same-gender and opposite-gender relations. The point is that the gender of the interviewer can affect interviews in particular ways, not that one type of interviewer is necessarily better than another. As I've said frequently in this book, the impact of the observer, whether in experiments, surveys, or field research, often cannot be avoided, but we can be conscious of it and take it into account in understanding what we've observed.

This discussion of the field researcher's relations to subjects flies in the face of the conventional view of "scientific objectivity." Before concluding this section, let's take the issue one step further.

In the conventional view of science, there are implicit differences of power and status separating the researcher from the subjects of research. When we discussed experimental designs in Chapter 8, for example, who was in charge was obvious: The experimenter organized things and told the subjects what to do. Often the experimenter was the only person who even knew what the research was really about. Something similar might be said about survey research. The person running the survey designs the questions, decides who will be selected for questioning, and is responsible for making sense out of the data collected.

Sociologists often look at these sorts of relationships as power or status relationships. In experimental and survey designs, the researcher clearly has more power and a higher status than do the people being studied. The researchers have a special knowledge that the subjects do not enjoy. They are not so crude as to say they are superior to their subjects, but there is a sense in which that's implicitly assumed. (Notice that there is a similar, implicit assumption about the writers and readers of textbooks.)

In field research, such assumptions can be problematic. When the early European anthropologists set out to study what were originally called "primitive" societies, there was no question that the anthropologists knew best. Whereas the natives "believed" in witchcraft, for example, the anthropologists "knew" it wasn't really true. And whereas the natives said some of their rituals would appease the gods, the anthropologists explained that the "real" functions of these rituals were the creation of social identity, the establishment of group solidarity, and so on.

Giampietro Gobo (2011) sensitizes us to the cultural roots (and limits) of commonly used social research techniques. For the most part, those roots are embedded in American and European cultures. By contrast, he notes that:

> If one wanted to undertake an ethnomethodological study in the Maya villages of Zincatela and San Juan Chamula, in Chiapas (Mexico), one should be careful about the following: no photos are allowed in the church or of people in this area due to the belief that photos will capture the soul, and if someone leaves the village with the photo then their soul goes with them.

(2011: 417)

Gobo also points out (2011: 424) some of the implicit assumptions that lie behind the use of standardized or unstructured interviews, such as "the sense that it is acceptable to have conversations with strangers (interviewers)," "the

ability on the part of the interviewee to speak for himself, and an awareness of himself as an autonomous and independent individual," and "experience in giving information in telephone interviews without seeing the face of the interviewer." These are foreign concepts for many who were raised outside Europe and America.

The more social researchers have gone into the field to study their fellow humans face to face, however, the more they have become conscious of these implicit assumptions about researcher superiority, and the more they have considered alternatives. As we turn now to the various paradigms of field research, we'll see some of the ways in which that ongoing concern has worked itself out.

Some Qualitative Field Research Paradigms

Although I've described field research as simply going where the action is and observing it, there are actually many different approaches to this research method. This section examines several field research paradigms: naturalism, ethnomethodology, grounded theory, case studies and the extended case method, institutional ethnography, and participatory action research. Although this survey won't exhaust the variations on the method, it should give you a broad appreciation of the possibilities.

There are no specific methods attached to each of these paradigms. You could do ethnomethodology or institutional ethnography by analyzing court hearings or conducting group interviews, for example. The important distinctions of this section are *epistemological*, that is, having to do with what data mean, regardless of how they were collected (see Chapter 1).

Naturalism

Naturalism is an old tradition in qualitative research. The earliest field researchers operated on the positivist assumption that social reality

---

naturalism An approach to field research based on the assumption that an objective social reality exists and can be observed and reported accurately.

was "out there," ready to be naturally observed and reported by the researcher as it "really is" (Gubrium and Holstein 1997). This tradition started in the 1930s and 1940s at the University of Chicago's sociology department, whose faculty and students fanned out across the city to observe and understand local neighborhoods and communities. The researchers of that era and their research approach are now often referred to as the Chicago School.

One of the earliest and best-known studies that illustrates this research tradition is William Foote Whyte's ethnography of Cornerville, an Italian American neighborhood, in his book *Street Corner Society* (1943). An **ethnography** is a study that focuses on detailed and accurate description rather than explanation. Like other naturalists, Whyte believed that in order to fully learn about social life on the streets, he needed to become more of an insider. He made contact with "Doc," his key informant, who appeared to be one of the street-gang leaders. Doc let Whyte enter his world, and Whyte got to participate in the activities of the people of Cornerville. His study offered something that surveys could not: a richly detailed picture of life among the Italian immigrants of Cornerville.

An important feature of Whyte's study is that he reported the reality of the people of Cornerville on their terms. The naturalist approach is based on telling "their" stories the way they "really are," not the way the ethnographer understands "them." The narratives collected by Whyte are taken at face value as the social "truth" of the Cornerville residents.

About 40 years later, David Snow and Leon Anderson (1987) conducted exploratory field research into the lives of homeless people in Austin, Texas. Their main task was to understand how the homeless construct and negotiate their identity while knowing that the society they live in attaches a stigma to homelessness. Snow and Anderson believed that, to achieve this goal, the collection of data had to arise naturally. Like Whyte in *Street Corner Society*, they found some

key informants whom they followed in their everyday journeys, such as at their day-labor pickup sites or under bridges. Snow and Anderson chose to memorize the conversations they participated in or the "talks" that homeless people had with each other. At the end of the day, the two researchers debriefed and wrote detailed field notes about all the "talks" they encountered. They also recorded in-depth interviews with their key informants to improve the accuracy of their research notes.

Snow and Anderson reported "hanging out" with homeless people over the course of 12 months for a total of 405 hours in 24 different settings. Out of these rich data, they identified three related patterns in homeless people's conversations. First, the homeless showed an attempt to "distance" themselves from other homeless people, from the low-status job they currently had, or from the Salvation Army they depended on. Second, they "embraced" their street-life identity—their group membership or a certain belief about why they are homeless. Third, they told "fictive stories" that always contrasted with their everyday life. For example, they would often say that they were making much more money than they really were, or even that they were "going to be rich."

Richard Mitchell (2002) offers another, timely illustration of the power of ethnographic reporting. Recent U.S. history has raised the specter of violence from secretive survivalist groups, dramatized by the 1992 siege at Ruby Ridge, Idaho, which left the wife and son of the white supremacist Randy Weaver dead; the 1993 shootout with David Koresh and his Branch Davidians in Waco, Texas; and Timothy McVeigh's 1995 bombing, which left 168 dead under the rubble of the nine-story Murrah Federal Building in Oklahoma City.

Mitchell describes a variety of survivalist individuals and groups, seeking to understand their reasoning, their plans, and the threat they may pose for the rest of us. Although he finds the survivalists disillusioned with and uncertain about the future of U.S. society, most are more interested in creating alternative lives and cultures for themselves than in blowing up the mainstream society. That's not to suggest none of the survivalists is a threat, but Mitchell's examination moves beyond the McVeighs, Koreshes,

ethnography A report on social life that focuses on detailed and accurate description rather than explanation.

and Weavers to draw a broader picture of the whole phenomenon.

While ethnographers seek to discover and understand the patterns of living among those they are studying, Mitchell Duneier (1999) has warned against what he calls the "ethnographic fallacy." This refers to an overgeneralization and oversimplification of the patterns observed. Despite the existence of patterns within groups, there is also diversity, and you need to be wary of broad assertions suggesting that "the poor," "the French," or "cheerleaders" act or think in certain ways as though all members of the group do so.

In Chapter 9, we saw how the Internet is affecting survey research. In a qualitative study of menopausal symptoms, Im and Chee (2012) report:

> *Some participants … tended not to discuss their menopausal symptom experience with strangers, including researchers, which might have prevented their participation in face-to-face discussions….*

The solution was to invite subjects to participate in online discussions. Each participant was asked to choose a pseudonym to use, and their reluctance to discuss their symptoms decreased.

Whereas this chapter aims at introducing you to some of the different approaches available to you in qualitative field research, please realize that this discussion of ethnography merely sketches some of the many avenues social researchers have established. If you're interested in this general approach, you might want to explore the idea of *virtual ethnography*, which uses ethnographic techniques for inquiry into online social networks or communities. Or, in a different direction, *autoethnography* intentionally assumes a personal stance, breaking with the general proscription against the researcher getting involved at that level. Lest it should seem a simple and/or trivial undertaking, you might look at "Easier Said Than Done: Writing an Autoethnography," Sarah Wall's 2008 article on the subject.

You can learn more about these variants on ethnography by searching the Web or your campus library. A later section of this chapter examines *institutional ethnography*, which links individuals and organizations.

Ethnomethodology

Ethnomethodology, which I introduced as a research paradigm in Chapter 2, is a unique approach to qualitative field research. It has its roots in the philosophical tradition of *phenomenology*, which can explain why ethnomethodologists are skeptical about the way people report their experience of reality (Gubrium and Holstein 1997). Alfred Schutz (1967, 1970), who introduced phenomenology, argued that reality was socially constructed rather than being "out there" for us to observe. People describe their world not "as it is" but "as they make sense of it." Thus, phenomenologists would argue that Whyte's street-corner men were describing their gang life as it made sense to them. Their reports, however, would not tell us how and why it made sense to them. For this reason, researchers cannot rely on their subjects' stories to depict social realities accurately.

Whereas traditional ethnographers believe in immersing themselves in a particular culture and reporting their informants' stories as if they represent reality, phenomenologists see a need to "make sense" out of the informants' perceptions of the world. Following in this tradition, some field researchers have tried to devise techniques that reveal how people make sense of their everyday world. As we saw in Chapter 2, the sociologist Harold Garfinkel suggested that researchers "break the rules" so that people's taken-for-granted expectations would become apparent. This is the technique that Garfinkel called *ethnomethodology*.

Garfinkel became known for engaging his students in performing a series of "breaching experiments" designed to break away from the ordinary (Heritage 1984). For instance, Garfinkel (1967) asked his students to do a "conversation clarification experiment." Students were told to engage in an ordinary conversation with an

---

ethnomethodology An approach to the study of social life that focuses on the discovery of implicit—usually unspoken—assumptions and agreements; this method often involves the intentional breaking of agreements as a way of revealing their existence.

acquaintance or a friend and to ask for clarification about any of this person's statements. Through this technique, they uncovered elements of conversation that are normally taken for granted. Here are two examples of what Garfinkel's students reported (1967: 42):

> *Case 1*
> The subject (S) was telling the experimenter (E), a member of the subject's car pool, about having had a flat tire while going to work the previous day.
> (S) I had a flat tire.
> (E) What do you mean, you had a flat tire?
> She appeared momentarily stunned. Then she answered in a hostile way: "What do you mean, 'What do you mean?' A flat tire is a flat tire. That is what I meant. Nothing special. What a crazy question."

> *Case 6*
> The victim waved his hand cheerily.
> (S) "How are you?"
> (E) "How I am in regard of what? My health, my finances, my school work, my peace of mind, my...?"
> (S) (Red in the face and suddenly out of control.) "Look I was just trying to be polite. Frankly, I don't give a damn how you are."

By setting aside or "bracketing" their expectations from these everyday conversations, the experimenters made visible the subtleties of mundane interactions. For example, although "How are you?" has many possible meanings, none of us have any trouble knowing what it means in casual interactions, as the unsuspecting subject revealed in his final comment.

Ethnomethodologists, then, are not simply interested in subjects' perceptions of the world. In these cases, we could imagine that the subjects may have thought that the experimenters were rude, stupid, or arrogant. The conversation itself, not the informants, becomes the object of ethnomethodological studies. In general, in ethnomethodology the focus is on the "underlying patterns" of interactions that regulate our everyday lives.

Ethnomethodologists believe that researchers who use a naturalistic analysis "[lose] the ability to analyze the commonsense world and its culture if [they use] analytical tools and insights that are themselves part of the world or culture being studied" (Gubrium and Holstein 1997: 43). D. L. Wieder has provided an excellent example of how much a naturalistic approach differs from an ethnomethodological approach (Gubrium and Holstein 1997). In his study, *Language and Social Reality: The Case of Telling the Convict Code* (1988), Wieder started to approach convicts in a halfway house in a traditional ethnographic style: He was going to become an insider by befriending the inmates and by conducting participant observations. He took careful notes and recorded interactions among inmates and between inmates and staff. His first concern was to describe the life of the convicts of the halfway house the way it "really was" for them. Wieder's observations allowed him to report on a "convict code" that he thought was the source of the deviant behavior expressed by the inmates toward the staff. This code, which consisted of a series of rules such as "Don't kiss ass," "Don't snitch," and "Don't trust the staff," was followed by the inmates who interfered with the staff members' attempts to help them make the transition from prison to the community.

It became obvious to Wieder that the code was more than an explanation for the convicts' deviant behavior; it was a "method of moral persuasion and justification" (Wieder 1988: 175). At this point he changed his naturalistic approach to an ethnomethodological one. Recall that whereas naturalistic field researchers aim to understand social life as the participants understand it, ethnomethodologists are more intent on identifying the methods through which understanding occurs. In the case of the convict code, Wieder came to see that convicts used the code to make sense of their own interactions with other convicts and with the staff. The ethnography of the halfway house thus shifted to an ethnography of the code. For instance, the convicts would say, "You know I won't snitch," referring to the code as a way to justify their refusal to answer Wieder's question (1988: 168). According to Wieder, the code "operated as a device for stopping or changing the topic of conversation" (1988: 175). Even the staff would refer to the code to justify their reluctance to help the convicts. Although the code was something that constrained behavior, it also functioned as a tool for the control of interactions.

Grounded Theory

Grounded theory originated from the collaboration of Barney Glaser and Anselm Strauss, sociologists who brought together two main traditions of research: positivism and interactionism. Essentially, **grounded theory** is the attempt to derive theories from an analysis of the patterns, themes, and common categories discovered in observational data. The first major presentation of this method can be found in Glaser and Strauss's book, *The Discovery of Grounded Theory* (1967). Grounded theory can be described as an approach that attempts to combine a naturalist approach with a positivist concern for a "systematic set of procedures" in doing qualitative research.

In *Basics of Qualitative Research: Techniques and Procedures for Developing Grounded Theory* (1998: 43–46), Anselm Strauss and Juliet Corbin suggested that grounded theory allows the researcher to be scientific and creative at the same time, as long as the researcher follows these guidelines:

- *Think comparatively:* The authors suggest that researchers must compare numerous incidents as a way of avoiding the biases that can arise from interpretations of initial observations.
- *Obtain multiple viewpoints:* In part, this refers to the different points of view of participants in the events under study, but Strauss and Corbin suggest that different observational techniques may also provide a variety of viewpoints.
- *Periodically step back:* As data accumulate, you'll begin to frame interpretations about what's going on, and it's important to keep checking your data against those interpretations. As Strauss and Corbin (1998: 45) say, "The data themselves do not lie."
- *Maintain an attitude of skepticism:* As you begin to interpret the data, you should regard all those interpretations as provisional, using new observations to *test* those interpretations, not just confirm them.
- *Follow the research procedures:* Grounded theory allows for flexibility in data collection as theories evolve, but Strauss and Corbin (1998: 46) stress that three techniques are essential: "making comparisons, asking questions, and sampling."

Although a review of the existing research literature is an early step in most research methods—toward the goal of learning what has been learned so far—the initial development of grounded theory argued specifically *against* this practice. Glaser and Strauss feared that grounding yourself in what has already been learned would create expectations that would constrain

Anselm Strauss was one of the most important figures in the rebirth of qualitative research and a founder of the Grounded Theory Method.

Lynda Koolish

what your research would look at, what you would see, and how you would interpret the data. Rather, they urged that categories and patterns be allowed to emerge from the new data. Once the data collection was ended, a review of the literature would provide another opportunity for comparison.

As Ciarán Dunne (2011) has detailed, this position regarding literature review is one now debated among grounded theorists. Strauss, himself, modified his own opinion on the matter, acknowledging that it might be appropriate to do an early literature review in some cases. Glaser, on the other hand, has maintained his original position. In any event, the initial concerns about preconceptions possibly blinding researchers to new discoveries can apply to any research method you might choose.

Grounded theory emphasizes research procedures. In particular, systematic coding is important for achieving validity and reliability in the data analysis. Because of this somewhat positivistic view of data, grounded theorists are quite open to

grounded theory An inductive approach to the study of social life that attempts to generate a theory from the constant comparing of unfolding observations. This differs greatly from hypothesis testing, in which theory is used to generate hypotheses to be tested through observations.

the use of qualitative studies in conjunction with quantitative ones. Here are two examples of the implementation of this approach.

Studying Academic Change

Clifton Conrad's (1978) study of academic change in universities is an early example of the grounded theory approach. Conrad hoped to uncover the major sources of changes in academic curricula and at the same time understand the process of change. Using the grounded theory idea of *theoretical sampling*—whereby groups or institutions are selected on the basis of their theoretical relevance—Conrad chose four universities for the purpose of his study. In two, the main vehicle of change was the formal curriculum committee; in the other two, the vehicle was an ad hoc group.

Conrad explained, step by step, the advantage of using the grounded theory approach in building his theory of academic change. He described the process of systematically coding data in order to create categories that must "emerge" from the data, and then assessing the fitness of these categories in relation to one other. Going continuously from data to theory and theory to data allowed him to reassess the validity of his initial conclusions about academic change.

For instance, it first seemed that academic change was caused mainly by an administrator who was pushing for it. By reexamining the data and looking for more-plausible explanations, Conrad found the pressure of interest groups to be a more convincing source of change. The emergence of these interest groups actually allowed the administrator to become an agent of change.

Assessing how data from each of the two types of universities fit with the other helped refine theory building. This refinement process stands in contrast to a naturalist approach, in which the process of theory building would have stopped with Conrad's first interpretation.

Conrad concluded that changes in university curricula are based on the following process: Conflict and interest groups emerge because of internal and external social structural forces; they push for administrative intervention and recommendation to make changes in the current academic program; these changes are then made by the most powerful decision-making body.

Shopping Romania

Much has been written about large-scale changes caused by the shift from socialism to capitalism in the former USSR and its Eastern European allies. Patrick Jobes and his colleagues (1997) wanted to learn about the transition on a smaller scale among average Romanians. They focused on the task of shopping.

Noting that shopping is normally thought of as a routine, relatively rational activity, the researchers suggested that it could become a social problem in a radically changing economy. They used the Grounded Theory Method to examine Romanian shopping as a social problem, looking for the ways in which ordinary people solved the problem.

Their first task was to learn something about how Romanians perceived and understood the task of shopping. The researchers—participants in a social problems class—began by interviewing forty shoppers and asking whether they had experienced problems in connection with shopping and what actions they had taken to cope with those problems.

Once the initial interviews were completed, the researchers reviewed their data, looking for categories of responses—the shoppers' most common problems and solutions. One of the most common problems was a lack of money. This led to the researchers' first working hypothesis: The "socio-economic position of shoppers would be associated with how they perceived problems and sought solutions" (Jobes et al. 1997: 133). This and other hypotheses helped the researchers focus their attention on more-specific variables in subsequent interviewing.

As they continued, they also sought to interview other types of shoppers. When they interviewed students, for example, they discovered that different types of shoppers were concerned with different kinds of goods, which in turn affected the problems faced and the solutions tried.

As additional hypotheses were developed in response to the continued interviewing, the researchers began to develop a more or less standardized set of questions to ask shoppers. Initially, all the questions were open-ended, but they eventually developed closed-ended items as well.

This study illustrates the key, inductive principles of grounded theory: Data are collected in the absence of hypotheses. The initial data are used to determine the key variables as perceived by those being studied, and hypotheses about relationships among the variables are similarly derived from the data collected. Continuing data collection yields refined understanding and, in turn, sharpens the focus of data collection itself.

Case Studies and the Extended Case Method

Social researchers often speak of **case studies**, which focus attention on one or a few instances of some social phenomenon, such as a village, a family, or a juvenile gang. As Charles Ragin and Howard Becker (1992) point out, there is little consensus on what constitutes a *case*, and the term is used broadly. The "case" being studied, for example, might be a period of time rather than a particular group of people. The limitation of attention to a particular instance of something is the essential characteristic of the case study.

The chief purpose of a case study can be descriptive, as when an anthropologist describes the culture of a preliterate tribe. Or the in-depth study of a particular case can yield explanatory insights, as when the community researchers Robert and Helen Lynd (1929, 1937) and W. Lloyd Warner (1949) sought to understand the structure and process of social stratification in small-town USA.

Case study researchers may seek only an idiographic understanding of the particular case under examination, or—as we've seen with grounded theory—case studies can form the basis for the development of more-general, nomothetic theories.

Racism is a persistent structural problem in America, manifest in many areas of life, including the criminal justice system. Racial discrimination in the justice system has been frequently documented statistically: minorities are more likely than whites to be arrested for a specific crime, more likely to be convicted, and likely to receive stiffer sentences. Some people are persuaded by statistical analyses, while others are not moved.

Nicole Gonzalez Van Cleave (2016) took a different approach. In her book, *Crook County*, she reported her experiences and observations during ten years in Chicago's Cook County courthouse, first as an employee and then as a researcher. She was able to put human faces on the statistics of prejudice and discrimination.

Her descriptive accounts detail the behavior and conversations of police officers, attorney, judges, and others involved in the often unjust criminal justice system. She wrote, for example, of the case of a black burglar who turned himself, remorseful for what he had done. The judge in his case set a very high bond amount. Yet the same judge, the same day, prescribed a much lower bail for a white burglar. Admittedly, many variables may be relevant to a particular case, but Van Cleave's research uncovered consistent patterns of discrimination.

Often the combination of quantitative and qualitative research on a particular topic is especially persuasive when the two methods yield the same conclusion.

Michael Burawoy and his colleagues (1991) have suggested a somewhat different relationship between case studies and theory. For them, the **extended case method** has the purpose of discovering flaws in, and then modifying, existing social theories. This approach differs importantly from some of the others already discussed.

Whereas traditional grounded theorists seek to enter the field with no preconceptions about what they'll find, Burawoy suggests just the opposite: to try "to lay out as coherently as possible what we expect to find in our site *before* entry" (1991: 9). Burawoy sees the extended case method as a way to rebuild or improve theory instead of approving or rejecting it. Thus, he looks for all the ways in which observations conflict with existing theories and what he calls "theoretical gaps and silences" (1991: 10). Burawoy and his colleagues' orientation to field research implies that knowing the literature beforehand is actually a must, whereas some grounded theorists would worry that knowing

case study The in-depth examination of a single instance of some social phenomenon, such as a village, a family, or a juvenile gang.

extended case method A technique developed by Michael Burawoy in which case study observations are used to discover flaws in, and to then improve, existing social theories.

what others have concluded might bias their observations and theories.

To illustrate the extended case method, I'll use two examples of studies by Burawoy's students.

Student–Teacher Negotiations

Leslie Hurst (1991) set out to study the patterns of interaction between teachers and students of a junior high school. She went into the field armed with existing, contradictory theories about the "official" functions of the school. Some theories suggested that the purpose of schools was to promote social mobility, whereas others suggested that schools mainly reproduced the status quo in the form of a stratified division of labor. The official roles assigned to teachers and students could be interpreted in terms of either view.

Hurst was struck, however, by the contrast between these theories and the types of interactions she observed in the classroom. In her own experiences as a student, teachers had total rights over the mind, body, and soul of their pupils. She observed something quite different at a school in a lower-middle-class neighborhood in Berkeley, California, where she volunteered as a tutor. She had access to several classrooms, the lunchroom, and the English department's meetings. She wrote field notes based on the negotiation interactions between students and teachers. She explained the nature of the student–teacher negotiations she witnessed by focusing on the separation of functions among the school, the teacher, and the family.

In Hurst's observation, the school fulfilled the function of controlling its students' "bodies"—for example, by regulating their general movements and activities within the school. The students' "minds" were to be shaped by the teacher, whereas students' families were held responsible for their "souls"; that is, families were expected to socialize students regarding personal values, attitudes, sense of property, and sense of decorum. When students don't come to school with these values in hand, the teacher, according to Hurst, "must first negotiate with the students some compromise on how the students will conduct themselves and on what will be considered classroom decorum" (1991: 185).

Hurst explained that the constant bargaining between teachers and students is an expression of the separation between "the body," which is the school's concern, and "the soul" as family domain. The teachers, who had limited sanctioning power to control their students' minds in the classroom, were using forms of negotiations with students so that they could "control...the student's body and sense of property" (1991: 185), or as Hurst defines it, "babysit" the student's body and soul.

Hurst says her view differs from the traditional sociological perspectives as follows:

> I do not approach schools with a futuristic eye. I do not see the school in terms of training, socializing, or slotting people into future hierarchies. To approach schools in this manner is to miss the negotiated, chaotic aspects of the classroom and educational experience. A futurist perspective tends to impose an order and purpose on the school experience, missing its day-to-day reality.
>
> (1991: 186)

In summary, what emerges from Hurst's study is an attempt to improve the traditional sociological understanding of education by adding the idea that classroom, school, and family have separate functions, which in turn can explain the emergence of "negotiated order" in the classroom.

The Fight against AIDS

Kathryn Fox (1991) set out to study an agency whose goal was to fight the AIDS epidemic by bringing condoms and bleach (for cleaning needles) to intravenous drug users. Her study offers a good example of finding the limitations of well-used models of theoretical explanation in the realm of understanding deviance—specifically, the "treatment model" that predicted that drug users would come to the clinic and ask for treatment. Fox's interactions with outreach workers—most of whom were part of the community of drug addicts or former prostitutes—contradicted that model.

To begin with, it was necessary to understand the drug users' subculture and to use that knowledge to devise more-realistic policies and programs. The target users had to be convinced, for example, that the program workers could be trusted, that they were really interested only in providing bleach and condoms. The target users needed to be sure that they were not going to be arrested.

Fox's field research didn't stop with an examination of the drug users. Fox also studied the agency workers, discovering that the outreach program meant different things to the research directors and the outreach workers. Some of the volunteers who were actually providing the bleach and condoms were frustrated about the minor changes they felt they could make. Many thought the program was just a bandage on the AIDS and drug-abuse problems. Some resented having to take field notes. Directors, on the other hand, needed reports and field notes so that they could validate their research in the eyes of the federal and state agencies that financed the project. Fox's study showed how the AIDS research project developed the bureaucratic inertia typical of established organizations: Its goal became that of sustaining itself.

Both of these studies illustrate how the extended case method can operate. The researcher enters the field with full knowledge of existing theories but aims to uncover contradictions that require the modification of those theories.

One criticism of the case study method is the limited generalizability of what is observed in a single instance of some phenomenon. This risk is reduced, however, when more than one case is studied in depth: the *comparative* case study method. You can find examples of this in the discussion of comparative and historical research methods in Chapter 11 of this book.

Institutional Ethnography

Institutional ethnography is an approach originally developed by Dorothy Smith (1978) to better understand women's everyday experiences by discovering the power relationships that shape those experiences. Today, this methodology has been extended to the ideologies that shape the experiences of any oppressed subjects.

Smith and other sociologists believe that if researchers ask women or other members of subordinated groups about "how things work," they can discover the institutional practices that shape their realities (Campbell 1998; Smith 1978). The goal of such inquiry is to uncover forms of oppression that more traditional types of research often overlook.

Courtesy of Dr. Dorothy E. Smith

Dorothy Smith, a pioneering social researcher and founder of institutional ethnography.

Smith's methodology is similar to ethnomethodology in the sense that the inquiry does not focus on the subjects themselves. The institutional ethnographer starts with the personal experiences of individuals but proceeds to uncover the institutional power relations that structure and govern those experiences. In this process, the researcher can reveal aspects of society that would have been missed by an inquiry that began with the official purposes of institutions.

This approach links the "microlevel" of everyday personal experiences with the "macrolevel" of institutions. As M. L. Campbell puts it:

> *Institutional ethnography, like other forms of ethnography, relies on interviewing, observations and document as data. Institutional ethnography departs from other ethnographic approaches by treating those data not as the topic or object of interest, but as "entry" into the social relations of the setting. The idea is to tap into people's expertise.*
>
> *(1998: 57)*

Here are two examples of this approach.

institutional ethnography A research technique in which the personal experiences of individuals are used to reveal power relationships and other characteristics of the institutions within which they operate.

Mothering, Schooling, and Child Development

Our first example of institutional ethnography is a study by Alison Griffith (1995), who collected data with Dorothy Smith on the relationship among mothering, schooling, and children's development. Griffith started by interviewing mothers from three cities of southern Ontario on their everyday work of creating a relationship between their families and the school. This was the starting point for other interviews with parents, teachers, school administrators, social workers, school psychologists, and central office administrators.

In her findings, Griffith explained how the discourse about mothering had shifted its focus over time from mother–child interactions to "child-centered" recommendations. She saw a distinct similarity in the discourse used by schools, the media (magazines and television programs), the state, and child development professionals.

Teachers and child development professionals saw the role of mothers in terms of a necessary collaboration between mothers and schools for the child's success not only in school but also in life. Because of unequal resources, all mothers do not participate in this discourse of "good" child development the same way. Griffith found that working-class mothers were perceived as weaker than middle-class mothers in the "stimulation" effort of schooling. Griffith argued that this child development discourse, embedded in the school institution, perpetuates the reproduction of class by making middle-class ideals for family–school relations the norm for everyone.

Compulsory Heterosexuality

The second illustration of institutional ethnography is taken from Didi Khayatt's (1995) study of the institutionalization of compulsory heterosexuality in schools and its effects on lesbian students. In 1990, Khayatt began her research by interviewing twelve Toronto lesbians, 15 to 24 years of age. Beginning with the young women's viewpoint, she then expanded her inquiry to other students, teachers, guidance counselors, and administrators.

Khayatt found that the school's administrative practices generated a compulsory heterosexuality, which produced a sense of marginality and vulnerability among lesbian students. For example, the school didn't punish harassment and name-calling against gay students. The issue of homosexuality was excluded from the curriculum lest it appear to students as an alternative to heterosexuality.

In both of the studies I've described, the inquiry began with the women's standpoint—mothers and lesbian students. However, instead of emphasizing the subjects' viewpoints, both analyses focused on the power relations that shaped these women's experiences and reality.

Participatory Action Research

Our final field research paradigm takes us further along in our earlier discussion of the status and power relationships linking researchers to the subjects of their research. Within the **participatory action research (PAR)** paradigm, the researcher's function is to serve as a resource to those being studied—typically, disadvantaged groups—as an opportunity for them to act effectively in their own interest. The disadvantaged subjects define their problems, define the remedies desired, and take the lead in designing the research that will help them realize their aims.

This approach began in Third World research development, but it spread quickly to Europe and North America (Gaventa 1991). It comes from a vivid critique of classical social science research. According to the PAR paradigm, traditional research is an "elitist model" (Whyte, Greenwood, and Lazes 1991) that reduces the "subjects" of research to "objects" of research. According to many advocates of this perspective, the distinction between the researcher and the researched should disappear. They argue that the subjects who will be affected by research should also be responsible for its design.

Implicit in this approach is the belief that research functions not only as a means of knowledge production but also as a "tool for the education and

---

participatory action research (PAR) An approach to social research in which the people being studied are given control over the purpose and procedures of the research; intended as a counter to the implicit view that researchers are superior to those they study.

development of consciousness as well as mobilization for action" (Gaventa 1991: 121–22). Advocates of PAR equate access to information with power and argue that this power has been kept in the hands of the dominant class, gender, ethnicity, or nation. Once people see themselves as researchers, they automatically regain power over knowledge.

By contrast, a participant in a Bangladeshi PAR study spoke of his community's previous experience with outside researchers (Datta 2015):

> There are a number of researchers who come to our community for research from various organizations (such as: the government, NGOs, research organizations, and universities) promising that they would bring many positive changes to us; but once they [outside researchers] are done with their research, they never come back. Even in most cases, we did not see our research results and/or research report. We do not know what information they have taken from us and for what.
>
> (592)

PAR poses a special challenge to researchers. On the one hand, a central intention is to empower participants to frame research relevant to their needs, as they define those needs. At the same time, the researcher brings special skills and insights that non-researchers lack. So, who should be in charge? Andrew Sense (2006: 1) suggests that this decision may have to be made in the moment, varying by particular circumstances: "Do I take the 'passenger' position on the bus or do I take the 'driver' seat and be a little more provocative to energise the session? My view at this moment is to judge it on the day."

Examples of the PAR approach include community power-structure research, corporate research, and "right-to-know" movements (Whyte, Greenwood, and Lazes 1991). Here are three more-detailed examples of research that used a PAR approach.

The Xerox Corporation

A participatory action research project took place at the Xerox Corporation at the instigation of leaders of both management and the union. Management's goal was to lower costs so that the company could thrive in an increasingly competitive market. The union suggested a somewhat broader scope: improving the quality of working life while lowering manufacturing costs and increasing productivity.

Company managers began by focusing attention on shop-level problems; they were less concerned with labor contracts and problematic managerial policies. At the time, management had a plan to start an "outsourcing" program that would lay off 180 workers, and the union had begun mobilizing to oppose the plan. Peter Lazes, a consultant hired by Xerox, spent the first month convincing management and the union to create a "cost study team" (CST) that included workers in the wire harness department.

Eight full-time workers were assigned to the CST for 6 months. Their task was to study the possibilities of making changes that would save the company $3.2 million and keep the 180 jobs. The team had access to all financial information and members were authorized to call on anyone within the company. This strategy allowed workers to make suggestions outside the realm usually available to them. According to Whyte and his colleagues, "reshaping the box enabled the CST to call upon management to explain and justify all staff services" (1991: 27). Because of the changes suggested by the CST and implemented by management, the company saved the targeted $3.2 million.

Management was so pleased by this result that it expanded the wire harness CST project to three other departments that were threatened by competition. Once again, management was happy about the money saved by the teams of workers.

The Xerox case study is interesting because it shows how the production of knowledge does not always have to be an elitist enterprise. The "experts" do not necessarily have to be the professionals. According to Whyte and his colleagues, "at Xerox, participatory action research created and guided a powerful process of organizational learning—a process whereby leaders of labor and management learned from each other and from the consultant/facilitator, while he learned from them" (1991: 30).

PAR and Welfare Policy

Participatory action research often involves poor people, as they are typically less able than others to influence the policies and actions that affect their lives. Bernita Quoss, Margaret Cooney, and Terri Longhurst (2000) report a research project involving welfare policy in Wyoming. University students, many of them welfare recipients, undertook research and lobbying efforts aimed at getting Wyoming to accept postsecondary education as "work" under the state's new welfare regulations.

This project began against the backdrop of the 1996 Personal Responsibility and Work Opportunity Reconciliation Act (PRWORA), which

> eliminated education waivers that had been available under the previous welfare law, the 1988 Family Support Act (FSA). These waivers had permitted eligible participants in the cash assistance AFDC program to attend college as an alternative to work training requirements. Empirical studies of welfare participants who received these waivers have provided evidence that education, in general, is the most effective way to stay out of poverty and achieve self-sufficiency.

> *(Quoss, Cooney, and Longhurst 2000: 47)*

The students began by establishing an organization, Empower, and making presentations on campus to enlist broad student and faculty support. They compiled existing research relevant to the issue and established relationships with members of the state legislature. By the time the 1997 legislative session opened, they were actively engaged in the process of modifying state welfare laws to take account of the shift in federal policy.

The students prepared and distributed fact sheets and other research reports that would be relevant to the legislators' deliberations. They attended committee meetings and lobbied legislators on a one-to-one basis. When erroneous or misleading data were introduced into the discussions, the student-researchers were on hand to point out the errors and offer corrections.

Ultimately, they were successful. Welfare recipients in Wyoming were allowed to pursue postsecondary education as an effective route out of poverty.

Participatory researchers at SaferSpaces (2014) in South Africa find the most effective way of identifying dangerous places in a community is to form a partnership with people who live there.

> Each participant draws a map of where they live and where they feel unsafe. By drawing these maps

> individually, participants have the opportunity to make a contribution and to express themselves. It also directs their thinking towards the physical environment within which they live and starts the process of thinking spatially.

> *(SaferSpaces, October 20, 2014)*

Small groups of participants then compare and combine their individual maps, creating a composite display of places people feel unsafe. Next, they visit the areas and make photographs to support analyses leading to solutions.

Some researchers speak of **emancipatory research**, which Ardha Danieli and Carol Woodhams (2005: 284) define as "first and foremost a process of producing knowledge which will be of benefit to oppressed people; a political outcome." Both qualitative and quantitative methods can be used to pursue this goal, but it goes well beyond simply learning what's so, even as seen from the subjects' point of view. The authors focus on the study of disability, and they note similarities in the development of emancipatory research and early feminist research.

Blocking a Demolition

In another example of researchers being directly involved in what they study, John Lofland (2003) detailed the demolition of a historic building in Davis, California, and community attempts to block the demolition. One thing that makes his book especially unusual is its reliance on photographs and facsimile news articles and government documents as raw data for the analysis (and for the reader): what Lofland refers to as "documentary sociology."

As Lofland explains, he was involved in the issue first as an active participant, joining with other community members in the attempt to block demolition of the Hotel Aggie (also known as the "Terminal Building" and "Terminal Hotel"). Built in 1924, in a town of around a thousand inhabitants, the hotel fell victim to population growth and urban development. Lofland says his role as researcher began on September 18, 2000, as the demolition of the building began.

emancipatory research Research conducted for the purpose of benefiting disadvantaged groups.

Before that, I was only and simply an involved citizen. Along with many other people, I was attempting to preserve the Terminal Building in some manner. This also explains why there are so few photographs in this book taken by me before that date, but many after that date. I had then begun seriously to document what was going on with a camera and field notes.

Therefore, questions of "informed consent" (now so often raised regarding research) were not pertinent before September 18. After that day, it was my practice to indicate to everyone I encountered that I was "writing a book" about the building.

(2003: 20)

Recall the discussion of informed consent in Chapter 3, a method of protecting research subjects. In this case, as Lofland notes elsewhere, explicit consent was not necessarily needed here because the behavior being studied was public. Still, his instincts as a social researcher were to ensure that he treat subjects appropriately.

One of Lofland's purposes was to study this failed attempt to secure "historic preservation" status for a building, thus providing useful information to activists in the future. This indicates that there can be many different forms of participatory action research.

At the same time, this is a valuable case for a study of research-methods, because Lofland, as the author of research-methods textbooks, is particularly sensitive to the methodological aspects of the study.

The depth and intensity of my involvement is a two-edged sword. On the one edge, my involvement provided me with a view closer than that of some other people. I was one type of "insider." This means I could gather data of certain sorts that were not available to the less involved.

On the other edge, my partisanship clearly poses the threat of bias. I have always been aware of this, and I have tried my best to correct for it. But, in the end, I cannot be the final judge. Each reader will have to form her or his own assessment. I can hope, however, that the "digital documentary" evidence I mention above helps the study tell itself, so to speak. It makes the reader less dependent on me than is the case with some other methods of representing what happened.

(2003: 20)

As you can see, the seemingly simple process of observing social action as it occurs has subtle though important variations. As we saw in Chapter 2, all our thoughts occur within, and are shaped by, paradigms, whether we're conscious of it or not. Qualitative field researchers have been unusually deliberate in framing a variety of paradigms to enrich the observation of social life.

The impact of researcher paradigms on the conduct of research is nowhere more explicitly recognized than in the case of *kaupapa Maori research,* a form of participatory action research developed within the indigenous Maori community of New Zealand. As Shayne Walker, Anaru Eketone, and Anita Gibbs (2006) report, an adherence to Maori culture shapes not only the purposes of research but its processes and practices as well. In a study of foster care, for example, the purpose of the study was established by those most directly concerned. The method of collecting data conformed to Maori practices, including public gatherings. The action implications derived from the analysis of data were tailored to Maori ways of doing things.

Conducting Qualitative Field Research

So far in this chapter we've considered the kinds of topics appropriate for qualitative field research, special considerations in doing this kind of research, and a sampling of paradigms that direct different types of research efforts. Along the way we've seen some examples that illustrate field research in action. To round out the picture, we turn now to specific ideas and techniques for conducting field research, beginning with how researchers prepare for work in the field.

Preparing for the Field

Suppose for the moment that you've decided to undertake field research into a campus political organization. Let's assume further that you're not a member of that group, that you do not know a great deal about it, and that you'll identify yourself to the participants as a researcher. To cover more of the activities common to research, we'll also assume that you've decided not to take a grounded theory or similar approach. This section will use this example and others to discuss some of the ways you might prepare yourself before undertaking direct observations.

As is true of most research methods (grounded theory being the most obvious exception), you would be well advised to begin with a search of the relevant literature, filling in your knowledge of the subject and learning what others have said about it (library research is discussed at length in Appendix A).

In the next phase of your research, you might wish to discuss the student political group with others who have already studied it or with anyone else likely to be familiar with it. In particular, you might find it useful to discuss the group with one or more informants (discussed in Chapter 7). Perhaps you have a friend who is a member, or you can meet someone who is. This aspect of your preparation is likely to be more effective if your relationship with the informant extends beyond your research role. In dealing with members of the group as informants, you should take care that your initial discussions do not compromise or limit later aspects of your research. Keep in mind that the impression you make on the informant, the role you establish for yourself, may carry over into your later efforts. For example, creating the initial impression that you may be an undercover FBI agent is unlikely to facilitate later observations of the group.

You should also be wary about the information you get from informants. Although they may have more-direct, personal knowledge of the subject under study than you do, what they "know" is probably a mixture of fact and point of view. Members of the political group in our example would be unlikely to provide completely unbiased information (as would members of opposing political groups). Before making your first contact with the student group, then, you should already be quite familiar with it, and you should understand its general philosophical context.

There are many ways to establish your initial contact with the people you plan to study. How you do it will depend, in part, on the role you intend to play. Especially if you decide to take on the role of complete participant, you must find a way to develop an identity with the people to be studied. If you wish to study people employed as dishwashers in a restaurant, the most direct method would be to get a job as a dishwasher. In the case of the student political group, you might simply join the group.

Many of the social processes appropriate for field research are open enough to make your contact with the people to be studied rather simple and straightforward. If you wish to observe a mass demonstration, just be there. If you wish to observe patterns in jaywalking, hang around busy streets.

Sometimes you will identify yourself as a researcher when you contact people you wish to study. You might contact a participant with whom you feel comfortable and gain that person's assistance. In studying a formal group, you might approach the group's leaders, or you may find that one of your informants can introduce you. Establishing rapport is particularly important in those cases. See "How to Do It: Establishing Rapport" for more details on this.

Whereas you'll probably have many options in making your initial contact with the group, realize that your choice can influence your subsequent observations. Suppose, for example, that you are studying a university and begin with high-level administrators. This choice is likely to have a couple of important consequences. First, your initial impressions of the university will be shaped to some extent by the administrators' views, which will differ significantly from those of students or faculty. This initial impression may influence the way you observe and interpret events subsequently—especially if you're unaware of the influence.

Second, if the administrators approve of your research project and encourage students and faculty to cooperate with you, the latter groups will probably look on you as somehow aligned with the administration, which can affect what they say to you. For example, faculty members might be reluctant to tell you about plans to organize through the teamsters' union.

In making a direct, formal contact with the people you want to study, you'll be required to give them some explanation of the purpose of your study. Here again, you face an ethical dilemma.

How to Do It

Establishing Rapport

In qualitative field research, it's almost always vital to establish rapport with those you're observing, especially if your observations include in-depth interviews and interactions. Rapport can be defined as an open and trusting relationship. But how do you establish that?

Let's assume that you've identified yourself as a researcher. You need to explain your research purpose in a nonthreatening way. Communicate that you're there to learn about them and understand them, not to judge them or cause them any problems. This will work best if you actually *have* a genuine interest in understanding the people you're observing and can communicate that interest to them. This gives them a sense of self-worth, which will increase their willingness to open up to you. *Pretending* to be interested is not the same as really being interested. In fact, if you're not interested in learning what things look like from the point of view of those you're observing, you might consider another activity and not waste their time or your own.

It follows that you'll function best as an attentive listener rather than a talker. You should not remain mute, of course, but you should talk primarily (1) to elicit more information from the other people or (2) to answer questions they may have about you and your research. While you don't have to agree with any points of view expressed by your subjects, you should never argue with them or try to change their minds. Keep reminding yourself that your genuine purpose is to understand their world and how it makes sense to them—whether it works for you or not. A little humility may help with this. You'll be able to hear and understand people better if you don't start with an implicit feeling of superiority to them.

Be relaxed and act appropriately for the setting. Some people are more formal or informal than others, and you'll do well to take on their general style and be comfortable with it. If you can get them to relax and enjoy the interaction, you'll have achieved the rapport you need, and you'll probably enjoy the interaction yourself.

Telling them the complete purpose of your research might eliminate their cooperation altogether or significantly affect their behavior. On the other hand, giving only what you believe would be an acceptable explanation might involve outright deception. Your decisions in this and other matters will probably be largely determined by the purpose of your study, the nature of what you're studying, the observations you wish to use, and similar factors, but ethical considerations must be taken into account as well.

Previous field research offers no fixed rule—methodological or ethical—to follow in this regard. Your appearance as a researcher, regardless of your stated purpose, may result in a warm welcome from people who are flattered that a scientist finds them important enough to study. Or it may result in your being totally ostracized—or worse. It probably wouldn't be a good idea, for example, to burst into a meeting of an organized crime syndicate and announce that you're writing a term paper on organized crime.

Qualitative Interviewing

In part, field research is a matter of going where the action is and simply watching and listening. As the baseball legend Yogi Berra said, "You can see a lot just by observing"—provided that you're paying attention. At the same time, as I've already indicated, field research can involve more-active inquiry. Sometimes it's appropriate to ask people questions and record their answers. Your on-the-spot observations of a full-blown riot will lack something if you don't know why people are rioting. Ask somebody.

When Cecilia Menjívar (2000) wanted to learn about the experiences of Salvadoran immigrants in San Francisco, she felt that in-depth interviews would be a useful technique, along with personal observations. Before she was done, she had discovered a much more complex system of social processes and structures than one would have imagined. It was important for new immigrants to have a support structure of family members already in the United States, as you might imagine, but Menjívar found that her interviewees were often reluctant to call on relatives for help. On the one hand, they might jeopardize those family members who were here without documentation and living in poverty. At the same time, asking for help would put the

---

rapport An open and trusting relationship; this is especially important in qualitative research between researchers and the people they're observing.

immigrants in debt to those helping them out. Menjívar also discovered that Salvadoran gender norms put women immigrants in an especially difficult situation because they were largely prohibited from seeking the help of men they weren't related to, lest they seem to obligate themselves sexually. These are the kinds of discoveries that can emerge from open-ended, in-depth interviewing.

We've already discussed interviewing in Chapter 9, and much of what was said there applies to qualitative field interviewing. The interviewing you'll do in connection with field observation, however, differs enough to demand separate treatment. In surveys, questionnaires are rigidly structured; however, less-structured interviews are more appropriate for field research. As Herbert and Riene Rubin describe the distinction, "Qualitative interviewing design is flexible, iterative, and continuous, rather than prepared in advance and locked in stone" (1995: 43). They elaborate in this way:

> Design in qualitative interviewing is iterative. That means that each time you repeat the basic process of gathering information, analyzing it, winnowing it, and testing it, you come closer to a clear and convincing model of the phenomenon you are studying....
>   The continuous nature of qualitative interviewing means that the questioning is redesigned throughout the project.
>
> (46, 47)

Unlike a survey, a **qualitative interview** is an interaction between an interviewer and a respondent in which the interviewer has a general plan of inquiry including the topics to be covered, but not a set of questions that must be asked with particular words and in a particular order. At the same time, the qualitative interviewer, like the survey interviewer, must be fully familiar with the questions to be asked. This allows the interview to proceed smoothly and naturally.

qualitative interview Contrasted with survey interviewing, the qualitative interview is based on a set of topics to be discussed in depth rather than the use of standardized questions.

A qualitative interview is essentially a conversation in which the interviewer establishes a general direction for the conversation and pursues specific topics raised by the respondent. Ideally, the respondent does most of the talking. If you're talking more than 5 percent of the time, that's probably too much.

Steinar Kvale offers two metaphors for interviewing: the interviewer as a "miner" or as a "traveler." The first model assumes that the subject possesses specific information and that the interviewer's job is to dig it out. By contrast, in the second model, the interviewer

> wanders through the landscape and enters into conversations with the people encountered. The traveler explores the many domains of the country, as unknown territory or with maps, roaming freely around the territory.... The interviewer wanders along with the local inhabitants, asks questions that lead the subjects to tell their own stories of their lived world.
>
> (1996: 3–5)

Asking questions and noting answers is a natural human process, and it seems simple enough to add it to your bag of tricks as a field researcher. Be a little cautious, however. Recall that wording questions is a tricky business. All too often, the way we ask questions subtly biases the answers we get. Sometimes we put our respondent under pressure to look good. Sometimes we put the question in a particular context that omits altogether the most relevant answers.

Suppose, for example, that you want to find out why a group of students is rioting and pillaging on campus. You might be tempted to focus your questioning on how students feel about the dean's recent ruling that requires students always to carry *The Basics of Social Research* with them on campus. (Makes sense to me.) Although you may collect a great deal of information about students' attitudes toward the infamous ruling, they may be rioting for some other reason. Perhaps most are simply joining in for the excitement. Properly done, field research interviewing enables you to find out.

In both qualitative and quantitative research, we tend to think of using face-to-face or telephone interviews. When Nicole Ison (2009) set out to conduct in-depth interviews with young people with cerebral palsy, their speech difficulties created a special problem. Her solution was to conduct e-mail interviews. Even for participants for whom typing may have been difficult, the subjects could work at their own pace, avoiding the frustration that would probably have attended spoken interviews. Subjects could create their responses and review them to be sure they had accurately expressed their intended communications.

Although you may set out with a reasonably clear idea of what you want to ask in your interviews, one of the special strengths of field research is its flexibility. In particular, the answers evoked by your initial questions should shape your subsequent ones. It doesn't work merely to ask preestablished questions and record the answers. Instead, you need to ask a question, listen carefully to the answer, interpret its meaning for your general inquiry, and then frame another question either to dig into the earlier answer or to redirect the person's attention to an area more relevant to your inquiry. In short, you need to be able to listen, think, and talk almost at the same time.

The discussion of probes in Chapter 9 provides a useful guide to getting answers in more depth without biasing later answers. More generally, field interviewers need to be good listeners. This means being more interested than interesting; learning to use prompts like "How is that?" "In what ways?" "How do you mean that?" "What would be an example of that?" and learning to look and listen expectantly and letting the person you're interviewing fill in the silence.

At the same time, you can't afford to be a totally passive receiver. You'll go into your interviews with some general (or specific) questions you want answered and some topics you want addressed. At times you'll need the skill of subtly directing the flow of conversation.

The martial arts offer a useful analogy in this regard. The aikido master never resists an opponent's blow but instead accepts it, joins with it, and then subtly redirects it in a more appropriate direction. So, instead of trying to halt your respondent's line of discussion, learn to take what he or she has just said and branch that comment back in the direction appropriate for your purposes. Most people love to talk to anyone who's really interested. Stopping their line of conversation tells them that you are not interested; asking

them to elaborate in a particular direction tells them that you are.

Consider this hypothetical example in which you're interested in why college students chose their majors.

You: What are you majoring in?
Resp: Engineering.
You: I see. How did you come to choose engineering?
Resp: I have an uncle who was voted the best engineer in Arizona in 1981.
You: Gee, that's great.
Resp: Yeah. He was the engineer in charge of developing the new civic center in Tucson. It was written up in most of the engineering journals.
You: I see. Did you talk to him about your becoming an engineer?
Resp: Yeah. He said that he got into engineering by accident. He needed a job when he graduated from high school, so he went to work as a laborer on a construction job. He spent eight years working his way up from the bottom, until he decided to go to college and come back nearer the top.
You: So is your main interest civil engineering, like your uncle, or are you more interested in some other branch of engineering?
Resp: Actually, I'm leaning more toward electrical engineering—computers, in particular. I started messing around with microcomputers when I was in high school, and my long-term plan is . . .

Notice how the interview first begins to wander off into a story about the respondent's uncle. The first attempt to focus things back on the student's own choice of major ("Did you talk to your uncle . . . ?") fails. The second attempt ("So is your main interest . . . ?") succeeds. Now the student is providing the kind of information you're looking for. It's important for field researchers to develop the ability to "control" conversations in this fashion. At the same time, of course, you need to be on the alert for "distractions" that point to unexpectedly important aspects of your research interest.

Herbert and Riene Rubin offer several ways to control a "guided conversation," such as the following:

If you can limit the number of main topics, it is easier to maintain a conversational flow from one topic to another. Transitions should be smooth and logical. "We have been talking about mothers, now let's talk about fathers," sounds abrupt. A smoother transition might be, "You mentioned your mother did not care how you performed in school—was your father more involved?" The more abrupt the transition, the more it sounds like the interviewer has an agenda that he or she wants to get through, rather than wanting to hear what the interviewee has to say.

(1995: 123)

Because field research interviewing is so much like normal conversation, researchers must keep reminding themselves that they are not having a normal conversation. In normal conversations, each of us wants to come across as an interesting, worthwhile person. If you watch yourself the next time you chat with someone you don't know too well, you'll probably find that much of your attention is spent on thinking up interesting things to say— contributions to the conversation that will make a good impression. Often, we don't really hear each other, because we're too busy thinking of what we'll say next. As an interviewer, the desire to appear interesting is counterproductive. The interviewer needs to make the other person seem interesting and can do so by being interested—and by listening more than talking. (Do this in ordinary conversations, and people will actually regard you as a great conversationalist.)

Lofland and colleagues (2006: 69–70) suggest that investigators adopt the role of the "socially acceptable incompetent" when interviewing. That is, you act as though you do not understand the situation you find yourself in and must be helped to grasp even the most basic and obvious aspects of that situation: "A naturalistic investigator, almost by definition, is one who does not understand. She or he is 'ignorant' and needs to be 'taught.' This role of watcher and asker of questions is the quintessential *student* role" (2006: 69).

Interviewing needs to be an integral part of the entire field research process. Later, I'll stress the need to review your observational notes every night—making sense out of what you've observed, getting a clearer feel for the situation you're studying, and finding out what you should pay more attention to in further observations. In the same fashion, you'll need to review your notes on interviews, recording especially effective questions, and detecting all those questions you should have asked but didn't. Start asking such questions the next time you interview. If you've recorded the interviews, replay them as a useful preparation for future interviews.

Steinar Kvale (1996: 88) details seven stages in the complete interviewing process:

1. *Thematizing:* Clarifying the purpose of the interviews and the concepts to be explored
2. *Designing:* Laying out the process through which you'll accomplish your purpose, including a consideration of the ethical dimension
3. *Interviewing:* Doing the actual interviews
4. *Transcribing:* Creating a written text of the interviews
5. *Analyzing:* Determining the meaning of gathered materials in relation to the purpose of the study
6. *Verifying:* Checking the reliability and validity of the materials
7. *Reporting:* Telling others what you've learned

As with all other aspects of field research, interviewing improves with practice. Fortunately, it's something you can practice any time you want. Practice on your friends.

As in other aspects of qualitative research, data collection can be supported by a variety of technologies: pencils and paper, computers, and audio and video recorders, among others. The editors of the *International Journal of Social Research Methodology* (Editorial 2011) have provided an excellent overview of the use of video cameras in qualitative data collection. The remainder of the issue consists of articles dealing with special aspects of the topic.

Focus Groups

Although our discussions of field research so far have focused on studying people in the process of living their lives, researchers sometimes bring people into the laboratory for qualitative interviewing and observation. The focus-group method, which is also called group interviewing, is essentially a qualitative method. It's based on structured, semi-structured, or unstructured interviews. It allows the researcher/interviewer to question several individuals systematically and simultaneously. The focus-group data technique is frequently used in political and market research, but it is also used for other purposes as well. In *Silent Racism* (2006), for example, Barbara Trepagnier used focus groups to examine the persistence of racism among "well-meaning white people."

Imagine that you're thinking about introducing a new product. Let's suppose that you've invented a new computer that not only does word processing, spreadsheets, data analysis, and the like, but also contains a GPS, MP3, climate change calculator, automobile diagnostic system, microwave oven, denture cleaner, and coffee maker. To highlight its computing and coffee-making features, you're thinking of calling it the "Compulator." You figure the new computer will sell for about $28,000, and you want to know whether people are likely to buy it. Your prospects might be well served by focus groups.

In a **focus group**, typically five to fifteen people are brought together in a private, comfortable environment to engage in a guided discussion of some topic—in this case, the acceptability and salability of the Compulator. The subjects are selected on the basis of relevance to the topic under study. Given the likely cost of the Compulator, your focus-group participants would probably be limited to upper-income groups, for example. Other, similar considerations might figure into the selection.

Participants in focus groups are not likely to be chosen through rigorous probability-sampling methods. This means that the participants do not statistically represent any meaningful population. However, the purpose of the study is to explore rather than to describe or explain in any definitive sense. Nevertheless, typically more than one focus group is convened in a given study because of the serious danger that a single group of five to fifteen people will be too atypical to offer any generalizable insights.

William Gamson (1992) used focus groups to examine how U.S. citizens frame their views of political issues. Having picked four issues—affirmative action, nuclear power, troubled industries, and the Arab–Israeli conflict—Gamson

undertook a content analysis of press coverage to get an idea of the media context within which we think and talk about politics. Then the focus groups were convened for a firsthand observation of the process of people discussing issues with their friends.

Richard Krueger (1988: 47) points to five advantages of focus groups:

1. The technique is a socially oriented research method capturing real-life data in a social environment.
2. It has flexibility.
3. It has high face validity.
4. It has speedy results.
5. It is low in cost.

In addition to these advantages, group dynamics frequently bring out aspects of the topic that would not have been anticipated by the researcher and would not have emerged from interviews with individuals. In a side conversation, for example, a couple of the participants might start joking about the results of leaving out one letter from a product's name. This realization might save the manufacturer great embarrassment later on.

Krueger (1988: 44–45) also notes some disadvantages of the focus-group method, however:

1. Focus groups afford the researcher less control than individual interviews.
2. Data are difficult to analyze.
3. Moderators require special skills.
4. Difference between groups can be troublesome.
5. Groups are difficult to assemble.
6. The discussion must be conducted in a conducive environment.

In a focus-group interview, more than in any other type of interview, the interviewer has to be a skilled moderator. Controlling the dynamics within the group is a major challenge. Letting one person dominate the focus-group interview reduces the likelihood that the other subjects will

---

focus group A group of subjects interviewed together, prompting a discussion. The technique is frequently used by market researchers, who ask a group of consumers to evaluate a product or discuss a type of commodity, for example.

participate. This can generate the problem of group conformity or what is sometimes called "groupthink," which is the tendency for people in a group to conform to the opinions and decisions of the most outspoken members of the group. This danger is compounded by the possibility that one or two people may sometimes dominate the conversation. Interviewers need to be aware of this phenomenon and try to get everyone to participate fully on all the issues brought up in the interview. In addition, interviewers must resist bringing their own views into play by overdirecting the interview and the interviewees.

Although focus group research differs from other forms of qualitative field research, it further illustrates the possibilities for doing social research face to face with those we wish to understand. In addition, David Morgan (1993) suggested that focus groups are an excellent device for generating questionnaire items for a subsequent survey.

In their typical form—centered on a particular topic and taking a limited amount of time—focus groups would not be regarded as an in-depth research technique. However, Carolina Överlien, Karin Aronsson, and Margareta Hydén (2005) have used the technique successfully for extended discussions of sexuality among Swedish teenagers in a youth detention home.

Like other social research techniques, focus groups are adapting to new communication modalities. George Silverman (2005), for example, offers a discussion of focus groups that are conducted online or via telephone.

Recording Observations

The greatest advantage of the field research method is the presence of an observing, thinking researcher at the scene of the action. Not even audio recorders and cameras can capture all the relevant aspects of social processes, although both of those devices can be very useful to the field researcher. Consequently, in both direct observation and interviewing, making full and accurate notes of what goes on is vital. If possible, take notes on your observations while you observe. When that's not feasible, write down your notes as soon as possible afterward.

In your notes, include both your empirical observations and your interpretations of them. In other words, record what you "know" has happened and what you "think" has happened. Be sure to identify these different kinds of notes for what they are. For example, you might note that Person X spoke out in opposition to a proposal made by a group leader (an observation), that you *think* this represents an attempt by Person X to take over leadership of the group (an interpretation), and that you *think* you heard the leader comment to that effect in response to the opposition (a tentative observation).

Of course, you cannot hope to observe everything, nor can you record everything you do observe. Just as your observations will represent a sample of all possible observations, your notes will represent a sample of your observations. The idea, of course, is to record the most pertinent ones. "Issues and Insights: Interview Transcript Annotated with Researcher Memos" provides an example given by Sandrine Zerbib from an in-depth interview with a woman film director. Notice that the illustration contains a portion of an in-depth interview, along with some of Zerbib's memos, written during her review of the interview later on.

Some of the most important observations can be anticipated before you begin the study; others will become apparent as your observations progress. Sometimes you can make note taking easier by preparing standardized recording forms in advance. In a study of jaywalking, for example, you might anticipate the characteristics of pedestrians that are most likely to be useful for analysis—age, gender, social class, ethnicity, and so forth—and prepare a form in which observations of these variables can be recorded easily. Alternatively, you might develop a symbolic shorthand in advance to speed up recording. For studying audience participation at a mass meeting, you might want to construct a numbered grid representing the different sections of the meeting room; then you could record the location of participants easily, quickly, and accurately.

None of this advance preparation should limit your recording of unanticipated events and aspects of the situation. Quite the contrary, the speedy handling of anticipated observations

Issues and Insights

Interview Transcript Annotated with Researcher Memos

Thursday August 26, 12:00–1:00

R: What is challenging for women directors on a daily experience, on a daily life?

J: Surviving.

R: OK. Could you develop a little bit on that? [I need to work on my interview schedule so that my interviewee answers with more elaboration without having to probe.]

J: Yeah, I mean it's all about trying to get, you know, in, trying to get the job, and try, you know, to do a great job so that you are invited back to the next thing. And particularly since they are so many, you know, difficulties in women directing. It makes it twice as hard to gain into this position where you do an incredible job, because … you can't just do an average job, you have to [347] do this job that just knocks your socks off all the time, and sometimes you don't get the opportunity to do that, because either you don't have a good producer or you have so many pressures that you can't see straight or your script is lousy, and you have to make a silk purse out of a sow's ear. You know, you have a lot of extra strikes against you than the average guy who has similar problems, because you are a woman and they look at it, and women are more visible than men … in unique positions. [It seems that Joy is talking about the particularities of the film industry. There are not that many opportunities and in order to keep working, she needs to build a certain reputation. It is only by continuing to direct that she can maintain or improve her reputation. She thinks that it is even harder for women but does not explain it.]

R: Hum … what about on the set did you experience, did it feel … did people make it clear that you were a woman, and you felt treated differently? [I am trying to get her to speak about more specific and more personal experiences without leading her answer.]

J: Yeah, oh yeah, I mean … a lot of women have commiserated about, you know when you have to walk on the set for the first time, they're all used to working like a well-oiled machine and they say, "Oh, here is the woman, something different" and sometimes they can be horrible, they can resist your directing and they can, they can sabotage you, by taking a long time to light, or to move sets, or to do something … and during that time you're wasting time, and that goes on a report, and the report goes to the front [368] office, and, you know, and so on and so on and so on and so forth. And people upstairs don't know what the circumstances are, and they are not about to fire a cinematographer that is on their show for ever and ever … nor do they want to know that this guy is a real bastard, and making your life a horror. They don't want to know that, so therefore, they go off, because she's a woman let's not hire any more women, since he has problems with women. You know, so, there is that aspect.

[I need to review the literature on institutional discrimination. It seems that the challenges that Joy is facing are not a matter of a particular individual. She is in a double bind situation where whether she complains or not, she will not be treated equal to men. Time seems to be one quantifiable measurement of how well she does her job and, as observed in other professions, the fact that she is a woman is perceived as a handicap. Review literature on women in high management positions. I need to keep asking about the dynamics between my interviewees and the crewmembers on the set. The cinematographer has the highest status on the set under the director. Explore other interviews about reasons for conflict between them.]

[Methods (note to myself for the next interviews): try to avoid phone interviews unless specific request from the interviewee. It is difficult to assess how the interviewee feels with the questions. Need body language because I become more nervous about the interview process.]

Note: A number in brackets represents a word that was inaudible from the interview. It is the number that appeared on the transcribing machine, with each interview starting at count 0. The numbers help the researcher locate a passage quickly when he or she reviews the interview.

---

can give you more freedom to observe the unanticipated.

You're already familiar with the process of taking notes, just as you already have at least informal experience with field research in general. Like good field research, however, good note taking requires careful and deliberate attention and involves specific skills. Some guidelines follow. (You can learn more from Lofland et al. 2006: 110–7.)

First, don't trust your memory any more than you have to; it's untrustworthy. To illustrate this point, try this experiment. Recall the last three or four movies you saw that you really liked. Now, name five of the actors or actresses. Who had the longest hair? Who was the most likely to start conversations? Who was the most likely to make suggestions that others followed? Now, if you didn't have any trouble answering any of those questions, how sure are you of your answers? Would you be willing to bet $100 that a panel of impartial judges would observe what you recall?

Even if you pride yourself on having a photographic memory, it's a good idea to take notes either during the observation or as soon afterward as possible. If you take notes during observation, do it unobtrusively, because people will tend to behave differently if they see you taking down everything they say or do.

Second, it's usually a good idea to take notes in stages. In the first stage, you may need to take sketchy notes (words and phrases) in order to keep abreast of what's happening. Then go off by yourself and rewrite your notes in more detail. If you do this soon after the events you've observed, the sketchy notes should allow you to recall most of the details. The longer you delay, the less likely you'll be able to recall things accurately and fully.

In his study of bike messengers in New York City, mentioned earlier, Jeffrey Kidder reports on this process:

> I obtained the vast majority of data for this article through informal interviews. I unobtrusively took notes throughout the day and at social events. Upon returning home, these data were compiled into my field notes. During the workday and during races, parties, and other social gatherings, casual conversations provided the truest glimpses into messenger beliefs, ideologies, and opinions. To this end, I avoided formal interviews and instead allowed my questions to be answered by normal talk within the social world.
>
> *(2005: 349)*

I know this method sounds logical, but it takes self-discipline to put it into practice. Careful observation and note taking can be tiring, especially if it involves excitement or tension and if it extends over a long period. If you've just spent eight hours observing and making notes on how people have been coping with a disastrous flood, your first desire afterward will likely be to get some sleep, change into dry clothes, or get a bite to eat. You may need to take some inspiration from newspaper reporters who undergo the same sorts of hardships then write to meet their deadlines.

Third, you'll inevitably wonder how much you should record. Is it really worth the effort to write out all the details you can recall right after the observation session? The general guideline is yes. Generally, in field research you can't be really sure of what's important and what's unimportant until you've had a chance to review and analyze a great volume of information, so you should record even things that don't seem important at the outset. They may turn out to be significant after all. Also, the act of recording the details of something "unimportant" may jog your memory about something important.

Realize that your final report on the project will not reflect most of your field notes. Put more harshly, most of your notes will be "wasted." But take heart: Even the richest gold ore yields only about 30 grams of gold per metric ton, meaning that 99.997 percent of the ore is wasted. Yet, that 30 grams of gold can be hammered out to cover an area 18 feet square—the equivalent of about 685 book pages. So take a ton of notes, and plan to select and use only the gold.

Like other aspects of field research (and all research for that matter), proficiency comes with practice. The nice thing about field research is that you can begin practicing now and can continue practicing in almost any situation. You don't have to be engaged in an organized research project to practice observation and recording. You might start by volunteering to take the minutes at committee meetings, for example. Or just pick a sunny day on campus, find a shady spot, and try observing and recording some specific characteristics of the people who pass by. You can do the same thing at a shopping mall or on a busy street corner. Remember that observing and recording are professional skills, and, like all worthwhile skills, they improve with practice.

Strengths and Weaknesses of Qualitative Field Research

Like all research methods, qualitative field research has distinctive strengths and weaknesses. As I've already indicated, field research is especially effective for studying subtle nuances in attitudes and behaviors and for examining social processes over time. As such, the chief strength of this method lies in the depth of understanding it permits. Whereas other research methods may be challenged as "superficial," field research seldom receives this criticism.

Flexibility is another advantage of field research. As discussed earlier, you can modify your field research design at any time. Moreover, you're always prepared to engage in field research, whenever the occasion should arise, whereas you could not as easily initiate a survey or an experiment.

Field research can be relatively inexpensive as well. Other social research methods may require costly equipment or an expensive research staff, but field research typically can be undertaken by one researcher with a notebook and a pencil. This is not to say that field research is never expensive. A particular project may require a large number of trained observers, for instance. Expensive recording equipment may be needed. Or you may wish to undertake participant observation of interactions in pricey nightclubs.

Field research has several weaknesses as well. First, being qualitative rather than quantitative, it is not an appropriate means for arriving at statistical descriptions of a large population. Observing casual political discussions in laundromats, for example, would not yield trustworthy estimates of the future voting behavior of the total electorate. Nevertheless, the study could provide important insights into how political attitudes are formed.

To assess field research further, we should focus on the issues of validity and reliability. Recall that validity concerns whether measurements actually measure what they're intended to, rather than something else. Reliability, on the other hand, is a matter of dependability: If you made the same measurement again and again, would you get the same result? Let's see how field research stacks up in these respects.

Validity

Field research seems to provide measures with greater validity than do survey and experimental measurements, which are often criticized as superficial and not really valid. Let's review a couple of field research examples to see why this is so.

"Being there" is a powerful technique for gaining insights into the nature of human affairs in all their rich complexity. Listen, for example, to what one nurse reports about the impediments to patients' coping with cancer:

Common fears that may impede the coping process for the person with cancer can include the following:

- *Fear of death—for the patient, and the implications his or her death will have for significant others.*
- *Fear of incapacitation—because cancer can be a chronic disease with acute episodes that may result in periodic stressful periods, the variability of the person's ability to cope and constantly adjust may require a dependency upon others for activities of daily living and may consequently become a burden.*
- *Fear of alienation—from significant others and health care givers, thereby creating helplessness and hopelessness.*
- *Fear of contagion—that cancer is transmissible and/or inherited.*
- *Fear of losing one's dignity—losing control of all bodily functions and being totally vulnerable.*

(Garant 1980: 2167)

Observations and conceptualizations such as these are valuable in their own right. In addition, they can provide the basis for further research—both qualitative and quantitative.

Here's what Joseph Howell has to say about "toughness" as a fundamental ingredient of life on Clay Street, a white, working-class neighborhood in Washington, D.C.:

Most of the people on Clay Street saw themselves as fighters in both the figurative and literal sense. They considered themselves strong, independent people who would not let themselves be pushed around. For Bobbi, being a fighter meant battling the welfare department and cussing out social workers and doctors upon occasion. It meant spiking Barry's beer with sleeping pills and bashing him over the head with a broom. For Barry it meant telling off his boss and refusing to hang the door, an act that led to his being fired. It meant going through the ritual of a duel with Al. It meant pushing Bubba around and at times getting rough with Bobbi.

June and Sam had less to fight about, though if pressed they both hinted that they, too, would fight. Being a fighter led Ted into near conflict with Peg's brothers, Les into conflict with Lonnie, Arlene into conflict with Phyllis at the bowling alley, etc.

(1973: 292)

Even without having heard the episodes Howell refers to in this passage, we get the distinct impression that Clay Street is a tough place to live. That "toughness" shows far more powerfully through these field observations than it would in a set of statistics on the median number of fistfights occurring during a specified period.

These examples point to the superior validity of field research, as compared with surveys and experiments. The kinds of comprehensive measurements available to the field researcher tap a depth of meaning in concepts such as common fears of cancer patients and "toughness" (or such as liberal and conservative) that is generally unavailable to surveys and experiments. Instead of specifying concepts, field researchers commonly give detailed illustrations.

Reliability

Field research has, however, a potential problem with reliability. Suppose you were to characterize your best friend's political orientations according to everything you know about him or her. Your assessment of your friend's politics would appear to have considerable validity; certainly it's unlikely to be superficial. We couldn't be sure, however, that another observer would characterize your friend's politics the same way you did, even with the same amount of observation.

Similarly, in-depth, field research measurements are often quite personal. How I judge your friend's political orientation depends very much on my own, just as your judgment depends on your political orientation. Conceivably, then, you could describe your friend as middle-of-the-road, although I might feel that I've been observing a fire-breathing radical.

As I've suggested earlier, researchers who use qualitative techniques are conscious of this issue and take pains to address it. Individual researchers often sort out their own biases and points of view, and the communal nature of science means that their colleagues will help them in that regard. Nevertheless, it's prudent to be wary of purely descriptive measurements in field research—your own or someone else's. If a researcher reports that the members of a club are fairly conservative, such a judgment is unavoidably linked to the researcher's own politics. You can be more trusting of *comparative* evaluations: identifying who is more conservative than whom, for example. Even if you and I had different political orientations, we would probably agree pretty much in ranking the relative conservatism of the members of a group.

As a means for both increasing and documenting the trustworthiness of qualitative research, Glenn Bowen (2009) illustrates the use of an "audit trail," which records the researcher's decisions throughout the conduct of the research and the analysis of data. Decisions on the coding of interview responses would be an example. Some computer programs for qualitative data analysis provide for the recording of an audit trail.

While the audit trail is suggested to counter concerns that qualitative analysis might lack rigor, a similar technique would be appropriate for quantitative research. While the results of measurement decisions in designing a quantitative survey are explicit in the actual wording of questionnaires, the reasoning behind those decisions is not always obvious.

As we've seen, field research is a potentially powerful tool for social scientists, one that provides a useful balance against the strengths and weaknesses of experiments and surveys. The remaining chapters of Part 3 present additional modes of observation available to social researchers.

Ethics in Qualitative Field Research

As I've noted repeatedly, all forms of social research raise ethical issues. By bringing researchers into direct and often intimate contact with their subjects, field research raises ethical concerns in a particularly dramatic way. Here are some of the issues mentioned by Lofland and colleagues (2006: 78–79):

- Is it ethical to talk to people when they do not know you will be recording their words?
- Is it ethical to get information for your own purposes from people you hate?
- Is it ethical to see a severe need for help and not respond to it directly?
- Is it ethical to be in a setting or situation but not commit yourself wholeheartedly to it?
- Is it ethical to develop a calculated stance toward other humans, that is, to be strategic in your relations?
- Is it ethical to take sides or to avoid taking sides in a factionalized situation?
- Is it ethical to "pay" people with trade-offs for access to their lives and minds?

What do you think?...Revisited

The impact of the observer affects most forms of social research. We've seen that experimenters can influence the way people behave in experiments, and survey researchers can affect how people respond to questionnaires. The problem is also present when participant-observers set out to study human behavior in its natural setting. As we've seen, researchers sometimes conceal their research identity as a way of reducing their impact, but we've also seen that anything they do in a social setting will have some impact.

Ultimately, the solution to this problem is awareness of it, because this allows you to have some control over the impact you have. This approach is coupled with replication by other researchers. Different researchers affect research situations in different ways. If they nonetheless discover the same patterns of behavior, our confidence that we've learned something about social life—not just something about our role in it—increases.

● Is it ethical to "use" people as allies or informants in order to gain entry to other people or to elusive understandings?

Participation observation brings special ethical concerns with it. When you ask people to reveal their inner thoughts and actions to you, you may be opening them up to a degree of suffering: perhaps by having them recall troubling experiences, for example, as in the earlier example of interviewing cancer patients. Moreover, you're also asking them to risk the public disclosure of what they've confided in you, and you're strictly obligated to honor their confidences. We've seen cases of researchers going to jail rather than reveal the private matters they observed in confidence.

Geoff Pearson (2009) examines the sticky question of how participant-observers should behave when studying people routinely engaged in criminal activities. The researcher's refusal to join in such illegal behavior might very well alter what is being studied and, in some cases, risk the researcher's study and/or safety. On the other hand, are researchers justified in breaking the law in such cases? Obviously, the severity of the crimes would affect your decisions, but when you examine such ethical questions in depth, you are likely to find yourself entering numerous gray areas.

Planning and conducting field research in a responsible way requires attending to these and other ethical concerns.

MAIN POINTS

Introduction

● Field research involves the direct observation of social phenomena in their natural settings. Typically, field research is qualitative rather than quantitative.

● In field research, observation, data processing, and analysis are interwoven, cyclical processes.

Topics Appropriate for Field Research

● Field research is especially appropriate for topics and processes that are not easily quantifiable, that change over time, or that are best studied in natural settings. Among these topics are practices, episodes, encounters, roles, relationships, groups, organizations, settlements, and lifestyles or subcultures.

Special Considerations in Qualitative Field Research

● Among the special considerations involved in field research are the various possible roles of

the observer and the researcher's relations to subjects. As a field researcher, you must decide whether to observe as an outsider or as a participant, whether or not to identify yourself as a researcher, and how to negotiate your relationships with subjects.

Some Qualitative Field Research Paradigms

● Field research can be guided by any one of several paradigms, such as naturalism, ethnomethodology, grounded theory, case studies and the extended case method, institutional ethnography, and participatory action research.

Conducting Qualitative Field Research

● Preparing for the field involves doing background research, determining how to make contact with subjects, and resolving issues of what your relationship to your subjects will be.

● Field researchers often conduct in-depth interviews that are much less structured than those conducted in survey research. Qualitative interviewing is more of a guided conversation than a search for specific information. Effective interviewing involves active listening and the ability to direct conversations unobtrusively.

- To create a focus group, researchers bring subjects together and observe their interactions as they explore a specific topic.

- Whenever possible, field observations should be recorded as they are made; otherwise, they should be recorded as soon afterward as possible.

Strengths and Weaknesses of Qualitative Field Research

- Among the advantages of field research are the depth of understanding it can provide, its flexibility, and (usually) its lack of costs.

- Compared with surveys and experiments, field research measurements generally have more validity but less reliability. Also, field research is generally not appropriate for arriving at statistical descriptions of large populations.

- An audit trail records the researcher's decisions throughout the conduct of the research and the analysis of data.

Ethics in Qualitative Field Research

- Conducting field research responsibly involves confronting several ethical issues that arise from the researcher's direct contact with subjects.

KEY TERMS

case study	institutional ethnography
emancipatory research	naturalism
ethnography	participatory action
ethnomethodology	research (PAR)
extended case method	qualitative interview
focus group	rapport
grounded theory	reactivity

PROPOSING SOCIAL RESEARCH: FIELD RESEARCH

This chapter has laid out a large number of possibilities for conducting field research. You should indicate the kind of study you plan to do. Will you be the sole observer in the study? If not, how will you select and train the other observers?

Will you be a participant in the events you're observing and, if so, will you identify yourself as a researcher to the people being observed? You might indicate how these choices may affect what you observe—as well as the ethical issues involved.

In earlier exercises, you've dealt with the variables you wish to examine and the ways you'll select informants and/or people to observe, as well as the times and locations of your observations. As this chapter has demonstrated, there are other logistical issues to be worked out. It may be appropriate to describe your note-taking plans if that task is likely to be difficult (for instance, as a participant not identified as a researcher).

If you'll be conducting in-depth interviews, you should include an outline of the topics to be covered in those interviews. Are there some topics/questions that must be addressed in each interview and others that will be pursued only if appropriate?

Compared with experiments and surveys, field research allows more flexibility in the timing of the research. Depending on how things go, you may find yourself concluding earlier or later than you had planned. Nevertheless, you should comment on your proposed schedule.

REVIEW QUESTIONS

1. Think of some group or activity you participate in or are very familiar with. In two or three paragraphs, describe how an outsider might effectively go about studying that group or activity. What should he or she read, what contacts should be made, and so on?

2. Choose any two of the paradigms discussed in this chapter. How might your hypothetical study from item 1 be conducted if you followed each? Compare and contrast the way these paradigms might work in the context of your study.

3. To explore the strengths and weaknesses of experiments, surveys, and field research, choose a general research area (such as prejudice, political orientation, or education) and write brief descriptions of studies in that area that could be conducted using each of these three methods. In each case, why is the chosen method the most appropriate for the study you describe?

4. Return to the example you devised in response to item 1. What five ethical issues can you imagine having to confront if you were to undertake your study?

5. Using the Web, find a research report that uses the Grounded Theory Method. Briefly, what are the study design and main findings?

CHAPTER 11

Unobtrusive Research

Ermolaer Alexander/Shutterstock.com

Learning Objectives

After studying this chapter, you will be able to . . .

- Understand the use of content analysis for both qualitative and quantitative social research.
- Give examples of researchers analyzing existing statistics in order to illustrate the strengths and weaknesses of this method.
- Describe comparative and historical methods and give examples to illustrate.
- List some of the options for online unobtrusive research and discuss advantages and shortcomings.
- Understand the ethical issues involved in the various methods of unobtrusive research.

Introduction

With the exception of the complete observer in field research, each of the modes of observation discussed so far requires the researcher to intrude to some degree on whatever he or she is studying. This is most obvious in the case of experiments, followed closely by survey research. Even the field researcher, as we've seen, can change things in the process of studying them.

At least one previous example in this book, however, was totally exempt from that danger. Durkheim's analysis of suicide did nothing to affect suicides one way or the other (see Chapter 5). His study is an example of **unobtrusive research**, or methods of studying social behavior without affecting it. As you'll see, unobtrusive measures can be qualitative or quantitative.

This chapter examines three types of unobtrusive research methods: content analysis, analysis of existing statistics, and comparative and historical research. First, we'll discuss content analysis, in which researchers examine a class of social artifacts that usually are written documents such as newspaper editorials. Next, the Durkheim study is an example of the analysis of existing statistics. As you'll see, there are great masses of data all around you, awaiting your use in the understanding of social life. Finally, comparative and historical research, a form of

research with a venerable history in the social sciences, is currently enjoying a resurgence of popularity. Like field research, comparative and historical research is usually a qualitative method, one in which the main resources for observation and analysis are historical records. The method's name includes the word *comparative* because social scientists—in contrast to historians who may simply describe a particular set of events—seek to discover common patterns that recur in different times and places.

To set the stage for our examination of these three research methods, I want to draw your attention to an excellent book that should sharpen your senses about the potential for unobtrusive measures in general. It is, among other things, the book from which I take the term *unobtrusive measures.*

In 1966, Eugene Webb and three colleagues published an ingenious little book on social research (revised in 2000) that has become a classic. It focuses on the idea of unobtrusive or nonreactive research. Webb and his colleagues played freely with the task of learning about human behavior by observing what people inadvertently leave behind them. Do you want to know what exhibits are the most popular at a museum? You could conduct a poll, but people might tell you what they thought you wanted to hear or what might make them look intellectual and serious. You could stand by different exhibits and count the viewers that came by, but people might come over to see what you were doing. Webb and his colleagues suggest that you check the wear and tear on the floor in front of various exhibits. Those that have the most-worn tiles are probably the most popular.

---

unobtrusive research Methods of studying social behavior without affecting it. This includes content analysis, analysis of existing statistics, and comparative and historical research.

What do you think?

This chapter presents several research techniques that by definition have no impact on what's being studied. If the impact of the observer is such a problem in social research, why don't social scientists limit themselves to unobtrusive techniques?

See the *What do you think? ... Revisited* box toward the end of the chapter.

Picture 1/Earl Babbie

Want to know which exhibits are popular with little kids? Look for mucus on the glass cases. To get a sense of the most popular radio stations, you could arrange with an auto mechanic to check what radio stations are programmed in for cars brought in for repair.

The possibilities are limitless. Like a detective investigating a crime, the social researcher looks for clues. If you stop to notice, you'll find that clues of social behavior are all around you. In a sense, everything you see represents the answer to some important social science question—all you have to do is think of the question.

Although problems of validity and reliability crop up in unobtrusive measures, a little ingenuity can either handle them or put them in perspective.

Content Analysis

As I mentioned in the chapter introduction, **content analysis** is the study of recorded human communications. Among the forms suitable for study are books, magazines, Web pages, poems, newspapers, songs, paintings, speeches, letters, e-mail messages, bulletin board postings on the Internet, laws, and constitutions, as well as any components or collections thereof. Shulamit Reinharz points out that feminist researchers have used content analysis to study

> *children's books, fairy tales, billboards, feminist nonfiction and fiction books, children's art work, fashion, fat-letter postcards, Girl Scout handbooks, works of fine art, newspaper rhetoric, clinical records, research publications, introductory sociology textbooks, and citations, to mention only a few.*
>
> *(1992: 146–7)*

In another example, when William Mirola set out to discover the role of religion in the movements to establish the eight-hour working day in America, his data were taken "from Chicago's labor, religious, and secular presses, from pamphlets, and from speeches given by eight-hour proponents from three representative factions within the movement" (2003: 273).

Topics Appropriate for Content Analysis

Content analysis is particularly well suited to the study of communications and to answering the classic question of communications research: "Who says what, to whom, why, how, and with what effect?" Are popular French novels more concerned with love than novels in the United States are? Was the popular British music of the 1960s more politically cynical than the popular German music during that period? Do political candidates who primarily speak to "bread and butter" issues get elected more often than those who address issues of high principle? Each of these questions suggests a social science research topic: The first might address national character, the second political orientations, and the third political processes. Although you might study such topics by observing individual people, content analysis provides another approach.

An early example of content analysis is the work of Ida B. Wells. In 1891, Wells,

---

content analysis The study of recorded human communications, such as books, websites, paintings, and laws.

whose parents had been slaves, wanted to test the widely held assumption that African American men were being lynched in the South primarily for raping white women. As a research method, she examined newspaper articles on the 728 lynchings reported during the previous ten years. In only a third of the cases were the lynching victims even accused of rape, much less proved guilty. Primarily, they were charged with being insolent, not staying in "their place" (cited in Reinharz 1992: 146).

More recently, the best-selling book *Megatrends 2000* (Naisbitt and Aburdene 1990) used content analysis to determine the major trends in modern U.S. life. The authors regularly monitored thousands of local newspapers per month in order to discover local and regional trends for publication in a series of quarterly reports. Their book examines some of the trends they observed in the nation at large. In a follow-up book (Aburdene 2005), this kind of analysis pointed to such trends as "the power of spirituality" and "the rise of conscious capitalism."

Some topics are more appropriately addressed by content analysis than by any other method of inquiry. Suppose that you're interested in violence on television. Maybe you suspect that the manufacturers of men's products are more likely to sponsor violent TV shows than are sponsors of other products or services. Content analysis would be the best way of finding out.

Briefly, here's what you'd do. First, you'd develop operational definitions of the two key variables in your inquiry: *men's products* and *violence*. The section on coding, later in this chapter, will discuss some of the ways you could do that. Ultimately, you'd need a plan that would allow you to watch TV, classify sponsors, and rate the degree of violence on particular shows.

Next, you'd have to decide what to watch. Probably you'd decide (1) what stations to watch, (2) for what period, and (3) at what hours. Then, you'd stock up on beer and potato chips and start watching, classifying, and recording. Once you'd completed your observations, you'd be able to analyze the data you collected and determine whether men's product manufacturers sponsored more blood and gore than other sponsors did.

Gabriel Rossman (2002) had a somewhat different question regarding the mass media. Public concern had grown over the concentration of media into fewer and fewer corporate hands, so Rossman decided to ask the following question: If a newspaper is owned by the same conglomerate that owns a movie production company, can you trust that newspaper's movie reviews of its parent company's productions?

You can't, according to Rossman's findings. Because many newspapers rate movies somewhat quantitatively (for example, three stars out of four), he could perform a simple quantitative analysis. For each movie review, he asked two main questions: (1) Was the movie produced by the same company that owned the newspaper? and (2) What rating did the film receive? He found that, indeed, movies produced by the parent company received higher ratings than other movies did. Further, the ratings given to movies by newspapers with the same parent company were higher than the ratings those movies received from other newspapers. This discrepancy, moreover, was strongest in the case of big-budget movies in which the parent company had invested heavily.

As a mode of observation, content analysis requires a thoughtful handling of the "what" that is being communicated. The analysis of data collected in this mode, as in others, addresses the "why" and "with what effect."

Sampling in Content Analysis

In the study of communications, as in the study of people, you often can't observe directly all you would like to explore. In your study of TV violence and sponsorship, for example, I'd advise against attempting to watch everything that's broadcast. It wouldn't be possible, and your brain would probably short-circuit before you came close to drawing any conclusion. Usually, it's appropriate to sample. Let's begin by revisiting the idea of units of analysis. We'll then review some of the sampling techniques that might be applied to such units in content analysis.

Units of Analysis

As I discussed in Chapter 4, determining appropriate units of analysis—the individual units that we make descriptive and explanatory statements about—can be a complicated task. For example, if we wish to compute average family income, the individual family is the unit of analysis. But we'll have to ask individual members of families how much money they make. Thus, individuals will be the units of observation, even though the individual family remains the unit of analysis. Similarly, we may wish to compare crime rates of different cities in terms of their size, geographic region, racial composition, and other differences. Even though the characteristics of these cities are partly a function of the behaviors and characteristics of their individual residents, the cities would ultimately be the units of analysis.

The complexity of this issue is often more apparent in content analysis than in other research methods, especially when the units of observation differ from the units of analysis. A few examples should clarify this distinction.

Let's suppose we want to find out whether criminal law or civil law makes the most distinctions between men and women. In this instance, individual laws would be both the units of observation and the units of analysis. We might select a sample of a state's criminal and civil laws and then categorize each law by whether or not it makes a distinction between men and women. In this fashion, we could determine whether criminal or civil law distinguishes by sex the most.

Somewhat differently, we might wish to determine whether states that enact laws distinguishing between different racial groups are also more likely than other states to enact laws distinguishing between men and women. Although the examination of this question would also involve the coding of individual acts of legislation, the unit of analysis in this case is the individual state, not the law.

Or, changing topics radically, let's suppose we're interested in representationalism in paintings. If we wish to compare the relative popularity of representational and nonrepresentational paintings, the individual paintings will be our units of analysis. If, on the other hand, we wish to discover whether representationalism in paintings is more characteristic of wealthy or impoverished painters, of educated or uneducated painters, of capitalist or socialist painters, the individual painters will be our units of analysis.

It's essential that this issue be clear, because sample selection depends largely on what the unit of analysis is. If individual writers are the units of analysis, the sample design should select all or a sample of the writers appropriate to the research question. If books are the units of analysis, we should select a sample of books, regardless of their authors. Bruce Berg (1989: 112–3) points out that even if you plan to analyze some body of textual materials, the units of analysis might be words, themes, characters, paragraphs, items (such as a book or letter), concepts, semantics, or combinations of these. Figure 11-1 illustrates some of those possibilities.

I'm not suggesting that sampling should be based solely on the units of analysis. Indeed, we may often subsample—select samples of subcategories—for each individual unit of analysis. Thus, if writers are the units of analysis, we might (1) select a sample of writers from the total population of writers, (2) select a sample of books written by each writer selected, and (3) select portions of each selected book for observation and coding.

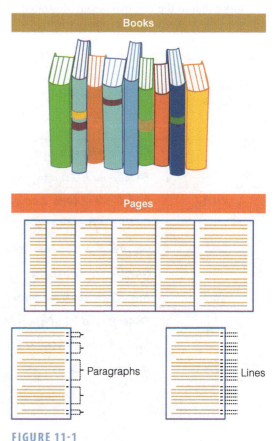

FIGURE 11-1

A Few Possible Units of Analysis for Content Analysis.

Finally, let's look at a trickier example: the study of TV violence and sponsors. What's the unit of analysis for the research question "Are the manufacturers of men's products more likely to sponsor violent shows than other sponsors are?" Is it the TV show? The sponsor? The instance of violence?

In the simplest study design, it would be none of these. Though you might structure your inquiry in various ways, the most straightforward design would be based on the commercial as the unit of analysis. You would use two kinds of observational units: the commercial and the program (the show that gets squeezed in between commercials). You would want to observe both units. You would classify commercials by whether they advertised men's products and the programs by their violence. The program classifications would be transferred to the commercials occurring near them. Figure 11-2 provides an example of the kind of record you might keep.

Notice that in the research design illustrated in Figure 11-2, all the commercials occurring in the same program break are grouped and get the same scores. Also, the number of violent instances recorded as following one commercial break is the same as the number preceding the next break. This simple design allows us to classify each commercial by its sponsorship and the degree of violence associated with it. Thus, for example, the first Grunt Aftershave commercial is coded as being a men's product and as having 10 instances of violence associated with it. The Buttercup Bra commercial is coded as not being a men's product and as having no violent instances associated with it.

In the illustration, we have four men's product commercials with an average of 7.5 violent instances each. The four commercials classified as definitely not men's products have an average of 1.75, and the two that might or might not be considered men's products have an average of 1 violent instance each. If this pattern of differences persisted across a much larger number of observations, we'd probably conclude that manufacturers of men's products are more likely to sponsor TV violence than other sponsors are.

Commercial Break	Sponsor	Men's Product?			Number of Instances of Violence	
		Yes	No	?	Before the Commercial Break	After the Commercial Break
1st	Grunt Aftershave	✓			6	4
	Brute Jock Straps	✓			6	4
2nd	Bald-No-More Lotion	✓			4	3
3rd	Grunt Aftershave	✓			3	0
	Snowflake Toothpaste		✓		3	0
	Godliness Cleanser		✓		3	0
4th	Big Thumb Hammers			✓	0	1
5th	Snowflake Toothpaste		✓		1	0
	Big Thumb Hammers			✓	1	0
6th	Buttercup Bras		✓		0	0

FIGURE 11-2

Example of Recording Sheet for TV Violence.

The point of this illustration is to demonstrate how units of analysis figure into the data collection and analysis. You need to be clear about your unit of analysis before planning your sampling strategy, but in this case you can't simply sample commercials. Unless you have access to the stations' broadcasting logs, you won't know when the commercials are going to occur. Moreover, you need to observe the programming as well as the commercials. As a result, you must set up a sampling design that will include everything you need in order to observe enough instances of violence and men's product commercials.

In designing the sample, you'd need to establish the universe to be sampled from. In this case, which TV stations will you observe? What will be the period of the study—the number of days? And during which hours of each day will you observe? Then, how many commercials do you want to observe and code for analysis? Watch television for a while and find out how many commercials occur each hour; then you can figure out how many hours of observation you'll need (and can stand).

Now you're ready to design the sample selection. As a practical matter, you wouldn't have to sample among the different stations if you had assistants—each of you could watch a different channel during the same period. But let's suppose you're working alone. Your final sampling frame, from which a sample will be selected and watched, might look something like this:

Jan. 7, Channel 2, 7–9 P.M.
Jan. 7, Channel 4, 7–9 P.M.
Jan. 7, Channel 9, 7–9 P.M.
Jan. 7, Channel 2, 9–11 P.M.
Jan. 7, Channel 4, 9–11 P.M.
Jan. 7, Channel 9, 9–11 P.M.
Jan. 8, Channel 2, 7–9 P.M.
Jan. 8, Channel 4, 7–9 P.M.
Jan. 8, Channel 9, 7–9 P.M.
Jan. 8, Channel 2, 9–11 P.M.
Jan. 8, Channel 4, 9–11 P.M.
Jan. 8, Channel 9, 9–11 P.M.
Jan. 9, Channel 2, 7–9 P.M.
Jan. 9, Channel 4, 7–9 P.M.
etc.

Notice that I've made several decisions for you in the illustration. First, I've assumed that channels 2, 4, and 9 are the ones appropriate to your study. I've assumed that you found the 7–11 P.M. prime-time hours to be the most relevant and that two-hour periods will do the job. I picked January 7 out of the hat for a starting date. In practice, of course, all these decisions should be based on your careful consideration of what would be appropriate to your particular study.

Once you have become clear about your units of analysis and the observations best suited to those units and have created a sampling frame like the one I've illustrated, sampling is simple and straightforward. The alternative procedures available to you are the same ones described in Chapter 7: random, systematic, stratified, and so on.

Sampling Techniques

As we've seen, in the content analysis of written prose, sampling may occur at any or all of several levels, including the contexts relevant to the works. Other forms of communication may also be sampled at any of the conceptual levels appropriate to them.

In content analysis, we could employ any of the conventional sampling techniques discussed in Chapter 7. We might select a random or systematic sample of French and U.S. novelists, of laws passed in the state of Mississippi, or of Shakespearean soliloquies. We might select (with a random start) every 23rd paragraph in Tolstoy's *War and Peace*. Or we might number all of the songs recorded by the Beatles and select a random sample of 25.

Stratified sampling is also appropriate for content analysis. To analyze the editorial policies of U.S. newspapers, for example, we might first group all newspapers by the region of the country or size of the community in which they are published, frequency of publication, or average circulation. We might then select a stratified random or systematic sample of newspapers for analysis. Having done so, we might select a sample of editorials from each selected newspaper, perhaps stratified chronologically.

Cluster sampling is equally appropriate to content analysis. Indeed, if individual editorials are our units of analysis, then the selection of newspapers at the first stage of sampling would be a cluster sample. In an analysis of political speeches, we might begin by selecting a sample of politicians; each politician would represent a cluster of political speeches. The TV commercial study described previously is another example of cluster sampling.

It should be repeated that sampling need not end when we reach the unit of analysis. If novels are the unit of analysis in a study, we might select a sample of novelists, a subsample of novels written by each selected author, and a subsample of paragraphs within each novel. We would then analyze the content of the paragraphs for the purpose of describing the novels themselves. (If you haven't realized this yet, researchers speak of samples within samples as "subsamples.")

Let's turn now to the coding or classification of the material being observed. Part 4 discusses the manipulation of such classifications to draw descriptive and explanatory conclusions.

Coding in Content Analysis

Content analysis is essentially a coding operation. **Coding** is the process of transforming raw data into a standardized form. In content analysis, communications—oral, written, or other—are coded or classified according to some conceptual framework. Thus, for example, newspaper editorials may be coded as liberal or conservative. Radio broadcasts may be coded as propagandistic or not, novels as romantic or not, paintings as representational or not, and political speeches as containing character assassinations or not. Recall that because terms such as these are subject to many interpretations, the researcher must specify definitions clearly.

Coding in content analysis involves the logic of conceptualization and operationalization, which I discussed in Chapter 5. As in other research methods, you must refine your conceptual framework and develop specific methods for observing in relation to that framework.

coding The process whereby raw data are transformed into standardized form suitable for machine processing and analysis.

manifest content In connection with content analysis, the concrete terms contained in a communication, as distinguished from *latent content*.

Manifest and Latent Content

In the earlier discussions of field research, we found that the researcher faces a fundamental choice between depth and specificity of understanding. Often, this represents a choice between validity and reliability, respectively. Typically, field researchers opt for depth, preferring to base their judgments on a broad range of observations and information, even at the risk that another observer might reach a different judgment for the same situation. Survey research—through the use of standardized questionnaires—represents the other extreme: total specificity, even though the specific measures of variables may not be adequately valid reflections of those variables. The content analyst has some choice in this matter, however.

Coding the **manifest content**—the visible, surface content—of a communication is analogous to using a standardized questionnaire. To determine, for example, how erotic certain novels are, you might simply count the number of times the word *love* appears in each novel or the average number of appearances per page. Or, you might use a list of words, such as *love, kiss, hug,* and *caress,* each of which might serve as an indicator of the erotic nature of the novel. This method would have the advantage of ease and reliability in coding and of letting the reader of the research report know precisely how eroticism was measured. It would have a disadvantage, on the other hand, in terms of validity. Surely the phrase *erotic novel* conveys a richer and deeper meaning than the number of times the word *love* is used.

While content analysts in the past have needed to engage in hand counts of terms being scored, the computer has made this task easier. If you are coding a digital-format document, for example, you could use a search function to jump from one appearance of a term to the next, counting as you went along. However, computer programs such as Wordscores are streamlining this task even further. Let's suppose you would like to code political documents according to political orientations. First, Wordscores can be used to analyze documents of known orientations (e.g., liberal, conservative, etc.), noting what terms tend to be used frequently. Those patterns can then be used to analyze documents you wish to code on the basis of political orientations.

Alternatively, you could code the communication's underlying meaning, or its **latent content**. In the present example, you might read an entire novel or a sample of paragraphs or pages and make an overall assessment of how erotic the novel was. Although your total assessment might very well be influenced by the appearance of words such as *love* and *kiss*, it would not depend fully on their frequency.

Clearly, this second method seems better designed for capturing the underlying meaning of communications, but its advantage comes at a cost to reliability and specificity. Especially if more than one person is coding the novel, somewhat different definitions or standards may be employed. A passage that one coder regards as erotic may not seem erotic to another. Even if you do all of the coding yourself, there is no guarantee that your definitions and standards will remain constant throughout the enterprise. Moreover, the reader of your research report will likely be uncertain about the definitions you've employed. See Figure 11-3 to compare manifest and latent coding.

Mother and daughter researchers, Carol J. Auster and Lisa A. Auster-Gussman (2014) set about analyzing American cultural differences regarding motherhood and fatherhood. They chose as source materials a sample of Mother's Day and Father's Day cards displayed on the Hallmark website. The researchers regarded the greeting cards as part of a recursive process in which cultural norms conditioned the nature of the cards, while the cards themselves served to reinforce those norms.

The analysis of themes illustrated differences in the traditional views of motherhood and fatherhood, reflecting the traditionally nurturing roles of mothers and instrumental roles of fathers. In addition to these themes, the researchers examined patterns relating to the pictures and objects included in the cards, as well as differences in the colors typically used. The authors note that the Hallmark cards available for analysis featured very few mothers and fathers of color, and they suggested data sources useful for an expanded analysis of racial and ethnic differences.

Manifest Coding of Materials (Objective)

Manifest coding involves the counting of specific elements, such as the word *love*, to determine whether and to what degree the passage should be judged "erotic."

Latent Coding of Materials (Subjective)

Latent coding calls for the researcher to view the entire unit of analysis (a paragraph in this case) and make a subjective assessment regarding whether and to what degree it is "erotic."

FIGURE 11-3
Manifest and Latent Coding.

Conceptualization and the Creation of Code Categories

For all research methods, conceptualization and operationalization typically involve the interaction of theoretical concerns and empirical observations. If, for example, you believe some news editorials on the Web to be liberal and others to be conservative, ask yourself why you

---

latent content In connection with content analysis, the underlying meaning of communications, as distinguished from their *manifest content*.

think so. Read several editorials, asking yourself which points of view are liberal and which ones are conservative. Was the political orientation of a particular editorial most clearly indicated by its manifest content or by its tone? Was your decision based on the use of certain terms (for example, *leftist*, *fascist*, and so on) or on the support or opposition given to a particular issue or political personality?

Both inductive and deductive methods should be used in this activity. If you're testing theoretical propositions, your theories should suggest empirical indicators of concepts. If you begin with specific empirical observations, you should attempt to derive general principles relating to them and then apply those principles to the other empirical observations.

Bruce Berg (1989: 111) places code development in the context of grounded theory and likens it to solving a puzzle:

> *Coding and other fundamental procedures associated with grounded theory development are certainly hard work and must be taken seriously, but just as many people enjoy finishing a complicated jigsaw puzzle, many researchers find great satisfaction in coding and analysis. As researchers . . . begin to see the puzzle pieces come together to form a more complete picture, the process can be downright thrilling.*

Throughout this activity, remember that the operational definition of any variable is composed of the attributes included in it. Such attributes, moreover, should be mutually exclusive and exhaustive. A political website, for example, should not be described as both liberal and conservative, though you should probably allow for some to be middle-of-the-road. It may be sufficient for your purposes to code novels as erotic or non-erotic, but you may also want to consider that some could be anti-erotic. Paintings might be classified as representational or not, if that satisfies your research purpose, or you might wish to classify them as impressionistic, abstract, allegorical, and so forth.

Realize further that different levels of measurement can be used in content analysis. You might, for example, use the nominal categories of liberal and conservative for characterizing political websites, or you might wish to use a more refined ordinal ranking, ranging from extremely liberal to extremely conservative. Bear in mind, however, that the level of measurement

implicit in your coding methods—nominal, ordinal, interval, or ratio—does not necessarily reflect the nature of your variables. If the word *love* appeared 100 times in Novel A and 50 times in Novel B, you would be justified in saying that the word *love* appeared twice as often in Novel A, but not that Novel A was twice as erotic as Novel B. Similarly, agreeing with twice as many anti-Semitic statements in a questionnaire as someone else does not necessarily make one twice as anti-Semitic as that other person.

Counting and Record Keeping

If you plan to evaluate your content analysis data quantitatively, your coding operation must be amenable to data processing. This means, first, that the end product of your coding must be numerical. If you're counting the frequency of certain words, phrases, or other manifest content, the coding is necessarily numerical. But even if you're coding latent content on the basis of overall judgments, it will be necessary to represent your coding decision numerically: 1 = very liberal, 2 = moderately liberal, 3 = moderately conservative, and so on.

Second, your record keeping must clearly distinguish between units of analysis and units of observation, especially if these two are different. The initial coding, of course, must relate to the units of observation. If novelists are the units of analysis, for example, and you wish to characterize them through a content analysis of their novels, your primary records will represent novels as the units of observation. You may then combine your scoring of individual novels to characterize each novelist, the unit of analysis.

Third, while you're counting, it will normally be important to record the base from which the counting is done. It would probably be useless to know the number of realistic paintings produced by a given painter without knowing the number he or she has painted altogether; the painter would be regarded as realistic if a high percentage of paintings were of that genre. Similarly, it would tell us little that the word *love* appeared 87 times in a novel if we did not know about how many words there were in the entire novel. The issue of observational base is most easily resolved if every observation is coded in terms of one of the attributes making up a variable. Rather than simply counting the number of liberal editorials in a given collection, for example, code each editorial by its political orientation, even if it must be coded "no apparent orientation."

Let's suppose we want to describe and explain the editorial policies of different newspapers. Figure 11-4 presents part of a tally sheet that might result from the coding of newspaper editorials. Note that newspapers are the units of analysis. Each newspaper has been assigned an identification number to facilitate mechanized processing. The second column has a space for the number of editorials coded for each newspaper. This will be an important piece of information, because we want to be able to say, for example, "Of all the editorials, 22 percent were pro–United Nations," not just "There were eight pro–United Nations editorials."

One column in Figure 11-4 is for assigning a subjective overall assessment of each newspaper's editorial policies. (Such assignments might later be compared with the several objective measures.) Other columns provide space for recording numbers of editorials reflecting specific editorial positions. In a real content analysis, there would be spaces for recording other editorial positions plus noneditorial information about each newspaper, such as the region in which it is published, its circulation, and so forth.

The type of content analysis just described is sometimes referred to as *conceptual analysis*, to distinguish it from *relational analysis*. The latter goes beyond observing the frequency of a particular concept in a sample of texts to examining the relationships among concepts. For example, you might look for references to "discrimination" in letters to the editor and also note the kind of discrimination being discussed: racial, religious, gender, and so forth. In fact, you could examine the change in that relationship over time.

Qualitative Data Analysis

Not all content analysis results in counting. Sometimes a qualitative assessment of the materials is most appropriate, as in Carol Auster's examination of Disney toys (2012).

Bruce Berg (1989: 123–5) discusses "negative case testing" as a technique for qualitative hypothesis testing. First, in the grounded theory tradition, you begin with an examination of the data, which may yield a general hypothesis. Let's say that you're examining the leadership of a new community association by reviewing the minutes of meetings to see who made motions that were subsequently passed. Your initial examination of the data suggests that the wealthier members are the most likely to assume this leadership role.

Newspaper ID	Number of editorials evaluated	SUBJECTIVE EVALUATION 1. Very liberal 2. Moderately liberal 3. Middle-of-the-road 4. Moderately conservative 5. Very conservative	Number of "isolationist" editorials	Number of "pro–United Nations" editorials	Number of "anti–United Nations" editorials
001	37	2	0	8	0
002	26	5	10	0	6
003	44	4	2	1	2
004	22	3	1	2	3
005	30	1	0	6	0

FIGURE 11-4

Sample Tally Sheet (Partial).

The second stage in the analysis is to search your data to find all the cases that contradict the initial hypothesis. In this instance, you would look for poorer members who made successful motions and wealthy members who never did. Third, you must review each of the disconfirming cases and either (1) give up the hypothesis or (2) see how it needs to be fine-tuned.

Let's say that in your analysis of disconfirming cases, you notice that each of the nonwealthy leaders has a graduate degree, whereas each of the wealthy nonleaders has very little formal education. You may revise your hypothesis to consider both education and wealth as routes to leadership in the association. Perhaps you'll discover some threshold for leadership (a white-collar job, a level of income, and a college degree) beyond which those with the most money, education, or both are the most active leaders.

This process is an example of what Barney Glaser and Anselm Strauss (1967) called *analytic induction*. It is inductive in that it begins primarily with observations, and it is analytic because it goes beyond description to find patterns and relationships among variables.

There are, of course, dangers in this form of analysis, as in all others. The chief risk is misclassifying observations so as to support an emerging hypothesis. For example, you may erroneously conclude that a nonleader didn't graduate from college or you may decide that the job of factory foreman is "close enough" to being white-collar.

Berg (1989: 124) offers techniques for avoiding these errors:

1. If there are sufficient cases, select some at random from each category in order to avoid merely picking those that best support the hypothesis.
2. Give at least three examples in support of every assertion you make about the data.
3. Have your analytic interpretations carefully reviewed by others uninvolved in the research project to see whether they agree.
4. Report whatever inconsistencies you do discover—any cases that simply do not fit your hypotheses. Realize that few social patterns are 100 percent consistent, so you may have discovered something important even if it doesn't apply to absolutely all of social life. However, you should be honest with your readers in that regard.

There are computer programs now available for content analysis. For example, you can try out MAXQDA online. Also, T-LAB provides for some interesting qualitative analyses, such as mapping word associations in a political speech. Some of the programs appropriate for content analysis are discussed in Chapter 13 in connection with other kinds of qualitative data analysis.

Illustrations of Content Analysis

Several studies have indicated that, historically, women have been stereotyped on television. R. Stephen Craig (1992) took this line of inquiry one step further to examine the portrayal of both men and women during different periods of television programming.

To study sex stereotyping in television commercials, Craig selected a sample of 2,209 network commercials during several periods between January 6 and 14, 1990.

> *The weekday day part (in this sample, Monday–Friday, 2–4 P.M.) consisted exclusively of soap operas and was chosen for its high percentage of women viewers. The weekend day part (two consecutive Saturday and Sunday afternoons during sports telecasts) was selected for its high percentage of men viewers. Evening "prime time" (Monday–Friday, 9–11 P.M.) was chosen as a basis for comparison with past studies and the other day parts.*

> *(1992: 199)*

Each of the commercials was coded in several ways. "Characters" were coded as:

All male adults
All female adults
All adults, mixed gender
Male adults with children or teens (no women)
Female adults with children or teens (no men)
Mixture of ages and genders

In addition, Craig's coders noted which character was on the screen longest during the commercial—the "primary visual character"—as well as the roles played by the characters (such as spouse, celebrity, parent), the type of product advertised (such as body product, alcohol), the setting (such as kitchen, school, business), and the voice-over narrator.

Table 11-1 indicates the differences in the times when men and women appeared in commercials. Women appeared most during the daytime (with its soap operas), men predominated

TABLE 11-1

Percentages of Adult Primary Visual Characters by Sex Appearing in Commercials at Different Times

	Daytime	Evening	Weekend
Adult male	40	52	80
Adult female	60	48	20

Source: R. Stephen Craig, "The Effect of Television Day Part on Gender Portrayals in Television Commercials: A Content Analysis," *Sex Roles* 26 5/6 (1992): 204.

during the weekend commercials (with its sports programming), and men and women were equally represented during evening prime time.

Craig found other differences in the ways men and women were portrayed.

> *Further analysis indicated that male primary characters were proportionately more likely than females to be portrayed as celebrities and professionals in every day part, while women were proportionately more likely to be portrayed as interviewer/demonstrators, parent/spouses, or sex object/models in every day part. . . . Women were proportionately more likely to appear as sex object/ models during the weekend than during the day.*

> *(1992: 204)*

The research also showed that different products were advertised during different time periods. As you might imagine, almost all the daytime commercials dealt with body, food, or home products. These products accounted for only one in three commercials on the weekends. Instead, weekend commercials stressed automotive products (29 percent), business products or services (27 percent), or alcohol (10 percent). There were virtually no alcohol ads during evenings and daytime.

As you might suspect, women were most likely to be portrayed in home settings, men most likely to be shown away from home. Other findings dealt with the different roles played by men and women.

> *The women who appeared in weekend ads were almost never portrayed without men and seldom as the commercial's primary character. They were generally seen in roles subservient to men (e.g., hotel receptionist, secretary, or stewardess), or as sex objects or models in which their only function seemed to be to lend an aspect of eroticism to the ad.*

> *(1992: 208)*

Although some of Craig's findings may seem unsurprising, remember that "common knowledge" does not always correspond with reality. It's always worthwhile to check out widely held assumptions. And even when we think we know about a given situation, it's often useful to know specific details such as those provided by a content analysis like this one.

In another content analysis that drew on popular culture for content, Charis Kubrin (2005) chose a primarily qualitative approach. Kubrin was interested in the themes put forth in rap music, particularly in gangsta rap, and the relationship of those themes to neighborhood culture and the "street code."

> *In response to societal and neighborhood conditions, black youth in disadvantaged communities have created a substitute social order governed by their own code—a street code—and rituals of authenticity. . . . This social order reflects the subcultural locus of interests that emerges from pervasive race and class inequality and the social isolation of poor black communities.*

> *(2005: 439)*

She began her study by identifying all the platinum rap albums released between 1992 and 2000: 130 albums containing a total of 1,922 songs. She then drew a simple random sample of one-third of the songs (632) and set about the task of listening to each. She did this twice with each song.

> *First, I listened to a song in its entirety while reading the printed lyrics to determine what the song was about. Second, I listened to the song again and coded each line to determine whether the street code elements described earlier were present: (1) respect, (2) willingness to fight or use violence, (3) material wealth, (4) violent retaliation, (5) objectification of women, and (6) nihilism.*

> *(2005: 443)*

Kubrin was particularly interested in the theme of nihilism—the rejection of traditional moral principles and a fundamental skepticism about the meaning of life. She was interested in how that theme was portrayed in gangsta rap and how it fit into the street code.

Though she began with a sample of 632 songs, she found that no new themes appeared to be showing up after about 350 songs had been analyzed. To be safe, she coded another 50 songs and found no new themes, completing her coding process at that point.

Kubrin noted that rap music is typically regarded as antisocial and resistant to organized society, but her in-depth analysis of lyrics suggested something different:

> *Rap music does not exist in a cultural vacuum; rather it expresses the cultural crossing, mixing, and engagement of black youth culture with the values, attitudes and concerns of the white majority. Many of the violent (and patriarchical, materialistic, sexist, etc.) ways of thinking that are glorified in gangsta rap are a reflection of the prevailing values created and sustained in the larger society.*
>
> *(2005: 454)*

She traces the implications of this for understanding street life as well as for the likely success of various crime-control strategies.

Strengths and Weaknesses of Content Analysis

Probably the greatest advantage of content analysis is its economy in terms of both time and money. A college student might undertake a content analysis, whereas undertaking a survey, for example, might not be feasible. There is no requirement for a large research staff; no special equipment is needed. As long as you have access to the material to be coded, you can undertake content analysis.

Content analysis also has the advantage of allowing the correction of errors. If you discover you've botched a survey or an experiment, you may be forced to repeat the whole research project with all its attendant costs in time and money. If you botch your field research, it may be impossible to redo the project; the event under study may no longer exist. In content analysis, it's usually easier to repeat a portion of the study than it is with other research methods. You might be required, moreover, to recode only a portion of your data rather than all of it.

A third advantage of content analysis is that it permits the study of processes occurring over a long time. You might focus on the imagery of Irish Americans conveyed in U.S. novels written between 1850 and 1860, for example, or you might examine how such imagery has changed from 1850 to the present.

Finally, content analysis has the advantage of all unobtrusive measures, namely, that the content analyst seldom has any effect on the subject being studied. Because the novels have already been written, the paintings already painted, the speeches already presented, content analyses can have no effect on them.

Content analysis has disadvantages as well. For one thing, it's limited to the examination of recorded communications. Such communications may be oral, written, or graphic, but they must be recorded in some fashion to permit analysis.

As we've seen, content analysis has both advantages and disadvantages in terms of validity and reliability. Problems of validity are likely unless you happen to be studying communication processes per se.

On the other side of the ledger, the concreteness of materials studied in content analysis strengthens the likelihood of reliability. You can always code your data and then recode the original documents from scratch. And you can repeat the process as many times as you want. In field research, by contrast, there's no way to return to the original events that were observed, recorded, and categorized.

Let's turn now from content analysis to a related research method: the analysis of existing data. Although numbers rather than communications are analyzed in this case, I think you'll see the similarity to content analysis.

Analyzing Existing Statistics

Frequently you can or must undertake social science inquiry through the use of official or quasi-official statistics. This differs from secondary analysis, in which you obtain a copy of someone else's data and undertake your own statistical analysis. In this section, we're going to look at ways of using the data analyses that others have already done.

This method is particularly significant because existing statistics should always be considered as at least a supplemental source of data. If you were planning a survey of political attitudes, for example, you would do well to examine and present your findings within a context of voting patterns, rates of voter turnout, or similar statistics relevant to your research interest. Or, if you were doing evaluation research on an experimental morale-building program on an assembly line, then statistics on absenteeism, sick leave, and so on would probably be interesting and revealing in connection with the data from your own research. Existing statistics, then, can often provide a historical or conceptual context within which to locate your original research.

Existing statistics can also provide the main data for a social science inquiry. An excellent example is the classic study mentioned at the beginning of this chapter, Emile Durkheim's *Suicide* ([1897] 1951). Let's take a closer look at Durkheim's work before considering some of the special problems this method presents.

Durkheim's Study of Suicide

Why do people kill themselves? Undoubtedly, every suicide case has a unique history and explanation, yet all such cases could no doubt be grouped according to certain common causes: financial failure, trouble in love, disgrace, and other kinds of personal problems. The French sociologist Emile Durkheim had a slightly different question in mind when he addressed the matter of suicide, however. He wanted to discover the environmental conditions that encouraged or discouraged it, especially social conditions.

The more Durkheim examined the available records, the more patterns of differences became apparent to him. One of the first things to attract his attention was the relative stability of suicide rates. Looking at several countries, he found suicide rates to be about the same year after year. He also discovered that a disproportionate number of suicides occurred in summer, leading him to hypothesize that temperature might have something to do with suicide. If this were the case, suicide rates should be higher in the southern European countries than in the temperate ones. However, Durkheim discovered that the highest rates were found in countries in the central latitudes, so temperature couldn't be the answer.

He explored the role of age (35 was the most common suicide age), sex (men outnumbered women around four to one), and numerous other factors. Eventually, a general pattern emerged from different sources.

In terms of the stability of suicide rates over time, for instance, Durkheim found that the pattern was not totally stable. There were spurts in the rates during times of political turmoil, which occurred in several European countries around 1848. This observation led him to hypothesize that suicide might have something to do with "breaches in social equilibrium." Put differently, social stability and integration seemed to be a protection against suicide.

This general hypothesis was substantiated and specified through Durkheim's analysis of a different set of data. The different countries of Europe had radically different suicide rates. The rate in Saxony, for example, was about ten times that of Italy, and the relative ranking of various countries persisted over time. As Durkheim considered other differences among the various countries, he eventually noticed a striking pattern: Predominantly Protestant countries had consistently higher suicide rates than Catholic ones did. The predominantly Protestant countries had 190 suicides per million population; mixed Protestant–Catholic countries, 96; and predominantly Catholic countries, 58 (Durkheim [1897] 1951: 152).

Although suicide rates thus seemed to be related to religion, Durkheim reasoned that some other factor, such as level of economic and cultural development, might explain the observed differences among countries. If religion had a genuine effect on suicide, then the religious difference would have to be found *within* given countries as well. To test this idea, Durkheim first noted that the German state of Bavaria had both the most Catholics and the lowest suicide rates in that country, whereas heavily Protestant Prussia had a much higher suicide rate. Not content to stop there, however, Durkheim examined the provinces composing each of those states.

Table 11-2 shows what he found. As you can see, in both Bavaria and Prussia, provinces with the highest proportion of Protestants also had the highest suicide rates. Durkheim became increasingly confident that religion played a significant role in the matter of suicide.

TABLE 11-2

Suicide Rates in Various German Provinces, Arranged in Terms of Religious Affiliation

Religious Character of Province	Suicides per Million Inhabitants
Bavarian Provinces (1867–1875)*	
Less than 50% Catholic	
Rhenish Palatinate	167
Central Franconia	207
Upper Franconia	204
Average	**192**
50% to 90% Catholic	
Lower Franconia	157
Swabia	118
Average	**135**
More than 90% Catholic	
Upper Palatinate	64
Upper Bavaria	114
Lower Bavaria	19
Average	**75**
Prussian Provinces (1883–1890)	
More than 90% Protestant	
Saxony	309.4
Schleswig	312.9
Pomerania	171.5
Average	**264.6**
68% to 89% Protestant	
Hanover	212.3
Hesse	200.3
Brandenburg and Berlin	296.3
East Prussia	171.3
Average	**220.0**
40% to 50% Protestant	
West Prussia	123.9
Silesia	260.2
Westphalia	107.5
Average	**163.6**
28% to 32% Protestant	
Posen	96.4
Rhineland	100.3
Hohenzollern	90.1
Average	**95.6**

* Note: The population below 15 years of age has been omitted.

Source: Adapted from Emile Durkheim, *Suicide* (Glencoe, IL: Free Press, [1897] 1951), 153.

Returning eventually to a more general theoretical level, Durkheim combined the religious findings with the earlier observation about increased suicide rates during times of political turmoil. As we've seen, Durkheim suggested that many suicides are a product of *anomie*, that is, "normlessness," or a general sense of social instability and disintegration. During times of political strife, people may feel that the old ways of society are collapsing. They become demoralized and depressed, and suicide is one answer to the severe discomfort. Seen from the other direction, social integration and solidarity—reflected in personal feelings of being part of a coherent, enduring social whole—offer protection against depression and suicide. That was where the religious difference fit in. Catholicism, as a far more structured and integrated religious system, gave people a greater sense of coherence and stability than did the more loosely structured Protestantism.

From these theories, Durkheim created the concept of *anomic suicide*. More importantly, as you may know, he added the concept of *anomie* to the lexicon of the social sciences.

This account of Durkheim's classic study is greatly simplified, of course. However, anyone studying social research would profit from examining the original. For our purposes, Durkheim's approach provides a good illustration of the possibilities for research contained in the masses of data regularly gathered and reported by government agencies and other organizations.

In a more recent examination of suicide rates, Steven Barkan, Michael Rocque, and Jason Houle (2013) try to explain the relatively higher rates of suicide in the American West. Reminiscent of Durkheim's conclusion regarding social solidarity, the researchers found that residential stability was a strong force for lowering suicide rates.

The Consequences of Globalization

The notion of "globalization" has become increasingly controversial in the United States and around the world, with reactions ranging from scholarly debates to violent confrontations in the streets. One point of view sees the spread of U.S.-style capitalism to developing countries as economic salvation for those regions. A very different point of view sees globalization as essentially neocolonial exploitation, in which multinational conglomerates exploit the resources and people of poor countries. And, of course, there are numerous variations on these contradictory views.

Jeffrey Kentor (2001) wanted to bring data to bear on the question of how globalization affects the developing countries that host the process. To that end, he used data available from the World Bank's "World Development Indicators." Noting past variations in the way globalization was measured, Kentor used the amount of foreign investment in a country's economy as a percentage of that country's whole economy. He reasoned that dependence on foreign investments was more important than the amount of the investment.

In his analysis of 88 countries with a per capita gross domestic product (the total goods and services produced in a country) of less than $10,000, Kentor found that dependence on foreign investment tended to increase income inequality among the citizens of a country. The greater the degree of dependence, the greater the income inequality. Kentor reasoned that globalization produced well-paid elites who, by working with the foreign corporations, maintained a status well above that of the average citizen. But because the profits derived from the foreign investments tended to be returned to the investors' countries instead of enriching the poor countries, the great majority of the population in the latter reaped little or no economic benefit.

Income inequality, in turn, was found to increase birth rates and, hence, population growth, in a process too complex to summarize here. Population growth, of course, brings a wide range of problems to countries already too poor to provide for the current basic needs of their people.

This research example, along with our brief look at Durkheim's studies, should broaden your understanding of the kinds of social phenomena that we can study through data already collected and compiled by others.

Units of Analysis

The unit of analysis involved in the analysis of existing statistics is often not the individual. Durkheim, for example, was required to work with political-geographic units: countries, regions, states, and cities. The same situation would probably appear if you were to undertake a study of crime rates, accident rates, or disease. By their nature, most existing statistics are aggregated: They describe groups.

The aggregate nature of existing statistics can present a problem, though not an insurmountable one. As we saw, for example, Durkheim wanted to determine whether Protestants or Catholics were more likely to commit suicide. The difficulty was that none of the records available to him indicated the religion of those people who committed suicide. Ultimately, then, it was not possible for him to say whether Protestants committed suicide more often than Catholics did, though he inferred as much. Because Protestant countries, regions, and states had higher suicide rates than did Catholic countries, regions, and states, he drew the obvious conclusion.

There's danger in drawing this kind of conclusion, however. It's always possible that patterns of behavior at a group level do not reflect corresponding patterns on an individual level. Such errors are due to an ecological fallacy, which was discussed in Chapter 4. In the case of Durkheim's study, it was altogether possible, for example, that it was Catholics who committed suicide in the predominantly Protestant areas. Perhaps Catholics in those predominantly Protestant areas were so badly persecuted that they were led into despair and suicide. In that case, it would be possible for Protestant countries to have high suicide rates without any Protestants committing suicide.

Durkheim avoided the danger of the ecological fallacy in two ways. First, his general conclusions were based as much on rigorous theoretical deductions as on the empirical facts. The correspondence between theory and fact made a counterexplanation, such as the one I just made up, less likely. Second, by extensively retesting his conclusions in a variety of ways, Durkheim further strengthened the likelihood that they were correct. Suicide rates were higher in Protestant countries than in Catholic ones; higher in Protestant regions of Catholic countries than in Catholic regions of Protestant countries; and so forth. The replication of findings added to the weight of evidence in support of his conclusions.

Problems of Validity

Whenever we base research on an analysis of data that already exist, we're obviously limited to what exists. Often, the existing data do not cover exactly what we're interested in, and our measurements may not be altogether valid representations of the variables and concepts we want to make conclusions about.

Two characteristics of science are used to handle the problem of validity in analysis of existing statistics: *logical reasoning* and *replication*. Durkheim's strategy provides an example of logical reasoning. Although he could not determine the religion of people who committed suicide, he reasoned that most of the suicides in a predominantly Protestant region would be Protestants.

Replication can be a general solution to problems of validity in social research. Recall the earlier discussion of the interchangeability of indicators (Chapter 5). Crying in sad movies isn't necessarily a valid measure of compassion, nor is putting little birds back in their nests or giving money to charity. None of these things, taken alone, would prove that one group (women, say) was more compassionate than another (men). But if women appeared more compassionate than men by all these measures, that would create a weight of evidence in support of the conclusion. In the analysis of existing statistics, a little ingenuity and reasoning can usually turn up several independent tests of a given hypothesis. If all the tests seem to confirm the hypothesis, then the weight of evidence supports the validity of the measure.

Problems of Reliability

The analysis of existing statistics depends heavily on the quality of the statistics themselves: Do they accurately report what they claim to report? This can be a substantial problem sometimes, because the weighty tables of government statistics, for example, are sometimes grossly inaccurate.

Consider research into crime. Because a great deal of this research depends on official crime statistics, this body of data has come under critical evaluation. The results have not been too encouraging. As an illustration, suppose you were interested in tracing long-term trends in marijuana use in the United States. Official statistics on the numbers of people arrested for selling or possessing marijuana would seem to be a reasonable measure of use, right? Not necessarily.

To begin, you face a hefty problem of validity. Before the passage of the Marihuana Tax Act in 1937, "grass" was legal in the United States,

so arrest records would not give you a valid measure of use. But even if you limited your inquiry to the times after 1937, you would still have problems of reliability, stemming from the nature of law enforcement and crime record keeping, not to mention the states, such as Washington, Oregon, Colorado, Alaska, and Massachusetts, that have legalized recreational uses of it as of this writing.

Law enforcement, for example, is subject to various pressures. A public outcry against marijuana, led perhaps by a vocal citizens' group, often results in a police crackdown on drug trafficking—especially during an election or budget year. A sensational story in the press can have a similar effect. In addition, the volume of other crimes facing the police can affect marijuana arrests.

In tracing the pattern of drug arrests in Chicago between 1942 and 1970, Lois DeFleur (1975) demonstrated that the official records present a far less accurate history of drug use than of police practices and political pressure on police. On a different level of analysis, Donald Black (1970) and others have analyzed the factors influencing whether an offender is actually arrested by police or let off with a warning. Ultimately, official crime statistics are influenced by whether specific offenders are well or poorly dressed, whether they are polite or abusive to police officers, and so forth. When we consider unreported crimes, sometimes estimated to be as much as ten times the number of crimes known to police, the reliability of crime statistics gets even shakier.

These comments concern crime statistics at a local level. Often it's useful to analyze national crime statistics, such as those reported in the FBI's annual *Uniform Crime Reports.* Additional problems are introduced at the national level. For example, different local jurisdictions define crimes differently. Also, participation in the FBI program is voluntary, so the data are incomplete.

Finally, the process of record keeping affects the data available to researchers. Whenever a law-enforcement unit improves its record-keeping system—computerizes it, for example—the apparent crime rates increase dramatically. This can happen even if the number of crimes committed, reported, and investigated does not increase.

Researchers' first protection against the problems of reliability in the analysis of existing statistics is knowing that the problem may exist. Investigating the nature of the data collection and tabulation may enable you to assess the nature and degree of unreliability so that you can judge its potential impact on your research interest. If you also use logical reasoning and replication, you can usually cope with the problem.

Sources of Existing Statistics

It would take a whole book just to list the sources of data available for analysis. In this section, I want to mention a few sources and point you in the direction of finding others relevant to your research interest.

Undoubtedly, one of the more important resources of data about the United States is the annual *Statistical Abstract of the United States*, published by the Department of Commerce from 1878 to 2012. It includes statistics on the individual states and (less extensively) cities, as well as for the nation as a whole. Where else can you find the number of work stoppages in the country year by year, the residential property taxes of major cities, the number of water-pollution discharges reported around the country, the number of business proprietorships in the nation, and hundreds of other such handy bits of information? Best of all, you can access past editions of the *Statistical Abstract* on the Web for free.

While you are probably most familiar with the U.S. Census in terms of the decennial enumeration of the whole population, as prescribed by the Constitution, the Census Bureau conducts numerous other studies. The American Community Survey is another useful source, employing more-frequent sample surveys of the nation. You should be able to learn about where you live, although the extent and accuracy of the data will depend on the size of your community. You can also use the online program, Census Explorer, to examine the American Community Survey data.

Suppose you were interested in the issue of income discrimination by sex. You could examine this rather easily through the *Statistical Abstract* data. The following table, for example, provides a look at sex, education, and income. As you can see, as of 2016, women still had not reached a parity with men, even when they have the same level of education.

Median Weekly Earnings of Full-Time Workers, 2016

	Men	Women	Ratio of Women/Men Earnings
All workers	$969	$784	0.81
Less than high school graduates	$551	$423	0.77
High school graduates	$769	$599	0.78
Some college	$896	$688	0.77
Bachelor's degree	$1,348	$994	0.74
Advanced degree	$1,707	$1,257	0.74

Note: These data point to a persistent difference between the incomes of men and women, even when both groups have achieved the same levels of education.

Source: Bureau of Labor Statistics, Usual Weekly Earnings of Wage and Salary Workers, Fourth Quarter 2016. January 27, 2017, Table 9. Quartiles and selected deciles of usual weekly earnings of full-time wage and salary workers by selected characteristics, 2016 annual averages.

As we've seen before, a graphic presentation can sometimes communicate data more clearly than tables of numbers. You could enter the above incomes into a spreadsheet program and have it create a graphic display as shown Figure 11-5.

These data point to a persistent difference between the incomes of men and women, even when both groups have achieved the same levels of education. Other variables could explain the differences, however; we'll return to this issue in Chapter 14.

World statistics are available through the United Nations. Its *Demographic Yearbook* presents annual vital statistics (births, deaths, and other data relevant to population) for the individual nations of the world. Other publications report a variety of other kinds of data. Again, utilizing the resources at your library or on the Web may be the best introduction to what's available.

The amount of data provided by nongovernment agencies is as staggering as the amount your taxes buy. Chambers of Commerce often publish data reports on businesses, as do private consumer groups. Common Cause covers politics and government. The Gallup Organization

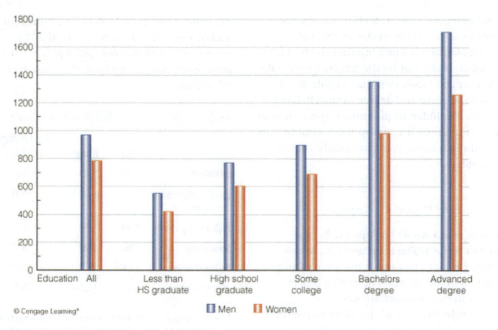

FIGURE 11-5

Graphic Display of Gender, Education, and Income.

publishes reference volumes on public opinion as measured by Gallup Polls since 1935.

Organizations such as the Population Reference Bureau publish a variety of demographic data, U.S. and international, that a secondary analyst could use. Their *World Population Data Sheet* and *Population Bulletin* are resources heavily used by social scientists. Social indicator data can be found in the journal *SINET: A Quarterly Review of Social Reports and Research on Social Indicators, Social Trends, and the Quality of Life.*

A new guide to Population Action International's mapping website shows how climate change and population dynamics will change the world over time. High rates of population growth and climate-change consequences overlap in many countries. Interactive maps illustrate how climate-change impacts, demographic trends, and the need for contraception are likely to affect countries' abilities to adapt to the effects of climate change.

The maps identify 33 population and climate-change hot spots—countries that are experiencing rapid population growth, low resilience to climate change, and high projected declines in agricultural production. Many hot spots are currently experiencing water stress or scarcity, a condition that will worsen with continued rapid

population growth. In many countries, a high proportion of women lack access to reproductive health services and contraceptives; investments in family-planning programs in these hot spots could improve health and well-being, slow population growth, and reduce vulnerability to climate-change impacts.

The sources I've mentioned here represent only a tiny fraction of the thousands that are available. With so much data already collected, the lack of funds to support expensive data collection is no reason for not doing good and useful social research. Moreover, as we've seen, this research method need not be limited to tables of numbers. There are graphic resources available as well, such as the *Social Explorer*. A wide range of data about the United States can be represented on a map of congressional districts or census tracts. You can examine aspects of population, religion, economy, and many other variables. For example, you can easily find the geographic concentrations of unmarried partners: male/female, male/male, and female/female.

You can do similar kinds of map-based examinations through the Census Bureau by clicking on "Maps" at their website. Once you've displayed a variable such as multiracial marriages state-by-state, you can click on a particular state and get a detailed graph of the interracial marriages in that state.

Google Public Data offers another convenient resource for a variety of social research analyses: particularly international data. Built-in software allows you to create graphic illustrations from your analyses.

Social media provides another source of data for social analysis, and Topsy Social Analytics (http://topsy.com) offers a tool for analyzing Twitter communications. You can enter a search term (for example, Ebola) and see tweets on that topic—and plot the frequency of such tweets over time.

Finally, the Association of Religious Data Archives (ARDA) (http://thearda.com) provides a substantial variety of both qualitative and quantitative religious data, including surveys of religious beliefs, attitudes, and practices. Moreover, ARDA's Measurement Wizard allows you to zero in on precisely the data you are interested in.

Let's move now from an inherently quantitative method to one that is typically qualitative: comparative and historical research.

Comparative and Historical Research

Comparative and historical research differs substantially from the methods discussed so far, though it overlaps somewhat with field research, content analysis, and the analysis of existing statistics. It involves the use of historical methods by sociologists, political scientists, and other social scientists to examine societies (or other social units) over time and in comparison with one another.

The discussion of longitudinal research designs in Chapter 4 notwithstanding, our examination of research methods so far has focused primarily on studies anchored in one point in time and in one locale, whether a small group or a nation. Although accurately portraying the main thrust of contemporary social science research, this focus conceals the fact that social scientists are also interested in tracing the development of social forms over time and comparing those developmental processes across cultures. James Mahoney and Dietrich

Rueschemeyer (2003: 4) suggest that current comparative and historical researchers "focus on a wide range of topics, but they are united by a commitment to providing historically grounded explanations of large-scale and substantively important outcomes." Thus, you find comparative and historical studies dealing with topics of social class, capitalism, religion, revolution, and the like.

After describing some major instances of comparative and historical research, past and present, this section discusses some of the key elements of this method.

Examples of Comparative and Historical Research

Auguste Comte, who coined the term *sociologie*, saw that new discipline as the final stage in a historical development of ideas. With his broadest brush, he painted an evolutionary picture that took humans from a reliance on religion to metaphysics to science. With a finer brush, he portrayed science as evolving from the development of biology and the other natural sciences to the development of psychology and, finally, to the development of scientific sociology.

A great many later social scientists have also turned their attention to broad historical processes. Several have examined the historical progression of social forms from the simple to the complex, from rural-agrarian to urban-industrial societies. The U.S. anthropologist Lewis Morgan, for example, saw a progression from "savagery" to "barbarism" to "civilization" (1870). Robert Redfield, another anthropologist, wrote more recently of a shift from "folk society" to "urban society" (1941). Emile Durkheim saw social evolution largely as a process of ever-greater division of labor ([1893] 1964). In a more specific analysis, Karl Marx examined economic systems progressing historically from primitive to feudal to capitalistic forms ([1867] 1967). All history, he wrote in this context, was a history of class struggle—the "haves" struggling to maintain

---

comparative and historical research The examination of societies (or other social units) over time and in comparison with one another.

their advantages and the "have-nots" struggling for a better lot in life. Looking beyond capitalism, Marx saw the development of socialism and finally communism.

Not all historical studies in the social sciences have had this evolutionary flavor, however. Some social science readings of the historical record, in fact, point to grand cycles rather than to linear progressions. No scholar better represents this view than Pitirim A. Sorokin. A participant in the Russian Revolution of 1917, Sorokin served as secretary to Prime Minister Kerensky. Both Kerensky and Sorokin fell from favor, however, and Sorokin began his second career—as a U.S. sociologist.

Whereas Comte read history as a progression from religion to science, Sorokin (1937–1940) suggested that societies alternate cyclically between two points of view, which he called "ideational" and "sensate." Sorokin's sensate point of view defined reality in terms of sense experiences. The ideational, by contrast, placed a greater emphasis on spiritual and religious factors. Sorokin's reading of the historical record further indicated that the passage between the ideational and sensate was through a third point of view, which he called the "idealistic." This third view combined elements of the sensate and ideational in an integrated, rational view of the world.

These examples indicate some of the topics comparative and historical researchers have examined. To get a better sense of what comparative and historical research entails, let's look at a few examples in somewhat more detail.

Weber and the Role of Ideas

In his analysis of economic history, Karl Marx put forward a view of economic determinism. That is, he postulated that economic factors determined the nature of all other aspects of society. For example, Marx's analysis showed that a function of European churches was to justify and support the capitalist status quo—religion was a tool of the powerful in maintaining their dominance over the powerless. "Religion is the sigh of the oppressed creature," Marx wrote in a famous passage, "the sentiment of a heartless world, and the soul of soulless conditions. It is the opium of the people" (Bottomore and Rubel [1843] 1956: 27).

Max Weber, a German sociologist, disagreed. Without denying that economic factors could and did affect other aspects of society, Weber argued that economic determinism did not explain everything. Indeed, Weber said, economic forms could come from noneconomic ideas. In his research in the sociology of religion, Weber examined the extent to which religious institutions were the source of social behavior rather than mere reflections of economic conditions. His most noted statement of this side of the issue is found in *The Protestant Ethic and the Spirit of Capitalism* ([1905] 1958). Here's a brief overview of Weber's thesis.

John Calvin (1509–1564), a French theologian, was an important figure in the Protestant reformation of Christianity. Calvin taught that the ultimate salvation or damnation of every individual had already been decided by God; this idea is called *predestination*. Calvin also suggested that God communicated his decisions to people by making them either successful or unsuccessful during their earthly existence. God gave each person an earthly "calling"—an occupation or profession—and manifested their success or failure through that medium. Ironically, this point of view led Calvin's followers to seek proof of their coming salvation by working hard, saving their money, and generally striving for economic success.

In Weber's analysis, Calvinism provided an important stimulus for the development of capitalism. Rather than "wasting" their money on worldly comforts, the Calvinists reinvested it in their economic enterprises, thus providing the capital necessary for the development of capitalism. In arriving at this interpretation of the origins of capitalism, Weber researched the official doctrines of the early Protestant churches, studied the preaching of Calvin and other church leaders, and examined other relevant historical documents.

In three other studies, Weber conducted detailed historical analyses of Judaism ([1934] 1952) and the religions of China ([1934] 1951) and India ([1934] 1958). Among other things, Weber wanted to know why capitalism had not developed in the ancient societies of China, India, and Israel. In none of the three religions of these countries did he find any teaching that would have supported the accumulation and reinvestment of capital—strengthening his conclusion about the role of Protestantism in that regard.

Fair Trade

If you buy coffee at a grocery store or coffeehouse, you may have noticed that some of the packages are labeled "fair trade." As you might know, the fair trade certification reflects an international, social/ecological/economic movement formed to support farmers and laborers in developing countries. The fair trade movement seeks equity in international trade and aims to ensure that these workers receive a higher price for the products they grow and export. In a free-market economy, it is common that growers of products like coffee, chocolate, and bananas actually receive very little of the money that you, a consumer in a developed country, might pay for it. In practice, fair trade reflects economic reorganization. It may include local farmer co-ops working with international nonprofit organizations, such as the Institute for Agriculture and Trade Policy, to cut out the "middlemen" and thus deliver more money and price stability to those doing the work. Fair trade practices are also focused on improving environmental standards and sustainability practices.

Daniel Jaffee (2007) came in contact with that movement in 2003 while attending a meeting of the World Trade Organization in Mexico. A group for the delegates staged a demonstration on behalf of fair trade and walked out of the WTO meeting to move into a smaller conference of their own. Jaffee followed them and began his extended study of fair trade economics.

> Over two years, I lived, worked, and talked with these farmers, as well as with their neighbors who know a very difference coffee market—the conventional market represented by local coyotes, middlemen who often pay them less than it costs to produce their coffee in the first place.
>
> (2007: xiv)

Jaffee's research involved participant observation, as his description indicates, but also the collection and analysis of quantitative data about production, prices, income, and the like. In part, he was interested in placing the new movement within the larger context of world coffee production and marketing. (Fair trade represents roughly 1 percent of the total.)

He was also interested in the evolution of the movement over time, as fair trade became better known and more popular. He examined the development of the organizations involved and looked at the adjustments required when large distributors such as Starbucks began offering fair trade coffee as an option for its customers. Whereas we have seen that some research methods offer a snapshot of social life at one point in time, Jaffee's analysis offers a motion picture of an ongoing social process.

Here are a few brief examples to illustrate some of the topics that are of interest to comparative and historical scholars today.

- *The Rise of Christianity:* Rodney Stark (1997) lays out his research question in the book's subtitle: *How the Obscure, Marginal Jesus Movement Became the Dominant Religious Force in the Western World in a Few Centuries.* For many people, the answer to this puzzle is a matter of faith in the miraculous destiny of Christianity. Without debunking Christian faith, Stark looks for a scientific explanation, undertaking an analysis of existing historical records that sketch out the population growth of Christianity during its early centuries. He notes, among other things, that the early growth rate of Christianity, rather than being unaccountably rapid, was very similar to the contemporary growth of Mormonism. He then goes on to examine elements in early Christian practice that gave it growth advantages over the predominant paganism of the Roman Empire. For example, the early Christian churches were friendlier to women than paganism was, and much of the early growth occurred among women—who often converted their husbands later on. And in an era of deadly plagues, the early Christians were more willing to care for stricken friends and family members, which not only enhanced the survival of Christians but also made it a more attractive conversion prospect. At every turn in the analysis, Stark makes rough calculations of the demographic impact of cultural factors. This study is an illustration of how social research methods can shed light on nonscientific realms such as faith and religion.
- *Policing World Society:* Mathieu Deflem (2002) set out to learn how contemporary systems of international cooperation among police agencies came about. All of us have heard

movie and TV references to the international police organization, Interpol. Deflem went back to the middle of the nineteenth century and traced its development through World War II. In part, his analysis examines the strains between the bureaucratic integration of police agencies in their home governments and the need for independence from those governments.

- *Organizing America:* Charles Perrow (2002) wanted to understand the roots of the uniquely American form of capitalism. Compared with European nations, the United States has shown less interest in providing for the needs of average citizens and has granted greater power to gigantic corporations. Perrow feels the die was pretty much cast by the end of the nineteenth century, resting primarily on Supreme Court decisions in favor of corporations and the experiences of the textile and railroad industries.

- *Diminished Democracy:* Theda Skocpol (2003) turns her attention to something that fascinated Alexis de Tocqueville in his 1840 *Democracy in America*: the grassroots commitment to democracy, which appeared in all aspects of American community life. It almost seemed as though democratic decision making was genetic in the New World, but what happened? Skocpol's analysis of contemporary U.S. culture suggests a "diminished democracy" that cannot be easily explained by the ideologies of either the right or the left.

These examples of comparative and historical research should give you some sense of the potential power of the method. Let's turn now to an examination of the sources and techniques used in this method.

Sources of Comparative and Historical Data

As we saw in the case of existing statistics, there is no end of data available for analysis in historical research. To begin, historians may have already reported on whatever it is you want to examine, and their analyses can give you an initial grounding in the subject, a jumping-off point for more in-depth research.

Most likely you'll ultimately want to go beyond others' conclusions and examine some "raw data" to draw your own conclusions. These data vary, of course, according to the topic under study. When W. I. Thomas and Florian Znaniecki

(1918) studied the adjustment process for Polish peasants coming to the United States early in the twentieth century, they examined letters written by the immigrants to their families in Poland. (They obtained the letters by placing newspaper advertisements.) Other researchers have analyzed old diaries. Such personal documents only scratch the surface, however. In discussing procedures for studying the history of family life, Ellen Rothman points to the following sources:

> In addition to personal sources, there are public records which are also revealing of family history. Newspapers are especially rich in evidence on the educational, legal, and recreational aspects of family life in the past as seen from a local point of view. Magazines reflect more general patterns of family life; students often find them interesting to explore for data on perceptions and expectations of mainstream family values. Magazines offer several different kinds of sources at once: visual materials (illustrations and advertisements), commentary (editorial and advice columns), and fiction. Popular periodicals are particularly rich in the last two. Advice on many questions of concern to families—from the proper way to discipline children to the economics of wallpaper—fills magazine columns from the early nineteenth century to the present. Stories that suggest common experiences or perceptions of family life appear with the same continuity.
>
> (1981: 53)

Organizations generally document themselves, so if you're studying the development of some organization you should examine its official documents: charters, policy statements, speeches by leaders, and so on. Once, when I was studying the rise of a contemporary Japanese religious group—Soka Gakkai—I discovered not only weekly newspapers and magazines published by the group but also a published collection of all the speeches given by the original leaders. With these sources, I could trace changes in recruitment patterns over time. At the outset, followers were enjoined to enroll the entire world. Later, the emphasis shifted specifically to Japan. Once a sizable Japanese membership had been established, an emphasis on enrolling the entire world returned (Babbie 1966).

Often, official government documents provide the data needed for analysis. To better appreciate the history of race relations in the United States, A. Leon Higginbotham Jr. (1978) examined

200 years of laws and court cases involving race. As the first African American appointed as a federal judge, Higginbotham found that, rather than protecting African Americans, the law embodied bigotry and oppression. In the earliest court cases, there was considerable ambiguity over whether African Americans were indentured servants or, in fact, slaves. Later court cases and laws clarified the matter—holding African Americans to be something less than human.

The sources of data for historical analysis are too extensive to cover even in outline here, although the examples we've looked at should suggest some ideas. Whatever resources you use, however, a couple of cautions are in order.

As we saw in the case of existing statistics, you can't trust the accuracy of records—official or unofficial, primary or secondary. Your protection lies in replication: In the case of historical research, that means corroboration. If several sources point to the same set of "facts," your confidence in them might reasonably increase.

At the same time, you need always be wary of bias in your data sources. If all your data on the development of a political movement are taken from the movement itself, you're unlikely to gain a well-rounded view of it. The diaries of well-to-do gentry of the Middle Ages may not give you an accurate view of life in general during those times. Where possible, obtain data from a variety of sources representing different points of view.

As Ron Aminzade and Barbara Laslett indicate in "How to Do It: Reading and Evaluating Documents," there is an art to knowing how to regard such documents and what to make of them.

Incidentally, the critical review that Aminzade and Laslett urge for the reading of historical documents is useful in many areas of your life besides the pursuit of comparative and historical research. Consider applying some of their questions to presidential press conferences, advertising, or (gasp) college textbooks. None of these offers a direct view of reality; all have human authors and human subjects.

Analytic Techniques

The analysis of comparative and historical data is another vast subject that I can't cover exhaustively here. Moreover, because comparative and

---

How to Do It

Reading and Evaluating Documents

by Ron Aminzade and Barbara Laslett
University of Minnesota

The purpose of the following comments is to give you some sense of the kind of interpretive work historians do and the critical approach they take toward their sources. It should help you to appreciate some of the skills historians develop in their efforts to reconstruct the past from residues, to assess the evidentiary status of different types of documents, and to determine the range of permissible inferences and interpretations. Here are some of the questions historians ask about documents:

1. Who composed the documents? Why were they written? Why have they survived all these years? What methods were used to acquire the information contained in the documents?

2. What are some of the biases in the documents and how might you go about checking or correcting them? How inclusive or representative is the sample of individuals, events, and so on, contained in the document? What were the institutional constraints and the general organizational routines under which the document was prepared? To what extent does the document provide more of an index of institutional activity than of the phenomenon being studied? What is the time lapse between the observation of the events documented and the witnesses' documentation of them? How confidential or public was the document meant to be? What role did etiquette, convention, and custom play in the presentation of the material contained within the document? If you relied solely upon the evidence contained in these documents, how might your vision of the past be distorted? What other kinds of documents might you look at for evidence on the same issues?

3. What are the key categories and concepts used by the writer of the document to organize the information presented? What selectivities or silences result from these categories of thought?

4. What sorts of theoretical issues and debates do these documents cast light on? What kinds of historical and/or sociological questions do they help to answer? What sorts of valid inferences can one make from the information contained in these documents? What sorts of generalizations can one make on the basis of the information contained in these documents?

historical research is usually a qualitative method, there are no easily listed steps to follow in the analysis of historical data. Nevertheless, a few comments are in order.

Max Weber used the German term *verstehen*—"understanding"—in reference to an essential quality of social research. He meant that the researcher must be able to take on, mentally, the circumstances, views, and feelings of those being studied, so that the researcher can interpret their actions appropriately. Certainly this concept applies to comparative and historical research. The researcher's imaginative understanding is what breathes life and meaning into the evidence being analyzed.

The comparative and historical researcher must find patterns among the voluminous details describing the subject matter under study. Often,

this takes the form of what Weber called "ideal types": conceptual models composed of the essential characteristics of social phenomena. For example, Weber himself did considerable research on bureaucracy. Having observed numerous actual bureaucracies, Weber ([1925] 1946) detailed those qualities essential to bureaucracies in general: jurisdictional areas, hierarchically structured authority, written files, and so on. Weber did not merely list those characteristics common to all the actual bureaucracies he observed. Rather, to create a theoretical model of the "perfect" (ideal type) bureaucracy, he needed to understand fully the essentials of bureaucratic operation. Figure 11-6 offers a more recent illustration of some positive and negative aspects of bureaucracy as a general social phenomenon.

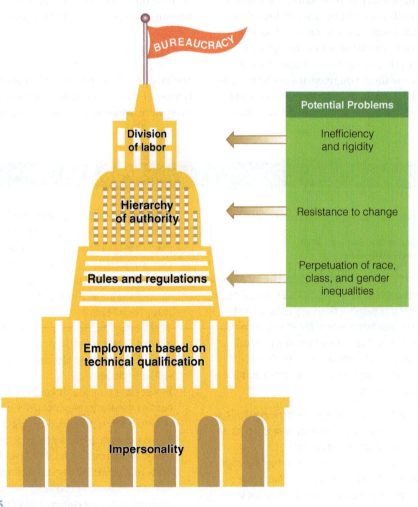

FIGURE 11-6

Some Positive and Negative Aspects of Bureaucracy.

Source: Adapted from Diana Kendall, *Sociology in Our Times*, 8th ed. (Belmont, CA: Wadsworth, 2011). Used by permission.

Often, comparative and historical research is informed by a particular theoretical paradigm. Thus, Marxist scholars may undertake historical analyses of particular situations—such as the history of Latinos and Latinas in the United States—to determine whether they can be understood in terms of the Marxist version of conflict theory. Sometimes, comparative and historical researchers attempt to replicate prior studies in new situations—for example, doing follow-up replications of Weber's studies of religion and economics.

Although comparative and historical research is often regarded as a qualitative rather than quantitative technique, this is by no means necessary. Historical analysts sometimes use time-series data to monitor changing conditions over time, such as data on population, crime rates, unemployment, infant mortality rates, and so forth. The analysis of such data sometimes requires sophistication, however. For example, Larry Isaac and Larry Griffin (1989) discuss the uses of a variation on regression techniques (see Chapter 16) in determining the meaningful breaking points in historical processes, as well as for specifying the periods within which certain relationships occur among variables. Criticizing the tendency to regard history as a steadily unfolding process, the authors focus their attention on the statistical relationship between unionization and the frequency of strikes, demonstrating that the relationship has shifted importantly over time.

Isaac and Griffin raise several important issues regarding the relationship among theory, research methods, and the "historical facts" they address. Their analysis, once again, warns against the naive assumption that history, as documented, necessarily coincides with what actually happened.

Unobtrusive Online Research

Since this is the final chapter on methods of data collection, it might be useful to review some of the ways in which online data are being used for unobtrusive social research. We've seen the wealth of data sources online, but online processes themselves can be the subject of study.

Big data is a term you are likely to hear increasingly in this context. It refers to the gigantic data sets being automatically compiled from online activity. The most notorious of these in recent times has been the National Security Agency's (NSA) compilation and analysis of phone, e-mail, and other communications. Prior to the development of massive data-storage capacities, this would have been unimaginable.

Big data are also in common use by commercial enterprises such as Google, Amazon, and many retailers. Have you ever noticed that after you've read an article online about cocker spaniels, you begin receiving e-mails advertising pet products, and the pop-up ads on websites you visit in the future may feature similar commodities. Social media such as Facebook and Twitter are another source of big data.

The rise of social media has both impacted society and opened new avenues for unobtrusive social research. For example, Rachel Gong (2011) examined the extent to which the use of social media might affect political success. Noting the general conclusion that President Barack Obama's electoral success rested heavily on his use of then-new electronic media, Gong tested the breadth of this effect—examining the utility of social media in a society such as Malaysia, where the conventional media were state controlled. She found that parliamentary candidates whose campaigns maintained a blog were far more successful than those who did not. This was found to be especially true for candidates running in opposition to the status quo.

Noah Smith and colleagues at Carnegie Mellon University are developing several computer programs for analyzing natural language. One of their special interests is the analysis of Twitter communications, and they have used those analyses to explain political positions in Congress and to predict NFL outcomes among other things. Katy Steinmetz (2013) has reviewed their work and you can learn more at Noah's Ark (http://www.ark.cs.washington.edu).

Social research journals are increasingly addressing the emerging methodologies. The May 2013 issue of the *International Journal of Social Research Methodology*, for example, is devoted to the topic of digital social research, examining new techniques and their implications for social science.

Robert Ackland and Rachel Gibson (2013) examined differences in the way 100 political parties in six countries used their websites for different kinds of communications. Some of the hyperlinks on political websites direct users to a candidate or to issue websites that reflect the values of the party, while others aim at discrediting opponents. Still other hyperlinks sought to enhance the party by associating it with larger or more-established groups. Among other things, the researchers found left-wing parties using hyperlinks to demonstrate an international orientation and an affinity for the nonprofit sector. Rightwing parties, by contrast, used hyperlinks to show their affinity for business and the commercial sector, and for same-country groups (2013: 241).

Rob Procter, Farida Vis, and Alex Voss (2013) set out to analyze the 2011 London anti-austerity riots by using 54 Twitter hashtags to identify 2.6 million tweets relating to various aspects of the riots. This enormous mass of data allowed them to monitor the rise and fall of a rumor that the London Eye and Big Ben were on fire, for example. The sheer volume of data required them to invent new techniques for content analysis.

Adam Edwards and colleagues (2013) point to an interesting distinction for social media research in comparison with other social science techniques. Some methods, such as survey research, permit a broad viewing (e.g., societywide) of social life at a specified time. Other methods, such as participant observation, allow for an immediate scrutiny of social processes but with a much narrower view. Some of the digital research into social media allows a dynamic observation of process—as it is happening—on a scale as broad as a national survey. Overall, the authors do not foresee digital research replacing conventional methods, but they do anticipate new potential for studying human social behavior with these emerging methodologies.

The variety of online sources and techniques opens new possibilities for approaching research questions from several directions. If each independent approach produces the same conclusion, we can be more confident of our findings. Seth Stephens-Davidowitz (2013) offered an intriguing example when he sought to estimate the percentage of American men who are gay. Noting surveys that may ask for this information, he worried that many gay men would not report their sexuality, especially in states and communities that were intolerant of gays.

Some of the data sources he examined online were Facebook profiles indicating preference for same-sex partners. He found fewer such profiles for Facebook members in states intolerant of gays. To accommodate the possibility that gay men might move to more-tolerant states, Stephens-Davidowitz focused his attention on high school students, reasoning that they would be less able to pack up and move.

Stephens-Davidowitz also obtained data from match.com, an online dating site, and Craigslist. Using Google, he was able to examine searches for gay porn sites and married women searching for answers to the question: "Is my husband gay?" Each of these approaches offered estimates of the percentage of openly gay men and the percentage still in the closet.

The possibilities for big data analysis are, well, big. There are also problems. First, the data available for analysis may not correspond exactly with the concepts you'd like to explore. Tyler Vigen (2014) has drawn attention to another, sometimes humorous, problem. With the volume of analytic possibilities, you may find correlations between variables that have no meaning: There is no causal relationship. The correlation is merely a matter of chance. For example, Vigen presents a persuasive graph showing the close correlation between "U.S. spending on science, space, and technology" on the one hand and "suicides by hanging, strangulation, and suffocation" on the other.

Ethics and Unobtrusive Measures

The use of unobtrusive measures avoids many of the ethical issues we've discussed in connection with other data-collection techniques, but if you reflect on the general principles we've discussed, I think you'll see that there are potential risks to guard against.

The general principle of confidentiality may be relevant in some projects, for example. Let's suppose you want to examine an immigrant subculture through a content analysis of letters written back to the old country, as was the case in the Thomas and Znaniecki (1918) study of Polish peasants, mentioned earlier in the chapter. To begin, you should obtain those letters legally and ethically (no getting a government agency

What do you think?...Revisited

Unobtrusive research techniques allow researchers to avoid having an effect on what is being studied. Given that, why don't social scientists limit themselves to these techniques?

As we've seen, each of the unobtrusive techniques presented in this chapter has shortcomings of its own. The most general is that we may not be able to find existing statistics, recorded communications, or historical records that provide valid and reliable data relevant to the topic we

wish to study. Other techniques, such as experiments, surveys, and field research, allow us to generate original data to fill such voids.

The ideal approach is to use multiple techniques, including unobtrusive ones. The use of multiple approaches to conduct research can substantiate our findings when they agree, and they can broaden our understanding of the subject matter when they do not agree.

---

to intercept the letters for you), and you need to protect the privacy of the letter writers and recipients.

As with all other research techniques, you're obliged to collect data, analyze them, and report your findings honestly, with the purpose of discovering what is so, rather than attempting to

support a favored hypothesis or personal agenda. While it may be easy to agree with such a principle, you're likely to find it somewhat more difficult to apply when you actually conduct research. Your ethical sensibilities will be more challenged by the vast gray areas than by the black-and-white ones.

MAIN POINTS

Introduction
- Unobtrusive measures are ways of studying social behavior without affecting it in the process.

Content Analysis
- Content analysis is a social research method appropriate for studying human communications. Researchers can use it to study not only communication processes but other aspects of social behavior as well.
- Common units of analysis in content analysis include units of communication—words, paragraphs, books, and so forth. Standard probability sampling techniques are sometimes appropriate in content analysis.
- Content analysis involves coding—transforming raw data into categories based on some conceptual scheme. Coding may attend to both manifest and latent content. The determination of latent content requires judgments on the part of the researcher.
- Both quantitative and qualitative techniques are appropriate for interpreting content analysis data.
- The advantages of content analysis include economy, safety, and the ability to study processes occurring over a long time. Also, it is unobtrusive. Its disadvantages are that it is limited to recorded communications and can raise issues of reliability and validity.

Analyzing Existing Statistics
- A variety of government and nongovernment agencies provide aggregate statistical data for studying aspects of social life.
- Problems of validity in the analysis of existing statistics can often be handled through logical reasoning and replication.
- Because existing statistics often generate problems of reliability, researchers must use them with caution.

Comparative and Historical Research
- Social scientists use comparative and historical methods to discover patterns in the histories of different cultures.
- Although often regarded as a qualitative method, comparative and historical research can make use of quantitative techniques, such as the analysis of time-series data.
- Social media and other online activities have opened up a new realm of possibilities for unobtrusive research.
- Researchers are developing new techniques for analyzing online content, such as big data.

Unobtrusive Online Research
- *Big data* refers to the gigantic data sets being automatically compiled from online activity.
- This is very much a method under development at present.

Ethics and Unobtrusive Measures
- Sometimes even unobtrusive measures can raise the possibility of violating subjects' privacy.

- The general principles of honest observation, analysis, and reporting apply to all research techniques.

KEY TERMS

coding

comparative and
  historical research

content analysis

latent content

manifest content

unobtrusive research

PROPOSING SOCIAL RESEARCH: UNOBTRUSIVE MEASURES

This chapter has provided an overview of three major types of unobtrusive measures: content analysis, existing statistics, and historical/comparative. While existing statistics represent, by their nature, a quantitative method, the other two can be done with a qualitative and/or quantitative approach. In this exercise, you'll want to identify which method and orientation you plan to use. If you're doing these exercises for the purpose of better understanding the topics of the book, you could try your hand at each of these methods.

You'll need to describe the data you'll use and anything special about your access to those data. Whether you're studying newspaper editorials, infant mortality rates, or accounts of political revolutions, you'll likely face potential problems of validity and reliability. Unobtrusive methods involve the use of available data, which often offer approximations of the observations we might ideally like to make. For example, you may need to use drug-arrest rates as an approximation of drug-use rates. You should discuss how you plan to deal with any such approximations.

REVIEW QUESTIONS

1. Is the Republican or the Democratic party the more supportive of free speech? In two or three paragraphs, outline a content analysis design to answer this question. Be sure to specify units of analysis, sampling methods, and relevant measurements.

2. Social scientists often contrast the sense of "community" in villages, small towns, and neighborhoods with life in large, urban societies. What, in your opinion, are the essential qualities of an ideal type of community?

3. How might you compare lifestyles in different societies around the world, using pictures on the World Wide Web? What cultural features would you look for? How would you identify differences and similarities?

4. How old is the college or university you're attending? When you decide on an age, specify *what* is that old. Is it people, buildings, or something else? Cite the sources you might use in arriving at your conclusion. Discuss any ambiguities that might exist in determining the age of your college or university.

CHAPTER 12

Evaluation Research

CHAPTER OVERVIEW

Now you're going to see one of the most rapidly growing uses of social research: the evaluation of social interventions. You'll come away from this chapter able to judge whether social programs have succeeded or failed.

Robert Kneschke/Shutterstock.com

Introduction

You may not be familiar with *Twende na Wakati* (*Let's Go with the Times*), but it was the most popular radio show in Tanzania a few years ago. It's a serial drama. The main character, Mkwaju, was a truck driver with some pretty traditional ideas about gender roles and sexuality. By contrast, Fundi Mitindo, a tailor, and his wife, Mama Waridi, had more modern ideas regarding the roles of men and women, particularly in relation to the issues of overpopulation and family planning.

Twende na Wakati was the creation of Population Communications International (PCI) and other organizations working in conjunction with the Tanzanian government in response to two problems facing that country: (1) a population growth rate over twice that of the rest of the world and (2) an AIDS epidemic particularly heavy along the international truck route, where more than a fourth of the truck drivers and over half the commercial sex workers were found to be HIV-positive in 1991. The prevalence of contraceptive use was 11 percent (Rogers et al. 1996: 5–6).

The purpose of the serial drama was to bring about a change in knowledge, attitudes, and practices (KAP) relating to contraception and family planning. Rather than instituting a conventional educational campaign, PCI felt it would be more effective to illustrate the message through entertainment.

Between 1993 and 1995, there were 208 episodes of *Twende na Wakati* aired, aiming at the 67 percent of Tanzanians who listen to the radio. Eighty-four percent of the radio listeners reported listening to the PCI serial drama,

making it the most popular show in the country. Ninety percent of the show's listeners recognized Mkwaju, the sexist truck driver, and only 3 percent regarded him as a positive role model. Over two-thirds identified Mama Waridi, a businesswoman, and her tailor husband as positive role models.

Surveys conducted to measure the impact of the show indicated that it had affected knowledge, attitudes, and behavior. For example, 49 percent of the married women who listened to the show said they now practiced family planning, compared with only 19 percent of the nonlisteners. There were other impacts:

> Some 72 percent of the listeners in 1994 said that they adopted an HIV/AIDS prevention behavior because of listening to "Twende na Wakati," and this percentage increased to 82 percent in our 1995 survey. Seventy-seven percent of these individuals adopted monogamy, 16 percent began using condoms, and 6 percent stopped sharing razors and/or needles.
>
> (Rogers et al. 1996: 21)

We can judge the effectiveness of the serial drama because of a particular form of social science. *Evaluation research* refers to a research purpose rather than a specific research method. This purpose is to evaluate the impact of social interventions: new teaching methods, innovations in parole, and a host of others. (See "Applying Concepts in Everyday Life: Serial Drama Research Success" to see some of the practical functions of evaluation research in our example.) Many methods—surveys, experiments, and so on—can be used in evaluation research.

What do you think?

Why is there so much continuing debate over issues that straightforward social research would likely resolve? For example, people still debate whether the threat of the death penalty successfully deters murderers. Can't that outcome be tested once and for all? Can't we tell yes from no, black from white, up from down?

See the *What do you think? ... Revisited* box toward the end of the chapter.

Earl Babbie

Peter Rossi and colleagues have defined *evaluation research* as follows:

> *Program evaluation is the use of social research procedures to systematically investigate the effectiveness of social intervention programs. More specifically, evaluation researchers [evaluators] use social research methods to study, appraise, and help improve social programs in all their important aspects, including the diagnosis of the social problems they address, their conceptualization and design, their implementation and administration, their outcomes, and their efficiency.*

> *(Rossi, Lipsey, and Freeman 2002: 4)*

Evaluation research is probably as old as social research itself. Whenever people have instituted a social reform for a specific purpose, they have paid attention to its actual consequences, even if they have not always done so in a conscious, deliberate, or sophisticated fashion. In recent years, however, the field of evaluation research has become an increasingly popular and active research specialty, as reflected in textbooks, courses, and projects. Moreover, the growth of evaluation research points to a more general trend in the social sciences. As a researcher, you'll likely be asked to conduct evaluations of your own.

In part, the growth of evaluation research reflects social scientists' increasing desire to make a difference in the world. At the same time, we can't discount the influence of (1) an increase in federal requirements that program evaluations must accompany the implementation of new programs and (2) the availability of research funds to fulfill those requirements. In any case, it seems clear that social scientists will be bringing their skills into the real world now more than ever before.

This chapter looks at some of the key elements in this form of social research. After considering the kinds of topics commonly subjected to evaluation, we'll move through some of its main operational aspects: measurement, study design, and execution. As you'll see, formulating questions is as important as answering them. Because it occurs within real life, evaluation research has special problems, some of which we'll examine. Besides logistical problems, special ethical issues arise from evaluation research generally and in its specific, technical procedures. As you review reports of program evaluations, you should be especially sensitive to these problems.

Evaluation is a form of applied research— that is, it's intended to have some real-world effect. It will be useful, therefore, to consider whether and how it's actually applied. As you'll see, the obvious implications of an evaluation

Applying Concepts in Everyday Life

Serial Drama Research Success

The research evaluating the serial drama produced in Tanzania served many practical functions. To begin with, it told the producers whether they had been successful in delivering each of their messages. These data helped them fine-tune their presentations and make it easier to promote similar programs. Serial dramas promoting small families, safe sex, and the liberation of women have been produced in several other countries in Africa as well as in Asia and Latin America, and the list is still growing.

Today, this work is primarily carried on by the Population Media Center (populationmedia.org).

research project do not necessarily affect real life. They may become the focus of ideological, rather than scientific, debates. They may simply be denied out of hand, for political or other reasons. Perhaps most typically, they may merely be ignored and forgotten, left to collect dust in bookcases across the land.

Toward the end of this chapter, we'll look at a particular resource for large-scale evaluation—social indicators research. This type of research is also a rapidly growing specialty. Essentially, it involves the creation of aggregated indicators of the "health" of society, similar to the economic indicators that give diagnoses and prognoses of economies.

Topics Appropriate for Evaluation Research

Evaluation research is appropriate whenever some social intervention occurs or is planned. A *social intervention* is an action taken within a social context for the purpose of producing some intended result. In its simplest sense, **evaluation research** is a process of determining whether a social intervention has produced the intended result.

The topics appropriate for evaluation research are limitless. When the federal government abolished the selective service system (military draft), military researchers began paying special attention to the impact on enlistment. As individual states have liberalized their marijuana laws, researchers have sought to learn the consequences, both for marijuana use and for other forms of social behavior. Do no-fault divorce reforms increase the number of divorces, and do related social problems decrease? Has no-fault automobile insurance really brought down insurance policy premiums? Agencies providing foreign aid conduct evaluations to determine whether the desired effects were produced. Government programs are also reviewed for their effectiveness.

> **evaluation research** Research undertaken for the purpose of determining the impact of some social intervention, such as a program aimed at solving a social problem.
>
> **program evaluation/outcome assessment** The determination of whether a social intervention is producing the intended result.

Has the "No Child Left Behind" program improved the quality of education in America? Have "Just Say No" abstinence programs reduced rates of sexual activity and pregnancies among young people? These are the kinds of issues that evaluation research can address.

The intent of evaluation research takes many forms. Needs assessment studies aim at determining the existence and extent of problems, typically among a segment of the population, such as the elderly. Cost–benefit studies determine whether the results of a program can be justified by its expense (both financial and other). Monitoring studies provide a steady flow of information about something of interest, such as crime rates or the outbreak of an epidemic. Sometimes the monitoring involves incremental interventions. Read the following description of "adaptive management" by the Nature Conservancy, a public-interest group seeking to protect natural areas:

> *First, partners assess assumptions and set management goals for the conservation area. Based on this assessment, the team takes action, then monitors the environment to see how it responds. After measuring results, partners refine their assumptions, goals and monitoring regimen to reflect what they've learned from past experiences. With refinements in place, the entire process begins again.*
>
> *(2005: 3)*

Much of evaluation research is referred to as **program evaluation** or **outcome assessment**: the determination of whether a social intervention is producing the intended result. Here's an example.

Some years ago, a project evaluating the nation's drivers' education programs, conducted by the National Highway and Transportation Safety Administration (NHTSA), stirred up a controversy. Philip Hilts (1981: 4) reported on the study's findings:

> *For years the auto insurance industry has given large insurance discounts for children who take drivers' education courses, because statistics show that they have fewer accidents.*
>
> *The preliminary results of a new major study, however, indicate that drivers' education does not prevent or reduce the incidence of traffic accidents at all.*

Based on an analysis of 17,500 young people in DeKalb County, Georgia (including Atlanta),

Solutions without Problems

As you will see in this chapter, the bare-bones outline of evaluation research is:

1. Identify a potential problem.
2. Measure the extent of the problem.
3. Devise and implement a potential solution.
4. Later, remeasure the extent of the problem to see whether the solution worked.

As you will also see, a common failure in this procedure is the omission of Step 4. Once a solution has been put in place, generating vested interests, no one bothers to check whether it did the job.

But, wait, sometimes there is a worse failure: the omission of Step 2. Often, people imagine there to be a problem, and a solution is put in place, but there was actually no problem to solve. However, the "solution" can cause problems of its own. I call these Solutions without Problems or soluprobs®. Here are just a few examples.

- Voter ID laws designed to prevent impersonation at polling places
  - Studies of the "problem" show that it almost never happens. Many eligible voters are prevented from voting by the Voter ID laws.

- The 2003 invasion of Iraq by the United States
  - Iraq did not have weapons of mass destruction, nor did they have plans to attack America.
  - That war still continues today.

- The Salem Witch Trials of 1692
  - There were no witches.
  - Suspects were imprisoned and/or executed.

- Banning same-sex marriage
  - Allowing gays to marry has not destroyed the institution of marriage or turned everyone gay.
  - Many committed, loving relationships were denied recognition.

- Ban the use of marijuana
  - With its growing legalization, we can see that none of the horrendous "dangers" has come true.
  - Hundreds of thousands of (mostly young) people have been imprisoned and their lives destroyed.

Perhaps you can think of other examples of solutions to "problems" that didn't actually exist.

---

the preliminary findings indicated that students who took drivers' education had just as many accidents and traffic violations as those who didn't. The study also seemed to reveal some subtle aspects of driver training.

First, it suggested that the apparent impact of drivers' education was largely a matter of self-selection. The kinds of students who took drivers' education were less likely to have accidents and traffic violations—with or without driver training. Students with high grades, for example, were more likely to sign up for driver training, and they were also less likely to have accidents.

More startling, however, was the suggestion that driver-training courses may have actually increased traffic accidents! The existence of drivers' education may encourage some students to get their licenses earlier than if there were no such courses. In a study of ten Connecticut towns that discontinued driver training, about three-fourths of those who probably would have been licensed through their classes delayed getting licenses until they were 18 or older (Hilts 1981: 4).

As you might imagine, these results were not well received by those most closely associated

with driver training. This matter was complicated, moreover, by the fact that the NHTSA study was also evaluating a new, more intensive training program—and the preliminary results showed that the new program was effective.

Here's a very different example of evaluation research. Rudolf Andorka, a Hungarian sociologist, had been particularly interested in his country's shift to a market economy. Even before the dramatic events in Eastern Europe in 1989, Andorka and his colleagues had been monitoring the nation's "second economy"—jobs pursued outside the socialist economy. Their surveys followed the rise and fall of such jobs and examined their impact within Hungarian society. One conclusion was that "the second economy, which earlier probably tended to diminish income inequalities or at least improved the standard of living of the poorest part of the population, in the 1980s increasingly contributed to the growth of inequalities" (Andorka 1990: 111).

Because evaluation research is basically a matter of discovering whether social interventions make a difference, it's sometimes coupled with the intentions of participatory action research (PAR), discussed in Chapter 10. PAR has been particularly strong among Australian researchers, so it's not surprising to find Australians Wayne Miller and June Lennie (2005) speaking of "empowerment evaluation" to characterize their assessment of a national school breakfast program. This approach, they say,

> is distinguished by its clearly articulated underlying principles that allow for the extensive participation of program management and staff, funders, community members and other stakeholders in all stages of the evaluation. This approach can build evaluation capacities, give voice to a diversity of people involved, and enable open and honest discussion about the strengths and weaknesses of key program activities. It also enables collaborative planning and identification of the documentation or evidence required to assess the goals and strategies that participants develop to improve key program activities. The ultimate aim is for evaluation to become a normal part of planning and managing programs, resulting in ongoing improvement and learning.

> (2005: 18)

As you can see, the questions appropriate for evaluation research are of great practical significance: Jobs, programs, and investments as well as beliefs and values are at stake. Let's now examine how these questions are answered— how evaluations are conducted.

Formulating the Problem: Issues of Measurement

Several years ago, I headed an institutional research office that conducted research directly relevant to the operation of the university. Often, we were asked to evaluate new programs in the curriculum. The following description shows the problem that arose in that context, and it points to one of the key barriers to good evaluation research.

Faculty members would appear at my office to say they'd been told by the university administration to arrange for an evaluation of the new program they had permission to try. This points to a common problem: Often the people whose programs are being evaluated aren't thrilled at the prospect. For them, an independent evaluation threatens the survival of the programs and perhaps even their jobs.

The main problem I want to introduce, however, has to do with the purpose of the

A Delicate Balance

Evaluation research shows up in many corners of society. For example, it is important to nonprofit organizations attempting to make a difference in the world: ending hunger, poverty, or war. Anyone involved in such enterprises must confront a delicate balance between (1) making a difference and (2) getting credit for making a difference.

While the noble choice would be to actually make a difference rather than trying to take credit, the matter is more complex than that. If your organization is unable to *demonstrate* that you are making a difference, you are unlikely to get financial support for continuing. So you won't be able to actually make a difference.

Demonstrating your impact may be a complex matter as well. For example, I am on the board of Population Media Center, which is focused on the problem of overpopulation and a range of related issues (e.g., the status of girls and women). For the most part, we produce and broadcast radio serial dramas in over 50 developing countries, having reached up to 500 million listeners. In addition, PMC also originated the Emmy-nominated TV series, *East Los High*, on Hulu for the U.S. audience. We do a lot of evaluation research, both to improve our effectiveness and to document it.

For example, we measure the percentage of the population listening to our radio dramas. It's typically very high (50% or 75% is not unusual). That's nice, but it doesn't reduce population growth by itself. So, does listening to the shows change behavior? Surveys compare the attitudes and behavior of listeners and nonlisteners. We work with family-planning centers to measure increases in clients seeking their services. We also ask what prompted them to come to the family-planning centers. Typically, 25 to 50% volunteer that the radio show led them to come.

But does a visit to a family-planning center produce changes in childbearing behavior? Ultimately, the question is whether fertility rates and population growth change in the society. If the population stops growing or grows more slowly in a target society, that's a mark of success, but at the global level, many other factors (e.g., economics, epidemics, politics, religion, etc.) might also have played a role in the change.

Thus, you can see that evaluation research is very important but can also be elusive. Effective evaluation research requires both experience and ingenuity.

intervention to be evaluated. The question "What is the intended result of the new program?" often produced a vague response such as "Students will get an in-depth and genuine understanding of mathematics, instead of simply memorizing methods of calculations." Fabulous! And how could we measure that "in-depth and genuine understanding"? Often, I was told that the program aimed at producing something that could not be measured by conventional aptitude and achievement tests. No problem there; that's to be expected when we're innovating and being unconventional. What would be an unconventional measure of the intended result? Sometimes this discussion came down to an assertion that the effects of the program would be "unmeasurable."

There's the common rub in evaluation research: measuring the "unmeasurable." Evaluation research is a matter of finding out whether something is there or not there, whether something happened or didn't happen. To conduct evaluation research, we must be able to operationalize, observe, and recognize the presence or absence of what is under study.

Often, outcomes can be derived from published program documents. Thus, when Edward Howard and Darlene Norman (1981) evaluated the performance of the Vigo County Public Library (VCPL) in Indiana, they began with the statement of purpose previously adopted by the library's board of trustees:

> To acquire by purchase or gift, and by recording and production, relevant and potentially useful information that is produced by, about, or for the citizens of the community;
>
>   To organize this information for efficient delivery and convenient access, furnish the equipment necessary for its use, and provide assistance in its utilization; and
>
>   To effect maximum use of this information toward making the community a better place in which to live through aiding the search for understanding by its citizens.
>
> (1981: 306)

As the researchers said, "Everything that VCPL does can be tested against the Statement of Purpose." They then set about creating operational measures for each of the purposes.

Although "official" purposes of interventions are often the key to designing an evaluation, these

may not always suffice. Anna-Marie Madison (1992), for example, warns that programs designed to help disadvantaged minorities do not always reflect what the proposed recipients of the aid may need and desire:

> The cultural biases inherent in how middle-class white researchers interpret the experiences of low-income minorities may lead to erroneous assumptions and faulty propositions concerning causal relationships, to invalid social theory, and consequently to invalid program theory. Descriptive theories derived from faulty premises, which have been legitimized in the literature as existing knowledge, may have negative consequences for program participants.
>
> (1992: 38)

In setting up an evaluation, then, researchers must pay careful attention to issues of measurement. Let's take a closer look at the types of measurements that evaluation researchers must deal with.

Specifying Outcomes

As I've already suggested, a key variable for evaluation researchers to measure is the *outcome*, or what is called the *response variable*. If a social program is intended to accomplish something, we must be able to measure that something. If we want to reduce prejudice, we need to be able to measure prejudice. If we want to increase marital harmony, we need to be able to measure that.

Achieving agreements on definitions in advance is essential:

> The most difficult situation arises when there is disagreement as to standards. For example, many parties may disagree as to what defines serious drug abuse—is it defined best as 15% or more of students using drugs weekly, 5% or more using hard drugs such as cocaine or PCP monthly, students beginning to use drugs as young as seventh grade, or some combination of the dimensions of rate of use, nature of use, and age of user?...Applied researchers should, to the degree possible, attempt to achieve consensus from research consumers in advance of the study (e.g., through advisory groups) or at least ensure that their studies are able to produce data relevant to the standards posited by all potentially interested parties.
>
> (Hedrick, Bickman, and Rog 1993: 27)

In some cases, you may find that the problem and a sufficient solution are defined by law or agency regulations; if so, you must be aware

Evaluation research involves the identification of a problem, designing a possible solution, and seeing if the solution worked. Problem: a dented right front panel. Solution: a bandage. Evaluation: the panel is still attached and functional, but it's ugly; the solution was cost-effective in terms of function.

Picture 3/Earl Babbie

of such specifications and accommodate them. Moreover, whatever the agreed-on definitions, you must also achieve agreement on how the measurements will be made. Because there are different possible methods for estimating the percentage of students "using drugs weekly"; for example, you would have to be sure that all the parties involved understood and accepted the method(s) you chose.

On the other side of the coin, Yuet Wah Cheung (2009) used "drug-free weeks" as the dependent variable in his evaluation of drug-treatment programs in Hong Kong. This longitudinal study examined the role of positive and negative "social capital" in determining success or failure. *Positive social capital* included degree of family support and support from non–drug-using friends, while *negative social capital* included stressful events and association with drug-using friends. Cheung found, for example, that if recovering drug users were able to establish networks of supportive, non–drug-using friends, this made it less likely that they would revert to associating with their old network of drug users.

In the case of the Tanzanian soap opera, there were several outcome measures. In part, the purpose of the program was to improve knowledge about both family planning and AIDS. Thus, for example, one show debunked the belief that the AIDS virus was spread by mosquitoes and could be avoided by the use of insect repellant. Studies of listeners showed a reduction in that belief (Rogers et al. 1996: 21).

PCI also wanted to change Tanzanian attitudes toward family size, gender roles, HIV/AIDS, and other related topics; the research indicated that the show had affected these areas as well. Finally, the program aimed at affecting behavior. We've already seen that radio listeners reported changing their behavior with regard to AIDS prevention. They reported a greater use of family planning as well. However, because there's always the possibility of a gap between what people say they do and what they actually do, the researchers sought independent data to confirm their conclusions.

Tanzania's national AIDS control program had been offering condoms free of charge to citizens. In the areas covered by the soap opera, the number of condoms given out increased sixfold between 1992 and 1994. This far exceeded the increase of 1.4 times in the control area, where broadcasters did not carry the soap opera.

Measuring Experimental Contexts

Measuring the dependent variables directly involved in the experimental program is only a beginning. As Henry Riecken and Robert Boruch (1974: 120–1) point out, it's often appropriate and important to measure the aspects of the context of an experiment that researchers think might affect the experiment. Though external to the experiment itself, some variables may affect it.

Suppose, for example, that you were conducting an evaluation of a program aimed at training unskilled people for employment. The primary outcome measure would be their success at gaining employment after completing the program. You would, of course, observe and calculate the subjects' employment rate, but you should also determine what has happened to the employment/unemployment rates of society at large during the evaluation. A general slump in the job market should be taken into account in assessing what might otherwise seem to be a relatively low employment rate for subjects. Or, if all the experimental subjects get jobs following the program, you should consider any general increase in available jobs. Combining complementary measures with proper control-group designs should allow you to pinpoint the effects of the program you're evaluating.

Specifying Interventions

Besides making measurements relevant to the outcomes of a program, researchers must measure the program intervention—the experimental stimulus. In part, this measurement will be handled by the assignment of subjects to experimental and control groups, if that's the research design. Assigning a person to the experimental group is the same as scoring that person "yes" on the stimulus, and assignment to the control group represents a score of "no." In practice, however, it's seldom that simple.

Let's stick with the job-training example. Some people will participate in the program; others will not. But imagine for a moment what job-training programs are probably like. Some subjects will participate fully; others will miss a lot of sessions or fool around when they are present. So we may need measures of the extent or quality of participation in the program. If the program is effective, you should find that those who participated fully have higher employment rates than do those who participated less.

Other factors may further confound the administration of the experimental stimulus. Suppose we're evaluating a new form of psychotherapy that's designed to cure sexual impotence. Several therapists administer it to subjects composing an experimental group. We plan to compare the recovery rate of the experimental group with that of a control group, which receives some other therapy or none at all. It may be useful to include the names of the therapists treating specific subjects in the experimental group, because some may be more effective than others. If this turns out to be the case, we must find out why the treatment worked better for some therapists than for others. What we learn will further develop our understanding of the therapy itself.

"How to Do It: Positive Deviance" offers an alternative view for designing an intervention.

How to Do It

Positive Deviance

In his examination of "positive deviance," Arvind Singhal (2011) points to an implicit idea about how we design social interventions. He uses an example of a rural village in Vietnam, where juvenile malnutrition is a chronic problem. Suppose we want to design a program to solve the problem. The usual approach, Singhal suggests, is to identify examples of the problem—malnourished kids—and use what we learn about their plight to design a solution.

As an alternative, Singhal suggests that we might also look for cases that deviate from the norm positively—children who are not malnourished—and ask why not. Some cases will have obvious explanations. For example, children in the wealthiest family in the village will probably not suffer from malnutrition. There may be other cases with obvious explanations, but some will represent more of a puzzle. He reports one poor family in which the children were quite healthy.

When researchers, Jerry and Monica Sternin, studied the family, they learned several ways in which the positive deviance [PD] family differed from others in the village. Singhal (2011: 198–99) summarizes:

- Family members collected tiny shrimps and crabs from paddy fields, adding them to their children's meals. These foods are rich in protein and minerals.

- Family members added greens of sweet potato plants to their children's meals. These greens are rich in beta carotene, and other essential micronutrients, e.g., iron and calcium.

- Interestingly, these foods were accessible to everyone, but most community members believed the foods were inappropriate for young children.

Further,

- PD mothers were feeding their children three to four times a day, rather than the customary twice a day.

- PD mothers were actively feeding their children, making sure there was no food wasted.

- PD mothers washed the hands of the children before and after they ate.

This approach to social change fits well within the context of our discussion of participant action research. Just as physicians feel they know more about their patients' bodies than the patients do themselves, social researchers can fall into the trap of discounting what the subjects of their study know about their own situations. When we do that, we may miss a powerful resource for understanding and improving social life.

Sources: Arvind Singhal. 2011. "Turning Diffusion of Innovations Paradigm on Its Head," pp. 193–205 in *The Diffusion of Innovations: A Communication Science Perspective*, edited by Arun Vishwanath and George A. Barnett. (New York: Peter Lang); Arvind Singhal. 2013. "The Value of Positive Deviations." *Monthly Developments Magazine*, June, pp. 17–20.

Specifying the Population

In evaluating an intervention, it's important to define the population of subjects for whom the program is appropriate. Ideally, all or a sample of appropriate subjects will then be assigned to experimental and control groups as warranted by the study design. Defining the population, however, can itself involve specifying measurements. If we're evaluating a new form of psychotherapy, it's probably appropriate to study people with mental problems. But how will "mental problems" be defined and measured? The job-training program mentioned previously is probably intended for people who are having trouble finding work, but what counts as "having trouble"?

Beyond defining the relevant population, then, the researcher should make fairly precise measurements on the variables considered in the definition. For example, even though the randomization of subjects in the psychotherapy study would ensure an equal distribution of those with mild and severe mental problems into the experimental and control groups, we'd need to keep track of the relative severity of different subjects' problems in case the therapy turns out to be effective for only those with mild disorders. Similarly, we should measure such demographic variables as gender, age, race, and so forth in case the therapy works only for women, the elderly, or some other group.

New versus Existing Measures

In providing for the measurement of these different kinds of variables, the researcher must continually choose whether to create new measures or to use ones already devised by others. If a study addresses something that's never been measured before, the choice is easy. If it addresses something that others have tried to measure, the researcher will need to evaluate the relative worth of various existing measurement devices in terms of her or his specific research situations and purpose. Recall that this is a general issue in social research that applies well beyond evaluation research. Let's examine briefly the advantages and disadvantages of creating new measures versus using existing ones.

Creating measurements specifically for a study can offer greater relevance and validity.

If the psychotherapy we're evaluating aims at a specific aspect of recovery, we can create measures that pinpoint that aspect. We might not be able to find any standardized psychological measures that hit that aspect right on the head. However, creating our own measure will cost us the advantages to be gained from using preexisting measures. Creating good measures takes time and energy, both of which could be saved by adopting an existing technique. Of greater scientific significance, measures that have been used frequently by other researchers carry a body of possible comparisons that might be important to our evaluation. If the experimental therapy raises scores by an average of ten points on a standardized test, we'll be in a position to compare that therapy with others that have been evaluated using the same measure. Finally, measures with a long history of use usually have known degrees of validity and reliability, but newly created measures will require pretesting or will be used with considerable uncertainty.

Operationalizing Success/Failure

Potentially, one of the most taxing aspects of evaluation research is determining whether the program under review succeeded or failed. The purpose of a foreign-language program may be to help students better learn the language, but how much better is enough? The purpose of a conjugal-visit program at a prison may be to raise morale, but how high does morale need to be raised to justify the program?

As you may anticipate, clear-cut answers to questions like these almost never materialize. This problem has surely led to what is generally called cost–benefit analysis. How much does the program cost in relation to what it returns in benefits? If the benefits outweigh the cost, keep the program going. If the reverse, junk it. That's simple enough, and it seems to work in straightforward economic situations: If it costs you $20 to produce something and you can't sell it for over $18, there's no way you can make up the difference by increasing volume.

Unfortunately, the situations usually faced by evaluation researchers are seldom amenable to straightforward economic accounting. The foreign-language program may cost the school district $100 per student, and it may raise students' performances on tests by an average of 15 points. Because the test scores can't be converted into dollars, though, no obvious ground for weighing the costs and benefits exists.

Sometimes, as a practical matter, the criteria of success and failure can be handled through competition among programs. If a different foreign-language program costs only $50 per student and produces an increase of 20 points in test scores, it will undoubtedly be considered more successful than the first program—assuming that test scores are seen as an appropriate measure of the purpose of both programs and that the less-expensive program has no unintended negative consequences.

When Scott Connolly, Katie Elmore, and Wendi Stein (2008) undertook a qualitative evaluation of a Jamaican radio drama, *Outta Road*, designed for youth, they utilized focus groups, in-depth interviews, and exercises in which respondents drew sketches to illustrate their answers. The researchers described their aims thusly:

> The purpose of the study was to assess how listeners to the program engaged with the program and to what extent they found personal meaning and were influenced by the educational messages and themes in the drama.
>
> Unlike a quantitative evaluation, this report does not attempt to generalize the findings to all Outta Road *youth listeners in Jamaica. The findings do, however, provide rich verbal and visual insights into how the program was incorporated into the lives of participants, what personal meaning they derived from the content, and through reflection how youth listeners internalized the key messages from the drama.*
>
> *(2008: 2)*

Ultimately, the criteria of success and failure are often a matter of agreement. The people responsible for the program may commit themselves in advance to a particular outcome that will be regarded as an indication of success. If that's the case, all you need to do is make absolutely certain that the research design will measure the specified outcome. I mention this obvious requirement simply because researchers sometimes fail to meet it, and there's little or nothing more embarrassing than that.

In summary, researchers must take measurement quite seriously in evaluation research,

carefully determining all the variables to be measured and getting appropriate measures for each. As I've implied, however, such decisions are typically not purely scientific ones. Evaluation researchers often must work out their measurement strategy with the people responsible for the program being evaluated. It usually doesn't make sense to determine whether a program achieves Outcome X when its purpose is to achieve Outcome Y. (Realize, however, that evaluation designs sometimes have the purpose of testing for unintended consequences.)

There is a political aspect to these choices, also. Because evaluation research often affects other people's professional interests—their pet program may be halted, or they may be fired or lose professional standing—the results of evaluation research are often contested.

Let's turn now to some of the evaluation designs that researchers commonly employ.

Types of Evaluation Research Designs

As I noted at the start of this chapter, evaluation research is not a method itself, but rather it is one application of social research methods. As such, it can involve any of several research designs. Here, we'll consider three main types of research design that are appropriate for evaluations: experimental designs, quasi-experimental designs, and qualitative evaluations.

Experimental Designs

Many of the experimental designs introduced in Chapter 8 can be used in evaluation research. By way of illustration, let's see how the classical experimental model might be applied to our evaluation of the new psychotherapeutic treatment for sexual impotence.

In designing our evaluation, we should begin by identifying a population of patients appropriate for the therapy. This identification might be made by researchers experimenting with the new therapy. Let's say we're dealing with a clinic that already has 100 patients being treated for sexual impotence. We might take that group and the clinic's definition of sexual impotence as a starting point, and we should maintain any existing assessments of the severity of the problem for each specific patient.

For purposes of the evaluation research, however, we would need to develop a more specific measure of impotence. Maybe it would involve whether patients have sexual intercourse at all (within a specified time), how often they have intercourse, or whether and how often they reach orgasm. Alternatively, the outcome measure might be based on the assessments of independent therapists not involved in the therapy who interview the patients later. In any event, we would need to agree on the measures to be used.

In the simplest design, we would assign the 100 patients randomly to experimental and control groups; the former would receive the new therapy, and the latter would be taken out of therapy altogether during the experiment. Because ethical practice would probably prevent withdrawing therapy altogether from the control group, it's more likely that the control group would continue to receive their conventional therapy.

Having assigned subjects to the experimental and control groups, we would need to agree on the length of the experiment. Perhaps the designers of the new therapy feel that it ought to be effective within two months, and an agreement could be reached. The duration of the study doesn't need to be rigid, however. One purpose of the experiment and evaluation might be to determine how long it actually takes for the new therapy to be effective. Conceivably, then, an agreement could be struck to measure recovery rates weekly, say, and let the ultimate length of the experiment rest on a continual review of the results.

Let's suppose the new therapy involves showing pornographic movies to patients. We'd need to specify that stimulus. How often would patients see the movies, and how long would each session be? Would they see the movies in private or in groups? Should therapists be present? Perhaps we should observe the patients while the movies are being shown and include our observations as well as the measurements of the experimental stimulus. Do some patients watch the movies eagerly but others look away from the screen? We'd have to ask these kinds of questions and create specific measurements to address them.

Having thus designed the study, all we have to do is "roll 'em." The study is set in motion, the observations are made and recorded, and the mass of data is accumulated for analysis. Once the study has run its course, we can determine whether the new therapy had its intended—or perhaps some unintended—consequences. We can tell whether the movies were most effective for patients with mild problems or severe ones, whether they worked for young subjects but not older ones, and so forth.

This simple illustration should show you how the standard experimental designs presented in Chapter 8 can be used in evaluation research. Many, perhaps most, of the evaluations reported in the research literature don't look exactly like this illustration, however. Because it's nested in real life, evaluation research often calls for *quasi-experimental designs*. Let's see what this means.

Quasi-Experimental Designs

Quasi experiments are distinguished from "true" experiments primarily by the lack of random assignment of subjects to an experimental and a control group. In evaluation research, it's often impossible to achieve such an assignment of subjects. Rather than forgo evaluation altogether, researchers sometimes create and execute research designs that give some evaluation of the program in question. This section describes some of these designs.

Time-Series Designs

To illustrate the **time-series design**—a research design that involves measurements taken over time—I'll begin by asking you to assess the meaning of some hypothetical data. Suppose

---

quasi experiments Nonrigorous inquiries somewhat resembling controlled experiments but lacking key elements such as pretesting and posttesting and/or control groups.

time-series design A research design that involves measurements made over some period, such as the study of traffic-accident rates before and after lowering the speed limit.

I come to you with what I say is an effective technique for getting students to participate in classroom sessions in a course I'm teaching. To prove my assertion, I tell you that on Monday only four students asked questions or made a comment in class; on Wednesday I devoted the class time to an open discussion of a controversial issue raging on campus; and on Friday, when we returned to the subject matter of the course, eight students asked questions or made comments. In other words, I contend, the discussion of a controversial issue on Wednesday has doubled classroom participation. This simple set of data is presented graphically in Figure 12-1.

Have I persuaded you that the open discussion on Wednesday has had the consequence I claim for it? Probably you'd object that my data do not prove the case. Two observations (Monday and Friday) aren't really enough to prove anything. Ideally, I should have had two classes, with students assigned randomly to each, held an open discussion in only one, and then compared the two on Friday. But I don't have two classes with random assignment of students. Instead, I've been keeping a record of class participation throughout the semester for the one class. This record allows you to conduct a *time-series* evaluation.

Figure 12-2 presents three possible patterns of class participation over time, both before and after the open discussion on Wednesday.

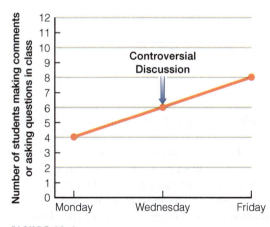

FIGURE 12-1

Two Observations of Class Participation: Before and After an Open Discussion.

Which of these patterns would give you some confidence that the discussion had the impact I contend it had?

If the time-series results looked like the first pattern in Figure 12-2, you'd probably conclude that the process of greater class participation had begun on the Wednesday before the discussion and had continued, unaffected, after the day devoted to the discussion. The long-term data suggest that the trend would have occurred even without the discussion on Wednesday. The first pattern, then, contradicts my assertion that the special discussion increased class participation.

The second pattern contradicts my assertion by indicating that class participation has been bouncing up and down in a regular pattern throughout the semester. Sometimes it increases from one class to the next, and sometimes it decreases; the open discussion on that Wednesday simply came at a time when the level of participation was about to increase. More to the point, we note that class participation decreased again in the class following the alleged postdiscussion increase.

Only the third pattern in Figure 12-2 supports my contention that the open discussion mattered. As depicted there, the level of discussion before that Wednesday had been a steady four students per class. Not only did the level of participation double following the day of discussion, but it continued to increase afterward. Although these data do not protect us against the possible influence of some extraneous factor (I might also have mentioned that participation would figure into students' grades), they do exclude the possibility that the increase results from a process of maturation (indicated in the first pattern) or from regular fluctuations (indicated in the second).

Nonequivalent Control Groups

The time-series design just described involves only an "experimental" group; it doesn't provide the value to be gained from having a control group. Sometimes, when researchers can't create experimental and control groups by random assignment from a common pool, they can find an existing "control" group that appears similar to the experimental group. Such a group is called

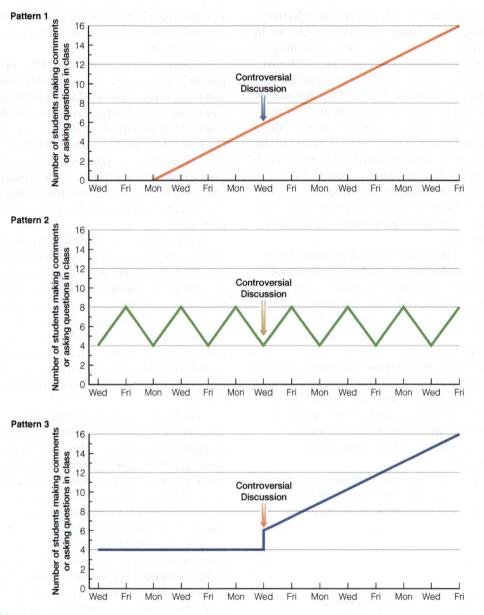

FIGURE 12-2

Three Patterns of Class Participation in a Longer Historical Period.

a **nonequivalent control group**. If an innovative foreign-language program is being tried in one class

> **nonequivalent control group** A control group that is similar to the experimental group but is not created by the random assignment of subjects. This sort of control group does differ significantly from the experimental group in terms of the dependent variable or variables related to it.

in a large high school, for example, you may be able to find another foreign-language class in the same school that has a very similar student population: one that has about the same composition in terms of grade in school, gender, ethnicity, IQ, and so forth. The second class, then, could provide a point of comparison. At the end of the semester, both classes could be given the same foreign-language test, and you could compare performances.

Here's how two junior high schools were selected for purposes of evaluating a program aimed at discouraging tobacco, alcohol, and drug use:

> The pairing of the two schools and their assignment to "experimental" and "control" conditions was not random. The local Lung Association had identified the school where we delivered the program as one in which administrators were seeking a solution to admitted problems of smoking, alcohol, and drug abuse. The "control" school was chosen as a convenient and nearby demographic match where administrators were willing to allow our surveying and breath-testing procedures. The principal of that school considered the existing program of health education to be effective and believed that the onset of smoking was relatively uncommon among his students. The communities served by the two schools were very similar. The rate of parental smoking reported by the students was just above 40 percent in both schools.
>
> (McAlister et al. 1980: 720)

In the initial set of observations, the experimental and control groups reported virtually the same (low) frequency of smoking. Over the 21 months of the study, smoking increased in both groups, but it increased less in the experimental group than in the control group, suggesting that the program had had an impact on students' behavior.

Multiple Time-Series Designs

Sometimes the evaluation of processes occurring outside of "pure" experimental controls can be made easier by the use of more than one time-series analysis. **Multiple time-series designs** are an improved version of the nonequivalent control group design just described. Carol Weiss (1972: 69) presents a useful example:

> An interesting example of multiple time series was the evaluation of the Connecticut crackdown on highway speeding. Evaluators collected reports of traffic fatalities for several periods before and after the new program went into effect. They found that fatalities went down after the crackdown, but since the series had had an unstable up-and-down pattern for many years, it was not certain that the drop was due to the program. They then compared the statistics with time-series data from four neighboring states where there had been no changes in traffic enforcement. Those states registered no equivalent drop in fatalities. The comparison lent credence to the conclusion that the crackdown had had some effect.

Although this study design is not as good as one in which subjects are assigned randomly, it's nonetheless an improvement over assessing the experimental group's performance without any comparison. That's what makes these designs quasi experiments instead of just fooling around. The key in assessing this aspect of evaluation studies is comparability, as the following example illustrates.

A growing concern in the poor countries of the world, rural development, has captured the attention and support of many rich countries. Through national foreign-assistance programs and through international agencies such as the World Bank, the developed countries are in the process of sharing their technological knowledge and skills with the developing countries. Such programs have had mixed results, however. Often, modern techniques do not produce the intended results when applied in traditional societies.

Rajesh Tandon and L. Dave Brown (1981) undertook an experiment in which instruction in village organization would accompany technological training. They felt it was important for poor farmers to learn how to organize and exert collective influence within their villages—getting needed action from government officials, for example. Only then would their new technological skills bear fruit.

Both intervention and evaluation were attached to an ongoing program in which twenty-five villages had been selected for technological training. Two poor farmers from each village had been trained in new agricultural technologies. Then they had been sent home to share their new knowledge with their

---

multiple time-series designs The use of more than one set of data that were collected over time, as in accident rates over time in several states or cities, so that comparisons can be made.

fellow villagers and to organize other farmers into "peer groups" who would assist in spreading that knowledge. Two years later, the authors randomly selected two of the twenty-five villages (subsequently called Group A and Group B) for special training and eleven others as controls. A careful comparison of demographic characteristics showed the experimental and control groups to be strikingly similar to each other, suggesting that they were sufficiently comparable for the study.

The peer groups from the two experimental villages were brought together for special training in organization building. The participants were given some information about organizing and making demands on the government; they were also given opportunities to act out dramas similar to the situations they faced at home. The training took three days.

The outcome variables considered by the evaluation all had to do with the extent to which members of the peer groups initiated group activities designed to improve their situation. Six types of initiative were studied. "Active initiative," for example, was defined as "active effort to influence persons or events affecting group members versus passive response or withdrawal" (Tandon and Brown 1981: 180). The data for evaluation came from the journals that the peer-group leaders had been keeping since their initial technological training. The researchers read through the journals and counted the number of initiatives taken by members of the peer groups. Two researchers coded the journals independently and compared their work to test the reliability of the coding process.

Figure 12-3 compares the number of active initiatives by members of the two experimental

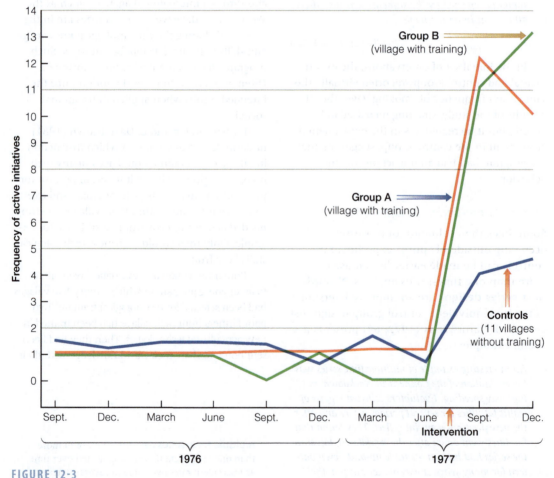

FIGURE 12-3

Active Initiatives over Time.

Source: Rajesh Tandon and L. Dave Brown, "Organization-Building for Rural Development: An Experiment in India," *Journal of Applied Behavioral Science,* April–June 1981, p. 182.

groups with those coming from the control groups. Similar results were found for the other outcome measures.

Notice two things about the graph. First, there's a dramatic difference in the number of initiatives by the two experimental groups as compared with the eleven controls. This seems to confirm the effectiveness of the special training program. The orange line in the graph represents Group A, which received training in active initiatives, while the green line represents Group B, which also received the training. We see that the number of instances of active initiatives greatly increased for these two groups.

Second, notice that the number of initiatives also increased among the control groups. The researchers explain this latter pattern as a result of contagion. Because all the villages were near each other, the lessons learned by peer-group members in the experimental groups were communicated in part to members of the control villages. Although the control groups showed an increase in initiatives, the key evaluation here lies in comparing the experimental (A and B) groups with the controls.

This example illustrates the strengths of multiple time-series designs in situations in which true experiments are inappropriate for the program being evaluated.

Qualitative Evaluations

I've laid out the steps involved in tightly structured, mostly quantitative evaluation research, but evaluations can also be less structured and more qualitative. For example, Pauline Bart and Patricia O'Brien (1985) wanted to evaluate different ways to stop rape, so they undertook in-depth interviews with both rape victims and women who had successfully fended off rape attempts. As a general rule, they found that resistance (such as yelling, kicking, running away) was more likely to be successful than to make the situation worse, as women sometimes fear it will.

Sometimes even structured, quantitative evaluations can yield unexpected, qualitative results. Paul Steel is a social researcher specializing in the evaluation of programs aimed at pregnant drug users. One program he evaluated involved counseling by public-health nurses,

who warned pregnant drug users that continuing to use drugs would likely result in underweight babies whose skulls would be an average of 10 percent smaller than normal. In his in-depth interviews with program participants, however, he discovered that the program omitted one important piece of information: that undersized babies were a bad thing. Many of the young women Steel interviewed thought that smaller babies would mean easier deliveries.

In another program, a local district attorney had instituted what would generally be regarded as a progressive, enlightened program. If a pregnant drug user were arrested, she could avoid prosecution if she would (1) agree to stop using drugs and (2) successfully complete a drug-rehabilitation program. Again, in-depth interviews suggested that the program did not always operate on the ground the way it did in principle. Specifically, Steel discovered that whenever a young woman was arrested for drug use, the other inmates would advise her to get pregnant as soon as she was released on bail. That way, she could avoid prosecution (Paul Steel: personal communication, November 22, 1993).

The most effective evaluation research is one that combines qualitative and quantitative components. Making statistical comparisons is useful, but so is gaining an in-depth understanding of the processes producing the observed results—or preventing the expected results from appearing.

The evaluation of the Tanzanian soap opera, presented earlier in this chapter, employed several research techniques. I've already mentioned the listener surveys and data obtained from clinics. In addition, the researchers conducted numerous focus groups to probe more deeply into the impact the shows had on listeners. Further, content analyses were done on the soap opera episodes themselves and on the many letters received from listeners. Both quantitative and qualitative analyses were undertaken (Swalehe et al. 1995).

The soap opera research also gave me a firsthand opportunity to see how different cultures affect the conduct of research. Not long ago I consulted on the evaluation of soap operas being planned in Ethiopia. In contrast to the Western concern for confidentiality in social research, respondents selected for interviews in rural Ethiopian villages often took a special pride at being selected and wanted their answers broadly known in the community.

Sometimes, local researchers' desires to please the client got in the way of the evaluation. For example, some pilot episodes were tested in focus groups to determine whether listeners would recognize any of the social messages being communicated. The results were more encouraging than could have been expected. When I asked how the focus group subjects had been selected, the researcher described his introductory conversation, "We would like you to listen to some radio programs designed to encourage people to have small families, and we'd like you to tell us whether we've been successful." Not surprisingly, the small-family theme came through clearly to the focus-group subjects.

These examples, along with earlier comments in previous sections, have hinted at the possibility of problems in the actual execution of evaluation research projects. Of course, all forms of research can run into problems, but evaluation research has a special propensity for it.

Logistical Problems

In a military context, *logistics* refers to moving supplies around—making sure people have food, guns, and tent pegs when they need them. Here, I use it to refer to getting subjects to do what they're supposed to do, getting research instruments distributed and returned, and other seemingly unchallenging tasks. These tasks pose more challenges than you might guess.

Motivating Sailors

When Kent Crawford, Edmund Thomas, and Jeffrey Fink (1980) set out to find a way to motivate "low performers" in the U.S. Navy, they found out just how many problems can occur. The purpose of the research was to test a three-pronged program for motivating sailors who were chronically poor performers and often in trouble aboard ship. First, a workshop was to be held for supervisory personnel, training them in effective leadership of low performers. Second, a few supervisors would be selected and trained as special counselors and role models—people the low performers could turn to for advice or just as sounding boards. Finally, the low performers themselves would participate in workshops aimed at

training them to be more motivated and effective in their work and in their lives. The project was to be conducted aboard a particular ship, with a control group selected from sailors on four other ships.

To begin, the researchers reported that the supervisory personnel were not exactly thrilled with the program.

> *Not surprisingly, there was considerable resistance on the part of some supervisors toward dealing with these issues. In fact, their reluctance to assume ownership of the problem was reflected by "blaming" any of several factors that can contribute to their personnel problem. The recruiting system, recruit training, parents, and society at large were named as influencing low performance—factors that were well beyond the control of the supervisors.*
>
> (Crawford, Thomas, and Fink 1980: 488)

Eventually, the reluctant supervisors came around and "this initial reluctance gave way to guarded optimism and later to enthusiasm" (1980: 489).

The low performers themselves presented even more of a problem, however. The research design called for pretesting and posttesting of attitudes and personalities, so that changes brought about by the program could be measured and evaluated.

> *Unfortunately, all of the LPs (Low Performers) were strongly opposed to taking these so-called personality tests and it was therefore concluded that the data collected under these circumstances would be of questionable validity. Ethical concerns also dictated that we not force "testing" on the LPs.*
>
> (Crawford, Thomas, and Fink 1980: 490)

As a consequence, the researchers had to rely on interviews with the low performers and on the judgments of supervisors for their measures of attitude change. The subjects continued to present problems, however.

Initially, the ship's command ordered fifteen low performers to participate in the experiment. Of the fifteen, however, one went into the hospital, another was assigned duties that prevented participation, and a third went "over the hill" (absent without leave). Thus, the experiment began with twelve subjects. But before it was completed, three more subjects completed their enlistment requirements and left the Navy, and another was thrown out for disciplinary reasons. The experiment concluded, then, with eight subjects. Although the evaluation pointed to positive results, the very small number of subjects warranted caution in any generalizations from the experiment.

The special, logistical problems of evaluation research grow out of the fact that it occurs within the context of real life. Although evaluation research is modeled after the experiment—which suggests that the researchers have control over what happens—it takes place within frequently uncontrollable daily life. Of course, the participant-observer in field research doesn't have control over what's observed either, but that method doesn't strive for control. Given the objectives of evaluation research, lack of control can create real dilemmas for the researcher.

Administrative Control

As suggested in the previous example, the logistical details of an evaluation project often fall to program administrators. Let's suppose you're evaluating the effects of a conjugal-visit program on the morale of married prisoners. The program allows inmates periodic visits from their spouses during which they can have sexual relations. On the fourth day of the program, a male prisoner dresses up in his wife's clothes and escapes. Although you might be tempted to assume that his morale was greatly improved by escaping, that turn of events would complicate your study design in many ways. Perhaps the warden will terminate the program altogether, and where's your evaluation then? Or, if the warden is brave, he or she may review the files of all those prisoners you selected randomly for the experimental group and veto the "bad risks." There goes the comparability of your experimental and control groups. As an alternative, stricter security measures may be introduced to prevent further escapes, and the security measures may have a dampening effect on morale. So the experimental stimulus has changed in the middle of your research project. Some of the data will reflect the original stimulus; other data will reflect the modification. Although you'll probably be able to sort it all out, your carefully designed study has become a logistical snake pit.

Or suppose you've been engaged to evaluate the effect of race-relations lectures on prejudice

in the Army. You've carefully studied the soldiers available to you for study, and you've randomly assigned some to attend the lectures and others to stay away. The rosters have been circulated weeks in advance, and at the appointed day and hour, the lectures begin. Everything seems to be going smoothly until you begin processing the files: The names don't match. Checking around, you discover that military field exercises, KP duty, and a variety of emergencies required some of the experimental subjects to be elsewhere at the time of the lectures. That's bad enough, but then you learn that helpful commanding officers sent others to fill in for the missing soldiers. And whom do you suppose they picked to fill in? Soldiers who didn't have anything else to do or who couldn't be trusted to do anything important. You might learn this bit of information a week or so before the deadline for submitting your final report on the impact of the race-relations lectures.

These are some of the logistical problems confronting evaluation researchers. You need to be familiar with the problems to understand why some research procedures may not measure up to the design of the classical experiment. As you read reports of evaluation research, however, you'll find that—my earlier comments notwithstanding—it is possible to carry out controlled social research in conjunction with real-life experiments.

"Issues and Insights: Testing Soap Operas in Tanzania" describes some of the logistical problems involved in the research discussed at the outset of this chapter.

Use of Research Results

One more facts-of-life aspect of evaluation research concerns how evaluations are used. Because the purpose of evaluation research is to determine the success or failure of social interventions, you might think it reasonable that a program would automatically be continued or terminated based on the results of the research.

Reality isn't that simple and reasonable, however. Other factors intrude on the assessment of evaluation research results, sometimes blatantly and sometimes subtly. As president, Richard Nixon appointed a blue-ribbon national commission to study the consequences of pornography.

Issues and Insights

Testing Soap Operas in Tanzania

by William N. Ryerson

President and Founder, Population Media Center

Twende na Wakati (*Let's Go with the Times*) has been broadcast on Radio Tanzania since mid-1993 with support from the United Nations Population Fund. The program was designed to encourage family-planning use and AIDS-prevention measures.

There were many different elements to the research. One was a nationwide, random-sample survey given prior to the first airing of the soap opera in June 1993 and then annually after that. Many interviewers faced particularly interesting challenges. For example, one interviewer, Fridolan Banzi, had never been in or on water in his life and couldn't swim. He arranged for a small boat to take him through the rough waters of Lake Victoria so he could carry out his interviews at a village that had no access by road. He repeated this nerve-wracking trip each year afterward in order to measure the change in that village.

Another interviewer, Mr. Tende, was invited to participate in a feast that the villagers held to welcome him and to indicate their enthusiasm about having been selected for the study. They served him barbequed rats. Though they weren't part of his normal diet, he ate them anyway to be polite and to ensure that the research interviews could be carried out in that village.

Still another interviewer, Mrs. Masanja, was working in a village in the Pwani region along the coast of the Indian Ocean when cholera broke out in that village. She wisely chose to abandon the interviews there, which reduced the 1993 sample size by one ward. The unsung heroes of this research, the Tanzanian interviewers, deserve a great deal of credit for carrying out this important work under difficult circumstances.

After a diligent, multifaceted evaluation, the commission reported that pornography didn't appear to have any of the negative social consequences often attributed to it. Exposure to pornographic materials, for example, didn't increase the likelihood of sex crimes. You might have expected liberalized legislation to follow from the research. Instead, the president said the commission was wrong.

Less dramatic examples of the failure to follow the implications of evaluation research could be listed endlessly. Undoubtedly, every evaluation researcher can point to studies he or she conducted—studies providing clear research results and obvious policy implications—that were ignored.

The 1990s saw the passage of three-strikes laws at the federal level and in numerous states. The intention was to reduce crime rates by locking up "career criminals." Under the 1994 California law, for example, having a past felony conviction would double your punishment when you were convicted of your second felony, and the third felony conviction would bring a mandatory sentence of 25 years to life. Over the years, only California has enforced such laws with any vigor.

Those who supported the passage of three-strikes legislation have been quick to link the dramatic drop in crime rates during the 1990s to the new policy of getting tough with career criminals. Other observers have looked for additional evidence to support the impact of three-strikes laws. Some critics of these laws, for example, have noted that crime rates have been dropping dramatically across the country, not only in California but in states that have no three-strikes laws and in those where the courts have not enforced the three-strikes laws that exist. In fact, crime rates have dropped in those California counties that have tended to ignore that state's law. Moreover, the drop in California crime rates began before the three-strikes law went into effect.

In 1994, Peter Greenwood and his colleagues at the Rand Corporation estimated that implementation of the law would cost California's criminal justice system approximately $5.5 billion more per year, especially in prison costs as "career criminals" were sentenced to longer terms. Although the Rand group did not deny that the three-strikes legislation would have some impact on crime—those serving long terms in prison can't commit crimes on the streets—a follow-up study (Greenwood, Rydell, and Model 1996) suggested that it was an inefficient way of attacking crime. The researchers estimated that a million dollars spent on "three strikes" would prevent 60 crimes, whereas the same amount spent on programs encouraging high school students to stay in school and graduate would prevent 258 crimes.

Criminologists have long recognized that most crimes are committed by young men. Focusing attention on older "career criminals" has little or no effect on the youthful offenders. In fact, three-strikes sentences disproportionately fall on those approaching the end of their criminal careers by virtue of growing older.

In a more general critique, John Irwin and James Austin (1997) suggest that people in the United States tend to overuse prisons as a solution to crime, ignoring other, more effective, solutions. Often, imprisonment causes problems more serious than those it was intended to remedy.

As with many other social interventions, however, much of the support for three-strikes laws in California and elsewhere has mostly to do with public emotions about crime and the political implications of such emotions. Thus, evaluation research on these laws may eventually bring about changes, but it will do so much more slowly than you might logically expect.

There are three important reasons why the implications of evaluation research results are not always put into practice. First, the implications may not always be presented in a way that nonresearchers can understand. Second, evaluation results sometimes contradict deeply held beliefs. That was certainly true in the case of the pornography commission. If everybody knows that pornography is bad, that it causes all manner of sexual deviance, then it's likely that research results to the contrary will have little immediate impact. By the same token, people thought Copernicus was crazy when he said the earth revolved around the sun. Anybody could tell that the earth was standing still. The third barrier to the use of evaluation results is vested interests. If I've devised a new rehabilitation program that I'm convinced will keep ex-convicts from returning to prison, and if people have taken to calling it the "Babbie Plan," how do you think I'm going to feel when your evaluation suggests that the program doesn't work? I might apologize for misleading people, fold up my tent, and go into another line of work. More likely, I'd call your research worthless and begin intense lobbying with the appropriate authorities to have my program continue.

In the earlier example of the evaluation of drivers' education, Philip Hilts reported some of the reactions to the researchers' preliminary results:

> *Ray Burneson, traffic safety specialist with the National Safety Council, criticized the study, saying that it was a product of a group (NHTSA) run by people who believe "that you can't do anything to train drivers. You can only improve medical facilities and build stronger cars for when the accidents happen.... This knocks the whole philosophy of education."*
>
> *(1981: 4)*

By its nature, evaluation research takes place in the midst of real life, affecting it and being affected by it. Here's another example, well known to social researchers.

Rape Reform Legislation

For years, many social scientists and other observers have noted certain problems with the prosecution of rape cases. All too often, it is felt, the victim's time on the witness stand severely prolongs the suffering. Frequently, defense lawyers portray her as having encouraged the sex act and being of shady moral character; other personal attacks are intended to deflect responsibility from the accused rapist.

Criticisms such as these have resulted in a variety of state laws aimed at remedying the problems. Cassie Spohn and Julie Horney (1990) were interested in tracking the impact of such legislation. These researchers summarized the ways in which new laws were intended to make a difference:

> *The most changes are: (1) redefining rape and replacing the single crime of rape with a series of graded offenses defined by the presence or absence of aggravating conditions; (2) changing the consent standard by eliminating the requirement that the victim physically resist her attacker; (3) eliminating the requirement that the victim's testimony be corroborated; and (4) placing restrictions on the introduction of evidence of the victim's prior sexual conduct.*
>
> *(1990: 2)*

It was generally expected that such legislation would encourage women to report being raped and would increase convictions when the cases

TABLE 12-1

Analysis of Rape Cases Before and After Legislation

	Rape	
	Before (N = 2,252)	After (N = 2,369)
Outcome of case (%)		
Convicted of original charge	45.8	45.4
Convicted of another charge	20.6	19.4
Not convicted	33.6	35.1
Median prison sentence (months)		
For those convicted of original charge	96.0	144.0
For those convicted of another charge	36.0	36.0

were brought to court. To examine the latter expectation, the researchers focused on the period from 1970 to 1985 in Cook County, Illinois: "Our data file includes 4,628 rape cases, 405 deviant sexual assault cases, 745 aggravated criminal sexual assault cases, and 37 criminal sexual assault cases" (1990: 4). Table 12-1 shows some of what they discovered.

Spohn and Horney summarized these findings as follows:

> The only significant effects revealed by our analyses were increases in the average maximum prison sentences; there was an increase of almost 48 months for rape and of almost 36 months for sex offenses. Because plots of the data indicated an increase in the average sentence before the reform took effect, we modeled the series with the intervention moved back one year earlier than the actual reform date. The size of the effect was even larger and still significant, indicating that the effect should not be attributed to the legal reform.
>
> (1990: 10)

Notice in the table that there was virtually no change in the percentages of cases ending in conviction for rape or some other charge (such as assault). Hence, the change in laws had no effect on the likelihood of conviction. As the researchers note, the one change that is evident—an increase in the length of sentences—cannot be attributed to the reform legislation itself.

In addition to the analysis of existing statistics, Spohn and Horney interviewed judges and lawyers to determine what they felt about the impact of the laws. Their responses were somewhat more encouraging.

> Judges, prosecutors and defense attorneys in Chicago stressed that rape cases are taken more seriously and rape victims treated more humanely as a result of the legal changes. These educative effects clearly are important and should please advocates of rape reform legislation.
>
> (1990: 17)

Thus, the study found other effects besides the qualitative results the researchers had looked for. This study demonstrates the importance of following up on social interventions to determine whether, in what ways, and to what degree they accomplish their intended results.

Preventing Domestic Violence

In a somewhat similar study, researchers in Indianapolis focused their attention on the problem of wife battering, with a special concern for whether prosecuting the batterers can lead to subsequent violence. David Ford and Mary Jean Regoli (1992) set about studying the consequences of various options for prosecution allowed within the "Indianapolis Prosecution Experiment" (IPE).

Wife-battering cases can follow a variety of patterns, as Ford and Regoli summarize:

> After a violent attack on a woman, someone may or may not call the police to the scene. If the police are at the scene, they are expected to investigate for evidence to support probable cause for a warrantless arrest. If it exists, they may arrest at their discretion. Upon making such an on-scene arrest, officers fill out a probable cause affidavit and slate the suspect into court for an initial hearing. When the police are not called, or if they are called but do not arrest, a victim may initiate charges on her own by going to the prosecutor's office and swearing out a probable cause affidavit with her allegation against the man. Following a judge's approval, the alleged batterer may either be summoned to court or be arrested on a warrant and taken to court for his initial hearing.
>
> (1992: 184)

What if a wife brings charges against her husband and then reconsiders later on? Many courts have a policy of prohibiting such actions, in the belief that they are serving the interests of the victim by forcing the case to be pursued to completion. In the IPE, however, some victims are offered the possibility of dropping the charges if they so choose later in the process. In addition, the court offers several other options. Because wife battering is largely a function of sexism, stress, and an inability to deal with anger, some of the innovative possibilities in the IPE involve educational classes with anger-control counseling.

If the defendant admits his guilt and is willing to participate in an anger-control counseling program, the judge may postpone the trial for that purpose and can later dismiss the charges if the defendant successfully completes the program. Alternatively, the defendant may be tried and, if found guilty, be granted probation provided he participates in the anger-control program. Finally, the defendant can be tried and, if found guilty, given a conventional punishment such as imprisonment.

Which of these possibilities most effectively prevents subsequent wife battering? That's the question Ford and Regoli addressed. Here are some of their findings.

First, men who are brought to court for a hearing are less likely to continue beating their wives, no matter what the outcome of the hearing. Simply being brought into the criminal justice system has an impact.

Second, women who have the right to drop charges later on are less likely to be abused subsequently than those who do not have that right. In particular, the combined policies of arresting defendants by warrant and allowing victims to drop charges provide victims with greater security from subsequent violence than do any of the other prosecution policies.

However, giving victims the right to drop charges has a somewhat strange impact. Women who exercise that right are more likely to be abused later than are those who insist on the prosecution proceeding to completion. The researchers interpret this as showing that future violence can be decreased when victims have a sense of control supported by a clear and consistent alliance with criminal justice agencies.

A decisive system response to any violation of conditions for pretrial release, including of course new violence, should serve notice that the victim-system alliance is strong. It tells the defendant that the victim is serious in her resolve to end the violence and that the system is unwavering in its support of her interest in securing protection.

(Ford and Regoli 1992: 204)

The effectiveness of anger-control counseling cannot be assessed simply. Policies aimed at getting defendants into anger-control counseling seem to be relatively ineffective in preventing new violence. The researchers note, however, that the policy effects should not be confused with actual counseling outcomes. Some defendants scheduled for treatment never received it. Considerably more information on implementing counseling is needed for a proper evaluation.

Moreover, the researchers caution that the results of their research point to general patterns, and that individual battered wives must choose courses of action appropriate for their particular situations and should not act blindly on the basis of the overall patterns. The research is probably more useful in what it says about ways of structuring the criminal justice system (giving victims the right to drop charges, for example) than in guiding the actions of individual victims.

Finally, the IPE offers an example of a common problem in evaluation research. Often, actual practices differ from what might be expected in principle. For example, the researchers considered the impact of different alternatives for bringing suspects into court: Specifically, the court can issue either a summons ordering the husband to appear in court or a warrant to have the husband arrested. The researchers were concerned that having the husband arrested might actually add to his anger over the situation. They were somewhat puzzled, therefore, to find no difference in the anger of husbands summoned or arrested.

The solution of the puzzle lay in the discrepancy between principle and practice:

Although a warrant arrest should in principle be at least as punishing as on-scene arrest, in practice it may differ little from a summons. A man usually knows about a warrant for his arrest and often elects to turn himself in at his convenience, or he is contacted by the warrant service agency and invited to turn himself in. Thus, he may not experience the obvious punishment of, say, being arrested, hand-cuffed, and taken away from a workplace.

(Ford 1989: 9–10)

In summary, many factors besides the scientific quality of evaluation research affect how its results are used. And, as we saw earlier, factors outside the evaluator's control can affect the quality of the study itself. But this "messiness" is balanced by the potential contributions that evaluation research can make toward the betterment of human life.

The Sabido Methodology

One of the clearest illustrations of the uses of evaluation research results comes from the omnibus methodology developed by Miguel Sabido for the use of "Entertainment-Education (E-E)" to promote social programs. The example of *Twende na Wakati* at the onset of this chapter illustrated the methods initially developed by Miguel Sabido in the 1970s when he was vice president for research at the Mexican broadcasting company Televisa. Sabido's first projects used television novellas to promote literacy and family planning. They were so successful that those methods have been used to promote a variety of social issues in the subsequent decades.

In part, the Sabido methodology concerns the nature of the radio or television dramas, particularly the kinds of characters portrayed. Some characters represent traditional points of view, some represent the modern views that the programming is designed to promote, and a third type of "transitional" character begins with traditional views and eventually shifts to the modern views. Typically, when a transitional character signs up for literacy classes, thousands of audience members do the same shortly thereafter. When the transitional character begins using condoms for family planning or safe sex, family-planning clinics are mobbed the next day by men wanting condoms.

The Sabido methodology extends beyond character definitions and plot structures. An E-E project begins with thorough research into the society where the change is being planned. A project in Ethiopia by the Population Media

Center, for example, aimed at lowering the birth rate, encouraging safe-sex practices, and enhancing the status of women. The production of radio serial dramas was preceded by extensive research into relevant existing conditions. What was the birth rate? How did it vary in different regions of the country and among different ethnic groups? What were the existing attitudes toward family planning? These questions were answered in part through national surveys. At the same time, qualitative researchers went into the countryside to observe rural villages, talking with residents and sometimes recording the sounds of village life.

This formative research provided the writers with ideas about which issues they should raise and how they should raise them. For example, the research indicated that in some regions, abduction was still a common method of mate selection: A man would kidnap a young woman, sexually assaulting her and holding her prisoner until she would consent to be his wife. The formative research also revealed a widespread belief that condoms were infected with HIV—that is, people believed that condom use increased the risk of AIDS rather than preventing it.

The initial research also provided a baseline for subsequent evaluations. By knowing public opinion toward family planning prior to the radio programs, researchers could determine how much it had changed afterward. Pre-programming measures of the use of family-planning centers could be compared with levels of use afterward. Many of these evaluation efforts ran concurrently with the radio programming. For example, regular focus groups were used to monitor public reactions to each of the serial installments, examining whether people were reacting as intended.

The Sabido methodology provides an excellent illustration of how research methods can be used to construct and evaluate social action programs aimed at resolving social problems. To learn more about the Sabido methodology, see Barker and Sabido (2005).

As you can see, evaluation research can provide a unique and powerful tool for effecting social change. However, it can also be useful on a personal level, in everyday situations, such as improving your grades, losing weight, making

Issues and Insights

Chinese Public Opinion

One of the consequences of the 1949 revolution in China was the cancellation of sociology as a field of study in Chinese universities. Some Chinese sociologists, like the respected Fei Xiao Tung, continued to conduct social research, monitoring the quality of life among citizens, especially those living in rural villages. Trained by the renowned Polish-British anthropologist, Bronislaw Malinowski, Fei primarily used qualitative, ethnographic methods. Whenever Chairman Mao would invite criticism of his regime, Professor Fei would be one of the first in line to report his research on what wasn't working and how to improve life in China. His findings were not always well received by the government, but he was persistent.

Finally, in 1979, the People's Republic of China reinstated sociology as a field of study, and Professor Fei was given the responsibility of creating the institutions necessary for the emergence of new generations of Chinese social scientists. Today, Chinese scholars are actively examining all aspects of social life. For example, the Canton Public Opinion Research Center (C-POR) in Guangzhou seeks to uncover problems in the quality of life of both urban and rural people, and draw the government's attention to those problems.

You can explore this research at the center's website: www.c-por .org/. As you will discover, most of the content is in Chinese characters, but a short detour to Google Translator, www.google.com/language_tools, will solve your problem. Here are only a few report titles from late 2011:

"Guangzhou City Public Health Assessment Survey";
"Guangzhou Environmental Protection Status Follow-Up Survey of Public Evaluation";
"Guangzhou Food Safety Assessment Survey of Public";
"Guangzhou Government 'Window Service' Satisfaction Follow-Up Investigation";
"Evaluation of Public Health in Rural Village";
"Rural Education, Supervision, Valuation of the Low Mountains of Northern Guangdong Villagers";
"Education and Supervision by the Villagers Recognized Quality Control to Be Improved";
"Beijing People's Life Experience Is Best."

When I had the opportunity to meet with some of the C-POR researchers in Shanghai in 2010, I was struck by their enthusiasm for using social research methods to improve the lot of the people. They are worthy successors to Professor Fei Xiao Tung.

friends, and influencing people. Moreover, evaluation research is hardly limited to the United States, as "Issues and Insights: Chinese Public Opinion" illustrates.

Social Indicators Research

I want to continue our discussion with a type of research that combines evaluation research with the analysis of existing data. A rapidly growing field in social research involves the development and monitoring of **social indicators**, aggregated statistics that reflect the social condition of a society or social subgroup. Researchers use social indicators to monitor aspects of social life in much the way that economists use indexes such as gross national product (GNP) per capita as an indicator of a nation's economic development.

Suppose we wanted to compare the relative health conditions in different societies. One strategy would be to compare their death rates (number of deaths per 1,000 population). More specifically, we could look at infant mortality: the number of infants who die during their first year of life among every 1,000 births. Depending on the particular aspect of health conditions we were interested in, we could devise any number of other measures: physicians per capita, hospital beds per capita, days of hospitalization per capita, and so forth. Notice that intersocietal comparisons are facilitated by calculating per capita rates (dividing by the size of the population).

Before we go further, recall from Chapter 11 the problems involved in using existing statistics. In a word, they're often unreliable, reflecting their modes of collection, storage, and calculation. With this in mind, we'll look at some of the ways we can use social indicators for evaluation research on a large scale.

social indicators Measurements that reflect the quality or nature of social life, such as crime rates, infant mortality rates, number of physicians per 100,000 population, and so forth. Social indicators are often monitored to determine the nature of social change in a society.

The Death Penalty and Deterrence

Does the death penalty deter capital crimes such as murder? This question is hotly debated every time a state considers eliminating or reinstating capital punishment and every time someone is executed. Those supporting capital punishment often argue that the threat of execution will keep potential murderers from killing people. Opponents of capital punishment often argue that it has no effect in that regard. Social indicators can help shed some light on the question.

If capital punishment actually deters people from committing murder, then we should expect to find murder rates lower in those states that have the death penalty than in those that do not. The relevant comparisons in this instance are not only possible, they've been compiled and published. William Bailey (1975) compiled data that directly contradicted the view that the death penalty deters murderers. In both 1967 and 1968, those states with capital punishment had dramatically higher murder rates than those without capital punishment. Some people criticized the interpretation of Bailey's data, saying that most states had not used the death penalty in recent years, even when they had it on the books. That could explain why it hadn't seemed to work as a deterrent. Further analysis, however, contradicts this explanation. When Bailey compared those states that hadn't used the death penalty with those that had, he found no real difference in murder rates.

A counterexplanation is possible, however. It could be the case that the interpretation given Bailey's data was backward. Maybe the existence of the death penalty as an option was a consequence of high murder rates: Those states with high rates instituted it; those with low rates didn't institute it or repealed it if they had it on the books. It could be the case, then, that instituting the death penalty would bring murder rates down. Not so, however. Analyses over time do not show an increase in murder rates when a state repeals the death penalty nor a decrease in murders when one is instituted. A more recent examination by Bailey and colleague Ruth Peterson (1994) confirmed the earlier findings and also indicated that law enforcement officials doubted the deterrent effect. Further, the pattern observed by Bailey in 1967 and 1968 has persisted over time, even when we take into account the substantial increase in the overall murder rate. In 2015, for example, the 31 death penalty states had a combined murder rate of 5.1 per 100,000, compared with a combined murder rate of 4.1 among the 19 states that lack the death penalty (Death Penalty Information Center 2017).

Notice from the preceding discussion that researchers can use social indicators data for comparison across groups either at one time or across some period. Often, doing both sheds the most light on the subject.

At present, work on the use of social indicators is proceeding on two fronts. On the one hand, researchers are developing more-refined indicators—finding which indicators of a general variable are the most useful in monitoring social life. At the same time, research is being devoted to discovering the relationships among variables within whole societies.

Computer Simulation

One of the most exciting prospects for social indicators research lies in the area of computer simulation. As researchers begin compiling mathematical equations describing the relationships that link social variables to one another (for example, the relationship between growth in population and the number of automobiles), those equations can be stored and linked in a computer. With a sufficient number of adequately accurate equations on hand, researchers one day will be able to test the implications of specific social changes by computer rather than in real life.

Suppose a state contemplated doubling the size of its tourism industry, for example. We could enter that proposal into a computer-simulation model and receive in seconds or minutes a description of all the direct and indirect consequences of the increase in tourism. We could know what new public facilities would be required; which public agencies, such as police and fire departments, would have to be increased and by how much; what the labor force would look like; what kind of training would be required to provide it; how much new income and tax revenue would be produced; and so forth, through all the intended and unintended consequences of the action. Depending on the results, the public planners might say, "Suppose we increased the industry only by half," and have a new printout of consequences immediately.

An excellent illustration of computer simulation linking social and physical variables is to be found in the research of Donella and Dennis Meadows and their colleagues at Dartmouth and the Massachusetts Institute of Technology (Meadows et al. 1972; Meadows, Meadows, and Randers 1992). They've taken as input data some known and estimated reserves of various nonreplaceable natural resources (for example, oil, coal, and iron), past patterns of population and economic growth, and the relationships between growth and use of resources. Using a complex computer-simulation model, they've been able to project, among other things, the probable number of years various resources will last in the face of alternative usage patterns in the future. Going beyond the initially gloomy projections, such models also make it possible to chart out less-gloomy futures, specifying the actions required to achieve them. Clearly, the value of computer simulation is not limited to evaluation research, though it can serve an important function in that regard.

This potentiality points to the special value of evaluation research in general. Throughout human history, we've been tinkering with our social arrangements, seeking better results. Evaluation research provides a means for us to learn right away whether a particular tinkering really makes things better. Social indicators allow us to make that determination on a broad scale; coupling them with computer simulation opens up the possibility of knowing how much we would like a particular intervention without having to experience its risks.

Ethics and Evaluation Research

Because it's embedded in the day-to-day events of real life, evaluation research entails special ethical problems. Evaluating the impact of busing school children to achieve educational

integration, for example, will throw the researchers directly into the political, ideological, and ethical issues of busing itself. It's not possible to evaluate a sex-education program in elementary schools without becoming involved in the heated issues surrounding sex education itself, and the researcher will find it difficult to remain impartial. The evaluation study design will require that some children receive sex education—in fact, you may very well be the one who decides which children do. (From a scientific standpoint, you *should* be in charge of selection.) This means that when parents become outraged that their child is being taught about sex, you'll be directly responsible.

Now let's look on the bright side. Maybe the experimental program is of great value to those participating in it. Let's say that the new industrial-safety program being evaluated reduces injuries dramatically. What about the control-group members who were deprived of the program by the research design? The evaluators' actions could be an important part of the reason that a control-group subject suffered an injury.

By its very nature, then, evaluation research is interwoven with real-world issues. We only evaluate programs when it matters whether or not they make a difference, and that means the results of the evaluation matter to people. This brings up another potential problem. Almost always, some people will want the results to turn out a certain way, and other people may want a different result. Often, as in the case of pharmaceutical testing, for example, those paying for the research may want a particular result. Further, the researchers themselves may have personal motives for wanting a given result. Evaluation researchers, therefore, often find themselves under internal or external pressure to produce a particular finding.

Of course, researchers must not be swayed by either personal desires or sponsors' demands in the design, execution, and analysis of results. This is true in all kinds of social research; however, unethical actions in evaluation research can produce particularly severe consequences. For example, the results of evaluation research may determine whether people are subjected to medical or social remedies. Imagine a medical researcher slanting drug-testing results to suggest

What do you think?...Revisited

The purpose of evaluation research is to determine whether social interventions or programs have had their desired effects. No matter how much research is done, however, debates tend to persist.

As we've seen in this chapter, the political and ideological viewpoints that inform positions on certain issues often are deeply ingrained and withstand contrary information. Moreover, evaluation research occurs in the "real world," where such assessments often affect people's self-interests. Research that says a program is ineffective threatens the jobs of those employed by the program, not to mention the reputations of those who created it.

Evaluation research typically examines studies with multiple variables. This means that researchers can easily argue about which variable caused an apparent effect. They may even argue that some other, previously unthought of variable is responsible (recall the discussion of internal and external invalidity in Chapter 8). Further, many of the variables evaluated tend to be somewhat ambiguous: What is "happiness," "motivation," and so forth? In short, there is usually ample "wiggle room" for people who disagree with the findings of evaluation research. As such, we haven't settled once and for all on whether the death penalty prevents murders.

that a new drug is more effective than it is or covering up the negative side effects of the drug—with the consequence of more patients being given the drug.

Or imagine an evaluation of a prison rehabilitation program being slanted to make the program seem more effective than it is. Limited resources might be diverted to support the ineffective program and possibly even harm the prisoners subjected to it. That's not the worst example, however.

Recall from Chapter 3 that in 1932 researchers in Tuskegee, Alabama, began a program that presumably provided treatment for syphilis to poor, African American men suffering from the disease. Over the years that followed, several hundred men participated in the program. What they didn't know was that they were not actually receiving any treatment at all; the physicians conducting the study merely wanted to observe the natural progress of the disease. Even after penicillin was found to be an effective cure, the researchers still withheld the treatment. Although there is unanimous agreement today that the study was inherently unethical, this was not the case at the time. Even when the study began being reported in research publications, the researchers refused to acknowledge they had done anything wrong (Jones 1981).

My purpose in these comments has not been to cast a shadow on evaluation research. Rather, I want to bring home the real-life consequences of the evaluation researcher's actions. Ultimately, as we saw in Chapter 3, all social research has ethical components.

I will close this discussion with a somewhat different observation made by Donald T. Campbell. In what has come to be known as Campbell's law, he observed, "The more any quantitative social indicator is used for social decision-making, the more subject it will be to corruption pressures and the more apt it will be to distort and corrupt the social processes it is intended to monitor" (1976: 54). One example of this is what educators refer to as "teaching to the test." If teachers are to be evaluated on the basis of how well their students perform on a standardized test, instruction tends to focus on that test rather than on the subject matter more generally. Similarly, when those managing stock portfolios are compensated on the basis of how many stocks have been traded, there is a temptation to trade stocks that might more wisely be held. Or when police departments are judged as to their ability to lower assault rates in the city, there will be a temptation to categorize and report incidents as lesser offenses.

Thus, we see that evaluation research is sometimes a part of the process it seeks to evaluate and that it can have unintended consequences. This is another example of the recursive nature of social research, discussed in Chapter 1.

MAIN POINTS

Introduction

- Evaluation research is a form of applied research that studies the effects of social interventions.

Topics Appropriate for Evaluation Research

- Evaluation research is appropriate whenever some social intervention occurs or is planned, so the potential for topics is limitless.

- Much of evaluation research is referred to as program evaluation or outcome assessment: the determination of whether a social intervention is producing the intended result.

Formulating the Problem: Issues of Measurement

- A careful formulation of the problem, including relevant measurements and criteria of success or failure, is essential in evaluation research. In particular, evaluators must carefully specify outcomes, measure experimental contexts, specify the intervention being studied and the population targeted by the intervention, decide whether to use existing measures or devise new ones, and assess the potential cost-effectiveness of an intervention.

Types of Evaluation Research Designs

- Evaluation researchers typically use experimental or quasi-experimental designs. Examples of quasi-experimental designs include time-series studies and the use of nonequivalent control groups.

- Evaluators can also use qualitative methods of data collection. Both quantitative and qualitative data analyses can be appropriate in evaluation research, sometimes in the same study.

- The special, logistical problems of evaluation research grow out of the fact that it occurs within the context of real life.

- The implications of evaluation research won't necessarily be put into practice, especially if they conflict with official points of view.

Social Indicators Research

- Social indicators can provide an understanding of broad social processes.

- Computer-simulation models hold the promise of allowing researchers to study the possible results of social interventions without having to incur those results in real life.

Ethics and Evaluation Research

- Evaluation research entails special ethical problems because it's embedded in the day-to-day events of real life.

- Evaluation research may bring added pressure to produce specific results, as desired by interested parties.

- Unethical actions in an evaluation study can have more-severe consequences than such actions in other types of research.

KEY TERMS

evaluation research

multiple time-series designs

nonequivalent control group

program evaluation/ outcome assessment

quasi experiments

social indicators

time-series design

PROPOSING SOCIAL RESEARCH: EVALUATION RESEARCH

Evaluation research represents a research purpose rather than a particular method. In the proposal, you need to spell out the type of evaluation you're conducting and perhaps the implications of various possible outcomes.

In earlier assignments, you'll have spelled out the data-collection and measurement methods to be used in your study. If your study is designed to determine the success or failure of a program, it may be appropriate to add a specification of the research results that will be deemed a positive or negative assessment in that regard. This may not always be appropriate or possible, but it adds integrity to the evaluation process when it can be done.

REVIEW QUESTIONS

1. Review the evaluation of the Navy low-performer program discussed in this chapter. How would you redesign the program and the evaluation to avoid the problems that appeared in the actual study?

2. Take a minute to think of the many ways our society has changed during your own lifetime. How would you specify those changes as social indicators that could be used in monitoring the quality of life in your society?

3. Identify at least three deliberate social interventions, such as lowering the voting age to 18. For each, how would you (1) specify the perceived problem and (2) describe the kind of research that would evaluate whether the intervention was successful in solving the perceived problem?

4. Think of something at your college that you feel could be improved. Now think of something that could be done to solve the problem you've identified. Pursue this line of thought until you've developed a clear operational definition of how you would know when the problem had been solved through some intervention. What future measurement would represent success?

CHAPTER 13

Qualitative Data Analysis

CHAPTER OVERVIEW

Qualitative data analysis is the nonnumerical assessment of observation made through participant observation, content analysis, in-depth interviews, and other qualitative research techniques. Although qualitative analysis is as much an art as a science, it has its own logic and techniques, some of which are enhanced by special computer programs.

Image Source/Corbis

Learning Objectives

After studying this chapter, you will be able to . . .

- Identify and explain the three methods for linking theory and analysis.
- Describe the use of coding in the analysis of qualitative data.
- Recognize some computer programs that can be used for qualitative analysis and explain how they work.
- Provide examples of how quantitative data can be analyzed qualitatively.
- Explain some of the ways researchers can assess the quality of qualitative data.
- Summarize some of the ethical issues that can occur in qualitative research.

Introduction

Chapter 14 will deal with the *quantitative* analysis of social research data, sometimes called *statistical analysis*. Recent decades of social science research have tended to focus on quantitative data-analysis techniques. This focus, however, sometimes conceals another approach to making sense of social observations: **qualitative analysis**—methods for examining social research data without converting them to a numerical format. This approach predates quantitative analysis. It remains useful in data analysis and is even enjoying a resurgence of interest among social scientists.

Although statistical analyses may intimidate some students, the steps involved can sometimes be learned by rote. That is, with practice, the rote exercise of quantitative skills can produce an increasingly sophisticated understanding of the logic that lies behind those techniques.

It's much more difficult to teach qualitative analysis as a series of rote procedures. In this case, understanding must precede practice. In this chapter, we begin with the links between research and theory in qualitative analysis. Then we examine some procedures that have proved useful in pursuing the theoretical aims. After considering some simple manual techniques, we'll take some computer programs out for a spin.

Linking Theory and Analysis

As suggested in Chapter 10 and elsewhere in this book, qualitative research methods involve a continuing interplay between data collection and theory. In quantitative research, it's sometimes easy to get caught up in the logistics of data collection and in the statistical analysis of data, thereby losing sight of theory for a time. This occurs less in qualitative research, where data collection, analysis, and theory intertwine more intimately.

In the discussions to follow, we'll use the image of theory offered by Anselm Strauss and Juliet Corbin (1994: 278) as consisting of "*plausible* relationships proposed among *concepts* and *sets of concepts*." They stress "plausible" to indicate that theories represent our best understanding of how life operates. The more our research confirms a particular set of relationships among particular concepts, however, the more confident we become that our understanding corresponds to social reality.

Although qualitative research is sometimes undertaken for purely descriptive purposes—such as the anthropologist's ethnography detailing the ways of life in a previously unknown tribe—the rest of this chapter focuses primarily on the search for explanatory patterns. As we'll see, sometimes the patterns occur over time, and sometimes they take the form of causal relations among variables. Let's look at some of the ways qualitative researchers uncover such patterns.

qualitative analysis The nonnumerical examination and interpretation of observations, for the purpose of discovering underlying meanings and patterns of relationships. This approach is most typical of field research and historical research.

What do you think?

Why do researchers sometimes use qualitative analyses when they might have used statistics? Isn't a statistical, or quantitative, analysis a more "scientific" way to study poverty, discrimination, and so forth?

See the *What do you think?...Revisited* box toward the end of the chapter.

Earl Babbie

Discovering Patterns

John Lofland and colleagues (2006: 149–65) suggest six different ways of looking for patterns in a particular research topic. Let's suppose you're interested in analyzing child abuse in a certain neighborhood. Here are some questions you might ask yourself, in an effort to make sense out of your data:

1. *Frequencies:* How often does child abuse occur among families in the neighborhood under study? (Realize that there may be a difference between the frequency and what people are willing to tell you.)
2. *Magnitudes:* What are the levels of abuse? How brutal are they?
3. *Structures:* What are the different types of abuse—physical, mental, sexual? Are they related in any particular manner?
4. *Processes:* Is there any order among the elements of structure? Do abusers begin with mental abuse and move on to physical and sexual abuse, or does the order of elements vary?
5. *Causes:* What are the causes of child abuse? Is it more common in particular social classes or among different religious or ethnic groups? Does it occur more often during good times or bad?
6. *Consequences:* How does child abuse affect the victims, in both the short and the long term? What changes does it cause in the abusers?

---

cross-case analysis An analysis that involves an examination of more than one case, either a variable-oriented or case-oriented analysis.

variable-oriented analysis An analysis that describes and/or explains a particular variable.

For the most part, in examining your data you'll look for patterns appearing across several observations that typically represent different cases under study. This approach is called **cross-case analysis**. A. Michael Huberman and Matthew Miles (1994) offer two strategies for cross-case analysis: variable-oriented and case-oriented analysis. **Variable-oriented analysis** is similar to a model we've already discussed from time to time in this book. If we were trying to predict the decision to attend college, Huberman and Miles suggest, we might consider variables such as "gender, socioeconomic status, parental expectations, school performance, peer support, and decision to attend college" (1994: 435). Thus, we would determine whether men or women were more likely to attend college. The focus of our analysis would be on interrelations among variables, and the people observed would primarily be the carriers of those variables.

Variable-oriented analysis may remind you of the discussion in Chapter 1 that introduced the idea of nomothetic explanation. The aim here is to achieve a partial, overall explanation using relatively few variables. The political pollster who attempts to explain voting intentions on the basis of two or three key variables is using this approach. There is no pretense that the researcher can predict every individual's behavior or even explain any one person's motivations in full. Sometimes, though, it's useful to have even a partial explanation of overall orientations and actions.

You may also recall the introduction to idiographic explanation in Chapter 1, wherein we attempt to understand a particular case fully. In the voting example, we would endeavor to learn everything we could about all the factors that came into play in determining one person's decision on how to vote. This orientation lies at the base of what Huberman and Miles call a **case-oriented analysis**.

> *In a case-oriented analysis, we would look more closely into a particular case, say, Case 005, who is female, middle-class, has parents with high expectations, and so on. These are, however, "thin" measures. To do a genuine case analysis, we need to look at a full history of Case 005; Nynke van der Molen, whose mother trained as a social worker but is bitter over the fact that she never worked outside the home, and whose father wants Nynke to work in the family florist shop. Chronology is also important: two years ago, Nynke's closest friend decided to go to college, just before Nynke began work in a stable and just before Nynke's mother showed her a scrapbook from social work school. Nynke then decided to enroll in veterinary studies.*
>
> *(1994: 436)*

This abbreviated commentary should give some idea of the detail involved in this type of analysis. Of course, an entire analysis would be more extensive and pursue issues in greater depth. This full, idiographic examination, however, tells us nothing about people in general. It offers no theory about why people choose to attend college.

Even so, in addition to understanding one person in great depth, the researcher sees the critical elements of the subject's experiences as instances of more-general social concepts or variables. For example, Nynke's mother's social work training can also be seen as "mother's education." Her friend's decision can be seen as "peer influence." More specifically, these could be seen as independent variables having an impact on the dependent variable of attending college.

Of course, one case does not a theory make—hence, Huberman and Miles's reference to cross-case analysis, in which the researcher turns to other subjects, looking into the full details of their lives as well but paying special note to the variables that seemed important in the first case. How much and what kind of education did other subjects' mothers have? Is there any evidence of close friends attending college?

Some subsequent cases will closely parallel the first one in the apparent impact of particular variables. Other cases will bear no resemblance to the first. These latter cases may require the identification of other important variables, which may invite the researcher to explore why some cases seem to reflect one pattern, whereas others reflect another.

Grounded Theory Method

The cross-case analysis just described should sound somewhat familiar. In the discussion of grounded theory in Chapter 10, we saw how qualitative researchers sometimes attempt to establish theories on a purely inductive basis. This approach begins with observations rather than hypotheses and seeks to discover patterns and develop theories from the ground up, with no preconceptions, although some research may build and elaborate on earlier grounded theories.

Recall that grounded theory was first developed by the sociologists Barney Glaser and Anselm Strauss (1967), in an attempt to come to grips with their clinical research in medical sociology. Since then, it has evolved as a method, with the cofounders taking it in slightly different directions. The following discussion will deal with the basic concepts and procedures of the **Grounded Theory Method (GTM)**.

In addition to the fundamental, inductive tenet of building theory from data, GTM employs the **constant comparative method**. As Glaser and Strauss originally described this method, it involved four stages (1967: 105–13):

1. "Comparing incidents applicable to each category." As Glaser and Strauss researched the reactions of nurses to the possible death of patients in their care, the researchers found that the nurses were assessing the "social loss" attendant

---

case-oriented analysis An analysis that aims to understand a particular case or several cases by looking closely at the details of each.

Grounded Theory Method (GTM) An inductive approach to research introduced by Barney Glaser and Anselm Strauss in which theories are generated solely from an examination of data rather than being derived deductively.

constant comparative method A component of the Grounded Theory Method in which observations are compared with one another and with the evolving inductive theory.

on a patient's death. Once this concept arose in the analysis of one case, they looked for evidence of the same phenomenon in other cases. When they found the concept arising in the cases of several nurses, they compared the different incidents. This process is similar to conceptualization as described in Chapter 5—specifying the nature and dimensions of the many concepts arising from the data.

2. "Integrating categories and their properties." Here the researcher begins to note relationships among concepts. In the assessment of social loss, for example, Glaser and Strauss found that nurses took special notice of a patient's age, education, and family responsibilities. For these relationships to emerge, however, it was necessary for the researchers to have noticed all these concepts.

3. "Delimiting the theory." Eventually, as the patterns of relationships among concepts become clearer, the researcher can ignore some of the concepts that were initially noted but are evidently irrelevant to the inquiry. In addition to the number of categories being reduced, the theory itself may become simpler. In the examination of social loss, for example, Glaser and Strauss found that the assessment processes could be generalized beyond nurses and dying patients: They seemed to apply to the ways all staff dealt with all patients (dying or not).

4. "Writing theory." Finally, the researcher must put his or her findings into words to be shared with others. As you may have already experienced for yourself, the act of communicating your understanding of a topic actually modifies and even improves your own grasp of it. In GTM, the writing stage is regarded as a part of the research process. A later section of this chapter (on memoing) elaborates on this point.

This brief overview should give you an idea of how grounded theory proceeds. The many techniques associated with GTM can be found both in print and on the Web. One of the key publications is Anselm Strauss and Juliet Corbin's *Basics of Qualitative Research* (1998), which elaborates on and extends many of the concepts and techniques found in the original 1967 Glaser and Strauss volume.

---

semiotics The study of signs and the meanings associated with them. This is commonly associated with content analysis.

GTM is only one analytic approach to qualitative data. In the remainder of this section, we'll look at some other specialized techniques.

Semiotics

Semiotics is commonly defined as the "science of signs" and has to do with symbols and meanings. It's often associated with content analysis, which was discussed in Chapter 11, though it can be applied in a variety of research contexts.

Peter Manning and Betsy Cullum-Swan (1994: 466) offer some sense of the applicability of semiotics, as follows: "Although semiotics is based on language, language is but one of the many sign systems of varying degrees of unity, applicability, and complexity. Morse code, etiquette, mathematics, music, and even highway signs are examples of semiotic systems."

There is no meaning inherent in any sign, however. Meanings reside in minds. So, a particular sign means something to a particular person. However, the agreements we have about the meanings associated with particular signs make semiotics a *social* science. As Manning and Cullum-Swan point out,

> For example, a lily is an expression linked conventionally to death, Easter, and resurrection as a content. Smoke is linked to cigarettes and to cancer, and Marilyn Monroe to sex. Each of these connections is social and arbitrary, so that many kinds of links exist between expression and content.
>
> *(1994: 466)*

To explore this contention, see if you can link the signs with their meanings in Figure 13-1. I'm confident enough that you know all the "correct" associations that there's no need for me to give the answers. (OK, you should have said 1c, 2a, 3b, 4e, 5d.) The point is this: What do any of these

SIGN	MEANING
1. Poinsettia	a. Good luck
2. Horseshoe	b. First prize
3. Blue ribbon	c. Christmas
4. "Say cheese"	d. Acting
5. "Break a leg"	e. Smile for a picture

FIGURE 13-1
Matching Signs and Their Meanings.

signs have to do with their "meanings"? Draft an e-mail message to a Martian social scientist explaining the logic at work here. (You might want to include some "emoticons" like :-)—another example of semiotics.)

While there is no doubt a story behind each of the linkages in Figure 13-1, the meanings you and I "know" today are socially constructed. Semiotic analysis involves a search for the meanings intentionally or unintentionally attached to signs.

Consider the sign shown in Figure 13-2, from a hotel lobby in Portland, Oregon. What's being communicated by the rather ambiguous sign? The first sentence seems to be saying that the hotel is up to date with the current move away from tobacco in the United States. Guests who want a smoke-free environment need look no further: This is a healthy place to stay. At the same time, says the second sentence, the hotel would not like to be seen as inhospitable to smokers. There's room for everyone under this roof. No one need feel excluded. This sign is more easily understood within a marketing paradigm than one of logic.

The "signs" examined in semiotics, of course, are not limited to this kind of sign. Most are quite different, in fact. *Signs* are any things that are assigned special meanings. They can include such things as logos, animals, people, and consumer products. Sometimes the symbolism is subtle. You can find a classic analysis in Erving Goffman's *Gender Advertisements* (1979). Goffman focused on advertising pictures found in magazines and newspapers. The overt purpose of the

ads, of course, was to sell specific products. But what else was communicated? What in particular did the ads say about men and women?

Analyzing pictures containing both men and women, Goffman was struck by the fact that men were almost always bigger and taller than the women accompanying them. (In many cases, in fact, the picture managed to convey the distinct impression that the women were merely accompanying the men.) Although the most obvious explanation is that men are, on average, heavier and taller than women, Goffman suggested the pattern had a different meaning: that size and placement implied *status*. Those who were larger and taller presumably had higher social standing—more power and authority (1979: 28). Goffman suggested that the ads communicated that men were more important than women.

In the spirit of Freud's comment that "sometimes a cigar is just a cigar" (he was a smoker), how would you decide whether the ads simply reflected the biological differences in the average sizes of men and women or whether they sent a message about social status? In part, Goffman's conclusion was based on an analysis of the exceptional cases: those in which the women appeared taller than the men. In these cases, the men were typically of a lower social status—the chef beside the society matron, for example. This confirmed Goffman's main point that size and height indicated social status.

The same conclusion could be drawn from pictures with men of different heights. Those of higher status were taller, whether it was the gentleman speaking to a waiter or the boss guiding the work of his younger assistants. Where actual height was unclear, Goffman noted the placement of heads in the picture. The assistants were crouching down while the boss leaned over them. The servant's head was bowed so it was lower than that of the master.

The latent message conveyed by the ads, then, was that the higher a person's head appeared in the ad, the more important that person was. And in the great majority of ads containing men and women, the former were clearly portrayed as more important. The subliminal message in the ads, whether intended or not, was that men are more powerful and enjoy a higher status than do women.

THIS ENTIRE ESTABLISHMENT IS NON-SMOKING. SMOKING IS ONLY PERMITTED IN DESIGNATED AREAS.

Earl Babbie

FIGURE 13-2
Mixed Signals?

Goffman examined several differences besides physical size in the portrayal of men and women. As another example, men were typically presented in active roles, women in passive ones. The (male) doctor examined the child while the (female) nurse or mother looked on, often admiringly. A man guided a woman's tennis stroke (all the while keeping his head higher than hers). A man gripped the reins of his galloping horse, while a woman rode behind him with her arms wrapped around his waist. A woman held the football, while a man kicked it. A man took a photo, which contained only women.

Goffman suggested that such pictorial patterns subtly perpetuated a host of gender stereotypes. Even as people spoke publicly about gender equality, these advertising photos established a quiet backdrop of men and women in their "proper roles."

Conversation Analysis

Ethnomethodology, as you'll recall, aims to uncover the implicit assumptions and structures in social life. **Conversation analysis (CA)** seeks to pursue that aim through an extremely close scrutiny of the way we converse with one another. In the examination of ethnomethodology in Chapter 10, you saw some examples of conversation analysis. Here we'll look a little more deeply into that technique.

David Silverman (1993: 125ff), reviewing the work of other CA theorists and researchers, speaks of three fundamental assumptions. First, conversation is a socially structured activity. Like other social structures, it includes established rules of behavior. For example, we're expected to take turns, with only one person speaking at a time. In telephone conversations, the person answering the call is expected to speak first (as in "Hello"). You can verify the existence of this rule, incidentally, by picking up the phone without speaking. You may recall that this is the sort of thing ethnomethodologists tend to do.

conversation analysis (CA) A meticulous analysis of the details of conversation, based on a complete transcript that includes pauses, "hems," and also "haws."

Second, Silverman points out that conversations must be understood contextually. The same utterance will have totally different meanings in different contexts. For example, notice how the meaning of "Same to you!" varies if preceded by "I don't like your looks" or by "Have a nice day."

Third, CA aims to understand the structure and meaning of conversation through excruciatingly accurate transcripts of conversations. Not only are the exact words recorded, but all the "uhs," "ers," bad grammar, and pauses are also noted. Pauses, in fact, are measured to the nearest tenth of a second.

The practical uses of this type of analysis are many. Ann Marie Kinnell and Douglas Maynard (1996), for example, analyzed conversations between staff and clients at an HIV-testing clinic to examine how information about safe sex was communicated. Among other things, they found that the staff tended to provide standard information rather than try to speak directly to a client's specific circumstances. Moreover, they seemed reluctant to give direct advice about safe sex, settling for information alone.

These discussions should give you some sense of the variety of qualitative analysis methods available to researchers. Now let's look at some of the data-processing and data-analysis techniques commonly used in qualitative research.

Qualitative Data Processing

Let me begin this section with a warning. The activity we're about to examine is as much art as science. At the very least, there are no cut-and-dried steps that guarantee success.

It's a lot like learning how to paint with watercolors or compose a symphony. You can certainly gain education in such activities; you can even take university courses in both. Each has its own conventions and techniques as well as tips you may find useful as you set out to create art or music. However, instruction can carry you only so far. The final product must come from you. Much the same can be said of qualitative data processing.

At the same time, researchers have developed systematic and rigorous techniques for this type of research. We'll examine some of those here. You can gain a more in-depth view of these techniques from *Constructing Grounded Theory: A Practical Guide through Qualitative Research*, an excellent book by social researcher Kathy Charmaz (2006).

This section presents some ideas on coding qualitative data, writing memos, and mapping concepts graphically. Although far from a "how-to" manual, these ideas give a useful starting point for finding order in qualitative data.

Coding

Whether you engage in participant observation, in-depth interviewing, collecting biographical narratives, doing content analysis, or doing some other form of qualitative research, you'll eventually possess a growing mass of data—most typically in the form of textual materials. What do you do next?

The key process in the analysis of qualitative social research data is *coding*—classifying or categorizing individual pieces of data—coupled with some kind of retrieval system. Together, these procedures allow you to retrieve materials you may later be interested in.

Let's say you're chronicling the growth of a social movement. You recall writing up some notes about the details of the movement's earliest beginnings. Now you need that information. If all your notes have been catalogued by topic, retrieving those you need should be straightforward. As a simple format for coding and retrieval, you might have created a set of file folders labeled with various topics, such as "History." Data retrieval in this case means pulling out the "History" folder and rifling through the notes contained therein until you find what you need.

As you'll see later in this chapter, several sophisticated computer programs allow for a faster, more certain, and more precise retrieval process. Rather than looking through a "History" file, you can go directly to notes dealing with the "Earliest History" or the "Founding" of the movement.

Coding has another, even more important purpose. As discussed earlier, the aim of data analysis is the discovery of patterns among the data, patterns that point to a theoretical understanding of social life. The coding and relating of concepts is key to this process and requires a more refined system than a set of manila folders. In this section, we'll assume that you'll be doing your coding manually. A later section of the chapter will illustrate the use of computer programs for qualitative data analysis.

Coding Units

As you may recall from the earlier discussion of content analysis, for statistical analysis it's important to identify a standardized unit of analysis prior to coding. If you were comparing American and French novels, for example, you might evaluate and code sentences, paragraphs, chapters, or whole books. It would be important, however, to code the same units for each novel analyzed. This uniformity is necessary in a quantitative analysis, as it allows us to report something like "Twenty-three percent of the paragraphs contained metaphors." This is possible only if we've coded the same unit—paragraphs—in each of the novels.

Coding data for a qualitative analysis, however, is quite different. The *concept* is the organizing principle for qualitative coding. Here the units of text appropriate for coding will vary within a given document. Thus, in a study of organizations, "Size" might require only a few words per coding unit, whereas "Mission" might take a few pages. Or a lengthy description of a heated stockholders' meeting might be coded as "Internal Dissent."

Realize also that a given code category may be applied to textual materials of quite different lengths. For example, some references to the organization's mission may be brief, others lengthy. Whereas standardization is a key principle in quantitative analysis, this is not the case in qualitative analysis.

Coding as a Physical Act

Before continuing with the logic of coding, let's take a moment to see what it actually looks like. Lofland and colleagues offer this description of manual filing:

> Prior to the widespread availability of personal computers beginning in the late 1980s, coding frequently took the specific physical form of filing. The researcher established an expanding set of file folders with code names on the tabs and physically placed either the item of data itself or a note that referenced its location in another file folder. Before photocopying was easily available and cheap, some fieldworkers typed their fieldnotes with carbon paper, wrote codes in the margins of the copies of the notes, and cut them up with scissors. They then placed the resulting slips of paper in corresponding file folders.

(2006: 203)

As these researchers point out, computers have greatly simplified this task. However, the image of slips of paper that contain text and are put in folders representing code categories is useful for understanding the process of coding. In the next section, when I suggest that we categorize a textual passage with a certain code, imagine that we have the passage typed on a slip of paper and that we place it in a file folder bearing the name of the code. Whenever we assign two codes to a passage, imagine placing duplicate copies of the passage in two different folders representing the two codes.

Creating Codes

So, what should your code categories be? Glaser and Strauss (1967: 101ff) allow for the possibility of coding data for the purpose of testing hypotheses that have been generated by prior theory. In that case, then, the theory would suggest the codes, in the form of variables.

In this section, however, we're going to focus on the more common process of **open coding**. Strauss and Corbin define it as follows:

> *To uncover, name, and develop concepts, we must open up the text and expose the thoughts, ideas, and meanings contained therein. Without the first analytic step, the rest of the analysis and the communication that follows could not occur. Broadly speaking, during open coding, data are broken down into discrete parts, closely examined, and compared for similarities and differences. Events, happenings, objects, and actions/interactions that are found to be conceptually similar in nature or related in meaning are grouped under more abstract concepts termed "categories."*

> *(1998: 102)*

---

open coding The initial classification and labeling of concepts in qualitative data analysis. In open coding, the codes are suggested by the researchers' examination and questioning of the data.

axial coding A reanalysis of the results of open coding in Grounded Theory Method, aimed at identifying the important, general concepts.

selective coding In Grounded Theory Method, this analysis builds on the results of open coding and axial coding to identify the central concept that organizes the other concepts that have been identified in a body of textual materials.

Open coding is the logical starting point for GTM qualitative coding. Beginning with some body of text (part of an interview, for example), you read and reread a passage, seeking to identify the key concepts contained within it. Any particular piece of data may be given several codes, reflecting as many concepts. For example, notice all the concepts contained in this comment by a student interviewee:

> *I thought the professor should have given me at least partial credit for the homework I turned in.*

Some obvious codes are "Professor," "Homework," and "Grading." The result of open coding is the identification of numerous concepts relevant to the subject under study. The open coding of more and more text will lengthen the list of codes.

Besides open coding, two other types of coding take place in this method. **Axial coding** aims to identify the core concepts in the study. Although axial coding uses the results of open coding, more concepts can be identified through continued open coding after the axial coding has begun. Axial coding involves a regrouping of the data, in which the researcher uses the open-code categories and looks for more-analytic concepts. For example, the passage just given also carries the concept of "Perceptions of Fairness," which might appear frequently in the student interviews, thereby suggesting that it's an important element in understanding students' concerns. Another axial code reflected in the student comment might be "Power Relationships," because the professor is seen to exercise power over the student.

The last kind of coding, **selective coding**, seeks to identify *the* central code in the study: the one that all the other codes relate to. Both of the axial codes just mentioned might be restructured as aspects of a more general concept: "Professor–Student Relationships." Of course, in a real data analysis, decisions such as the ones we've been discussing would arise from masses of textual data, not from a single quotation. The basic notion of the Grounded Theory Method is that patterns of relationships can be teased out of an extensive, in-depth examination of a large body of observations.

Here's a concrete example to illustrate how you might engage in this form of analysis. Suppose you're interested in the religious bases for homophobia. You've interviewed some people opposed to homosexuality who cite a religious basis for their feelings. Specifically, they refer you to these passages in the Book of Leviticus (Revised Standard Version):

> 18:22 *You shall not lie with a male as with a woman; it is an abomination.*
>
> 20:13 *If a man lies with a male as with a woman, both of them have committed an abomination; they shall be put to death, their blood is upon them.*

Although the point of view expressed here seems unambiguous, you might decide to examine it in more depth. Perhaps a qualitative analysis of Leviticus can yield a fuller understanding of where these injunctions against homosexuality fit into the larger context of Judeo-Christian morality.

Let's start our analysis by examining the two passages just quoted. We might begin by coding each passage with the label "Homosexuality." This is clearly a key concept in our analysis. Whenever we focus on the issue of homosexuality in our analysis of Leviticus, we want to consider these two passages.

Because homosexuality is such a key concept, let's look more closely into what it means within the data under study. We first notice the way *homosexuality* is identified: a man lying with a man "as with a woman." Although we can imagine a lawyer seeking admission to heaven saying, "But here's my point; if we didn't actually lie down…"; it seems safe to assume that the passage refers to having sex, though it is not clear what specific acts might or might not be included.

Notice, however, that the injunctions appear to concern *male* homosexuality only; lesbianism is not mentioned. In our analysis, then, each of these passages might also be coded "Male Homosexuality." This illustrates two more aspects of coding: (1) Each unit can have more than one code and (2) hierarchical codes (one included within another) can be used. Now each passage has two codes assigned to it.

An even more general code might be introduced at this point: "Prohibited Behavior." This is important for two reasons. First, homosexuality is not inherently wrong, from an analytic standpoint. The purpose of the study is to examine the way it's made wrong by the religious texts in question. Second, our study of Leviticus may turn up other behaviors that are prohibited.

There are at least two more critical concepts in the passages: "Abomination" and "Put to Death." Notice that whereas these are clearly related to "Prohibited Behavior," they are hardly

the same. Parking without putting money in the meter is prohibited, but few would call it an abomination and fewer still would demand the death penalty for that transgression. Let's assign these two new codes to our first two passages.

At this point, we want to branch out from the two key passages and examine the rest of Leviticus. We therefore examine and code each of the remaining chapters and verses. In our subsequent analyses, we'll use the codes we have already and add new ones as appropriate. When we do add new codes, it will be important to review the passages already coded to see whether the new codes apply to any of them.

Here are the passages we decide to code "Abomination." (The abominations are in bold.)

> 7:18 *If any of the flesh of the sacrifice of **his peace offering is eaten on the third day**, he who offers it shall not be accepted, neither shall it be credited to him; it shall be an abomination, and he who eats of it shall bear his iniquity.*
>
> 7:21 *And if any one **touches an unclean thing**, whether the uncleanness of man or an unclean beast or any unclean abomination, **and then eats of the flesh of the sacrifice** of the LORD's peace offerings, that person shall be cut off from his people.*
>
> 11:10 *But **anything in the seas or the rivers that has not fins and scales**, of the swarming creatures in the waters and of the living creatures that are in the waters, is an abomination to you.*
>
> 11:11 *They shall remain an abomination to you; **of their flesh you shall not eat, and their carcasses you shall have in abomination**.*
>
> 11:12 ***Everything in the waters that has not fins and scales** is an abomination to you.*
>
> 11:13 *And these you shall have in abomination among the birds, they **shall not be eaten**, they are an abomination: the **eagle**, the **vulture**, the **osprey**,*
>
> 11:14 *the **kite**, the **falcon** according to its kind,*
>
> 11:15 *every **raven** according to its kind,*
>
> 11:16 *the **ostrich**, the **nighthawk**, the **sea gull**, the **hawk** according to its kind,*
>
> 11:17 *the **owl**, the **cormorant**, the **ibis**,*
>
> 11:18 *the **water hen**, the **pelican**, the **carrion vulture**,*

11:19   the **stork**, the **heron** according to its kind, the **hoopoe**, and the **bat**.

11:20   **All winged insects that go upon all fours** are an abomination to you.

11:41   **Every swarming thing** that swarms upon the earth is an abomination; it shall not be eaten.

11:42   Whatever goes on its belly, and whatever goes on all fours, or whatever has many feet, all the **swarming things** that swarm upon the earth, you shall not eat; for they are an abomination.

11:43   You shall not make yourselves abominable with **any swarming thing that swarms**; and you shall not defile yourselves with them, lest you become unclean.

18:22   You shall not **lie with a male as with a woman**; it is an abomination.

19:6   It shall be eaten the same day you offer it, or on the morrow; and anything left over until the third day shall be burned with fire.

19:7   **If it is eaten at all on the third day**, it is an abomination; it will not be accepted,

19:8   **and every one who eats it shall bear his iniquity**, because he has profaned a holy thing of the LORD; and that person shall be cut off from his people.

20:13   **If a man lies with a male as with a woman**, both of them have committed an abomination; they shall be put to death, their blood is upon them.

20:25   You shall therefore make a distinction between the clean beast and the unclean, and between the unclean bird and the clean; **you shall not make yourselves abominable by beast or by bird or by anything with which the ground** teems, which I have set apart for you to hold unclean.

Male homosexuality, then, isn't the only abomination identified in Leviticus. As you compare these passages, looking for similarities and differences, it will become apparent that most of the abominations have to do with dietary rules—specifically those potential foods deemed "unclean." Other abominations flow from the mishandling of ritual sacrifices. "Dietary Rules" and "Ritual Sacrifices" thus represent additional codes to be used in our analysis.

Earlier, I mentioned the death penalty as another concept to be explored in our analysis. When we take this avenue, we discover that many behaviors besides male homosexuality warrant the death penalty. Among them are these:

20:2   Giving your children to Molech (human sacrifice)

20:9   Cursing your father or mother

20:10   Adultery with your neighbor's wife

20:11   Adultery with your father's wife

20:12   Adultery with your daughter-in-law

20:14   Taking a wife and her mother also

20:15   Men having sex with animals (the animals are to be killed, also)

20:16   Women having sex with animals

20:27   Being a medium or wizard

24:16   Blaspheming the name of the Lord

24:17   Killing a man

As you can see, the death penalty is broadly applied in Leviticus: everything from swearing to murder, including male homosexuality somewhere in between.

An extended analysis of prohibited behavior, short of abomination and death, also turns up a lengthy list. Among them are slander, vengeance, grudges, cursing the deaf, and putting stumbling blocks in front of blind people. In chapter 19, verse 19, Leviticus quotes God as ordering, "You shall not let your cattle breed with a different kind; you shall not sow your field with two kinds of seed; nor shall there come upon you a garment of cloth made of two kinds of stuff." Shortly thereafter, he adds, "You shall not eat any flesh with the blood in it. You shall not practice augury or witchcraft. You shall not round off the hair on your temples or mar the edges of your beard." Tattoos were prohibited, though Leviticus is silent on body piercing. References to all of these practices would be coded "Prohibited Acts" and perhaps given additional codes as well (recall "Dietary Rules").

I hope this brief glimpse into a possible analysis will give you some idea of the process by which codes are generated and applied. You should also have begun to see how such coding would allow you to understand better the messages being put forward in a text and to retrieve data appropriately as you need them.

Memoing

In the Grounded Theory Method, the coding process involves more than simply categorizing chunks of text. As you code data, you should also be using the technique of **memoing**—writing memos or notes to yourself and others involved in the project. Some of what you write during analysis may end up in your final report; much of it will at least stimulate what you write.

In GTM, these memos have a special significance. Strauss and Corbin (1998: 217) distinguish three kinds of memos: code notes, theoretical notes, and operational notes.

Code notes identify the code labels and their meanings. This is particularly important because, as in all social science research, most of the terms we use with technical meanings also have meanings in everyday language. It's essential, therefore, to write down a clear account of what you mean by the codes used in your analysis. In the Leviticus analysis, for example, you would want a code note regarding the meaning of "abomination" and how you've used that code in your analysis of text.

Theoretical notes cover a variety of topics: reflections of the dimensions and deeper meanings of concepts, relationships among concepts, theoretical propositions, and so on. All of us occasionally ruminate over the nature of something, try to think it out, to make sense out of it. In qualitative data analysis, it's vital to write down these thoughts, even those you'll later discard as useless. They will vary greatly in length, but you should limit each to a single main thought so that you can sort and organize them all later. In the Leviticus analysis, one theoretical note might discuss the way that most of the injunctions implicitly address the behavior of *men*, with women being mostly incidental.

Operational notes deal primarily with methodological issues. Some will draw attention to data-collection circumstances that may be relevant to understanding the data later on. Others will consist of notes directing future data collection.

These memos are written throughout the data-collection and analysis process. Thoughts demanding memos will come to you as you reread notes or transcripts, code chunks of text, or discuss the project with others. It's a good idea to get in the habit of writing out your memos as soon as possible after the thoughts come to you.

Notice that whereas we often think of writing as a linear process, starting at the beginning and moving through to the conclusion, memoing does not follow this pattern. It might be characterized as a process of creating chaos and then finding order within it.

To explore this process further, you can refer to the works cited in this discussion. You'll also find a good deal of information on the Web. Ultimately, the best education in this process comes from practice. Even if you don't have a research project underway, you can practice now on class notes. Or start a journal and code it.

Concept Mapping

It should be clear by now that qualitative data analysts spend a lot of time committing thoughts to paper (or to a computer file) and figuring out how they relate to one another. Often, we can think out relationships among concepts even more clearly by putting the concepts in a graphical format, a process called **concept mapping**. Some researchers find it useful to put all their major concepts on a single sheet of paper, whereas others spread their thoughts across several sheets of paper, whiteboards, magnetic boards, computer pages, or other media. Figure 13-3 shows how we might think out some of the concepts of Goffman's examination of gender and advertising. (This image was created through the use of Inspiration, a concept-mapping computer program.)

Incidentally, many of the topics discussed in this section have useful applications in quantitative

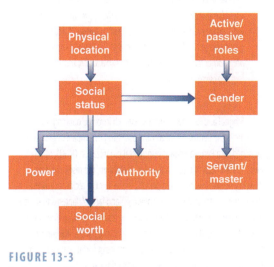

FIGURE 13-3

An Example of Concept Mapping.

---

memoing Writing memos that become part of the data for analysis in qualitative research such as grounded theory. Memos can describe and define concepts, deal with methodological issues, or offer initial theoretical formulations.

concept mapping The graphical display of concepts and their interrelations, useful in the formulation of theory.

Issues and Insights

Pencils and Photos in the Hands of Research Subjects

How would you go about studying the life conditions of Peruvian Indians living in the Amazon rain forest? With minimal telecommunications infrastructure and a slow ferry-based postal service in the vast region, a mail or telephone survey wouldn't be the best approach. It might occur to you to conduct in-depth interviews in which you would work from an outline of topics to be covered. Arvind Singhal and Elizabeth Rattine-Flaherty (2006) opted for a very different approach, which put the subjects of study more in control of the research and allowed for important but unexpected discoveries. They derived their inspiration from the work of the renowned Brazilian educator, Paulo Freire, who once set out to measure exploitation among street children. Instead of interviewing them, he gave them cameras and asked them to bring back photographs of exploitation. As Singhal and Rattine-Flaherty report in their own findings:

> One child took a photo of a nail on a wall. It made no sense to adults, but other children were in strong agreement. The ensuing discussions showed that many young boys of that neighborhood worked in the shoe-shine business. Their clients were mainly in the city, not in the barrio where they lived. As their shoe-shine boxes were too heavy for them to carry, these boys rented a nail on a wall (usually in a shop), where they could hang their boxes for the night. To them, that nail on the wall represented "exploitation." The "nail on the wall" photograph spurred widespread discussions in the Peruvian barrio about other forms of institutionalized exploitation, including ways to overcome them.

> (2006: 314)

Singhal and Rattine-Flaherty's research involved gauging the quality of life in the Peruvian Amazon and assessing the impact of programs launched by a Peruvian nongovernmental organization (NGO), Minga Peru. To view society through the eyes of children, the researchers set up drawing sessions with colored pencils. In the spirit of reciprocity, one of the authors sketched pictures of snowmen and jack-o'-lanterns that were a part of her growing up in the Midwest. In addition to depicting life in their villages and their close relationship with the natural environment, the children's sketches often featured examples of social change being brought about by the NGO's developmental programs.

> These include sketches of chicken coops, fish farms, and agro-forestry projects. These enterprises, all launched by Minga Peru, began in the Peruvian Amazon only in the past few years. For children to sketch these "new" initiatives in their pictures on their own, without prompts, is noteworthy.

> (2006: 322)

The photographs taken by the adult women were equally revealing. Several drew attention to the patriarchal social structure. As the authors report:

> Several photographs depicted the subservient position of the Amazonian women relative to men, a situation that Minga Peru seeks to address. For instance, Adela's picture shows a middle-aged Amazonian woman and her husband sitting on their porch and having a conversation. The woman, sporting a forlorn expression, sits with her legs crossed while her husband stares directly into the camera, squatting with his arms and feet spread in an open position. Especially noticeable is the physical distance of about 10 feet that separates the woman and the man. When Adela was asked why she took the picture and why were the man and woman sitting so far apart, she noted: "The woman is sitting at one side of the house and he is on the other and this was not anything unusual." Upon probing, we learned that Amazonian men determine how close the couple sits. If they are sitting closer, and if the man has his arm around his partner, it is his decision to do so. This authority also applies to initiation of sex: The man determines if and when sex will happen.

> (2006: 323–24)

This research not only illustrates some unusual data-collection techniques, it also represents the spirit of participatory action research, discussed in Chapter 10. With a very different setting and purpose—looking for evidence of the impact of globalization in Ireland—Pat O'Connor (2006) asked Irish adolescents to write essays about themselves and about Ireland, including drawings, poems, and songs. Both studies demonstrate that qualitative field research can involve a lot more than just observing and interviewing.

M. Morgan and colleagues (2009) used this technique in the examination of a very sensitive topic: chronic vaginal infections among Australian women. In addition to in-depth interviews in which the female interviewers often spoke of their own experiences, the subject-women were asked to draw pictures to illustrate their feelings in relation to the medical condition.

Sources: Arvind Singhal and Elizabeth Rattine-Flaherty. 2006. "Pencils and Photos as Tools of Communicative Research and Praxis: Analyzing Minga Peru's Quest for Social Justice in the Amazon." *International Communication Gazette* 68 (4): 313–30; Pat O'Connor. 2006. "Globalization, Individualization and Gender in Adolescents' Texts." *International Journal of Social Research Methodology* 9 (4): 261–77; M. Morgan, F. McInerney, J. Rumbold, and P. Liamputtong. 2009. "Drawing the Experience of Chronic Vaginal Thrush and Complementary and Alternative Medicine." *International Journal of Social Research Methodology* 12 (2): 127–46.

as well as qualitative analyses. Certainly, concept mapping is appropriate in both types of analysis. The several types of memos would also be useful in both. And the discussion of coding readily applies to the coding of open-ended questionnaire responses for the purpose of quantification and statistical analysis. (We'll look at coding again in the next chapter, on quantifying data.)

The use of visual portrayals can profit data collection as well as the organization of data analysis. For examples of this, see "Issues and Insights: Pencils and Photos in the Hands of Research Subjects."

Having noted the overlap of qualitative and quantitative techniques, it seems fitting now to address an instrument that is primarily associated with quantitative research but that is proving quite valuable for qualitative analysts as well—the computer.

Computer Programs for Qualitative Data

Not so many decades ago, the recording and analysis of qualitative research data was paper-bound: Interview transcripts, observational notes, official documents, and the like were all compiled and evaluated by hand. The work of writing documents by hand was eventually streamlined by the advent of typewriters. Copying technologies, first carbon paper and later photocopying, made it possible to easily reproduce duplicate pages of research information. These printed data could then be cut into slips of paper, with each strip displaying an individual, coded item. This procedure allowed researchers to categorize collected data according to different themes or concepts. Recalling the discussion of Leviticus, one pile of paper slips might contain passages referencing homosexuality, another pile could contain references to abominations, and so forth. And, as our earlier discussion mentioned, a given passage might show up in more than one pile. Once this coding and sorting was done, researchers could manually review all the materials within a particular category, allowing them to look for and identify common patterns and important distinctions.

As you can imagine, computers changed all that. Once information has been entered into a computer, copying whole documents or pieces is a trivial matter. Simulating the earlier paperbound method, you can copy an interview comment relevant to, say, discrimination against women and paste it into another document created to hold anything relevant to that theme. With nothing but the simplest word processor or text editor, you can streamline the coding process in any number of ways. Imagine this paragraph was part of textual materials you were analyzing in a study of technological progress. You could augment this document by adding coding notations such as: <computer> <qualitative> <coding>.

You could go through this whole chapter, adding these and other notations wherever appropriate. And once you were finished, you could use a simple "Search" function to review all the materials marked <coding>, for example. You could augment this process by searching the chapter for words such as *code*, *coding*, *category*, *classify*, or other applicable terms.

I'd like these short paragraphs to draw attention to and to honor the researchers who could make sense out of social life with little more than paper and pencil. And I hope that a review of past practices will demonstrate what a powerful tool the simplest, bare-bones computer can represent for the analysis of qualitative data.

Now, let's turn 180 degrees from these simple computer manipulations. As you will soon see, those methods may seem like crude tools from the Dark Ages of social research. We are now going to spend some time with a few of the computer programs devised precisely for the analysis of qualitative data.

QDA Programs

Today, qualitative data analysis (QDA) programs abound. Where the analyst's problem used to be merely finding any such program, the problem now lies in choosing one of so many. Here are a few commonly used QDA programs with online sites where you can learn more about them and, in many cases, download demo copies.

- **Atlas.ti:** www.atlasti.com/index.php
- **Dedoose:** http://www.dedoose.com
- **Ethno:** www.indiana.edu/%7Esocpsy/ESA/
- **Ethnograph:** www.qualisresearch.com/
- **HyperResearch:** www.researchware.com/
- **HyperTranscribe:** www.researchware.com/
- **MAXQDA:** www.maxqda.com/
- **NVivo:** www.qsrinternational.com/products_nvivo.aspx
- **QDA Miner:** http://provalisresearch.com/products/qualitative-data-analysis-software/
- **Qualrus:** www.qualrus.com/
- **TAMS:** sourceforge.net/projects/tamsys
- **Weft:** www.pressure.to/qda/

Another excellent resource is "Choosing a CAQDAS Package" by Ann Lewins and Christina

Silver (2009). This will familiarize you with some of the key features in such computer programs and help you choose which one is best suited to your purposes.

We'll turn now to a couple of illustrations of QDA programs at work. Although the available programs differ somewhat from one another, I think these illustrations will give you a good sense of the general use of computers to analyze qualitative data. We'll begin by returning to the earlier example of the Book of Leviticus, with its sexual and other prohibitions. Let's see how a computer analysis might assist the search for patterns.

Leviticus as Seen through Qualrus

We'll first consider one of the programs just listed: Qualrus. Although the text to be coded can be typed directly into Qualrus, usually materials already in existence—such as field notes or, in this case, the verses of Leviticus—are *imported* into the program. For purposes of this illustration, I simply copied chapter 20 of Leviticus from the Web and pasted it into the document section of the layout.

Figure 13-4 shows how the text is presented within Qualrus. As you can see, although the

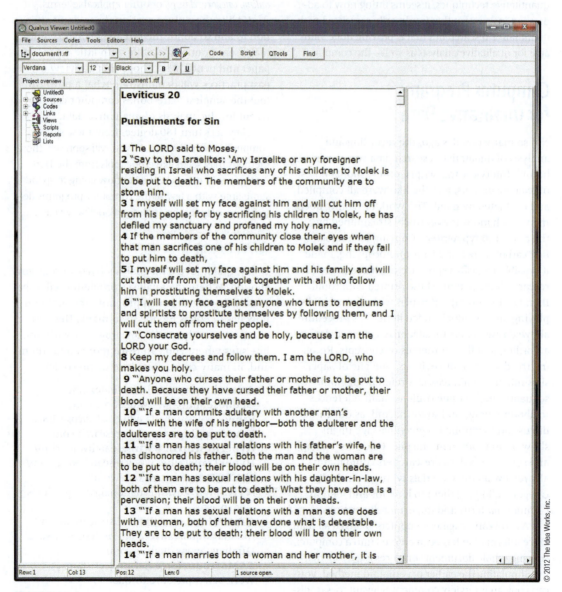

FIGURE 13-4

Leviticus Text Displayed in Qualrus.

text inserted cannot all be displayed in the window at once, you can easily scroll up and down through the document.

One of the themes we discussed earlier concerned the various actions that could result in death, so let's create that as a code in Qualrus. This is most easily done by highlighting it with the cursor to select a passage we wish to code: verse 2, for example. Then click the "Code" button at the top of the page, and you will be presented with another window for creating codes, as illustrated in Figure 13-5.

Click the top, leftmost icon in the window and you will be asked to enter a new code in the lower left portion of the window, where I have typed "death." This is the first in what will become a list of codes. Notice also that the software has automatically entered the code "death" in the upper right portion of the window, indicating the designated code for this particular passage. Once you have decided on and inserted additional codes into the program, you will be able to select from that list and click the arrow to assign codes to the passage. Once you've assigned the proper code, click "OK" to return to the document.

Having created the code, we can proceed through the text, identifying all the passages that specify the death penalty for some action. Figure 13-6 illustrates what this coding looks like.

As you build your coding structure, Qualrus begins to anticipate and suggest appropriate codes for some of the passages you have selected. Figure 13-7 gives an example of this. Qualrus has examined a passage we've selected and has guessed that perhaps it should be coded "death."

The suggested code appears in the upper left panel of the window. If you choose to accept that suggestion, you must select the code and transfer it to the upper right panel.

Notice that in Figure 13-7 I have added several additional codes, which might be useful in our analysis of Leviticus. I've provided for passages that mention adultery, bestiality, homosexuality, incest, the occult, and sex. In addition, in my review of the document, I noticed some passages specifying that God would punish the sinners, while others indicated that the community should mete out the punishment. Let's see what the document looks like when all these codes have been employed. See Figure 13-8.

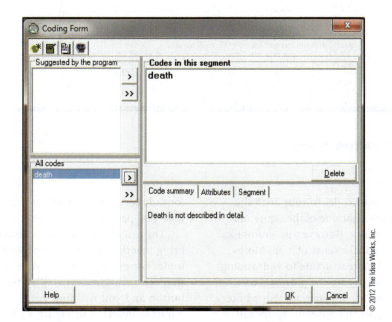

FIGURE 13-5

Creating a Code for Death.

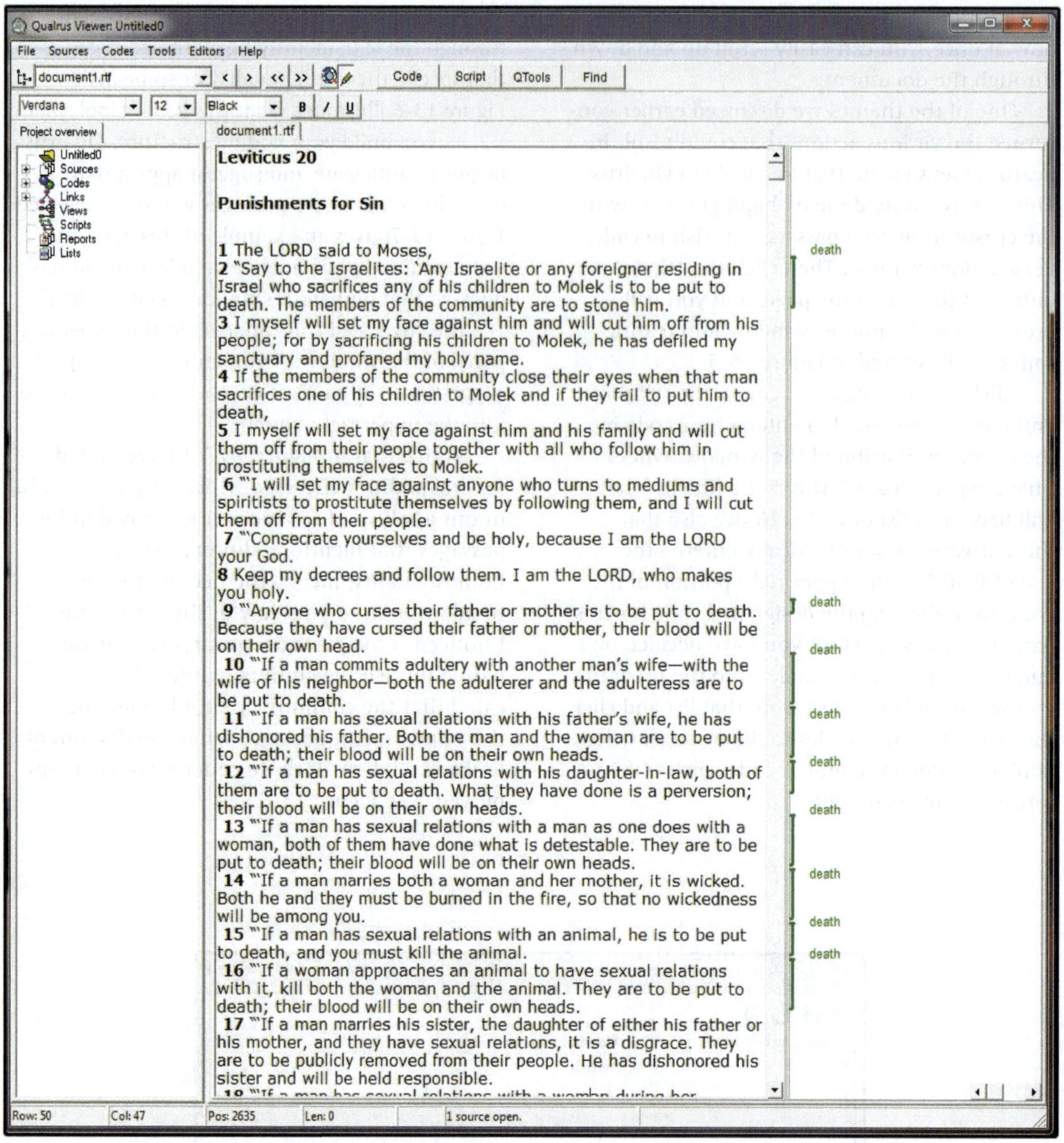

FIGURE 13-6

Displaying the Coding of "Death" Passages.

As you add more codes to your analysis, the display on the right side of the figure may become very complex. However, programs like Qualrus help you make sense of it all. Moreover, as you add more structure to your coding scheme, Qualrus can ease the process. For example, you can tell the program that adultery, bestiality, incest, and homosexuality are subsets of the code "sex." Once you have informed Qualrus of that, every time you code a passage

for "adultery," for example, it will automatically code that passage as "sex," also.

The purpose of all this coding is to let you bring together all the instances of a particular topic. For example, let's review all the actions that can result in death. Click the "QTools" button and you will find yourself in a new window, depicted in Figure 13-9. I have instructed Qualrus to locate all the passages that were coded "death."

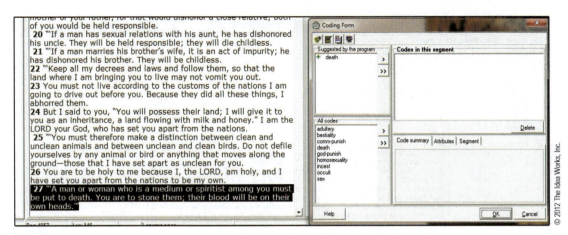

FIGURE 13-7
Suggesting a Code.

By scrolling up and down through the main panel that contains the passages, you can review all the relevant passages and look for patterns. Let's say you are particularly interested in cases in which sexual behavior warrants death. Figure 13-10 illustrates that search. Now we have a shorter, more focused set of passages to study.

This has been a brief look into some of the ways you could use the Qualrus program to analyze the prohibitions detailed in Book of Leviticus. I have not attempted to offer a theoretical framework for your analysis, I merely want to illustrate the most elementary techniques that could be used in your research into the topic. Moreover, as you can see at the top of the last two figures, the QTools that are available go well beyond reviewing one or two codes. To learn more about the additional analytic tools, you should visit the Qualrus Web page.

At this point, I will shift to another of the listed QDA programs. Although NUD*IST was one of the first such programs to be popularly employed by social researchers, it has evolved through many forms over time. At this writing, QSR International's latest tool was NVivo 10, and it is the program we will discuss next.

NVivo

To get started in our introduction to this program for the analysis of qualitative data, Figure 13-11 shows a sample of how NVivo software might display the data in Figure 13-8 (analysis of the Book of Leviticus using Qualrus). NVivo refers to established codes as *nodes*.

While the formats of different programs vary, of course, the fundamental process of coding and analyzing the coded data is the bedrock of all QDA programs. However, we are going to shift gears at this point.

Thus far, I have limited discussion to the coding of a written document, because it offers a clear example of coding. Written documents are hardly the appropriate materials for such an analysis. You've probably already anticipated that these programs would be especially valuable for the analysis of in-depth interviews, and you would be right. At this point, I want to make use of a research example used to introduce NVivo that is found on their website.

"Environmental Change Down East" is a sample project provided by NVivo that explores environmental change in a coastal area of North Carolina. Part of the research for this project involved qualitative interviews with residents of the communities in the area. Figure 13-12 shows a coded portion of one such interview.

Notice the colored coding strips to the right of the interview transcript. The red strip, for example, represents the node, "Jobs and cost of living." If you read the passages marked with red strips, you can see why the researchers coded those passages as they did. Note also that a given passage may reflect more than one of the themes being examined, and so it has been assigned more than one node.

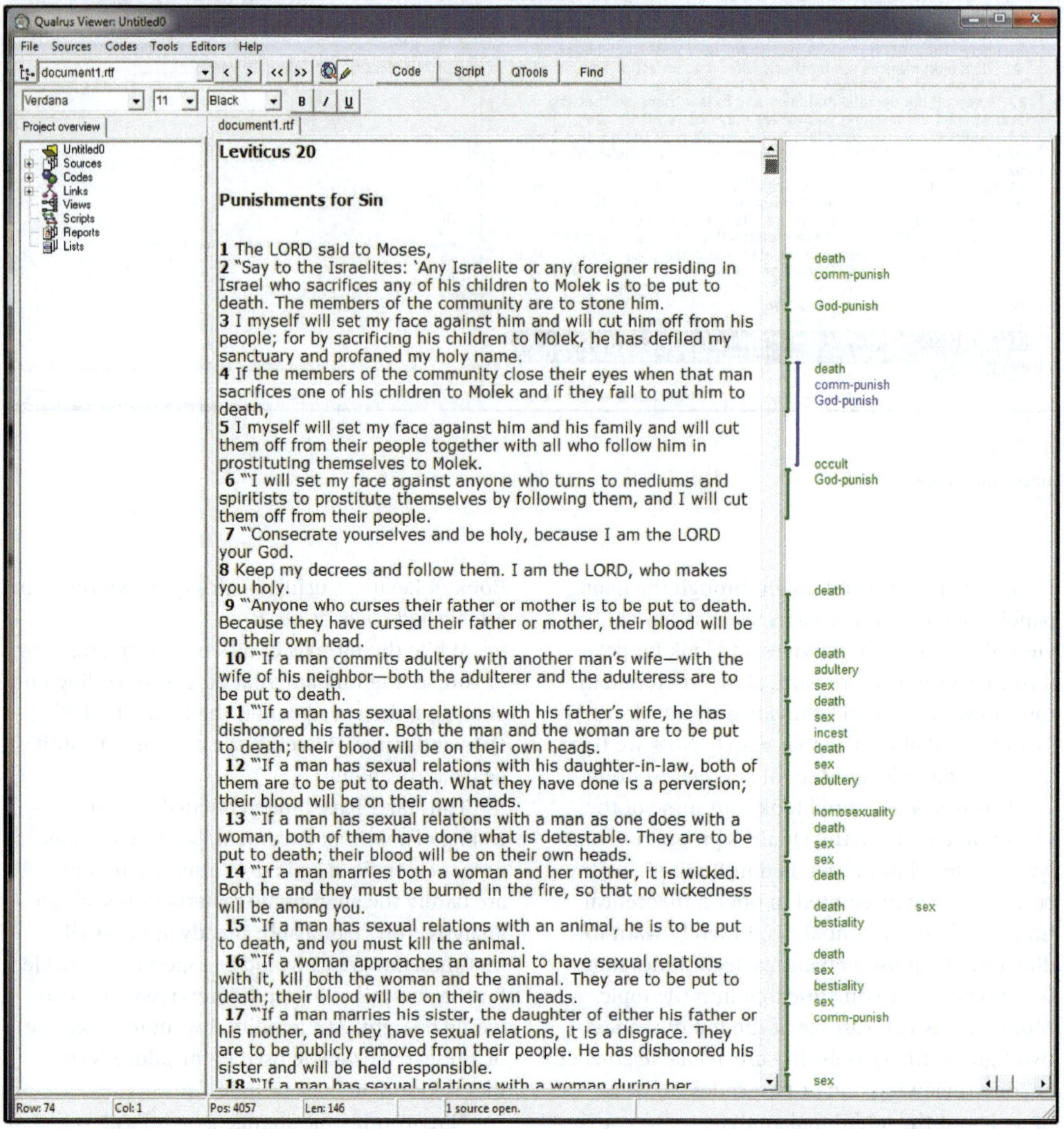

FIGURE 13-8

Display of Many Codes.

Thus far, we have seen that QDA programs can be used to organize documents and interview transcripts in preparation for analysis. That's only the beginning, however. Whereas the Down East project we are now examining contains the interviews of a number of residents, the researchers also collected data about physical locations in the study area, such as communities. Figure 13-13 illustrates some of those data.

In a record such as this one, the textual material and/or the photograph (or a portion of the photograph) can be coded by assigning nodes. Later on, a review of materials relating to a particular theme will bring together the relevant interviews, documents, photographs, and anything else that has been assigned the node in question.

Visual data for analysis are not limited to photographs. Perhaps your project has collected some of its data in a video format. Figure 13-14 illustrates this possibility.

Once the video files have been imported into the NVivo software, they can be coded, retrieved,

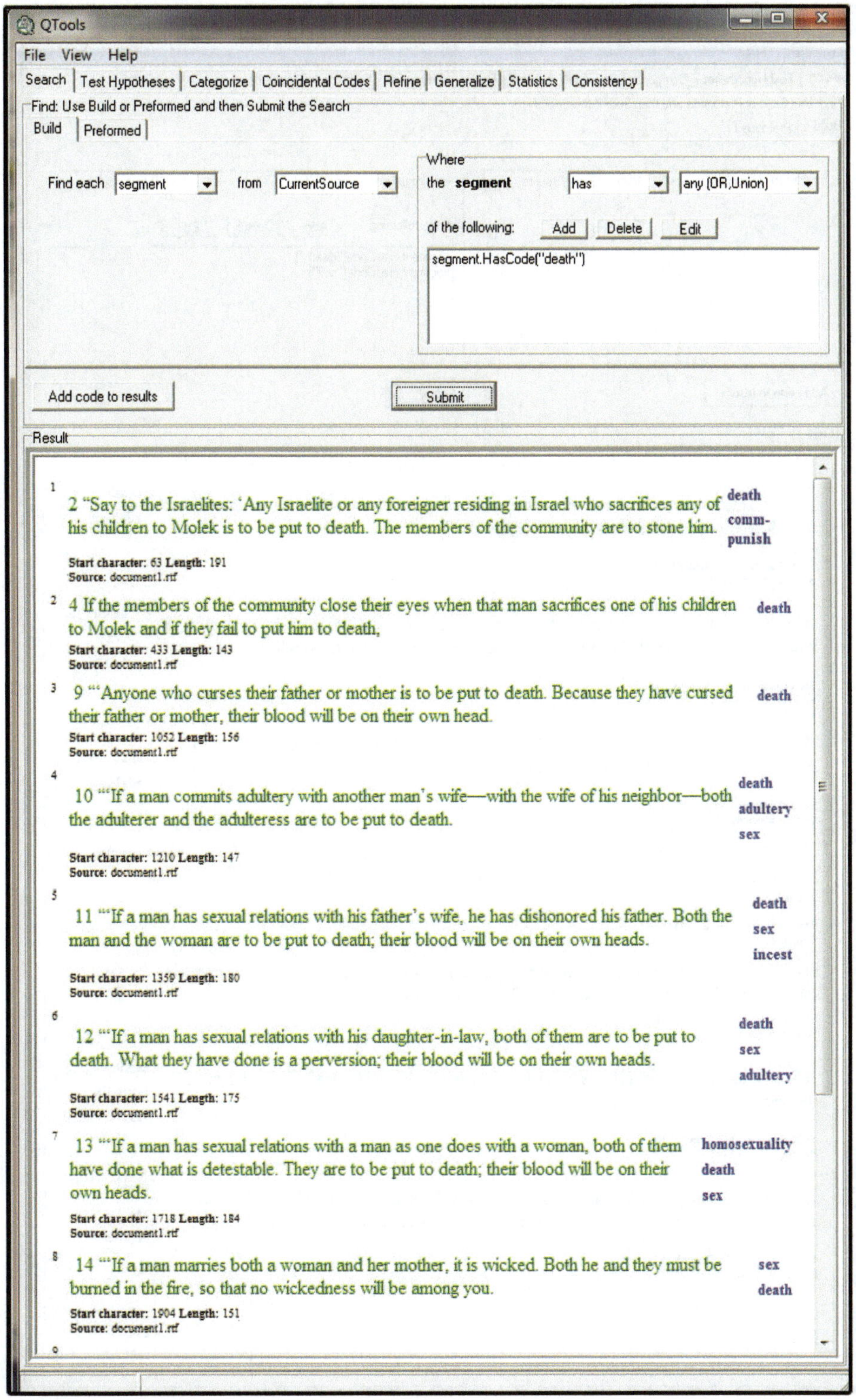

FIGURE 13-9
Reviewing All the Passages Coded "Death."

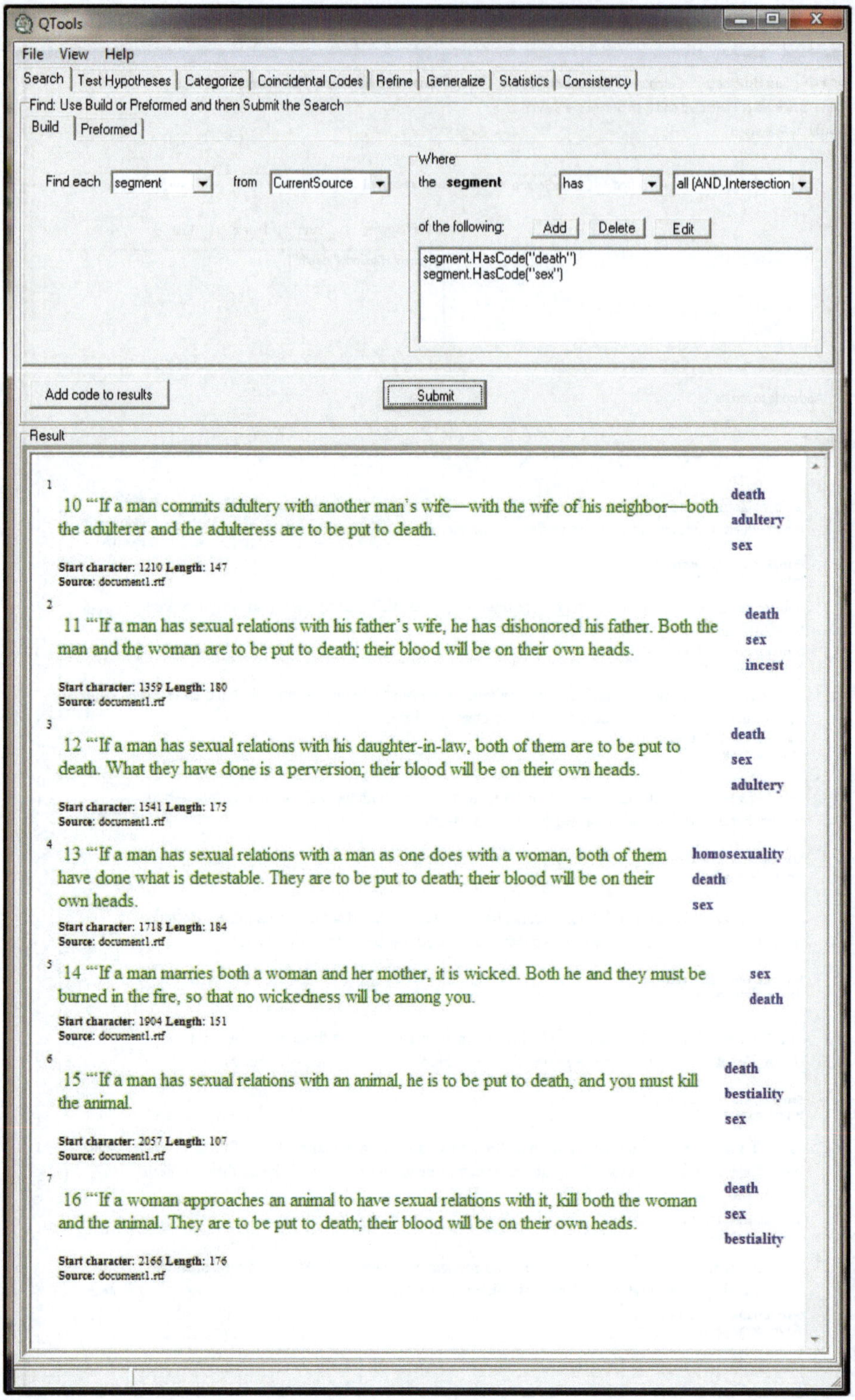

FIGURE 13-10

Reviewing the Passages Coded "Death" and "Sex."

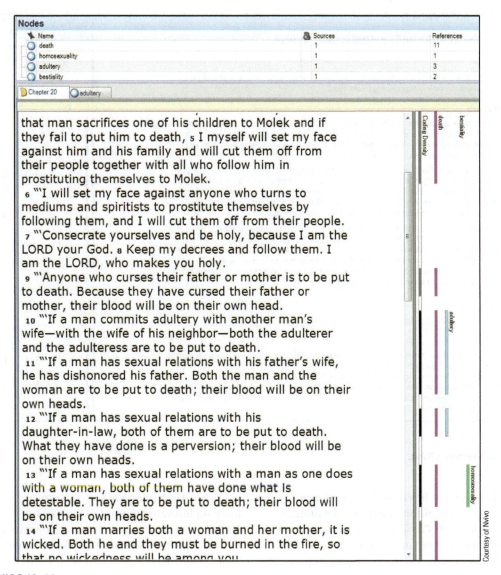

FIGURE 13-11

Looking at Leviticus with NVivo.

and played within the program. Audio recording can, of course, be treated similarly.

In these few illustrations, I have intended to give you an idea of the variety of data that can be amassed and organized within NVivo. If you have used a bibliographic program such as EndNote, you can import those materials into the project as well. Your own imagination will expand the list of possibilities further.

I'll conclude this discussion by touching on one further element that we mentioned in our earlier discussions of qualitative research:

memos. At every turn, NVivo allows you to record your own notes in the context of other materials—assigning those memos to the appropriate nodes. As you begin to recognize a certain pattern emerging in the intersection of two themes, for example, you can jot down notes about what you've observed so that you don't forget the details.

In the discussion of Qualrus, we saw a couple of examples of preliminary analyses. You'll recall that we retrieved all the passages from Leviticus that dealt with the death penalty; then we searched for passages prescribing

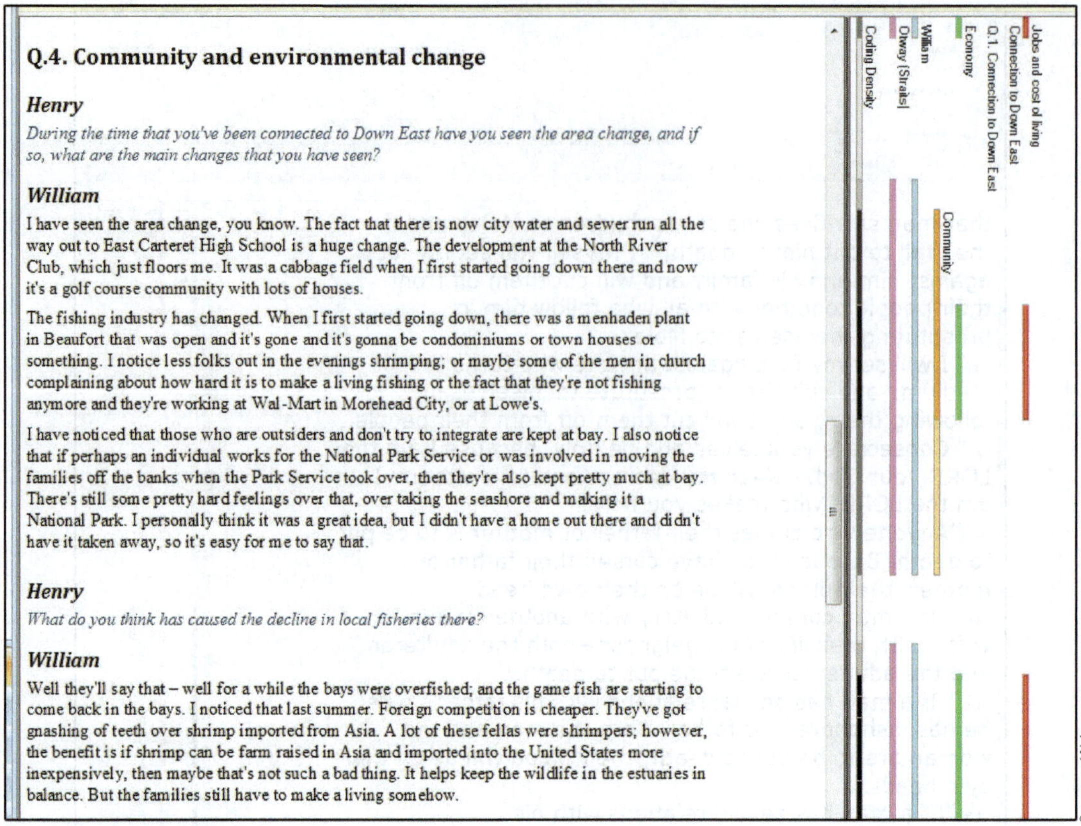

Q.4. Community and environmental change

Henry

During the time that you've been connected to Down East have you seen the area change, and if so, what are the main changes that you have seen?

William

I have seen the area change, you know. The fact that there is now city water and sewer run all the way out to East Carteret High School is a huge change. The development at the North River Club, which just floors me. It was a cabbage field when I first started going down there and now it's a golf course community with lots of houses.

The fishing industry has changed. When I first started going down, there was a menhaden plant in Beaufort that was open and it's gone and it's gonna be condominiums or town houses or something. I notice less folks out in the evenings shrimping; or maybe some of the men in church complaining about how hard it is to make a living fishing or the fact that they're not fishing anymore and they're working at Wal-Mart in Morehead City, or at Lowe's.

I have noticed that those who are outsiders and don't try to integrate are kept at bay. I also notice that if perhaps an individual works for the National Park Service or was involved in moving the families off the banks when the Park Service took over, then they're also kept pretty much at bay. There's still some pretty hard feelings over that, over taking the seashore and making it a National Park. I personally think it was a great idea, but I didn't have a home out there and didn't have it taken away, so it's easy for me to say that.

Henry

What do you think has caused the decline in local fisheries there?

William

Well they'll say that – well for a while the bays were overfished; and the game fish are starting to come back in the bays. I noticed that last summer. Foreign competition is cheaper. They've got gnashing of teeth over shrimp imported from Asia. A lot of these fellas were shrimpers; however, the benefit is if shrimp can be farm raised in Asia and imported into the United States more inexpensively, then maybe that's not such a bad thing. It helps keep the wildlife in the estuaries in balance. But the families still have to make a living somehow.

FIGURE 13-12

A Coded Interview in NVivo.

death for sexual behavior. NVivo offers a variety of analytical tools under the heading of "Queries." The NVivo "Getting Started Guide" is available as a pdf on the QSR website. This guide offers some examples of queries posed in the Down East project:

- Gather material coded at combinations of nodes—for example, gather content coded at *water quality* and *recreational fishing* and explore the associations.
- Gather material from classified nodes with specific attribute values—for example, *what do fishermen think about the rise of tourism*?
- Search for content coded at multiple nodes and use operators to further refine the query—for example, gather content coded at *community change* where it overlaps with content coded at *real estate development*.
- Search for content that is not coded at a specific node—find content coded at *environmental impacts* but not coded at *negative attitude*. (Qsr 2011: 31.)

Today, QDA programs can also provide for far more sophisticated displays of analytical results. Figure 13-15, for example, offers a summary of positive, negative, and mixed comments made about commercial fishing in several Down East communities.

This bar graph might make you curious about the positive attitudes toward commercial fishing among residents of Harkers Island. Using QDA software, it would be a simple matter to retrieve those specific comments in full.

The QDA programs also offer a graphic presentation of your coding structure, supporting the development of a coherent, theoretical understanding of the subject under study. See Figure 13-16 for an example.

I hope this brief overview of two QDA programs, Qualrus and NVivo, will give you a good idea of the powerful tools available for qualitative data analysts. Please understand that I have selected only a few of the features available in the two programs I've illustrated

Courtesy of QSR International

FIGURE 13-13

Data about a Physical Location.

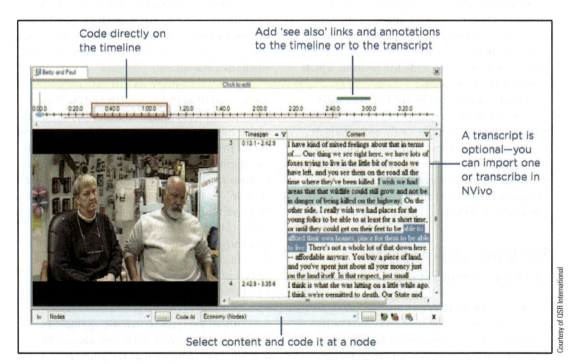

Courtesy of QSR International

FIGURE 13-14

Using Video Data with NVivo.

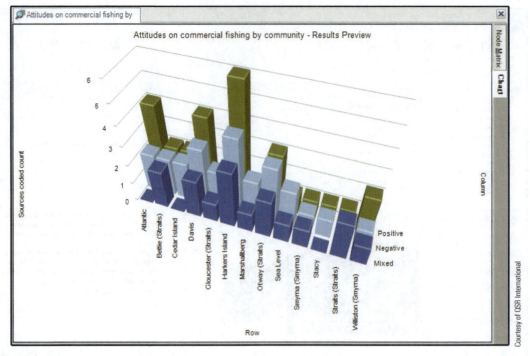

FIGURE 13-15

Attitudes toward Commercial Fishing by Community.

here and that the other programs I listed earlier in the chapter have many of the same or similar capabilities. Most programs provide comprehensive tutorials online, so you can explore, compare, and decide for yourself which program fits your needs.

Finally, you should be warned that such programs may be more expensive than the average student budget can easily accommodate. However, you may find that your school has licensed copies available for your use. The same situation pertains for *quantitative* data analysis programs such as Statistical Package for the Social Sciences (SPSS).

The Qualitative Analysis of Quantitative Data

Although it's important and appropriate to distinguish between qualitative and quantitative research and to discuss them separately, they are by no means incompatible or in competition. You need to operate in both modes to explore your full potential as a social researcher.

Chapter 14 explores some ways in which quantitative analyses can strengthen qualitative studies. Before we move on, however, let's look at an example of how quantitative data demand qualitative assessment.

Figure 13-17 presents FBI data on homicides committed in the United States. These data are often presented in a tabular form, but notice how clearly the patterns of crime appear in this three-dimensional graph. Even though the graph is based on statistical data, it conveys its meaning quite clearly. Although summarizing it in the form of equations may be useful for certain purposes, it would add nothing to the clarity of the picture itself. Thus, the qualitative assessment of the graph clarifies the quantitative data in a way that no other representation could. Here's a case in which a picture is truly worth a thousand words, or perhaps numbers.

With an even more striking example of how to understand quantitative data qualitatively, Mark Newman at the University of Michigan begins with a map showing 2012 U.S. presidential election results. You've

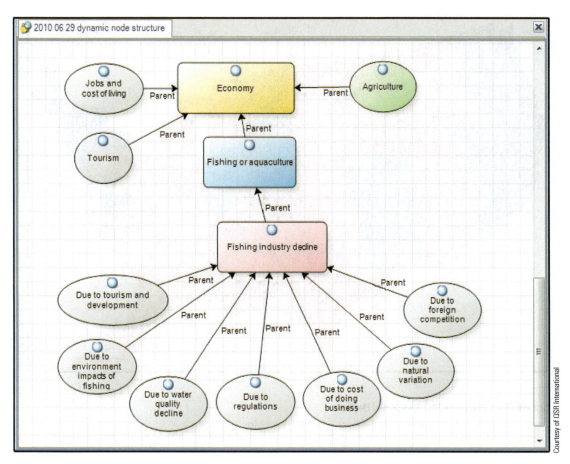

FIGURE 13-16

Mapping Down East Concepts.

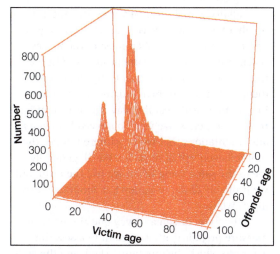

FIGURE 13-17

Number of One-on-One Homicides by Age of Victim and Age of Offender, Raw Data.

Source: Michael D. Maltz, "Visualizing Homicide: A Research Note," *Journal of Quantitative Criminology* 15, no. 4 (1998): 401.

probably seen something like this before. (See Figure 13-18.)

The general pattern shows Democrats winning the coastal and Northeast states, and Republicans winning the Midwest and South. In terms of the simple number of states, it looks as though the Republican, Governor Mitt Romney, surely must have beaten the Democrat, incumbent president Barack Obama. When Newman adjusts the map to take account of population—stretching or shrinking states to reflect their population (and electoral college votes)—the Democratic dominance becomes more apparent. (See Figure 13-19.)

In the extended analysis on his website, Newman also shifts his unit of analysis from states to counties for a different look, and he goes beyond the dichotomous blue/red labeling of states and counties to use purple for those where the results were close.

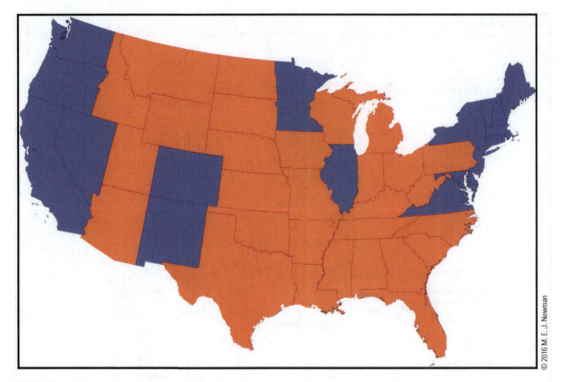

FIGURE 13-18

Red and Blue States in the 2012 Presidential Election.

Source: Figure by Mark Newman, used with permission, http://www-personal.umich.edu/~mejn/election/2012/.

Evaluating the Quality of Qualitative Research

As you have seen in earlier chapters and will see in the next chapter, there are often clear guidelines for evaluating the quality of quantitative research. In the case of survey research, for example, we can note the size of the sample, the manner in which it was selected, and the completion rate achieved. The questionnaire items are standardized and open to scrutiny. You can also use statistical tests to assess research findings.

Judging the quality of qualitative research is more elusive, though no less important. Given that there are many different ways of conducting qualitative research, we'll examine some general guidelines you can use in distinguishing first-rate qualitative investigations from those not so well done.

In Chapter 5, we examined two aspects of measurement quality: validity and reliability.

That's a reasonable point to start our look at assessing qualitative research.

Validity, you'll recall, involves the question of whether you are measuring what you say you are measuring. Remember, most of the things social scientists measure are products of human thought and agreement, not things that exist independently of human judgment. Prejudice, for example, isn't real the way age or weight is real. Nonetheless, we've all observed behaviors and orientations that we've gathered under the umbrella concept of *prejudice*. We all generally mean the same thing when we use the term, but we probably disagree somewhat about its meaning as well.

When you design a survey questionnaire to measure prejudice, it's important to assess the extent to which the questions asked, and the answers received, actually reflect what we can agree to mean by the term. The same logic applies in qualitative research projects such as field observations or historical studies. If field researchers

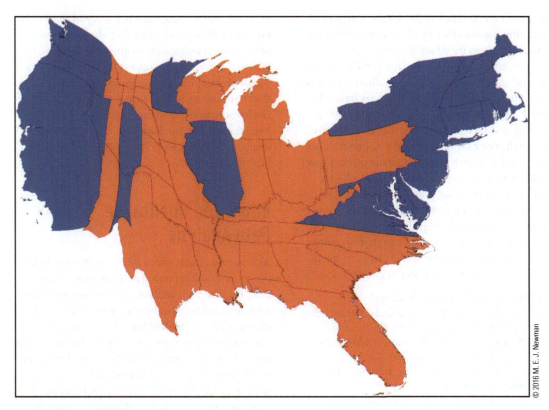

FIGURE 13-19

Adjusting State Sizes to Reflect Population.

Source: Figure by Mark Newman, used with permission, http://www-personal.umich.edu/~mejn/election/2016/.

characterize a subject of observation as prejudiced, you should examine their basis for saying that. Qualitative researchers, more than quantitative researchers, pay special attention to understanding life as the participants see it, so you may find the researchers in this case reporting that people who knew the subject in question also mentioned that he or she was prejudiced.

Some qualitative researchers prefer to use the term *credibility* in place of *validity* in this context. This is done as a caution against the older, positivistic view that social concepts represent real phenomena, existing objectively, independent of human thought. Be warned, however, that some researchers use the term with other meanings, quite distant from that of *validity*. Also, in this textbook, my use of the term *validity* explicitly denies objective reality for the concepts we use and study.

Reliability is also a reasonable criterion of quality with regard to qualitative research,

though it needs to be applied appropriately. Recall that this is a question of whether a measurement or observational technique would yield the same data if it were possible to measure or observe the same thing several times independently. In the case of categorizing raw data, such as those produced by in-depth interviews or by open-ended answers to survey questions, we can ask more than one person to undertake the coding or categorizing process independently to see whether they produce the same results. In most aspects of social research, however, the concept of reliability is more elusive, since what we are observing may be constantly changing, and the act of measuring (such as asking a question) may have an effect on the person being studied. Still, the basic concept of reliability, which some qualitative researchers prefer to call *dependability*, is meaningful for qualitative research. Yvonna Lincoln and Egon Guba (1985), for example, proposed an "inquiry audit" for the purpose of assessing the consistency of what was observed and the process by which it was observed.

Lincoln and Guba's research (1985) laid out several classic ways in which qualitative research could be assessed. Since then, others have offered somewhat modified schemes for both assessing and increasing the quality of qualitative research. For example, Britain's National Centre for Social Research sought to assist cabinet-level officials in assessing the quality of qualitative research projects that evaluated government programs. Although the study focused on the use of qualitative methods for purposes of evaluation research, the eighteen questions that organized such assessments can be applied to most forms of qualitative research (Spencer et al. 2003: 22–28):

1. How credible are the findings?
2. How has knowledge or understanding been extended by the research?
3. How well does the evaluation address its original aims and purpose?
4. How well is the scope for drawing wider inference explained?
5. How clear is the basis of evaluative appraisal?
6. How defensible is the research design?
7. How well-defended are the same design/ target selection of cases/documents?
8. How well is the eventual sample composition and coverage described?
9. How well was the data collection carried out?
10. How well has the approach to, and formulation of, analysis been conveyed?
11. How well are the contexts of data sources retained and portrayed?
12. How well has diversity of perspective and content been explored?
13. How well has detail, depth, and complexity (i.e., richness) of the data been conveyed?
14. How clear are the links between data, interpretation, and conclusions—i.e., how well can the route to any conclusions be seen?
15. How clear and coherent is the reporting?
16. How clear are the assumptions/theoretical perspectives/values that have shaped the form and output of the evaluation?
17. What evidence is there of attention to ethical issues?
18. How adequately has the research process been documented?

The attempt to settle on criteria for evaluating qualitative social research is far from over. For example, some researchers are wary of the British effort just delineated. Worrying about the implications of having a government body specify research criteria, they suggest that the list grows out of philosophical and political orientations that have not been made clear (J. Smith and Hodkinson: personal communication, 2005). Nonetheless, for our purposes, asking these or similar questions can help uncover problems in qualitative research.

Ethics and Qualitative Data Analysis

At least two ethical issues raise special concern in the analysis and reporting of qualitative research. First, because such analysis depends so directly on subjective judgments, there is an obvious risk of seeing what you are looking for or want to find. The risk increases in the case of participatory action research or other projects involving an element of social justice. Researcher bias is hardly an inevitable outcome, however. Experienced qualitative analysts avoid this pitfall in at least two ways: by cultivating a deliberate awareness of their own values and preferences and by adhering to established techniques for data collection and analysis. And as an additional protection, the peer-review process in scientific research encourages colleagues to point out any failings in this regard.

Second, qualitative research makes protecting subjects' privacy particularly important. The qualitative researcher will often analyze and report data collected from identifiable individuals. Throughout the book, I've indicated the importance of not revealing what we learn about subjects, as in the case of data collection. When writing up the results of your analyses, you'll often need to make concerted efforts to conceal identities. To achieve this end, individuals, organizations, and communities are often given pseudonyms. Sometimes, you may need to suppress details that would let outsiders figure out who you're talking about. Thus, it may be appropriate to speak about interviewing "a church leader" rather than "the head deacon." You may also need to suppress or alter age, race, or gender references if that would give away a subject's identity. The key principle is to respect the privacy of those you study.

What do you think?...Revisited

Quantification requires a simplification of data through a loss of detail. Sometimes those details are critical to understanding the "whole picture." You've experienced this if you've ever found yourself being categorized by someone else. Let's say you express some political opinion. Someone then asks what your major is, and you reply, "sociology." Then that same person says, "Well, of course!"—implying that they now "know" a long list of things about you (some true, some false) that will now shape their "understanding" of the political opinion you expressed. You may have experienced being similarly categorized in terms of your religion, race, place of birth, or gender. A similar loss can occur in the quantification of data, in which a limited number of categories takes the place of varied details. Qualitative analysis aims at staying closer to the original details, even through the coding and categorizing process.

MAIN POINTS

Introduction

- Qualitative analysis is the nonnumerical examination and interpretation of observations.

Linking Theory and Analysis

- Qualitative analysis involves a continual interplay between theory and analysis. In analyzing qualitative data, we seek to discover patterns such as changes over time or possible causal links between variables.

- Examples of approaches to the discovery and explanation of such patterns are Grounded Theory Method (GTM), semiotics, and conversation analysis (CA).

Qualitative Data Processing

- The processing of qualitative data is as much art as science. Three key tools for preparing data for analysis are coding, memoing, and concept mapping.

- In contrast to the standardized units used in coding for statistical analyses, the units to be coded in qualitative analyses may vary within a document. Although codes may be derived from the theory being explored, more often researchers use open coding, in which codes are suggested by the researchers' examination and questioning of the data.

- Memoing is appropriate at several stages of data processing and serves to capture code meanings, theoretical ideas, preliminary conclusions, and other thoughts that will be useful during analysis.

- Concept mapping uses diagrams to explore relationships in the data graphically.

Computer Programs for Qualitative Data

- Many computer programs, such as Qualrus and NVivo, are specifically designed to assist researchers in the analysis of qualitative data.

The Qualitative Analysis of Quantitative Data

- Researchers need both qualitative and quantitative analysis for the fullest understanding of social science data.

Evaluating the Quality of Qualitative Research

- Validity (credibility) and reliability (dependability) are ways to assess the quality of qualitative research.

Ethics and Qualitative Data Analysis

- The subjective element in qualitative data analysis provides an added challenge to avoiding bias in the interpretation of data.

- Because the qualitative data analyst knows the identity of subjects, taking special steps to protect their privacy is crucial.

KEY TERMS

axial coding	Grounded Theory Method (GTM)
case-oriented analysis	
concept mapping	memoing
constant comparative method	open coding
	qualitative analysis
conversation analysis (CA)	selective coding
	semiotics
cross-case analysis	variable-oriented analysis

PROPOSING SOCIAL RESEARCH: QUALITATIVE DATA ANALYSIS

In this chapter we've seen some of the approaches to qualitative data analysis that are available to social researchers. Since you won't have analyzed your data when you write this portion of the proposal, of course, you can't say anything about the conclusions you'll draw. However, you can describe your initial plans for the analysis. I say "initial" plans because you may change directions somewhat as the data accumulate and patterns begin to emerge. In some cases, your analysis will begin as observations are being made and/or other data are being gathered, or you may plan to complete the data-collection phase before starting your data analysis.

This is the place to indicate whether you plan to employ a particular method of analysis, such as Grounded Theory, semiotics, conversation analysis, and so forth. If you're going to use a particular QDA program, mention that here as well.

REVIEW QUESTIONS

1. Review Goffman's examination of gender advertising, and collect and analyze a set of advertising photos from magazines or newspapers or online. What is the relationship between gender and status in the materials you found?

2. Review the discussion of homosexuality in the Book of Leviticus. How might the examination be structured as a cross-case analysis?

3. Imagine that you're conducting a cross-case analysis of revolutionary documents such as the Declaration of Independence and the Declaration of the Rights of Man and of the Citizen (from the French Revolution). What key concepts might you code in the following sentence from the Declaration of Independence?

When in the Course of human events, it becomes necessary for one people to dissolve the political bands which have connected them with another, and to assume among the Powers of the earth, the separate and equal station to which the Laws of Nature and of Nature's God entitle them, a decent respect to the opinions of mankind requires that they should declare the causes which impel them to the separation.

4. Go to the World Press Review online (www .worldpress.org) and pick a controversial news topic discussed by several newspapers. See if you can identify characteristics of those newspapers (such as political leaning, region) that might explain the different points of view expressed on the topic.

CHAPTER 14

Quantitative Data Analysis

CHAPTER OVERVIEW

Often, social data are converted to numerical form for statistical analyses. In this chapter, we'll begin with the process of quantifying data, then turn to analysis. Quantitative analysis may be descriptive or explanatory; it may involve, one, two, or several variables. We begin our examination of how quantitative analyses are done with some simple but powerful ways of manipulating data in order to attain research conclusions.

conejota/Shutterstock.com

Introduction

In Chapter 13, we saw some of the logic and techniques by which social researchers analyze the qualitative data they've collected. This chapter will examine **quantitative analysis**, or the techniques by which researchers convert data to a numerical form and subject it to statistical analyses.

To begin, we'll look at *quantification*—the process of converting data to a numerical format. This involves converting social science data into a *machine-readable form*—a form that can be read and manipulated by computers and similar machines used in quantitative analysis.

The rest of the chapter will present the logic and some of the techniques of quantitative data analysis—starting with the simplest case, univariate analysis, which involves one variable, then discussing bivariate analysis, which involves two variables. We'll move on to a brief introduction to multivariate analysis, or the examination of several variables simultaneously, such as *age, education,* and *prejudice,* and then we'll move to a discussion of sociological diagnostics. Finally, we'll look at the ethics of quantitative data analysis.

Before we can do any sort of analysis, we need to quantify our data. Let's turn now to the basic steps involved in converting data into machine-readable forms amenable to computer processing and analysis.

Some students take to statistics more readily than do others.

Aaron Babbie

Quantification of Data

Today, quantitative analysis is almost always done by computer programs such as Statistical Package for the Social Sciences (SPSS) and MicroCase. For those programs to work their magic, they must be able to read the data you've collected in your research. If you've conducted a survey, for example, some of your data are inherently numerical: age or income, for instance. Whereas the writing and check marks on a questionnaire are qualitative in nature, a scribbled age is easily converted to quantitative data.

Other data are also easily quantified: Transforming male and female into "1" and "2" is hardly rocket science. Researchers can also easily assign numerical representations to such variables as *religious affiliation, political party,* and *region of the country.*

quantitative analysis The numerical representation and manipulation of observations for the purpose of describing and explaining the phenomena that those observations reflect.

What do you think?

In Chapter 13, we saw several inherent shortcomings in quantitative data. These shortcomings centered primarily on standardization and superficiality in the face of a social reality that is varied and deep. Can anything meaningful be learned from data that sacrifice meaningful detail in order to permit numerical manipulations?

See the *What do you think?... Revisited* box toward the end of the chapter.

Picture 6/Earl Babbie

Some data are more challenging, however. If a survey respondent tells you that he or she thinks the biggest problem facing Woodbury, Vermont, is "the disintegrating ozone layer," the computer can't process that response numerically. You must translate by coding the responses. We've already discussed coding in connection with content analysis (Chapter 11) and again in connection with qualitative data analysis (Chapter 13). Now we look at coding specifically for quantitative analysis, which differs from the other two primarily in its goal of converting raw data into numbers.

As with content analysis, the task of quantitative coding is to reduce a wide variety of idiosyncratic items of information to a more limited set of attributes composing a variable. Suppose, for example, that a survey researcher asks respondents, "What is your occupation?" The responses to such a question will vary considerably. Although it will be possible to assign each reported occupation a separate numerical code, this procedure will not facilitate analysis, which typically depends on several subjects having the same attribute.

The variable *occupation* has many preestablished coding schemes. One such scheme distinguishes professional and managerial occupations, clerical occupations, semiskilled occupations, and so forth. Another scheme distinguishes different sectors of the economy: manufacturing, health, education, commerce, and so forth. Still others combine these two schemes. Using an established coding scheme gives you the advantage of being able to compare your research results with those of other studies.

The occupational coding scheme you choose should be appropriate for the theoretical concepts being examined in your study. For some studies, coding all occupations as either white-collar or blue-collar might suffice. For others, self-employed and not self-employed might do. Or a peace researcher might wish to know only whether the occupation depended on the defense establishment or not.

Although the coding scheme should be tailored to meet particular requirements of the analysis, you should keep one general guideline in mind. If the data are coded to maintain a great deal of detail, code categories can always be combined during an analysis that does not require such detail. If the data are coded into relatively few, gross categories, however, you'll have no way during analysis to re-create the original detail. To keep your options open, it's a good idea to code your data in greater detail than you plan to use in the analysis.

Developing Code Categories

There are two basic approaches to the coding process. First, you may begin with a relatively well-developed coding scheme, derived from your research purpose. Thus, as suggested previously, the peace researcher might code occupations in terms of their relationship to the defense establishment. You might also use an existing coding scheme so that you can compare your findings with those of previous research.

The alternative method is to generate codes from your data, as discussed in Chapter 13. Let's say we've asked students in a self-administered campus survey to say what they believe is the biggest problem facing their college today.

Here are a few of the answers they might have written in.

Tuition is too high

Not enough parking spaces

Faculty don't know what they are doing

Advisors are never available

Not enough classes offered

Cockroaches in the dorms

Too many requirements

Cafeteria food is infected

Books cost too much

Not enough financial aid

Take a minute to review these responses and see whether you can identify some categories represented. Realize that there is no right answer; several coding schemes might be generated from these answers.

Let's start with the first response: "Tuition is too high." What general areas of concern does that response reflect? One obvious possibility is "Financial Concerns." Are there other responses that would fit into that category? Table 14-1 shows which of the questionnaire responses could fit.

In more-general terms, the first answer can also be seen as reflecting nonacademic concerns. This categorization would be relevant if your research interest included the distinction between academic and nonacademic concerns. If that were the case, the responses might be coded as shown in Table 14-2.

TABLE 14-1

Student Responses That Can Be Coded "Financial Concerns"

	Financial Concerns
Tuition is too high	X
Not enough parking spaces	
Faculty don't know what they are doing	
Advisors are never available	
Not enough classes offered	
Cockroaches in the dorms	
Too many requirements	
Cafeteria food is infected	
Books cost too much	X
Not enough financial aid	X

TABLE 14-2

Student Concerns Coded as "Academic" and "Nonacademic"

	Academic	Nonacademic
Tuition is too high		X
Not enough parking spaces		X
Faculty don't know what they are doing	X	
Advisors are never available	X	
Not enough classes offered	X	
Cockroaches in the dorms		X
Too many requirements	X	
Cafeteria food is infected		X
Books cost too much		
Not enough financial aid		X

Notice that I didn't code the response "Books cost too much" in Table 14-2, because this concern could be seen as representing both of the categories. Books are part of the academic program, but their cost is not. This signals the need to refine the coding scheme we're developing. Depending on our research purpose, we might be especially interested in identifying any problems that had an academic element; hence we'd code this one "Academic." Just as reasonably, however, we might be more interested in identifying nonacademic problems and would code the response accordingly. Or, as another alternative, we might create a separate category for responses that involved both academic and nonacademic matters.

As yet another alternative, we might want to separate nonacademic concerns into those involving administrative matters and those dealing with campus facilities. Table 14-3 shows how the ten responses would be coded in that event.

As these few examples illustrate, there are many possible schemes for coding a set of data. Your choices should match your research purposes and reflect the logic that emerges from the data themselves. Often, you'll find yourself modifying the code categories as the coding process proceeds. Whenever you change the list of categories, however, you must review the data already coded to see whether changes are in order.

Like the set of attributes composing a variable, and like the response categories in a closed-ended questionnaire item, code categories should be both exhaustive and mutually exclusive. Every

TABLE 14-3

Nonacademic Concerns Coded as "Administrative" or "Facilities"

	Academic	Administrative	Facilities
Tuition is too high		X	
Not enough parking spaces			X
Faculty don't know what they are doing	X		
Advisors are never available	X		
Not enough classes offered	X		
Cockroaches in the dorms			X
Too many requirements	X		
Cafeteria food is infected			X
Books cost too much	X		
Not enough financial aid		X	

piece of information being coded should fit into one and only one category. Problems arise whenever a given response appears to fit equally into more than one code category or whenever it fits into no category: Both signal a mismatch between your data and your coding scheme.

If you're fortunate enough to have assistance in the coding process, you'll need to train your coders in the definitions of code categories and show them how to use those categories properly. To do so, explain the meaning of the code categories and give several examples of each. To make sure your coders fully understand what you have in mind, code several cases ahead of time. Then ask your coders to code the same cases without knowing how you coded them. Finally, compare your coders' work with your own. Any discrepancies will indicate an imperfect communication of your coding scheme to your coders. Even with perfect agreement between you and your coders, however, it's best to check the coding of at least a portion of the cases throughout the coding process.

If you're not fortunate enough to have assistance in coding, you should still obtain some verification of your own reliability as a coder. Nobody is perfect, especially a researcher hot on the trail of a finding. Suppose that you're studying an emerging cult and that you have the impression that people who do not have a regular family will be the most likely to regard the new cult as a family substitute. The danger is that whenever you discover a subject who reports

no family, you'll unconsciously try to find some evidence in the subject's comments that the cult is a substitute for family. If at all possible, then, get someone else to code some of your cases to see whether that person makes the same assignments you made.

Codebook Construction

The end product of the coding process in quantitative analysis is the conversion of data items into numerical codes. These codes represent attributes composing variables, which, in turn, are assigned locations within a data file. A **codebook** is a document that describes the locations of variables and lists the assignments of codes to the attributes composing those variables.

A codebook serves two essential functions. First, it is the primary guide used in the coding process. Second, it is your guide for locating variables and interpreting codes in your data file during analysis. If you decide to correlate two variables as a part of your analysis of your data, the codebook tells you where to find the variables and what the codes represent.

Figure 14-1 is a partial codebook created from two variables from the General Social Survey. Though there is no one right format for a codebook, this example presents some of the common elements.

Notice first that each variable is identified by an abbreviated variable name: POLVIEWS, ATTEND. We can determine the religious service attendance of respondents, for example, by referencing ATTEND. This example uses the format established by the General Social Survey, which has been carried over into SPSS. Other data sets and/or analysis programs might format variables differently. Some use numerical codes in place of abbreviated names, for example. You must, however, have some identifier that will allow you to locate and use the variable in question.

Next, every codebook should contain the full definition of the variable. In the case of a

---

codebook The document used in data processing and analysis that tells the location of different data items in a data file. Typically, the codebook identifies the locations of data items and the meaning of the codes used to represent different attributes of variables.

POLVIEWS	ATTEND
We hear a lot of talk these days about liberals and conservatives. I'm going to show you a seven-point scale on which the political views that people might hold are arranged from extremely liberal—point 1—to extremely conservative—point 7. Where would you place yourself on this scale? 1. Extremely liberal 2. Liberal 3. Slightly liberal 4. Moderate, middle of the road 5. Slightly conservative 6. Conservative 7. Extremely conservative 8. Don't know 9. No answer	How often do you attend religious services? 0. Never 1. Less than once a year 2. About once or twice a year 3. Several times a year 4. About once a month 5. 2–3 times a month 6. Nearly every week 7. Every week 8. Several times a week 9. Don't know, No answer

FIGURE 14-1

Partial Codebook.

questionnaire, the definition consists of the exact wordings of the questions asked, because, as we've seen, the wording of questions strongly influences the answers returned. In the case of POLVIEWS, you know that respondents were given the several political categories and asked to pick the one that best fit them.

The codebook also indicates the attributes composing each variable. In POLVIEWS, for example, the political categories just mentioned serve as these attributes: "Extremely liberal," "Liberal," "Slightly liberal," and so forth.

Finally, notice that each attribute also has a numeric label. Thus, in POLVIEWS, "Extremely liberal" is code category 1. These numeric codes are used in various manipulations of the data. For example, you might decide to combine categories 1 through 3 (all the "liberal" responses). It's easier to do this with code numbers than with lengthy names.

Data Entry

In addition to transforming data into quantitative form, researchers interested in quantitative analysis also need to convert data into a machine-readable format so that computers can read and manipulate the data. There are many ways of accomplishing this step, depending on the original form of your data and on the computer program you'll use for analyzing the data. I'll simply introduce you to the process here. If you find yourself undertaking this

task, you should be able to tailor your work to the particular data source and program you're using.

If your data have been collected by questionnaire, you might do your coding on the questionnaire itself. Then, data-entry specialists (including yourself) could enter the data into, say, an SPSS data matrix or into an Excel spreadsheet that would later be imported into SPSS.

Sometimes, social researchers use optical scan sheets for data collection. These sheets can be fed into machines that will convert the black marks into data, which can be imported into the analysis program. This procedure works only with subjects who are comfortable using such sheets, and it's usually limited to closed-ended questions.

Sometimes, data entry occurs in the process of data collection. In computer-assisted telephone interviewing (CATI), for example, the interviewer keys responses directly into the computer, where the data are compiled for analysis (see Chapter 9). Even more effortlessly, online surveys can be constructed so that the respondents enter their own answers directly into the accumulating database, without the need for an intervening interviewer or data-entry person.

Once data have been fully quantified and entered into the computer, researchers can begin quantitative analysis. Let's look at the three cases mentioned at the start of this chapter: univariate, bivariate, and multivariate analyses.

Univariate Analysis

The simplest form of quantitative analysis, **univariate analysis**, involves describing a case in terms of a single variable—specifically, the distribution of attributes that compose it. For example, if *gender* were measured, we would look at how many of the subjects were men and how many were women.

Distributions

The most basic format for presenting univariate data is to report all individual cases, that is, to list the attribute for each case under study in terms of the variable in question. Let's take as an example the General Social Survey (GSS) data on attendance at religious services, ATTEND.

Figure 14-2 shows how you could request these data, using the Berkeley SDA online analysis program introduced earlier in the book. You can access this program at http://sda.berkeley .edu/sdaweb/analysis/?dataset=gss14.

In the figure you'll see that ATTEND has been entered as the Row variable, and I have specified a Selection Filter to limit the analysis to the data collected in the 2014 GSS. The consequence of this will be apparent shortly.

Figure 14-3 represents the tabular response to our request. We see, for example, that 26.2 percent say they never attend religious services. As we move down the table, we see that 17.6 percent say they attend every week. To simplify the results, we might want to combine the last three categories and say that 28.8 percent attend "About weekly."

A description of the number of times that the various attributes of a variable are observed in a

Frequency Distribution		
Cells contain: -Column percent -Weighted N		Distribution
ATTEND	0: NEVER	26.2 660.6
	1: LT ONCE A YEAR	7.7 193.4
	2: ONCE A YEAR	12.9 326.9
	3: SEVRL TIMES A YR	10.4 263.7
	4: ONCE A MONTH	5.8 146.8
	5: 2-3X A MONTH	8.2 207.0
	6: NRLY EVERY WEEK	4.3 109.1
	7: EVERY WEEK	17.6 443.4
	8: MORE THN ONCE WK	6.9 175.1
	COL TOTAL	100.0 2,526.0

FIGURE 14-3

Attendance at Religious Services, 2014.

sample is called a **frequency distribution**. Sometimes it's easiest to see a frequency distribution in a graph. Figure 14-4 was created by SDA based on the specifications in the chart options section of Figure 14-2. The vertical scale on the left side of the graph indicates the percentages selecting each of the answers that are displayed along the horizontal axis of the graph. Take a minute to notice how the percentages in Figure 14-3 correspond to the heights of the bars in Figure 14-4.

This program also offers other graphical possibilities. In Figure 14-2, you could have specified "Pie Chart" instead of "Bar Chart" as the type of chart desired.

SDA Frequencies/Crosstabulation Program
Help: General / Recoding Variables

Row:	ATTEND	(Required)
Column:		
Control:		
Selection Filter(s):	YEAR(2014)	
Weight:	COMPWT - Composite weight: WTSSALL * OVERSAMP * FOF ▾	

▸ Output Options
▸ Chart Options
▸ Decimal Options

Run the Table Clear Fields

© 2012 The Idea Works, Inc.

FIGURE 14-2

Requesting a Univariate Analysis of ATTEND, 2014.

univariate analysis The analysis of a single variable, for purposes of description. Frequency distributions, averages, and measures of dispersion are examples of univariate analysis, as distinguished from *bivariate* and *multivariate analysis.*

frequency distribution A description of the number of times the various attributes of a variable are observed in a sample. The report that 53 percent of a sample were men and 47 percent were women would be a simple example of a frequency distribution. Another example would be the report that 15 of the cities studied had populations under 10,000, 23 had populations between 10,000 and 25,000, and so forth.

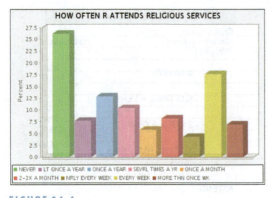

FIGURE 14-4

Bar Chart of GSS ATTEND, 2014.

Since pie charts are particularly appropriate for nominal variables, Figure 14-5 is drawn from a different study. In his examination of religion in Navi Mumbia, India, demographer Prem Saxena (2012: 20) felt that a pie chart best communicated the predominance of Hindu temples among the various religious institutions.

Central Tendency

Beyond simply reporting the overall distribution of values, sometimes called the *marginal frequencies* or just the *marginals*, you may choose to present your data in the form of an **average**, or measure of central tendency. You're already familiar with the concept of central tendency from the many kinds of averages you use in everyday life to express the "typical" value of a variable. For instance, in baseball a batting average of .300 says that a batter gets

---

average An ambiguous term generally suggesting typical or normal—a central tendency. The *mean*, *median*, and *mode* are specific examples of mathematical averages.

mean An average computed by summing the values of several observations and dividing by the number of observations. If you now have a grade point average of 4.0 based on 10 courses, and you get an F in this course, your new grade point (mean) average will be 3.6.

mode An average representing the most frequently observed value or attribute. If a sample contains 1,000 Protestants, 275 Catholics, and 33 Jews, "Protestant" is the modal category.

median An average representing the value of the "middle" case in a rank-ordered set of observations. If the ages of five men are 16, 17, 20, 54, and 88, the median would be 20. (The mean would be 39.)

---

a hit three out of every ten opportunities, on average. Over the course of a season, a hitter might go through extended periods without getting any hits at all and go through other periods when he or she gets a bunch of hits all at once. Over time, though, the central tendency of the batter's performance can be expressed as getting three hits in every ten chances. Similarly, your grade point average expresses the "typical" value of all your grades taken together, even though some of them might be A's, others B's, and one or two might be C's (I know you never get anything lower than a C).

Averages like these are more properly called the *arithmetic mean* (the result of dividing the sum of the values by the total number of cases). The **mean** is only one way to measure central tendency or "typical" values. Two other options are the **mode** (the most frequently occurring attribute) and the **median** (the middle attribute in the ranked distribution of observed attributes). Here's how the three averages would be calculated from a set of data.

Suppose you're conducting an experiment that involves teenagers as subjects. They range in age from 13 to 19, as indicated in the following table:

Age	Number
13	3
14	4
15	6
16	8
17	4
18	3
19	3

Now that you've seen the actual ages of the 31 subjects, how old would you say they are in general, or "on average"? Let's look at three different ways you might answer that question.

The easiest average to calculate is the mode, the most frequent value. As you can see, there were more 16-year-olds (eight of them) than any other age, so the modal age is 16, as indicated in Figure 14-6. Technically, the modal age is the category "16," which may include some people who are closer to 17 than 16 but who haven't yet reached that birthday.

Figure 14-6 also demonstrates the calculation of the mean. There are three steps: (1) multiply each age by the number of subjects who have

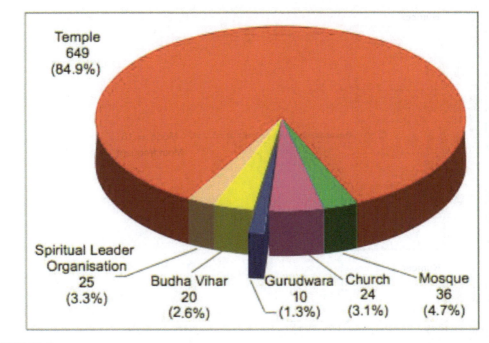

FIGURE 14-5

Pie Chart Showing Number and Percentage of Religious Institutions by Type.

Source: Prem Saxena. 2012. Social, Religious and Regional Institutions in Navi Mumbai. (Navi Mumbai, India: City and Industrial Development Corporation of Maharashtra). This illustration is taken from Figure 2.3, p. 20.

that age, (2) total the results of all those multiplications, and (3) divide that total by the number of subjects.

In the case of age, a special adjustment is needed. As indicated in the discussion of the mode, those who call themselves "13" actually range from exactly 13 years old to those just short of 14. It is reasonable to assume, moreover, that as a group the "13-year-olds" in the country are evenly distributed within that one-year span, making their average age 13.5 years. This is true for each of the age groups. Hence, it's appropriate to add 0.5 years to the final calculation, making the mean age 16.37, as indicated in Figure 14-6.

The third measure of central tendency, the median, represents the "middle" value: Half are above it, half below. If we had the precise ages of each subject (for example, 17 years 124 days), we'd be able to arrange all 31 subjects in order by age, and the median for the whole group would be the age of the middle subject.

As you can see, however, we do not know precise ages; our data constitute "grouped data" in this regard. For example, three people who are not precisely the same age have been grouped in the category "13-year-olds."

Figure 14-6 illustrates the logic of calculating a median for grouped data. Because there are 31 subjects altogether, the "middle" subject would be subject number 16 if they were arranged by age—15 teenagers would be younger and 15 older. Look at the bottom portion of Figure 14-6, and you'll see that the middle person is one of the eight 16-year-olds. In the enlarged view of that group, we see that number 16 is the third from the left.

Because we do not know the precise ages of the subjects in this group, the statistical convention here is to assume that they are evenly spread along the width of the group. In this instance, the possible ages of the subjects go from 16 years 0 days to 16 years 364 days. Strictly speaking, the range, then, is 364/365 days. As a practical matter, it's sufficient to call it one year.

If the eight subjects in this group were evenly spread from one limit to the other, they would be one-eighth of a year apart from each other—a 0.125-year interval. Look at the illustration and you'll see that if we place the first subject at half the interval from the lower limit and add a full interval to the age of each successive subject, the final one is half an interval from the upper limit.

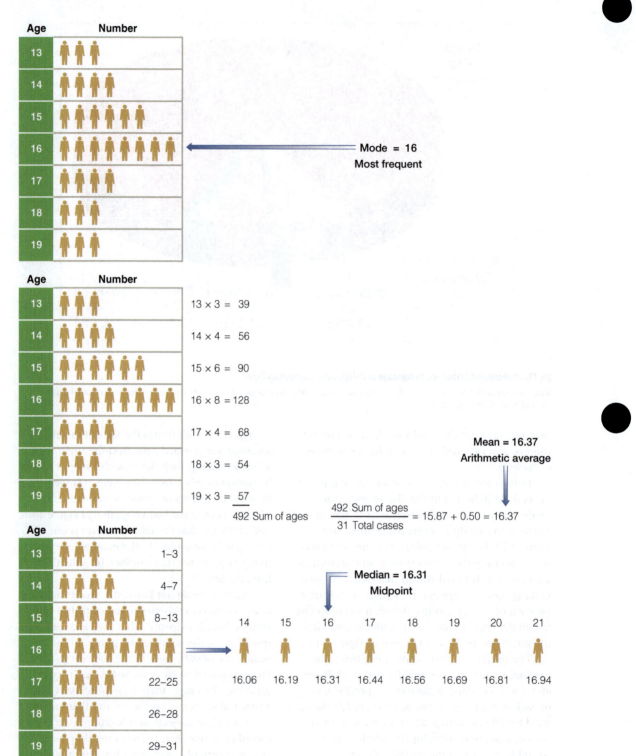

What we've done is calculate, hypothetically, the precise ages of the eight subjects, assuming their ages were spread out evenly. Having done this, we merely note the age of the middle subject—16.31—and that is the median age for the group.

Whenever the total number of subjects is an even number, of course, there is no middle case. To get the median, you merely calculate the mean of the two values on either side of the midpoint in the ranked data. Suppose, for example, that there was one more 19-year-old in our sample, giving us a total of 32 cases. The midpoint would then fall between subjects 16 and 17. The median would therefore be calculated as $(16.31 + 16.44)/2 = 16.38$.

As you can see in Figure 14-6, the three measures of central tendency produce three different values for our set of data, which is often (but not necessarily) the case. Which measure, then, best represents the "typical" value? More generally, which measure of central tendency should we prefer? The answer depends on the nature of your data and the purpose of your analysis. For example, whenever means are presented, you should be aware that they are susceptible to extreme values—a few very large or very small numbers. As only one example, the (mean) average person in Medina, Washington, has a net worth in excess of a million dollars. If you were to visit Medina, however, you might not find that the "average" resident lives up to your idea of a millionaire. The very high mean reflects the influence of one extreme case among Medina's 3,000 or so residents—Bill Gates of Microsoft, who has a net worth of tens of billions of dollars. Clearly, the median wealth would give you a more accurate picture of the residents of Medina as a whole.

This example should illustrate the need to choose carefully among the various measures of central tendency. A course or textbook in statistics will give you a broader understanding of the variety of situations in which each is appropriate.

Dispersion

Averages offer readers the advantage of reducing the raw data to the most manageable form: A single number (or attribute) can represent all the detailed data collected in regard to the variable. This advantage comes at a cost, of course, because the reader cannot reconstruct the original data from an average. Summaries of the dispersion of responses can somewhat alleviate this disadvantage.

Dispersion refers to the way values are distributed around some central value, such as an average. The simplest measure of dispersion is the range: the distance separating the highest from the lowest value. Thus, besides reporting that our subjects have a mean age of 15.87, we might also indicate that their ages range from 13 to 19.

A more sophisticated measure of dispersion is the **standard deviation**. This measure was briefly mentioned in Chapter 7 as the standard error of a sampling distribution. Essentially, the standard deviation is an index of the amount of variability in a set of data. A higher standard deviation means that the data are more dispersed; a lower standard deviation means that they are more bunched together. Figure 14-7 illustrates the basic idea. Notice that the professional golfer not only has a lower mean score but is also more consistent—represented by the lower standard deviation. The duffer, on the other hand, has a higher average and is also less consistent: sometimes doing much better, sometimes much worse.

There are many other measures of dispersion. In reporting intelligence test scores, for example, researchers might determine the interquartile range, the range of scores for the middle 50 percent of subjects. If the top one-fourth had scores ranging from 120 to 150, and if the bottom one-fourth had scores ranging from 60 to 90, the

---

dispersion The distribution of values around some central value, such as an average. The range is a simple example of a measure of dispersion. Thus, we may report that the mean age of a group is 37.9, and the range is from 12 to 89.

standard deviation A measure of dispersion around the mean, calculated so that approximately 68 percent of the cases will lie within plus or minus one standard deviation from the mean, 95 percent will lie within plus or minus two standard deviations, and 99.9 percent will lie within three standard deviations. Thus, for example, if the mean age in a group is 30 and the standard deviation is 10, then 68 percent have ages between 20 and 40. The smaller the standard deviation, the more tightly the values are clustered around the mean; if the standard deviation is high, the values are widely spread out.

a. High standard deviation = spread-out values

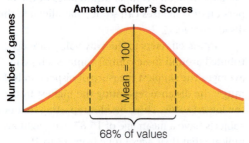

b. Low standard deviation = tightly clustered values

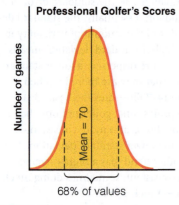

FIGURE 14-7
High and Low Standard Deviations.

report might say that the interquartile range was from 90 to 120 (or 30 points) with a mean score of, let's say, 102.

Continuous and Discrete Variables

The preceding calculations are not appropriate for all variables. To understand this point, we must distinguish between two types of variables: continuous and discrete. A **continuous variable** (or ratio variable) increases steadily in tiny fractions.

> **continuous variable** A variable whose attributes form a steady progression, such as *age* or *income*. Thus, the ages of a group of people might include 21, 22, 23, 24, and so forth and could even be broken down into fractions of years.
>
> **discrete variable** A variable whose attributes are separate from one another, or discontinuous, as in the case of *gender* or *religious affiliation*. In other words, there is no progression from male to female in the case of *gender*.

An example is *age*, which increases steadily with each increment of time. A **discrete variable** jumps from category to category without intervening steps. Examples include *gender*, *military rank*, or *year in college* (you go from being a sophomore to a junior in one step).

In analyzing a discrete variable—a nominal or ordinal variable, for example—some of the techniques discussed previously do not apply. Strictly speaking, modes should be calculated for nominal data, medians for interval data, and means for ratio data, not for nominal data (see Chapter 5). If the variable in question is *gender*, for example, raw numbers (23 of the cross-dressing outlaw bikers in our sample are women) or percentages (7 percent are women) can be appropriate and useful analyses, but neither a median nor a mean would make any sense. Calculating the mode would be legitimate, though not very revealing, because it would only tell us "most were men." However, the mode for data on *religious affiliation* might be more interesting, as in "most people in the United States are Protestant."

Detail versus Manageability

In presenting univariate and other data, you'll be constrained by two goals. On the one hand, you should attempt to provide your reader with the fullest degree of detail regarding those data. On the other hand, the data should be presented in a manageable form. As these two goals are often in direct conflict, you'll find yourself continually seeking the best compromise between them. One useful solution is to report a given set of data in more than one form. In the case of age, for example, you might report the distribution of ungrouped ages plus the mean age and standard deviation.

As you can see from this introductory discussion of univariate analysis, this seemingly simple matter can be rather complex. In any event, the lessons of this section pave the way for a consideration of subgroup comparisons and bivariate analyses.

Subgroup Comparisons

Univariate analyses describe the units of analysis of a study and, if they are a sample drawn from some larger population, allow us to make descriptive inferences about the larger population. Bivariate and

TABLE 14-4

Marijuana Legalization by Age of Respondents, 2014

	Under 25	25–39	40–54	55 or Older
Should be legalized	71%	60%	55%	47%
Should not be legalized	29	40	45	53
100% =	(161)	(441)	(429)	(533)

Source: General Social Survey, 2014, National Opinion Research Center.

multivariate analyses are aimed primarily at explanation. Before turning to explanation, however, we should consider the case of subgroup description.

Often it's appropriate to describe subsets of cases, subjects, or respondents. Here's a simple example from the GSS. In 2014, respondents were asked, "Should marijuana be made legal?" In response, 55.5 percent said it should and 44.5 percent said it shouldn't. Table 14-4 presents the responses given to this question by respondents in different age categories.

Notice that the subgroup comparisons tell us how different groups in the population responded to this question. You can undoubtedly see a pattern in the results, though possibly not exactly what you expected; we'll return to that in a moment. First, let's see how another set of subgroups answered this question.

Table 14-5 presents different political subgroups' attitudes toward legalizing marijuana, based on whether respondents characterized themselves as conservative or liberal. Before looking at the table, you might try your hand at hypothesizing what the results are likely to be and why. Notice that I've changed the direction of

percentaging this table, to make it easier to read. To compare the subgroups in this case, you would read down the columns, not across the rows.

Before examining the logic of causal analysis, let's consider another example of subgroup comparisons—one that will let us address some table-formatting issues.

"Collapsing" Response Categories

"Textbook examples" of tables are often simpler than you'll typically find in published research reports or in your own analyses of data, so this section and the next address two common problems and suggest solutions.

Let's begin with examining Table 14-6, which reports data from a hypothetical political poll. As you'll see, respondents were asked whom they would vote for if an election were held today: candidate Smith or Jones. They were further asked to indicate the strength of their commitment to their chosen candidate. The results have been reported in terms of the political party preferences of the respondents.

While you would probably have no trouble guessing which candidate was the Republican and which was the Democrat, you might feel the table had a few too many numbers for an easy interpretation. We can simplify matters a bit by collapsing the extreme and moderate categories for each candidate. I've done this in Table 14-7.

Handling "Don't Knows"

While it is useful to know what portion of the electorate is undecided in an election campaign, it is

TABLE 14-5

Marijuana Legalization by Political Orientation, 2014

	Should Legalize	Should Not Legalize	100% =
Extremely liberal	84%	16	(58)
Liberal	76%	24	(172)
Slightly liberal	69%	31	(168)
Moderate	54%	46	(597)
Slightly conservative	54%	46	(219)
Conservative	33%	67	(222)
Extremely conservative	34%	66	(72)

Source: General Social Survey, 2014, National Opinion Research Center.

TABLE 14-6

A Hypothetical Political Poll

"If the election were held today, whom do you think you would vote for: Smith or Jones?"

	Political Party Affiliation of Respondents		
	Democrat	Republican	None
Definitely Smith	40%	10%	10%
Probably Smith	30	10	20
Probably Jones	10	25	20
Definitely Jones	5	35	15
Don't know	15	20	35
Total	100%	100%	100%

TABLE 14-7

A Hypothetical Political Poll with Collapsed Categories

"If the election were held today, whom do you think you would vote for: Smith or Jones?"

| | Political Party Affiliation of Respondents | | |
	Democrat	Republican	None
Definitely or probably:			
Smith	70%	20%	30%
Jones	15	60	35
Don't know	15	20	35
Total	100%	100%	100%

worth noting that no one gets to vote "undecided" on Election Day, so you may want to examine the relative strength of the two candidates with the "don't knows" excluded from the calculation. That result is shown in Table 14-8.

This recalculation is fairly straightforward. Because Democrats gave a total of 85 percent of their intended votes to the two candidates, we need only divide each of the original percentages by 0.85. Thus, Smith's 70 percent among Democrats becomes 82 percent when the "don't know" responses are excluded. Republicans gave 80 percent of their votes to the two candidates, so we divide each of their percentages in Table 14-8 by 0.80. And thus, the percentages of those with no party affiliation are divided by 0.65.

At this point, having seen three versions of the data, you may be asking yourself, "Which is the right one?" The answer depends on your purpose in analyzing and interpreting the data. For example, if it is not essential for you to distinguish "definitely" from "probably," it makes

TABLE 14-8

A Hypothetical Political Poll with Collapsed Categories

"If the election were held today, whom do you think you would vote for: Smith or Jones?"

| | Political Party Affiliation of Respondents | | |
	Democrat	Republican	None
Definitely or probably:			
Smith	82%	25%	46%
Jones	18	75	54
Total	100%	100%	100%

sense to combine them, because it's easier to read the table.

Whether to include or exclude the "don't knows" is harder to decide in the abstract. It may be a very important finding that such a large percentage of voters have not made up their minds. On the other hand, it might be more appropriate to exclude the "don't knows" on the assumption that they wouldn't vote or that ultimately their votes would most likely be divided equally between the two candidates.

In any event, the truth contained within your data is that a certain percentage said they didn't know and the remainder divided their responses in whatever manner they did. Often, it's appropriate to report your data in both forms— with and without the "don't knows"—so your readers can also draw their own conclusions. Of course, you yourself will be a reader of such tables, drawn up by others, and knowing the logic behind constructing them will help you be a savvy consumer of quantitative data.

Numerical Descriptions in Qualitative Research

Although this chapter deals primarily with quantitative research, the discussions are also relevant to qualitative studies. Numerical testing can often verify the findings of in-depth, qualitative studies. Thus, for example, when David Silverman wanted to compare the cancer treatments received by patients in private clinics with those in Britain's National Health Service, he primarily chose in-depth analyses of the interactions between doctors and patients:

> My method of analysis was largely qualitative and ... I used extracts of what doctors and patients had said as well as offering a brief ethnography of the setting and of certain behavioural data. In addition, however, I constructed a coding form which enabled me to collate a number of crude measures of doctor and patient interactions.
>
> (1993: 163)

Not only did the numerical data fine-tune Silverman's impressions based on his qualitative observations, but his in-depth understanding of the situation allowed him to craft an increasingly appropriate quantitative analysis. Listen to the interaction between qualitative and quantitative approaches in this lengthy discussion:

My overall impression was that private consultations lasted considerably longer than those held in the NHS clinics. When examined, the data indeed did show that the former were almost twice as long as the latter (20 minutes as against 11 minutes) and that the difference was statistically highly significant. However, I recalled that, for special reasons, one of the NHS clinics had abnormally short consultations. I felt a fairer comparison of consultations in the two sectors should exclude this clinic and should only compare consultations taken by a single doctor in both sectors. This subsample of cases revealed that the difference in length between NHS and private consultations was now reduced to an average of under 3 minutes. This was still statistically significant, although the significance was reduced. Finally, however, if I compared only new patients seen by the same doctor, NHS patients got 4 minutes more on the average—34 minutes as against 30 minutes in the private clinic.

(1993: 163–64)

This example further demonstrates the special power that can be gained from a combination of approaches in social research. The combination of qualitative and quantitative analyses can be especially potent.

Bivariate Analysis

In contrast to univariate analysis, subgroup comparisons involve two variables. In this respect, subgroup comparisons constitute a kind of **bivariate analysis**—that is, an analysis of two variables simultaneously. However, as with univariate analysis, the purpose of subgroup comparisons is largely descriptive. Most bivariate analysis in social research adds another element: determining relationships between the variables themselves. Thus, univariate analysis and subgroup comparisons focus on describing the people (or other units of analysis) under study, whereas bivariate analysis focuses on the variables and their empirical relationships.

Table 14-9 could be regarded as an instance of subgroup comparison: It independently describes the attendance of men and women at religious services, as reported in the 2014 GSS. It shows—comparatively and descriptively—that the women under study attended religious services more often than did the men. However, the same table, seen as an explanatory bivariate analysis, tells a somewhat different story. It suggests that the variable

TABLE 14-9

Religious Attendance Reported by Men and Women, 2014

	Men	Women
Weekly	24%	33%
Less often	76	67
100% =	(1,148)	(1,378)

Source: General Social Survey, 2014, National Opinion Research Center.

gender has an effect on the variable *religious service attendance*. That is, we can view the behavior as a dependent variable that is partially determined by the independent variable, *gender*.

Explanatory bivariate analyses, then, involve the "variable language" introduced in Chapter 1. In a subtle shift of focus, we are no longer talking about men and women as different subgroups but about *gender* as a variable: one that has an influence on other variables. The theoretical interpretation of Table 14-9 might be taken from Charles Glock's comfort hypothesis as discussed in Chapter 2:

1. Women are still treated as second-class citizens in U.S. society.
2. People denied status gratification in the secular society may turn to religion as an alternative source of status.
3. Hence, women should be more religious than men.

The data presented in Table 14-9 confirm this reasoning. Thirty-three percent of the women attend religious services weekly, as compared with 24 percent of the men.

Adding the logic of causal relationships among variables has an important implication for the construction and reading of percentage tables. One of the chief bugaboos for new data analysts is deciding on the appropriate "direction of percentaging" for any given table. In Table 14-9, for example, I've divided the group of subjects into two subgroups—men and women—and then described the behavior of each subgroup. That is the correct method for

bivariate analysis The analysis of two variables simultaneously, for the purpose of determining the empirical relationship between them. The construction of a simple percentage table or the computation of a simple correlation coefficient are examples of bivariate analyses.

constructing this table. Notice, incidentally, that we could—however inappropriately—construct the table differently. We could first divide the subjects into different degrees of religious attendance and then describe each of those subgroups in terms of the percentage of men and women in each. This method would make no sense in terms of explanation, however. Table 14-9 suggests that your gender will affect your frequency of religious service attendance. Had we used the other method of construction, the table would suggest that your religious service attendance affects whether you are a man or a woman—which makes no sense. Your behavior cannot determine your gender.

A related problem complicates the lives of new data analysts. How do you read a percentage table? There is a temptation to read Table 14-9 as follows: "Of the women, only 33 percent attended religious services weekly, and 67 percent said they attended less often; therefore, being a woman makes you less likely to attend religious services frequently." This is, of course, an incorrect reading of the table. Any conclusion that *gender*—as a variable—has an effect on religious service attendance must hinge on a comparison between men and women. Specifically, we compare the 33 percent with the 24 percent and note that women are more likely than men to attend religious services weekly. The comparison of subgroups, then, is essential in reading an explanatory bivariate table.

In constructing and presenting Table 14-9, I've used a convention called *percentage down.* This term means that you can add the percentages down each column to total 100 percent. You read this form of table across a row. For the row labeled "Weekly," what percentage of the men attends weekly? What percentage of the women attend weekly?

The direction of percentaging in tables is arbitrary, and some researchers prefer to percentage across. They would organize Table 14-9 so that "Men" and "Women" were shown on the left side of the table, identifying the two rows, and "Weekly" and "Less often" would appear at the top to identify the columns. The actual numbers in the table would be moved around accordingly, and each row of percentages would total 100 percent. In that case, you would read the table down a column, still asking what percentage of men and

women attended frequently. The logic and the conclusion would be the same in either case; only the form would differ.

In reading a table that someone else has constructed, you need to find out in which direction it has been percentaged. Usually this will be labeled or be clear from the logic of the variables being analyzed. As a last resort, however, you should add the percentages in each column and each row. If each of the columns totals 100 percent, the table has been percentaged down. If the rows total 100 percent each, it has been percentaged across. The rule, then, is as follows:

1. If the table is percentaged down, read across.
2. If the table is percentaged across, read down.

Percentaging a Table

Figure 14-8 reviews the logic by which we create percentage tables from two variables. I've used as variables *gender* and *attitude toward equality for men and women.*

Here's another example. Suppose we're interested in learning something about newspaper editorial policies regarding the legalization of marijuana. We undertake a content analysis of editorials on this subject that have appeared during a given year in a sample of daily newspapers across the nation. Each editorial has been classified as favorable, neutral, or unfavorable toward the legalization of marijuana. Perhaps we wish to examine the relationship between editorial policies and the types of communities in which the newspapers are published, thinking that rural newspapers might be more conservative in this regard than urban ones. Thus, each newspaper (hence, each editorial) has been classified in terms of the population of the community in which it is published.

Table 14-10 presents hypothetical data describing the editorial policies of rural and urban newspapers. Note that the unit of analysis in this example is the individual editorial. Table 14-10 tells us that there were 127 editorials about marijuana in our sample of newspapers published in communities with populations under 100,000. (Note that this cutting point is chosen for simplicity of illustration and does not mean that *rural* refers to a community of less than 100,000 in any absolute sense.) Of these, 11 percent (14 editorials divided by a

a. Some men and women who either favor (+) gender equality or don't (−) favor it.

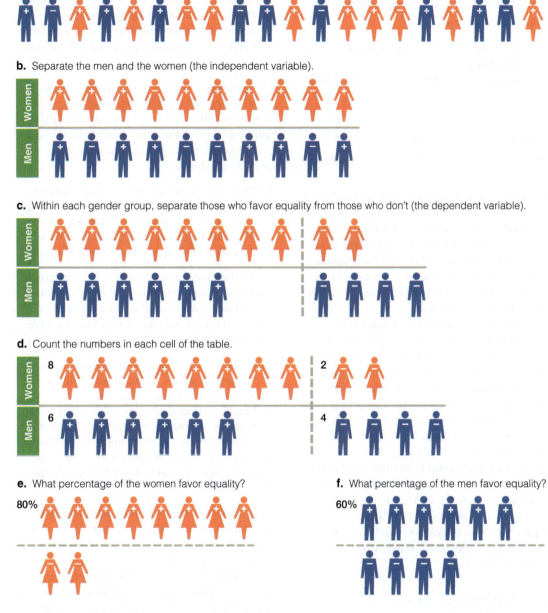

b. Separate the men and the women (the independent variable).

c. Within each gender group, separate those who favor equality from those who don't (the dependent variable).

d. Count the numbers in each cell of the table.

e. What percentage of the women favor equality?

80%

f. What percentage of the men favor equality?

60%

g. Conclusions.

While a majority of both men and women favored gender equality, women were more likely than men to do so.

Thus, gender appears to be one of the causes of attitudes toward sexual equality.

	Women	Men
Favor equality	80%	60%
Don't favor equality	20	40
Total	**100%**	**100%**

FIGURE 14-8

Percentaging a Table.

TABLE 14-10

Hypothetical Data Regarding Newspaper Editorials on the Legalization of Marijuana

Editorial Policy toward Legalizing Marijuana	Community Size	
	Under 100,000	Over 100,000
Favorable	11%	32%
Neutral	29	40
Unfavorable	60	28
100% =	(127)	(438)

base of 127) were favorable toward legalization of marijuana, 29 percent were neutral, and 60 percent were unfavorable. Of the 438 editorials that appeared in our sample of newspapers published in communities of more than 100,000 residents, 32 percent (140 editorials) were favorable toward legalizing marijuana, 40 percent were neutral, and 28 percent were unfavorable.

When we compare the editorial policies of rural and urban newspapers in our imaginary study, we find—as expected—that rural newspapers are less favorable toward the legalization of marijuana than are urban newspapers. We determine this by noting that a larger percentage (32 percent) of the urban editorials were favorable than the percentage of rural ones (11 percent). We might note as well that more rural than urban editorials were unfavorable (60 percent compared with 28 percent). Note that this table assumes that the size of a community might affect its newspapers' editorial policies on this issue, rather than that editorial policy might affect the size of communities.

Constructing and Reading Bivariate Tables

Let's now review the steps involved in the construction of explanatory bivariate tables:

1. The cases are divided into groups according to the attributes of the independent variable.

> **contingency table** A format for presenting the relationships among variables as percentage distributions; typically used to reveal the effects of the independent variable on the dependent variable.

2. Each of these subgroups is then described in terms of attributes of the dependent variable.
3. Finally, the table is read by comparing the independent variable subgroups with each other in terms of a given attribute of the dependent variable.

Let's repeat the analysis of *gender* and attitude on gender equality following these steps. For the reasons outlined previously, *gender* is the independent variable; *attitude toward gender equality* constitutes the dependent variable. Thus, we proceed as follows:

1. The cases are divided into men and women.
2. Each gender subgrouping is described in terms of approval or disapproval of gender equality.
3. Men and women are compared in terms of the percentages approving of gender equality.

In the example of editorial policies regarding the legalization of marijuana, *size of community* is the independent variable and a *newspaper's editorial policy* the dependent variable. The table would be constructed as follows:

1. Divide the editorials into subgroups according to the sizes of the communities in which the newspapers are published.
2. Describe each subgroup of editorials in terms of the percentages favorable, neutral, or unfavorable toward the legalization of marijuana.
3. Compare the two subgroups in terms of the percentages favorable toward the legalization of marijuana.

Bivariate analyses typically have an explanatory causal purpose. These two hypothetical examples have hinted at the nature of causation as social scientists use it.

Tables such as the ones we've been examining are commonly called **contingency tables**: Values of the dependent variable are contingent on (depend on) values of the independent variable. Although contingency tables are common in social science, their format has never been standardized. As a result, you'll find a variety of formats in the research literature. As long as a table is easy to read and interpret, there's probably no reason to strive for standardization. However, there are several guidelines that you should follow in the presentation of most tabular data:

1. A table should have a heading or a title that succinctly describes what is contained in the table.

2. The original content of the variables should be clearly presented—in the table itself if at all possible or in the text with a paraphrase in the table. This information is especially critical when a variable is derived from responses to an attitudinal question, because the meaning of the responses will depend largely on the wording of the question.

3. The attributes of each variable should be clearly indicated. Though complex categories will have to be abbreviated, their meaning should be clear in the table and, of course, the full description should be reported in the text.

4. When percentages are reported in the table, the base on which they are computed should be indicated. It's redundant to present all the raw numbers for each category, because these could be reconstructed from the percentages and the bases. Moreover, the presentation of both numbers and percentages is often confusing in a table and makes it more difficult to read.

5. If any cases are omitted from the table because of missing data ("no answer," for example), their numbers should be indicated in the table.

Although I have introduced the logic of causal, bivariate analysis in terms of percentage tables, many other formats are appropriate for this topic. Scatterplot graphs are one possibility, providing a visual display of the relationship between two variables. For an engaging example of this, you might check out the GapMinder software available on the Web. Using countries as the unit of analysis, you can examine the relationship between birth rate and infant mortality, for example. In fact, you can watch the relationship develop over time.

Introduction to Multivariate Analysis

The logic of **multivariate analysis**, or the analysis of more than two variables simultaneously, can be seen as an extension of bivariate analysis. Specifically, we can construct multivariate tables on the basis of a more complicated subgroup description by following essentially the same steps outlined for bivariate tables. Instead of one independent variable and one dependent variable, however, we'll have more than one independent variable. Instead of understanding the dependent variable on the basis of a single independent variable, we'll seek an understanding through the use of more than one independent variable.

Let's return to the example of religious attendance. Suppose we believe that age would also affect such behavior (Glock's comfort hypothesis suggests that older people are more religious than younger people). As the first step in table construction, we would divide the total sample into subgroups based on the attributes of both independent variables simultaneously: younger men, older men, younger women, and older women. Then the several subgroups would be described in terms of the dependent variable, *religious service attendance*, and comparisons would be made. Table 14-11, from an analysis of the 2006 GSS data, is the result.

Table 14-11 has been percentaged down and therefore should be read across. The interpretation of this table warrants several conclusions:

1. Among both men and women, older people attend religious services more often than younger people do. Among women, 27 percent of those under 40 years old and 41 percent of those 40 or older attend religious services weekly. Among men, the respective figures are 19 and 31 percent.

2. Within each age group, women attend slightly more frequently than men. Among those respondents under 40 years old, 27 percent

TABLE 14-11

Multivariate Relationship: Religious Service Attendance, Gender, and Age, 2006

"How often do you attend religious services?"

	Under 40		40 or Older	
	Men	*Women*	*Men*	*Women*
About weekly*	19%	27%	31%	41%
Less often	81	73	69	59
100% =	(832)	(958)	(1,211)	(1,477)

*About weekly = "More than once a week," "Weekly," and "Nearly every week."

Source: General Social Survey, 2006, National Opinion Research Center.

multivariate analysis The analysis of the simultaneous relationships among several variables. Examining simultaneously the effects of *age, gender,* and *social class* on *religiosity* would be an example of multivariate analysis.

of the women attend weekly, compared with 19 percent of the men. Among those 40 or over, 41 percent of the women and 31 percent of the men attend weekly.

3. As measured in the table, age appears to have a greater effect on attendance at religious services than does gender.

4. Age and gender have independent effects on religious service attendance. Within a given attribute of one independent variable, different attributes of the second variable still affect behaviors.

5. Similarly, the two independent variables have a cumulative effect on behaviors. Older women attend the most often (41 percent), and younger men attend the least often (19 percent).

6. Note also that it would make sense to say there is a *positive* relationship between age and religious service attendance. An increase in age is associated with an increase in attendance. If older people went less often, we would say there was a *negative* relationship. We can't make similar statements in the case of gender, however, since it is a nominal variable, with no "higher" or "lower" intrinsic order to distinguish women and men.

Before I conclude this section, it will be useful to note an alternative format for presenting such data. Several of the tables presented in this chapter are somewhat inefficient. When the dependent variable *religious attendance* is dichotomous (having exactly two attributes), knowing one attribute permits the reader to reconstruct the other easily. Thus, if we know that 27 percent of the women under 40 attend religious services weekly, then we know automatically that 73 percent attend less often. So, reporting the percentages who attend less often is unnecessary.

On the basis of this recognition, Table 14-11 could be presented in the alternative format shown in Table 14-12. In this table, the percentages of people saying they attend religious services about weekly are reported in the cells representing the intersections of the two independent variables. The numbers presented in parentheses below each percentage represent the number of cases on which the percentages are based. Thus, for example, the reader knows that there are 958 women under 40 years of age in the sample and that 27 percent of them attend religious services weekly. We can calculate from this that 259 of those 958 women attend weekly and that the other 699 younger women (or 73 percent) attend less frequently. This

TABLE 14-12
A Simplification of Table 14-11

	Percent Who Attend about Weekly	
	Men	Women
Under 40	19%	27%
	(832)	(958)
40 or older	31%	41%
	(1,211)	(1,477)

Source: General Social Survey, 2006, National Opinion Research Center.

new table is easier to read than the former one, and it does not sacrifice any detail.

Sociological Diagnostics

The multivariate techniques we are now exploring can serve as powerful tools for diagnosing social problems. They can be used to replace opinions with facts and to settle ideological debates with data analysis.

For an example, let's return to the issue of gender and income. Many explanations have been advanced to account for the long-standing pattern of women in the labor force earning less than men. One explanation is that, because of traditional family patterns, women as a group have participated less in the labor force and many only begin working outside the home after completing certain child-rearing tasks. Thus, women as a group will probably have less seniority at work than will men, and income increases with seniority. A landmark 1984 study by the Census Bureau showed this reasoning to be partly true, as Table 14-13 shows.

Table 14-13 indicates, first of all, that job tenure did indeed affect income. Among both men and women, those with more years on the job earned more. This is seen by reading down the first two columns of the table.

The table also indicates that women earned less than men, regardless of job seniority. This can be seen by comparing average wages across the rows of the table, and the ratio of women's-to-men's wages is shown in the third column. Thus, years on the job was an important determinant of earnings, but seniority did not adequately explain the pattern of women earning

TABLE 14-13

Gender, Job Tenure, and Income, 1984 (Full-time workers 21–64 years of age)

Years Working with Current Employer	Average Hourly Income		Women/Men Ratio
	Men	Women	
Less than 2 years	$8.46	$6.03	0.71
2 to 4 years	$9.38	$6.78	0.72
5 to 9 years	$10.42	$7.56	0.73
10 years more	$12.38	$7.91	0.64

Source: U.S. Bureau of the Census, Current Population Reports, Series P-70, No. 10, *Male–Female Differences in Work Experience, Occupation, and Earning, 1984* (Washington, DC: U.S. Government Printing Office, 1987), 4.

less than men. In fact, we see that women with 10 or more years on the job earned substantially less ($7.91/hour) than men with less than 2 years ($8.46/hour).

Although years on the job did not fully explain the difference between men's and women's pay, there are other possible explanations: level of education, child-care responsibilities, and so forth. The researchers who calculated Table 14-13 also examined some of the other variables that might reasonably explain the differences in pay without representing gender discrimination, including these:

- Number of years in the current occupation
- Total years of work experience (any occupation)
- Whether they have usually worked full time
- Marital status
- Size of city or town they live in
- Whether covered by a union contract
- Type of occupation
- Number of employees in the firm
- Whether private or public employer
- Whether they left previous job involuntarily
- Time spent between current and previous job
- Race
- Whether they have a disability
- Health status
- Age of children
- Whether they took an academic curriculum in high school
- Number of math, science, and foreign language classes in high school
- Whether they attended private or public high school
- Educational level achieved
- Percentage of women in the occupation
- College major

Each of the variables listed here might reasonably affect earnings and, if women and men differ in these regards, could help to account for male and female income differences. When all these variables were taken into account, the researchers were able to account for 60 percent of the discrepancy between the incomes of men and women. The remaining 40 percent, then, is a function of other "reasonable" variables and/or prejudice. This kind of conclusion can be reached only by examining the effects of several variables at the same time—that is, through multivariate analysis.

I hope this example shows how the logic implicit in day-to-day conversations can be represented and tested in a quantitative data analysis like this. Let's review some data introduced in Chapter 11.

In 2016, the median weekly income for a full-time, year-round male worker was $969. The average full-time, year-round female worker earned $784, or about 81 percent as much as her male counterpart (U.S. Bureau of Labor Statistics 2017, Table 9). But does that difference represent gender discrimination or does it reflect legitimate factors?

Some argue that education, for example, affects income and that in the past, women have gotten less education than men. We might start, therefore, by checking whether educational differences explain why women today earn less, on average, than men. Table 14-14 offers data to test this hypothesis.

TABLE 14-14

Median Weekly Earnings of Full-Time Workers, 2016

	Men	Women	Ratio of Women/Men Earnings
All workers	$969	$784	0.81
Less than H.S. graduate	$551	$423	0.77
H.S. graduates	$769	$599	0.81
Some college	$896	$688	0.77
Bachelor's degree	$1,348	$994	0.74
Advanced degree	$1,707	$1,257	0.74

Note: These data point to a persistent difference between the incomes of men and women, even when both groups have achieved the same levels of education.

Source: Bureau of Labor Statistics, Usual Weekly Earnings of Wage and Salary Workers, Fourth Quarter 2016. January 27, 2017, Table 9. Quartiles and selected deciles of usual weekly earnings of full-time wage and salary workers by selected characteristics, 2016 annual averages.

As the table shows, at each level of comparable education, women earn substantially less than men do. Clearly, education does not explain the discrepancy.

Sex and gender are generally regarded as a simple and distinct division between men and women. Social researchers, however, recognize that the matter is more complex than that. For example, transsexuals are individuals who choose to change their biological sex permanently through surgery and hormones. Clearly, such a radical change brings many adjustments and challenges that would make for interesting studies, but Kristen Schilt (2006) has taken an unusual tack.

While many kinds of research point to the disadvantaged status of women in the workplace, Schilt's research on transsexuals reveals the impact of gender on a personal level. In many of the cases, the subjects changed their sex while maintaining the same job in their employing organization. Following their sex change, female-to-male transsexuals were likely to enjoy pay raises and increased authority. In other studies, male-to-female transsexuals reported just the opposite experiences. Personal accounts such as these flesh out statistical studies that consistently show women earning less than men, even when they do the same work.

As another example of multivariate data analysis in real life, consider the common observation that minority group members are more likely to be denied bank loans than are white applicants. A counterexplanation might be that the minority applicants in question were more likely to have had a prior bankruptcy or that they had less collateral to guarantee the requested loan—both reasonable bases for granting or denying loans. However, the kind of multivariate analysis we've just examined could easily resolve the disagreement.

Let's say we look only at those who have not had a prior bankruptcy and who have a certain level of collateral. Are whites and minorities equally likely to get the requested loan? We could conduct the same analysis in subgroups determined by level of collateral. If whites and minorities were equally likely to get their loans in each of the subgroups, we would need to conclude that there was no ethnic discrimination. If minorities were still less likely to get

their loans, however, that would indicate that bankruptcy and collateral differences were not the explanation—strengthening the case that discrimination was at work.

All this should make clear that social research can play a powerful role in serving the human community. It can help us determine the current state of affairs and can often point the way to where we want to go.

Welcome to the world of sociological diagnostics!

Ethics and Quantitative Data Analysis

In Chapter 13, I pointed out that the subjectivity present in qualitative data analysis increases the risk of biased analyses, which experienced researchers learn to avoid. Some think, however, that quantitative analyses are not susceptible to subjective biases. Unfortunately, this isn't so. Even the most mathematically explicit analysis yields ample room for defining and measuring variables in ways that encourage one finding over another, and quantitative analysts need to guard against this. Sometimes, the careful specification of hypotheses in advance can offer protection, although this can also inhibit a full exploration of what data can tell us.

The quantitative analyst has an obligation to report any formal hypotheses and other expectations that didn't pan out. Suppose that you think that a particular variable will prove to be a powerful cause of gender prejudice, but your data analysis contradicts that expectation. You should report the lack of correlation, because such information is useful to others who conduct research on this topic. Although it would be more satisfying to discover what causes prejudice, it's quite important to know what doesn't cause it.

The protection of subjects' privacy is as important in quantitative analysis as in qualitative analysis. However, with quantitative methods it's often easier to collect and record data in ways that make subject identification more difficult. However, the first time public officials demand that you reveal the names of student subjects who reported using illegal drugs in a survey, this issue will take on more salience. (Don't reveal the names, by the way. If necessary, burn the questionnaires—"accidentally.")

What do you think?...Revisited

This chapter began with a question about whether anything meaningful or useful could be learned from the analysis of data that have been stripped of many details in order to permit statistical manipulation. The answer, we've seen, is an unqualified "yes."

Quantitative analysis can be a tool for social change. For instance, calculating the average incomes of men and women or of whites and minorities can demonstrate the inequalities that exist for people doing exactly the same job. Such quantitative analyses can overpower anecdotal evidence about particular women or minorities who earn large salaries. We've also seen that quantitative analyses of qualitative phenomena, such as voting intentions, can be done with precision and utility.

The key lesson is that both qualitative and quantitative research are legitimate and powerful approaches to understanding social life. They are particularly useful, moreover, when used together.

MAIN POINTS

Introduction

- Most data are initially qualitative: They must be quantified to permit statistical analysis.
- Quantitative analysis involves the techniques by which researchers convert data to a numerical form and subject it to statistical analyses.

Quantification of Data

- Some data, such as age and income, are intrinsically numerical.
- Often, quantification involves coding into categories that are then given numerical representations.
- Researchers may use existing coding schemes, such as the Census Bureau's categorization of occupations, or develop their own coding categories. In either case, the coding scheme must be appropriate for the nature and objectives of the study.
- A codebook is the document that describes the identifiers assigned to different variables and the codes assigned to represent the attributes of those variables.

Univariate Analysis

- Univariate analysis is the analysis of a single variable. Because univariate analysis does not involve the relationships between two or more variables, its purpose is descriptive rather than explanatory.
- Several techniques allow researchers to summarize their original data to make them more manageable while maintaining as much of the original detail as possible. Frequency distributions, averages, grouped data, and measures of dispersion are all ways of summarizing data concerning a single variable.

Subgroup Comparisons

- Subgroup comparisons can be used to describe similarities and differences among subgroups with respect to some variable.
- Collapsing response categories and handling "don't knows" are two techniques for presenting and interpreting data.

Bivariate Analysis

- Bivariate analysis focuses on relationships between variables rather than comparisons of groups. Bivariate analysis explores the statistical association between the independent variable and the dependent variable. Its purpose is usually explanatory rather than merely descriptive.
- The results of bivariate analyses often are presented in the form of contingency tables, which are constructed to reveal the effects of the independent variable on the dependent variable.

Introduction to Multivariate Analysis

- Multivariate analysis is a method of analyzing the simultaneous relationships among several variables. It may also be used to understand the relationship between two variables more fully.
- The logic and techniques of quantitative research can be valuable to qualitative researchers.

Sociological Diagnostics

- Sociological diagnostics is a quantitative analysis technique for determining the nature of social problems such as ethnic or gender discrimination.

Ethics and Quantitative Data Analysis

- Unbiased analysis and reporting is as much an ethical concern in quantitative analysis as it is in qualitative analysis.
- Subjects' privacy must be protected in quantitative data analysis and reporting.

KEY TERMS

average	mean
bivariate analysis	median
codebook	mode
contingency table	multivariate analysis
continuous variable	quantitative analysis
discrete variable	standard deviation
dispersion	univariate analysis
frequency distribution	

PROPOSING SOCIAL RESEARCH: QUANTITATIVE DATA ANALYSIS

In this exercise, you should outline your plans for analysis. In earlier exercises, you'll have specified the variables to be analyzed, including precisely how you'll measure those variables.

Now you'll report how you plan to conduct your analysis. Are your aims primarily descriptive or explanatory? If explanatory, are you planning a simple bivariate analysis or a multivariate one? Here's where you should say whether you're planning a tabular analysis or something more complex than what has been discussed in this chapter. It doesn't really matter which computer program you use [SPSS, Statistical Analysis System (SAS), and so forth] unless it's a specialized program or one that is not commonly used.

If you've derived precise hypotheses, you may want to specify levels of statistical significance that will determine the meaning of the outcomes. This is not always necessary, however.

REVIEW QUESTIONS

1. How might the various majors at your college be classified into categories? Create a coding system that would allow you to categorize them according to some meaningful variable. Then create a different coding system, using a different variable.

2. How many ways could you describe these in numerical terms? What are some of your intrinsically numerical attributes? Could you express some of your qualitative attributes in quantitative terms?

3. How would you construct and interpret a contingency table from the following information: 150 Democrats favor raising the minimum wage, and 50 oppose it; 100 Republicans favor raising the minimum wage, and 300 oppose it?

4. Using the hypothetical data in the following table, how would you construct and interpret tables showing these three relationships?

   a. The bivariate relationship between age and attitude toward abortion

   b. The bivariate relationship between political orientation and attitude toward abortion

   c. The multivariate relationship linking age, political orientation, and attitude toward abortion

Age	Political Orientation	Attitude toward Abortion	Frequency
Young	Liberal	Favor	90
Young	Liberal	Oppose	10
Young	Conservative	Favor	60
Young	Conservative	Oppose	40
Old	Liberal	Favor	60
Old	Liberal	Oppose	40
Old	Conservative	Favor	20
Old	Conservative	Oppose	80

CHAPTER 15

The Logic of Multivariate Analysis

CHAPTER OVERVIEW

We'll use the elaboration model to examine the fundamental logic of multivariate and causal analysis. Exploring applications of this logic in the form of simple percentage tables provides a foundation for making sense of more-complex analytic methods.

Rawpixel.com/Shutterstock.com

Introduction

In this chapter, we'll use the elaboration model to examine the fundamental logic of multivariate and causal analysis. Exploring applications of this logic in the form of simple percentage tables provides a foundation for making sense of more complex analytic methods. We will also delve more deeply into the logic of multivariate analysis in quantitative social research, introduced in Chapter 14. It builds on earlier discussions of causation among variables. In Chapter 4, we looked at the criteria for causation, and I introduced the idea of spuriousness. As we saw, sometimes there appears to be a causal relationship between two variables (e.g., number of storks and birth rates), but a more careful analysis shows that apparent relationship to be caused by the influence of a third variable (e.g., rural/urban). Rural communities have higher birth rates and also more storks than urban areas do. As we will see in this chapter, there are a number of other possible multivariate relationships.

To explore this topic, we are going to utilize a social science analysis perspective that is referred to variously as the **elaboration model**, the interpretation method, the Lazarsfeld method, or the Columbia school. Its many names reflect the fact that it aims at elaborating on an empirical relationship among variables in order to interpret

that relationship, in the manner developed by Paul Lazarsfeld while he was a professor at Columbia University. As such, the elaboration model is one method for doing multivariate analysis.

Researchers use the logic of the elaboration model to understand the relationship between two variables through the simultaneous introduction of additional variables, though they may not always refer to the model by name. It was developed primarily through the medium of percentage tables, but it can be used with other statistical techniques, as Chapter 16 will show.

I firmly believe that the elaboration model offers the clearest available picture of the logic of causal analysis in social research. Especially through the use of contingency tables, this method portrays the logical process of scientific analysis. Moreover, if you can comprehend fully the use of the elaboration model using contingency tables, you should greatly improve your ability to use and understand more-sophisticated statistical techniques, such as partial regressions and log-linear models, for example.

In a sense, this discussion of elaboration analysis is an extension of our examination of spuriousness in Chapter 4. As you'll recall, one of the criteria of causal relations in social research is that the observed relationship between two variables not be an artifact caused by some other variable. In the case of the positive relationship between the number of fire trucks responding to a fire and the amount of damage done, for example, we saw that the size of the fire explained away the apparent relationship between trucks and damage. The bigger the fire, the more trucks responding to it; and the bigger the fire, the more damage was done.

elaboration model A logical model for understanding the relationship between two variables by controlling for the effects of a third. Principally developed by Paul Lazarsfeld. The various outcomes of an elaboration analysis are replication, explanation, interpretation, and specification.

What do you think?

A constant dilemma for researchers concerns the determination of causes. What causes voting choices, religious preferences, prejudice, etc.? Usually we can find variables related to these variations—women attend church more often than men—but how do we know we have actually found the *cause*? How can we be sure the observed correlation isn't, in fact, caused by something else?

See the *What Do You Think?... Revisited* box toward the end of the chapter.

Earl Babbie

The logic used in that hypothetical example was the same as the logic of the elaboration model. As the early examples that gave birth to the elaboration model will illustrate, social research often reveals a counterintuitive understanding of social life.

Using both hypothetical and real examples, we'll see that the testing of an observed relationship may result in a variety of discoveries and logical interpretations. Spuriousness is only one of the possibilities.

The "Tips and Tools: Why Do Elaboration?" by one of the elaboration model's creators,

Patricia Kendall, provides another powerful justification for using this model.

The Origins of the Elaboration Model

The historical origins of the elaboration model provide a good illustration of how scientific research works in practice. As I mentioned in Chapter 1, during World War II Samuel Stouffer organized and headed a special social research

How to Do It

Why Do Elaboration?

Patricia L. Kendall

Department of Sociology, Queens College, CUNY

There are several aspects of a true controlled experiment. The most crucial are (1) creating experimental and control groups that are identical within limits of chance (this is done by assigning individuals to the two groups through processes of randomization: using tables of random numbers, flipping coins, etc.); (2) making sure that it is the experimenter who introduces the stimulus, not external events; and (3) waiting to see whether the stimulus has had its presumed effect.

We may have the hypothesis, for example, that attending Ivy League colleges leads to greater success professionally than attending other kinds of colleges and universities does. How would we study this through a true experiment? Suppose you said, "Take a group of people in their 40s, find out which ones went to Ivy League colleges, and see whether they are more successful than those who went to other kinds of colleges." If that is your answer, you are wrong.

A true experiment would require the investigator to select several classes of high school seniors, divide each class at random into experimental and control groups, send the experimental groups to Ivy

League colleges (regardless of their financial circumstances or academic qualifications and regardless of the desire of the colleges to accept them) and the control group to other colleges and universities, wait 20 years or so until the two groups have reached professional maturity, and then measure the relative success of the two groups. Certainly a bizarre process.

Sociologists also investigate the hypothesis that coming from a broken home leads to juvenile delinquency. How would we go about studying this experimentally? If you followed the example above, you would see that studying this hypothesis through a true experiment would be totally impossible. Just think of what the experimenter would have to do!

The requirements of true experiments are so unrealistic in sociological research that we are forced to use other, and less ideal, methods in all but the most trivial situations. We can study experimentally whether students learn more from one type of lecture than another, or whether a film changes viewers' attitudes. But these are not always the sorts of questions in which we are truly interested.

We therefore resort to approximations—generally surveys—that have their own shortcomings. However, the elaboration model allows us to examine survey data, take account of their possible shortcomings, and draw rather sophisticated conclusions about important issues.

branch within the U.S. Army. Throughout the war, this group conducted a large number and variety of surveys among U.S. servicemen. Although the objectives of these studies varied somewhat, they generally focused on the factors affecting soldiers' combat effectiveness.

Several of the studies examined morale in the military. Because morale seemed to be related positively to combat effectiveness, improving morale would make the war effort more effective. Stouffer and his research staff sought to uncover some of the variables that affected morale. In part, the group sought to confirm empirically some commonly accepted propositions, including the following:

1. Promotions surely affect soldiers' morale, so soldiers serving in units with low promotion rates should have relatively low morale.
2. Given racial segregation and discrimination in the South, African American soldiers being trained in northern training camps should have higher morale than should those being trained in the South.
3. Soldiers with more education should be more likely to resent being drafted into the army as enlisted men than should those with less education.

Each of these propositions made sense logically, and common wisdom held each to be true. Stouffer decided to test each empirically. To his surprise, none of the propositions was confirmed.

We discussed the first proposition in Chapter 1. As you may recall, Stouffer found that soldiers serving in the Military Police (where promotions were the slowest in the army) had fewer complaints about the promotion system than did those serving in the Army Air Corps (where promotions were the fastest in the army). The other propositions fared just as badly. African American soldiers serving in northern training camps and those serving in southern training camps seemed to differ little, if at all, in their general morale. And less-educated soldiers were more likely to resent being drafted into the army than those with more education were.

Rather than trying to hide the findings or just running tests of statistical significance and publishing the results, Stouffer asked,

"Why?" He found the answer to this question within the concepts of reference group and relative deprivation. Put simply, Stouffer suggested that soldiers did not evaluate their positions in life according to absolute, objective standards, but rather on the basis of their position relative to others around them. The people they compared themselves with were in their reference group, and they felt relative deprivation if they didn't compare favorably in that regard.

Following this logic, Stouffer found an answer to each of the anomalies in his empirical data. Regarding promotion, he suggested that soldiers judged the fairness of the promotion system based on their own experiences relative to others around them. In the Military Police, where promotions were few and slow, few soldiers knew of a less-qualified buddy who had been promoted faster than they had. In the Army Air Corps, however, the rapid promotion rate meant that many soldiers knew of less-qualified buddies who had been promoted faster than seemed appropriate. Thus, ironically, the MPs said the promotion system was generally fair, and the air corpsmen said it was not.

A similar analysis seemed to explain the case of the African American soldiers. Rather than comparing conditions in the North with those in the South, African American soldiers compared their own status with the status of the African American civilians around them. In the South, where discrimination was at its worst, they found that being a soldier insulated them somewhat from adverse cultural norms in the surrounding community. Whereas southern African American civilians were grossly discriminated against and denied self-esteem, good jobs, and so forth; African American soldiers had a slightly better status. In the North, however, many of the African American civilians they encountered held well-paying defense jobs. And with discrimination being less severe, being a soldier did not help one's status in the community.

Finally, the concepts of reference group and relative deprivation seemed to explain the anomaly of highly educated draftees accepting their induction more willingly than those with less education did. Stouffer reasoned as follows:

1. *A person's friends, on the whole, have about the same educational status as that person does.*
2. *Draft-age men with less education are more likely to engage in semi-skilled production-line occupations and farming than more educated men.*
3. *During wartime, many production-line industries and farming are vital to the national interest; workers in those industries and farmers are exempted from the draft.*
4. *A man with little education is more likely to have friends in draft-exempt occupations than a man with more education.*
5. *When each compares himself with his friends, a less educated draftee is more likely to feel discriminated against than a draftee with more education.*

(Stouffer et al. 1949–1950: 122–27)

Stouffer's explanations unlocked the mystery of the three anomalous findings. Because they were not part of a preplanned study design, however, he lacked empirical data for testing them. Nevertheless, Stouffer's logical exposition provided the basis for the later development of the elaboration model: understanding the relationship between two variables through the controlled introduction of other variables.

Paul Lazarsfeld and his associates at Columbia University formally developed the elaboration model in 1946. In a methodological review of Stouffer's Army studies, Lazarsfeld and Patricia Kendall used the logic of the elaboration model to present hypothetical tables that would have proved Stouffer's contention regarding education and acceptance of induction had the empirical data been available (Kendall and Lazarsfeld 1950).

The central logic of the elaboration model begins with an observed relationship between two variables and the possibility that one variable may be causing the other. In the Stouffer example, the initial two variables were *educational level* and *acceptance of being drafted as fair.* Because the soldiers' educational levels were set before they were drafted (and thus having an opinion about being drafted) it would seem that *educational level* was the cause, or independent variable, and *acceptance of induction* was the effect, or dependent variable. As we

just saw, however, the observed relationship countered what the researchers had expected.

The elaboration model examines the impact of other variables on the relationship first observed. Sometimes, this analysis reveals the mechanisms through which the causal relationship occurs. Other times, an elaboration analysis disproves the existence of a causal relationship altogether.

In the present example, the additional variable was whether or not a soldier's friends were deferred or drafted. In Stouffer's speculative explanation, this variable showed how it was actually logical that soldiers with more education would be the more accepting of being drafted: because it was likely that their friends would have been drafted. Those with the least education were likely to have been in occupations that often brought deferments from the draft, leading those drafted to feel they had been treated unfairly.

Kendall and Lazarsfeld began with Stouffer's data showing the positive association between education and acceptance of induction (see Table 15-1). In this and the following tables, "should have been deferred" and "should not have been deferred" represent inductees' judgments of their own situation, with the latter group feeling it was fair for them to have been drafted.

Then, Kendall and Lazarsfeld created some hypothetical tables to represent what the analysis might have looked like had soldiers been asked whether most of their friends had been drafted or deferred. In Table 15-2, 19 percent of those

TABLE 15-1

Summary of Stouffer's Data on Education and Acceptance of Induction

	High Ed.	Low Ed.
Should not have been deferred	88%	70%
Should have been deferred	12%	30%
Total	100%	100%
	(1,761)	(1,876)

Source: Tables 15-1, 15-2, 15-3, and 15-4 are modified with permission of Macmillan Publishing Co., Inc., from *Continuities in Social Research: Studies in the Scope and Method of "The American Soldier"* by Robert K. Merton and Paul F. Lazarsfeld (eds.). 1950 by The Free Press, a Corporation, renewed 1978 by Robert K. Merton.

TABLE 15-2

Hypothetical Relationship between Education and Deferment of Friends

Friends Deferred?	High Ed.	Low Ed.
Yes	19%	79%
No	81%	21%
Total	100%	100%
	(1,761)	(1,876)

with high education hypothetically said their friends were deferred, compared with 79 percent of the soldiers with less education.

Notice that the numbers of soldiers with high and low education are the same as in Stouffer's real data. In later tables, you'll see that the numbers who accepted or resented being drafted remain true to the original data. Only the numbers saying that friends were or were not deferred were made up.

Stouffer's explanation next assumed that soldiers with friends who had been deferred would be more likely to resent their own induction than those would who had no deferred friends. Table 15-3 presents the hypothetical data that would have supported that assumption.

The hypothetical data in Tables 15-2 and 15-3 would confirm linkages that Stouffer had specified in his explanation. First, soldiers with low education were more likely to have friends who were deferred than were soldiers with more education. Second, having friends who were deferred made a soldier more likely to think he should have been deferred. Stouffer had suggested that these two relationships would clarify the original

relationship between education and acceptance of induction. Kendall and Lazarsfeld created a hypothetical table that would confirm Stouffer's explanation (see Table 15-4).

Recall that the original finding was that draftees with high education were more likely to accept their induction into the army as fair than were those with less education. In Table 15-4, however, we note that level of education has no effect on the acceptance of induction among those who report having friends deferred: 63 percent among both educational groups indicate that they accept their induction (that is, they say they should not have been deferred). Similarly, educational level has no significant effect on acceptance of induction among those who reported having no friends deferred: 94 and 95 percent say they should not have been deferred.

On the other hand, among those with high education the acceptance of induction is strongly related to whether or not friends were deferred: 63 percent versus 94 percent. And the same is true among those with less education. The hypothetical data in Table 15-4, then, would support Stouffer's contention that education affected acceptance of induction only through the medium of having friends deferred. Highly educated draftees were less likely to have friends deferred and, by virtue of that fact, were more likely to accept their own induction as fair. Those with less education were more likely to have friends deferred and, by virtue of that fact, were less likely to accept their own induction.

TABLE 15-3

Hypothetical Relationship between Deferment of Friends and Acceptance of One's Own Induction

	Friends Deferred?	
	Yes	No
Should not have been deferred	63%	94%
Should have been deferred	37%	6%
Total	100%	100%
	(1,819)	(1,818)

TABLE 15-4

Hypothetical Data Relating Education to Acceptance of Induction through the Factor of Having Friends Who Were Deferred

	Friends Deferred		No Friends Deferred	
	High Ed.	Low Ed.	High Ed.	Low Ed.
Should not have been deferred	63%	63%	94%	95%
Should have been deferred	37%	37%	6%	5%
Total	100%	100%	100%	100%
	(335)	(1,484)	(1,426)	(392)

Recognize that neither Stouffer's explanation nor the hypothetical data denied the reality of the original relationship. As educational level increased, acceptance of one's own induction also increased. The nature of this empirical relationship, however, was interpreted through the introduction of a third variable. The variable, *deferment of friends,* did not deny the original relationship; it merely clarified the mechanism through which the original relationship occurred.

This, then, is the heart of the elaboration model and of multivariate analysis. Having observed an empirical relationship between two variables (such as *level of education* and *acceptance of induction*), we seek to understand the nature of that relationship through the effects produced by introducing other variables (such as *having friends who were deferred*). Mechanically, we accomplish this by first dividing our sample into subsets on the basis of the **test variable**, also called the *control variable.* In our example, having friends deferred or not is the test variable, and the sample is divided into those who have deferred friends and those who do not. The relationship between the original two variables (*acceptance of induction* and *level of education*) is then recomputed separately for each of the subsamples. The tables produced in this manner are called the *partial tables,* and the relationships found in the partial tables are called the **partial relationships**, or *partials.* The partial relationships are then compared with the initial relationship discovered in the total sample, often referred to as the **zero-order relationship** to indicate that no test variables have been controlled for.

Although the elaboration was first demonstrated through the use of hypothetical data, it laid out a logical method for analyzing relationships among variables that have been actually measured. As we'll see, our first, hypothetical example describes only one possible outcome in the elaboration model. There are others.

The Elaboration Paradigm

This section presents guidelines for understanding an elaboration analysis. To begin, we must know whether the test variable is antecedent (prior in time) to the other two variables or whether it is intervening between them, because these positions suggest different logical relationships in the multivariate model. If the test variable is intervening, as in the case of *education, deferment of friends,* and *acceptance of induction,* then the analysis is based on the model shown in Figure 15-1. The logic of this multivariate relationship is that the independent variable (*educational level*) affects the intervening test variable (*having friends deferred or not*), which in turn affects the dependent variable (*accepting induction*).

If the test variable is antecedent to both the independent and dependent variables, a different model must be used (see Figure 15-2). Here, the test variable affects both the "independent" and "dependent" variables. Realize, of course, that the terms *independent variable* and *dependent variable* are, strictly speaking, used incorrectly in the diagram. In fact, we have one independent variable (the test variable) and two dependent variables. The incorrect terminology has been used only to provide continuity with the

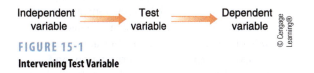

FIGURE 15-1

Intervening Test Variable

© Cengage Learning®

---

test variable A variable that is held constant in an attempt to clarify further the relationship between two other variables. Having discovered a relationship between *education* and *prejudice,* for example, we might hold *sex* constant by examining the relationship between education and prejudice among men only and then among women only. In this example, *sex* would be the test variable.

partial relationship In the elaboration model, this is the relationship between two variables when examined in a subset of cases defined by a third variable. Beginning with a zero-order relationship between *political party* and *attitudes toward abortion,* for example, we might want to see whether the relationship held true among both men and women (i.e., controlling for *sex*). The relationship found among men and the relationship found among women would be the partial relationships, sometimes simply called the partials.

zero-order relationship In the elaboration model, this is the original relationship between two variables, with no test variables controlled for.

FIGURE 15-2

Antecedent Test Variable

preceding example. Because of their individual relationships to the test variable, the "independent" and "dependent" variables are empirically related to each other, but there is no causal link between them. Their empirical relationship is merely a product of their coincidental relationships to the test variable. (Subsequent examples will further clarify this relationship.)

Table 15-5 provides a guide to understanding an elaboration analysis. The two columns in the table indicate whether the test variable is antecedent or intervening in the sense described previously. The left side of the table shows the nature of the partial relationships as compared with the original relationship between the independent and dependent variables. The body of the table gives the technical notations—replication, explanation, interpretation, and specification—assigned to each case. We'll discuss each in turn.

TABLE 15-5

The Elaboration Paradigm

Partial Relationships Compared with Original	Test Variable	
	Antecedent	Intervening
Same Relationship	Replication	Replication
Less or none	Explanation	Interpretation
Split*	Specification	Specification

*One partial is the same or greater, and the other is less or none.

> **replication** A technical term used in connection with the elaboration model, referring to the elaboration outcome in which the initially observed relationship between two variables persists when a control variable is held constant, thereby supporting the idea that the original relationship is genuine.

Replication

Whenever the partial relationships are essentially the same as the original relationship, the term **replication** is assigned to the result, regardless of whether the test variable is antecedent or intervening. This means that the original relationship has been replicated under test conditions. If, in our previous example, education still affected acceptance of induction both among those who had friends deferred and those who did not, then we would say the original relationship had been replicated. Note, however, that this finding would not confirm Stouffer's explanation of the original relationship. Having friends deferred or not would *not* be the mechanism through which education affected the acceptance of induction.

To see what a replication looks like, turn back to Tables 15-3 and 15-4. Imagine that our initial discovery was that having friends deferred strongly influenced how soldiers felt about being drafted, as shown in Table 15-3. Had we first discovered this relationship, we might have wanted to see whether it was equally true for soldiers of different educational backgrounds. To find out, we would have made *education* our control or test variable.

Table 15-4 contains the results of such an examination, though it is constructed somewhat differently from what we would have done had we used *education* as the test variable. Nevertheless, we see in the table that having friends deferred or not still influences attitudes toward being drafted among both soldiers with high education and those with low education. (Compare columns 1 and 3, then 2 and 4.) This result represents a replication of the relationship between having friends deferred and attitude toward being drafted.

Researchers frequently use the elaboration model rather routinely in the hope of replicating their findings among subsets of the sample. If we discovered a relationship between education and prejudice, for example, we might introduce such test variables as *age, region of the country, race, religion,* and so forth to test the stability of the original relationship. If the relationship were replicated among young and old, among people from different parts of the country, and so forth, we would have grounds for concluding that the original relationship was a genuine and general one.

Explanation

Explanation is the term used to describe a *spurious relationship*: an original relationship shown to be false through the introduction of a test variable. This requires two conditions: (1) The test variable must be antecedent to both the independent and dependent variables. (2) The partial relationships must be zero or significantly less than those found in the original. Several examples will illustrate this situation.

Let's look at an example we touched on in Chapter 4. There is an empirical relationship between the number of storks in different areas and the birth rates for those areas. The more storks in an area, the higher the birth rate. This empirical relationship might lead one to assume that the number of storks affects the birth rate. An antecedent test explains away this relationship, however. Rural areas have both more storks and higher birth rates than urban areas do. Within rural areas, there is no relationship between the number of storks and the birth rate; nor is there a relationship within urban areas.

Figure 15-3 illustrates how the rural/urban variable causes the apparent relationship between storks and birth rates. Part I of the figure shows the original relationship. Notice that all but one of the entries in the box for towns and cities with many storks have high birth rates and that all but one of those in the box for towns and cities with few storks have low birth rates. In percentage form, we say that 93 percent of the towns and cities with many storks also had high birth rates, contrasted with 7 percent of those with few storks. That's quite a large difference and represents a strong association between the two variables.

Part II of the figure separates the towns from the cities (the rural from urban areas) and examines storks and babies in each type of place separately. Now we can see that all the rural places have high birth rates and all the urban places have low birth rates. Also notice that only one rural place had few storks and only one urban place had lots of storks.

Here's a similar example, also mentioned in Chapter 4 and at the beginning of this chapter. There is a positive relationship between the number of fire trucks responding to a fire and

I. Birth Rates in Towns and Cities Having Few or Many Storks

II. Birth Rates in Towns and Cities Having Few or Many Storks, Controlling for Rural (Towns) and Urban (Cities)

H = Town or city with high birth rate
L = Town or city with low birth rate

FIGURE 15-3

The Facts of Life about Storks and Babies

the amount of damage done. If more trucks respond, more damage is done. One might assume from this fact that the fire trucks themselves cause the damage. However, an antecedent test variable, the *size of the fire,* explains away the original relationship. Large fires do more damage than small ones do, and more fire trucks show up at large fires than at small ones. Looking only at large fires, we would see that the original relationship vanishes (or perhaps reverses itself); and the same would be true in looking only at small fires.

Finally, let's take a real research example. Years ago, I found an empirical relationship between the region of the country in which medical school faculty members attended medical school and their attitudes toward Medicare (Babbie 1970). To simplify matters, only the East and the South will be examined.

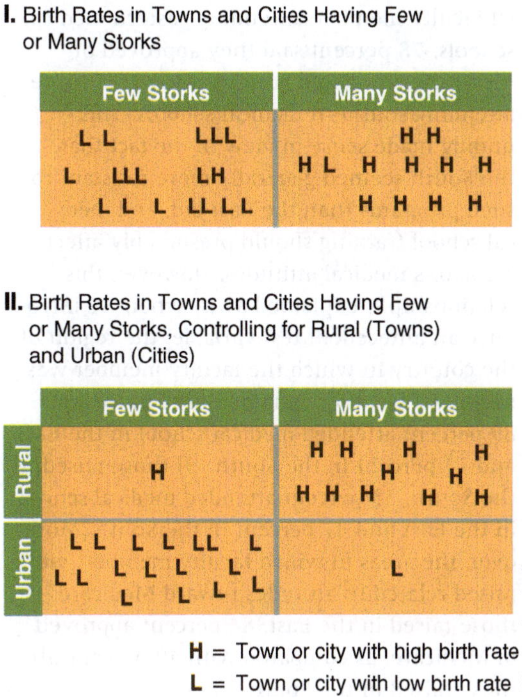

> **explanation** An elaboration model outcome in which the original relationship between two variables is revealed to have been spurious, because the relationship disappears when an antecedent test variable is introduced.

Of faculty members attending eastern medical schools, 78 percent said they approved of Medicare, compared with 59 percent of those attending southern medical schools. This finding made sense in view of the fact that the South seemed generally more resistant to such programs than the East did, and medical school training should presumably affect a doctor's medical attitudes. However, this relationship is explained away when we introduce an antecedent test variable: the region of the country in which the faculty member was raised. Of faculty members raised in the East, 89 percent attended medical school in the East and 11 percent in the South. Of those raised in the South, 53 percent attended medical school in the East and 47 percent in the South. Moreover, the areas in which faculty members were raised related to attitudes toward Medicare. Of those raised in the East, 84 percent approved of Medicare, as compared with 49 percent of those raised in the South.

Table 15-6 presents the three-variable relationship among (1) region in which raised, (2) region of medical school training, and (3) attitude toward Medicare. Faculty members raised in the East are quite likely to approve of Medicare, regardless of where they attended medical school. Those raised in the South are relatively less likely to approve of Medicare,

TABLE 15-6

Region of Origin, Region of Medical School Training, and Attitude toward Medicare

	Percent Who Approve of Medicare	
	Raised in East	Raised in South
Medical School in East	84	50
Medical School in South	80	47

Source: Earl R. Babbie, *Science and Morality in Medicine* (Berkeley: University of California Press, 1970), 181.

interpretation A technical term used in connection with the elaboration model. It represents the research outcome in which a control variable is discovered to be the mediating factor through which an independent variable has its effect on a dependent variable.

but, again, the region of their medical school training has little or no effect. These data indicate, therefore, that the original relationship between region of medical training and attitude toward Medicare was spurious; it was due only to the coincidental effect of region of origin on both region of medical training and attitude toward Medicare. When region of origin is held constant, as in Table 15-6, the original relationship disappears in the partials.

In "Applying Concepts in Everyday Life: Attending an Ivy League College and Success in Later Professional Life," Patricia Kendall, one of the founders of the elaboration model, recalls a study in which the researcher suspected an explanation but found a replication. Though the data are no longer current, the topic is still of vital interest to students: To what extent does your professional success depend on attending the "right" school?

Interpretation

Interpretation is similar to explanation, except for the time placement of the test variable and the implications that follow from that difference. **Interpretation** represents the research outcome in which a test or control variable is discovered to be the mediating factor through which an independent variable has its effect on a dependent variable. The earlier example of education, friends deferred, and acceptance of induction is an excellent illustration of interpretation. In terms of the elaboration model, the effect of education on acceptance of induction is not explained away; it is still a genuine relationship. In a real sense, educational differences cause differential acceptance of induction. The intervening variable, *deferment of friends,* merely helps to interpret the mechanism through which the relationship occurs. Thus, an interpretation does not deny the validity of the original causal relationship but simply clarifies the process through which that relationship functions.

Here's another example of interpretation. Researchers have observed that children from broken homes are more likely to become delinquent than those from intact homes are. This relationship may be interpreted, however, through the introduction of *supervision* as a test variable. Among children who are supervised, delinquency rates are not affected by whether or not their parents are divorced. The same is true among those who are not supervised. It is the relationship between broken homes and the lack of supervision that produced the original relationship.

Specification

Sometimes the elaboration model produces partial relationships that differ significantly from each other. For example, one partial relationship is the same as or stronger than the original two-variable relationship, and the second partial relationship is less than the original and may be reduced to zero. In the elaboration paradigm, this situation is referred to as **specification**: We have specified the conditions under which the original relationship occurs.

Now recall the study, cited earlier in this book, of the sources of religious involvement (Glock, Ringer, and Babbie 1967: 92). It was discovered that among Episcopal church members, involvement decreased as social class increased. This finding is reported in Table 15-7, which examines mean levels of church involvement among women parishioners at different levels of social class.

Glock interpreted this finding in the context of others in the analysis and concluded that church involvement provides an alternative form of gratification for people who are denied gratification in the secular society. This conclusion explained why women were more religious than men, why old people were more religious than young people, and so forth. Glock reasoned that people of lower social class (measured by income and education) had fewer chances to gain self-esteem from the secular society than people of higher social class did. To illustrate this idea, he noted that social class was strongly related to the likelihood that a woman had ever held an office in a secular organization (see Table 15-8).

Glock then reasoned that if social class were related to church involvement only by virtue of the fact that lower-class women would be denied opportunities for gratification in the secular society, the original relationship should not hold among women who were getting gratification. As a rough indicator of the receipt of gratification from the secular society, he used as a variable the holding of secular office. In this test, social class should be unrelated to church involvement among those who had held such an office.

Table 15-9 presents an example of a specification. Among women who have held office in secular organizations, there is essentially no relationship between social class and church involvement. In effect, the table specifies the conditions under which the original relationship holds: among the women lacking gratification in the secular society.

TABLE 15-7
Social Class and Mean Church Involvement among Episcopal Women

| | Social Class Levels | | | | |
	Low 0	1	2	3	High 4
Mean involvement	0.63	0.58	0.49	0.48	0.45

Note: Mean scores rather than percentages have been used here.

Source: Tables 15-7, 15-8, and 15-9 are from Charles Y. Glock, Benjamin B. Ringer, and Earl R. Babbie, *To Comfort and to Challenge* (Berkeley: University of California Press, 1967). Regents of the University of California.

TABLE 15-8
Social Class and the Holding of Office in Secular Organizations

| | Social Class Levels | | | | |
	Low 0	1	2	3	High 4
Percent who have held office in a secular organization	46	47	54	60	83

specification A technical term used in connection with the elaboration model, representing the elaboration outcome in which an initially observed relationship between two variables is replicated among some subgroups created by the control variable but not among others. In such a situation, you will have specified the conditions under which the original relationship exists: for example, among men but not among women.

Attending an Ivy League College and Success in Later Professional Life

by Patricia L. Kendall

Department of Sociology, Queens College, CUNY

Probably the main danger for survey analysts is that a relationship they hope is causal will turn out to be spurious. That is, the original relationship between *X* and *Y* is explained by an antecedent test factor. More specifically, the partial relationships between *X* and *Y* reduce to 0 when that antecedent test factor is held constant.

This was a distinct possibility in a major finding from a study carried out several decades ago. One of my fellow graduate students at Columbia University, Patricia Salter West, based her dissertation on questionnaires obtained by *Time Magazine* from 10,000 of its male subscribers. Among many of the hypotheses developed by West was that male graduates of Ivy League schools (Brown, Columbia, Cornell, Dartmouth, Harvard, University of Pennsylvania, Princeton, and Yale) were more successful in their later professional careers, as defined by their annual earnings, than those who graduated from other colleges and universities.

The initial fourfold table (Table 1) supported West's expectation. Although I made up the figures, they conform closely to what West actually found in her study. Having attended an Ivy League school seems to lead to considerably greater professional success than does being a graduate of some other kind of college or university.

But wait a minute. Isn't this a relationship that typically could be spurious? Who can afford to send their sons to Ivy League schools? Wealthy families, of course.[†] And who can provide the business and professional connections that could help sons become successful in their careers? Again, wealthy or well-to-do families.

In other words, the socioeconomic status of the student's family may explain away the apparent causal relationship. In fact, some of West's findings suggest that this might indeed be the case.

TABLE 1*

Later Professional success (Y)	College Attended (X)	
	Ivy League College	Other College or University
Successful	1,300 (65%)	2,000 (25%)
Unsuccessful	1,700 (35%)	6,000 (75%)
Total	3,000 (100%)	8,000 (100%)

*I have had to invent relevant figures because the only published version of West's study contained no totals. See Ernest Havemann and Patricia Salter West, *They Went to College* (New York: Harcourt, Brace, 1952).

TABLE 2

Attendance at Ivy League Colleges According to Family Socioeconomic Status (SES)

College Attended (X)	Family SES (T)	
	High SES	Low SES
Ivy League colleges	1,500 (33%)	500 (9%)
Other colleges and universities	3,000 (67%)	5,000 (91%)
Total	4,500 (100%)	5,500 (100%)

According to Table 2, a third of those coming from families defined as wealthy, compared with 1 in 11 coming from less well-to-do backgrounds, attended Ivy League colleges. Thus, there is a very high correlation between the two variables, *X* and *T.* (There is a similarly high correlation between family socioeconomic status [*T*] and later professional success [*Y*].)

TABLE 15-9

Church Involvement by Social Class and Holding Secular Office

	Mean Church Involvement for Social Class Levels				
	Low 0	1	2	3	High 4
Have held office	0.46	0.53	0.46	0.46	0.46
Have not held office	0.62	0.55	0.47	0.46	0.40

The term *specification* is used in the elaboration paradigm regardless of whether the test variable is antecedent or intervening. In either case,

the meaning is the same. We have specified the particular conditions under which the original relationship holds.

Refinements to the Paradigm

The preceding sections have presented the primary logic of the elaboration model as developed by Lazarsfeld and his colleagues. Here, we look at some logically possible variations, some of which can be found in a book by Morris Rosenberg (1968).

First, the basic paradigm assumes an initial relationship between two variables. It might be useful, however, in a more comprehensive

TABLE 3

Partial Relationships between *X* and *Y* with *T* Held Constant

	High Family SES (T)		Low Family SES (T)	
Later Success (Y)	Ivy League College (X)	Other College (X)	Ivy League College (X)	Other College
Successful	1,000 (67%)	1,000 (33%)	300 (60%)	1,000 (20%)
Not successful	500 (33%)	2,000 (67%)	200 (40%)	4,000 (80%)
Total	1,500 (100%)	3,000 (100%)	500 (100%)	5,000 (100%)

The magnitude of these so-called marginal correlations suggest that West's hypothesis regarding the causal nature of having attended an Ivy League college might be incorrect; it suggests instead that the socioeconomic status of the students' families accounted for the original relationship she observed.

We are not done yet, however. The crucial question is what happens to the partial relationships once the test factor is controlled. These are shown in Table 3.

These partial relationships show that, even when family socioeconomic status is held constant, there is still a marked relationship between having attended an Ivy League college and success in later professional life. As a result, West's initial hypothesis received support from the analysis she carried out.

Despite this, West had in no way proved her hypothesis. There are almost always additional antecedent factors that might explain the original relationship. Consider, for example, the intelligence of the students

(as measured by IQ tests or SAT scores). Ivy League colleges pride themselves on the excellence of their student bodies. They may therefore be willing to award merit scholarships to students with exceptional qualifications but not enough money to pay tuition and board. Once admitted to these prestigious colleges, bright students may develop the skills—and connections—that will lead to later professional success. Since West had no data on the intelligence of the men she studied, she was unable to study whether the partial relationships disappeared once this test factor was introduced.

In sum, the elaboration paradigm permits the investigator to rule out certain possibilities and to gain support for others. It does not permit us to prove anything.

†Since she had no direct data on family socioeconomic status, West defined as wealthy or having high socioeconomic status those who supported their sons completely during all four years of college. She defined as less wealthy or having low socioeconomic status those whose sons worked their way through college, in part or totally

model to differentiate between positive and negative relationships. Moreover, Rosenberg suggests using the elaboration model even with an original relationship of zero. He cites as an example a study of union membership and attitudes toward having Jews on the union staff (see Table 15-10). The initial analysis indicated that length of union membership did not relate to the attitude: Those who had belonged to the union for less than four years were just as willing to accept Jews on the staff as were those who had belonged for more than four years. The *age* of union members, however, was found to *suppress* the relationship between length of union membership and attitude toward Jews.

Overall, younger members were more favorable toward Jews than older members were. At the same time, of course, younger members were not likely to have been in the union as long as the old members. Within specific age groups, however, those in the union longest were the most supportive of having Jews on the staff. Age, in this case, was a **suppressor variable**, concealing the relationship between length of membership and attitude toward Jews.

suppressor variable In the elaboration model, a test variable that prevents a genuine relationship from appearing at the zero-order level.

TABLE 15-10

Example of a Suppressor Variable

I: No Apparent Relationship between Attitudes toward Jews and Length of Time in the Union

	Length of Time in the Union	
	Less than four years	*Four years or more*
Percent who don't care if there are Jews on the union staff	49.2	50.5
	(126)	(256)

II: In Each Age Group, Length of Time in Union Increases Willingness to Have Jews on Union Staff

		Length of Time in the Union	
		Less than four years	*Four years or more*
Percent who don't care if there are Jews on the union staff, by age	29 years or under	56.4	62.7
	100% =	(78)	(51)
	30–49 years	37.1	48.3
	100% =	(35)	(116)
	50 years or older	38.4	56.1
	100% =	(13)	(89)

Source: Adapted from Morris Rosenberg, *The Logic of Survey Analysis* (New York: Basic Books, 1968), 88–89.

Second, the basic paradigm focuses on partials being the same as or weaker than the original relationship but does not provide guidelines for specifying what constitutes a significant difference between the original and the partials. When you use the elaboration model, you'll frequently find yourself making an arbitrary decision about whether a given partial is significantly weaker than the original. This, then, suggests another dimension that could be added to the paradigm.

Third, the limitation of the basic paradigm to partials that are the same as or weaker than the original neglects two other possibilities. A partial relationship might be stronger than the original. Or, on the other hand, a partial relationship might be the reverse of the original—for example, negative whereas the original was positive.

Rosenberg provides a hypothetical example of the latter possibility by first suggesting that

---

distorter variable In the elaboration model, a test variable that reverses the direction of a zero-order relationship.

---

a researcher might find that working-class respondents in his study are more supportive of the civil rights movement than middle-class respondents are (see Table 15-11). He further suggests that race might be a **distorter variable** in this instance, reversing the true relationship between class and attitudes. Presumably, African American respondents would be more supportive of the movement than whites would, but African Americans would also be overrepresented among working-class respondents and underrepresented among the middle class. Middle-class African American respondents might be more supportive than working-class African Americans, however; and the same relationship might be found among whites. Holding race constant, then, the researcher would conclude that support for the civil rights movement was greater among the middle class than among the working class.

Here's another example of a distorter variable at work. When Michel de Seve set out to examine the starting salaries of men and

TABLE 15-11

Example of a Distorter Variable (Hypothetical)

I: Working-Class Subjects Appear More Liberal on Civil Rights than Middle-Class Subjects

	Civil Rights Score	
	Middle Class	Working Class
High	37%	45%
Low	63%	55%
	100%	100%
100% =	(120)	(120)

II: Controlling for Race Shows the Middle Class to Be More Liberal than the Working Class

	Race and Social Class			
	Blacks		Whites	
Civil Rights Score	Middle Class	Working Class	Middle Class	Working Class
High	70%	50%	30%	20%
Low	30%	50%	70%	80%
	100%	100%	100%	100%
100% =	(20)	(100)	(100)	(20)

Source: Morris Rosenberg, *The Logic of Survey Analysis* (New York: Basic Books, 1968), 94–95.

women in the same organization, she was surprised to find the women were receiving higher starting salaries, on the average, than their male counterparts were. The distorter variable was *time of first hire.* Many of the women had been hired relatively recently, when salaries were higher overall than in the earlier years when many of the men had been hired (reported in E. Cook 1995).

All these new dimensions further complicate the notion of specification. If one partial is the same as the original, and the other partial is even stronger, how should you react to that situation? You've specified one condition under which the original relationship holds up, but you've also specified another condition under which it holds even more clearly.

Finally, the basic paradigm focuses primarily on dichotomous test variables. In fact, the elaboration model is not so limited—either in theory or in use—but the basic paradigm becomes more complicated when the test variable divides the sample into three or more subsamples. And the paradigm becomes more complicated yet when more than one test variable is used simultaneously.

I'm not saying all this to fault the basic elaboration paradigm. To the contrary, I want to emphasize that the elaboration model is not a simple algorithm—a set of procedures through which to analyze research. Rather, it's primarily a logical device for assisting the researcher in understanding his or her data. A firm understanding of the elaboration model will make a sophisticated analysis easier. However, this model suggests neither which variables should be introduced as controls nor definitive conclusions about the nature of elaboration results. For all these things, you must look to your own ingenuity. Such ingenuity, moreover, will come only through extensive experience. By pointing to oversimplifications in the basic elaboration paradigm, I've sought to bring home the point that the model provides only a logical framework. You'll find sophisticated analyses far more complicated than the examples I've used to illustrate the basic paradigm.

At the same time, if you fully understand the basic model, you'll understand other techniques such as correlations, regressions, and factor analyses a lot more easily. Chapter 16 places techniques such as partial correlations and partial regressions in the context of the elaboration model.

Elaboration and Ex Post Facto Hypothesizing

Before we leave the discussion of the elaboration model, we should look at it in connection with a form of fallacious reasoning called **ex post facto hypothesizing**. Although the social science literature presents a host of references warning against it, inexperienced researchers can sometimes be confused about its implications.

ex post facto hypothesis A hypothesis created after confirming data have already been collected. It is a meaningless construct because there is no way for it to be disconfirmed.

"Ex post facto" means "after the fact." When you observe an empirical relationship between two variables and then simply suggest a reason for that relationship, that is sometimes called ex post facto hypothesizing. You've generated a hypothesis linking two variables after their relationship is already known. You'll recall, from an early discussion in this book, that all hypotheses must be subject to disconfirmation in order to be meaningful. Unless you can specify empirical findings that would *disprove* your hypothesis, it's not really a *hypothesis* as researchers use that term. You might reason, therefore, that once you've observed a relationship between two variables, any hypothesis regarding that relationship cannot be disproved.

This is a fair assessment if you're doing nothing more than dressing up your empirical observations with deceptive hypotheses after the fact. Having observed that women are more religious than men, you should not simply assert that women will be more religious than men because of some general dynamic of social behavior and then rest your case on the initial observation.

The unfortunate spin-off of the injunction against ex post facto hypothesizing is its inhibition of good, honest hypothesizing after the fact. Inexperienced researchers are often led to believe that they must make all their hypotheses before examining their data—even if that process means making a lot of poorly reasoned hypotheses. Furthermore, they're led to ignore any empirically observed relationships that do not confirm some prior hypothesis.

Surely, few researchers would now wish that Samuel Stouffer had hushed up his anomalous findings regarding morale among soldiers in the army. Stouffer noted peculiar empirical observations and set about hypothesizing the reasons for those findings. And his reasoning has proved invaluable to researchers ever since. The key is that his "after the fact" hypotheses could themselves be tested.

There is another, more sophisticated point to be made here, however. Anyone can generate hypotheses to explain observed empirical relationships in a body of data, but the elaboration model provides the logical tools for *testing* those hypotheses within the same body of data. A good example of this testing may be found in the earlier discussion of social class and church involvement. Glock explained the original relationship in terms of social deprivation theory. If he had stopped at that point, his comments would have been interesting but hardly persuasive. He went beyond that point, however. He noted that if the hypothesis was correct, then the relationship between social class and church involvement should disappear among the women who were receiving gratification from the secular society—those who had held office in a secular organization. This hypothesis was then subjected to an empirical test. Had the new hypothesis not been confirmed by the data, he would have been forced to reconsider.

These additional comments should further illustrate the point that data analysis is a continuing process, demanding all the ingenuity and perseverance you can muster. The image of a researcher carefully laying out hypotheses and then testing them in a ritualistic fashion results only in ritualistic research.

In case you're concerned that the strength of ex post facto proofs seems to be less than that of the traditional kinds, let me repeat the earlier assertion that "scientific proof" is a contradiction in terms. Nothing is ever *proved* scientifically. Hypotheses, explanations, theories, or hunches can all escape a stream of attempts at disproof, but none can be proved in any absolute sense. The acceptance of a hypothesis, then, is really a function of the extent to which it has been tested and not disconfirmed. No hypothesis, therefore, should be considered sound on the basis of one test—whether the hypothesis was generated before or after the observation of empirical data. With this in mind, you should not deny yourself some of the most fruitful avenues available to you in data analysis. You should always try to reach an honest understanding of your data, develop meaningful theories for more general understanding, and not worry about the manner of reaching that understanding.

What do you think?...Revisited

Earlier in this chapter, we noted the problem of knowing whether we have discovered a genuine cause, when two variables are found to be correlated with one another. For example, it is often asserted that gender is an important cause of income differences: on average, women earn less than men. This difference is supported by the data.

A counterargument is that the real cause is seniority. The data verify that because of traditional family responsibilities, the average female employee has less seniority than the average male employee. However,

when we hold seniority constant—examining groups of men and women with the same seniority—the women still earn less.

Another counterargument is that men tend to occupy higher-status occupations: men are doctors, women are nurses; men are business executives, women are secretaries. However, when we hold occupation constant, men still earn more: male doctors earn more than female doctors, etc.

The elaboration model offers a protection against identifying the wrong cause for observed variations

MAIN POINTS

Introduction

- The elaboration model is a method of multivariate analysis appropriate for social research. It is primarily a logical model that can illustrate the basic logic of other multivariate methods.

The Origins of the Elaboration Model

- Paul Lazarsfeld and Patricia Kendall used the logic of the elaboration model to present hypothetical tables regarding Samuel Stouffer's work regarding education and acceptance of induction in the Army.

- A partial relationship (or "partial") is the observed relationship between two variables within a subgroup of cases based on some attribute of the test or control variable.

- A zero-order relationship is the observed relationship between two variables without a third variable being held constant or controlled.

The Elaboration Paradigm

- The basic steps in elaboration are as follows: (1) A relationship is observed to exist between two variables, (2) a third variable (the test variable) is held constant in the sense that the cases under study are subdivided according to the attributes of that third variable, (3) the original two-variable relationship is recomputed within each of the subgroups, and (4) the comparison of the original relationship with the relationships found within each subgroup (the partial relationships) provides a fuller understanding of the original relationship itself.

- The logical relationships of the variables differ depending on whether the test variable is antecedent to the other two variables or intervening between them.

- The outcome of an elaboration analysis may be replication (whereby a set of partial relationships is essentially the same as the

corresponding zero-order relationship), explanation (whereby a set of partial relationships is reduced essentially to zero when an antecedent variable is held constant), interpretation (whereby a set of partial relationships is reduced essentially to zero when an intervening variable is held constant), or specification (whereby one partial relationship is reduced, ideally to zero, and the other remains about the same as the original relationship or is stronger).

- A suppressor variable conceals the relationship between two other variables; a distorter variable causes an apparent reversal in the relationship between two other variables (from negative to positive or vice versa).

Elaboration and Ex Post Facto Hypothesizing

- Ex post facto hypothesizing, or the development of hypotheses "predicting" relationships that have already been observed, is invalid in science, because disconfirming such hypotheses is impossible. Although nothing prevents us from suggesting reasons that observed relationships may be the way they are, we should not frame those reasons in the form of "hypotheses." More important, one observed relationship and possible reasons for it may suggest hypotheses about other relationships that have not been examined. The elaboration model is an excellent logical device for this kind of unfolding analysis of data.

KEY TERMS

distorter variable	replication
elaboration model	specification
ex post facto hypothesis	suppressor variable
explanation	test variable
interpretation	zero-order relationship
partial relationship	

PROPOSING SOCIAL RESEARCH: THE ELABORATION MODEL

See the exercise for Chapter 16.

REVIEW QUESTIONS

1. Review the Stouffer–Kendall–Lazarsfeld example of education, friends deferred, and attitudes toward being drafted. Suppose they had begun with an association between friends deferred and attitudes toward being drafted, and then they had controlled for education. What conclusion would they have reached?

2. In your own words describe the elaboration logic of (a) replication, (b) interpretation, (c) explanation, and (d) specification.

3. Review "Applying Concepts in Everyday Life: Attending an Ivy League College and Success in Later Professional Life." In your own words, explain what Patricia Kendall means when she says, "Despite this [support from the analysis of partial relationships], West had in no way proved her hypothesis." What conclusions can one reasonably draw from West's study?

4. Construct hypothetical examples of suppressor and distorter variables.

5. Search the Web for a research report on the discovery of a spurious relationship. Give the Web address of the document and quote or paraphrase what was discovered.

CHAPTER 16

Social Statistics

CHAPTER OVERVIEW

Statistics allow researchers to summarize data, measure associations between variables, and draw inferences from samples to populations. Getting acquainted with a few simple statistics frequently used in social research is less painful (and less threatening to your social life) than you might believe.

alphaspirit/Shutterstock.com

Learning Objectives

After studying this chapter, you will be able to . . .

- Understand the logic and techniques of descriptive statistics.
- Understand the logic and techniques of inferential statistics.
- Learn the logic of several multivariate techniques in social research.

Introduction

It has been my experience over the years that many students are intimidated by statistics. Sometimes statistics makes them feel they're

- A few clowns short of a circus
- Dumber than a box of hair
- A few feathers short of a duck
- All foam, no beer
- Missing a few buttons on their remote control
- A few beans short of a burrito
- As screwed up as a football bat
- About as sharp as a bowling ball
- About four cents short of a nickel
- Not running on full thrusters*

Many people are intimidated by quantitative research because they feel uncomfortable with mathematics and statistics. And indeed, many research reports are filled with unspecified computations. The role of statistics in social research is often important, but it's equally important to see this role in its proper perspective.

Empirical research is first and foremost a logical rather than a mathematical operation. Mathematics is merely a convenient and efficient language for accomplishing the logical operations inherent in quantitative data analysis. *Statistics* is the applied branch of mathematics especially appropriate for a variety of research analyses. This textbook is not intended to teach you statistics or torture you with them. Rather, I want to sketch

out a logical context within which you might learn and understand statistics. There is a good chance (i.e., probability) that you will need to take a statistics course as part of your program of study, and I want the discussions of this chapter to give you a running start on that course if you do need (or want) to take it.

We'll be looking at two types of statistics: descriptive and inferential. *Descriptive statistics* is a medium for describing data in manageable forms. *Inferential statistics,* on the other hand, assists researchers in drawing conclusions from their observations; typically, this involves drawing conclusions about a population from the study of a sample drawn from it. After that discussion, I'll briefly introduce you to some of the analytic techniques you may come across in your reading of the social science literature.

Descriptive Statistics

As I've already suggested, **descriptive statistics** present quantitative descriptions in a manageable form. Sometimes we want to describe single variables, and sometimes we want to describe the associations that connect one variable with another. Let's look at some of the ways to do these things.

Data Reduction

Scientific research often involves collecting large masses of data. Suppose we surveyed 2,000 people, asking each of them 100 questions—not an unusually large study. We would then have a staggering 200,000 answers! No one could possibly read all those answers and reach any meaningful conclusion about them. Thus, much scientific analysis

> **descriptive statistics** Statistical computations describing either the characteristics of a sample or the relationship among variables in a sample. Descriptive statistics merely summarize a set of sample observations, whereas inferential statistics move beyond the description of specific observations to make inferences about the larger population from which the sample observations were drawn.

*Thanks to the many contributors to humor lists on the Internet.

What do you think?

Societies around the globe differ greatly from one another in terms of economics, religion, family, and other cultural practices. Can the same research techniques be used to study such different kinds of living experiences? Are different techniques required for the study of different societies?

> By the end of this chapter, you should be able to answer these questions. See the *What do you think? … Revisited* box toward the end of the chapter.

Earl Babbie

involves the reduction of data from unmanageable details to manageable summaries.

To begin our discussion, let's look briefly at the raw-data matrix created by a quantitative research project. Table 16-1 presents a partial data matrix. Notice that each row in the matrix represents a person (or other unit of analysis), each column represents a variable, and each cell represents the coded attribute or value a given person has on a given variable. The first column in Table 16-1 represents a person's sex. Let's say a 1 represents male and a 2 represents female. This means that persons A and B are male, person C is female, and so forth.

In the case of age, person A's "3" might mean 30–39 years old, person B's "4" might mean 40–49. However age has been coded (see Chapter 14), the code numbers shown in Table 16-1 describe each of the people represented there.

Notice that the data have already been reduced somewhat by the time a data matrix like this one has been created. If age has been coded

as suggested previously, the specific answer "33 years old" has already been assigned to the category "30–39." The people responding to our survey may have given us 60 or 70 different ages, but we've now reduced them to 6 or 7 categories.

Chapter 14 discussed some of the ways of further summarizing univariate data: averages such as the mode, median, and mean and measures of dispersion such as the range, the standard deviation, and so forth. It's also possible to summarize the associations among variables.

Measures of Association

The association between any two variables can also be represented by a data matrix, this time produced by the joint frequency distributions of the two variables. Table 16-2 presents such a matrix. It provides all the information needed to determine the nature and extent of the relationship between education and prejudice.

TABLE 16-1
Partial Raw-Data Matrix

	Sex	Age	Education	Income	Occupation	Political Affiliation	Political Orientation	Religious Affiliation	Importance of Religion
Person A	1	3	2	4	1	2	3	0	4
Person B	1	4	2	4	4	1	1	1	2
Person C	2	2	5	5	2	2	4	2	3
Person D	1	5	4	4	3	2	2	2	4
Person E	2	3	7	8	6	1	1	5	1
Person F	2	1	3	3	5	3	5	1	1

TABLE 16-2

Hypothetical Raw Data on Education and Prejudice

		Educational Level			
Prejudice	None	Grade School	High School	College	Graduate Degree
High	23	34	156	67	16
Medium	11	21	123	102	23
Low	6	12	95	164	77

Notice, for example, that 23 people (1) have no education and (2) scored high on prejudice; 77 people (1) had graduate degrees and (2) scored low on prejudice.

Like the raw-data matrix in Table 16-1, this matrix provides more information than can easily be comprehended. A careful study of the table shows that as education increases from "None" to "Graduate Degree," there is a general tendency for prejudice to decrease, but no more than a general impression is possible. For a more precise summary of the data matrix, we need one of several types of descriptive statistics. Selecting the appropriate measure depends initially on the nature of the two variables.

We'll turn now to some of the options available for summarizing the association between two variables. Each of these measures of association is based on the same model—**proportionate reduction of error (PRE)**.

To see how this model works, let's assume that I asked you to guess respondents' attributes on a given variable: for example, whether they answered yes or no to a given questionnaire item. To assist you, let's first assume you know the overall distribution of responses in the total sample—say, 60 percent said yes and 40 percent said no. You would make the fewest errors in this process if you always guessed the modal (most frequent) response: yes.

---

proportionate reduction of error (PRE) A logical model for assessing the strength of a relationship by asking how much knowing values of one variable would reduce our errors in guessing values of the other. For example, if we know how much education people have, we can improve our ability to estimate how much they earn, thus indicating that there is a relationship between the two variables.

Second, let's assume you also know the empirical relationship between the first variable and some other variable: say, *gender*. Now, each time I ask you to guess whether a respondent said yes or no, I'll tell you whether the respondent is a man or a woman. If the two variables are related, you should make fewer errors the second time. It's possible, therefore, to compute the PRE by knowing the relationship between the two variables: the greater the relationship, the greater the reduction of error.

This basic PRE model is modified slightly to take account of different levels of measurement—nominal, ordinal, or interval. The following sections will consider each level of measurement and present one measure of association appropriate for each. Bear in mind that the three measures discussed are only an arbitrary selection from among many appropriate measures.

Nominal Variables

If the two variables consist of nominal data (for example, gender, religious affiliation, race), lambda (λ) would be one appropriate measure. (Lambda is a letter in the Greek alphabet corresponding to *l* in our alphabet. Greek letters are used for many concepts in statistics, which perhaps helps to account for the number of people who say of statistics, "It's all Greek to me.") Lambda is based on your ability to guess values on one of the variables: the PRE achieved through knowledge of values on the other variable.

Imagine this situation: I tell you that a room contains 100 people and I would like you to guess the gender of each person, one at a time. If half are men and half women, you'll probably be right half the time and wrong half the time.

But suppose I tell you each person's occupation before you guess that person's sex. What sex would you guess if I said the person was a truck driver? You would probably be wise to guess "male"; although there are now plenty of women truck drivers, most are still men. If I said the next person was a nurse, you'd probably be wisest to guess "female," following the same logic. Although you would still make errors in guessing "sexes," you would clearly do better than you would if you didn't know their occupations. The extent to which you did better (the proportionate reduction of error) would be an indicator of the association that exists between sex and occupation.

Here's another simple hypothetical example that illustrates the logic and method of lambda. Table 16-3 presents hypothetical data relating sex to employment status. Overall, we note that 1,100 people are employed and 900 are not employed. If you were to predict whether people were employed, and if you knew only the overall distribution on that variable, you would always predict "employed," because that would result in fewer errors than always predicting "not employed." Nevertheless, this strategy would result in 900 errors out of 2,000 predictions.

Let's suppose that you had access to the data in Table 16-3 and that you were told each person's sex before making your prediction of employment status. Your strategy would change in that case. For every man you would predict "employed," and for every woman you would predict "not employed." In this instance, you would make 300 errors—the 100 men who were not employed and the 200 employed women—or 600 fewer errors than you would make without knowing the person's sex.

Lambda, then, represents the reduction in errors as a proportion of the errors that would have been made on the basis of the overall distribution. In this hypothetical example, lambda would equal 0.67; that is, 600 fewer errors divided by the 900 total errors based on employment status alone. In this fashion, lambda measures the statistical association between sex and employment status.

If sex and employment status were statistically independent, we would find the same distribution of employment status for men and women. In this case, knowing each person's sex would not affect the number of errors made in predicting employment status, and the resulting lambda would be zero. If, on the other hand, all men were employed and none of the women were employed, by knowing sex you would avoid errors in predicting employment status. You would make 900 fewer errors (out of 900), so lambda would be 1.0—representing a perfect statistical association.

Lambda is only one of several measures of association appropriate for the analysis of two nominal variables. You could look at any statistics textbook for a discussion of other appropriate measures.

Ordinal Variables

If the variables being related are ordinal (for example, social class, religiosity, alienation), gamma (γ) is one appropriate measure of association. Like lambda, gamma is based on our ability to guess values on one variable by knowing values on another. However, whereas lambda is based on guessing exact values, gamma is based on guessing the ordinal arrangement of values. For any given pair of cases, we guess that their ordinal ranking on one variable will correspond (positively or negatively) to their ordinal ranking on the other.

Let's say we have a group of elementary school students. It's reasonable to assume that there is a relationship between their ages and their heights. We can test this by comparing every pair of students: Brett and Sophia, Brett and Terrell, Sophia and Terrell, and so forth. Then we ignore all the pairs in which the students are the same age and/or the same height. We then classify each of the remaining pairs (those who differ in both age and height) into one of two categories: those in which the older child is also the taller ("same" pairs) and those in which the older child is the shorter ("opposite" pairs). So, if Brett is older and taller than Sophia, the Brett–Sophia pair is counted as a "same." If Brett is older but shorter than Sophia, then that pair is an "opposite."

To determine whether age and height are related to each other, we compare the number of same and opposite pairs. If the same pairs outnumber the opposite pairs, we can conclude that there is a positive association between the two variables—as one increases, the other increases. If there are more opposites than sames, we can conclude that the relationship is negative. If there are about as many sames as opposites, we can conclude

TABLE 16-3

Hypothetical Data Relating Sex to Employment Status

	Men	Women	Total
Employed	900	200	1,100
Unemployed	100	800	900
Total	1,000	1,000	2,000

that age and height are not related to each another, that they're independent of one another.

Here's a social science example to illustrate the simple calculations involved in gamma. Let's say you suspect that religiosity is positively related to political conservatism, and if Person A is more religious than Person B, you guess that A is also more conservative than B. Gamma is the proportion of paired comparisons that fits this pattern.

Table 16-4 presents hypothetical data relating social class to prejudice. The general nature of the relationship between these two variables is that as social class increases, prejudice decreases. There is a negative association between social class and prejudice.

Gamma is computed from two quantities: (1) the number of pairs having the same ranking on the two variables and (2) the number of pairs having the opposite ranking on the two variables. The pairs having the same ranking are computed as follows: The frequency of each cell in the table is multiplied by the sum of all cells appearing below and to the right of it—with all these products being summed. In Table 16-4, the number of pairs with the same ranking would be 200(900 + 300 + 400 + 100) + 500(300 + 100) + 400(400 + 100) + 900(100), or 340,000 + 200,000 + 200,000 + 90,000 = 830,000.

The pairs having the opposite ranking on the two variables are computed as follows: The frequency of each cell in the table is multiplied by the sum of all cells appearing below and to the left of it—with all these products being summed. In Table 16-4, the numbers of pairs with opposite rankings would be 700(500 + 800 + 900 + 300) + 400(800 + 300) + 400(500 + 800) + 900(800), or 1,750,000 + 440,000 + 520,000 + 720,000 = 3,430,000. Gamma is computed from the numbers of same-ranked pairs and opposite-ranked pairs as follows:

$$gamma = \frac{same - opposite}{same + opposite}$$

TABLE 16-4

Hypothetical Data Relating Social Class to Prejudice

Prejudice	Lower Class	Middle Class	Upper Class
Low	200	400	700
Medium	500	900	400
High	800	300	100

In our example, gamma equals (830,000 − 3,430,000) divided by (830,000 + 3,430,000), or −0.61. The negative sign in this answer indicates the negative association suggested by the initial inspection of the table. Social class and prejudice, in this hypothetical example, are negatively associated with one another. The numerical figure for gamma indicates that 61 percent more of the pairs examined had the opposite ranking than had the same ranking.

Note that whereas values of lambda vary from 0 to 1, values of gamma vary from −1 to +1, representing the direction as well as the magnitude of the association. Because nominal variables have no ordinal structure, it makes no sense to speak of the direction of the relationship. (A negative lambda would indicate that you made more errors in predicting values on one variable while knowing values on the second than you made in ignorance of the second, and that's not logically possible.)

Table 16-5 is an example of the use of gamma in social research. To study the extent to which widows sanctified their deceased husbands, Helena Lopata (1981) administered a questionnaire to a probability sample of 301 widows. In part, the questionnaire asked the respondents to characterize their deceased husbands in terms of the following semantic differentiation scale:

	Characteristic							
Positive Extreme								Negative Extreme
Good	1	2	3	4	5	6	7	Bad
Useful	1	2	3	4	5	6	7	Useless
Honest	1	2	3	4	5	6	7	Dishonest
Superior	1	2	3	4	5	6	7	Inferior
Kind	1	2	3	4	5	6	7	Cruel
Friendly	1	2	3	4	5	6	7	Unfriendly
Warm	1	2	3	4	5	6	7	Cold

Respondents were asked to describe their deceased spouses by circling a number for each pair of opposing characteristics. Notice that the series of numbers connecting each pair of characteristics is an ordinal measure.

Next, Lopata wanted to discover the extent to which the several measures were related to one another. Appropriately, she chose gamma

TABLE 16-5

Gamma Associations among the Semantic Differentiation Items of the Sanctification Scale

	Useful	Honest	Superior	Kind	Friendly	Warm
Good	0.79	0.88	0.80	0.90	0.79	0.83
Useful		0.84	0.71	0.77	0.68	0.72
Honest			0.83	0.89	0.79	0.82
Superior				0.78	0.60	0.73
Kind					0.88	0.90
Friendly						0.90

Source: Helena Znaniecki Lopata, "Widowhood and Husband Sanctification," *Journal of Marriage and the Family* (May 1981): 439–50.

as the measure of association. Table 16-5 shows how she presented the results of her investigation.

The format presented in Table 16-5 is called a *correlation matrix*. For each pair of measures, Lopata has calculated the gamma. Good and Useful, for example, are related to each other by a gamma equal to 0.79. The matrix is a convenient way of presenting the intercorrelations among several variables, and you'll find it frequently in the research literature. In this case, we see that all the variables are quite strongly related to one another, though some pairs are more strongly related than others.

Gamma is only one of several measures of association appropriate for ordinal variables. Again, any introductory statistics textbook will give you a more comprehensive treatment of this subject.

Interval or Ratio Variables

If interval or ratio variables (for example, *age, income, grade point average,* and so forth) are being associated, one appropriate measure of association is Pearson's product-moment correlation (*r*). The derivation and computation of this measure of association are complex enough to lie outside the scope of this book, so I'll make only a few general comments here.

Like both gamma and lambda, *r* is based on guessing the value of one variable by knowing another. For continuous interval or ratio variables, however, it's unlikely that you could predict the precise value of the variable. On the

other hand, predicting only the ordinal arrangement of values on the two variables would not take advantage of the greater amount of information conveyed by an interval or ratio variable. In a sense, *r* reflects how closely you can guess the value of one variable through your knowledge of the value of another.

To understand the logic of *r*, consider the way you might hypothetically guess values that particular cases have on a given variable. With nominal variables, we've seen that you might always guess the modal value. But for interval or ratio data, you would minimize your errors by always guessing the mean value of the variable. Although this practice produces few if any perfect guesses, the extent of your errors will be minimized. Imagine the task of guessing peoples' incomes and how much better you would do if you knew how many years of education they had as well as the mean incomes for people with 0, 1, 2 (and so forth) years of education.

In the computation of lambda, we noted the number of errors produced by always guessing the modal value. In the case of *r*, errors are measured in terms of the sum of the squared differences between the actual value and the mean. This sum is called the *total variation*.

To understand this concept, we must expand the scope of our examination. Let's look at the logic of regression analysis and discuss correlation within that context.

Regression Analysis

The general formula for describing the association between two variables is $Y = f(X)$. This formula is read "*Y* is a function of *X*," meaning that values of *Y* can be explained in terms of variations in the values of *X*. Stated more strongly, we might say that *X* causes *Y*, so the value of *X* determines the value of *Y*. **Regression analysis** is a method of determining the specific function relating *Y* to *X*. There are several forms of regression analysis, depending on the complexity of the relationships being studied. Let's begin with the simplest.

> **regression analysis** A method of data analysis in which the relationships among variables are represented in the form of an equation, called a regression equation.

Linear Regression

The regression model can be seen most clearly in the case of a **linear regression analysis**, in which a perfect linear association between two variables exists or is approximated. Figure 16-1 is a scattergram presenting in graphic form the values of X and Y as produced by a hypothetical study. It shows that for the four cases in our study, the values of X and Y are identical in each instance. The case with a value of 1 on X also has a value of 1 on Y, and so forth. The relationship between the two variables in this instance is described by the equation $Y = X$; this is called the *regression equation*. Because all four points lie on a straight line, we could superimpose that line over the points; this is the *regression line*.

The linear regression model has important descriptive uses. The regression line offers a graphic picture of the association between X and Y, and the regression equation is an efficient form for summarizing that association. The regression model has inferential value as well. To the extent that the regression equation correctly describes the general association between the two variables, it may be used to predict other sets of values. If, for example, we know that a new case has a value of 3.5 on X, we can predict the value of 3.5 on Y as well.

In practice, of course, studies are seldom limited to four cases, and the associations between variables are seldom as clear as the one presented in Figure 16-1.

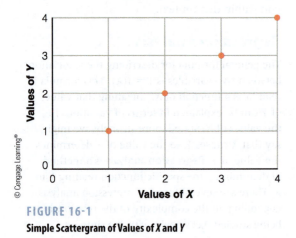

FIGURE 16-1

Simple Scattergram of Values of X and Y

© Cengage Learning®

---

linear regression analysis A form of statistical analysis that seeks the equation for the straight line that best describes the relationship between two ratio variables.

A somewhat more realistic example is presented in Figure 16-2, which represents a hypothetical relationship between population and crime rate in small- to medium-size cities. Each dot in the scattergram is a city, and its placement reflects that city's population and its crime rate. As was the case in our previous example, the values of Y (crime rates) generally correspond to those of X (populations), and as values of X increase, so do values of Y. However, the association is not nearly as clear as it is in Figure 16-1.

In Figure 16-2 we can't superimpose a straight line that will pass through all the points in the scattergram. But we can draw an approximate line showing the best possible linear representation of the several points. I've drawn that line on the graph.

You may (or may not) recall from algebra that any straight line on a graph can be represented by an equation of the form $Y = a + bX$, where X and Y are values of the two variables. In this equation, a equals the value of Y when X is 0, and b represents the slope of the line. If we know the values of a and b, we can calculate an estimate of Y for every value of X.

We can now say more formally that regression analysis is a technique for establishing the regression equation representing the geometric line that comes closest to the distribution of points on a graph. The regression equation provides a mathematical *description* of the relationship between the variables, and it allows us to *infer* values of Y when we have values of X. Recalling Figure 16-2, we could estimate crime rates of cities if we knew their populations.

To improve your guessing, you construct a *regression line*, stated in the form of a regression equation that permits the estimation of values on one variable from values on the other. The general format for this equation is $Y' = a + b(X)$, where a and b are computed values, X is a given value on one variable, and Y' is the estimated value on the other. The values of a and b are computed to minimize the differences between actual values of Y and the corresponding estimates (Y') based on the known value of X. The sum of squared differences between actual and estimated values of Y is called the *unexplained variation* because it represents errors that still exist even when estimates are based on known values of X.

The *explained variation* is the difference between the total variation and the unexplained

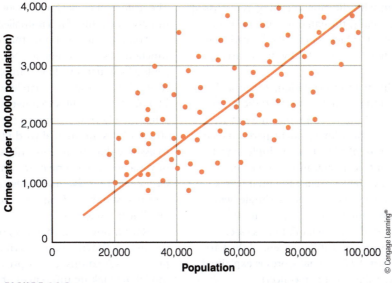

FIGURE 16-2

A Scattergram of the Values of Two Variables with Regression Line Added (Hypothetical)

variation. Dividing the explained variation by the total variation produces a measure of the *proportionate reduction of error* corresponding to the similar quantity in the computation of lambda. In the present case, this quantity is the *correlation squared*: r^2. Thus, if $r = 0.7$, then $r^2 = 0.49$, meaning that about half the variation has been explained. In practice, we compute r rather than r^2, because the product-moment correlation can take either a positive or a negative sign, depending on the direction of the relationship between the two variables. (Computing r^2 and taking a square root would always produce a positive quantity.) You can consult any standard statistics textbook for the method of computing r, although there are many data analysis programs available to do this.

Unfortunately—or perhaps fortunately—social life is so complex that the simple linear regression model often does not sufficiently represent the state of affairs. As we saw in Chapter 14, it's possible, using percentage tables, to analyze more than two variables. As the number of variables increases, such tables become increasingly complicated and hard to read. The regression model offers a useful alternative in such cases.

Multiple Regression

Very often, social researchers find that a given dependent variable is affected simultaneously by several independent variables. **Multiple regression analysis** provides a means of analyzing such

situations. This was the case when Beverly Yerg (1981) set about studying teacher effectiveness in physical education. She stated her expectations in the form of a multiple regression equation:

$$F = b_0 + b_1 I + b_2 X_1 + b_3 X_2 + b_4 X_3 + b_5 X_4 + e,$$

where

F = Final pupil-performance score

I = Initial pupil-performance score

X_1 = Composite of guiding and supporting practice

X_2 = Composite of teacher mastery of content

X_3 = Composite of providing specific, task-related feedback

X_4 = Composite of clear, concise task presentation

b = Regression weight

e = Residual

(Adapted from Yerg 1981: 42)

Notice that in place of the single X variable in a linear regression, there are several X's, and there are also several b's instead of just one. Also, Yerg has chosen to represent a as b_0 in this

multiple regression analysis A form of statistical analysis that seeks the equation representing the impact of two or more independent variables on a single dependent variable.

equation but with the same meaning as discussed previously. Finally, the equation ends with a residual factor (e), which represents the variance in Y that is not accounted for by the X variables analyzed.

Beginning with this equation, Yerg calculated the values of the several b's to show the relative contributions of the several independent variables in determining final student-performance scores. She also calculated the multiple-correlation coefficient as an indicator of the extent to which all six variables predict the final scores. This follows the same logic as the simple bivariate correlation discussed earlier, and it's traditionally reported as a capital R. In this case, $R = 0.877$, meaning that 77 percent of the variance ($0.877^2 = 0.77$) in final scores is explained by the six variables acting in concert.

Partial Regression

In exploring the elaboration model in Chapter 15, we paid special attention to the relationship between two variables when a third test variable was held constant. Thus, we might examine the effect of education on prejudice with age held constant, testing the independent effect of education. To do so, we would compute the tabular relationship between education and prejudice separately for each age group.

Partial regression analysis is based on this same logical model. The equation summarizing the relationship between variables is computed on the basis of the test variables remaining constant. As in the case of the elaboration model, the result may then be compared with the uncontrolled relationship between the two variables to clarify further the overall relationship.

Curvilinear Regression

Up to now, we've been discussing the association among variables as represented by a straight line.

> **partial regression analysis** A form of regression analysis in which the effects of one or more variables are held constant, similar to the logic of the elaboration model.
>
> **curvilinear regression analysis** A form of regression analysis that allows relationships among variables to be expressed with curved geometric lines instead of straight ones.

The regression model is even more general than our discussion thus far has implied.

You may already know that curvilinear functions, as well as linear ones, can be represented by equations. For example, the equation $X^2 + Y^2 = 25$ describes a circle with a radius of 5. Raising variables to powers greater than 1 has the effect of producing curves rather than straight lines. In the real world there is no reason to assume that the relationship among every set of variables will be linear. In some cases, then, **curvilinear regression analysis** can provide a better understanding of empirical relationships than any linear model can.

Recall, however, that a regression line serves two functions. It describes a set of empirical observations, and it provides a general model for making inferences about the relationship between two variables in the general population that the observations represent. A very complex equation might produce an erratic line that would indeed pass through every individual point. In this sense, it would perfectly describe the empirical observations. There would be no guarantee, however, that such a line could adequately predict new observations or that it in any meaningful way represented the relationship between the two variables in general. Thus, it would have little or no inferential value.

Earlier in this book, we discussed the need for balancing detail and utility in data reduction. Ultimately, researchers attempt to provide the most faithful, yet also the simplest, representation of their data. This practice also applies to regression analysis. Data should be presented in the simplest fashion that best describes the actual data; as such, linear regressions are the ones most frequently used. Curvilinear regression analysis adds a new option to the researcher in this regard, but it does not solve the problems altogether. Nothing does that.

Cautions in Regression Analysis

The use of regression analysis for statistical inferences is based on the same assumptions made for correlational analysis: simple random sampling, the absence of nonsampling errors, and continuous interval data. Because social science research seldom completely satisfies these assumptions, you should use caution in assessing the results in regression analyses.

Also, regression lines—linear or curvilinear—can be useful for *interpolation* (estimating cases lying between those observed), but they are less trustworthy when used for *extrapolation* (estimating cases that lie beyond the range of observations). This limitation on extrapolations is important in two ways. First, you're likely to come across regression equations that seem to make illogical predictions. An equation linking population and crimes, for example, might seem to suggest that small towns with, say, a population of 1,000 should have 123 crimes a year. This failure in predictive ability does not disqualify the equation but dramatizes that its applicability is limited to a particular range of population sizes. Second, researchers sometimes overstep this limitation, drawing inferences that lie outside their range of observation, and you'd be right in criticizing them for that.

The preceding sections have introduced some of the techniques for measuring associations among variables at different levels of measurement. Matters become slightly more complex when the two variables represent different levels of measurement. Though we aren't going to pursue this issue in this textbook, UCLA provides an excellent resource online at http://www.ats.ucla.edu/stat /mult_pkg/whatstat/default.htm, adapting the work of Dr. James Leeper at the University of Alabama.

Inferential Statistics

Many, if not most, social science research projects involve the examination of data collected from a sample drawn from a larger population. A sample of people may be interviewed in a survey; a sample of divorce records may be coded and analyzed; a sample of newspapers may be examined through content analysis. Researchers seldom, if ever, study samples just to describe the samples per se; in most instances, their ultimate purpose is to make assertions about the larger population from which the sample has been selected. Frequently, then, you'll wish to interpret your univariate and multivariate sample findings as the basis for inferences about some population.

This section examines **inferential statistics**—the statistical measures used for making inferences

from findings based on sample observations to a larger population. We'll begin with univariate data and move to multivariate.

Univariate Inferences

Chapter 14 dealt with methods of presenting univariate data. Each summary measure was intended as a method of describing the sample studied. Now we'll use such measures to make broader assertions about a population. This section addresses two univariate measures: percentages and means.

If 50 percent of a sample of people say they had colds during the past year, 50 percent is also our best estimate of the proportion of colds in the total population from which the sample was drawn. (This estimate assumes a simple random sample, of course.) It's rather unlikely, however, that precisely 50 percent of the population had colds during the year. If a rigorous sampling design for random selection has been followed, however, we'll be able to estimate the expected range of error when the sample finding is applied to the population.

Chapter 7, on sampling theory, covered the procedures for making such estimates, so I'll only review them here. In the case of a percentage, the quantity

$$\sqrt{\frac{p \times q}{n}}$$

where *p* is a proportion, *q* equals $(1 - p)$, and *n* is the sample size, is called the *standard error*. As noted in Chapter 7, this quantity is very important in the estimation of sampling error. We may be 68 percent confident that the population figure falls within plus or minus one standard error of the sample figure; we may be 95 percent confident that it falls within plus or minus two standard errors; and we may be 99.9 percent confident that it falls within plus or minus three standard errors.

Any statement of sampling error, then, must contain two essential components: the *confidence level* (for example, 95 percent) and the *confidence interval* (for example, ±2.5 percent). If 50 percent of a sample of 1,600 people say they had

---

inferential statistics The body of statistical computations relevant to making inferences from findings based on sample observations to some larger population.

colds during the year, we might say we're 95 percent confident that the population figure is between 47.5 percent and 52.5 percent.

In this example we've moved beyond simply describing the sample into the realm of making estimates (inferences) about the larger population. In doing so, we must take care in several ways.

First, the sample must be drawn from the population about which inferences are being made. A sample taken from a telephone directory cannot legitimately be the basis for statistical inferences about the population of a city, but only about the population of telephone subscribers with listed numbers.

Second, the inferential statistics assume several things. To begin with, they assume simple random sampling, which is virtually never the case in sample surveys. The statistics also assume sampling with replacement, which is almost never done—but this is probably not a serious problem. Although systematic sampling is used more frequently than random sampling, it, too, probably presents no serious problem if done correctly. Stratified sampling, because it improves representativeness, clearly presents no problem. Cluster sampling does present a problem, however, because the estimates of sampling error may be too small. Quite clearly, street-corner sampling does not warrant the use of inferential statistics. Finally, the calculation of standard error in sampling assumes a 100 percent completion rate—that is, that everyone in the sample completed the survey. The seriousness of this problem increases as the completion rate decreases.

Third, inferential statistics are addressed to sampling error only, not **nonsampling error** such as coding errors or misunderstandings of questions by respondents. Thus, although we might state correctly that between 47.5 and 52.5 percent of the population (95 percent confidence) would *report* having colds during the previous year, we couldn't so confidently guess the percentage who had actually *had* them. Because nonsampling errors are probably larger than sampling errors in a respectable sample design, we need to be especially cautious in generalizing from our sample findings to the population.

Tests of Statistical Significance

There is no scientific answer to the question of whether a given association between two variables is significant, strong, important, interesting, or worth reporting. Perhaps the ultimate test of significance rests in your ability to persuade your audience (present and future) of the association's significance. At the same time, there is a body of inferential statistics to assist you in this regard called *parametric tests of significance.* As the name suggests, parametric statistics are those that make certain assumptions about the parameters describing the population from which the sample is selected. They allow us to determine the **statistical significance** of associations. "Statistical significance" does not imply "importance" or "significance" in any general sense. It refers simply to the likelihood that relationships observed in a sample could be attributed to sampling error alone. Researchers often distinguish between statistical significance and *substantive significance* in this regard, with the latter referring to whether the relationship between variables is big enough to make a meaningful difference. Whereas statistical significance can be calculated, substantive significance is always a judgment call.

Although **tests of statistical significance** are widely reported in the social science literature, the logic underlying them is rather subtle and often misunderstood. Tests of significance are based on the same sampling logic discussed elsewhere in this book. To understand that logic, let's return for a moment to the concept of sampling error in regard to univariate data.

Recall that a sample statistic normally provides the best single estimate of the corresponding population parameter, but the statistic and the parameter seldom correspond precisely. Thus, we report the probability that the parameter falls within a certain range (confidence interval). The degree of uncertainty within that range is due to normal sampling error. The corollary of such a

---

nonsampling error Those imperfections of data quality that are a result of factors other than sampling error. Examples include misunderstandings of questions by respondents and erroneous recordings by interviewers and coders.

statistical significance A general term referring to the likelihood that relationships observed in a sample could be attributed to sampling error alone.

tests of statistical significance A class of statistical computations that indicate the likelihood that the relationship observed between variables in a sample can be attributed to sampling error only.

statement is, of course, that it is improbable that the parameter would fall outside the specified range *only* as a result of sampling error. Thus, if we estimate that a parameter (99.9 percent confidence) lies between 45 percent and 55 percent, we say, by implication, that it is extremely improbable that the parameter is actually, say, 90 percent if our *only* error of estimation is due to normal sampling. This is the basic logic behind tests of statistical significance.

The Logic of Statistical Significance

I think I can illustrate the logic of statistical significance best in a series of diagrams representing the selection of samples from a population. Here are the elements in the logic:

1. Assumptions regarding the independence of two variables in the population study
2. Assumptions regarding the representativeness of samples selected through conventional probability-sampling procedures
3. The observed joint distribution of sample elements in terms of the two variables

Figure 16-3 represents a hypothetical population of 256 people; half are women, half are men. The diagram also indicates how each person feels about seeing women as equal to men. In the diagram, those favoring equality have open circles, those opposing it have their circles filled in.

The question we'll be investigating is whether there is any relationship between sex and feelings about equality for men and women. More specifically, we'll see if women are more likely to favor equality than men are, because women would presumably benefit more from it. Take a moment to look at Figure 16-3 and see what the answer to this question is.

This illustration indicates no relationship between sex and attitudes about equality. Exactly half of each group favors equality and half opposes it. Recall the earlier discussion of proportionate reduction of error. In this instance, knowing a person's sex would not reduce the "errors" we'd make in guessing his or her attitude toward equality. Figure 16-3 also provides a tabular view of what you can observe in the diagram.

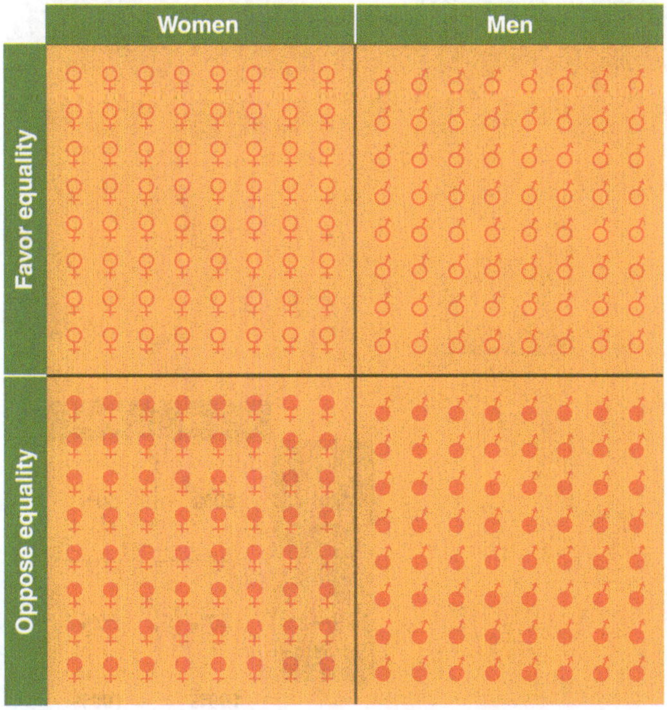

	Women	Men
Favor equality	50%	50%
Oppose equality	50%	50%
	100%	100%

Legend
♀ = Woman who favors equality
♂ = Man who favors equality
♀ = Woman who opposes equality
♂ = Man who opposes equality

© Cengage Learning®

FIGURE 16-3

A Hypothetical Population of Men and Women Who Either Favor or Oppose Sexual Equality

Figure 16-4 represents the selection of a one-fourth sample from the hypothetical population. In terms of the graphic illustration, a "square" selection from the center of the population provides a representative sample. Notice that our sample contains 16 of each type of person: Half are men and half are women; half of each sex favors equality, and the other half opposes it.

The sample selected in Figure 16-4 would allow us to draw accurate conclusions about the relationship between sex and equality in the larger population. Following the sampling logic we saw in Chapter 7, we'd note that there was no relationship between sex and equality in the sample; thus, we'd conclude that there was similarly no relationship in the larger population—because we've presumably selected a sample in accordance with the conventional rules of sampling.

Of course, real-life samples are seldom such perfect reflections of the populations from which they are drawn. It would not be unusual for us to have selected, say, one or two extra men who opposed equality and a couple of extra women who favored it—even if there was no

relationship between the two variables in the population. Such minor variations are part and parcel of probability sampling, as we saw in Chapter 7.

Figure 16-5, however, represents a sample that falls far short of the mark in reflecting the larger population. Notice that it includes far too many supportive women and opposing men. As the table shows, three-fourths of the women in the sample support equality, but only one-fourth of the men do so. If we had selected this sample from a population in which the two variables were unrelated to each other, we'd be sorely misled by our sample.

As you'll recall, it's unlikely that a properly drawn probability sample would ever be as inaccurate as the one shown in Figure 16-5. In fact, if we actually selected a sample that gave us the results this one does, we'd look for a different explanation. Figure 16-6 illustrates the more likely situation.

Notice that the sample selected in Figure 16-6 also shows a strong relationship between sex and equality. The reason is quite different this time.

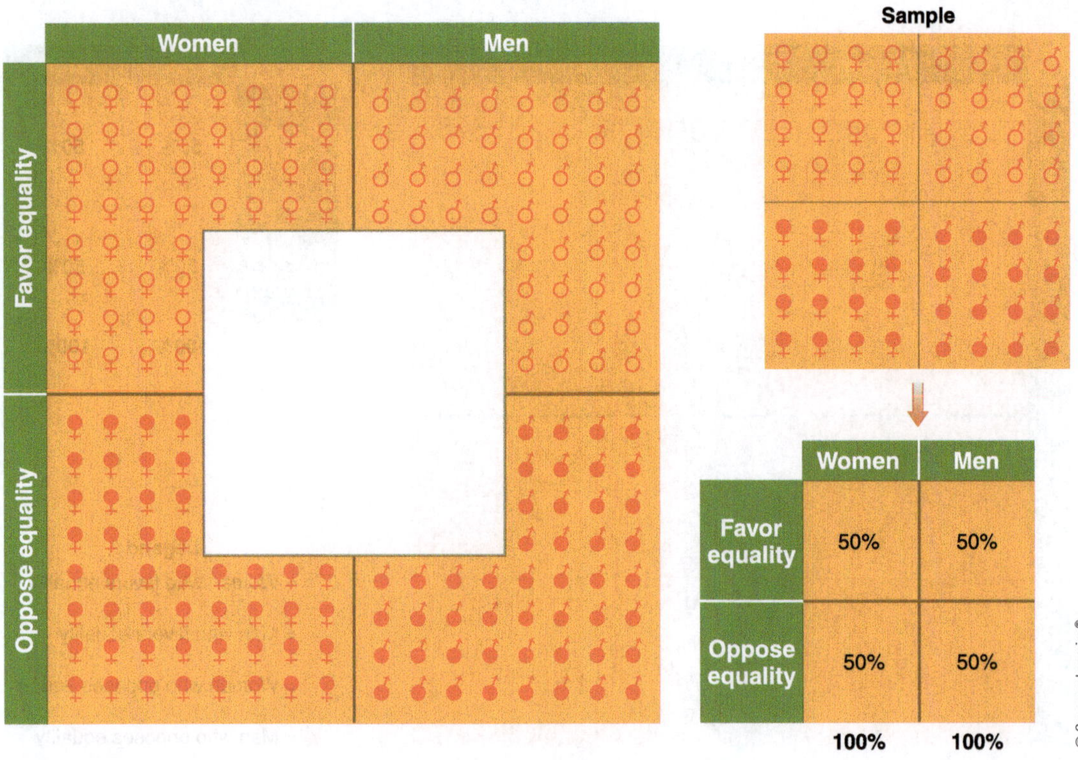

FIGURE 16-4

A Representative Sample

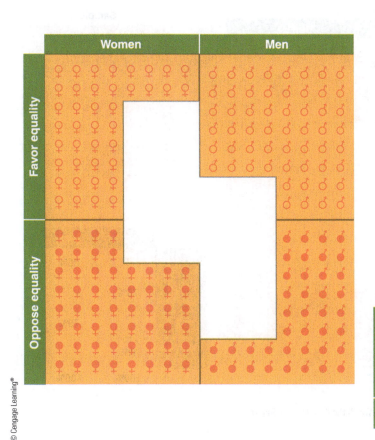
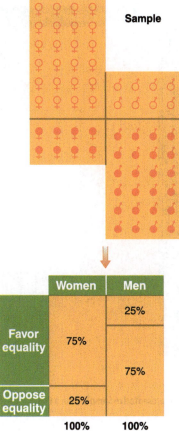

FIGURE 16-5

An Unrepresentative Sample

We've selected a perfectly representative sample, but we see that there is actually a strong relationship between the two variables in the population at large. In this latest figure, women are more likely to support equality than men are: That's the case in the population, and the sample reflects it.

In practice, of course, we never know what's so for the total population; that's why we select samples. So if we selected a sample and found the strong relationship presented in Figures 16-5 and 16-6, we'd need to decide whether that finding accurately reflected the population or was simply a product of sampling error.

The fundamental logic of tests of statistical significance, then, is this: Faced with any discrepancy between the assumed independence of variables in a population and the observed distribution of sample elements, we may explain that discrepancy in either of two ways: (1) we may

attribute it to an unrepresentative sample, or (2) we may reject the assumption of independence. The logic and statistics associated with probability sampling methods offer guidance about the varying probabilities of varying degrees of unrepresentativeness (expressed as sampling error). Most simply put, there is a *high* probability of a small degree of unrepresentativeness and a *low* probability of a large degree of unrepresentativeness.

The statistical significance of a relationship observed in a set of sample data, then, is always expressed in terms of probabilities. "Significant at the 0.05 level ($p \leq 0.05$)" simply means that the probability that a relationship as strong as the observed one can be attributed to sampling error alone is no more than 5 in 100. Put somewhat differently, if two variables are independent of each other in the population, and if 100 probability samples are selected from that population, no more than 5 of those samples should provide

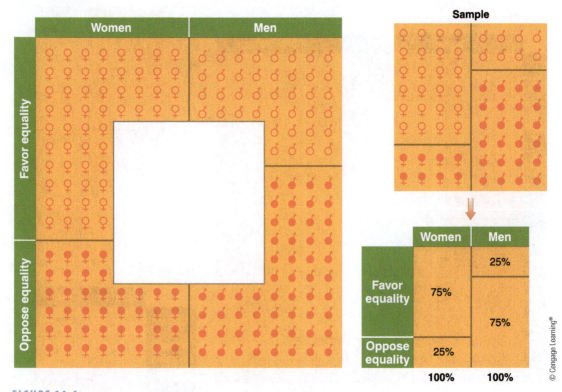

FIGURE 16-6

A Representative Sample from a Population in Which the Variables Are Related

a relationship as strong as the one that has been observed.

There is, then, a corollary to confidence intervals in tests of significance, which represents the probability of the measured associations being due *only* to sampling error. This is called the **level of significance**. Like confidence intervals, levels of significance are derived from a logical model in which several samples are drawn from a given population. In the present case, we assume that there is no association between the variables in the population, and then we ask what proportion of the samples drawn from that population would produce associations at least

as great as those measured in the empirical data. Three levels of significance are frequently used in research reports: 0.05, 0.01, and 0.001. These mean, respectively, that the chances of obtaining the measured association as a result of sampling error are 5/100, 1/100, and 1/1,000.

Researchers who use tests of significance normally follow one of two patterns. Some specify in advance the level of significance they'll regard as sufficient. If any measured association is statistically significant at that level, they'll regard it as representing a genuine association between the two variables. In other words, they're willing to discount the possibility of its resulting from sampling error only.

Other researchers prefer to report the specific level of significance for each association, disregarding the conventions of 0.05, 0.01, and 0.001. Rather than reporting that a given association is significant at the 0.05 level, they might report significance at the 0.023 level, indicating the chances of its having resulted from sampling error as 23 out of 1,000.

> **level of significance** In the context of tests of statistical significance, the degree of likelihood that an observed, empirical relationship could be attributable to sampling error. A relationship is significant at the 0.05 level if the likelihood of its being only a function of sampling error is no greater than 5 out of 100.

Chi Square

Chi square (χ^2) is a frequently used test of significance in social science. It's based on the *null hypothesis*: the assumption that there is no relationship between two variables in the total population (as you may recall from Chapter 2). Given the observed distribution of values on the two separate variables, we compute the conjoint distribution that would be expected if there were no relationship between the two variables. The result of this operation is a set of *expected frequencies* for all the cells in the contingency table. We then compare this expected distribution with the distribution of cases actually found in the sample data, and we determine the probability that the discovered discrepancy could have resulted from sampling error alone. An example will illustrate this procedure.

Let's assume we're interested in the possible relationship between church attendance and sex for the members of a particular church. To test this relationship, we select a sample of 100 church members at random. We find that our sample is made up of 40 men and 60 women and that 70 percent of our sample say they attended church during the preceding week, whereas the remaining 30 percent say they did not.

If there is no relationship between sex and church attendance, then 70 percent of the men in the sample should have attended church during the preceding week and 30 percent should have stayed away. Moreover, women should have attended in the same proportion. Table 16-6 (part I) shows that, based on this model, 28 men and 42 women would have attended church, with 12 men and 18 women not attending.

Part II of Table 16-6 presents the observed attendance for the hypothetical sample of 100 church members. Note that 20 of the men report having attended church during the preceding week, and the remaining 20 say they did not. Among the women in the sample, 50 attended church and 10 did not. Comparing the expected and observed frequencies (parts I and II), we note that somewhat fewer men attended church than expected, whereas somewhat more women attended than expected.

Chi square is computed as follows. For each cell in the tables, the researcher (1) subtracts the expected frequency for that cell from the observed frequency, (2) squares this quantity, and (3) divides the squared difference by the expected frequency. This procedure is carried out for each cell in the tables; part III of Table 16-6 presents the cell-by-cell computations. The several results are then added together to find the value of chi square: 12.70 in the example.

This value is the overall discrepancy between the observed conjoint distribution in the sample and the distribution we would expect if the two variables were unrelated to each other. Of course, the mere discovery of a discrepancy does not prove that the two variables are related, because normal sampling error might produce discrepancies even when there is no relationship in the total population. The magnitude of the value of chi square, however, permits us to estimate the probability of that having happened.

To determine the statistical significance of the observed relationship, we must use a standard set of chi square values. This will require the computation of the *degrees of freedom*, which refer to the possibilities for variation within a statistical model. Suppose I challenge you to find three numbers whose mean is 11. There are infinite solutions to this problem: (11, 11, 11), (10, 11, 12), (−11, 11, 33), and so on. Now, suppose I require that one of the numbers be 7. There would still be an infinite number of possibilities for the other two numbers.

TABLE 16-6

A Hypothetical Illustration of Chi Square

I. Expected Frequencies	Men	Women	Total
Attended church	28	42	70
Did not attend church	12	18	30
Total	40	60	100

II. Observed Cell Frequencies	Men	Women	Total
Attended church	20	50	70
Did not attend church	20	10	30
Total	40	60	100

III. (Observed − Expected)^2 ÷ Expected	Men	Women	
Attended church	2.29	1.52	$\chi^2 = 12.70$
Did not attend church	5.33	3.56	$p < 0.001$

If I told you one number had to be 7 and another 10, however, there would be only one possible value for the third. If the average of three numbers is 11, their sum must be 33. If two of the numbers total 17, the third must be 16. In this situation, we say there are two degrees of freedom. Two of the numbers could have any values we choose, but once they are specified, the third number is determined.

More generally, whenever we're examining the mean of N values, we can see that the degrees of freedom equal $N - 1$. Thus, in the case of the mean of 23 values, we could make 22 of them anything we like, but the 23rd would then be determined.

A similar logic applies to bivariate tables, such as those analyzed by chi square. Consider a table reporting the relationship between two dichotomous variables: *sex* (men/women) and *abortion attitude* (approve/disapprove). Notice that the table provides the marginal frequencies of both variables.

Abortion Attitude	Men	Women	Total
Approve			500
Disapprove			500
Total	500	500	1,000

Despite the conveniently round numbers in this hypothetical example, notice that there are numerous possibilities for the cell frequencies. For example, it could be the case that all 500 men approve and all 500 women disapprove, or it could be just the reverse. Or there could be 250 cases in each cell. Notice there are numerous other possibilities.

Now the question is, How many cells could we fill in pretty much as we choose before the remainder are determined by the marginal frequencies? The answer is only one. If we know that 300 men approved, for example, then 200 men would have had to disapprove, and the distribution would need to be just the opposite for the women.

In this instance, then, we say the table has one degree of freedom. Now, take a few minutes to construct a three-by-three table. Assume you know the marginal frequencies for each variable, and see if you can determine how many degrees of freedom it has.

For chi square, the degrees of freedom are computed as follows: the number of rows in the table of observed frequencies, minus 1, is multiplied by the number of columns, minus 1. This may be written as $(r - 1)(c - 1)$. For a three-by-three table, then, there are four degrees of freedom: $(3 - 1)(3 - 1) = (2)(2) = 4$.

In the example of sex and church attendance, we have two rows and two columns (discounting the totals), so there is one degree of freedom. Turning to a table of chi square values (see Appendix D), we find that for one degree of freedom and random sampling from a population in which there is no relationship between two variables, 10 percent of the time we should expect a chi square of at least 2.7. Thus, if we selected 100 samples from such a population, we should expect about 10 of those samples to produce chi squares equal to or greater than 2.7. Moreover, we should expect chi square values of at least 6.6 in only 1 percent of the samples and chi square values of 10.8 in only one tenth of a percent (0.001) of the samples. The higher the chi square value, the less probable it is that the value could be attributed to sampling error alone.

In our example, the computed value of chi square is 12.70. If there were no relationship between sex and church attendance in the church-member population and a large number of samples had been selected and studied, then we would expect a chi square of this magnitude in fewer than 1/10 of 1 percent (0.001) of those samples. Thus, the probability of obtaining a chi square of this magnitude is less than 0.001, if random sampling has been used and there is no relationship in the population. We report this finding by saying that the relationship is statistically significant at the 0.001 level. Because it is so improbable that the observed relationship could have resulted from sampling error alone, we're likely to reject the null hypothesis and assume that there is a relationship between the two variables in the population of church members.

t-Test

Chi square is appropriate for testing the statistical association of relations found in typically nominal or ordinal tabular data, as in the example just discussed. Suppose your data represent a high level of measurement such as interval or ratio data.

Let's say you want to know if men and women have significantly different weights. To determine this, you measure the weights of a sample of men and women and then calculate the mean average for each sex. Let's say that the average weight for men is 170 and for women 135. That seems like a pretty substantial difference on the face of it. But what if your "sample" consists of two men and two women. Intuitively, you can see that even a difference of the observed magnitude could have resulted from your picking two big men and two small women, just by chance. We wouldn't want to conclude that we had discovered something about men and women in general, based simply on four people who might not be typical.

The *t*-test, sometimes known as Student's *t*, is a commonly used measure for judging the statistical significance of differences in group means. The formula for calculating *t* involves some statistics we haven't discussed in this book, so let me give you a sense of the logic involved in this measure.

First, it makes sense that the value of *t* will increase with the size of the difference between the means.

The value of *t* will also increase with the size of the sample involved; hence, differences found in larger samples—as we saw in regard to chi square—are more likely to be judged statistically significant.

Finally, the value of *t* will be larger when variations of values within each group are smaller. In the case of sex and weight, the value of *t* will be greatest when

- The difference between the average weight of men and that of women is large.
- We've examined a large sample.
- Most women's weights are clustered around the mean weight for women and most men's weights around the mean for men. In the extreme case, the heaviest woman would weigh less than the lightest man, though this is unlikely in any substantial sample.

Once you calculate a value for *t* in your data, you look that value up in a *t*-test table, found in any statistics textbook. This gives you the significance of that value, expressed as the probability that the observed difference might have been due to sampling error alone—the same logic used in the case of chi square.

Most measures of association can be tested for statistical significance in a similar manner. Standard tables of values permit us to determine whether a given association is statistically significant and at what level. Any standard statistics textbook provides instructions on the use of such tables.

There are several possible outcomes of hypothesis testing in relation to the truth. To begin, you might accept the null hypothesis (concluding that the variables under study are unrelated to one another), or you may reject it (concluding that the variables are related to one another).

In reality, there are two situations in which you draw the correct conclusion. You can accept the null hypothesis when there really is no relationship between the variables in the whole population. Or you can reject the null hypothesis when there really is a relationship between the two variables.

Statisticians speak of two kinds of errors in this regard. *Type I error*, refers to the incorrect rejection of the null hypothesis: concluding that there is a relationship between the two variables, where there is no relationship in the whole population. In other words, the relationship discovered in the sample is a product of sampling error, not indicative of circumstances in the whole population. On the other hand, *Type II error* refers to the incorrect acceptance of the null hypothesis: concluding that there is no relationship between the variables when, in fact, there is.

Here's a simple table to illustrate this terminology.

Situation in the Real World

		Are the variables related?	
		Related	Unrelated
Conclusion drawn from a sample about the variables	**Related**	Correct	Type I Error
	Unrelated	Type II Error	Correct

Suppose you are testing whether an innovative educational program will reduce delinquency rates. Suppose further that the program would be very expensive to implement. In that situation, you would be especially concerned about avoiding the Type I error: concluding that the program works when it really doesn't. If the cost of the program was low and the potential payoff great, you would especially want to avoid the Type II error: missing a genuine solution.

Some Words of Caution

Tests of significance provide an objective yardstick that we can use to estimate the statistical significance of associations between variables. They help us rule out associations that may not represent genuine relationships in the population under study. However, the researcher who uses or reads reports of significance tests should remain wary of several dangers in their interpretation.

First, we've been discussing tests of statistical significance; there are no objective tests of *substantive* significance. Thus, we may be legitimately convinced that a given association is not due to sampling error, but we may be in the position of asserting without fear of contradiction that two variables are only slightly related to each other. Recall that sampling error is an inverse function of sample size—the larger the sample, the smaller the expected error. Thus, a correlation of, say, 0.1 might very well be significant (at a given level) if discovered in a large sample, whereas the same correlation between the same two variables would not be significant if found in a smaller sample. This makes perfectly good sense given the basic logic of tests of significance: In the larger sample, there is less chance that the correlation could be simply the product of sampling error. In both samples, however, it might represent an essentially zero correlation.

The distinction between statistical and substantive significance is perhaps best illustrated by cases in which there is *absolute certainty* that observed differences cannot be a result of sampling error. This would be the case when we observe an entire population. Suppose we were able to learn the ages of every public official in the United States and of every public official in Russia. For argument's sake, let's assume further that the average age of U.S. officials was 45 years, compared with, say, 46 for the Russian officials. Because we would have the ages of all officials, there would be no question of sampling error. We would know with certainty that the Russian officials were older than their U.S. counterparts. At the same time, we would say that the difference was of no substantive significance. We would conclude, in fact, that they were essentially the same age.

Second, lest you be misled by this hypothetical example, realize that statistical significance should not be calculated on relationships observed in data collected from whole populations. Remember, tests of statistical significance measure the likelihood of relationships between variables being only a product of sampling error; if there's no sampling, there's no sampling error.

Third, tests of significance are based on the same sampling assumptions we used in computing confidence intervals. To the extent that these assumptions are not met by the actual sampling design, the tests of significance are not strictly legitimate.

We've examined statistical significance here in the form of chi square and *t*-tests, but social scientists commonly use several other measures as well. Analysis of variance is one example you may run across in your studies.

As is the case for most matters covered in this book, I have a personal prejudice. In this instance, it's against tests of significance. I don't object to the statistical logic of those tests because the logic is sound. Rather, I'm concerned that such tests seem to mislead more than they enlighten. Here are my principal reservations:

1. Tests of significance make sampling assumptions that are virtually never satisfied by actual sampling designs.
2. They depend on the absence of nonsampling errors, a questionable assumption in most actual empirical measurements.
3. In practice, they are too often applied to measures of association that have been computed in violation of the assumptions made by those measures (for example, product-moment correlations computed from ordinal data).
4. Statistical significance is too easily misinterpreted as "strength of association," or substantive significance.

These concerns are underscored by a study (Sterling, Rosenbaum, and Weinkam 1995) examining the publication policies of nine psychology and three medical journals. As the researchers discovered, the journals were quite unlikely to publish articles that did not report statistically significant correlations among variables. They quote the following from a rejection letter:

> *Unfortunately, we are not able to publish this manuscript. The manuscript is very well written and the study was well documented. Unfortunately, the negative results translate into a minimal contribution to the field. We encourage you to continue your work in this area and we will be glad to consider additional manuscripts that you may prepare in the future.*

(Sterling et al. 1995: 109)

Let's suppose a researcher conducts a scientifically excellent study to determine whether *X* causes *Y*. The results indicate no statistically significant correlation. That's good to know. If we're interested in what causes cancer, war, or juvenile delinquency, it's good to know that a possible cause actually does *not* cause it. That knowledge would free researchers to look elsewhere for causes.

As we've seen, however, journals might very well reject such a study. Other researchers would likely continue testing whether *X* causes *Y*, not knowing that previous studies found no causal relationship. This would produce many wasted studies, none of which would see publication and draw a close to the analysis of *X* as a cause of *Y*.

From what you've learned about probabilities, however, you can understand that if enough studies are conducted, one will eventually measure a statistically significant correlation between *X* and *Y*. If there is absolutely no relationship between the two variables, we would expect a correlation significant at the 0.05 level five times out of a hundred, because that's what the 0.05 level of significance means. If a hundred studies were conducted, therefore, we could expect five to suggest a causal relationship where there was actually none—and those five studies would be published!

There are, then, serious problems inherent in too much reliance on tests of statistical significance. At the same time (perhaps paradoxically) I would suggest that tests of significance *can* be a valuable asset to the researcher—useful tools for understanding data. Although many of my comments suggest an extremely conservative approach to tests of significance—that you should use them only when all assumptions are met—my general perspective is just the reverse.

I encourage you to use any statistical technique—any measure of association or test of significance—if it will help you understand your data. If the computation of product-moment correlations among nominal variables and the testing of statistical significance in the context of uncontrolled sampling will meet this criterion, then I encourage such activities. I say this in the spirit of what Hanan Selvin, another pioneer in developing the elaboration model, referred to as "data-dredging techniques." Anything goes, if it leads ultimately to the understanding of data and of the social world under study.

The price of this radical freedom, however, is the giving up of strict, statistical interpretations.

You will not be able to base the ultimate importance of your finding solely on a significant correlation at the 0.05 level. Whatever the avenue of discovery, empirical data must ultimately be presented in a legitimate manner, and their importance must be argued logically.

Other Multivariate Techniques

For the most part, this book has focused on rather rudimentary forms of data manipulation, such as the use of contingency tables and percentages. The elaboration model of analysis was presented in this form, as well as many of the examples of data analysis throughout the book.

This section of the chapter presents a cook's tour of several other multivariate techniques from the logical perspective of elaborating the relationships among social variables. This discussion is intended not to teach you how to use these techniques but rather to present sufficient information so that you can understand them if you run across them in a research report. The methods of analysis that we'll examine—path analysis, time-series analysis, factor analysis, analysis of variance, log-linear models, odds-ratio analysis, Geographic Information Systems, and Demographic Analyses—are only a few of the many multivariate techniques used by social scientists.

Path Analysis

Path analysis is a causal model for understanding relationships between variables. Though based on regression analysis, it can provide a more useful graphic picture of relationships among several variables than other means can. Path analysis assumes that the values of one variable are caused by the values of another, so distinguishing independent and dependent variables is essential. This requirement is not unique to path analysis, of course, but path analysis provides a unique way of displaying explanatory results for interpretation.

Recall for a moment one of the ways I represented the elaboration model in Chapter 15

---

path analysis A form of multivariate analysis in which the causal relationships among variables are presented in a graphic format.

(Figure 15-1). Here's how we might diagram the logic of interpretation:

Independent variable → Intervening variable → Dependent variable

The logic of this presentation is that an independent variable has an impact on an intervening variable, which in turn has an impact on a dependent variable. The path analyst constructs similar patterns of relationships among variables, but the typical path diagram contains many more variables than shown in this diagram.

Besides diagramming a network of relationships among variables, path analysis also shows the strengths of those several relationships. The strengths of relationships are calculated from a regression analysis that produces numbers analogous to the partial relationships in the elaboration model. These *path coefficients*, as they're called, represent the strengths of the relationships between

pairs of variables, with the effects of all other variables in the model held constant.

The analysis in Figure 16-7, for example, focuses on the religious causes of anti-Semitism among Christian church members. The variables in the diagram are, from left to right, (1) orthodoxy, or the extent to which the subjects accept conventional beliefs about God, Jesus, biblical miracles, and so forth; (2) particularism, the belief that one's religion is the "only true faith"; (3) acceptance of the view that the Jews crucified Jesus; (4) religious hostility toward contemporary Jews, such as believing that God is punishing them or that they will suffer damnation unless they convert to Christianity; and (5) secular anti-Semitism, such as believing that Jews cheat in business, are disloyal to their country, and so forth.

To start with, the researchers who conducted this analysis proposed that secular anti-Semitism was produced by moving through the

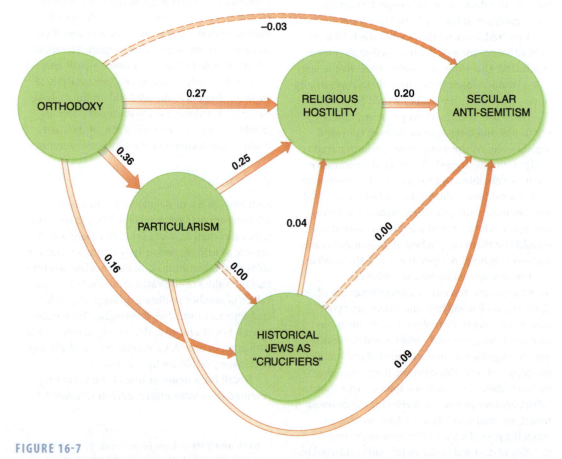

FIGURE 16-7

Diagramming the Religious Sources of Anti-Semitism

Source: Rodney Stark, Bruce D. Foster, Charles Y. Glock, and Harold E. Quinley, *Wayward Shepherds—Prejudice and the Protestant Clergy*. 1971 by Anti-Defamation League of B'nai Brith. Harper & Row, Publishers, Inc.

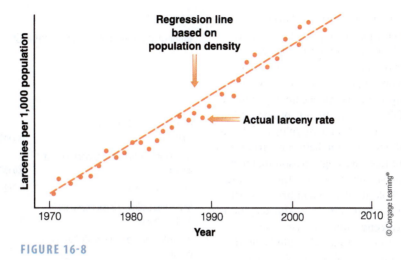

FIGURE 16-8

The Larceny Rates over Time in a Hypothetical City

five variables: Orthodoxy caused particularism, which caused the view of the historical Jews as crucifiers, which caused religious hostility toward contemporary Jews, which resulted, finally, in secular anti-Semitism.

The path diagram tells a different story. The researchers found, for example, that belief in the historical role of Jews as the crucifiers of Jesus doesn't seem to matter in the process that generates anti-Semitism. And, although particularism is a part of one process resulting in secular anti-Semitism, the diagram also shows that anti-Semitism is created more directly by orthodoxy and religious hostility. Orthodoxy produces religious hostility even without particularism, and religious hostility generates secular hostility in any event.

One last comment on path analysis is in order. Although it's an excellent way of handling complex causal chains and networks of variables, path analysis itself does not tell the causal order of the variables. Nor was the path diagram in Figure 16-7 generated by computer. The researcher decided the structure of relationships among the variables and used computer analysis merely to calculate the path coefficients that applied to the structure.

Time-Series Analysis

The various forms of regression analysis are often used to examine time-series data, representing changes in one or more variables over time. As I'm sure you know, U.S. crime rates have generally increased over the years. A **time-series analysis** of crime rates could express the long-term trend in

a regression format and provide a way of testing explanations for the trend—such as population growth or economic fluctuations—and could permit forecasting of future crime rates.

In a simple illustration, Figure 16-8 graphs the larceny rates of a hypothetical city over time. Each dot on the graph represents the number of larcenies reported to police during the year indicated.

Suppose we feel that larceny is partly a function of overpopulation. You might reason that crowding would lead to psychological stress and frustration, resulting in increased crimes of many sorts. Recalling the discussion of regression analysis, we could create a regression equation representing the relationship between larceny and population density—using the actual figures for each variable, with years as the units of analysis. Having created the best-fitting regression equation, we could then calculate a larceny rate for each year, based on that year's population density rate. For the sake of simplicity, let's assume that the city's population size (and hence density) has been steadily increasing. This would lead us to predict a steadily increasing larceny rate as well. These regression estimates are represented by the dashed regression line in Figure 16-8.

Time-series relationships are often more complex than this simple illustration suggests. For one thing, there can be more than one causal variable. For example, we might find that unemployment

time-series analysis An analysis of changes in a variable (such as *crime rates*) over time.

rates also had a powerful impact on larceny. We might develop an equation to predict larceny on the basis of both of these causal variables. As a result, the predictions might not fall along a simple, straight line. Whereas population density was increasing steadily in the first model, unemployment rates rise and fall. As a consequence, our predictions of the larceny rate would similarly go up and down.

Pursuing the relationship between larceny and unemployment rates, we might reason that people do not begin stealing as soon as they become unemployed. Typically, they might first exhaust their savings, borrow from friends, and keep hoping for work. Larceny would be a last resort.

Time-lagged regression analysis could be used to address this more complex case. Thus, we might create a regression equation that predicted a given year's larceny rate based, in part, on the previous year's unemployment rate or perhaps on an average unemployment rates of the 2 years. The possibilities are endless.

If you think about it, a great many causal relationships are likely to involve a time lag. Historically, many of the world's poor countries have maintained their populations by matching high death rates with equally high birth rates. It has been observed repeatedly, moreover, that when a society's death rate is drastically reduced—through improved medical care, public sanitation, and improved agriculture, for example—that society's birth rate drops sometime later on, but with an intervening period of rapid population growth. Or, to take a very different example, a crackdown on speeding on a state's highways would likely reduce the average speed of cars. Again, however, the causal relationship would undoubtedly involve a time lag—days, weeks, or months, perhaps—as motorists began to realize the seriousness of the crackdown.

In all such cases, the regression equations generated might take many forms. In any event, the criterion for judging success or failure is the extent to which the researcher can account for the actual values observed for the dependent variable.

factor analysis A complex algebraic method for determining the general dimensions or factors that exist within a set of concrete observations.

TABLE 16-7
Modern and Traditional Orientations in Shanghai

	Factors	
	1	2
My main goal in life is to become a millionaire	0.6544	0.0742
I pursue jobs with high remuneration and high risks	0.6568	−0.1174
To get rich is glorious	0.3727	0.1977
Respecting authority is not important in modern society	0.3574	−0.0744
It is better not to disagree with those in power	0.0347	0.4968
Go with the flow even when natural disasters and social trouble occur	0.0070	0.4890
Family background and personal relationships are most important to personal status	0.0139	0.3570

Source: Jiaming Sun. 2008. *Global Connectivity and Local Transformation: A Micro Approach to Studying the Effect of Globalization in Shanghai.* Lanham, MD: University Press of America, p. 110.

Factor Analysis

Factor analysis is a unique approach to multivariate analysis. Its statistical basis is complex enough and different enough from the foregoing discussions to suggest a general discussion here.

Factor analysis is a complex algebraic method used to discover patterns among the variations in values of several variables. This is done essentially through the generation of artificial dimensions (*factors*) that correlate highly with several of the real variables and that are independent of one another. A computer must be used to perform this complex operation.

Here's a simple example of factor analysis used in a study of social change in Shanghai, China. Jiaming Sun (2008) used factor analysis to detect whether a series of attitudes reflected some overall orientations to life. Table 16-7 presents an extract of his analysis.

As you can see, the first four statements correlate highly with the first factor, while the final three statements correlate highly with the second factor. If you read through the first four statements, you can see that the factor analysis has identified a common orientation Sun labeled "secular-rational," whereas the last three statements reflect a more traditional point of view.

Here's a more complex example of the use of factor analysis. Many social researchers have studied the problem of delinquency. If you look deeply into the problem, however, you'll discover that there are many different types of delinquents. In a survey of high school students in a small Wyoming town, Morris Forslund (1980) set out to create a typology of delinquency. His questionnaire asked students to report whether they had committed a variety of delinquent acts. He then submitted their responses to factor analysis. The results are shown in Table 16-8.

As you can see in this table, the various delinquent acts are listed on the left. The numbers shown in the body of the table are the factor loadings on the four factors constructed in the analysis. You'll notice that after examining the dimensions, or factors, Forslund labeled them. I've bracketed the items on each factor that led to his choice of labels. Forslund summarized the results as follows:

For the total sample four fairly distinct patterns of delinquent acts are apparent. In order of variance explained, they have been labeled: 1) Property Offenses, including both vandalism and

TABLE 16-8

Factor Analysis: Delinquent Acts, Whites

Delinquent Act	Property Offenses Factor I	Incorrigibility Factor II	Drugs/Truancy Factor III	Fighting Factor IV
Broke street light, etc.	0.669	0.126	0.119	0.167
Broke windows	0.637	0.093	0.077	0.215
Broke down fences, clotheslines, etc.	0.621	0.186	0.186	0.186
Taken things worth $2 to $50	0.616	0.187	0.233	0.068
Let air out of tires	0.587	0.243	0.054	0.156
Taken things worth over $50	0.548	−0.017	0.276	0.034
Thrown eggs, garbage, etc.	0.526	0.339	−0.023	0.266
Taken things worth under $2	0.486	0.393	0.143	0.077
Taken things from desks, etc., at school	0.464	0.232	−0.002	0.027
Taken car without owner's permission	0.461	0.172	0.080	0.040
Put paint on something	0.451	0.237	0.071	0.250
Disobeyed parents	0.054	0.642	0.209	0.039
Marked on desk, wall, etc.	0.236	0.550	−0.061	0.021
Said mean things to get even	0.134	0.537	0.045	0.100
Disobeyed teacher, school official	0.240	0.497	0.223	0.195
Defied parents to their face	0.232	0.458	0.305	0.058
Made anonymous telephone calls	0.373	0.446	0.029	0.135
Smoked marijuana	0.054	0.064	0.755	−0.028
Used other drugs for kicks	0.137	0.016	0.669	0.004
Signed name to school excuse	0.246	0.249	0.395	0.189
Drank alcohol, parents absent	0.049	0.247	0.358	0.175
Skipped school	0.101	0.252	0.319	0.181
Beat up someone in a fight	0.309	0.088	0.181	0.843
Fought—hit or wrestled	0.242	0.266	0.070	0.602
Percent of variance	67.2	13.4	10.9	8.4

Source: Morris A. Forslund, *Patterns of Delinquency Involvement: An Empirical Typology,* paper presented at the Annual Meeting of the Western Association of Sociologists and Anthropologists, Lethbridge, Alberta, February 8, 1980. The table above is adapted from page 10.

theft; 2) Incorrigibility; 3) Drugs/Truancy; and 4) Fighting. It is interesting, and perhaps surprising, to find both vandalism and theft appear together in the same factor. It would seem that those high school students who engage in property offenses tend to be involved in both vandalism and theft. It is also interesting to note that drugs, alcohol and truancy fall in the same factor.

(1980: 4)

Having determined this overall pattern, Forslund reran the factor analysis separately for boys and for girls. Essentially the same patterns emerged in both cases.

This example shows that factor analysis is an efficient method of discovering predominant patterns among a large number of variables. Instead of being forced to compare countless correlations—simple, partial, and multiple—to discover those patterns, researchers can use factor analysis for this task. Incidentally, this is a good example of a helpful use of computers.

Factor analysis also presents data in a form that the reader or researcher can interpret. For a given factor, the reader can easily discover the variables loading highly on it, thus noting clusters of variables. Or the reader can easily discover which factors a given variable is or is not loaded highly on.

But factor analysis also has disadvantages. First, as noted previously, factors are generated with no regard to substantive meaning. Often researchers will find factors producing very high loadings for a group of substantively disparate variables. They might find, for example, that prejudice and religiosity have high positive loadings on a given factor, with education having an equally high negative loading. Surely the three variables are highly correlated, but what does the factor represent in the real world? All too often, inexperienced researchers will be led into naming such factors as "religio-prejudicial lack of education" or something similarly nonsensical.

Second, factor analysis is often criticized on basic philosophical grounds. Recall that to be useful, a hypothesis must be disprovable. If the researcher

cannot specify the conditions under which the hypothesis would be disproved, the hypothesis is either a tautology or useless. In a sense, factor analysis suffers this defect. No matter what data are input, factor analysis produces a solution in the form of factors. Thus, if the researcher were asking, "Are there any patterns among these variables?" the answer always would be yes. This fact must also be taken into account in evaluating the results of factor analysis. The generation of factors does not ensure meaning.

My personal view of factor analysis is the same as that for other complex modes of analysis. It can be an extremely useful tool for the social science researcher. Its use should be encouraged whenever such activity might assist researchers in understanding a body of data. As in all cases, however, such tools are just that—tools; they are never magical solutions.

Let me reiterate that the analytic techniques we've touched on are only a few of the many techniques commonly used by social scientists. As you pursue your studies, you may very well want to study this subject in more depth later.

Analysis of Variance

Analysis of variance (ANOVA) applies the logic of statistical significance, discussed earlier. Fundamentally, the cases under study are combined into groups representing an independent variable, and the extent to which the groups differ from one another is analyzed in terms of some dependent variable. The extent to which the groups differ is compared with the standard of random distribution: Could we expect to obtain such differences if we had assigned cases to the various groups through random selection?

We'll look briefly now at two common forms of ANOVA: one-way analysis of variance and two-way analysis of variance.

One-Way Analysis of Variance

Suppose we want to compare income levels of Republicans and Democrats to see if Republicans are really richer. We select a sample of individuals for our study, and we ask them (1) which political party they identify with and (2) their total income for the past year. We calculate the mean or median incomes of each political group, finding that the Republicans in our sample have a mean income of, say, $52,000, compared with $46,000 for the Democrats. Clearly, our

---

analysis of variance (ANOVA) Method of analysis in which cases under study are combined into groups representing an independent variable, and the extent to which the groups differ from one another is analyzed in terms of some dependent variable. Then, the extent to which the groups differ is compared with the standard of random distribution.

Republicans are richer than our Democrats, but is the difference "significant"? Would we have been likely to get a $6,000 difference if we had created two groups by way of random selection?

ANOVA answers this question through the use of variance. Simply put, the variance of a distribution (or incomes, for example) is a measurement of the extent to which a set of values are clustered close to the mean or range high and low away from it.

Figure 16-9 illustrates these two possibilities. Notice that in both distributions the Republicans have a mean income of $52,000 and the Democrats have $46,000. In panel a, most Republicans have incomes relatively close to the mean of $52,000, and most Democrats have incomes close to their party's mean of $46,000. Panel b,

however, presents quite a different picture. Although the group means are the same as in panel a, both Republicans and Democrats have incomes ranging from very high to very low, with considerable overlap in the parties' distributions. In technical terms, there is a higher degree of variance in panel b than in panel a. On the face of it, we'd conclude that panel a of Figure 16-9 indicates a genuine difference in the incomes of Republicans and Democrats. With data like those presented in panel b, we wouldn't be so sure; in this case, there seems more likelihood that the normal variations produced by random sampling error could have produced means of $52,000 and $46,000.

In an actual ANOVA, statistical calculations rather than impressions are used to make this

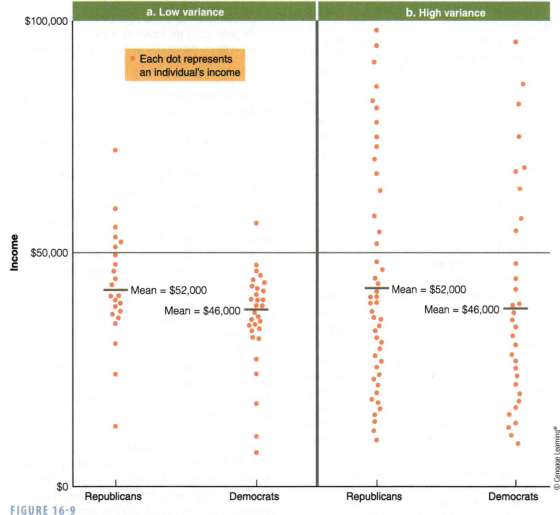

FIGURE 16-9

Two Distribution Patterns of the Incomes of Republicans and Democrats

decision. The observed difference in means is expressed as standardized multiples and fractions of the observed variance. Because the variance in panel a of Figure 16-9 is smaller than the variance in panel b, $6,000 would represent a larger difference in panel a than in panel b. The resulting difference of means—standardized by the variance—would then be checked against a standard statistical table showing the theoretical distribution of such values, as in our earlier discussion of statistical significance. Ultimately, we'd conclude that the difference was significant at some level of significance. We might discover, for example, that sampling error would have produced a difference as large as the one observed only one time in a thousand. Thus, we would say that difference was "significant at the 0.001 level."

In the example just given, I've glossed over the actual calculations in favor of the basic logic of the procedure.

This simplest case is often referred to as a *t*-test for the difference between two means. With more than two groups, the calculations become more complex, because more comparisons must be made. Basically, it's necessary to compare the differences separating group means with the variations found within each group. The end result of the analysis, as discussed in the simplest case, is expressed in terms of statistical significance—the likelihood of the observed differences resulting from sampling error in random selection.

Two-Way Analysis of Variance

One-way ANOVA represents a form of bivariate analysis (*political party* and *income* were the two variables in our example). As we've seen, however, social researchers often engage in multivariate analysis. Two-way ANOVA permits the simultaneous examination of more than two variables. Suppose, for example, that we suspect that the income differences between Republicans and Democrats are a function of education. Our hypothesis is that Republicans are better educated than Democrats and that educated people—regardless of party—earn more, on average, than people with less education do. A two-way ANOVA would sort out the effects

> **log-linear models** Data-analysis technique based on specifying models that describe the interrelationships among variables and then comparing expected and observed table-cell frequencies.

of the two explanatory variables in a manner similar to that of the elaboration model discussed in Chapter 15 and following the same logic discussed in the case of partial correlations and regressions.

Log-Linear Models

The **log-linear model**, which involves rather elaborate logarithmic calculations, is based on specifying models that describe the interrelationships among variables and then comparing expected and observed table-cell frequencies. (The logic here is similar to that for chi square, discussed earlier.) H. T. Reynolds describes the process:

> *At the outset of log-linear analysis, as in most statistical procedures, the investigator proposes a model that he feels might fit the data. The model is a tentative statement about how a set of variables are interrelated. After choosing the model, he next estimates the frequencies expected in a sample of the given size if the model were true. He then compares these estimates, F, with the observed values.*

> *(1977: 76–77)*

In specifying the models to be tested in a log-linear analysis, the researcher will consider direct relationships between the dependent variable and each independent variable, relationships between pairs of independent variables, and three-variable (and more, depending on the total number of variables) relationships similar to those already discussed in the elaboration model (Chapter 15). We'll consider a three-variable case taken from the preceding example.

We might suspect that a person's political party affiliation ("party") is a function of political orientation ("philosophy") and race. The components of this model, then, include (1) the direct effect of philosophy on party, (2) the direct effect of race on party, (3) the effect of race on philosophy, (4) the effect of race on the relationship between philosophy and party (as in the elaboration model), and (5) the effect of philosophy on the relationship between race and party. Though each of these components will have some explanatory power, log-linear analysis provides a means of identifying which are the most important and which can, as a practical matter, be ignored. Although the calculations involved in log-linear analysis are many and complex, computer programs can perform them all handily. If you find references in the research literature to logit, probit, or multi-way frequency analysis (MFA), those analyses are using this model.

Log-linear analysis has two main shortcomings. First, its logic makes certain mathematical assumptions that a particular set of data might not satisfy, but this issue is far too complex to be pursued here. Second, as with other summary techniques discussed, the results of log-linear analysis do not permit the immediate, intuitive grasp possible in simple comparisons of percentages or means. Because of this, log-linear methods would not be appropriate—even if statistically justified—in cases in which the analysis can be managed through simple percentage tables. It's best reserved for complex situations in which tabular analyses are not powerful enough.

Odds-Ratio Analysis

Another popular technique for analyzing relationships is based on the familiar notion of the odds of things happening. For example, when you roll a pair of dice, there are 36 possible outcomes, but the various numerical possibilities have different odds. There is only one possibility for rolling a two ("snake eyes"), so that means the odds of doing so is 35 to 1 against it. By contrast, there are six ways of rolling a seven (1-6, 6-1, 2-5, 5-2, 3-4, 4-3), so the odds are only 30 to 6 against rolling a seven. While the difference between 35 and 30 doesn't seem that impressive, notice it is also the case that the chance of rolling a seven is 6 times better than rolling a two. A similar logic can be used to examine the relationship between social research variables.

Suppose you are interested in juvenile delinquency. The National Center for Juvenile Justice (2009) reports that around 9 percent (0.086422) of males 10 to 17 years of age had been arrested in 2009, including arrests for suspicion of criminal behavior. The comparable figure for females was around 4 percent (0.038782). The sex difference is 5 percent, which doesn't seem like much, perhaps. However, we note that being arrested is, thankfully, a reasonably rare event. You might notice, however, that males are over twice as likely as females to be arrested. Here's how an **odds ratio** would be calculated from these data.

	Arrested	Not Arrested
Male	0.086422	0.913578
Female	0.038782	0.961218

$$\text{Odds Ratio} = \frac{(.086422 \times .961218)}{(.038782 \times .913578)} = 2.344608703$$

If the result had been 1.00, we would conclude that there was no difference in the odds of a male or a female being arrested. If the result is above 1.00, we would conclude that the first-listed group (males, in this case) is more likely to be arrested. With a result between 0.00 and 1.00, we conclude that the second-listed group is more likely to be arrested. You can discover this for yourself by reversing the males and females and recalculating the odds ratio.

Geographic Information Systems (GIS)

Finally, let's examine a very different analytic technique: the **Geographic Information Systems (GIS)**. Much of the aggregated data of interest to social scientists describes geographic units: countries, states, counties, cities, census tracts, and the like. Whereas such data can and often are presented in statistical tables, the patterns they represent can often be grasped more readily in a graphic format. With this in mind, U.S. Census data are increasingly being made available in a mappable format.

Much of the analysis of recent presidential elections in the United States was couched in terms of red (Republican) and blue (Democratic) states, and I'm sure you've seen maps of the distribution of the two. Some researchers have pointed out that no state was completely red or blue, and they added purple for those fairly evenly divided in their support for the two major parties.

Other researchers have pointed out that counties are a more appropriate unit of analysis in this case, displaying the political diversity within a given state. As a general pattern, Republicans did better in rural counties, Democrats did better in the urban ones. In the 2016

odds ratio A statistical technique for expressing the relationship between variables by comparing the odds of different occurrences.

Geographic Information Systems (GIS) Analytic technique in which researchers map quantitative data that describe geographic units for a graphic display.

election, for example, Robert Vanderbei (2016) used GIS mapping to display 2016 presidential voting patterns in a way that reflected all these concerns. The unit of analysis in Figure 16-10 is the county. The relative tilt toward the Republicans or Democrats is indicated by shading, and the height of each county column reflects the number of voters per square mile: the higher the column, the more urban the county.

If you're interested in pursuing the possibilities of this analytic technique, you might try a Web search for "GIS" or "Geographic Information Systems." By the time you read this paragraph, newer applications of the technique will have appeared. And you'll find that its use is hardly limited to the United States.

This completes our discussion of some of the analytic techniques commonly used by social scientists. I've merely brushed the surface of each, and there are many other techniques that I haven't touched on at all. My purpose has been to give you a preview of some of the techniques you might want to study in more depth later on, as well as to familiarize you with them in case you run across them in reading the research reports of others.

Demographic Analyses

Demography is the study of population. Demographers study the size and composition of populations, as well as changes over time. Specifically, some demographers are concerned with population growth and sometimes refer to *overpopulation*. Just to set the stage for this discussion, Figure 16-11 presents a graph of world population growth during the past 2,000 years.

As I write this, the world population is nearly 7.5 billion people and is growing by roughly a quarter of a million people a day. Here are some of the tools demographers use to track and understand population.

Population size and growth are a function of only two factors: births and deaths. The simplest

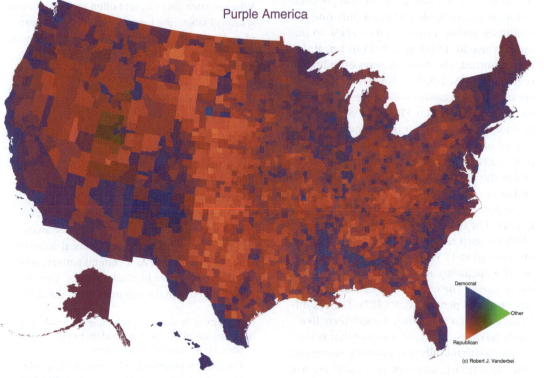

2016 Presidential Election

Purple America

FIGURE 16-10

GIS Display of 2016 U.S. Presidential Election

Source: Robert Vanderbei, "Election 2016 Results" (http://www.princeton.edu/~rvdb/JAVA/election2016/).

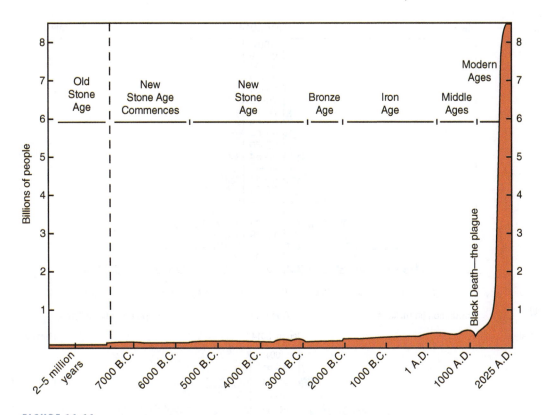

FIGURE 16-11

World Population Growth through History

Source: http://www.susps.org/overview/numbers.html

measure of births is the **crude birth rate (CBR)**: the number of births in a year, divided by the population at the middle of that year. Typically, this result is multiplied by 1,000 and is reported as X births per thousand. For example, in the West African country of Niger, the 2015 crude birth rate was 50/1,000. In Japan, by contrast, it was 8/1,000 (Population Reference Bureau [PRB] 2015).

The **crude death rate (CDR)** is a comparable calculation for mortality. Continuing our earlier comparison, Niger's CDR is 11/1,000, while Japan's was a similar 10/1,000. But notice the stark difference in these two countries: Niger's population was growing by 39 per thousand, while Japan's population was actually declining by 2 per thousand. Combining the CBR and the CDR produces a **rate of natural increase** in the world or in a smaller subdivision. Niger is growing by nearly 4 percent a year (PRB 2015).

Knowing a country's rate of increase permits a simple, but useful, calculation. Dividing 70 by the percentage increase yields the *doubling time*

for the population. Thus, at its present rate of growth, Niger's population will double every 18 years, doubling the need for food, water, housing, schools, hospitals, and so forth.

Another measure of growth commonly used by demographers is the **total fertility rate**: the number of children born during an average woman's lifetime. Currently in Niger, that number is 7.6 births. In Japan, by contrast, the average woman currently has 1.4 children during her lifetime (PRB 2015).

Beyond the sheer size of a population, demographers are also interested in its composition:

crude birth rate (CBR) The number of births in a year, divided by the population at the beginning of that year.

crude death rate (CDR) The number of deaths in a year, divided by the population at the beginning of that year.

rate of natural increase The CBR plus the CDR.

total fertility rate The number of children born during an average woman's lifetime.

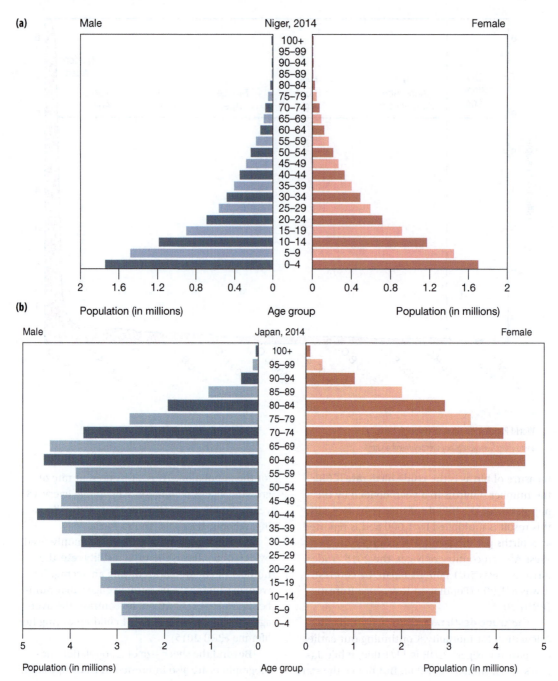

FIGURE 16-12

Age/Sex Pyramids for Niger (a) and Japan (b).

Sources: http://www.indexmundi.com/niger/age_structure.html and http://www.indexmundi.com/japan/age_structure.html

age and sex, for example. One observation to take away from these dramatically different **age/ sex pyramids** is the large portion of the Niger

population who will soon be entering the child-bearing years, contrasted to the relatively smaller portion in Japan. (See Figure 16-12.)

Finally, demographers use several measures reflecting on the quality of life, such as life expectancy at birth and maternal deaths

> **age/sex pyramids** A graphic display of the age and sex distributions of a population.

in childbirth. One measure that seems a useful index of the quality of life in general is the **infant mortality rate (IMR)**: the number of children per thousand births who die during their first year of life. Worldwide, the figure is around 50, or 5 percent of all births. In Niger, it is 60, while it is 2.1 in Japan.

These are just a few of the basic concepts and calculations that allow demographers to examine whole societies or, indeed, the whole human race.

> **infant mortality rate (IMR)** The number of children per thousand births who die during their first year of life.

What do you think?...Revisited

The most recent examination of demographic analyses gave some examples of how a specific research method can be used not only in different societies but it can be used to compare and contrast societies as well. Each of the other techniques covered in this chapter has also been used in varied settings. This is not to say that there are no problems involved. In the case of demographic analyses, for example, the quality of available data varies from country to country. Some totalitarian regimes may be unwilling to make data public for the use of external researchers.

MAIN POINTS

Introduction

- Statistics is the applied branch of mathematics especially appropriate for a variety of research analyses.

Descriptive Statistics

- Descriptive statistics are used to summarize data under study. Some descriptive statistics summarize the distribution of attributes on a single variable; others summarize the associations between variables.

- Descriptive statistics summarizing the relationships between variables are called measures of association.

- Many measures of association are based on a proportionate reduction of error (PRE) model.

  - This model is based on a comparison of (1) the number of errors we would make in attempting to guess the attributes of a given variable for each of the cases under study—if we knew nothing but the distribution of attributes on that variable—and (2) the number of errors we would make if we knew the joint distribution overall and were told for each case the attribute of one variable each time we were asked to guess the attribute of the other.

  - These measures include lambda (λ), which is appropriate for the analysis of two nominal variables; gamma (γ), which is appropriate for the analysis of two ordinal variables; and Pearson's product-moment correlation (r), which is appropriate for the analysis of two interval or ratio variables.

- Regression analysis represents the relationships between variables in the form of equations, which can be used to predict the values of a dependent variable on the basis of values of one or more independent variables.

- Regression equations are computed on the basis of a regression line: that geometric line representing, with the least amount of discrepancy, the actual location of points in a scattergram.

- Types of regression analysis include linear regression analysis, multiple regression analysis, partial regression analysis, and curvilinear regression analysis.

Inferential Statistics

- Inferential statistics are used to estimate the generalizability of findings arrived at through the analysis of a sample to the larger population from which the sample was selected. Some inferential statistics estimate the single-variable characteristics of the population; others—tests of statistical significance—estimate the relationships between variables in the population.

- Inferences about some characteristic of a population must indicate a confidence interval and a confidence level. Computations of confidence levels and intervals are based on probability theory and assume that conventional probability-sampling techniques have been employed in the study.

- Inferences about the generalizability, to a population, of the associations discovered between variables in a sample involve tests of statistical significance, which estimate the likelihood that an association as large as the observed one could result from normal sampling error if no such association exists between the variables in the larger population. Tests of statistical significance are also based on probability theory and assume that conventional probability-sampling techniques have been employed in the study.

- The level of significance of an observed association is reported in the form of the probability that the association could have been produced merely by sampling error. To say that an association is significant at the 0.05 level is to say that an association as large as the observed one could not be expected to result from sampling error more than 5 times out of 100.

- Social researchers tend to use a particular set of levels of significance in connection with tests of statistical significance: 0.05, 0.01, and 0.001. This is merely a convention, however.

- A frequently used test of statistical significance in tabular data is chi square (χ^2).

- The *t*-test is a frequently used test of statistical significance for comparing means.

- Statistical significance must not be confused with substantive significance, the latter meaning that an observed association is strong, important, meaningful, or worth writing home to your mother about.

- Tests of statistical significance, strictly speaking, make assumptions about data and methods that are almost never satisfied completely by real social research. Despite this, the tests can serve a useful function in the analysis and interpretation of data.

Other Multivariate Techniques

- Path analysis is a method of presenting graphically the networks of causal relationships among several variables. It illustrates the primary "paths" of variables through which independent variables cause dependent ones. Path coefficients are standardized regression coefficients that represent the partial relationships between variables.

- Time-series analysis is an analysis of changes in a variable (such as *crime rates*) over time.

- Factor analysis, feasible only with a computer, is an analytic method of discovering the general dimensions represented by a collection of actual variables. These general dimensions, or factors, are calculated hypothetical dimensions that are not perfectly represented by any of the empirical variables under study but are highly associated with groups of empirical variables. A factor loading indicates the degree of association between a given empirical variable and a given factor.

- Analysis of variance (ANOVA) is based on comparing variations between and within groups and determining whether between-group differences could reasonably have occurred in simple random sampling or whether they likely represent a genuine relationship between the variables involved.

- Log-linear models offer a method for analyzing complex relationships among several nominal variables having more than two attributes each.

- Odds-ratio analysis expresses the relationship between variables in terms of the odds of different occurrences.

- Geographic Information Systems (GIS) map quantitative data that describe geographic units for a graphic display.

- Demographic analyses employ a number of calculations to examine human population.

KEY TERMS

age/sex pyramid

analysis of variance (ANOVA)

crude birth rate (CBR)

crude death rate (CDR)

curvilinear regression analysis

descriptive statistics

factor analysis

Geographic Information Systems (GIS)

infant mortality rate (IMR)

inferential statistics

level of significance

linear regression analysis

log-linear models

multiple regression analysis

nonsampling error

odds ratio

partial regression analysis

path analysis

proportionate reduction of error (PRE)

rate of natural increase

regression analysis

statistical significance

tests of statistical significance

time-series analysis

total fertility rate

PROPOSING SOCIAL RESEARCH: QUANTITATIVE DATA ANALYSIS

Chapters 14–16 all discuss different aspects of a quantitative data analysis. In this exercise, you should outline your plans for analysis.

In earlier exercises, you specified the variables to be analyzed, including precisely how you'll measure those variables. Now you need to present how you'll conduct your analysis. Here's where you should say whether you're planning a tabular analysis, multiple regression, factor analysis, or something else. It doesn't really matter which computer program you're using (e.g., Statistical Package for the Social Sciences [SPSS], Statistical Analysis System [SAS]) unless it's a specialized program or one that is not commonly used.

If you've derived precise hypotheses, you may want to specify levels of statistical significance that will determine the meaning of the outcomes. This is not always necessary, however.

REVIEW QUESTIONS

1. In your own words, explain the logic of proportionate reduction of error (PRE) measures of associations.

2. In your own words, explain the purpose of regression analyses.

3. In your own words, distinguish between measures of association and tests of statistical significance.

4. Find a study that reports the statistical significance of its findings and critique the clarity with which it is reported.

CHAPTER 17

Reading and Writing Social Research

CHAPTER OVERVIEW

Social research is useless unless communicated effectively to others. Special skills are involved in reading the research of others and writing about your own.

Dragon Images/Shutterstock.com

Introduction

Meaningful scientific research is inextricably wedded to communication, but it's not always an easy or comfortable marriage. Scientists—social and other—are not necessarily good at communicating their methods and findings. Thus, it's often hard to read and understand the research of others. You may also find it difficult to write up your own research in ways that communicate your ideas effectively. This final chapter addresses these two problems.

We'll begin with reading social research, then we'll turn to writing it. Although I'll offer guidance on both topics, you'll find that practicing both is key. The more you read social science research, the easier it gets, and the same is true of writing it.

Reading Social Research

"Reading" is not as simple a task as it may seem, especially when it involves social research. First, you need to organize a review of the literature in order to focus on the resources that will help you the most. Then, when you actually sit down to read them, you'll need certain skills for doing so efficiently. Finally, you should know how to find and assess sources on the Internet.

Organizing a Review of the Literature

With the exception of some grounded theory methodologists, most social researchers begin the design of a research project with a review of the literature. Most original research is seen as an extension of what has previously been learned about a particular topic. A review of the literature is the way we learn what's already known and not known.

In most cases, you should organize your search of the literature around the key concepts you wish to study; alternatively, you may want to study a certain population: Iraqi War veterans, computer hackers, Catholic priests, gay athletes, and so forth. In any case, you'll identify a set of terms that represent your core interests.

Your college or university library will probably have several search programs you can use at the library or online. Let's say you're interested in designing a study of attitudes toward capital punishment. If your library provides access to InfoTrac College Edition or a similar program, you might discover, as I just did, 8,735 newspaper references and 5,489 periodical references to capital punishment. In such situations, InfoTrac College Edition is indexed to allow users to narrow the search, so I soon discovered 249 entries for "public opinion" on capital punishment. Some of the entries were bibliographic citations and some were full-text articles I could read online.

When reading or accessing an article online, you should see whether you can download it as a pdf version. This format replicates the document with the original pagination, which will be useful if you wish to quote or cite specific portions of the article. Sometimes you can read abstracts or even full articles for free, and sometimes you may need to buy the articles.

At the top of the page, you'll see a place to enter a name, title, topic, or other search term. When I entered the term, "racism," I received a list of 8,082 items, I turned to the "sort" options and asked for "Newest to Oldest." Here's some of what I was given.

- Leonard, David J.,
  Playing While White: Privilege and Power on and off the Field/Dr. David J. Leonard. Seattle: University Washington Press, 2017.

What do you think?

The Internet seems like a great place to get information for term papers, but some of your professors may be hesitant about Web sources, saying the quality of data on the Internet can't be trusted. What should you do? First, read this chapter. Then . . .

See the *What do you think? ... Revisited* box toward the end of the chapter.

Earl Babbie

- Corthron, Kara Lee,
   Truth of right now/Kara Lee Corthron.
   New York: Simon Pulse, 2017.
   PZ7.1.C673 Tr 2017
- Cloud Tapper, Suzanne,
   Views on slavery: in the words of enslaved Africans, merchants, owners, and abolitionists/ Suzanne Cloud Tapper.
   New York, NY: Enslow Publishing, 2017.
   E441.C595 2017
   Request in Jefferson or Adams Building Reading Rooms

Sometimes a simple Web search is a useful way to begin. Use a search engine such as Google, Bing, or Yahoo to look for Web resources on "capital punishment" or "death penalty." Be sure to use quotation marks to look for a phrase rather than using two separate words. You might also add "public opinion" to the request to narrow the field of possible resources. In general, online searches tend to turn up huge numbers of entries, most of which will not help you much. You'll need some time to separate the wheat from the chaff. Later in this chapter, I'll give you further guidelines for searching the Web.

No matter how you start the literature-review process, you should always consider a technique akin to snowball sampling, discussed in Chapter 7. Once you identify a particularly useful book or article, note which publications its author cites. Some of these will likely be useful. In fact, you'll probably discover some citations that appear again and again, suggesting that they're core references within the subject matter area you're exploring. This last point is important, because the literature review is not about providing "window dressing" in the form of a few citations. Rather,

it's about digging into the body of knowledge that previous researchers have generated—and taking advantage of that knowledge as you design your own inquiry.

Once you've identified some potential resources, you must read them and find anything of value to your project. Here are some guidelines for reading research publications.

Reading Journals versus Books

As you might have guessed, you don't read a social research report the way you'd read a novel. You can, of course, but it's not the most effective approach. Journal articles and books are laid out somewhat differently, so here are some initial guidelines for reading each.

Reading a Journal Article

In most journals, each article begins with an **abstract**, or a summary of the article. Read it first. It should tell you the purpose of the research, the methods used, and the major findings.

In a good detective or spy novel, the suspense builds throughout the book and is resolved in some kind of surprise ending. This is not the effect most scholarly writers are going for. Social research is purposely anticlimactic. Rather than stringing the reader along, dragging out the

---

abstract A summary of a research article. The abstract usually begins the article and states the purpose of the research, the methods used, and the major findings.

suspense over whether *X* causes *Y*, social researchers willingly give away the punch line in the abstract.

The abstract serves two main functions. First, it gives you a good idea as to whether you'll want to read the rest of the article. If you're reviewing the literature for a paper you're writing, the abstract tells you whether that particular article is relevant. Second, the abstract establishes a framework within which to read the rest of the article. It may raise questions in your mind regarding methods or conclusions, thereby creating an agenda to pursue in your reading. (It's not a bad idea to jot those questions down, to be sure you get answers to them.)

After you've read the abstract, you might go directly to the summary and/or conclusions at the end of the article. That will give you a more detailed picture of what the article is all about. (You can also do this with detective and spy novels; it makes reading them a lot faster, but maybe not as much fun.) Jot down any new questions or observations that occur to you.

Next, skim the article, noting the section headings and any tables or graphs. You don't need to study any of these things in your skimming, though it's okay to dally with anything that catches your attention. By the end of this step, you should start feeling familiar with the article. You should be pretty clear on the researcher's conclusions and have a general idea of the methods used in reaching them.

Now, when you carefully read the whole article, you'll have a good idea of where it's heading and how each section fits into the logic of the whole article. Keep taking notes. Mark any passages you think you might like to quote later on.

After carefully reading the article, it's a good idea to skim it quickly one more time. This way you get back in touch with the forest after having focused on the trees.

If you want to fully grasp what you've just read, find someone else to explain it to. If you're doing the reading in connection with a course,

research monograph A book-length research report, either published or unpublished. This is distinguished from a textbook, a book of essays, a novel, and so forth.

Earl Babbie

There's nothing like sinking your teeth into a good book.

you should have no trouble finding someone willing to listen. If you can explain it coherently to someone who has had no prior contact with the subject matter, however, you'll have an absolute lock on the material.

Reading a Book

The approach for reading articles can be adapted to reading a book-length report, sometimes also called a **research monograph**. These longer research reports cover the same basic terrain and structure. Instead of an abstract, the preface and opening chapter of the book should lay out the purpose, method, and main findings of the study. The preface tends to be written more informally and is usually easier to understand than an abstract.

As with an article, it's useful to skim through the book, getting a sense of its organization, such as its use of tables, graphs, or other visuals. You should come away from this step feeling somewhat familiar with the book. And, as I suggested in connection with reading an article, you should take notes as you go along, writing down things you observe and questions that are raised.

As you settle in to read the book more carefully, you should repeat this same process with each chapter. Read the opening paragraphs to get a sense of what's to come, and then skip to the concluding paragraphs for the summary. Skim the chapter to increase your familiarity with it, and then read more deliberately, taking notes as you go.

It's sometimes OK to skip portions of a scholarly book, unlike the way you were taught to read and appreciate literature. This all depends on your purpose in reading it in the first place. Perhaps there are only a few portions of the book that are relevant to your purposes. However, realize that if you're interested in the researcher's findings, you must pay some attention to the methods used (for instance, who was studied? How? When?) in order to judge the quality of the conclusions offered by the author.

Evaluation of Research Reports

In this section, I provide sets of questions you might ask in reading and evaluating a research report. I've organized these questions to parallel some of the preceding chapters in this book, to facilitate your getting more details on a topic if necessary. Although they're hardly exhaustive, I hope these questions will help you grasp the meanings of research reports you read and alert you to potential problems in them.

Theoretical Orientations

- Is there a theoretical aspect to the study, or do no references to theory appear?
- Can you identify the researcher's chief paradigm or theoretical orientation? (Authors quoted in the report's review of the literature and elsewhere may offer a clue.)
- On the other hand, is the author attempting to refute some paradigm or theory?
- Is a theory or hypothesis being tested?
- In what way has the theoretical orientation shaped the methodology used in the study, such as the data-collection technique and the choice of which data were collected and which were ignored?
- Is the methodology used appropriate for the theoretical issues involved?

Research Design

- What was the purpose of the study: exploration, description, explanation, or some combination of these?
- Who conducted the research? Who paid for it, if anyone? What motivated the study? If the study's conclusions happen to correspond to the interests of the sponsor or researcher, this doesn't disqualify the conclusions, but you'll want to be especially wary.
- What was the unit of analysis? Was it appropriate for the purpose of the study? Are the conclusions drawn from the research appropriate for the unit of analysis? For example, have the researchers studied cities and ended up with assertions about individuals?
- Is this a cross-sectional or a longitudinal study? Be especially wary of longitudinal assertions being made on the basis of cross-sectional observations.
- If longitudinal data have been collected, have comparable measurements been made at each point in time? In the case of survey data, have the same questions been asked each time? If the report compares, say, crime or poverty rates, are they defined the same way each time? (Definitions of poverty, for example, change frequently.)
- If a panel study has been conducted, how many people dropped out over the course of the study?

Measurement

- What are the names of the concepts under study?
- Has the researcher delineated different dimensions of the variables? Do the analysis and reporting maintain those distinctions?
- What indicators—either qualitative or quantitative—have been chosen as measures of those dimensions and concepts? Is each indicator a valid measure of what it's intended to measure? What else could the indicator be a measure of? Is it a reliable measure? Has the reliability been tested?
- What is the level of measurement of each variable: nominal, ordinal, interval, or ratio? Is it the appropriate level?
- Have composite measurements (indexes, scales, or typologies) been used? If so, are they appropriate for the purpose of the study? Have they been constructed correctly?

Sampling

- Was it appropriate to study a sample, or should all elements have been studied? Remember, it's not always feasible to select a random sample.
- If sampling was called for, were probability-sampling methods appropriate, or would a purposive, snowball, or quota sample have been better? Has the appropriate sampling design been used?
- What population does the researcher want to draw conclusions about?

- What is the researcher's purpose? If it's statistical description, then rigorous probability-sampling methods are called for.
- If a probability sample has been selected, what sampling frame has been used? Does it appropriately represent the population that interests the researcher? What elements of the population have been omitted from the sampling frame, and what extraneous elements have been included?
- What specific sampling techniques have been employed: simple random sampling, systematic sampling, or cluster sampling? Has the researcher stratified the sampling frame prior to sampling? Have the stratification variables been chosen wisely? That is, are they relevant to the variables under study?
- How large a sample was selected? What percentage of the sample responded? Are there any likely differences between those who responded and those who didn't?
- Even assuming that the respondents are representative of those selected in the sample, what sampling error do you expect from a sample of this size?
- Has the researcher tested for representativeness: comparing the gender distribution of the population and of respondents, for example, or their ages, ethnicity, education, or income?
- Ultimately, do the studied individuals (or other units of analysis) represent the larger population from which they were chosen? That is, do conclusions drawn about the sample tell us anything meaningful about populations or about life in general?
- If probability sampling and statistical representation were not appropriate for the study—in a qualitative study, for example—have subjects and observations been selected in such a way as to provide a broad overview of the phenomenon being examined? Has the researcher paid special attention to deviant or disconfirming cases?

Experiments

- What is the primary dependent variable in the experiment? For example, what effect is the experimenter trying to achieve?
- What is the experimental stimulus?
- What other variables are relevant to the experiment? Have they been measured?
- How has each variable been defined and measured? What potential problems of validity and reliability do these definitions and measurements raise?

- Has a proper control group been used? Have subjects been assigned to the experimental and control groups through random selection or by matching? Has it been done properly? Has the researcher provided any evidence of the initial comparability of experimental and control-group subjects?
- Have there been pretest and posttest measurements of the dependent variable?
- What is the chance of a placebo (or "Hawthorne") effect in the experiment? Has any attention been given to the problem? Does the study employ a double-blind design, for example?
- Are there any problems of internal invalidity: history, maturation, testing, instrumentation, statistical regression, selection bias, experimental mortality, or demoralization?
- Are there issues of external invalidity? How has the experimenter ensured that the laboratory findings will apply to life in the real world?

Survey Research

- Does the study stand up to all the relevant questions regarding sampling?
- What questions were asked of respondents? What was the precise wording of the questions? Be wary of researcher reports that provide only paraphrases of the questions.
- If closed-ended questions were asked, were the answer categories provided appropriate, exhaustive, and mutually exclusive?
- If open-ended questions were asked, how have the answers been categorized? Has the researcher guarded against his or her own bias creeping in during the coding of open-ended responses?
- Are all the questions clear and unambiguous? Could respondents have misinterpreted them? If so, could the answers given mean something other than what the researcher has assumed?
- Were the respondents capable of answering the questions asked? If not, they may have answered anyway, but their answers might not mean anything.
- Are any of the questions double-barreled? Look for conjunctions (such as *and, or*). Are respondents being asked to agree or disagree with two ideas, when they might like to agree with one and disagree with the other?

- Do the questions contain negative terms? If so, respondents may have misunderstood them and answered inappropriately.
- Is there a danger of social desirability in any of the questions? Is any answer so right or so wrong that respondents may have answered on the basis of what people would think of them?
- How would you yourself answer each item? As a general rule, test all questionnaire items by asking yourself how you would answer. Any difficulty you might have in answering might also apply to others. Then, try to assume different points of view (for example, liberal and conservative, religious and unreligious) and ask how the questions might sound to someone with each point of view.
- Has the researcher conducted a secondary analysis of previously collected data? If so, determine the quality of the research that produced the data originally. Also, are the data available for analysis appropriate for the current purposes? Do the questions originally asked reflect adequately on the variables now being analyzed?

Field Research

- What theoretical paradigm has informed the researcher's approach to the study?
- Has the researcher set out to test hypotheses or generate theory from the observations? Or is there no concern for theory in the study?
- What are the main variables in this study? How have they been defined and measured? Do you see any problems of validity?
- How about reliability? Would another researcher, observing the same events, classify things the same way?
- Is there any chance that the classification of observations has been influenced by the way those classifications will affect the research findings and/or the researcher's hypotheses?
- If descriptive conclusions have been drawn— for example, "the group's standards were quite conservative"—what are the implicit standards being used?
- How much can the study's findings be generalized to a broader sector of society? What claims has the researcher made in this regard? What is the basis for such claims?
- If people have been interviewed, how were they selected? Do they represent all appropriate types?
- How much did the researcher participate in the events under study? How might

that participation have affected the events themselves?
- Did the researcher reveal his or her identity as a researcher? If so, what influence could that revelation have had on the behavior of those being observed?
- Does the research indicate any personal feelings—positive or negative—about those being observed? If so, what effect might these feelings have had on the observations that were made and the conclusions that were drawn from them?
- How has the researcher's own cultural identity or background affected the interpretation of what has been observed?

Content Analysis

- What are the key variables in the analysis? Are they appropriate for the research questions being asked?
- What is the source and form of data being analyzed? Are they appropriate for the research questions being asked?
- Is the time frame of the data being analyzed appropriate for the research questions?
- What is the unit of analysis?
- If a quantitative analysis has been conducted, (1) has an appropriate sample been selected from the data source and (2) have the appropriate statistical techniques been used?
- If a qualitative analysis has been conducted, (1) has an appropriate range of data been examined and (2) are the researcher's conclusions logically consistent with the data presented?

Analyzing Existing Statistics

- Who originally collected the data being reanalyzed? Were there any flaws in the data-collection methods? What was the original purpose of the data collection? Would that have affected the data that was collected?
- What was the unit of analysis of the data? Is it appropriate for the current research question and the conclusions being drawn? Is there a danger of the ecological fallacy?
- When were the data collected? Are they still appropriate for present concerns?
- What are the variables being analyzed in the present research? Were the definitions used by the original researchers appropriate for present interests?

Comparative and Historical Research

- Is this a descriptive or an explanatory study? Does it involve cross-sectional comparisons or changes over time?
- What is the unit of analysis in this study (for example, country, social movement)?
- What are the key variables under study? If it is an explanatory analysis, what causal relationships are examined?
- Does the study involve the use of other research techniques, such as existing statistics, content analysis, surveys, or field research? Use the guidelines elsewhere in this section to assess those aspects of the study.
- Is the range of data appropriate for the analysis: for example, the units being compared or the number of observations made for the purpose of characterizing units?
- If historical or other documents are used as a data source, who produced them and for what purposes? What biases might be embedded in them? Diaries kept by members of the gentry, for example, will not reflect the life of peasants of the same time and country.

Evaluation Research

- What is the social intervention being analyzed? How has it been measured? Are there any problems of validity or reliability?
- Have the appropriate people (or other units of analysis) been observed?
- How has "success" been defined? Where would the success be manifested—in individuals, in organizations, in crime rates? Has it been measured appropriately?
- Has the researcher judged the intervention a success or a failure? Is the judgment well founded?
- Who paid for the research, and who actually conducted it? Can you be confident of the researcher's objectivity? Did the sponsor interfere in any way?

Data Analysis

- Did the purpose and design of the study call for a qualitative or a quantitative analysis?
- How have nonstandardized data been coded? This question applies to both qualitative and quantitative analysis. To what extent were the codes (1) based on prior theory or (2) generated by the data?
- Has the researcher undertaken all relevant analyses? Have all appropriate variables

been identified and examined? Could the correlation observed between two variables have been caused by an antecedent third variable, making the observed relationship spurious?
- Does a particular research finding really matter? Is an observed difference between subgroups, for example, a large or meaningful one? Are there any implications for action?
- Has the researcher gone beyond the actual findings in drawing conclusions and implications?
- Are there logical flaws in the analysis and interpretation of data?
- Have the empirical observations of the study revealed new patterns of relationships, providing the bases for grounded theories of social life? Has the researcher looked for disconfirming cases that would challenge the new theories?
- Are the statistical techniques used in the analysis of data appropriate for the levels of measurement of the variables involved?
- If tests of statistical significance were used, have they been interpreted correctly? Has statistical significance been confused with substantive significance?

Reporting

- Has the researcher placed this particular project in the context of previous research on the topic? Does this research add to, modify, replicate, or contradict previous studies?
- In general, has the researcher reported the details of the study design and execution fully? Are there parts of the report that seem particularly vague or incomplete in the reporting of details?
- Has the researcher reported any flaws or shortcomings in the study design or execution? Are there any suggestions for improving research on the topic in the future?

I hope this section will prove useful to you in reading and understanding social research. The exercises at the end of this chapter will walk you through the reading of two journal articles: one qualitative and one quantitative. As I said earlier, you'll find that your proficiency in reading social research reports will mature with practice.

Before discussing how to go about creating social research reports for others to read, let's look at how to read and evaluate data from an increasingly popular source of information—the Internet.

Using the Internet Wisely

In the closing decade of the twentieth century, the World Wide Web developed into a profoundly valuable tool for social research. As it expands exponentially, the Web is becoming the mind of humanity, the repository of human knowledge, opinions, and beliefs—carrying with it intellectual insights but also misconceptions and outright bigotry. Clearly, it will continue to evolve as an increasingly powerful entity. As with gunpowder and television, the power of the technology does not guarantee that it will always be used wisely. I have opted to encourage use of the Web rather than opposing it, but I am mindful of the problems that make many of my colleagues more cautious.

In this section of the chapter, I share websites useful to social researchers and give some general advice on searching the Web. Then I address the major problems inherent in using the Web and suggest ways to avoid them.

Some Useful Websites

I want to mention a few key websites here and, more importantly, offer advice on how to search the Web.

As the World Wide Web has evolved, the online resources available to social researchers (and others) have expanded enormously. Moreover, I'll bet you are already able to search the Web for information of interest to you. To focus your interest for present purposes, let me mention just a few generally useful websites that you might like to check out:

- General Social Survey (GSS)
- U.S. Bureau of the Census
- USA Statistics in Brief
- Statistical Resources on the Web, University of Michigan
- IFLA—The Social Science Virtual Library
- QUALPAGE: Resources for Qualitative Research
- CAQDAS: Computer Assisted Qualitative Data Analysis Software, University of Surrey, England

Now, let's assume you need some information that you suspect is somewhere on the Web, but you don't know where to locate it. Here are some ideas to help you become a Web detective.

Searching the Web

There are millions and millions of pages of information on the Web. Estimating the number

"Don't believe everything you read on the Internet just because there's a picture with a quote next to it."

—Abraham Lincoln

of "facts" or pieces of data on the Web would be impossible, but most of the factual questions you might have can be answered by using the Internet. Finding answers, however, involves skill.

Go to Google or another **search engine** and search for "infant mortality rate." If you put your request inside quotation marks, as I just did, the search engine will look for that exact phrase instead of reporting websites that happen to have all three words. Figure 17-1 presents the initial results I received.

As you can see, our search yielded over half a million websites potentially relevant to our inquiry. Some of the websites shown in the figure focus on the definition of infant mortality rate; others provide data pertaining to the world or to specific countries. Some would let us compare different countries; others show changes in infant mortality rate (IMR) over time. Let's say we are specifically interested in infant mortality in Africa. To search for this, I have added the word *Africa* to the request: Africa "infant mortality rate."

Like many other search engines, Google interpreted this as a request to find websites that contain the word *Africa* plus the exact phrase *infant mortality rate*. Figure 17-2 presents the first set of results.

While some of the websites in Figure 17-2 are still quite general, we can assume that they contain information about African IMRs. Some in the list deal specifically with Africa. The first Web link is to Wikipedia, a free encyclopedia

---

search engine A computer program designed to locate where specified terms appear on websites throughout the World Wide Web.

Infant mortality rate - The World Factbook
https://www.cia.gov/.../2091rank.html ▼ Central Intelligence Agency (CIA) ▼
Infant mortality rate compares the number of deaths of infants under one year old in a
given year per 1,000 live births in the same year. This rate is often used as ...

List of countries by infant mortality rate - Wikipedia, the free ...
en.wikipedia.org/.../List_of_countries_by_**infant_mortality_rat**... ▼ Wikipedia ▼
The **infant mortality rate** (IMR) is the number of deaths of infants under one year old
per 1,000 live births. This rate is often used as an indicator of the level of ...

Infant mortality - Wikipedia, the free encyclopedia
en.wikipedia.org/wiki/Infant_mortality ▼ Wikipedia ▼
Infant mortality rate (**IMR**) is the number of deaths of children less than one year of age
per 1000 live births. The rate for a given region is the number of children ...
Infant mortality rate - Causes - Measuring IMR - Societal infrastructure

Infant Mortality - Centers for Disease Control and Prevention
www.cdc.gov/.../inf... ▼ United States Centers for Disease Control and Preve... ▼
Oct 1, 2012 - The death of a baby before his or her first birthday is called infant mortality.
The **infant mortality rate** is an estimate of the number of infant deaths ...

Mortality rate, infant (per 1,000 live births) | Data | Table
data.worldbank.org › Indicators ▼ World Bank ▼
Infant mortality rate is the number of infants dying before reaching one year of age, per
1,000 live births in a given year. Estimates developed by the UN ...

Images for "Infant Mortality Rate"
Report images

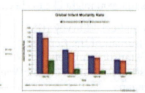

More images for "Infant Mortality Rate"

Infant Mortality Rate - The Henry J. Kaiser Family Foundation
kff.org/global-indicator/**infant-mortality-rate**/ ▼ Kaiser Family Foundation ▼
Infant Mortality Rate (Total Deaths per 1,000 Live Births) .... **Infant Mortality Rate**: The
number of infant deaths in a given year divided by the number of live births ...

FIGURE 17-1

Search for "Infant Mortality Rate."

Source: © 2014 Google. Downloaded May 9, 2014, 1:00 P.M. CST.

The infant mortality rate (IMR) is the number of deaths of infants under one year old per 1,000 live births. This rate is often used as an indicator of the level of health in a country. The infant mortality rate of the world is **49.4** according to the United Nations and 42.09 according to the CIA World Factbook.

List of countries by **infant mortality rate** - Wikipedia, the free ...
en.wikipedia.org/wiki/List_of_countries_by_**infant_mortality_rate**   Wikipedia ▾

Feedback

Infant mortality rate - The World Factbook
https://www.cia.gov/.../2091rank.html ▾ Central Intelligence Agency (CIA) ▾
Infant mortality rate compares the number of deaths of infants under one year old in a
given year per 1,000 live births in the same year. .... 51, South **Africa**. 41.61.

List of countries by **infant mortality rate** - Wikipedia, the free ...
en.wikipedia.org/.../List_of_countries_by_**infant_mortality_rat**... ▾ Wikipedia ▾
List of countries by **infant mortality rate** ... The **infant mortality rate** (IMR) is the
number of deaths of infants under one year old per 1,000 .... South **Africa**, 42.15, - >.

WHO | Infant mortality
www.who.int/gho/urban_health/.../en/ ▾ World Health Organization ▾
For example, the urban poorest 20% in **Africa** have witnessed a decrease in **infant
mortality rate** from 99 per 1000 live births in the 1990s to 70 per 1000 live ...

Africa's infant mortality rate threatens growth potential - SA...
www.sabc.co.za/.../**Africa's-infa**... ▾ South African Broadcasting Corporation ▾
Aug 1, 2013 - **Africa's** growth potential is at risk if the continent does not take care of its
new- borns who continue to die at birth.

Infant mortality rate by country - Thematic Map - **Africa** - M...
www.indexmundi.com/map/?v=29&r=af&l=en ▾
Reading Data. Please Wait. Rendering Map. Please Wait. ▢ 9 - 22 ▢ 22 - 34 ▢ 34 - 47 ▢
47 - 60 ▢ 60 - 74 ▢ 74 - 88 ▢ 88 - 102 ▢ 102 - 116. Mauritius, 11.2.

Images for **Africa "infant mortality rate"**   Report images

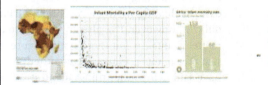

More images for **Africa "infant mortality rate"**

FIGURE 17-2

Search for "Africa 'Infant Mortality Rate'."

Source: © 2014 Google. Downloaded May 9, 2014, 1:05 P.M. CST.

How to Do It

Using Google Scholar and Other Online Resources

In searching the Web for research materials, you can narrow your focus with Google Scholar at scholar.google.com. Let's say you're interested in studying same-sex marriage and want to know what research has already been done on that topic. Enter that phrase in the box and click the "Search" button. Whereas a regular Google search would have turned up many websites that used the words "same-sex marriage" but were not much use in a research literature review, Google Scholar will provide you with richer pickings, although you'll still need to judge the quality of documents returned through your search.

You can also take advantage of the "Advanced Scholar Search" to specify a set of words, indicating that all must appear in an article—or just some of them. You can specify a particular author or journal, and you can indicate which scholarly field you're interested in, so that the search is limited to articles in that field.

There are some online resources that are especially useful for social science research. Your college library may have a license for JSTOR, an online repository for hundreds of academic journals. There are over a hundred in sociology alone. If you have access to this service, you can search for specific titles or authors, or you can do a broader search for articles on a particular topic. One limitation is that the most recent issues of journals are not available, due to agreements with the journal publishers, who, understandably, would like people to subscribe to their journals. The window of noncoverage is typically 3 to 5 years, so anything older than that is freely available for download and use.

Another useful resource is Sociological Abstracts, an online compendium of journal-article abstracts. ProQuest, the publisher, states that their site "abstracts and indexes the international literature in sociology and related disciplines in the social and behavioral sciences. The database provides abstracts of journal articles and citations to book reviews drawn from over 1,800+ serials publications, and also provides abstracts of books, book chapters, dissertations, and conference papers" (http://www.csa.com/factsheets/socioabs-set-c.php. Accessed December 14, 2019). While you will not be able to download the full text of articles, you will be able to locate those most relevant to your research interests, and then set about finding the actual articles in the journals where they were published.

These are only two examples of online resources that may be available at your school library.

---

compiled by the Web community. The rapid growth of Wikipedia has been a source of conversation and concern among academics. No one questions how extensive or user-friendly it is, but some worry that entries are not always accurate and that errors may go unnoticed. Rarely, true mischief has been perpetrated, with opposing political candidates maliciously altering each other's entries in the encyclopedia, for example. In one response to academic concerns, the history department at Middlebury College, one of the nation's most highly rated liberal arts colleges, told students in 2007 they could not cite Wikipedia as a source in term papers and exams. Lest this be seen as a condemnation of Wikipedia, however, Middlebury clarified:

> While the department did vote to restrict the use of the online encyclopedia as a source in course work, it did not suggest, as some reports had it, that students should be prevented from accessing Wikipedia or should not use it as a research tool. In fact, the department praised Wikipedia as "extraordinarily convenient and, for some general purposes, extremely useful."

(Middlebury College 2007)

Conducting this search on your own and visiting the Web links that result is a useful exercise. You'll find that some of the sites are discussions of the topic rather than tables of data. Others present a limited set of data ("selected countries"). Thus, compiling a list of Web links like this is a step along the way to obtaining relevant data, but it is not the final step. See "How to Do It: Using Google Scholar and Other Online Resources" for more.

Evaluating the Quality of Internet Materials

You now know enough about Web searches to begin learning through experience. You'll quickly learn that finding data on the Web is relatively easy. Evaluating what you've found is a bit more difficult, however. I've already alluded to the matter of quality, but there's much more to be said on the topic. In fact, many other people have said many other things about it. What do you suppose is your best source of such advice? If you said, "The Web," you got it.

Open up a search engine and ask it to find websites having to do with "evaluating websites." (Using alternative spellings can yield more results; for example, you could also enter "evaluating Web sites" and get a similar yet different set of entries.) Figure 17-3 gives you some idea of the extent of advice available to you.

As you can tell from the ".edu" in the addresses of most of these sites, this is a topic of concern for colleges and universities. Although each of the various sites approaches the topic differently, the guidance they offer has some elements in common. You would do well to study one or more of the sites in depth. In the meantime, here's an overview of the most common questions and suggestions for evaluating the data presented on websites.

1. *Who or what is the author of the website?* The two biggest risks you face in getting information from the Web are (1) bias and (2) sloppiness. The democratic beauty of the Web is its accessibility to such a large proportion of the population and the lack of censorship. These pluses also present dangers, in that just about anyone can put just about anything on the Web. The first thing you should note, therefore, is who the author of the website is—either an organization or an individual.

2. *Is the site advocating a particular point of view?* Many of the sites on the World Wide Web have been created to support a particular political, religious, nationalistic, social, or other point of view. This fact does not necessarily mean that the data they present are false, though that's sometimes the case. Beyond outright lying, however, you can be relatively sure that the website will present only data supporting its particular point of view. You can usually tell whether a website is reasonably objective or has an ax to grind, and you should be wary of those that go overboard to convince you of something.

3. *Does the website give accurate and complete references?* When data are presented, can you tell where they came from—how they were created? If the website is reporting data collected by someone else, are you given sufficient guidance to locate the original researchers? Or, if the data were compiled by the website authors, do they provide you with sufficiently detailed descriptions of their research methods? If data are presented without such clarifications, you should move on.

4. *Are the data up to date?* Another common problem on the Web is that materials may be posted and forgotten. Hence, you may find data reporting crime rates, chronicles of peace negotiations, and so forth that are out-of-date. Be sure that the data you obtain are timely for your purposes.

5. *Are the data official?* It's often a good idea to find data at official government research sites, such as the Bureau of the Census, the Bureau of Labor Statistics, the National Center for Health Statistics, and others. FedStats is a good launching point for finding data among some 100 federal research agencies. As we saw in Chapter 11, data presented by official agencies are not necessarily "The Truth," but they are grounded in a commitment to objectivity and have checks and balances to support them in achieving that goal.

6. *Is it a university research site?* Like government research agencies, university research centers and institutes are usually safe resources, committed to conducting professional research and having checks and balances (such as peer review) to support their achieving that. Throughout this book, I've mentioned the General Social Survey, conducted regularly by the National Opinion Research Center at the University of Chicago. You could use data presented here with confidence: confidence in the legitimacy of the data and confidence that your instructor will not question your use of that resource.

7. *Do the data seem consistent with data from other sites?* Verify (cross-check) data whenever possible. We've already seen that a Web search is likely to turn up more than one possible source of data. Take the time to compare what they present. If several websites present essentially the same data, you can use any of those sources with confidence.

As with so many things, your effective use of the Web will improve with practice. Moreover, the Web itself will be evolving alongside your use of it.

Citing Internet Materials

If you use materials from the Web, you must provide a bibliographic citation that allows your reader to locate the original materials—to see them in context. This also protects you from the serious problem of plagiarism, discussed a little later in this chapter.

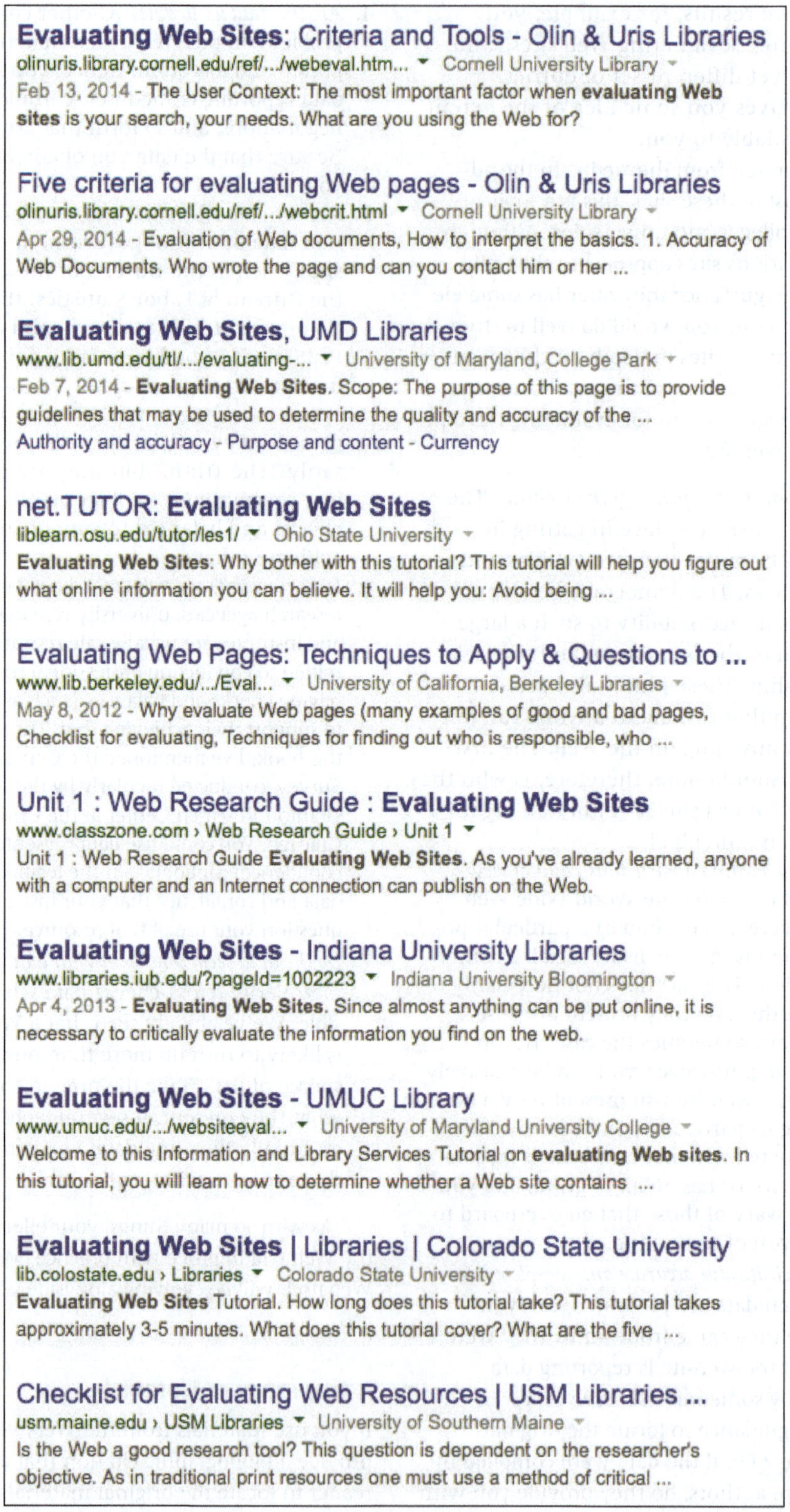

Evaluating Web Sites: Criteria and Tools - Olin & Uris Libraries
olinuris.library.cornell.edu/ref/.../webeval.htm... ▼ Cornell University Library ▼
Feb 13, 2014 - The User Context: The most important factor when **evaluating Web sites** is your search, your needs. What are you using the Web for?

Five criteria for evaluating Web pages - Olin & Uris Libraries
olinuris.library.cornell.edu/ref/.../webcrit.html ▼ Cornell University Library ▼
Apr 29, 2014 - Evaluation of Web documents, How to interpret the basics. 1. Accuracy of Web Documents. Who wrote the page and can you contact him or her ...

Evaluating Web Sites, UMD Libraries
www.lib.umd.edu/tl/.../evaluating-... ▼ University of Maryland, College Park ▼
Feb 7, 2014 - **Evaluating Web Sites**. Scope: The purpose of this page is to provide guidelines that may be used to determine the quality and accuracy of the ...
Authority and accuracy - Purpose and content - Currency

net.TUTOR: **Evaluating Web Sites**
liblearn.osu.edu/tutor/les1/ ▼ Ohio State University ▼
Evaluating Web Sites: Why bother with this tutorial? This tutorial will help you figure out what online information you can believe. It will help you: Avoid being ...

Evaluating Web Pages: Techniques to Apply & Questions to ...
www.lib.berkeley.edu/.../Eval... ▼ University of California, Berkeley Libraries ▼
May 8, 2012 - Why evaluate Web pages (many examples of good and bad pages, Checklist for evaluating, Techniques for finding out who is responsible, who ...

Unit 1 : Web Research Guide : **Evaluating Web Sites**
www.classzone.com › Web Research Guide › Unit 1 ▼
Unit 1 : Web Research Guide **Evaluating Web Sites**. As you've already learned, anyone with a computer and an Internet connection can publish on the Web.

Evaluating Web Sites - Indiana University Libraries
www.libraries.iub.edu/?pageId=1002223 ▼ Indiana University Bloomington ▼
Apr 4, 2013 - **Evaluating Web Sites**. Since almost anything can be put online, it is necessary to critically evaluate the information you find on the web.

Evaluating Web Sites - UMUC Library
www.umuc.edu/.../websiteeval... ▼ University of Maryland University College ▼
Welcome to this Information and Library Services Tutorial on **evaluating Web sites**. In this tutorial, you will learn how to determine whether a Web site contains ...

Evaluating Web Sites | Libraries | Colorado State University
lib.colostate.edu › Libraries ▼ Colorado State University ▼
Evaluating Web Sites Tutorial. How long does this tutorial take? This tutorial takes approximately 3-5 minutes. What does this tutorial cover? What are the five ...

Checklist for Evaluating Web Resources | USM Libraries ...
usm.maine.edu › USM Libraries ▼ University of Southern Maine ▼
Is the Web a good research tool? This question is dependent on the researcher's objective. As in traditional print resources one must use a method of critical ...

Google Inc.

FIGURE 17-3

Search for "Evaluating Websites."

Source: © 2014 Google. Downloaded May 9, 2014, 1:25 P.M. CST.

There are many standardized formats for bibliographic citations, illustrated later in this chapter. (See "How to Do It: Citing Bibliographic Sources.") Web materials, unfortunately, don't always fit those familiar formats.

Fortunately, each of these organizations—and many, many others—have risen to the challenge of Web citations. If you don't believe me, go to your favorite search engine and look for "Web citations." You'll find plenty of guidance.

Your instructor may prefer a specific format for Web citations. However, here are the elements commonly suggested for inclusion:

- The **URL** or Web address. For example, https://www.census.gov/newsroom/press -releases/2015/cb15-56.html provides data on the fastest-growing areas of Florida from mid-2013 to mid-2014. So if I tell you Miami-Fort Lauderdale-West Palm Beach grew by 66,000 residents in that one year, you can go directly to the source of my data.
- The date and time when the site was accessed. As you may have discovered, many online sites change from time to time. It may be useful for the reader to know when you visited the site in question. Some editing guides say to include this, whereas others say not to. When in doubt, check with your instructor or publisher. It's usually better to have too much information than too little. The data cited above were accessed March 30, 2015.
- If you're citing textual materials, there may very well be an author and title, as well as publishing information. These should be cited the same way you would cite printed materials, as in the following: Doe, John. 2015. "How I Learned to Love the Web." *Journal of Web Worship* 5 (3): 22–45.
- Sometimes, you'll use the Web to read a published journal article, locating it with InfoTrac College Edition or another vehicle. Such materials may be presented in a print format, with page numbers. If so, cite the appropriate page number. Lacking that, you may be able to cite the section where the materials in question appeared. The goal in all this is to help your reader locate the original Web materials you're using. Although you sometimes cannot give a precise location in an article posted to a website, most browsers allow users to search the site for a specified word or phrase and thus locate the materials being cited.

The fluidity of Web citations is a problem being addressed by the International DOI Foundation

(IDF), Inc., the abbreviation standing for "Digital Object Identifier." In this system, digital content, such as a journal article, is registered with the IDF and assigned a unique DOI number. The registered content is then locatable via a Web link that persists even if the article has been removed from its original Web location. The DOI number, when available, should be included when citing a source. You can learn more about this valuable citation tool at the Digital Object Identifier website.

Writing Social Research

Unless research is properly communicated, all the efforts devoted to the various procedures discussed throughout this book will go for naught. This means, first and foremost, that good social reporting requires good English or Spanish or whatever language you use. Whenever we ask the figures "to speak for themselves," they tend to remain mute. Whenever we use unduly complex terminology or construction, communication suffers.

My first advice to you is to read and reread (at approximately three-month intervals) an excellent small book by William Strunk Jr., and E. B. White, *The Elements of Style* (1999). If you do this faithfully, and if even 10 percent of the content rubs off, you'll stand a good chance of making yourself understood and your findings appreciated.

Next, you need to understand that scientific reporting has several functions. First, your report should communicate a body of specific data and ideas. You should provide those specifics clearly and with sufficient detail to permit an informed evaluation by others. Second, you should view your report as a contribution to the general body of scientific knowledge. While remaining appropriately humble, you should always regard your research report as an addition to what we know about social behavior. Finally, the report should stimulate and direct further inquiry. See "Applying Concepts in Everyday Life: Communication Is the Key" for more on the importance of knowing how to read and write well.

URL Web address, typically beginning with "http://"; stands for "uniform resource locator" or "universal resource locator."

Some Basic Considerations

Despite these general guidelines, different reports serve different purposes. A report appropriate for one purpose might be wholly inappropriate for another. This section deals with some of the contexts that affect choices in writing.

Audience

Before drafting your report, ask yourself who you hope will read it. Normally you should make a distinction between scientists and general readers. If the report is written for the former, you can make certain assumptions about their existing knowledge and therefore summarize certain points rather than explain them in detail. Similarly, you can use more-technical language than would be appropriate for a general audience.

At the same time, remain aware that any science has its factions and cults. Terms, assumptions, and special techniques familiar to your immediate colleagues might only confuse other scientists. The sociologist of religion writing for a general sociology audience, for example, should explain previous findings in more detail than he or she would if addressing an audience of sociologists of religion.

Form and Length of Report

My comments here apply to both written and oral reports. Each form, however, affects the nature of the report.

It's useful to think about the variety of reports that might result from a research project. To begin, you may wish to prepare a short *research note* for publication in an academic or technical journal. Such reports are approximately one to five pages long (typed, double-spaced) and should be concise and direct. In a small amount of space, you can't present the state of the field in any detail, so your methodological notes must be abbreviated. Basically, you should tell the reader why you feel your findings justify a brief note, then tell what those findings are.

Often researchers must prepare reports for the sponsors of their research. These reports can vary greatly in length. In preparing such a report, you should bear in mind your audience—scientific or lay—and their reasons for sponsoring the project in the first place. It's both bad politics and bad manners to bore the sponsors with research findings that have no interest or value to them. At the same time, it may be useful to summarize how the research has advanced basic scientific knowledge (if it has).

Working papers are another form of research reporting. In a large and complex project especially, you'll find comments on your analysis and the interpretation of your data useful. A working paper constitutes a tentative presentation with an implicit request for comments. Working papers can also vary in length, and they may present all of the research findings of the project or only a portion of them. Because your professional reputation is not at stake in a working paper, feel free to present tentative interpretations that you can't altogether justify—identifying them as such and asking for evaluations.

Many research projects result in papers delivered at professional meetings. These often serve the same purpose as working papers. You can present findings and ideas of possible interest to your colleagues and ask for their comments. Although the length of such professional papers varies, depending on the organization of the meetings, it's best to say too little rather than too much. Although a working paper may ramble somewhat through tentative conclusions, conference participants should not be forced to sit through an oral unveiling of the same. Interested listeners can always ask for more details later, and uninterested ones can gratefully escape.

Probably the most popular research report is the article published in an academic journal. Again, lengths vary, and you should examine the lengths of articles previously published by the journal in question. As a rough guide, however, 25 typed pages is a good length. A subsequent section on the organization of the report is based primarily on the structure of a journal article, so I'll say no more at this point except to indicate that student term papers should follow this model. As a general rule, a term paper that would make a good journal article also makes a good term paper.

A book, of course, represents the most prestigious form of research reporting. It has the length and detail of a working paper but is more polished. Because publishing research findings as a book lends them greater substance and worth, you have a special obligation to your audience. Although some colleagues may provide comments, possibly leading you to revise your ideas, other readers may be led to accept your findings uncritically.

Aim of the Report

Earlier in this book, we considered the different purposes of social research projects. In preparing your report, keep these different purposes in mind.

Some reports focus primarily on the exploration of a topic. As such, their conclusions are tentative and incomplete. If you're writing this sort of report, clearly indicate to your audience the exploratory aim of the study and present the shortcomings of the particular project. An exploratory report points the way to more-refined research on the topic.

Most research reports have a descriptive element, recounting their subject matter. In yours, carefully distinguish those descriptions that apply only to the sample and those that apply to the population. Give your audience some indication of the probable range of error in any inferential descriptions you make.

Many reports have an explanatory aim: pointing to causal relationships among variables. Depending on your probable audience, carefully delineate the rules of explanation that lie

behind your computations and conclusions. Also, as in the case of description, give your readers some guide to the relative certainty of your conclusions.

If your intention is to test a hypothesis based in theory, you should make that hypothesis clear and succinct. Specify what will constitute acceptance or rejection of the hypothesis and how either of those reflects on the theoretical underpinnings.

Finally, some research reports propose action. For example, if you've studied prejudice, you may suggest in your report how prejudice can be reduced on the basis of your research findings. This suggestion may become a knotty problem for you, however, because your values and orientations may have interfered with your proposals. Although it's perfectly legitimate for such proposals to be motivated by personal values, you must ensure that the data actually warrant the specific actions you've proposed. Thus, you should be especially careful to spell out the logic by which you move from empirical data to proposed action.

Organization of the Report

Although the various forms and purposes of reports somewhat affect the way they are organized, knowing a general format for presenting research data can be helpful. The following comments apply most directly to a journal article, but with some modification they apply to most forms of research reports as well.

Purpose and Overview

It's always helpful if you begin with a brief statement of the purpose of the study and the main findings of the analysis. In a journal article, as we've seen, this overview sometimes takes the form of an abstract.

Some researchers find this difficult to do. For example, your analysis may have involved considerable detective work, with important findings revealing themselves only as a result of imaginative deduction and data manipulation. You may wish, therefore, to lead the reader through the same exciting process, chronicling your discoveries with a degree of suspense and surprise. To the extent that this form of reporting gives an accurate picture of the research process, it has considerable instructional value. Nevertheless, many readers may not be interested in following your entire research account, and not knowing the purpose and general conclusions in advance may make it difficult for them to understand the significance of the study.

An old forensic dictum says, "Tell them what you're going to tell them; tell them; and tell them what you told them." You would do well to follow this dictum.

Review of the Literature

Next, you must indicate where your report fits into the general body of scientific knowledge. After presenting the general purpose of your study, you should bring the reader up to date on the previous research in the area, pointing to general agreements and disagreements among the previous researchers. Your review of the literature should lay the groundwork for your own study, showing why your research may have value in the larger scheme of things.

In some cases, you may wish to challenge previously accepted ideas. Carefully review the studies that have led to the acceptance of those ideas, then indicate the factors that have not been previously considered or the logical fallacies present in the previous research.

When you're concerned with resolving a disagreement among previous researchers, you should summarize the research supporting one view, then summarize the research supporting the other, and finally suggest the reasons for the disagreement.

Your review of the literature serves a bibliographic function for readers by indexing the previous research on a given topic. This can be overdone, however, and you should avoid an opening paragraph that runs three pages, mentioning every previous study in the field. The comprehensive bibliographic function can best be served by a reference list at the end of the report, and the review of the literature should focus on only the studies that have direct relevance to the present one.

Avoiding Plagiarism

Whenever you're reporting on the work of others, you must be clear about who said what. That is, you must avoid **plagiarism**: the theft of

> **plagiarism** Presenting someone else's words or thoughts as though they were your own, constituting intellectual theft.

another's words and/or ideas—whether intentional or accidental—and the presentation of those words and ideas as your own. Because this is a common and sometimes unclear problem for college students, especially in regard to the review of the literature, we'll consider the issue here. Realize, of course, that these concerns apply to everything you write.

The following are the ground rules regarding plagiarism:

- You cannot use another writer's exact words without using quotation marks and giving a complete citation, which indicates the source of the quotation so that your reader could locate the quotation in its original context. As a general rule, taking a passage of eight or more words without citation is a violation of federal copyright laws.
- It's also not acceptable to edit or paraphrase another's words and present the revised version as your own work.
- Finally, it's not even acceptable to present another's ideas as your own—even if you use totally different words to express those ideas.

The following examples should clarify what is or is not acceptable in the use of another's work.

The Original Work

Laws of Growth

Systems are like babies: once you get one, you have it. They don't go away. On the contrary, they display the most remarkable persistence. They not only persist; they grow. And as they grow, they encroach. The growth potential of systems was explored in a tentative, preliminary way by Parkinson, who concluded that administrative systems maintain an average growth of 5 to 6 percent per annum regardless of the work to be done. Parkinson was right so far as he goes, and we must give him full honors for initiating the serious study of this important topic. But what Parkinson failed to perceive, we now enunciate—the general systems analogue of Parkinson's Law.

*The System Itself Tends To Grow
At 5 To 6 Percent Per Annum*

Again, this Law is but the preliminary to the most general possible formulation, the Big-Bang Theorem of Systems Cosmology.

Systems Tend To Expand To Fill The Known Universe (Gall 1975: 12–14)

Now let's look at some of the acceptable ways you might make use of Gall's work in a term paper.

- *Acceptable:* John Gall, in his work *Systemantics*, draws a humorous parallel between systems and infants: "Systems are like babies: once you get one, you have it. They don't go away. On the contrary, they display the most remarkable persistence. They not only persist; they grow." *
- *Acceptable:* John Gall warns that systems are like babies. Create a system and it sticks around. Worse yet, Gall notes, systems keep growing larger and larger.**
- *Acceptable:* It has also been suggested that systems have a natural tendency to persist, even grow and encroach (Gall 1975: 12).

Note that the last format requires that you give a complete citation in your bibliography or reference list, as I do in this book. (See the References section toward the end of this book.) Complete footnotes or endnotes work as well. See the publication manuals of various organizations such as the APA or the ASA, as well as the *Chicago Manual of Style*, for appropriate citation formats. "How to Do It: Citing Bibliographic Sources" has some specific examples to get you started.

Here now are some unacceptable uses of the same material, reflecting some common errors.

- *Unacceptable:* In this paper, I want to look at some of the characteristics of the social systems we create in our organizations. First, systems are like babies: once you get one, you have it. They don't go away. On the contrary, they display the most remarkable persistence. They not only persist; they grow. [It's unacceptable to quote someone else's materials directly without using quotation marks and giving a full citation.]
- *Unacceptable:* In this paper, I want to look at some of the characteristics of the social systems we create in our organizations. First, systems are a lot like children: once you get one, it's yours. They don't go away; they persist. They not only persist, in fact: they grow.

*John Gall, *Systemantics: How Systems Work and Especially How They Fail* (New York: Quadrangle, 1975), 12.

** Ibid.

[It's unacceptable to edit another's work and present it as your own.]
- *Unacceptable:* In this paper, I want to look at some of the characteristics of the social systems we create in our organizations. One thing I've noticed is that once you create a system, it never seems to go away. Just the opposite, in fact: systems have a tendency to grow. You might say systems are a lot like children in that respect. [It's unacceptable to paraphrase someone else's ideas and present them as your own.] If you need more guidance on this, I've prepared an e-book, *Avoiding Plagiarism*, available on Amazon.

Each of the preceding unacceptable examples is an example of plagiarism and represents a serious offense. Admittedly, there are some "gray areas." Some ideas are more or less in the public domain, not "belonging" to any one person. Or you may reach an idea on your own that someone else has already put in writing. If you have a question about a specific situation, discuss it with your instructor in advance.

I've covered this topic in some detail because, although you must place your research in the context of what others have done and said, the improper use of their materials is a serious offense. Learning to avoid plagiarism is a part of your "coming of age" as a scholar.

Study Design and Execution

A research report containing interesting findings and conclusions will frustrate readers if they can't determine the methodological design and execution of the study. The worth of all scientific findings depends heavily on the manner in which the data were collected and analyzed.

In reporting the design and execution of a survey, for example, always include the following: the population, the sampling frame, the sampling method, the sample size, the data-collection method, the completion rate, and the methods of data processing and analysis. Comparable details should be given if other methods are used. The experienced researcher can report these details in a rather short space, without omitting anything required for the reader's evaluation of the study.

How to Do It

Citing Bibliographic Sources

Your review of the literature and other readings that figure in your paper all need to be cited properly. The good news is that proper citation isn't that hard to do. The bad news is that there are several formats in common use. I'll illustrate a few of the more common formats here, but you should ask your instructor what version you're expected to use. I'll illustrate citations of both a book and an article.

Book Information

Author: C. Wright Mills
Title: *The Power Elite*
City of publication: New York
Publisher: Oxford University Press
Year of publication: 1956

Article Information

Authors: Sharon Sassler and Anna Cunningham
Title: "How Cohabitors View Childbearing"
Journal name: *Sociological Perspectives*
Year of publication: 2008
Month/season of publication: Spring
Volume: 51
Number: 1
Pages: 3–28

   With such "raw data" in hand, you can format them by following any of the following bibliographic styles.

ASA Style Guide (American Sociological Association)

Mills, C. Wright. 1956. *The Power Elite*. New York: Oxford University Press.

Sassler, Sharon and Anna Cunningham. 2008. "How Cohabitors View Childbearing." *Sociological Perspectives* 51:3–28.

MLA Style Guide (Modern Language Association)

Mills, C. Wright. *The Power Elite*. New York: Oxford University Press, 1956.

Sassler, Sharon, and Anna Cunningham. "How Cohabitors View Childbearing." *Sociological Perspectives* 51.1 (2008): 3–28.

APSA Style Guide (American Political Science Association)

Mills, C. Wright. 1956. *The Power Elite*. New York: Oxford University Press.

Sassler, Sharon, and Anna Cunningham. 2008. "How Cohabitors View Childbearing." *Sociological Perspectives* 51(Spring): 3–28.

APA Style Guide (American Psychological Association)

Mills, C. Wright. (1956). *The power elite*. New York: Oxford University Press.

Sassler, S., & Cunningham, A. (2008). How cohabitors view childbearing. *Sociological Perspectives*, 51(1), 3–28.

Analysis and Interpretation

Having set the study in the perspective of previous research and having described the design and execution of it, you should then present your data. This chapter will provide further guidelines in this regard. For now, a few general comments are in order.

   The presentation of data, the manipulation of those data, and your interpretations should be integrated into a logical whole. It frustrates the reader to discover a collection of seemingly unrelated analyses and findings with a promise that all the loose ends will be tied together later in the report. Every step in the analysis should make sense at the time it is taken. You should present your rationale for a particular analysis, present the data relevant to it, interpret the results, and then indicate where that result leads next.

Summary and Conclusions

According to the forensic dictum mentioned earlier, summarizing the research report is essential. Avoid reviewing every specific finding, but review all the significant ones, pointing once more to their general significance.

   The report should conclude with a statement of what you've discovered about your subject matter and where future research might be directed. Many journal articles end with the statement "It is clear that much more research is needed." This conclusion is probably always true, but it has little value unless you can offer pertinent suggestions about the nature of that future research. You should review the particular shortcomings of your own study and suggest ways those shortcomings might be avoided.

Guidelines for Reporting Analyses

The presentation of data analyses should provide a maximum of detail without being cluttered. You can accomplish this best by continually examining your report to see whether it achieves the following aims.

If you're using quantitative data, present them so the reader can recompute them. In the case of percentage tables, for example, the reader should be able to collapse categories and recompute the percentages. Readers should receive sufficient information to permit them to compute percentages in the table in the direction opposite from that of your own presentation.

Describe all aspects of a quantitative analysis in sufficient detail to permit a secondary analyst to replicate the analysis from the same body of data. This means that he or she should be able to create the same indexes and scales, produce the same tables, arrive at the same regression equations, obtain the same factors and factor loadings, and so forth. This will seldom be done, of course, but if the report allows for it, the reader will be far better equipped to evaluate the report than if it does not.

Provide details. If you're doing a qualitative analysis, you must provide enough detail that your reader has a sense of having made the observations with you. Presenting only the data that support your interpretations is not sufficient; you must also share the data that conflict with the way you've made sense of things. Ultimately, you should provide enough information that the reader might reach a different conclusion than you did—though you can hope your interpretation will make the most sense. The reader, in fact, should be in position to replicate the entire study independently, whether it involves participant observation among heavy-metal groupies, an experiment regarding jury deliberation, or any other study format. Recall that replicability is an essential norm of science. A single study does not prove a point; only a series of studies can begin to do so. And unless studies can be replicated, there can be no meaningful series of studies.

Integrate supporting materials. I have previously mentioned the importance of integrating data and interpretations in the report. Here is a more specific guideline for doing this. Tables, charts, and figures, if any, should be integrated into the text of the report—appearing near that portion of the text discussing them. Sometimes students describe their analyses in the body of the report and place all the tables in an appendix. This procedure greatly impedes the reader, however. As a general rule, it is best to (1) describe the purpose for presenting the table, (2) present it, and (3) review and interpret it.

Draw explicit conclusions. Although research is typically conducted for the purpose of drawing general conclusions, you should carefully note the specific basis for such conclusions. Otherwise you may lead your reader into accepting unwarranted conclusions.

Point to any qualifications or conditions warranted in the evaluation of conclusions. Typically, you know best the shortcomings and tentativeness of your conclusions, and you should give the reader the advantage of that knowledge. Failure to do so can misdirect future research and result in a waste of research funds.

As I said at the outset of this discussion, research reports should be written in the best possible literary style. Writing lucidly is easier for some people than for others, and it's always harder than writing poorly. You are again referred to the Strunk and White book. Every researcher would do well to follow this procedure: Write. Read Strunk and White. Revise. Reread Strunk and White. Revise again. This will be a difficult and time-consuming endeavor, but so is science.

A perfectly designed, carefully executed, and brilliantly analyzed study will be altogether worthless unless you can communicate your findings to others. This chapter has attempted to provide some guidelines toward that end. The best guides are logic, clarity, and honesty. Ultimately, there is no substitute for practice.

Going Public

Though I have written this chapter with a particular concern for the research projects you may be called on to undertake in your research methods course, you should realize that graduate and even undergraduate students are increasingly presenting the results of their research as professional papers or published articles.

If you would like to explore these possibilities further, you may find state and regional associations to be more open to students than are national associations, although students may present papers to the American Sociological Association, for example. Some associations have special sessions and programs for student participants. You can learn more about these possibilities by visiting the associations' websites to learn of upcoming meetings and the topics for which papers are being solicited.

Typically, you'll submit your paper to someone who has agreed to organize a session with three to five papers on a particular topic. The organizer chooses which of the submissions will be accepted for presentation. Oral presentations at scholarly meetings are typically 15 to 20 minutes long, with the possibility of questions from the audience. Some presenters read a printed paper, whereas others speak from notes. Increasingly, presenters use PowerPoint presentations or the like.

To publish an article in a scholarly journal, you would do well to identify a journal that publishes articles on the topic of your research. Again, the journals published by state or regional associations may be the most accessible to student authors. Each journal will contain instructions for submitting articles, including instructions for formatting your article. Typically, articles submitted to a journal are circulated among three or so anonymous reviewers, who make comments and recommendations to the journal's editor. This is referred to as the "peer-review" process. Sometimes manuscripts are accepted pretty much as submitted, some are returned for revision and resubmission, and still others are rejected. The whole process from submission to a decision to publish or reject may take a few months, and there will be a further delay before the article is actually published.

The peer-review process is a distinguishing feature in academic publishing. The purpose is to help ensure that the book or article is considered a worthwhile addition to what is known about the topic under study. There is, to be sure, the possibility that peer review may favor established points of view over innovative ones, but the large number of publishing options makes it likely that a friendly journal or publisher might be found. Each would exercise peer judgment as to the scholarly quality of pieces submitted for publication. With the growth of online journals, you will find some that are peer-reviewed and others that are reviewed and judged by the editor in charge.

To meet the costs of publication, a journal will sometimes require that authors pay a small fee on acceptance. Typically, authors receive extra copies of their article—called "reprints"—to give to friends and family and to satisfy requests from professional colleagues.

The Ethics of Reading and Writing Social Research

I've already commented on some ethical issues involved in writing research reports. However, there are also some ethical issues at play in terms of reading the research literature. There has always been the risk of reviewing the literature with a special eye toward reports that support a point of view you may be fond of. Although wonderful in most respects, the power of the Internet to provide fast and expansive searches can allow even more "cherry picking" of supportive research literature. This places an ever-greater burden on researchers to exercise professional honesty in representing the history of research findings in a particular area.

Because this chapter concludes the main body of the book, I hope this final section makes clear that research ethics constitute not merely a nice thing to consider as long as it doesn't get in the way, but rather, a fundamental component of social science. Research ethics has not always been recognized in this fashion. When I first began writing this textbook, there was some objection to including this topic. It wasn't so much that researchers objected to the ethical treatment of subjects—ethics simply wasn't considered a proper topic for a book like this one. Attitudes have changed substantially over the years, however. I hope you benefit from understanding the crucial role of ethics in your work as well as in your life.

This chapter, and indeed this book, has provided what I hope will be a springboard for you to engage in and enjoy the practice of social research. The next time you find yourself pondering the cause of prejudice, observing a political rally, or being just plain curious about the latest trends in television, I trust you'll have the tools to explore your world with a social scientist's eye.

What do you think?...Revisited

There is a vast amount of information available on the Internet, but it's not all equally trustworthy and usable in scholarly research. This chapter suggested guidelines for sorting the wheat from the chaff. For example, data provided on government websites or on those of university research centers and institutes, although not perfect, are usually dependable. Clarity regarding how the data were collected is a good sign; so is the clear documentation of any other sources used. Be wary of websites that push a particular point of view or agenda. Their data may be valid and useful, but caution is in order. Never trust websites that are ambiguous about the methods used or about the exact meanings of variables reported on. Finally, look for agreement across several websites, if possible.

In some cases, you may find the website SourceWatch a useful tool to help you judge the trustworthiness of Web sources. Sometimes, you'll find that a "research team" is actually a public relations firm or that an individual "expert" always seems to report findings in support of a particular company or industry.

MAIN POINTS

Introduction

- Meaningful scientific research is inextricably wed to communication; knowing how to read and write it requires practice.

Reading Social Research

- Social researchers can access many resources, including the library and the Internet, for organizing a review of the literature.
- Reading scholarly literature is different from reading other works, such as novels.
- In reading scholarly literature, you should begin by reading the abstract, skimming the piece, and reading the conclusion to get a good sense of what it is about.
- When you read social science literature, you should form questions and take notes as you're reading.
- The key elements to note in reading a research report include theoretical orientation, research design, measurement methods, sampling (if any), and other considerations specific to the several data-collection methods discussed in this book.
- The Internet is a powerful tool for social researchers, but it also carries risks.
- Not everything you read on the Web is necessarily true.
- Original sources of data are preferred over those that take data from elsewhere.
- In evaluating a Web source, you should ask the following:

  Who or what is the author of the website?

  Is the site advocating a particular point of view?

  Does the site give accurate and complete references?

  Are the data up to date?

- Official data are usually a good source, although they are subject to error.
- The reader of a report should verify (cross-check) data whenever possible.
- Web citations, like other bibliographic references, should be complete in order to allow the reader to locate and review the materials cited. DOI numbers should be included when available.

Writing Social Research

- Good social research writing begins with good writing—period. Write to communicate rather than to impress.
- Being mindful of one's audience and one's purpose in writing the report is important.
- Plagiarism—presenting someone else's words or thoughts as though they were one's own—must be avoided. Whenever using someone else's exact words, you must be sure to use quotation marks or some other indication that you're quoting. In paraphrasing someone else's words or ideas, you must provide a full bibliographic citation of the source.
- The research report should include an account of the study design and execution.
- The analysis of a report should be clear at each step, and its conclusion should be specific but not overly detailed.
- To write good reports, researchers need to provide details, integrate supporting materials, and draw explicit conclusions.
- Increasingly, students are presenting papers at professional meetings and publishing articles in scholarly journals.

The Ethics of Reading and Writing Social Research

- A review of the literature should not be biased toward a particular point of view.

- Research ethics is a fundamental component of social science, not just a nice afterthought.

KEY TERMS

abstract

plagiarism

research monograph

search engine

URL

PROPOSING SOCIAL RESEARCH: PUTTING THE PROPOSAL TOGETHER

If you've been doing the Proposing Social Research exercises all through the book, you should have just about everything you need now to create the finished product. This chapter has given you some additional guidance on the review of the literature—both printed and online—and for writing social research, so you can review what you've written already and tidy it up. Appendix A will provide guidelines on gathering information for your bibliography.

Now you're ready to assemble the parts into a coherent whole. The following is the outline we discussed in Chapter 1. Use it as a quick reference to the social research topics discussed in each chapter.

Introduction (Chapter 1)
Review of the Literature (Chapters 2, 15;
    Appendix A)
Specifying the Problem/Question/Topic
    (Chapters 5, 6, 12)
Research Design (Chapter 4)
Data-Collection Method (Chapters 4, 8–11)
Selection of Subjects (Chapter 7)
Ethical Issues (Chapter 3)
Data Analysis (Chapters 13, 14)
Bibliography/References (Chapter 15;
    Appendix A)

Perhaps you'll be able to present this proposal as evidence that you've mastered the materials of the textbook. Or, something similar to this could be used to propose a senior thesis or graduate dissertation. If you go on to a career in social research, you could use a proposal like this to obtain funding to support the conduct of the research. In the last of these possibilities, you should include a project budget to indicate how much support you'll need and for what.

However you choose to use this kind of document, I wish you every success.

REVIEW QUESTIONS

1. Analyze the following quantitative research report: Stanley Lieberson, Susan Dumais, and Shyon Baumann, "The Instability of Androgynous Names: The Symbolic Maintenance of Gender Boundaries," *American Journal of Sociology* 105 (5): 1249. (This can be accessed in print or online through InfoTrac College Edition, for example.) Use the following questions as your guide:

   a. What are the theoretical underpinnings of the study?

   b. How are some of the key variables such as *androgynous, racial,* and *gender segregation* conceptualized and operationalized?

   c. On what data is this research based?

   d. Are there controlling variables?

   e. What is the unit of analysis?

   f. What type of analysis was done?

   g. What did the authors find?

   h. What are the strengths and weaknesses of this study?

2. Analyze the following qualitative research report: Dingxin Zhao, "State-Society Relations and the Discourses and Activities of the 1989 Beijing Student Movement," *American Journal of Sociology* 105 (6): 1592. (This can be accessed in print or online through InfoTrac College Edition, for example.) Use the following questions as your guide:

   a. What is the author's main research question?

   b. What theoretical frameworks are referred to, and which ones are used?

   c. What methodology is the author using? What type of data collection was chosen? What is the unit of analysis?

   d. Does the author have a hypothesis? If so, what is it?

   e. How does the author conceptualize key terms such as *state, state-society,* and *traditionalism*?

   f. What are the study's findings?

   g. What is the significance of this study? Are you convinced by the author, or do you see weaknesses in the study?

cigdem/ShutterStock.com

Using the Library

Introduction

We live in a world filled with social science research reports. Our daily newspapers, magazines, professional journals, alumni bulletins, club newsletters—virtually everything we pick up to read may carry reports dealing with a particular topic. For formal explorations of a topic, of course, the best place to start is still a good college or university library.

Getting Help

When you want to find something in the library, your best friend is the reference librarian, who is specially trained to find things in the library. Some libraries have specialized reference librarians—for the social sciences, humanities, government documents, and so forth. Find the librarian who specializes in your field. Make an appointment. Tell the librarian what you're interested in. He or she will probably put you in touch with some of the many available reference sources.

Reference Sources

You've probably heard the expression "information explosion." Your library is ground zero. Fortunately, a large number of reference volumes exist to offer a guide to the information that's available.

Books in Print

This volume lists all the books currently in print in the United States—listed separately by author and by title. Out-of-print books can often be found in older editions of *Books in Print*.

Readers' Guide to Periodical Literature

This annual volume with monthly updates lists articles published in many journals and magazines. Because the entries are organized by subject matter, this is an excellent source for organizing your reading on a particular topic.

In addition to these general reference volumes, you'll find a great variety of specialized references/databases, many of which are online. Here are just a few:

- *Sociological Abstracts*
- *Psychological Abstracts*
- *Social Science Index*
- *Social Science Citation Index*
- *Popular Guide to Government Publications*
- *New York Times Index*
- *Facts on File*
- *Editorial Research Reports*
- *Business Periodicals Index*
- *Monthly Catalog of Government Publications*
- *Public Affairs Information Service Bulletin*
- *Education Index*
- *Applied Science and Technology Index*
- *A Guide to Geographic Periodicals*
- *General Science Index*
- *Biological and Agricultural Index*
- *Nursing and Applied Health Index*

- *Nursing Studies Index*
- *Index to Little Magazines*
- *Popular Periodical Index*
- *Biography Index*
- *Congressional Quarterly Weekly Report*
- *Library Literature*
- *Bibliographic Index*

Using the Stacks

Serious research usually involves using the stacks, where most of the library's books are stored. This section provides information about finding books in the stacks.

The Electronic Catalog

Your library's catalog of holdings will be available electronically; you can access an electronic catalog through a computer search on either a library computer or your personal computer. Online catalog systems vary, but the following illustration from Chapman University's Leatherby Libraries will probably resemble what you'll find in your own library.

To start, let's look up a book I wrote: *The Sociological Spirit*. The library home page is shown in Figure A-1, and we begin our search by clicking on "Find Books . . ."

In this case, we're given several choices of libraries to search (Figure A-2). We'll choose the Leatherby Libraries, which are located on campus.

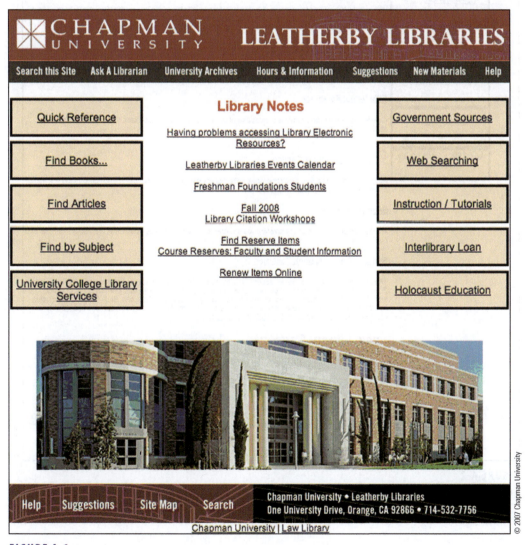

FIGURE A-1
Library Home Page

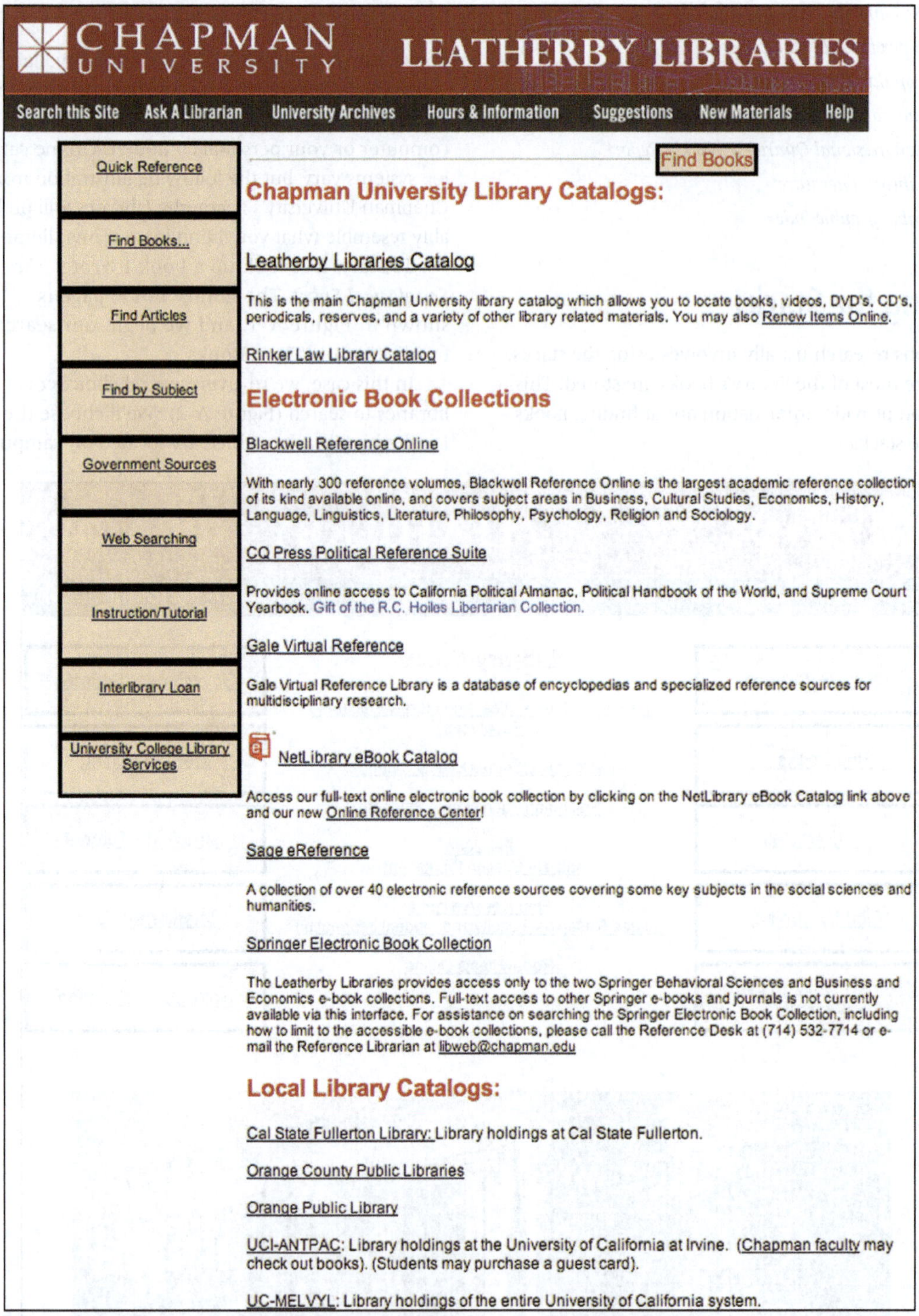

CHAPMAN UNIVERSITY **LEATHERBY LIBRARIES**

Search this Site   Ask A Librarian   University Archives   Hours & Information   Suggestions   New Materials   Help

Find Books

Quick Reference

Find Books...

Find Articles

Find by Subject

Government Sources

Web Searching

Instruction/Tutorial

Interlibrary Loan

University College Library Services

Chapman University Library Catalogs:

Leatherby Libraries Catalog

This is the main Chapman University library catalog which allows you to locate books, videos, DVD's, CD's, periodicals, reserves, and a variety of other library related materials. You may also Renew Items Online.

Rinker Law Library Catalog

Electronic Book Collections

Blackwell Reference Online

With nearly 300 reference volumes, Blackwell Reference Online is the largest academic reference collection of its kind available online, and covers subject areas in Business, Cultural Studies, Economics, History, Language, Linguistics, Literature, Philosophy, Psychology, Religion and Sociology.

CQ Press Political Reference Suite

Provides online access to California Political Almanac, Political Handbook of the World, and Supreme Court Yearbook. Gift of the R.C. Hoiles Libertarian Collection.

Gale Virtual Reference

Gale Virtual Reference Library is a database of encyclopedias and specialized reference sources for multidisciplinary research.

NetLibrary eBook Catalog

Access our full-text online electronic book collection by clicking on the NetLibrary eBook Catalog link above and our new Online Reference Center!

Sage eReference

A collection of over 40 electronic reference sources covering some key subjects in the social sciences and humanities.

Springer Electronic Book Collection

The Leatherby Libraries provides access only to the two Springer Behavioral Sciences and Business and Economics e-book collections. Full-text access to other Springer e-books and journals is not currently available via this interface. For assistance on searching the Springer Electronic Book Collection, including how to limit to the accessible e-book collections, please call the Reference Desk at (714) 532-7714 or e-mail the Reference Librarian at libweb@chapman.edu

Local Library Catalogs:

Cal State Fullerton Library: Library holdings at Cal State Fullerton.

Orange County Public Libraries

Orange Public Library

UCI-ANTPAC: Library holdings at the University of California at Irvine. (Chapman faculty may check out books). (Students may purchase a guest card).

UC-MELVYL: Library holdings of the entire University of California system.

FIGURE A-2

A Choice of Libraries to Search

Selecting the "Leatherby Libraries Catalog" presents us with the screen shown in Figure A-3, which provides for several ways of searching: by author, by title, by subject, and so forth.

Clicking "AUTHOR" will present a screen (not shown) that asks for the author's name. Searching for the name "Babbie" in this system eventually presents a list of books with that author name, including the one we're looking for, as shown in Figure A-4.

You'll notice two entries for the title we're looking for. These represent two editions of the same book. Let's click on the more recent edition, published in 1994. Figure A-5 is an electronic catalog record for the desired edition of this book.

Notice the adjoining bars marked "LOCATION," "CALL #," and "STATUS." Just below

the bar, we learn that the book is on the second floor, in the social science collection. More specifically, the Library of Congress number (or call number)—HM51.B164 1994—will help us locate the book on the shelves, which we see is, in fact, available rather than checked out.

Here's a useful strategy to use when you're researching a topic. Once you've identified the call number for a particular book in your subject area, go to the stacks, find that book, and look over the other books on the shelves near it. Because the books are arranged by subject, this method will help you locate relevant books you didn't know about.

Alternatively, you may want to go directly to the stacks and look at books in your subject area. In most libraries, books are arranged by the Library of Congress numbers. (Some follow

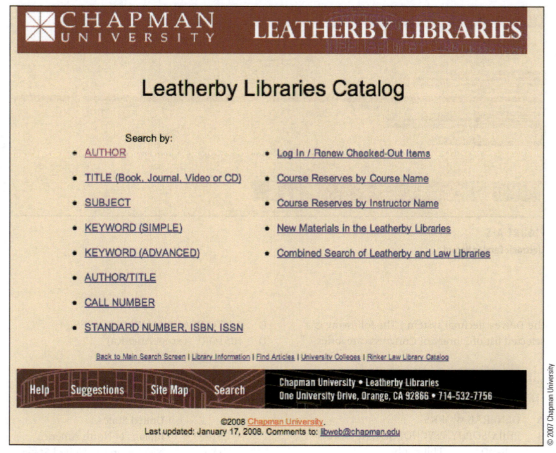

FIGURE A-3
Search Strategies

17	Adventures in social research : data analysis using SPSS for Windows / Earl Babbie, Fred Halley	
Save	Babbie, Earl R.	
☐	2nd FL Social Science Library Books HA32 .B283 1995	
	Thousand Oaks, Calif. : Pine Forge Press, c1995	
18	What is society? : reflections on freedom, order, and change / Earl R. Babbie	
Save	Babbie, Earl R.	
☐	2nd FL Social Science Library Books HM131 .B133 1994	
	Thousand Oaks, Calif. : Pine Forge Press, c1994	
19	The sociological spirit / Earl Babbie	
Save	Babbie, Earl R.	
☐	2nd FL Social Science Library Books HM51 .B164 1994	
	Belmont, Calif. : Wadsworth Pub. Co., c1994	
20	Research methods for social work / Allen Rubin, Earl Babbie	
Save	Rubin, Allen.	
☐	2nd FL Social Science Library Books HV11 .R84 1993	
	Pacific Grove, Calif. : Brooks/Cole, c1993	

FIGURE A-4
A List of Babbie Books

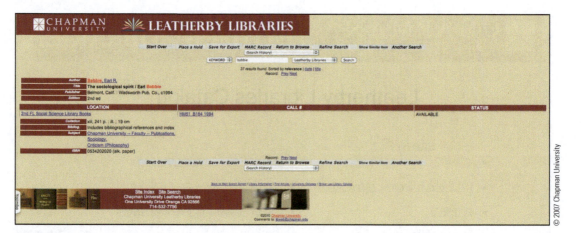

FIGURE A-5
Electronic Catalog Record

the Dewey decimal system.) The following is a selected list of Library of Congress categories.

Library of Congress Classifications (partial)

A	GENERAL WORKS	
B	PHILOSOPHY, PSYCHOLOGY, RELIGION	
	B-BD	Philosophy
	BF	Psychology
	BL-BX	Religion

C	HISTORY-AUXILIARY SCIENCES	
D	HISTORY (except America)	
	DA-DR	Europe
	DS	Asia
	DT	Africa
E–F	HISTORY (America)	
	E	United States
	E51–99	Indians of North America
	E185	Negroes in the United States
	F101–1140	Canada
	F1201–3799	Latin America

G   GEOGRAPHY-ANTHROPOLOGY
    G-GF   Geography
    GC   Oceanology and oceanography
    GN   Anthropology
    GV   Sports, amusements, games

H   SOCIAL SCIENCES
    H62.B2   *The Practice of Social Research*
    HB-HJ   Economics and business
    HM-HX   Sociology

J   POLITICAL SCIENCE
    JK   United States
    JN   Europe
    JQ   Asia, Africa
    JX   International relations

K   LAW

L   EDUCATION

M   MUSIC

N   FINE ARTS
    NA   Architecture
    NB   Sculpture
    NC   Graphic arts
    ND   Painting
    NE   Engraving
    NK   Ceramics, textiles

P   LANGUAGE AND LITERATURE
    PE   English language
    PG   Slavic language
    PJ-PM   Oriental language
    PN   Drama, oratory, journalism
    PQ   Romance literature
    PR   English literature
    PS   American literature
    PT   Germanic literature

Q   SCIENCE
    QA   Mathematics
    QB   Astronomy
    QC   Physics
    QD   Chemistry
    QE   Geology
    QH-QR   Biology

R   MEDICINE
    RK   Dentistry
    RT   Nursing

S   AGRICULTURE—PLANT AND ANIMAL
    INDUSTRY

T   TECHNOLOGY
    TA-TL   Engineering
    TR   Photography

U   MILITARY SCIENCE

V   NAVAL SCIENCE

Z   BIBLIOGRAPHY AND LIBRARY SCIENCE

Searching the Periodical Literature

Sometimes you will want to search the articles published in academic journals and other periodicals. Electronic library systems have a very powerful process for this.

Most college libraries now have access to the Education Resources Information Center (ERIC). This computer-based system allows you to search through hundreds of major journals to find articles published in the subject area of your interest. As a rule, each library website should have a list of the databases that you can visit; they also list them by discipline, which may help you limit the number of titles related to a specific keyword. Make sure you narrow your search by limiting, for instance, the language or period of the publication. Once you identify the articles you're interested in, the computer will print out their abstracts.

Of particular value to social science researchers, the publications *Sociological Abstracts* and *Psychological Abstracts* present summaries of books and articles—often prepared by the original authors—so that you can locate a great many relevant references easily and effectively. As you find relevant references, you can track down the original works and see the full details. The summaries are available in both written and computerized forms.

Figure A-6 contains the abstract of an article obtained in a computer search of *Sociological Abstracts*. I began by asking for a list of articles dealing with sociology textbooks. After reviewing the list, I asked to see the abstracts of each of the listed articles. Here's an example of what I received seconds later: an article by the sociologist Graham C. Kinloch, published in the *International Review of Modern Sociology*.

In case the meaning of the abbreviations in Figure A-6 isn't immediately obvious, I should explain that AU is author; TI is title; SO is the source or location of the original publication; DE indicates classification codes under which the abstract is referenced; and AB is the abstract. The computerized availability of resources such as *Sociological Abstracts* provides a powerful research tool for modern social scientists. You'll have the option to download or print, with or without the abstract, any title you find through the library's browsers.

AU   Kinloch-Graham-C.
TI   The Changing Definition and Content of Sociology in Introductory Textbooks, 1894–1981.
SO   International Review of Modern Sociology. 1984, 14, 1, spring, 89–103.
DE   Sociology-Education; (D810300). Textbooks; (D863400).
AB   An analysis of 105 introductory sociology textbooks published between 1894 & 1981 reveals historical changes in definitions of the discipline & major topics in relation to professional factors & changing societal contexts. Predominant views of sociology in each decade are discussed, with the prevailing view being that of a "scientific study of social structure in order to decrease conflict & deviance, thereby increasing social control." Consistencies in this orientation over time, coupled with the textbooks' generally low sensitivity to social issues, are explored in terms of their authors' relative homogeneity in age & educational backgrounds. 1 Table, 23 References. Modified HA.

FIGURE A-6

A Research Summary from *Sociological Abstracts* Used by Permission

If a document is not available in the library itself or via the Web, you always have the resource of interlibrary loans, which often are free. Libraries don't own every document or multimedia material (e.g., DVDs, videos), but many have loan agreements that can serve your needs. You need to be aware of how much time it will actually take for you to receive the book or article from the time you make your request. In the case of a book that is located in another library close by, for example, it may be faster for you to get it directly yourself. The key to a good library search is to become well informed. So start networking with librarians, faculty, and peers!

Random Numbers

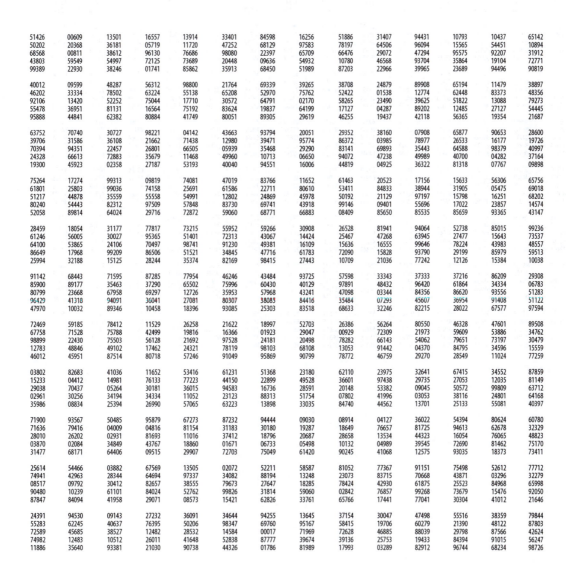

51426	00609	13501	16557	13914	33401	84598	16256	51886	31407	94431	10793	10437	65142
50202	20368	36181	05719	11720	47252	68129	97583	78197	64506	96094	15565	54451	10894
68568	00811	38612	96130	76686	98080	22397	65709	66476	29072	47294	95575	92207	31912
43803	59549	54997	72125	73689	20448	09636	54932	10780	46568	93704	35864	19104	72771
99389	22930	38246	01741	85862	35913	68450	51989	87203	22966	39965	23689	94496	90819
40012	09599	48287	56312	98800	21764	69339	39265	38708	24879	89908	65194	11479	38897
46202	33334	78502	63224	55138	65208	52970	75762	52422	01538	12774	62448	83373	48356
92106	13420	52252	75044	17710	30572	64791	02170	58265	23490	39625	51822	13088	79273
55478	36951	81131	16564	75192	83624	19837	64199	17127	04287	89202	12485	27127	54445
95888	44841	62382	80884	41749	80051	89305	29619	46255	19437	42118	56365	19354	21687
63752	70740	30727	98221	04142	43663	93794	20051	29352	38160	07908	65877	90653	28600
39706	31586	36108	21662	71438	12980	39471	95774	86372	03985	78977	26533	16177	19726
70394	94351	22457	26801	66505	05939	35468	29290	83141	69893	35443	64588	98379	40997
24328	66613	72883	35679	11468	49960	10713	06650	94072	47238	49989	40700	04282	37164
19300	45923	02358	27187	53193	40040	94551	16006	44819	04925	36322	81318	07767	09898
75264	17274	99313	09819	74081	47019	83766	11652	61463	20523	17156	15633	56306	65756
61801	25803	99036	74158	25691	61586	22711	80610	53411	84833	38944	31905	05475	69018
51217	44878	35559	55558	54991	12802	24869	45978	50192	21129	97197	15798	16251	68202
80240	54443	82312	97509	57848	83730	69741	43918	99146	09401	55696	17022	23857	14574
52058	89814	64024	29716	72872	59060	68771	66883	08409	85650	85535	85659	93365	43147
28459	18054	31177	77817	73215	55952	59266	30908	26528	81941	94064	52738	85015	99236
61246	56005	30027	95365	51401	72313	43067	14424	25467	47268	63945	27477	15643	73537
64100	53865	24106	70497	98741	91230	49381	16109	15636	16555	99646	78224	43983	48557
86649	17968	99209	86506	51521	34845	47716	61783	72090	15828	93790	29199	85979	59513
25994	32188	15125	28244	35374	82169	98415	27443	10709	21036	77242	12126	15384	10038
91142	68443	71595	87285	77954	46246	43484	93725	57598	33343	37333	37216	86209	29308
85900	89177	35463	37290	65502	75996	60430	40129	97891	48432	96420	61864	34334	06783
80799	23668	67958	69297	12726	35953	57968	43241	47098	03344	84356	86620	93556	51283
96429	41318	94091	36041	27081	80307	38083	84416	35484	07293	45607	36954	91408	51122
47970	10032	89346	10458	18396	93085	25303	83518	68633	32246	82215	28022	67577	97594
72469	59185	78412	11529	26258	21622	18997	52703	26386	56264	80550	46328	47601	89508
67758	71528	75788	42499	19816	16366	01923	29047	00929	72309	21973	59609	53886	34762
98899	22430	75503	56128	21692	97528	24181	20498	78282	66143	54062	79651	73197	30479
12783	48846	49102	17462	24321	78119	98103	68108	13053	91442	04370	84795	34596	15559
46012	45951	87514	80718	57246	91049	95869	90799	78772	46759	29270	28549	11024	77259
03802	82683	41036	11652	53416	61231	51368	23180	62110	23975	32641	67415	34552	87859
15233	04412	14981	76133	77223	44150	22899	49528	36601	97438	29735	27053	12035	81149
29038	70437	05264	30181	36015	94583	16736	28591	20148	53382	09045	50572	99809	63712
02961	30256	34194	34334	11052	23123	88313	51754	07802	41996	03053	38116	24801	64168
35986	08834	25394	26990	57065	63223	13898	33035	84740	44562	13701	25133	55081	40397
71900	93567	50485	95879	67273	87232	94444	09030	08914	04127	36022	54394	80624	60780
71636	79416	04009	04816	81154	31183	30180	19287	18649	76657	81725	94613	62678	32329
28010	26202	02931	81693	11016	37412	18796	20687	28658	13534	44323	16054	76065	48823
03870	02084	34849	43767	18860	01671	06733	05498	10132	04989	39545	72690	81462	75170
31477	68171	64406	09515	29907	72703	75049	61420	90245	41068	12575	93035	18373	73411
25614	54466	03882	67569	13505	02072	52211	58587	81052	77367	91151	75498	52612	77712
74941	42963	28344	64694	97337	34082	88194	13248	23073	83715	70668	43871	03296	32279
08517	09792	30412	82657	38555	79673	27647	18285	78424	42930	61875	25523	84968	65998
90480	10239	61101	84024	52762	99826	31814	59060	02842	76857	99268	73679	15476	92050
87847	84094	41958	29071	08573	15421	62826	33761	65766	17441	77041	30304	41012	21646
24391	94530	09143	27232	36091	34644	94255	13645	37154	30047	47498	55516	38359	79844
55283	62245	40637	76395	50206	98347	69760	95167	58415	19706	60279	21390	48122	87803
72589	45685	38527	12482	28532	14584	00017	71969	72628	46885	88039	29798	87566	42624
74982	12483	10512	26011	41648	52838	87777	39674	39136	25753	19433	84394	91015	56247
11886	35640	93381	21030	90738	44326	01786	81989	17993	03289	82912	96744	68234	98726

03777	40789	53138	84902	65517	67119	43016	45594	99378	26264	86455	41026	59843	91783
77434	88814	80202	35045	23861	66976	34547	62509	17306	39141	95877	56427	51271	21327
24031	79905	97125	99910	02012	98644	44131	90151	37559	85274	85888	83933	00595	05395
61411	26229	47339	39764	06203	89006	52147	66580	97816	46291	64695	19294	48456	65158
67605	28722	84993	40213	74203	17588	67884	65144	46757	76731	91814	82825	14921	09640
08399	54673	57424	52049	10022	80280	31618	30265	07223	01091	16857	12886	52200	25183
24661	01155	82608	14475	34709	18864	24666	80520	83407	31450	76563	13025	29970	71077
99946	90982	42196	89827	77686	97350	53420	85961	58836	63948	45483	41791	61909	92707
20479	03337	72012	68631	05734	66688	47879	27971	81284	71171	11497	16424	05229	33760
05878	82649	94108	04121	11154	97428	28550	50033	58366	16488	91552	31099	07497	37391
26765	43188	54789	34860	18404	87493	45808	69413	63670	65165	56961	36021	59176	45006
15069	48616	94053	66582	03240	19418	03006	92491	49077	15557	01484	40976	06847	95247
68642	49316	36286	17395	96578	75722	19864	04578	84155	99469	51186	12091	83697	21341
55787	36896	06645	76602	81478	47159	93149	92944	83403	35955	27043	82757	10447	95157
06646	80150	16643	42758	51005	50512	14497	88500	99547	95014	19788	52496	45661	82747
66780	49368	50975	19424	95851	61634	25554	87257	85018	02944	12394	02527	88003	77793
88208	75978	21467	13547	96535	86968	67281	08110	53381	54810	43583	96063	95300	86188
30356	70518	23955	20891	79713	56715	83046	79108	04521	41114	57848	31624	78945	25332
36986	65744	62601	56588	87171	27663	30679	25292	79814	97617	53606	77353	73868	87559
06312	95954	66193	09065	21998	34974	45725	00429	27951	58058	53538	71743	52870	53884
22033	94983	28160	08825	24088	23067	37465	08067	87338	13078	89357	05941	05270	82129
30452	89567	71981	33583	28892	18855	58394	39515	86250	42349	00832	41061	99545	26312
87612	59785	04514	02606	69365	28933	42218	91714	27058	40027	09691	54653	57232	56866
51590	24073	86172	55704	96959	73360	17055	66148	08078	20372	36932	71432	62588	72328
57163	99264	22454	55588	78458	91353	27547	27991	19627	63115	63099	59856	74653	69930
48780	80362	32025	83247	38147	91095	96062	50857	55831	62380	29003	80076	22990	60988
98912	00813	03183	15462	72115	02817	32788	30368	34305	07644	56157	65898	18561	37797
39402	14014	30652	27732	93899	78595	31964	36084	97566	34682	72458	98496	77969	14661
03842	24891	26115	26115	48801	26504	77741	82048	19748	83084	35668	23498	83585	31927
75596	26192	94550	45662	62572	34149	13402	85687	24250	65416	45033	48814	17003	76631
03591	98982	17722	51727	71369	62706	32211	09130	80850	19401	70052	37468	63436	82305
90888	86243	39999	02703	47268	00308	85152	09997	19070	04917	24351	48171	78505	66626
08573	24734	09760	51974	95354	16357	15969	12817	61896	31250	80066	38064	24088	98685
70629	30816	81429	71243	93048	43257	81387	95825	93165	20492	55200	56831	91286	79550
97402	96506	80817	15478	27808	04941	37273	69213	36638	85812	47422	16816	61468	53373
27250	03388	28225	96621	44165	59379	29178	87172	79478	58092	17710	63104	60684	97932
65971	48407	61392	11205	21776	03233	27068	46038	92918	25029	31686	15337	82092	17198
94424	24776	37573	52605	58251	42114	51162	20341	21685	79477	41030	33130	89819	35592
67035	34237	80576	98987	86458	05605	65635	39528	30420	75826	89077	50686	49972	97172
26583	89285	05050	72244	74086	97706	65120	16301	67917	21787	51785	59042	87324	30893
43684	96716	85263	40147	12867	00177	74088	19076	22915	72550	04976	52557	22961	71430
58977	02675	44573	43331	58957	22473	30080	35672	41973	70946	13049	35109	15024	45136
09507	92785	19629	14846	08127	93307	95036	78313	52446	01067	51465	62061	36698	04085
01738	05229	77024	21950	74783	42771	76450	63057	61615	34045	30701	18141	04768	00347
13403	24248	37469	17695	29452	29346	96446	72124	08531	06716	03668	98751	47708	03926
47813	95237	28518	84809	79497	25096	62922	86883	98553	32668	23650	12537	73446	80052
32411	26508	55034	61179	95124	83411	36322	87567	78589	69819	54656	09644	02350	65753
90886	05927	51880	67581	39310	01761	37345	36425	12883	77970	06829	65588	31084	04563
19712	56193	05978	74167	03347	36293	18145	39273	41897	64083	35547	67152	06188	94961
38191	90572	51923	10301	36802	90114	81194	55254	80329	49383	44090	15160	34222	23886
82520	77570	64671	06575	01907	54598	75591	12631	16676	49430	24133	66462	41574	16974
35050	44842	31469	43533	39343	79219	21618	89864	47156	13642	10654	88072	01650	18002
41269	69507	96835	61976	91903	54412	56619	65650	22130	25349	54952	08277	24992	53833
63840	22761	16566	18174	17073	15678	06395	72369	23714	69974	12838	71230	73589	55864
48616	17356	68349	30107	18604	60016	36241	30883	10979	28281	92015	73791	68528	54736

Distribution of Chi Square

	Probability					
	0.995	0.99	0.975	0.95	0.90	0.75
df	0.005	0.01	0.025	0.05	0.10	0.25
1	0.0000393	0.000157	0.000982	0.00393	0.0158	0.101
2	0.0100	0.0201	0.0506	0.103	0.211	0.575
3	0.0717	0.115	0.216	0.352	0.584	1.21
4	0.207	0.297	0.484	0.711	1.06	1.92
5	0.412	0.554	0.831	1.15	1.61	2.67
6	0.676	0.872	1.24	1.64	2.20	3.45
7	0.990	1.24	1.69	2.17	2.83	4.25
8	1.34	1.65	2.18	2.73	3.49	5.07
9	1.73	2.09	2.70	3.33	4.17·	5.90
10	2.16	2.56	3.25	3.94	4.87	6.74
11	2.60	3.05	3.82	4.57	5.58	7.58
12	3.07	3.57	4.40	5.23	6.30	8.44
13	3.57	4.11	5.01	5.89	7.04	9.30
14	4.07	4.66	5.63	6.57	7.79	10.2
15	4.60	5.23	6.26	7.26	8.55	11.0
16	5.14	5.81	6.91	7.96	9.31	11.9
17	5.70	6.41	7.56	8.67	10.1	12.8
18	6.26	7.01	8.23	9.39	10.9	13.7
19	6.84	7.63	8.91	10.1	11.7	14.6
20	7.43	8.26	9.59	10.9	12.4	15.5
21	8.03	8.90	10.3	11.6	13.2	16.3
22	8.64	9.54	11.0	12.3	14.0	17.2
23	9.26	10.2	11.7	13.1	14.8	18.1
24	9.89	10.9	12.4	13.8	15.7	19.0
25	10.5	11.5	13.1	14.6	16.5	19.9
26	11.2	12.2	13.8	15.4	17.3	20.8
27	11.8	12.9	14.6	16.2	18.1	21.7
28	12.5	13.6	15.3	16.9	18.9	22.7
29	13.1	14.3	16.0	17.7	19.8	23.6
30	13.8	15.0	16.8	18.5	20.6	24.5
40	20.7	22.2	24.4	26.5	29.1	33.7
50	28.0	29.7	32.4	34.8	37.7	42.9
60	35.5	37.5	40.5	43.2	46.5	52.3
70	43.3	45.4	48.8	51.7	55.3	61.7
80	51.2	53.5	57.2	60.4	64.3	71.1
90	59.2	61.8	65.6	69.1	73.3	80.6
100	67.3	70.1	74.2	77.9	82.4	90.1

continued

Probability

df	0.50	0.25	0.10	0.05	0.025	0.01	0.005
	0.50	0.25	0.10	0.05	0.025	0.01	0.005
1	0.455	1.32	2.71	3.84	5.02	6.63	7.88
2	1.39	2.77	4.61	5.99	7.38	9.21	10.6
3	2.37	4.11	6.25	7.82	9.35	11.3	12.8
4	3.36	5.39	7.78	9.49	11.1	13.3	14.9
5	4.35	6.63	9.24	11.1	12.8	15.1	16.8
6	5.35	7.84	10.6	12.6	14.5	16.8	18.6
7	6.35	9.04	12.0	14.1	16.0	18.5	20.3
8	7.34	10.2	13.4	15.5	17.5	20.1	22.0
9	8.34	11.4	14.7	16.9	19.0	21.7	23.6
10	9.34	12.5	16.0	18.3	20.5	23.2	25.2
11	10.34	13.7	17.3	19.7	21.9	24.7	26.8
12	11.34	14.8	18.5	21.0	23.3	26.2	28.3
13	12.34	16.0	19.8	22.4	24.7	27.7	29.8
14	13.34	17.1	21.1	23.7	26.1	29.1	31.3
15	14.34	18.2	22.3	25.0	27.5	30.6	32.8
16	15.34	19.4	23.5	26.3	28.8	32.0	34.3
17	16.34	20.5	24.8	27.6	30.2	33.4	35.7
18	17.34	21.6	26.0	28.9	31.5	34.8	37.2
19	18.34	22.7	27.2	30.1	32.9	36.2	38.6
20	19.34	23.8	28.4	31.4	34.2	37.6	40.0
21	20.34	24.9	29.6	32.7	35.5	38.9	41.4
22	21.34	26.0	30.8	33.9	36.8	40.3	42.8
23	22.34	27.1	32.0	35.2	38.1	41.6	44.2
24	23.34	28.2	33.2	36.4	39.4	43.0	45.6
25	24.34	29.3	34.4	37.7	40.6	44.3	46.9
26	25.34	30.4	35.6	38.9	41.9	45.6	48.3
27	26.34	31.5	36.7	40.1	43.2	47.0	49.6
28	27.34	32.6	37.9	41.3	44.5	48.3	51.0
29	28.34	33.7	39.1	42.6	45.7	49.6	52.3
30	29.34	34.8	40.3	43.8	47.0	50.9	53.7
40	39.34	45.6	51.8	55.8	59.3	63.7	66.8
50	49.33	56.3	63.2	67.5	71.4	76.2	79.5
60	59.33	67.0	74.4	79.1	83.3	88.4	92.0
70	69.33	77.6	85.5	90.5	95.0	100.0	104.0
80	79.33	88.1	96.6	102.0	107.0	112.0	116.0
90	89.33	98.6	108.0	113.0	118.0	124.0	128.0
100	99.33	109.0	118.0	124.0	130.0	136.0	140.0

Normal Curve Areas

z	0.00	0.01	0.02	0.03	0.04	0.05	0.06	0.07	0.08	0.09
0.0	0.0000	0.0040	0.0080	0.0120	0.0160	0.0199	0.0239	0.0279	0.0319	0.0359
0.1	0.0398	0.0438	0.0478	0.0517	0.0557	0.0596	0.0636	0.0675	0.0714	0.0753
0.2	0.0793	0.0832	0.0871	0.0910	0.0948	0.0987	0.1026	0.1064	0.1103	0.1141
0.3	0.1179	0.1217	0.1255	0.1293	0.1331	0.1368	0.1406	0.1443	0.1480	0.1517
0.4	0.1554	0.1591	0.1628	0.1664	0.1700	0.1736	0.1772	0.1808	0.1844	0.1879
0.5	0.1915	0.1950	0.1985	0.2019	0.2054	0.2088	0.2123	0.2157	0.2190	0.2224
0.6	0.2257	0.2291	0.2324	0.2357	0.2389	0.2422	0.2454	0.2486	0.2517	0.2549
0.7	0.2580	0.2611	0.2642	0.2673	0.2704	0.2734	0.2764	0.2794	0.2823	0.2852
0.8	0.2881	0.2910	0.2939	0.2967	0.2995	0.3023	0.3051	0.3078	0.3106	0.3133
0.9	0.3159	0.3186	0.3212	0.3238	0.3264	0.3289	0.3315	0.3340	0.3365	0.3389
1.0	0.3413	0.3438	0.3461	0.3485	0.3508	0.3531	0.3554	0.3577	0.3599	0.3621
1.1	0.3643	0.3665	0.3686	0.3708	0.3729	0.3749	0.3770	0.3790	0.3810	0.3830
1.2	0.3849	0.3869	0.3888	0.3907	0.3925	0.3944	0.3962	0.3980	0.3997	0.4015
1.3	0.4032	0.4049	0.4066	0.4082	0.4099	0.4115	0.4131	0.4147	0.4162	0.4177
1.4	0.4192	0.4207	0.4222	0.4236	0.4251	0.4265	0.4279	0.4292	0.4306	0.4319
1.5	0.4332	0.4345	0.4357	0.4370	0.4382	0.4394	0.4406	0.4418	0.4429	0.4441
1.6	0.4452	0.4463	0.4474	0.4484	0.4495	0.4505	0.4515	0.4525	0.4535	0.4545
1.7	0.4554	0.4564	0.4573	0.4582	0.4591	0.4599	0.4608	0.4616	0.4625	0.4633
1.8	0.4641	0.4649	0.4656	0.4664	0.4671	0.4678	0.4686	0.4693	0.4699	0.4706
1.9	0.4713	0.4719	0.4726	0.4732	0.4738	0.4744	0.4750	0.4756	0.4761	0.4767
2.0	0.4772	0.4778	0.4783	0.4788	0.4793	0.4798	0.4803	0.4808	0.4812	0.4817
2.1	0.4821	0.4826	0.4830	0.4834	0.4838	0.4842	0.4846	0.4850	0.4854	0.4857
2.2	0.4861	0.4864	0.4868	0.4871	0.4875	0.4878	0.4881	0.4884	0.4887	0.4890
2.3	0.4893	0.4896	0.4898	0.4901	0.4904	0.4906	0.4909	0.4911	0.4913	0.4916
2.4	0.4918	0.4920	0.4922	0.4925	0.4927	0.4929	0.4931	0.4932	0.4934	0.4936
2.5	0.4938	0.4940	0.4941	0.4943	0.4945	0.4946	0.4948	0.4949	0.4951	0.4952
2.6	0.4953	0.4955	0.4956	0.4957	0.4959	0.4960	0.4961	0.4962	0.4963	0.4964
2.7	0.4965	0.4966	0.4967	0.4968	0.4969	0.4970	0.4971	0.4972	0.4973	0.4974
2.8	0.4974	0.4975	0.4976	0.4977	0.4977	0.4978	0.4979	0.4979	0.4980	0.4981
2.9	0.4981	0.4982	0.4982	0.4983	0.4984	0.4984	0.4985	0.4985	0.4986	0.4986
3.0	0.4987	0.4987	0.4987	0.4988	0.4988	0.4989	0.4989	0.4989	0.4990	0.4990

Abridged from Table I of *Statistical Tables and Formulas,* by A. Hald (New York: John Wiley & Sons, Inc., 1952).

Estimated Sampling Error

How to use this table: Find the intersection between the sample size and the approximate percentage distribution of the binomial in the sample. The number appearing at this intersection represents the estimated sampling error, at the 95 percent confidence level, expressed in percentage points (plus or minus).

For example: In the sample of 400 respondents, 60 percent answer yes and 40 percent answer no.

The sampling error is estimated at plus or minus 4.9 percentage points. The confidence interval, then, is between 55.1 percent and 64.9 percent. We would estimate (95 percent confidence) that the proportion of the total population who would say yes is somewhere within that interval.

Sample Size	Binomial Percentage Distribution				
	50/50	60/40	70/30	80/20	90/10
100	10	9.8	9.2	8	6
200	7.1	6.9	6.5	5.7	4.2
300	5.8	5.7	5.3	4.6	3.5
400	5	4.9	4.6	4	3
500	4.5	4.4	4.1	3.6	2.7
600	4.1	4	3.7	3.3	2.4
700	3.8	3.7	3.5	3	2.3
800	3.5	3.5	3.2	2.8	2.1
900	3.3	3.3	3.1	2.7	2
1000	3.2	3.1	2.9	2.5	1.9
1100	3	3	2.8	2.4	1.8
1200	2.9	2.8	2.6	2.3	1.7
1300	2.8	2.7	2.5	2.2	1.7
1400	2.7	2.6	2.4	2.1	1.6
1500	2.6	2.5	2.4	2.1	1.5
1600	2.5	2.4	2.3	2	1.5
1700	2.4	2.4	2.2	1.9	1.5
1800	2.4	2.3	2.2	1.9	1.4
1900	2.3	2.2	2.1	1.8	1.4
2000	2.2	2.2	2	1.8	1.3

Glossary

abstract A summary of a research article. The abstract usually begins the article and states the purpose of the research, the methods used, and the major findings. See Chapter 17.

age/sex pyramids Graphic portrayals of the age and sex distributions of a population. See Chapter 16.

agreement reality Those things we "know" as part and parcel of the culture we share with those around us. See Chapter 1.

analysis of variance (ANOVA) Method of analysis in which cases under study are combined into groups representing an independent variable, and the extent to which the groups differ from one another is analyzed in terms of some dependent variable. Then, the extent to which the groups differ is compared with the standard of random distribution. See Chapter 16.

anonymity Anonymity is guaranteed in a research project when neither the researchers nor the readers of the findings can link a given response with a given respondent. See Chapter 3.

attribute A characteristic of a person or a thing. See *variable* and Chapter 1.

average An ambiguous term generally suggesting typical or normal—a central tendency. The *mean, median,* and *mode* are specific examples of mathematical averages. See Chapter 14.

axial coding A reanalysis of the results of open coding in Grounded Theory Method, aimed at identifying the important, general concepts. See Chapter 13.

bias (1) That quality of a measurement device that tends to result in a misrepresentation, in a particular direction, of what is being measured. For example, the questionnaire item "Don't you agree that the president is doing a good job?" would be biased in that it would generally encourage favorable responses. See Chapter 9.

(2) The thing inside you that makes other people or groups seem consistently better or worse than they really are. (3) What a nail looks like after you hit it crooked. (If you drink, don't drive.)

bivariate analysis The analysis of two variables simultaneously, for the purpose of determining the empirical relationship between them. The construction of a simple percentage table or the computation of a simple correlation coefficient are examples of bivariate analyses. See Chapter 14.

Bogardus social distance scale A measurement technique for determining the willingness of people to participate in social relations—of varying degrees of closeness—with other kinds of people. It is an especially efficient technique in that one can summarize several discrete answers without losing any of the original details of the data. See Chapter 6.

case study The in-depth examination of a single instance of some social phenomenon, such as a village, a family, or a juvenile gang. See Chapter 10.

case-oriented analysis (1) An analysis that aims to understand a particular case or several cases by looking closely at the details of each. See Chapter 13. (2) A private investigator's billing system.

closed-ended questions Survey questions in which the respondent is asked to select an answer from among a list provided by the researcher. These are popular in survey research because they provide a greater uniformity of responses and are more easily processed than *open-ended questions.* See Chapter 9.

cluster sampling (1) A multistage sampling in which natural groups (clusters) are sampled initially, with the members of each selected group being subsampled afterward. For example, you might select a sample of U.S. colleges and universities from a directory, get lists of the students at all the selected schools, then draw samples of students from each. This procedure is discussed in Chapter 7. (2) Pawing around in a box of macadamia nut clusters to take all the big ones for yourself.

codebook (1) The document used in data processing and analysis that tells the location of different data items in a data file. Typically, the codebook identifies the locations of data items and the meaning of the codes used to represent different attributes of variables. See Chapter 14. (2) The document that cost you 38 box tops just to learn that Captain Marvelous wanted you to brush your teeth and always tell the truth. (3) The document that allows CIA agents to learn that Captain Marvelous wants them to brush their teeth.

coding (1) The process whereby raw data are transformed into standardized form suitable for machine processing and analysis. See Chapter 11. (2) A strong drug you may take when you hab a bad code.

cognitive interviewing Testing potential questions in an interview setting, probing to learn how respondents understand or interpret the questions. See Chapter 5.

cohort study A study in which some specific subpopulation, or cohort, is studied over time, although data may be collected from different members in each set of observations. See Chapter 4 for more on this topic (if you want more). See also *longitudinal study, panel study,* and *trend study.*

comparative and historical research The examination of societies (or other social units) over time and in comparison with one another. See Chapter 11.

concept A family of conceptions, such as "chair," representing the whole class of actual chairs. See Chapter 5.

concept mapping (1) The graphical display of concepts and their interrelations, useful in the formulation of theory. See Chapter 13. (2) A masculine technique for finding locations by logic and will, without asking for directions.

conceptualization (1) The mental process whereby fuzzy and imprecise notions (concepts) are made more specific and precise. So you want to study prejudice. What do you mean by "prejudice"? Are there different kinds of prejudice? What are they? See Chapter 5, which is all about conceptualization and its pal, operationalization. (2) Sexual reproduction among intellectuals.

confidence interval (1) The range of values within which a population parameter is estimated to lie. A survey, for example, may show 40 percent of a sample favoring Candidate A (poor devil). Although the best estimate of the support existing among all voters would also be 40 percent, we would not expect it to be exactly that. We might, therefore, compute a confidence interval (such as from 35 to 45 percent) within which the actual percentage of the population probably lies. Note that we must specify a confidence level in connection with every confidence interval. See Chapter 7. (2) How close you dare to get to an alligator.

confidence level (1) The estimated probability that a population parameter lies within a given confidence interval. Thus, we might be 95 percent confident that between 35 and 45 percent of all voters favor Candidate A. See Chapter 7. (2) How sure you are that the ring you bought from a street vendor for $10 is really a three-carat diamond.

confidentiality A research project guarantees confidentiality when the researcher can identify a given person's responses but promises not to do so publicly. See Chapter 3.

constant comparative method (1) A component of the Grounded Theory Method in which observations are compared with one another and with the evolving inductive theory. See Chapter 13. (2) A blind-dating technique.

construct validity The degree to which a measure relates to other variables as expected within a system of theoretical relationships. See Chapter 5.

content analysis The study of recorded human communications, such as books, websites, paintings, and laws. See Chapter 11.

content validity The degree to which a measure covers the range of meanings included within a concept. See Chapter 5.

contingency question A survey question intended for only some respondents, determined by their responses to some other question. For example, all respondents might be asked whether they belong to the Cosa Nostra, and only those who said "yes" would be asked how often they go to company meetings and picnics. The latter would be a contingency question. See Chapter 9.

contingency table (1) A format for presenting the relationships among variables as percentage distributions; typically used to reveal the effects of the independent variable on the dependent variable. See Chapter 14 for several illustrations and guides to making such tables. (2) The card table you keep around in case your guests bring their seven kids to dinner with them.

continuous variable   A variable whose attributes form a steady progression, such as *age* or *income*. Thus, the ages of a group of people might include 21, 22, 23, 24, and so forth and could even be broken down into fractions of years. Contrast this with *discrete variables*, such as *sex* or *religious affiliation*, whose attributes form discontinuous chunks. See Chapter 14.

control group   (1) In experimentation, a group of subjects to whom no experimental stimulus is administered and who resemble the experimental group in all other respects. The comparison of the control group and the *experimental group* at the end of the experiment points to the effect of the experimental stimulus. See Chapter 8. (2) American Association of Managers.

control variable   See *test variable*.

conversation analysis (CA)   A meticulous analysis of the details of conversation, based on a complete transcript that includes pauses, "hems," and also "haws." See Chapter 13.

correlation   (1) An empirical relationship between two variables such that (a) changes in one are associated with changes in the other, or (b) particular attributes of one variable are associated with particular attributes of the other. Thus, for example, we say that *education* and *income* are correlated in that higher levels of education are associated with higher levels of income. Correlation in and of itself does not constitute a causal relationship between the two variables, but it is one criterion of causality. See Chapter 4. (2) Someone you and your friend are both related to.

criterion-related validity   The degree to which a measure relates to some external criterion. For example, the validity of the College Board exams is shown in their ability to predict the college success of students. Also called *predictive validity*. See Chapter 5.

critical realism   A paradigm that holds that things are real insofar as they produce effects. See Chapter 2.

cross-case analysis   An analysis that involves an examination of more than one case, either a *variable-oriented* or *case-oriented analysis*. See Chapter 13.

cross-sectional study   A study based on observations representing a single point in time. Contrasted with a *longitudinal study*. See Chapter 4.

crude birth rate (CBR)   The number of births in a year, divided by the population at the beginning of that year. See Chapter 16.

curvilinear regression analysis   A form of regression analysis that allows relationships among variables to be expressed with curved geometric lines instead of straight ones. See Chapter 16.

debriefing   (1) Interviewing subjects to learn about their experience of participation in the project and to inform them of any unrevealed purpose. This is especially important if there's a possibility that they have been damaged by that participation. See Chapter 3. (2) Pulling someone's shorts down. Don't do that. It's not nice.

deduction   (1) The logical model in which specific expectations of hypotheses are developed on the basis of general principles. Starting from the general principle that all deans are meanies, you might anticipate that this one won't let you change courses. This anticipation would be the result of deduction. See also *induction* and Chapter 1. (2) What the Internal Revenue Service said your good-for-nothing moocher of a brother-in-law technically isn't. (3) Of a duck.

dependent variable   (1) A variable assumed to depend on or be caused by another (called the *independent variable*). If you find that *income* is partly a function of *amount of formal education*, *income* is being treated as a dependent variable. See Chapter 1. (2) A wimpy variable.

descriptive statistics   Statistical computations describing either the characteristics of a sample or the relationship among variables in a sample. Descriptive statistics merely summarize a set of sample observations, whereas inferential statistics move beyond the description of specific observations to make inferences about the larger population from which the sample observations were drawn. See Chapter 16.

dimension   A specifiable aspect of a concept. "Religiosity," for example, might be specified in terms of a belief dimension, a ritual dimension, a devotional dimension, a knowledge dimension, and so forth. See Chapter 5.

discrete variable   A variable whose attributes are separate from one another, or discontinuous, as in the case of *gender* or *religious affiliation*. Contrast this with *continuous variables*, in which one attribute shades off into the next. Thus, in *age* (a continuous variable), the attributes progress steadily from 21 to 22 to 23, and so forth, whereas there is no progression from male to female in the case of *gender*. See Chapter 14.

dispersion   The distribution of values around some central value, such as an average. The range is a simple example of a measure of dispersion. Thus, we may report that the mean age of a group is 37.9, and the range is from 12 to 89. See Chapter 14.

distorter variable   In the elaboration model, a test variable that reverses the direction of a zero-order relationship. See Chapter 15.

double-blind experiment   An experimental design in which neither the subjects nor the experimenters know which is the experimental group and which is the control. See Chapter 8.

ecological fallacy   Erroneously basing conclusions about individuals solely on the observation of groups. See Chapter 4.

elaboration model   A logical model for understanding the relationship between two variables by controlling for the effects of a third. Principally developed by Paul Lazarsfeld. The various outcomes of an elaboration analysis are replication, explanation, interpretation, and specification. See Chapter 15.

element   That unit of which a population is composed and which is selected in a sample. Distinguished from *units of analysis,* which are used in data analysis. See Chapter 7.

emancipatory research   Research conducted for the purpose of benefiting disadvantaged groups. See Chapter 10.

epistemology   The science of knowing; systems of knowledge. See Chapter 1.

EPSEM (equal probability of selection method)   A sample design in which each member of a population has the same chance of being selected for the sample. See Chapter 7.

ethnography   A report on social life that focuses on detailed and accurate description rather than explanation. See Chapter 10.

ethnomethodology   An approach to the study of social life that focuses on the discovery of implicit—usually unspoken—assumptions and agreements; this method often involves the intentional breaking of agreements as a way of revealing their existence. See Chapter 10.

evaluation research   Research undertaken for the purpose of determining the impact of some social intervention, such as a program aimed at solving a social problem. See Chapter 12.

experimental group   In experimentation, a group of subjects to whom an experimental stimulus is administered. Contrast with *control group.* See Chapter 8.

explanation   An elaboration model outcome in which the original relationship between two variables is revealed to have been spurious, because the relationship disappears when an antecedent test variable is introduced. See Chapter 15.

ex post facto hypothesis   A hypothesis created after confirming data have already been collected. It is a meaningless construct because there is no way for it to be disconfirmed. See Chapter 15.

extended case method   A technique developed by Michael Burawoy in which case study observations are used to discover flaws in, and to then

improve, existing social theories. See Chapter 10.

external invalidity   Refers to the possibility that conclusions drawn from experimental results may not be generalizable to the "real" world. See Chapter 8 and also *internal invalidity.*

external validation   The process of testing the validity of a measure, such as an index or scale, by examining its relationship to other presumed indicators of the same variable. If the index really measures prejudice, for example, it should correlate with other indicators of prejudice. See Chapter 6 for a fuller discussion and illustrations.

face validity   (1) That quality of an indicator that makes it seem to be a reasonable measure of some variable. That the frequency of attendance at religious services is some indication of a person's religiosity seems to make sense without a lot of explanation. It has face validity. See Chapter 5. (2) When your face looks like your driver's license photo (rare and perhaps unfortunate).

factor analysis   A complex algebraic method for determining the general dimensions or factors that exist within a set of concrete observations. See Chapter 16.

field experiment   A formal experiment conducted outside the laboratory, in a natural setting. See Chapter 8.

focus group   A group of subjects interviewed together, prompting a discussion. The technique is frequently used by market researchers, who ask a group of consumers to evaluate a product or discuss a type of commodity, for example. See Chapter 10.

frequency distribution   (1) A description of the number of times the various attributes of a variable are observed in a sample. The report that 53 percent of a sample were men and 47 percent were women would be a simple example of a frequency distribution. Another example would be the report that 15 of the cities studied had populations under 10,000, 23 had populations between 10,000 and 25,000, and so forth. See Chapter 14. (2) A radio dial.

Geographic Information Systems (GIS)   Analytic technique in which researchers map quantitative data that describe geographic units for a graphic display. See Chapter 16.

grounded theory   (1) An inductive approach to the study of social life that attempts to generate a theory from the constant comparing of unfolding observations. This differs greatly from hypothesis testing, in which theory is used to generate hypotheses to be tested through observations. See Chapter 10. (2) A theory that is not allowed to fly.

Grounded Theory Method (GTM) An inductive approach to research introduced by Barney Glaser and Anselm Strauss in which theories are generated solely from an examination of data rather than being derived deductively. See Chapter 13.

Guttman scale (1) A type of composite measure used to summarize several discrete observations and to represent some more-general variable. See Chapter 6. (2) The device Louis Guttman weighed himself on.

hypothesis A specified testable expectation about empirical reality that follows from a more general proposition; more generally, an expectation about the nature of things derived from a theory. It is a statement of something that ought to be observed in the real world if the theory is correct. See Chapter 2.

idiographic An approach to explanation in which we seek to exhaust the idiosyncratic causes of a particular condition or event. Imagine trying to list all the reasons why you chose to attend your particular college. Given all those reasons, it's difficult to imagine your making any other choice. Contrast with *nomothetic*. See Chapter 1.

independent variable (1) A variable with values that are not problematical in an analysis but are taken as simply given. An independent variable is presumed to cause or determine a dependent variable. If we discover that religiosity is partly a function of sex—women are more religious than are men—*sex* is the independent variable and *religiosity* is the dependent variable. Note that any given variable might be treated as independent in one part of an analysis and as dependent in another part of it. *Religiosity* might become an independent variable in an explanation of crime. See Chapter 1. (2) A variable that refuses to take advice.

index A type of composite measure that summarizes and rank-orders several specific observations and represents some more-general dimension. Contrast with *scale*. See Chapter 6.

indicator An observation that we choose to consider as a reflection of a variable we wish to study. Thus, for example, attending religious services might be considered an indicator of *religiosity*. See Chapter 5.

induction (1) The logical model in which general principles are developed from specific observations. Having noted that Jews and Catholics are more likely to vote Democratic than are Protestants, you might conclude that religious minorities in the United States are more affiliated with the Democratic party, and then your task is to explain why. See also *deduction* and Chapter 1. (2) The culinary art of stuffing ducks.

infant mortality rate (IMF): The number of children per thousand births who die during their first year of life. See Chapter 16.

inferential statistics The body of statistical computations relevant to making inferences from findings based on sample observations to some larger population. Chapter 16.

informant Someone who is well versed in the social phenomenon that you wish to study and who is willing to tell you what he or she knows about it. Not to be confused with a respondent. If you were planning participant observation among the members of a religious sect, you would do well to make friends with someone who already knows about them—possibly a member of the sect—who could give you some background information. Not to be confused with a *respondent*. See Chapter 7.

informed consent A norm in which subjects base their voluntary participation in research projects on a full understanding of the possible risks involved. See Chapter 3.

institutional ethnography A research technique in which the personal experiences of individuals are used to reveal power relationships and other characteristics of the institutions within which they operate. See Chapter 10.

interest convergence The thesis that majority-group members will only support the interests of minorities when those actions also support the interests of the majority group. See Chapter 2.

internal invalidity Refers to the possibility that the conclusions drawn from experimental results may not accurately reflect what went on in the experiment itself. See also *external invalidity* and Chapter 8.

interpretation A technical term used in connection with the elaboration model. It represents the research outcome in which a control variable is discovered to be the mediating factor through which an independent variable has its effect on a dependent variable. See Chapter 15.

interval measure A level of measurement describing a variable whose attributes are rank-ordered and have equal distances between adjacent attributes. The Fahrenheit temperature scale is an example of this, because the distance between 17 and 18 is the same as that between 89 and 90. See Chapter 5 and *nominal measure, ordinal measure*, and *ratio measure*.

interview A data-collection encounter in which one person (an interviewer) asks questions of another (a respondent). Interviews may be conducted face to face or by telephone. See Chapter 9 for more information on interviewing as a method of survey research.

item analysis   An assessment of whether each of the items included in a composite measure makes an independent contribution or merely duplicates the contribution of other items in the measure. See Chapter 6.

judgmental sampling   (1) See *purposive sampling* and Chapter 7. (2) A sampling of opinionated people.

latent content   (1) In connection with content analysis, the underlying meaning of communications, as distinguished from their *manifest content*. See Chapter 11. (2) What you need to make a latent.

level of significance   In the context of tests of statistical significance, the degree of likelihood that an observed, empirical relationship could be attributable to sampling error. A relationship is significant at the .05 level if the likelihood of its being only a function of sampling error is no greater than 5 out of 100. See Chapter 16.

Likert scale   A type of composite measure developed by Rensis Likert in an attempt to improve the levels of measurement in social research through the use of standardized response categories in survey questionnaires to determine the relative intensity of different items. Likert items are those using such response categories as "strongly agree," "agree," "disagree," and "strongly disagree." Such items may be used in the construction of true Likert scales as well as other types of composite measures. See Chapter 6.

linear regression analysis   A form of statistical analysis that seeks the equation for the straight line that best describes the relationship between two ratio variables. See Chapter 16.

log-linear models   Data-analysis technique based on specifying models that describe the interrelationships among variables and then comparing expected and observed table-cell frequencies. See Chapter 16.

longitudinal study   A study design involving data collected at different points in time, as contrasted with a *cross-sectional study*. See also Chapter 4 and *cohort study, panel study,* and *trend study.*

macrotheory   A theory aimed at understanding the "big picture" of institutions, whole societies, and the interactions among societies. Karl Marx's examination of the class struggle is an example of macrotheory. Contrast with *microtheory*. See Chapter 2.

manifest content   (1) In connection with content analysis, the concrete terms contained in a communication, as distinguished from *latent content*. See Chapter 11. (2) What you have after a manifest bursts.

matching   In connection with experiments, the procedure whereby pairs of subjects are matched on the basis of their similarities on one or more variables, and one member of the pair is assigned to the experimental group and the other to the control group. See Chapter 8.

mean   (1) An average computed by summing the values of several observations and dividing by the number of observations. If you now have a grade point average of 4.0 based on 10 courses, and you get an F in this course, your new grade point (mean) average will be 3.6. See Chapter 14. (2) The quality of the thoughts you might have if your instructor did that to you.

median   (1) An average representing the value of the "middle" case in a rank-ordered set of observations. If the ages of five men are 16, 17, 20, 54, and 88, the median would be 20. (The mean would be 39.) See Chapter 14. (2) The dividing line between safe driving and exciting driving.

memoing   Writing memos that become part of the data for analysis in qualitative research such as grounded theory. Memos can describe and define concepts, deal with methodological issues, or offer initial theoretical formulations. See Chapter 13.

methodology   The science of finding out; procedures for scientific investigation. See Chapter 1.

microtheory   A theory aimed at understanding social life at the level of individuals and their interactions. Explaining how the play behavior of girls differs from that of boys is an example of microtheory. Contrast with *macrotheory*. See Chapter 2.

mode   (1) An average representing the most frequently observed value or attribute. If a sample contains 1,000 Protestants, 275 Catholics, and 33 Jews, "Protestant" is the modal category. See Chapter 14 for more thrilling disclosures about averages. (2) Better than apple pie à la median.

multiple regression analysis   A form of statistical analysis that seeks the equation representing the impact of two or more independent variables on a single dependent variable. See Chapter 16.

multiple time-series designs   The use of more than one set of data that were collected over time, as in accident rates over time in several states or cities, so that comparisons can be made. See Chapter 12.

multivariate analysis   The analysis of the simultaneous relationships among several variables. Examining simultaneously the effects of *age, gender,* and *social class* on *religiosity* would be an example of multivariate analysis. See Chapter 14.

naturalism   An approach to field research based on the assumption that an objective social reality exists and can be observed and reported accurately. See Chapter 10.

nominal measure   A variable whose attributes have only the characteristics of exhaustiveness and mutual exclusiveness—in other words, a level of measurement describing a variable that has attributes that are merely different, as distinguished from *ordinal, interval,* or *ratio measures. Gender* is an example of a nominal measure. See Chapter 5.

nomothetic   An approach to explanation in which we seek to identify a few causal factors that generally impact a class of conditions or events. Imagine the two or three key factors that determine which colleges students choose, such as proximity, reputation, and so forth. Contrast with *idiographic.* See Chapter 1.

nonequivalent control group   A control group that is similar to the experimental group but is not created by the random assignment of subjects. This sort of control group does differ significantly from the experimental group in terms of the dependent variable or variables related to it. See Chapter 12.

nonprobability sampling   Any technique in which samples are selected in some way not suggested by probability theory. Examples include reliance on available subjects as well as *purposive (judgmental), snowball,* and *quota sampling.* See Chapter 7.

nonsampling error   Those imperfections of data quality that are a result of factors other than sampling error. Examples include misunderstandings of questions by respondents and erroneous recordings by interviewers and coders. See Chapter 16.

null hypothesis   (1) In connection with hypothesis testing and tests of statistical significance, the hypothesis that suggests there is no relationship among the variables under study. You may conclude that the variables are related after having statistically rejected the null hypothesis. See Chapter 2. (2) An expectation about nulls.

odds ratio   A statistical technique for expressing the relationship between variables by comparing the odds of different occurrences. See Chapter 16.

open coding   The initial classification and labeling of concepts in qualitative data analysis. In open coding, the codes are suggested by the researchers' examination and questioning of the data. See Chapter 13.

open-ended questions   Questions for which the respondent is asked to provide his or her own answers. In-depth, qualitative interviewing relies almost exclusively on open-ended questions. See Chapter 9.

operational definition   The concrete and specific definition of something in terms of the operations by which observations are to be categorized. The operational definition of "earning an A in this course" might be "correctly answering at least 90 percent of the final exam questions." See Chapter 2.

operationalization   (1) One step beyond conceptualization. Operationalization is the process of developing operational definitions, or specifying the exact operations involved in measuring a variable. See Chapter 2. (2) Surgery on intellectuals.

ordinal measure   A level of measurement describing a variable with attributes we can rank-order along some dimension. An example is *socioeconomic status,* composed of the attributes high, medium, and low. See Chapter 5 and *nominal measure, interval measure,* and *ratio measure.*

panel mortality   The failure of some panel subjects to continue participating in the study. See Chapter 4.

panel study   A type of longitudinal study, in which data are collected from the same set of people (the sample or panel) at several points in time. See Chapter 4 and *cohort, longitudinal,* and *trend study.*

paradigm   (1) A model or framework for observation and understanding that shapes both what we see and how we understand it. The conflict paradigm causes us to see social behavior one way, the interactionist paradigm causes us to see it differently. See Chapter 2. (2) $0.20.

parameter   The summary description of a given variable in a population. See Chapter 7.

partial regression analysis   A form of regression analysis in which the effects of one or more variables are held constant, similar to the logic of the elaboration model. See Chapter 16.

partial relationship   In the elaboration model, this is the relationship between two variables when examined in a subset of cases defined by a third variable. Beginning with a zero-order relationship between *political party* and *attitudes toward abortion,* for example, we might want to see whether the relationship held true among both men and women (i.e., controlling for *sex*). The relationship found among men and the relationship found among women would be the partial relationships, sometimes simply called the partials. See Chapter 15.

participatory action research (PAR)   An approach to social research in which the people being studied are given control over the purpose and procedures of the research; intended as a counter to the implicit view that researchers are superior to those they study. See Chapter 10.

path analysis   A form of multivariate analysis in which the causal relationships among variables are presented in a graphic format. See Chapter 16.

plagiarism   Presenting someone else's words or thoughts as though they were your own, constituting intellectual theft. See Chapter 17.

population   The theoretically specified aggregation of the elements in a study. See Chapter 7.

posttesting   The remeasurement of a dependent variable among subjects after they've been exposed to a stimulus representing an independent variable. See *pretesting* and Chapter 8.

PPS (probability proportionate to size)   (1) This refers to a type of multistage cluster sample in which clusters are selected, not with equal probabilities (see *EPSEM*) but with probabilities proportionate to their sizes—as measured by the number of units to be subsampled. See Chapter 7. (2) The odds on who gets to go first: you or the 275-pound fullback.

pretesting   The measurement of a dependent variable among subjects before they are exposed to a stimulus representing an independent variable. See *posttesting* and Chapter 8.

probability sampling   The general term for samples selected in accordance with probability theory, typically involving some random-selection mechanism. Specific types of probability sampling include *EPSEM, PPS, simple random sampling,* and *systematic sampling*. See Chapter 7.

probe   A technique employed in interviewing to solicit a more complete answer to a question. It is a nondirective phrase or question used to encourage a respondent to elaborate on an answer. Examples include "Anything more?" and "How is that?" See Chapter 9 for a discussion of interviewing.

program evaluation/outcome assessment   The determination of whether a social intervention is producing the intended result. See Chapter 12.

proportionate reduction of error (PRE)   A logical model for assessing the strength of a relationship by asking how much knowing values of one variable would reduce our errors in guessing values of the other. For example, if we know how much education people have, we can improve our ability to estimate how much they earn, thus indicating that there is a relationship between the two variables. See Chapter 16.

purposive sampling   A type of nonprobability sampling in which the units to be observed are selected on the basis of the researcher's judgment about which ones will be the most useful or representative. Also called *judgmental sampling*. See Chapter 7.

qualitative analysis   (1) The nonnumerical examination and interpretation of observations, for the purpose of discovering underlying meanings and patterns of relationships. This approach is most typical of field research and historical research. See Chapter 13. (2) A classy analysis.

qualitative interview   Contrasted with survey interviewing, the qualitative interview is based on a set of topics to be discussed in depth rather than the use of standardized questions. See Chapter 10.

quantitative analysis   (1) The numerical representation and manipulation of observations for the purpose of describing and explaining the phenomena that those observations reflect. See Chapter 14. (2) A BIG analysis.

quasi experiments   Nonrigorous inquiries somewhat resembling controlled experiments but lacking key elements such as pretesting and posttesting and/or control groups. See Chapter 12.

questionnaire   A document containing questions and other types of items designed to solicit information appropriate for analysis. Questionnaires are used primarily in survey research but also in experiments, field research, and other modes of observation. See Chapter 9.

quota sampling   A type of nonprobability sampling in which units are selected for a sample on the basis of prespecified characteristics, so that the total sample will have the same distribution of characteristics assumed to exist in the population being studied. See Chapter 7.

random selection   A sampling method in which each element has an equal chance of selection independent of any other event in the selection process. See Chapter 7.

randomization   A technique for assigning experimental subjects to experimental and control groups randomly. See Chapter 8.

rapport   An open and trusting relationship; this is especially important in qualitative research between researchers and the people they're observing. See Chapter 10.

rate of natural increase   The CBR plus the CDR. See Chapter 16.

ratio measure   A level of measurement describing a variable with attributes that have all the qualities of nominal, ordinal, and interval measures and in addition are based on a "true zero" point. *Age* is an example of a ratio measure. See Chapter 5 and *interval measure, nominal measure,* and *ordinal measure*.

reactivity The problem that the subjects of social research may react to the fact of being studied, thus altering their behavior from what it would have been normally. See Chapter 10.

reductionism (1) A fault of some researchers: a strict limitation (reduction) of the kinds of concepts to be considered relevant to the phenomenon under study. See Chapter 4. (2) The cloning of ducks.

reliability (1) That quality of measurement methods that suggests that the same data would have been collected each time in repeated observations of the same phenomenon. In the context of a survey, we would expect that the question "Did you attend religious services last week?" would have higher reliability than the question "About how many times have you attended religious services in your life?" This is not to be confused with *validity*. See Chapter 5. (2) Quality of repeatability in untruths.

regression analysis A method of data analysis in which the relationships among variables are represented in the form of an equation, called a regression equation. See Chapter 16.

replication (1) Repeating an experiment to expose or reduce error. See Chapter 1. (2) A technical term used in connection with the elaboration model, referring to the elaboration outcome in which the initially observed relationship between two variables persists when a control variable is held constant, thereby supporting the idea that the original relationship is genuine. See Chapter 15.

representativeness (1) That quality of a sample of having the same distribution of characteristics as the population from which it was selected. By implication, descriptions and explanations derived from an analysis of the sample may be assumed to represent similar ones in the population. Representativeness is enhanced by probability sampling and provides for generalizability and the use of inferential statistics. See Chapter 7. (2) A noticeable quality in the presentation of self of some members of the U.S. Congress.

research monograph A book-length research report, either published or unpublished. This is distinguished from a textbook, a book of essays, a novel, and so forth. See Chapter 17.

respondent A person who provides data for analysis by responding to a survey questionnaire. See Chapter 9.

response rate The number of people participating in a survey divided by the number selected in the sample, in the form of a percentage. This is also called the *completion rate* or, in self-administered surveys, the return rate: the percentage of questionnaires sent out that are returned. See Chapter 9.

sampling error The degree of error to be expected in probability sampling. The formula for determining sampling error contains three factors: the parameter, the sample size, and the standard error. See Chapter 7.

sampling frame The list or quasi-list of units composing a population from which a sample is selected. If the sample is to be representative of the population, it is essential that the sampling frame include all (or nearly all) members of the population. See Chapter 7.

sampling interval The standard distance (k) between elements selected from a population for a sample. See Chapter 7.

sampling ratio The proportion of elements in the population that are selected to be in a sample. See Chapter 7.

sampling unit That element or set of elements considered for selection in some stage of sampling. See Chapter 7.

scale (1) A type of composite measure composed of several items that have a logical or empirical structure among them. Examples of scales include the *Bogardus social distance, Guttman, Likert,* and *Thurstone scales*. Contrast with *index*. See Chapter 6. (2) One of the less appetizing parts of a fish.

search engine A computer program designed to locate where specified terms appear on websites throughout the World Wide Web. See Chapter 17.

secondary analysis (1) A form of research in which the data collected and processed by one researcher are reanalyzed—often for a different purpose—by another. This is especially appropriate in the case of survey data. Data archives are repositories or libraries for the storage and distribution of data for secondary analysis. See Chapter 9. (2) Estimating the weight and speed of an opposing team's linebackers.

selective coding In Grounded Theory Method, this analysis builds on the results of open coding and axial coding to identify the central concept that organizes the other concepts that have been identified in a body of textual materials. See Chapter 13.

semantic differential A questionnaire format in which the respondent is asked to rate something in terms of two, opposite adjectives (e.g., rate textbooks as "boring" or "exciting"), using qualifiers such as "very," "somewhat," "neither," "somewhat," and "very" to bridge the distance between the two opposites. See Chapter 6.

semiotics The study of signs and the meanings associated with them. This is commonly associated with content analysis. See Chapter 13.

simple random sampling   (1) A type of probability sampling in which the units composing a population are assigned numbers. A set of random numbers is then generated, and the units having those numbers are included in the sample. Although probability theory and the calculations it provides assume this basic sampling method, it's seldom used, for practical reasons. An equivalent alternative is *systematic sampling* (with a random start). See Chapter 7. (2) A random sample with a low IQ.

snowball sampling   (1) A nonprobability-sampling method, often employed in field research, whereby each person interviewed may be asked to suggest additional people for interviewing. See Chapter 7. (2) Picking the icy ones to throw at your methods instructor.

social artifact   Any product of social beings or their behavior. It can be a unit of analysis. See Chapter 4.

social indicators   Measurements that reflect the quality or nature of social life, such as crime rates, infant mortality rates, number of physicians per 100,000 population, and so forth. Social indicators are often monitored to determine the nature of social change in a society. See Chapter 12.

sociobiology   A paradigm based on the view that social behavior can be explained solely in terms of genetic characteristics and behavior. See Chapter 4.

specification   A technical term used in connection with the elaboration model, representing the elaboration outcome in which an initially observed relationship between two variables is replicated among some subgroups created by the control variable but not among others. In such a situation, you will have specified the conditions under which the original relationship exists: for example, among men but not among women. See Chapter 15.

spurious relationship   A coincidental statistical correlation between two variables, shown to be caused by some third variable. For example, there is a positive relationship between the number of fire trucks responding to a fire and the amount of damage done: the more trucks, the more damage. The third variable is the size of the fire. They send lots of fire trucks to a large fire and a lot of damage is done because of the size of the fire. For a little fire, they just send a little fire truck, and not much damage is done because it's a small fire. Sending more fire trucks does not cause more damage. For a given size of fire, in fact, sending more trucks would reduce the amount of damage. See Chapter 4.

standard deviation   (1) A measure of dispersion around the mean, calculated so that approximately 68 percent of the cases will lie within plus or minus one standard deviation from the mean, 95 percent will lie within plus or minus two standard deviations, and 99.9 percent will lie within three standard deviations. Thus, for example, if the mean age in a group is 30 and the standard deviation is 10, then 68 percent have ages between 20 and 40. The smaller the standard deviation, the more tightly the values are clustered around the mean; if the standard deviation is high, the values are widely spread out. See Chapter 14. (2) Routine rule-breaking.

statistic   The summary description of a variable in a sample, used to estimate a population parameter. See Chapter 7.

statistical significance   A general term referring to the likelihood that relationships observed in a sample could be attributed to sampling error alone. See Chapter 16.

stratification   The grouping of the units composing a population into homogeneous groups (or strata) before sampling. This procedure, which may be used in conjunction with simple random, systematic, *or* cluster sampling, improves the representativeness of a sample, at least in terms of the variables used for stratification. See Chapter 7.

study population   That aggregation of elements from which a sample is actually selected. See Chapter 7.

suppressor variable   In the elaboration model, a test variable that prevents a genuine relationship from appearing at the zero-order level. See Chapter 15.

systematic sampling   (1) A type of probability sampling in which every *k*th unit in a list is selected for inclusion in the sample—for example, every 25th student in the college directory of students. You compute *k* by dividing the size of the population by the desired sample size; *k* is called the sampling interval. Within certain constraints, systematic sampling is a functional equivalent of simple random sampling and usually easier to do. Typically, the first unit is selected at random. See Chapter 7. (2) Picking every third one whether it's icy or not. See *snowball sampling* (2).

test variable   A variable that is held constant in an attempt to clarify further the relationship between two other variables. Having discovered a relationship between *education* and *prejudice*, for example, we might hold *sex* constant by examining the relationship between education and prejudice among men only and then among women only. In this example, *sex* would be the test variable. See Chapter 15.

tests of statistical significance A class of statistical computations that indicate the likelihood that the relationship observed between variables in a sample can be attributed to sampling error only. See Chapter 16.

theory A systematic explanation for the observations that relate to a particular aspect of life: juvenile delinquency, for example, or perhaps social stratification or political revolution. See Chapter 1.

Thurstone scale A type of composite measure, constructed in accordance with the weights assigned by "judges" to various indicators of some variables. See Chapter 6.

tolerance for ambiguity The ability to hold conflicting ideas in your mind simultaneously, without denying or dismissing any of them. See Chapter 1.

time-series analysis An analysis of changes in a variable (such as *crime rates*) over time. See Chapter 16.

time-series design A research design that involves measurements made over some period, such as the study of traffic-accident rates before and after lowering the speed limit. See Chapter 12.

trend study A type of longitudinal study in which a given characteristic of some population is monitored over time. An example would be the series of Gallup polls showing the electorate's preferences for political candidates over the course of a campaign, even though different samples were interviewed at each point. See Chapter 4 and *cohort, longitudinal,* and *panel study.*

typology (1) The classification (typically nominal) of observations in terms of their attributes on two or more variables. The classification of newspapers as liberal-urban, liberal-rural, conservative-urban, or conservative-rural would be an example. See Chapter 6. (2) Apologizing for your neckwear.

units of analysis The what or whom being studied. In social science research, the most typical units of analysis are individual people. See Chapter 4.

univariate analysis The analysis of a single variable, for purposes of description. Frequency distributions, averages, and measures of dispersion are examples of univariate analysis, as distinguished from *bivariate* and *multivariate analysis.* See Chapter 14.

unobtrusive research Methods of studying social behavior without affecting it. This includes *content analysis,* analysis of existing statistics, and *comparative and historical research.* See Chapter 11.

URL Web address, typically beginning with "http://"; stands for "uniform resource locator" or "universal resource locator." See Chapter 17.

validity A term describing a measure that accurately reflects the concept it is intended to measure. For example, your IQ would seem to be a more valid measure of your intelligence than would the number of hours you spend in the library. Though the ultimate validity of a measure can never be proved, we may agree to its relative validity on the basis of face validity, criterion-related validity, content validity, construct validity, internal validation, *and* external validation. This must not be confused with *reliability.* See Chapter 5.

variable A logical set of attributes. The variable *sex* is made up of the attributes *male* and *female.* See Chapter 1.

variable-oriented analysis An analysis that describes and/or explains a particular variable. See Chapter 13.

weighting Assigning different weights to cases that were selected into a sample with different probabilities of selection. In the simplest scenario, each case is given a weight equal to the inverse of its probability of selection. When all cases have the same chance of selection, no weighting is necessary. See Chapter 7.

zero-order relationship In the elaboration model, this is the original relationship between two variables, with no test variables controlled for.

Bibliography

Abdollahyan, Hamid, and Taghi Azadarmaki. 2001. "Sampling Design in a Survey Research: The Sampling Practice in Iran." Paper presented at the meeting of the American Sociological Association, Washington, DC, August 12–16.

Abdulhadi, Rabab. 1998. "The Palestinian Women's Autonomous Movement: Emergence, Dynamics, and Challenges." *Gender and Society* 12 (6): 649–73.

Aburdene, Patricia. 2005. *Megatrends 2010*. Charlottesville, VA: Hampton Roads.

Ackland, Robert, and Rachel Gibson. 2013. "Hyperlinks and Networked Communications: A Comparative Study of Political Parties Online." *International Journal of Social Research Methodology* 16 (3): 231–44. doi: 10.1080/13645579.2013.774179.

Akerlof, Karen, and Chris Kennedy. 2013. "Nudging toward a Healthy Natural Environment: How Behavioral Change Research Can Inform Conservation." Fairfax, VA: George Mason University Center for Climate Change, Communication . http://climatechangecommunication.org /report/nudging-toward-healthy-natural -environment-how-behavioral-change-research -can-inform.

American Association for Public Opinion Research (AAPOR). 2008. *Standard Definitions: Final Dispositions of Case Codes and Outcome Rates for Surveys*. 5th ed. Lenexa, KS: AAPOR.

American Association for Public Opinion Research (AAPOR). 2009, March. *An Evaluation of the Methodology of the 2008 Pre-Election Primary Polls*. Deerfield, IL: AAPOR.

American Association for Public Opinion Research (AAPOR) Cell Phone Task Force. 2010. *New Considerations for Survey Researchers When Planning and Conducting RDD Telephone Surveys in the U.S. with Respondents Reached via Cell Phone Numbers*. http://www.aapor.org/AAPORKentico /Education-Resources/Reports/Cell-Phone -Task-Force-Report.aspx.

American Sociological Association. "Teaching Ethics throughout the Curriculum." https:// www.asanet.org/teaching-learning/faculty /teaching-ethics-throughout-curriculum.

Andorka, Rudolf. 1990. "The Importance and the Role of the Second Economy for the Hungarian Economy and Society." *Quarterly Journal of Budapest University of Economic Sciences* 12 (2): 95–113.

Arbesman, Samuel. 2012. *The Half-Life of Facts: Why Everything We Know Has an Expiration Date*. New York: Penguin.

Asch, Solomon. 1958. "Effects of Group Pressure upon the Modification and Distortion of Judgments," pp. 174–83 in *Readings in Social Psychology*, 3rd ed., edited by Eleanor E. Maccoby et al. New York: Holt, Rinehart & Winston.

Asher, Ramona M., and Gary Alan Fine. 1991. "Fragile Ties: Sharing Research Relationships with Women Married to Alcoholics," pp. 196–205 in *Experiencing Fieldwork: An Inside View of Qualitative Research*, edited by William B. Shaffir and Roberta A. Stebbins. Thousand Oaks, CA: Sage.

Ashley, Amaya, and Stanley Presser. 2017. "Nonresponse Bias for Univariate and Multivariate Estimates of Social Activities and Roles," *Public Opinion Quarterly* 81 (1): 1–36. https://doi.org/10.1093/poq/nfw037.

Auster, Carol J., and Lisa A. Auster-Gussman. 2014. "Contemporary Mother's Day and Father's Day Greeting Cards: A Reflection of Traditional Ideologies of Motherhood and Fatherhood?" *Journal of Family Issues* 37(9), 1294–1326. https:// doi.org/10.1177/0192513X14528711.

Auster, Carol J., and Claire S. Mansbach. 2012. "The Gender Marketing of Toys: An Analysis of Color and Type of Toy on the Disney Store Website." *Sex Roles* 67 (7/8): 375–88. doi: 10.1007/s11199-012-0177-8.

Babbie, Earl. 1966. "The Third Civilization." *Review of Religious Research*, Winter, pp. 101–21.

Babbie, Earl. 1967. "A Religious Profile of Episcopal Churchwomen." *Pacific Churchman*, January, pp. 6–8, 12.

Babbie, Earl. 1970. *Science and Morality in Medicine*. Berkeley: University of California Press.

Babbie, Earl. 1982. *Social Research for Consumers*. Belmont, CA: Wadsworth.

Babbie, Earl. 1985. *You Can Make a Difference*. New York: St. Martin's Press.

Babbie, Earl. 2004. "Laud Humphreys and Research Ethics." *International Journal of Sociology and Social Policy* 24 (3/4/5): 12–18.

Bailey, William C. 1975. "Murder and Capital Punishment." In *Criminal Law in Action*, edited by William J. Chambliss. New York: Wiley.

Bailey, William C., and Ruth D. Peterson. 1994. "Murder, Capital Punishment, and Deterrence: A Review of the Evidence and an Examination on Police Killings." *Journal of Social Issues* 50: 53–74.

Ball-Rokeach, Sandra J., Joel W. Grube, and Milton Rokeach. 1981. "Roots: The Next Generation—Who Watched and with What Effect." *Public Opinion Quarterly* 45 (1): 58–68.

Barkan, Steven E., Michael Rocque, and Jason Houle. 2013. "State and Regional Suicide Rates: A New Look at an Old Puzzle." *Sociological Perspectives* 56 (2): 287–97. doi: 10.1525 /sop.2013.56.2.287.

Barker, Kriss, and Miguel Sabido, eds. 2005. *Soap Operas for Social Change to Prevent HIV/AIDS: A Training Guide for Journalists and Media Personnel*. Shelburne, VT: Population Media Center.

Bart, Pauline, and Patricia O'Brien. 1985. *Stopping Rape: Successful Survival Strategies*. New York: Pergamon.

Beatty, Paul C., and Gordon B. Willis. 2007. "Research Synthesis: The Practice of Cognitive Interviewing." *Public Opinion Quarterly* 71 (2): 287–311.

Becker, Penny Edgell. 1998. "Making Inclusive Communities: Congregations and the 'Problem' of Race." *Social Problems* 45 (4): 451–72.

Bednarz, Marlene. 1996. "Push Polls Statement." Report to the AAPORnet listserv, April 5. http:// www.aapor.org/ethics/pushpoll.html.

Belenky, Mary Field, Blythe McVicker Clinchy, Nancy Rule Goldberger, and Jill Mattuck Tarule.

1986. *Women's Ways of Knowing: The Development of Self, Voice, and Mind*. New York: Basic Books.

Bell, Derrick A. 1980. "*Brown v. Board of Education* and the Interest Convergence Dilemma." *Harvard Law Review* 93: 518–33.

Bellah, Robert N. 1970. "Christianity and Symbolic Realism." *Journal for the Scientific Study of Religion* 9: 89–96.

Bellah, Robert N. 1974. "Comment on the Limits of Symbolic Realism." *Journal for the Scientific Study of Religion* 13: 487–89.

Belmont Report. https://www.hhs.gov/ohrp /regulations-and-policy/belmont-report/index .html.

Berbrier, Mitch. 1998. "'Half the Battle': Cultural Resonance, Framing Processes, and Ethnic Affectations in Contemporary White Separatist Rhetoric." *Social Problems* 45 (4): 431–50.

Berg, Bruce L. 1989. *Qualitative Research Methods for the Social Sciences*. Boston: Allyn and Bacon.

Berkowitz, Dana, and Maura Ryan. Fall 2011. "Bathrooms, Baseball, and Bra Shopping: Lesbian and Gay Parents Talk about Engendering Their Children." *Sociological Perspectives* 54 (3): 329–50.

Beveridge, W. I. B. 1950. *The Art of Scientific Investigation*. New York: Vintage Books.

Bhattacharjee, Yudhijit. 2013. "The Mind of a Con Man Scientist," *New York Times Magazine*, April 26. http://www.nytimes.com/2013/04/28 /magazine/diederik-stapels-audacious-academic -fraud.html?pagewanted=all&_r=1&.

Bian, Yanjie. 1994. *Work and Inequality in Urban China*. Albany: State University of New York Press.

Biddle, Stuart J. H., David Markland, David Gilbourne, Nikos L. D. Chatzisarantis, and Andrew C. Sparkes. 2001. "Research Methods in Sport and Exercise Psychology: Quantitative and Qualitative Issues." *Journal of Sports Sciences* 19 (10): 777–809.

Bielby, William T., and Denise Bielby. 1999. "Organizational Mediation of Project-Based Labor Markets: Talent Agencies and the Careers of Screenwriters." *American Sociological Review* 64: 64–85.

Black, Donald. 1970. "Production of Crime Rates." *American Sociological Review* 35 (August): 733–48.

Blair, Johnny, Shanyang Zhao, Barbara Bickart, and Ralph Kuhn. 1995. *Sample Design for Household Telephone Surveys: A Bibliography 1949–1995*. College Park: Survey Research Center, University of Maryland.

Blaunstein, Albert, and Robert Zangrando, eds. 1970. *Civil Rights and the Black American*. New York: Washington Square Press.

Bogle, Kathleen A. 2008. *Hooking Up: Dating and Relationships on Campus*. New York: New York University Press.

Bolstein, Richard. 1991. "Comparison of the Likelihood to Vote among Preelection Poll Respondents and Nonrespondents." *Public Opinion Quarterly* 55 (4): 648–50.

Bottomore, T. B., and Maximilien Rubel, eds. [1843] 1956. *Karl Marx: Selected Writings in Sociology and Social Philosophy*. Translated by T. B. Bottomore. New York: McGraw-Hill.

Bowen, Glenn. 2009. "Supporting a Grounded Theory with an Audit Trail: An Illustration." *International Journal of Social Research Methodology* 12 (4): 305–16.

Boyle, John M., Faith Lewis, and Brian Tefft. 2010, December. "Segmented or Overlapping Dual Frame Samples in Telephone Surveys." *Survey Practice*. http://surveypractice.org/2010/12/08/segmented-or-overlapping-samples/.

Brennan, Mike, and Jan Charbonneau. 2009. "Improving Mail Survey Response Rates Using Chocolate and Replacement Questionnaires." *Public Opinion Quarterly* 73 (2): 368–78.

Brick, J. Michael. 2011. "The Future of Survey Sampling." *Public Opinion Quarterly* 75 (5): 872–88.

Brick, J. Michael, W. R. Andrews, Pat D. Brick, Howard King, Nancy. A. Mathiowetz, and Lynne Stokes. 2012. "Methods for Improving Response Rates in Two-Phase Mail Surveys." *Survey Practice* 5 (3). http://surveypractice.org/index.php/SurveyPractice/article/view/17/html.

Broom, Alex, Kelly Hand, and Philip Tovey. 2009. "The Role of Gender, Environment, and Individual Biography in Shaping Qualitative Interview Data." *International Journal of Social Research Methodology* 12 (1): 51–65.

Browne, Kath. 2005. "Snowball Sampling: Using Social Networks to Research Non-heterosexual Women." *International Journal of Social Research Methodology* 8 (1): 47–60.

Burawoy, M., A. Burton, A. A. Ferguson, K. J. Fox, J. Gamson, N. Gartrell, L. Hurst, C. Kurzman, L. Salzinger, J. Schiffman, and S. Ui, eds. 1991. *Ethnography Unbound: Power and Resistance in the Modern Metropolis*. Berkeley: University of California Press.

Byrne, Anne, John Canavan, and Michelle Millar. 2009. "Participatory Research and the Voice-Centered Relational Method of Data Analysis: Is It Worth It?" *International Journal of Social Research Methodology* 12 (1): 67–77.

Cahn, Naomi, and June Carbone. 2010. *Red Families v. Blue Families: Legal Polarization and the Creation of Culture*. New York: Oxford University Press.

Campbell, Donald T. 1976. *Assessing the Impact of Planned Social Change*. Hanover, NH: Public Affairs Center, Dartmouth College.

Campbell, Donald, and Julian Stanley. 1963. *Experimental and Quasi-Experimental Designs for Research*. Chicago: Rand McNally.

Campbell, M. L. 1998. "Institutional Ethnography and Experience as Data." *Qualitative Sociology* 21 (1): 55–73.

Carpini, Michael X. Delli, and Scott Keeter. 1991. "Stability and Change in the U.S. Public's Knowledge of Politics." *Public Opinion Quarterly* 55 (4): 583–612.

Carr, C. Lynn. 1998. "Tomboy Resistance and Conformity: Agency in Social Psychological Gender Theory." *Gender and Society* 12 (5): 528–53.

Carroll, Lewis. [1895] 2009. *Through the Looking-Glass and What Alice Found There*. Oxford: Oxford University Press.

Census Bureau. *See* U.S. Bureau of the Census.

Charmaz, Kathy. 2006. *Constructing Grounded Theory: A Practical Guide through Qualitative Research*. Thousand Oaks, CA: Sage.

Cheung, Yuet Wah. 2009. *A Brighter Side: Protective and Risk Factors in the Rehabilitation of Chronic Drug Abusers in Hong Kong*. Hong Kong: Chinese University Press.

Chirot, Daniel, and Jennifer Edwards. 2003. "Making Sense of the Senseless: Understanding Genocide." *Contexts* 2 (2): 12–19.

Choi, Bernard C. K., and Anita W. P. Pak. 2005. "A Catalog of Biases in Questionnaires." *Preventing Chronic Disease* 2 (1). http://www.cdc.gov/pcd/issues/2005/jan/04_0050.htm.

Chossudovsky, Michel. 1997. *The Globalization of Poverty: Impacts of IMF and World Bank Reforms*. London: Zed Books.

Christian, Leah Melani, Don A. Dillman, and Jolene D. Smyth. 2007. "Helping Respondents Get It Right the First Time: The Influence of Words, Symbols, and Graphics in Web Surveys." *Public Opinion Quarterly* (1): 113–25.

Christian, Leah, Scott Keeter, Kristen Purcell, and Aaron Smith. (2010, May 20). "Assessing the Cell Phone Challenge." Executive Summary. Pew Center Publications. http://pewresearch.org/pubs/1601/assessing-cell-phone-challenge-in-public-opinion-surveys.

Coates, Rodney D. 2006. "Towards a Simple Typology of Racial Hegemony." *Societies without Borders* 1: 69–91.

Coleman, James. 1966. *Equality of Educational Opportunity*. Washington, DC: U.S. Government Printing Office.

Collins, G. C., and Timothy B. Blodgett. 1981. "Sexual Harassment . . . Some See It . . . Some Won't." *Harvard Business Review*, March–April, pp. 76–95.

Comstock, Donald. 1980. "Dimensions of Influence in Organizations." *Pacific Sociological Review*, January, pp. 67–84.

Comte, Auguste. 1822. *Plan de Travaux Scientifiques Nécessaires pour Réorganiser la Société.* [Publisher city and name unknown].

Connolly, Scott, Katie Elmore, and Wendi Stein. 2008. *Qualitative Assessment of* Outta Road, *an Entertainment-Education Radio Soap Opera for Jamaican Adolescents.* Shelburne, VT: Population Media Center.

Conrad, Clifton F. 1978. "A Grounded Theory of Academic Change." *Sociology of Education* 51: 101–12.

Contemporary Sociology. 2008. *A Symposium on Public Sociology: Looking Back and Forward.* 37 (6, November).

Cook, Elizabeth. 1995. Communication to the METHODS listserv, April 25, from Michel de Seve (T120@music.ulaval.ca) to Cook (EC1645A@american.edu).

Cook, Thomas D., and Donald T. Campbell. 1979. *Quasi-Experimentation: Design and Analysis Issues for Field Settings.* Chicago: Rand McNally.

Cooper-Stephenson, Cynthia, and Athanasios Theologides. 1981. "Nutrition in Cancer: Physicians' Knowledge, Opinions, and Educational Needs." *Journal of the American Dietetic Association,* May, pp. 472–76.

Correll, Shelley J., Stephen Benard, and In Paik. 2007. "Getting a Job: Is There a Motherhood Penalty?" *American Journal of Sociology* 112 (5): 1297–338.

Couper, Mick P. 2001. "Web Surveys: A Review of Issues and Approaches." *Public Opinion Quarterly* 64 (4): 464–94.

Couper, Mick P. 2008. *Designing Effective Web Surveys.* New York: Cambridge University Press.

Couper, Mick P. 2011. "The Future of Modes of Data Collection." *Public Opinion Quarterly* 75 (5): 889–908.

Couper, Mick P., and Peter V. Miller. 2008. "Web Survey Methods: Introduction." *Public Opinion Quarterly* 72 (5): 831–5.

Craig, R. Stephen. 1992. "The Effect of Television Day Part on Gender Portrayals in Television Commercials: A Content Analysis." *Sex Roles* 26 (5/6): 197–211.

Crawford, Kent S., Edmund D. Thomas, and Jeffrey J. Fink. 1980. "Pygmalion at Sea: Improving the Work Effectiveness of Low Performers." *Journal of Applied Behavioral Science,* October–December, pp. 482–505.

Curtin, Richard, Stanley Presser, and Eleanor Singer. 2005. "Changes in Telephone Survey Nonresponse over the Past Quarter Century." *Public Opinion Quarterly* 69 (1): 87–98.

Danieli, Ardha, and Carol Woodhams. 2005. "Emancipatory Research Methodology and Disability: A Critique." *International Journal of Social Research* 8 (4): 281–96.

Datta, R., Nyojy U. Khyang, Hla Kray Prue Khyang, Hla Aung Prue Kheyang, Mathui Ching Khyang, and Jebunnessa Chapola. 2015. "Participatory Action Research and Researcher's Responsibilities: An Experience with an Indigenous Community." *International Journal of Social Research Methodology* 18 (6): 581–99, http://dx.doi.org/10.1080/13645579.2014.927492.

Davis, Fred. 1973. "The Martian and the Convert: Ontological Polarities in Social Research." *Urban Life* 2 (3): 333–43.

Day, Lemuel, Christopher Bader, and Ann Gordon. 2014. "The Chapman University Survey on American Fears." http://www.chapman.edu/wilkinson/research-centers/babbie-center/survey-american-fears.aspx.

Day, Lemuel, Christopher Bader, and Ann Gordon. 2018 "America's Top Fears 2018." https://blogs.chapman.edu/wilkinson/2018/10/16/americas-top-fears-2018/.

Death Penalty Information Center. 2017. http://www.deathpenaltyinfo.org/murder-rates-nationally-and-state.

De Coster, Stacy. 2005. "Depression and Law Violation: Gendered Responses to Gendered Stresses." *Sociological Perspectives* 48 (2): 155–87.

Deflem, Mathieu. 2002. *Policing World Society: Historical Foundations of International Police Cooperation.* New York: Oxford University Press.

DeFleur, Lois. 1975. "Biasing Influences on Drug Arrest Records: Implications for Deviance Research." *American Sociological Review* 40: 88–103.

de Leeuw, Edith D. 2010. "Mixed-Mode Surveys and the Internet." *Survey Practice* 3 (6). www.surveyresearch.org.

de Leeuw, E. D., and J. J. Hox. 2011. "Internet Surveys as Part of a Mixed-Mode Design," pp. 45–76 in *Social and Behavioral Research and the Internet: Advances in Applied Methods and Research Strategies,* edited by M. Das, P. Ester, and L. Kaczmirek. New York: Taylor & Francis.

Delgado, Richard. 2002. "Explaining the Rise and Fall of African American Fortunes—Interest Convergence and Civil Rights Gains." *Harvard Civil Rights–Civil Liberties Law Review* 37: 369–87.

Denscombe, Martyn. 2009. "Item Non-Response Rates: A Comparison of Online and Paper Questionnaires." *International Journal of Social Research Methodology* 12 (4): 281–91.

Dillman, Don A. 1978. *Mail and Telephone Surveys: The Total Design Method*. New York: Wiley.

Dillman, Don A. 2009. *Mail and Internet Surveys: The Tailored Design Method*. 3rd ed. Hoboken, NJ: Wiley.

Dillman, Don A. 2012. "Introduction to Special Issue of Survey Practice on Item Nonresponse." *Survey Practice* 5 (2). www.surveypractice.org.

Doyle, Sir Arthur Conan. [1891] 1892. "A Scandal in Bohemia." First published in *The Strand*, July 1891. Reprinted in *The Original Illustrated Sherlock Holmes*, pp. 11–25. Secaucus, NJ: Castle.

DuBois, W. E. B. 1903. *The Souls of Black Folk*. Chicago: McClurg.

Duneier, Mitchell. 1999. *Sidewalk*. New York: Farrar, Straus, & Giroux.

Dunne, Ciarán. 2011. "The Place of Literature Review in Grounded Theory Research." *International Journal of Social Research Methodology* 14 (2): 111–24.

Durkheim, Emile. [1893] 1964. *The Division of Labor in Society*. Translated by George Simpson. New York: Free Press.

Durkheim, Emile. [1897] 1951. *Suicide*. Glencoe, IL: Free Press.

Eastman, Crystal. 1910. "Work-Accidents and Employers' Liability." *The Survey*, September 3, pp. 788–94.

Edin, Kathryn, and Maria Kefalas. 2005. *Promises I Can Keep: Why Poor Women Put Motherhood before Marriage*. Berkeley: University of California Press.

Editorial. 2011, May. *International Journal of Social Research Methodology* 14 (3): 171–78.

Edwards, Adam, William Housley, Matthew Williams, Luke Sloan, and Malcolm Williams. 2013. "Digital Social Research, Social Media and the Sociological Imagination: Surrogacy, Augmentation, and Re-orientation." *International Journal of Social Research Methodology* 16 (3): 245–60. doi: 10.1080/13645579.2013.774185.

Ellison, Christopher G., and Darren E. Sherkat. 1990. "Patterns of Religious Mobility among Black Americans." *Sociological Quarterly* 31 (4): 551–68.

Emde, Matthias, and Marek Fuchs. 2012. "Exploring Animated Faces Scales in Web Surveys: Drawbacks and Prospects." *Survey Practice* 5 (1). http://surveypractice.org.

Emerson, Robert M., Kerry O. Ferris, and Carol Brooks Gardner. 1998. "On Being Stalked." *Social Problems* 45 (3): 289–314.

Farquharson, Karen. 2005. "A Different Kind of Snowball: Identifying Key Policymakers." *International Journal of Social Research* 8 (4): 345–53.

Farrall, Steven, Ben Hunter, Gilly Sharpe, and Adam Calverley. 2016. "What 'Works' When Retracing Sample Members in a Qualitative Longitudinal Study?" *International Journal of Social Research Methodology* 19 (3): 287–300. http://dx.doi.org /10.1080/13645579.2014.993839

Fausto-Sterling, Anne. 1992. "Why Do We Know So Little about Human Sex?" *Discover* Archives, June. http://cas.bellarmine.edu/tietjen/Human%20 Nature%20S%201999/why_do_we_know_so _little_about_h.htm.

Federal Communications Commission (FCC). 2015. Wireless Phones and the National Do-Not-Call List. http://www.fcc.gov/guides/truth-about -wireless-phones-and-national-do-not-call-list.

Festinger, Leon, Henry W. Reicker, and Stanley Schachter. 1956. *When Prophecy Fails*. Minneapolis: University of Minnesota Press.

Fevre, Ralph, Amanda Robinson, Trevor Jones, and Duncan Lewis. 2010. "Researching Workplace Bullying: The Benefits of Taking an Integrated Approach." *International Journal of Social Research Methodology* (13) 1: 71–85.

Fielding, Nigel. 2004. "Getting the Most from Archived Qualitative Data: Epistemological, Practical and Professional Obstacles." *International Journal of Social Research Methodology* 7 (1): 97–104.

Ford, David A. 1989. "Preventing and Provoking Wife Battery through Criminal Sanctioning: A Look at the Risks." September, unpublished manuscript.

Ford, David A., and Mary Jean Regoli. 1992. "The Preventive Impacts of Policies for Prosecuting Wife Batterers," pp. 181–208 in *Domestic Violence: The Changing Criminal Justice Response*, edited by E. S. Buzawa and C. G. Buzawa. New York: Auburn.

Fox, Kathryn J. 1991. "The Politics of Prevention: Ethnographers Combat AIDS among Drug Users," pp. 227–49 in *Ethnography Unbound: Power and Resistance in the Modern Metropolis*, edited by M. Burawoy, A. Burton, A. A. Ferguson, K. J. Fox, J. Gamson, N. Gartrell, L. Hurst, C. Kurzman, L. Salzinger, J. Schiffman, and S. Ui. Berkeley: University of California Press.

Frankel, Mark S., and Sanyin Siang. 1999 November. "Ethical and Legal Aspects of Human Subjects Research on the Internet: A Report of a Workshop." Washington, DC: American Association for the Advancement of Science. http://www.aaas.org/spp/dspp/sfrl /projects/intres/main.htm.

Galesic, Mirta, Roger Tourangeau, Mick P. Couper, and Frederick G. Conrad. 2008. "Eye-Tracking Data: New Insights on Response Order Effects and Other Cognitive Shortcuts in Survey Responding." *Public Opinion Quarterly* 72 (5): 892–913.

Gall, John. 1975. *Systemantics: How Systems Work and Especially How They Fail*. New York: Quadrangle.

Galton, Francis. 1889. *Natural Inheritance*. New York: Macmillan.

Gamson, William A. 1992. *Talking Politics*. New York: Cambridge University Press.

Gans, Herbert. 1971. "The Uses of Poverty: The Poor Pay All." *Social Policy*, July–August, pp. 20–24.

Gans, Herbert. 2002. "More of Us Should Become Public Sociologists." *Footnotes*, July–August. Washington, DC: American Sociological Association. http://www.asanet.org/footnotes /julyaugust02/fn10.html.

Garant, Carol. 1980. "Stalls in the Therapeutic Process." *American Journal of Nursing*, December, pp. 2166–67.

Gard, Greta, ed. 1993. *Ecofeminism: Women, Animals, Nature*. Philadelphia: Temple University Press. Quoted in Linda J. Rynbrandt and Mary Jo Deegan. 2002. "The Ecofeminist Pragmatism of Caroline Bartlett Crane, 1896–1935." *American Sociologist* 33 (3): 60.

Garfinkel, H. 1967. *Studies in Ethnomethodology*. Englewood Cliffs, NJ: Prentice-Hall.

Gaventa, J. 1991. "Towards a Knowledge Democracy: Viewpoints on Participatory Research in North America," pp. 121–31 in *Action and Knowledge: Breaking the Monopoly with Participatory Action-Research*, edited by O. Fals-Borda and M. A. Rahman. New York: Apex Press.

Geertz, Clifford. 1973. *The Interpretation of Cultures*. New York: Basic Books.

General Social Survey. http://gss.norc.org/Documents /other/GSS%20Response%20Period%20and%20 Field%20Period.pdf.

Glaser, Barney, and Anselm Strauss. 1967. *The Discovery of Grounded Theory: Strategies for Qualitative Research*. Chicago: Aldine.

Glock, Charles Y., Benjamin B. Ringer, and Earl R. Babbie. 1967. *To Comfort and to Challenge*. Berkeley: University of California Press.

Gobo, Giampietro. 2006. "Set Them Free: Improving Data Quality by Broadening the Interviewer's Tasks." *International Journal of Social Research Methodology* 9 (4): 279–301.

Gobo, Giampietro. 2011. "Glocalizing Methodology? The Encounter between Local Methodologies." *International Journal of Social Research Methodology* 14 (6): 417–37.

Goffman, Erving. 1961. *Asylums: Essays on the Social Situation of Mental Patients and Other Inmates*. Chicago: Aldine.

Goffman, Erving. 1963. *Stigma: Notes on the Management of a Spoiled Identity*. Englewood Cliffs, NJ: Prentice-Hall.

Goffman, Erving. 1974. *Frame Analysis*. Cambridge, MA: Harvard University Press.

Goffman, Erving. 1979. *Gender Advertisements*. New York: Harper & Row.

Gong, Rachel. 2011. "Internet Politics and State Media Control: Candidate Weblogs in Malaysia." *Sociological Perspectives* 54 (3): 307–28, doi: 10.1525/sop.2011.54.3.307.

Graham, Laurie, and Richard Hogan. 1990. "Social Class and Tactics: Neighborhood Opposition to Group Homes." *Sociological Quarterly* 31 (4): 513–29.

Greenwood, Peter W., C. Peter Rydell, Allan F. Abrahamse, Jonathan P. Caulkins, James Chiesa, Karyn E. Mode, and Stephen P. Klein. 1994. *Three Strikes and You're Out: Estimated Benefits and Costs of California's New Mandatory-Sentencing Law*. Santa Monica, CA: Rand.

Greenwood, Peter W., C. Peter Rydell, and Karyn Model. 1996. *Diverting Children from a Life of Crime: Measuring Costs and Benefits*. Santa Monica, CA: Rand.

Griffith, Alison I. 1995. "Mothering, Schooling, and Children's Development," pp. 108–21 in *Knowledge, Experience, and Ruling Relations: Studies in the Social Organization of Knowledge*, edited by M. Campbell and A. Manicom. Toronto, Canada: University of Toronto Press.

Gronvik, Lars. 2009. "Defining Disability: Effects of Disability Concepts on Research Outcomes." *International Journal of Social Research Methodology* 12 (1): 1–18.

Groves, Robert M. 2006. "Nonresponse Rates and Nonresponse Bias in Household Surveys." *Public Opinion Quarterly* 70 (5): 646–75.

Groves, Robert M. 2011. "Three Eras of Survey Research." *Public Opinion Quarterly* 75 (5): 861–71.

Gubrium, Jaber F., and James A. Holstein. 1997. *The New Language of Qualitative Method*. New York: Oxford University Press.

Hammersley, Martyn, and Anna Traianou. 2011. "Moralism and Research Ethics: A Machiavellian Perspective." *International Journal of Social Research Methodology* 14 (5): 379–90.

Hartsock, Nancy. 1983. "The Feminist Standpoint," pp. 283–310 in *Discovering Reality*, edited by S. Harding and M. B. Hintikka. Boston: D. Riedel.

Hawking, Stephen. 2001. *The Universe in a Nutshell*. New York: Bantam Books.

Hedrick, Terry E., Leonard Bickman, and Debra J. Rog. 1993. *Applied Research Design: A Practical Guide*. Thousand Oaks, CA: Sage.

Hempel, Carl G. 1952. "Fundamentals of Concept Formation in Empirical Science." *International Encyclopedia of United Science II*, no. 7.

Heritage, J. 1984. *Garfinkel and Ethnomethodology*. Cambridge: Polity Press.

Heritage, John, and David Greatbatch. 1992. "On the Institutional Character of Institutional Talk," pp. 99–137 in *Talk at Work*, edited by P. Drew and J. Heritage. Cambridge, England: Cambridge University Press.

Higginbotham, A. Leon, Jr. 1978. *In the Matter of Color: Race and the American Legal Process*. New York: Oxford University Press.

Hill, Lewis. 2000. *Yankee Summer: The Way We Were: Growing up in Rural Vermont in the 1930s*. Bloomington, IN: First Books Library.

Hilts, Philip J. 1981. "Values of Driving Classes Disputed." *San Francisco Chronicle*, June 25, p. 4.

Hogan, Richard, and Carolyn C. Perrucci. 1998. "Producing and Reproducing Class and Status Differences: Racial and Gender Gaps in U.S. Employment and Retirement Income." *Social Problems* 45 (4): 528–49.

Howard, Edward N., and Darlene M. Norman. 1981. "Measuring Public Library Performance." *Library Journal*, February, pp. 305–8.

Howell, Joseph T. 1973. *Hard Living on Clay Street*. Garden City, NY: Doubleday Anchor.

Huberman, A. Michael, and Matthew B. Miles. 1994. "Data Management and Analysis Methods," pp. 428–44 in *Handbook of Qualitative Research*, edited by Norman K. Denzin and Yvonna S. Lincoln. Thousand Oaks, CA: Sage.

Hughes, Michael. 1980. "The Fruits of Cultivation Analysis: A Reexamination of Some Effects of Television Watching." *Public Opinion Quarterly* 44 (3): 287–302.

Humphreys, Laud. 1970. *Tearoom Trade: Impersonal Sex in Public Places*. Chicago: Aldine.

Hurst, Leslie. 1991. "Mr. Henry Makes a Deal," pp. 183–202 in *Ethnography Unbound: Power and Resistance in the Modern Metropolis*, edited by M. Burawoy, A. Burton, A. A. Ferguson, K. J. Fox, J. Gamson, N. Gartrell, L. Hurst, C. Kurzman, L. Salzinger, J. Schiffman, and S. Ui. Berkeley: University of California Press.

Husser, Jason A., and Kenneth E. Fernandez. 2013. "To Click, Type, or Drag? Evaluating Speed of Survey Data Input Methods." *Survey Practice* 6 (2). http://www.surveypractice.org.

Iannacchione, Vincent G., Jennifer M. Staab, and David T. Redden. 2003. "Evaluating the Use of Residential Mailing Addresses in a Metropolitan Household Survey." *Public Opinion Quarterly* 67 (2): 202–10.

Ibrahim, Saad Eddin. 2003. "Letter from Cairo." *Contexts* 2 (2): 68–72.

Im, E.-O., and Wonshik Chee. 2012. "Practical Guidelines for Qualitative Research Using Online Forums," *Comput Inform Nurs*. 30(11): 604–11. doi: 10.1097/NXN.0b013e318266cade.

Irwin, John, and James Austin. 1997. *It's About Time: America's Imprisonment Binge*. Belmont, CA: Wadsworth.

Ison, Nicole L. 2009. "Having Their Say: Email Interviews for Research Data Collection with People Who Have Verbal Communication Impairment." *International Journal of Social Research Methodology* 12 (2): 161–72.

Isaac, Larry W., and Larry J. Griffin. 1989. "Ahistoricism in Time-Series Analyses of Historical Process: Critique, Redirections, and Illustrations from U.S. Labor History." *American Sociological Review* 54: 873–90.

Jackman, Molly R., and Lauri Kanerva. 2016. Evolving the IRB: Building Robust Review for Industry Research. *Washington and Lee Law Review Online* 72 (3). http://scholarlycommons .law.wlu.edu/cgi/viewcontent.cgi?article =1042&context=wlulr-online.

Jackman, Mary R., and Mary Scheuer Senter. 1980. "Images of Social Groups: Categorical or Qualified?" *Public Opinion Quarterly* 44 (3): 340–61.

Jackson, Jonathan. 2005. "Validating New Measures of the Fear of Crime." *International Journal of Social Research* 8 (4): 297–315.

Jacobs, Bruce A., and Jody Miller. 1998. "Crack Dealing, Gender, and Arrest Avoidance." *Social Problems* 45 (4): 550–69.

Jaffee, Daniel. 2007. *Brewing Justice: Fair Trade Coffee, Sustainability and Survival*. Berkeley: University of California Press.

Jasso, Guillermina. 1988. "Principles of Theoretical Analysis." *Sociological Theory* 6: 1–20.

Jensen, Arthur. 1969. "How Much Can We Boost IQ and Scholastic Achievement?" *Harvard Educational Review* 39: 273–74.

Jobes, Patrick C., Andra Aldea, Constantin Cernat, Ioana-Minerva Icolisan, Gabriel Iordache, Sabastian Lazeru, Catalin Stoica, Gheorghe Tibil, and Eugenia Udangiu. 1997. "Shopping as a Social Problem: A Grounded Theoretical Analysis of Experiences among Romanian Shoppers." *Journal of Applied Sociology* 14 (1): 124–46.

Johnson, Jeffrey C. 1990. *Selecting Ethnographic Informants*. Thousand Oaks, CA: Sage.

Johnston, Hank. 1980. "The Marketed Social Movement: A Case Study of the Rapid Growth of TM." *Pacific Sociological Review*, July, pp. 333–54.

Johnston, Hank, and David A. Snow. 1998. "Subcultures and the Emergence of the Estonian Nationalist Opposition 1945–1990." *Sociological Perspectives* 41 (3): 473–97.

Jones, James H. 1981. *Bad Blood: The Tuskegee Syphilis Experiments*. New York: Free Press.

Kaplan, Abraham. 1964. *The Conduct of Inquiry*. San Francisco: Chandler.

Kaplowitz, Michael D., Timothy D. Hadlock, and Ralph Levine. 2004. "A Comparison of Web and Mail Survey Response Rates." *Public Opinion Quarterly* 68 (1): 94–101.

Kasl, Stanislav V., Rupert F. Chisolm, and Brenda Eskenazi. 1981. "The Impact of the Accident at Three Mile Island on the Behavior and Well-Being of Nuclear Workers." *American Journal of Public Health*, May, pp. 472–95.

Kebede, Alemseghed, and J. David Knottnerus. 1998. "Beyond the Pales of Babylon: The Ideational Components and Social Psychological Foundations of Rastafari." *Sociological Perspectives* 42 (3): 499–517.

Keeter, Scott, and Kylie McGeeney. 2015. Pew Research Will Call More Cellphones in 2015. http://www.pewresearch.org/fact -tank/2015/01/07/pew-research-will-call -more-cellphones-in-2015/.

Kendall, Patricia L., and Paul Lazarsfeld. 1950. "Problems of Survey Analysis," pp. 133–96 in *Continuities in Social Research: Studies in the Scope and Methods of "The American Soldier,"* edited by Robert K. Merton and Paul Lazarsfeld. New York: Free Press.

Kentor, Jeffrey. 2001. "The Long Term Effects of Globalization on Income Inequality, Population Growth, and Economic Development." *Social Problems* 48 (4): 435–55.

Khayatt, Didi. 1995. "Compulsory Heterosexuality: Schools and Lesbian Students," pp. 149–63 in *Knowledge, Experience, and Ruling Relations: Studies in the Social Organization of Knowledge*, edited by M. Campbell and A. Manicom. Toronto, Canada: University of Toronto Press.

Kidder, Jeffrey L. 2005. "Style and Action: A Decoding of Bike Messenger Symbols." *Journal of Contemporary Ethnography* 34 (2): 344–67.

Kilburn, John C., Jr. 1998. "It's a Matter of Definition: Dilemmas of Service Delivery and Organizational Structure in a Growing Voluntary Organization." *Journal of Applied Sociology* 15 (1): 89–103.

Kinnell, Ann Marie, and Douglas W. Maynard. 1996. "The Delivery and Receipt of Safer Sex Advice in Pretest Counseling Sessions for HIV and AIDS." *Journal of Contemporary Ethnography* 24: 405–37.

Kinsey, Alfred C., et al. 1948. *Sexual Behavior in the Human Male*. Philadelphia: Saunders.

Kinsey, Alfred C., et al. 1953. *Sexual Behavior in the Human Female*. Philadelphia: Saunders.

Knowles, Caroline. 2006. "Handling Your Baggage in the Field: Reflections on Research Relationships." *International Journal of Social Research Methodology* 9 (5): 393–404.

Koopman-Boyden, Peggy, and Margaret Richardson. 2013. "An Evaluation of Mixed Methods (Diaries and Focus Groups) When Working with Older People." *International Journal of Social Research Methodology* 16 (5): 389–401. doi:10.1080 /13645579.2012.716971.

Koppel, Ross, Joshua P. Metlay, Abigail Cohen, Brian Abaluck, A. Russell Localio, Stephen E. Kimmel, and Brian L. Strom. 2005. "Role of Computerized Physician Order Entry Systems in Facilitating Medication Errors." *Journal of the American Medical Association* 293: 1197–203.

Kramer, A. D. I., Jamie E. Guillory, and Jeffrey T. Hancock. 2014. "Experimental Evidence of Massive-Scale Emotional Contagion through Social Networks," PNAS | June 17, 2014 | vol. 111 | no. 24, pp. 8788–8790.

Kreuter, Frauke, Stanley Presser, and Roger Tourangeau. 2008. "Social Desirability Bias in CATI, IVR, and Web Surveys: The Effects of Mode and Question Sensitivity." *Public Opinion Quarterly* 72 (5): 847–65.

Krueger, Richard A. 1988. *Focus Groups*. Thousand Oaks, CA: Sage.

Kubrin, Charis E. 2005. "I See Death around the Corner: Nihilism in Rap Music." *Sociological Perspectives* 48 (4): 433–59.

Kubrin, Charis E., and Ronald Weitzer. 2003. "Retaliatory Homicide: Concentrated Disadvantage and Neighborhood Culture." *Social Problems* 50 (2): 157–80.

Kuhn, Thomas. 1970. *The Structure of Scientific Revolutions*. Chicago: University of Chicago Press.

Kvale, Steinar. 1996. *InterViews: An Introduction to Qualitative Research Interviewing*. Thousand Oaks, CA: Sage.

Lakoff, George. 2002. *Moral Politics: How Liberals and Conservatives Think*. Chicago: University of Chicago Press.

Larcker, David F., and Anastasia A. Zakolyukina. (2010, July 29). "Detecting Deceptive Discussions in Conference Calls." Rock Center for Corporate Governance Working Paper No. 83. Palo Alto, CA: Stanford University Graduate School of Business. http://ssrn.com/abstract=1572705.

Laumann, Edward O., John H. Gagnon, Robert T. Michael, and Stuart Michaels. 1994. *The Social Organization of Sexuality*. Chicago: University of Chicago Press.

Lee, Motoko Y., Stephen G. Sapp, and Melvin C. Ray. 1996. "The Reverse Social Distance Scale." *Journal of Social Psychology* 136 (1): 17–24.

Lewins, Ann, and Christina Silver. 2009. "Choosing a CAQDAS Package." July. http://eprints.ncrm .ac.uk/791/1/2009ChoosingaCAQDASPackage .pdf.

Libin, A., and J. Cohen-Mansfield. 2000. "Individual versus Group Approach for Studying and Intervening with Demented Elderly Persons: Methodological Perspectives." *Gerontologist*, October 15, p. 105.

Lincoln, Yvonna S., and Egon G. Guba. 1985. *Naturalistic Inquiry*. Beverly Hills, CA: Sage.

Link, Michael W., Joe Murphy, Michael F. Schober, N., Trent D. Buskirk, Jennifer Hunter Childs, and Casey Langer Tesfaye. 2014. "Mobile Technologies for Conducting, Augmenting and Potentially Replacing Surveys: Report of the AAPOR Task Force on Emerging Technologies in Public Opinion Research." April 25. http://www .aapor.org/AAPORKentico/AAPOR_Main /media/MainSiteFiles/REVISED_Mobile_ Technology_Report_Final_revised10June14.pdf.

Linton, Ralph. 1937. *The Study of Man*. New York: D. Appleton-Century.

Literary Digest. 1936a. "Landon, 1,293,669: Roosevelt, 972,897." October 31, pp. 5–6.

Literary Digest. 1936b. "What Went Wrong with the Polls?" November 14, pp. 7–8.

Lofland, John. 2003. *Demolishing a Historic Hotel*. Davis, CA: Davis Research.

Lofland, John, David Snow, Leon Anderson, and Lyn H. Lofland. 2006. *Analyzing Social Settings: A Guide to Qualitative Observation and Analysis*. 4th ed. Belmont, CA: Wadsworth.

Lopata, Helena Znaniecki. 1981. "Widowhood and Husband Sanctification," *Journal of Marriage and the Family,* May: 439–50.

Lynd, Robert S., and Helen M. Lynd. 1929. *Middletown*. New York: Harcourt, Brace.

Lynd, Robert S., and Helen M. Lynd. 1937. *Middletown in Transition*. New York: Harcourt, Brace.

Lynn, Peter, ed. *Methodology of Longitudinal Surveys*. Hoboken, NJ: Wiley, 2009.

Madison, Anna-Marie. 1992. "Primary Inclusion of Culturally Diverse Minority Program Participants in the Evaluation Process." *New Directions for Program Evaluation* 53: 35–43.

Maeda, H. 2015. "Response Option Configuration of Online Administered Likert Scales," *International Journal of Social Research Methodology* 18 (1): 15–26. http://dx.doi.org/10.1080/13645579.20 14.885159.

Mahoney, James, and Dietrich Rueschemeyer, eds. 2003. *Comparative Historical Analysis in the Social Sciences*. New York: Cambridge University Press.

Malakhoff, Lawrence A., and Matt Jans. 2011. "Towards Usage of Avatar Interviewers in Web Surveys." *Survey Practice* 4 (3). http:// surveypractice.org.

Manning, Peter K., and Betsy Cullum-Swan. 1994. "Narrative, Content, and Semiotic Analysis," pp. 463–77 in *Handbook of Qualitative Research*, edited by Norman K. Denzin and Yvonna S. Lincoln. Thousand Oaks, CA: Sage.

Marshall, Catherine, and Gretchen B. Rossman. 1995. *Designing Qualitative Research*. Thousand Oaks, CA: Sage.

Marx, Karl. [1867] 1967. *Capital: A Critique of Political Economy*. New York: International Publishers.

McAlister, Alfred, Cheryl Perry, Joel Killen, Lee Ann Slinkard, and Nathan Maccoby. 1980. "Pilot Study of Smoking, Alcohol, and Drug Abuse Prevention." *American Journal of Public Health*, July, pp. 719–21.

McGeeney, Kylie. 2015. "Tips for Creating Web Surveys for Completion on a Mobile Device." July 15. http://www.pewresearch.org/2015/06/11 /tips-for-creating-web-surveys-for-completion -on-a-mobile-device/.

McGrane, Bernard. 1994. *The Un-TV and the 10 mph Car: Experiments in Personal Freedom and Everyday Life*. Fort Bragg, CA: Small Press.

Meadows, Donella H., Dennis L. Meadows, and Jørgen Randers. 1992. *Beyond the Limits: Confronting Global Collapse, Envisioning a Sustainable Future*. Post Mills, VT: Chelsea Green.

Meadows, Donella H., Dennis L. Meadows, Jørgen Randers, and William W. Behrens, III. 1972. *The Limits to Growth*. New York: Signet Books.

Meko, Tim, Denise Lu, and Lazaro Gamio. "How Trump Won the Presidency with Razor-Thin Margins in Swing States, *Washington Post,* November 11, 2016. https://www.washingtonpost.com/graphics /politics/2016-election/swing-state-margins/.

Menjívar, Cecilia. 2000. *Fragmented Ties: Salvadoran Immigrant Networks in America*. Berkeley: University of California Press.

Merton, Robert K. 1938. "Social Structure and Anomie." *American Sociological Review* 3: 672–82.

Merton, Robert K. 1957. *Social Theory and Social Structure*. New York: Free Press.

Merton, Robert K., and Alice Kitt. (1950). "Contributions to the Theory of Reference Group Behavior." Glencoe, IL: Free Press. Reprinted in part from *Studies in the Scope and Method of "The American Soldier,"* edited by R. K. Merton and Paul Lazarfeld. Glencoe, IL: Free Press.

Middlebury College. 2007. http://www.middlebury .edu/about/newsevents/archive/2007 /newsevents_633084484309809133.htm.

Milgram, Stanley. 1963. "Behavioral Study of Obedience." *Journal of Abnormal Social Psychology* 67: 371–78.

Milgram, Stanley. 1965. "Some Conditions of Obedience and Disobedience to Authority." *Human Relations* 18: 57–76.

Millar, Morgan M., and Don A. Dillman. 2012. "Encouraging Survey Response via Smartphones: Effects on Respondents' Use of Mobile Devices and Survey Response Rates." *Survey Practice* 5 (3). www.surveypractice.org.

Miller, Peter V. 2010. "The Road to Transparency in Survey Research." *Public Opinion Quarterly* 74 (3): 602–06.

Miller, Wayne, and June Lennie. 2005. "Empowerment Evaluation: A Practical Method for Evaluating a National School Breakfast Program." *Evaluation Journal of Australiasia* 5 (2): 18–26.

Milner, Murray, Jr. 2004. *Freaks, Geeks, and Cool Kids: American Teenagers, Schools, and the Culture of Consumption*. New York: Routledge.

Mirola, William A. 2003. "Asking for Bread, Receiving a Stone: The Rise and Fall of Religious Ideologies in Chicago's Eight-Hour Movement." *Social Problems* 50 (2): 273–93.

Mitchell, Richard G. Jr. 1991. "Secrecy and Disclosure in Fieldwork," pp. 97–108 in *Experiencing Fieldwork: An Inside View of Qualitative Research*, edited by William B. Shaffir and Robert A. Stebbins. Thousand Oaks, CA: Sage.

Mitchell, Richard G. Jr. 2002. *Dancing at Armageddon: Survivalism and Chaos in Modern Times*. Chicago: University of Chicago Press.

Morgan, David L., ed. 1993. *Successful Focus Groups: Advancing the State of the Art*. Thousand Oaks, CA: Sage.

Morgan, Lewis H. 1870. *Systems of Consanguinity and Affinity*. Washington, DC: Smithsonian Institution.

Moskowitz, Milt. 1981. "The Drugs That Doctors Order." *San Francisco Chronicle*, May 23, p. 33.

Moynihan, Daniel. 1965. *The Negro Family: The Case for National Action*. Washington, DC: U.S. Government Printing Office.

Myrdal, Gunnar. 1944. *An American Dilemma*. New York: Harper & Row.

Naisbitt, John, and Patricia Aburdene. 1990. *Megatrends 2000: Ten New Directions for the 1990's*. New York: Morrow.

National Center for Health Statistics, "Wireless Substitution: Early Release of Estimates from the National Health Interview Survey, July–December 2016, Released May 2017. https://www.cdc.gov /nchs/data/nhis/earlyrelease/wireless201705.pdf.

National Center for Juvenile Justice. 2009. "National Arrest Rates by Offense, Sex, and Race." October 31. http://ojjdp.ncjrs.org /ojstatbb/crime/excel/JAR_2008.xls.

Nature Conservancy. 2005. "The Nature of Science on the Sacramento River." *Nature Conservancy* [newsletter], Spring–Summer.

Neuman, W. Lawrence. 1998. "Negotiated Meanings and State Transformation: The Trust Issue in the Progressive Era." *Social Problems* 45 (3): 315–35.

Noy, Chaim. 2008. "Sampling Knowledge: The Hermeneutics of Snowball Sampling in Qualitative Research." *International Journal of Social Research Methodology* 11 (4): 327–44.

Olson, Kristen, and Andy Peytchev. 2007. "Effect of Interviewer Experience on Interview Pace and Interviewer Attitudes." *Public Opinion Quarterly* 71 (2): 273–86.

O'Neill, Harry W. 1992. "They Can't Subpoena What You Ain't Got." *AAPOR News* 19 (2): 4, 7.

Onwuegbuzie, Anthony J., and Nancy L. Leech. 2005. "On Becoming a Pragmatic Researcher: The Importance of Combining Quantitative and Qualitative Research Methodologies." *International Journal of Research Methodology* 8 (5): 375–87.

Överlien, Carolina, Karin Aronsson, and Margareta Hydén. 2005. "The Focus Group Interview as an In-Depth Method? Young Women Talking about Sexuality." *International Journal of Social Research* 8 (4): 331–44.

Park, Hyunjoo, M. Mandy Sha, and Yuling Pan. 2014. "Investigating Validity and Effectiveness of Cognitive Interviewing as a Pretesting Method for Non-English Questionnaires: Findings from Korean Cognitive Interviews." *International Journal of Social Research Methodology* 17 (6): 643–58. http://dx.doi.org/10.1080 /13645579.2013.823002.

Pearson, Geoff. 2009. "The Researcher as Hooligan: Where 'Participant' Observation Means Breaking the Law." *International Journal of Social Research Methodology* 12 (3): 243–55.

Perinelli, Phillip J. 1986. "Nonsuspecting Public in TV Call-in Polls." *New York Times*, February 14, letter to the editor.

Perrow, Charles. 2002. *Organizing America: Wealth, Power, and the Origins of Corporate Capitalism*. Princeton, NJ: Princeton University Press.

Petersen, Larry R., and Judy L. Maynard. 1981. "Income, Equity, and Wives' Housekeeping Role Expectations." *Pacific Sociological Review*, January, pp. 87–105.

Petrolia, Daniel R., and Sanjoy Bhattacharjee. 2009. "Revisiting Incentive Effects: Evidence from a Random-Sample Mail Survey on Consumer Preferences for Fuel Ethanol." *Public Opinion Quarterly* 73 (3): 537–50.

Picou, J. Steven. 1996a. "Compelled Disclosure of Scholarly Research: Some Comments on High Stakes Litigation." *Law and Contemporary Problems* 59 (3): 149–57.

Picou, J. Steven. 1996b. "Sociology and Compelled Disclosure: Protecting Respondent Confidentiality." *Sociological Spectrum* 16 (3): 207–38.

Plutzer, Eric, and Michael Berkman. 2005. "The Graying of America and Support for Funding the Nation's Schools." *Public Opinion Quarterly* 69 (1): 66–86.

Popper, Karl. 2002 [1934]. *The Logic of Scientific Discovery*. New York: Routledge Classics.

Population Communications International. 1996. *International Dateline* [February]. New York: Author.

Population Reference Bureau, World Population Data Sheet 2015. http://www.prb.org/pdf15/2015-world-population-data-sheet_eng.pdf.

Porter, Stephen R., and Michael E. Whitcomb. 2003. "The Impact of Contact Type on Web Survey Response Rates." *Public Opinion Quarterly* 67 (4): 579–88.

Powell, Elwin H. 1958. "Occupation, Status, and Suicide: Toward a Redefinition of Anomie." *American Sociological Review* 23 (4): 131–39.

Presser, Stanley, and Johnny Blair. 1994. "Survey Pretesting: Do Different Methods Produce Different Results?" pp. 73–104 in *Sociological Methodology 1994*, edited by Peter Marsden. San Francisco: Jossey-Bass.

Prewitt, Kenneth. 2003. "Partisan Politics in the 2000 U.S. Census." *Population Reference Bureau*, November. http://www.prb.org/Template.cfm?Section=PRB&template=/Content/ContentGroups/Articles/03/Partisan_Politics_in_the_2000_U_S__Census.htm.

Priede, Camilla, and Stephen Farrall. 2011. "Comparing Results from Different Styles of Cognitive Interviewing: 'Verbal Probing' vs 'Thinking Aloud.'" *International Journal of Social Research Methodology* 14 (4): 271–87.

Procter, Rob, Farida Vis, and Alex Voss. 2013. "Reading the Riots on Twitter: Methodological Innovation for the Analysis of Big Data." *International Journal of Social Research Methodology* 16 (3): 197–214. doi: 10.1080/13645579.2013.774172.

Public Opinion Quarterly. 2006. 70 (5). [Special issue: *Nonresponse Bias in Household Surveys*].

QSR International. 2011. "NVivo 9: Getting Started." http://download.qsrinternational.com/Document/NVivo9/NVivo9-Getting-Started-Guide.pdf.

Quoss, Bernita, Margaret Cooney, and Terri Longhurst. 2000. "Academics and Advocates: Using Participatory Action Research to Influence Welfare Policy." *Journal of Consumer Affairs* 34 (1): 47–61.

Ragin, Charles C., and Howard S. Becker. 1992. *What Is a Case? Exploring the Foundations of Social Inquiry*. Cambridge, England: Cambridge University Press.

Rasinski, Kenneth A. 1989. "The Effect of Question Wording on Public Support for Government Spending." *Public Opinion Quarterly* 53 (3): 388–94.

Rauch, Jonathan 2010 May. "Do 'Family Values' Weaken Families?" *Dallas Morning News*. http://www.dallasnews.com/opinion/commentary/2010/05/28/Jonathan-Rauch-Do-family-9089.

Redfield, Robert. 1941. *The Folk Culture of Yucatan*. Chicago: University of Chicago Press.

Reinharz, Shulamit. 1992. *Feminist Methods in Social Research*. New York: Oxford University Press.

Reynolds, H. T. 1977. Analysis of Nominal Data. Beverly Hills, CA: Sage.

Rhodes, Bryan B., and Ellen L. Marks. 2011. "Using Facebook to Locate Sample Members." *Survey Practice* 4 (5). http://36.22.354a.static.theplanet.com/index.php/SurveyPractice/article/view/83/pdf.

Riecken, Henry W., and Robert F. Boruch. 1974. *Social Experimentation: A Method for Planning and Evaluating Social Intervention*. New York: Academic Press.

Roethlisberger, F. J., and W. J. Dickson. 1939. *Management and the Worker*. Cambridge, MA: Harvard University Press.

Rogers, Everett M., Peter W. Vaughan, Ramadhan M. A. Swalehe, Nagesh Rao, and Suruchi Sood. 1996. "Effects of an Entertainment-Education Radio Soap Opera on Family Planning and HIV/AIDS Prevention Behavior in Tanzania." Report presented at a technical briefing on the Tanzania Entertainment-Education Project, Rockefeller Foundation, New York, March 27.

Roper, Burns. 1992. ". . . But Will They Give the Poll Its Due?" *AAPOR News* 19 (2): 5–6.

Rosenberg, Morris. 1968. *The Logic of Survey Analysis*. New York: Basic Books.

Rosenthal, Robert, and Lenore Jacobson. 1968. *Pygmalion in the Classroom*. New York: Holt, Rinehart & Winston.

Ross, Jeffrey Ian. 2004. "Taking Stock of Research Methods and Analysis on Oppositional Political Terrorism." *American Sociologist* 35 (2): 26–37.

Rossi, Peter H., Mark Lipsey, and Howard E. Freeman. 2002. *Evaluation: A Systematic Approach*. 6th ed. Thousand Oaks, CA: Sage.

Rossman, Gabriel. 2002. *The Qualitative Influence of Ownership on Media Content: The Case of Movie Reviews*. Paper presented to the American Sociological Association, Chicago. Reported in *Contexts* 2 (2, Spring 2003): 7.

Rothman, Ellen K. 1981. "The Written Record." *Journal of Family History*, Spring, pp. 47–56.

Rubin, Herbert J., and Riene S. Rubin. 1995. *Qualitative Interviewing: The Art of Hearing Data*. Thousand Oaks, CA: Sage.

Sacks, Jeffrey J., W. Mark Krushat, and Jeffrey Newman. 1980. "Reliability of the Health Hazard Appraisal." *American Journal of Public Health* 70: 730–2.

SaferSpaces. 2014. "Mapping and Spatial Design for Community Crime Prevention." http://www.saferspaces.org.za/be-inspired/entry/mapping-and-spatial-design-for-community-crime-prevention.

Sanders, William B. 1994. *Gangbangs and Drive-bys: Grounded Culture and Juvenile Gang Violence*. New York: Aldine De Gruyter.

Santos, Robert, Trent Buskirk, and Andrew Gelman. AAPOR Ad Hoc Committee on Credibility Interval. 2012. "AAPOR Statement: Understanding a 'Creditability Interval' and How it Differs from the 'Margin of Sampling Error' in a Public Opinion Poll." http://www.aapor.org/Understanding_a_credibility_interval_and_how_it_differs_from_the_margin_of_sampling_error_in_a_publi.htm.

Saunders, Benjamin, Jenny Kitzinger, and Celia Kitzinger. 2015. "Anonymising Interview Data: Challenges and Compromise in Practice." *Qualitative Research* 15 (5): 616–31.

Saxena, Prem. 2012. *Social, Religious and Regional Institutions in Navi Mumbai*. Navi Mumbai, India: City and Industrial Development Corporation of Mahrashtra.

Saxena, Prem C., Andrzej Kulczycki, and Rozzet Jurdi. 2004. "Nuptiality Transition and Marriage Squeeze in Lebanon—Consequences of Sixteen Years of Civil War." *Journal of Comparative Family Studies* 35 (2): 241–58. [Special issue on turbulence in the Middle East].

Saxena, Prem C., and Dhirendra Kumar. 1997. "Differential Risk of Mortality among Pensioners after Retirement in the State of Maharashtra, India." *Genus* 53 (1/2): 113–28.

Scarce, Rik. 1990. *Ecowarriors: Understanding the Radical Environmental Movement*. Chicago: Noble Press.

Scarce, Rik. 1999. "Good Faith, Bad Ethics: When Scholars Go the Distance and Scholarly Associations Do Not." *Law and Social Inquiry: Journal of the American Bar Foundation* 24 (4): 977–86.

Schiflett, Kathy L., and Mary Zey. 1990. "Comparison of Characteristics of Private Product Producing Organizations and Public Service Organizations." *Sociological Quarterly* 31 (4): 569–83.

Schilt, Kristen. 2006. "Just One of the Guys? How Transmen Make Gender Visible in the Workplace." *Gender and Society* 20 (4): 465–90.

Schmitt, Frederika E., and Patricia Yancey Martin. 1999. "Unobtrusive Mobilization by an Institutionalized Rape Crisis Center: 'All We Do Comes from Victims.'" *Gender and Society* 13 (3): 364–84.

Schutz, Alfred. 1967. *The Phenomenology of the Social World*. Evanston, IL: Northwestern University Press.

Schutz, Alfred. 1970. *On Phenomenology and Social Relations*. Chicago: University of Chicago Press.

Sense, Andrew J. 2006. "Driving the Bus from the Rear Passenger Seat: Control Dilemmas of Participative Action Research." *International Journal of Social Research Methodology* 9 (1): 1–13.

Shaffir, William B., and Robert A. Stebbins, eds. 1991. *Experiencing Fieldwork: An Inside View of Qualitative Research*. Thousand Oaks, CA: Sage.

Shea, Christopher. 2000. "Don't Talk to the Humans: The Crackdown on Social Science Research." *Lingua Franca* 10 (6): 27–34.

Sherif, Muzafer. 1935. "A Study of Some Social Factors in Perception." *Archives of Psychology* 27: 1–60.

Silver, Nate. 2010a. "The Broadus Effect? Social Desirability Bias and California Proposition 19." *FiveThirtyEight* (blog), *New York Times*, July 26. http://www.fivethirtyeight.com/2010/07/broadus-effect-social-desirability-bias.html.

Silver, Nate. 2010b. "'Robopolls' Significantly More Favorable to Republicans Than Traditional Surveys." *FiveThirtyEight* (blog) *New York Times*, October 28. http://fivethirtyeight.blogs.nytimes.com/2010/10/28/robopolls-significantly-more-favorable-to-republicans-than-traditional-surveys/.

Silverman, David. 1993. *Interpreting Qualitative Data: Methods for Analyzing Talk, Text, and Interaction*. Thousand Oaks, CA: Sage.

Silverman, George. 2005. "Qualitative Research: Face-to-Face Focus Groups, Telephone Focus Groups, Online Focus Groups." http://www.mnav.com/qualitative_research.htm.

Simi, P., and Robert Futrell. 2015. *American Swastika: Inside the White Power Movement's Hidden Spaces of Hate*. Lanham, MD: Rowman & Littlefield.

Singer, Eleanor, and Mick Couper. Fall 2014. "The Effect of Question Wording on Attitudes toward Prenatal Testing and Abortion." *Public Opinion Quarterly* 78 (3): 751–60.

Singer, Eleanor, Robert M. Groves, and Amy D. Corning. 1999. "Differential Incentives: Beliefs about Practices, Perceptions of Equity, and Effects on Survey Participation." *Public Opinion Quarterly* 63 (2): 251–60.

Skedsvold, Paula. 2002. "New Developments Concerning Public Use Data Files." *Footnotes* 30 (1): 3.

Skocpol, Theda. 2003. *Diminished Democracy: From Membership to Management in American Civic Life*. Norman: Oklahoma University Press.

Smith, Aaron. 2013. "Smartphone Ownership." June 5. Report from the Pew Research Center's Internet and American Life Project. www.pewinternet.org/reports/2013/Smaartphone-Ownership-2013.aspx.

Smith, Andrew E., and G. F. Bishop. 1992. *The Gallup Secret Ballot Experiments: 1944–1988*. Paper presented at the annual conference of the American Association for Public Opinion Research, St. Petersburg, FL, May.

Smith, Dorothy E. 1978. *The Everyday World as Problematic*. Boston: Northeastern University Press.

Smith, Vicki. 1998. "The Fractured World of the Temporary Worker: Power, Participation, and Fragmentation in the Contemporary Workplace." *Social Problems* 45 (4): 411–30.

Snow, David A., and Anderson, Leon. 1987. "Identity Work among the Homeless: The Verbal Construction and Avowal of Personal Identities." *American Journal of Sociology* 92:1336–71.

Sorokin, Pitirim A. 1937–1940. *Social and Cultural Dynamics*. 4 vols. Englewood Cliffs, NJ: Bedminster Press.

Special Issue. 2013, May. "Computational Social Science: Research Strategies, Design and Methods." *International Journal of Social Research Methodology* 16 (6).

Spencer, Liz, Jane Ritchie, Jane Lewis, and Lucy Dillon. 2003. *Quality in Qualitative Evaluation: A Framework for Assessing Research Evidence*. London: National Centre for Social Research. http://www.cabinetoffice.gov.uk/strategy/~/media/assets/www.cabinetoffice.gov.uk/strategy/qqe_rep%20pdf.ashx.

Spohn, Cassie, and Julie Horney. 1990. "A Case of Unrealistic Expectations: The Impact of Rape Reform Legislation in Illinois." *Criminal Justice Policy Review* 4 (1): 1–18.

Srole, Leo. 1956. "Social Integration and Certain Corollaries: An Exploratory Study." *American Sociological Review* 21: 709–16.

Stark, Rodney. 1997. *The Rise of Christianity: How the Obscure, Marginal Jesus Movement Became the Dominant Religious Force in the Western World in a Few Centuries*. San Francisco: HarperCollins.

Stearns, Cindy A. 1999. "Breastfeeding and the Good Maternal Body." *Gender and Society* 13 (3): 308–26.

Steinmetz, Katy. 2013."What Twitter Says to Linguists." *Time*, September 9. http://content.time.com/time/magazine/article/0,9171,2150609,00.html.

Stephens-Davidowitz, Seth. 2013. "How Many American Men Are Gay?" *New York Times Sunday Review*, December 7. http://www.nytimes.com/2013/12/08/opinion/sunday/how-many-american-men-are-gay.html?pagewanted=1&_r=0&hp&rref=opinion.

Sterling, T. D., W. L. Rosenbaum, and J. J. Weinkam. 1995. "Publication Decisions Revisited: The Effect of the Outcome of Statistical Tests on the Decision to Publish and Vice Versa." *American Statistician* 49 (1): 108–12.

Stouffer, Samuel, et al. 1949–1950. *The American Soldier*. 3 vols. Princeton, NJ: Princeton University Press.

Strang, David, and James N. Baron. 1990. "Categorical Imperatives: The Structure of Job Titles in California State Agencies." *American Sociological Review* 55: 479–95.

Strauss, Anselm. 1977. *Social Psychology of George Herbert Mead*. Chicago: University of Chicago Press.

Strauss, Anselm, and Juliet Corbin. 1994. "Grounded Theory Methodology: An Overview," pp. 273–85 in *Handbook of Qualitative Research*, edited by Norman K. Denzin and Yvonna S. Lincoln. Thousand Oaks, CA: Sage.

Strauss, Anselm, and Juliet Corbin. 1998. *Basics of Qualitative Research: Techniques and Procedures for Developing Grounded Theory*. Thousand Oaks, CA: Sage.

Strunk, William, Jr., and E. B. White. 1999. *The Elements of Style*. 4th ed. Boston: Allyn and Bacon.

Swalehe, Ramadhan, Everett M. Rogers, Mark J. Gilboard, Krista Alford, and Rima Montoya. 1995. *A Content Analysis of the Entertainment Education Radio Soap Opera 'Twende na Wakati' (Let's Go with the Times) in Tanzania*. Arusha,

Tanzania: Population Family Life and Education Programme (POFLEP), Ministry of Community Development, Women Affairs, and Children, November 15.

Takeuchi, David. 1974. "Grass in Hawaii: A Structural Constraints Approach." M.A. thesis, University of Hawaii.

Tan, Alexis S. 1980. "Mass Media Use, Issue Knowledge and Political Involvement." *Public Opinion Quarterly* 44 (2): 241–48.

Tandon, Rajesh, and L. Dave Brown. 1981. "Organization-Building for Rural Development: An Experiment in India." *Journal of Applied Behavioral Science*, April–June, pp. 172–89.

Taylor, Jerome. 2008. "An Illness That Has Defied Medical Science." *The Independent*, February 4. http://www.independent.co.uk/opinion /commentators/jerome-taylor-an-illness-that -has-defied-medical-science-777669.html.

Thomas, W. I., and Dorothy Swain Thomas. 1928. *The Child in America: Behavior Problems and Programs*. New York: Knopf.

Thomas, W. I., and Florian Znaniecki. 1918. *The Polish Peasant in Europe and America*. Chicago: University of Chicago Press.

Thompson, Paul. 2004. "Researching Family and Social Mobility with Two Eyes: Some Experiences of the Interaction between Qualitative and Quantitative Data." *International Journal of Social Research Methodology* 7 (3): 237–57.

Tocqueville, Alexis de (2000). *Democracy in America*. Chicago: The University of Chicago Press.

Tourangeau, Roger, Mick P. Couper, and Frederick G. Conrad. 2013. "'Up Means Good': The Effect of Screen Position on Evaluative Ratings in Web Surveys." *Public Opinion Quarterly* 77 (S1): 69–88. [Special issue: *Topics in Survey Measurement and Public Opinion*].

Tourangeau, Roger, and Cong Ye. 2009. "The Framing of the Survey Request and Panel Attrition." *Public Opinion Quarterly* 73 (2): 338–48.

Trepagnier, Barbara. 2006. *Silent Racism: How Well-Meaning White People Perpetuate the Racial Divide*. Boulder, CO: Paradigm.

Trials of War Criminals before the Nuremberg Military Tribunals under Control Council Law No. 10. Nuremberg, October 1946–April 1949. 1949–1953. Washington, DC: U.S. Government Printing Office. http://www.ushmm.org/research /doctors/indiptx.htm.

Tuckel, Peter, and Harry O'Neill. 2002. "The Vanishing Respondent in Telephone Surveys." *Journal of Advertising Research*, September–October, pp. 26–48.

Turk, Theresa Guminski. 1980. "Hospital Support: Urban Correlates of Allocation Based on Organizational Prestige." *Pacific Sociological Review*, July, pp. 315–32.

Turner, J. H., ed. 1989. *Theory Building in Sociology: Assessing Theoretical Cumulation*. Newbury Park, CA: Sage.

Union of Concerned Scientists. 2005. "Political Interference in Science." October 23. http:// ucsusa.org/scientific_integrity/interference/.

United Nations. 2015. "Human Development Report: Statistical Annex." http://hdr.undp.org /sites/default/files/hdr_2015_statistical_annex .pdf.

U.S. Bureau of the Census. 2012. Statistical Abstract of the United States. Washington, DC: U.S. Government Printing Office.

U.S. Bureau of Labor Statistics, Usual Weekly Earnings of Wage and Salary Workers, Fourth Quarter 2016. January 27, 2017, Table 9.

U.S. Department of Health and Human Services. 2002. http://www.grants.nih.gov/grants/policy /coc/index.htm/.

Van Cleave, Nicole G. 2016. *Crook County: Racism and Injustice in America's Largest Criminal Court*. Palo Alto, CA: Stanford University Press.

Vanderbei, Robert. "Election 2016 Results." http:// www.princeton.edu/~rvdb/JAVA/election2016/.

Veroff, Joseph, Shirley Hatchett, and Elizabeth Douvan. 1992. "Consequences of Participating in a Longitudinal Study of Marriage." *Public Opinion Quarterly* 56 (3): 325–27.

Vigen, Tyler. 2014. "Spurious Correlations." http://www.tylervigen.com.

Votaw, Carmen Delgado. 1979. *Women's Rights in the United States*. United States Commission on Civil Rights, Inter-American Commission on Women. Washington, DC: Clearinghouse Publications.

Walker, Shayne, Anaru Eketone, and Anita Gibbs. 2006. "An Exploration of Kaupapa Maori Research, Its Principles, Processes, and Applications." *International Journal of Social Research Methodology* 9 (4): 331–44.

Wall, Sarah. 2008. "Easier Said Than Done: Writing an Autoethnography." *International Journal of Qualitative Methods* 7 (1): 38–53. http://ejournals .library.ualberta.ca/index.php/IJQM/article /view/1621/1144.

Wallace, Walter. 1971. *The Logic of Science in Sociology*. Hawthorne, NY: Aldine.

Ward, Lester. 1906. *Applied Sociology*. Boston: Ginn.

Warner, W. Lloyd. 1949. *Democracy in Jonesville*. New York: Harper.

Waters, Jaime. 2015. "Snowball Sampling: A Cautionary Tale Involving a Study of Older Drug Users." *International Journal of Social Research Methodology* 18 (4): 367–80. http://dx.doi.org/10.1080/13645579.2014.953316.

Webb, Eugene J., Donald T. Campbell, Richard D. Schwartz, and Lee Sechrest. 2000. *Unobtrusive Measures*. Rev. ed. Thousand Oaks, CA: Sage.

Weber, Max. [1905] 1958. *The Protestant Ethic and the Spirit of Capitalism*. Translated by Talcott Parsons. New York: Scribner.

Weber, Max. [1925] 1946. "Science as a Vocation," pp. 129–56 in *From Max Weber: Essays in Sociology*, edited and translated by Hans Gerth and C. Wright Mills. New York: Oxford University Press.

Weber, Max. [1934] 1951. *The Religion of China*. Translated by Hans H. Gerth. New York: Free Press.

Weber, Max. [1934] 1952. *Ancient Judaism*. Translated by Hans H. Gerth and Don Martindale. New York: Free Press.

Weber, Max. [1934] 1958. *The Religion of India*. Translated by Hans H. Gerth and Don Martindale. New York: Free Press.

Weiss, Carol. 1972. *Evaluation Research*. Englewood Cliffs, NJ: Prentice-Hall.

Wells, T., J. T. Bailey, and M. W. Link. 2013. "Filling the Void: Gaining a Better Understanding of Tablet-based Surveys." *Survey Practice* 6 (1). www.surveypractice.org.

Wharton, Amy S., and James N. Baron. 1987. "So Happy Together? The Impact of Gender Segregation on Men at Work." *American Sociological Review* 52:574–87.

White, William S. 1997. Communication to APP-SOC listserv (appsoc@miagra.ucs.edu) from wwite@jaguar1.usouthal.edu, October 11.

Whyte, W. F. 1943. *Street Corner Society*. Chicago: University of Chicago Press.

Whyte, W. F., D. J. Greenwood, and P. Lazes. 1991. "Participatory Action Research: Through Practice to Science in Social Research," pp. 19–55 in *Participatory Action Research*, edited by W. F. Whyte. New York: Sage.

Wieder, D. L. 1988. *Language and Social Reality: The Case of Telling the Convict Code*. Lanham, MD: University Press of America.

Wiles, Rose, Amanda Coffey, Judy Robinson, and Sue Heath. 2012. "Anonymisation and Visual Images: Issues of Respect, 'Voice' and Protection." *International Journal of Social Research Methodology* 15 (1): 41–53.

Williams, Malcolm, and Kerryn Husk. 2013. "Can We, Should We, Measure Ethnicity?" *International Journal of Social Research Methodology* 16 (4): 285–300. doi: 10.1080/13645579.2012.682794.

Wilson, Camilo. 1999. Private e-mail, September 8.

Wolff, Kurt H., ed. and trans. 1950. *The Sociology of Georg Simmel*. Glencoe, IL: Free Press.

Worthy, M., and D. Mayclin. 2013. "A Comparison of Data Quality Across Modes in a Mixed-Mode Collection of Administrative Records." *Survey Practice* 6 (2).

Yeager, D. S., and Jon A. Krosnick. 2012. "Does Mentioning 'Some People' and 'Other People' in an Opinion Question Improve Measurement Quality?" *Public Opinion Quarterly* 76 (1): 131–41.

Yeager, D. S., and Jon A. Krosnick. 2011. "Does Mentioning 'Some People' and 'Other People' in a Survey Question Increase the Accuracy of Adolescents' Self-Reports?" *Developmental Psychology* 47 (6): 1674–79. doi:10.1037/a0025440.

Yerg, Beverly J. 1981. "Reflections on the Use of the RTE Model in Physical Education." *Research Quarterly for Exercise and Sport*, March, pp. 38–47.

Yinger, J. Milton, Kiyoshi Ikeda, Frank Laycock, and Stephen J. Cutler. 1977. *Middle Start: An Experiment in the Educational Enrichment of Young Adolescents*. London: Cambridge University Press.

Zimbardo, Philip. 1971. The Stanford Prison Experiment: A Simulation Study on the Psychology of Imprisonment. https://www.prisonexp.org.

Index

M